STUDENT SOLUTIONS MANUAL

Randy Gallaher ∞ Kevin Bodden
Lewis and Clark Community College

College Algebra
Enhanced with Graphing Utilities
Fourth Edition

Sullivan
Sullivan

Upper Saddle River, NJ 07458

Editor-in-Chief: Sally Yagan
Acquisitions Editor: Adam Jaworski
Supplement Editor: Christopher Truchan
Executive Managing Editor: Kathleen Schiaparelli
Assistant Managing Editor: Becca Richter
Production Editor: Donna Crilly
Supplement Cover Manager: Paul Gourhan
Supplement Cover Designer: Joanne Alexandris
Manufacturing Buyer: Ilene Kahn

© 2006 Pearson Education, Inc.
Pearson Prentice Hall
Pearson Education, Inc.
Upper Saddle River, NJ 07458

All rights reserved. No part of this book may be reproduced in any form or by any means, without permission in writing from the publisher.

Pearson Prentice Hall™ is a trademark of Pearson Education, Inc.

The author and publisher of this book have used their best efforts in preparing this book. These efforts include the development, research, and testing of the theories and programs to determine their effectiveness. The author and publisher make no warranty of any kind, expressed or implied, with regard to these programs or the documentation contained in this book. The author and publisher shall not be liable in any event for incidental or consequential damages in connection with, or arising out of, the furnishing, performance, or use of these programs.

> This work is protected by United States copyright laws and is provided solely for teaching courses and assessing student learning. Dissemination or sale of any part of this work (including on the World Wide Web) will destroy the integrity of the work and is not permitted. The work and materials from it should never be made available except by instructors using the accompanying text in their classes. All recipients of this work are expected to abide by these restrictions and to honor the intended pedagogical purposes and the needs of other instructors who rely on these materials.

Printed in the United States of America

10 9 8 7 6 5 4 3 2 1

ISBN 0-13-149107-5

Pearson Education Ltd., *London*
Pearson Education Australia Pty. Ltd., *Sydney*
Pearson Education Singapore, Pte. Ltd.
Pearson Education North Asia Ltd., *Hong Kong*
Pearson Education Canada, Inc., *Toronto*
Pearson Educación de Mexico, S.A. de C.V.
Pearson Education—Japan, *Tokyo*
Pearson Education Malaysia, Pte. Ltd.

Table of Contents

Preface

Chapter R Review
R.1	Real Numbers	1
R.2	Algebra Review	2
R.3	Geometry Review	6
R.4	Polynomials	8
R.5	Factoring Polynomials	11
R.6	Synthetic Division	13
R.7	Rational expressions	14
R.8	nth Roots; Rational Exponents	19
	Chapter Review	23
	Chapter Test	28

Chapter 1 Graphs, Equations, and Inequalities
1.1	Rectangular Coordinates; Graphing Utilities; Introduction to Graphing Equations	32
1.2	Solving Equations Using a Graphing Utility; Linear and Rational Equations	39
1.3	Quadratic Equations	46
1.4	Complex Numbers; Quadratic Equations in the Complex Number System	52
1.5	Radical Equations; Equations Quadratic in Form; Absolute Value Equations; Factorable Equations	56
1.6	Problem Solving: Interest, Mixture, Uniform Motion, Constant Rate Jobs	66
1.7	Solving Inequalities	69
1.8	Lines	75
1.9	Circles	82
	Chapter Review	87
	Chapter Test	99

Chapter 2 Functions and Their Graphs
2.1	Symmetry; Graphing Key Equations	105
2.2	Functions	109
2.3	The Graph of a Function	115
2.4	Properties of Functions	119
2.5	Linear Functions and Models	127
2.6	Library of Functions; Piecewise-defined Functions	132
2.7	Graphing Techniques: Transformations	137
2.8	Mathematical Models: Constructing Functions	145
	Chapter Review	149
	Chapter Test	158
	Cumulative Review	162

Chapter 3 Polynomial and Rational Functions
3.1	Quadratic Functions and Models	165
3.2	Polynomial Functions and Models	176
3.3	Properties of Rational Functions	187
3.4	The Graph of a Rational Function; Inverse and Joint Variation	190
3.5	Polynomial and Rational Inequalities	214
3.6	The Real Zeros of a Polynomial Function	222
3.7	Complex Zeros; Fundamental Theorem of Algebra	236
Chapter Review		240
Chapter Test		262
Cumulative Review		267

Chapter 4 Exponential and Logarithmic Functions
4.1	Composite Functions	270
4.2	One-to-One Functions; Inverse Functions	278
4.3	Exponential Functions	287
4.4	Logarithmic Functions	294
4.5	Properties of Logarithms	302
4.6	Logarithmic and Exponential Equations	305
4.7	Compound Interest	311
4.8	Exponential Growth and Decay; Newton's Law; Logistic Growth and Decay	313
4.9	Building Exponential, Logarithmic, and Logistic Models from Data	318
Chapter Review		321
Chapter Test		329
Cumulative Review		332

Chapter 5 Systems of Equations and Inequalities
5.1	Systems of Linear Equations: Substitution and Elimination	334
5.2	Systems of Linear Equations: Matrices	344
5.3	Systems of Linear Equations: Determinants	356
5.4	Matrix Algebra	361
5.5	Partial Fraction Decomposition	369
5.6	Systems of Linear Inequalities	375
5.7	Linear Programming	382
Chapter Review		389
Chapter Test		401
Cumulative Review		410

Chapter 6 Analytic Geometry
6.2	The Parabola	413
6.3	The Ellipse	420
6.4	The Hyperbola	429
6.5	Systems of Nonlinear Equations and Inequalities	437
Chapter Review		452
Chapter Test		458
Cumulative Review		462

Chapter 7 Sequences; Induction; the Binomial Theorem
7.1	Sequences	464
7.2	Arithmetic Sequences	469
7.3	Geometric Sequences; Geometric Series	472
7.4	Mathematical Induction	475
7.5	The Binomial Theorem	479
Chapter Review		481
Chapter Test		486
Cumulative Review		490

Chapter 8 Counting and Probability
8.1	Sets and Counting	491
8.2	Permutations and Combinations	492
8.3	Probability	494
Chapter Review		497
Chapter Test		498
Cumulative Review		501

Preface

This solution manual is to accompany *College Algebra Enhanced with Graphing Utilities, 4th Edition* by Michael Sullivan and Michael Sullivan, III. The Instructor Solutions Manual (ISM) contains detailed solutions to all exercises in the textbook and the chapter projects (both in the text and on the internet). The Student Solutions Manual (SSM) contains detailed solutions to all odd exercises in the textbook and all solutions to chapter tests. In both manuals, TI-83 Plus graphing calculator screenshots have been included in many solutions to demonstrate how technology can be used to solve problems and check solutions. A concerted effort has been made to make this manual as easy to read and as error free as possible. Please feel free to send us any suggestions or corrections.

We would like thank Dawn Murrin, Bob Walters, and Chris Truchan at Prentice Hall for their prompt and endless help with manuscript pages. Thanks for everything!

We would also like to thank our wives (Angie and Karen) and our children (Annie, Ben, Ethan, Logan, Payton, and Shawn) for their patient support and for enduring many late hours of typing.

A special thanks to Bill Bodden for his help with error-checking solutions and for the many great hours of math discussions.

Randy Gallaher and Kevin Bodden
Department of Mathematics
Lewis and Clark Community College
5800 Godfrey Road
Godfrey, IL 62035
rgallahe@lc.edu   kbodden@lc.edu

Chapter R
Review

Section R.1

1. rational

3. Distributive

5. True

7. False; 6 is the Greatest Common Factor of 12 and 18. The Least Common Multiple is the smallest value that both numbers will divide evenly. The LCM for 12 and 18 is 36.

9. a. $\{2,5\}$
   b. $\{-6,2,5\}$
   c. $\{-6, \frac{1}{2}, -1.333... = -1.\overline{3}, 2, 5\}$
   d. $\{\pi\}$
   e. $\{-6, \frac{1}{2}, -1.333... = -1.\overline{3}, \pi, 2, 5\}$

11. a. $\{1\}$
    b. $\{0,1\}$
    c. $\{0, 1, \frac{1}{2}, \frac{1}{3}, \frac{1}{4}\}$
    d. None
    e. $\{0, 1, \frac{1}{2}, \frac{1}{3}, \frac{1}{4}\}$

13. a. None
    b. None
    c. None
    d. $\{\sqrt{2}, \pi, \sqrt{2}+1, \pi+\frac{1}{2}\}$
    e. $\{\sqrt{2}, \pi, \sqrt{2}+1, \pi+\frac{1}{2}\}$

15. a. 18.953 b. 18.952

17. a. 28.653 b. 28.653

19. a. 0.063 b. 0.062

21. a. 9.999 b. 9.998

23. a. 0.429 b. 0.428

25. a. 34.733 b. 34.733

27. $3+2=5$

29. $x+2=3\cdot 4$

31. $3y=1+2$

33. $x-2=6$

35. $\frac{x}{2}=6$

37. $9-4+2=5+2=7$

39. $-6+4\cdot 3=-6+12=6$

41. $4+5-8=9-8=1$

43. $4+\frac{1}{3}=\frac{12+1}{3}=\frac{13}{3}$

45. $6-[3\cdot 5+2\cdot(3-2)]=6-[15+2\cdot(1)]$
    $=6-17$
    $=-11$

47. $2\cdot(3-5)+8\cdot 2-1=2\cdot(-2)+16-1$
    $=-4+16-1$
    $=12-1$
    $=11$

49. $10-[6-2\cdot 2+(8-3)]\cdot 2=10-[6-4+5]\cdot 2$
    $=10-[2+5]\cdot 2$
    $=10-[7]\cdot 2$
    $=10-14$
    $=-4$

51. $(5-3)\frac{1}{2}=(2)\frac{1}{2}=1$

53. $\dfrac{4+8}{5-3} = \dfrac{12}{2} = 6$

55. $\dfrac{3}{5} \cdot \dfrac{10}{21} = \dfrac{3 \cdot 2 \cdot 5}{5 \cdot 3 \cdot 7} = \dfrac{\cancel{3} \cdot 2 \cdot \cancel{5}}{\cancel{5} \cdot \cancel{3} \cdot 7} = \dfrac{2}{7}$

57. $\dfrac{6}{25} \cdot \dfrac{10}{27} = \dfrac{2 \cdot 3 \cdot 5 \cdot 2}{5 \cdot 5 \cdot 3 \cdot 9} = \dfrac{2 \cdot \cancel{3} \cdot \cancel{5} \cdot 2}{\cancel{5} \cdot 5 \cdot \cancel{3} \cdot 9} = \dfrac{4}{45}$

59. $\dfrac{3}{4} + \dfrac{2}{5} = \dfrac{15+8}{20} = \dfrac{23}{20}$

61. $\dfrac{5}{6} + \dfrac{9}{5} = \dfrac{25+54}{30} = \dfrac{79}{30}$

63. $\dfrac{5}{18} + \dfrac{1}{12} = \dfrac{10+3}{36} = \dfrac{13}{36}$

65. $\dfrac{1}{30} - \dfrac{7}{18} = \dfrac{3-35}{90} = -\dfrac{32}{90} = -\dfrac{16}{45}$

67. $\dfrac{3}{20} - \dfrac{2}{15} = \dfrac{9-8}{60} = \dfrac{1}{60}$

69. $\dfrac{\left(\dfrac{5}{18}\right)}{\left(\dfrac{11}{27}\right)} = \dfrac{5}{18} \cdot \dfrac{27}{11} = \dfrac{5 \cdot 9 \cdot 3}{9 \cdot 2 \cdot 11} = \dfrac{5 \cdot \cancel{9} \cdot 3}{\cancel{9} \cdot 2 \cdot 11} = \dfrac{15}{22}$

71. $6(x+4) = 6x + 24$

73. $x(x-4) = x^2 - 4x$

75. $(x+2)(x+4) = x^2 + 4x + 2x + 8$
$= x^2 + 6x + 8$

77. $(x-2)(x+1) = x^2 + x - 2x - 2$
$= x^2 - x - 2$

79. $(x-8)(x-2) = x^2 - 2x - 8x + 16$
$= x^2 - 10x + 16$

81. $(x+2)(x-2) = x^2 - 2x + 2x - 4$
$= x^2 - 4$

83. $2x + 3x = x(2+3)$
$= x(5)$
$= 5x$

85. $2(3 \cdot 4) = 2(12) = 24$
$(2 \cdot 3) \cdot (2 \cdot 4) = (6)(8) = 48$

87. Subtraction is not commutative; for example: $2 - 3 = -1 \neq 1 = 3 - 2$.

89. Division is not commutative; for example: $\dfrac{2}{3} \neq \dfrac{3}{2}$.

91. The Symmetric Property of Equality implies that if $2 = x$, then $x = 2$.

93. There are no real numbers that are both rational and irrational, since an irrational number, by definition, is a number that cannot be expressed as the ratio of two integers; that is, not a rational number

   Every real number is either a rational number or an irrational number, since the decimal form of a real number either involves an infinitely repeating pattern of digits or an infinite, non-repeating string of digits.

95. $x = 0.\overline{9}$
$10x = 9.\overline{9}$
Now compute $10x = 9.\overline{9}$
$\underline{ - x = 0.\overline{9}}$
$9x = 9 \Rightarrow x = \dfrac{9}{9} = 1$

So $0.999... = 1$.

Section R.2

1. variable

3. strict

5. 1.2345678×10^3

7. True, if the points are distinct. If the points are the same, the distance between them is 0.

9. False; a number in scientific notation is expressed as the product of a number, x, $1 \le x < 10$ or $-10 < x \le -1$, and a power of 10.

11.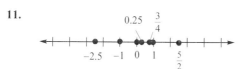

13. $\frac{1}{2} > 0$

15. $-1 > -2$

17. $\pi > 3.14$

19. $\frac{1}{2} = 0.5$

21. $\frac{2}{3} < 0.67$

23. $x > 0$

25. $x < 2$

27. $x \le 1$

29. Graph on the number line: $x \ge -2$

30.

31. Graph on the number line: $x > -1$

33. $d(C,D) = d(0,1) = |1-0| = |1| = 1$

35. $d(D,E) = d(1,3) = |3-1| = |2| = 2$

37. $d(A,E) = d(-3,3) = |3-(-3)| = |6| = 6$

39. $x + 2y = -2 + 2 \cdot 3 = -2 + 6 = 4$

41. $5xy + 2 = 5(-2)(3) + 2 = -30 + 2 = -28$

43. $\frac{2x}{x-y} = \frac{2(-2)}{-2-3} = \frac{-4}{-5} = \frac{4}{5}$

45. $\frac{3x+2y}{2+y} = \frac{3(-2)+2(3)}{2+3} = \frac{-6+6}{5} = \frac{0}{5} = 0$

47. $|x+y| = |3+(-2)| = |1| = 1$

49. $|x|+|y| = |3|+|-2| = 3+2 = 5$

51. $\frac{|x|}{x} = \frac{|3|}{3} = \frac{3}{3} = 1$

53. $|4x - 5y| = |4(3) - 5(-2)|$
$= |12 + 10|$
$= |22|$
$= 22$

55. $||4x| - |5y|| = ||4(3)| - |5(-2)||$
$= ||12| - |-10||$
$= |12 - 10|$
$= |2|$
$= 2$

57. $\frac{x^2 - 1}{x}$

Part (c) must be excluded. The value $x = 0$ must be excluded from the domain because it causes division by 0.

59. $\frac{x}{x^2 - 9} = \frac{x}{(x-3)(x+3)}$

Part (a) must be excluded. The values $x = -3$ and $x = 3$ must be excluded from the domain because they cause division by 0.

61. $\frac{x^2}{x^2 + 1}$

None of the given values are excluded. The domain is all real numbers.

63. $\frac{x^2 + 5x - 10}{x^3 - x} = \frac{x^2 + 5x - 10}{x(x-1)(x+1)}$

Parts (b), (c), and (d) must be excluded. The values $x = 0$, $x = 1$, and $x = -1$ must be excluded from the domain because they cause division by 0.

65. $\frac{4}{x-5}$

Domain $= \{x | x \ne 5\}$

67. $\dfrac{x}{x+4}$
    Domain $= \{x \mid x \neq -4\}$

69. $C = \dfrac{5}{9}(F - 32) = \dfrac{5}{9}(32 - 32) = \dfrac{5}{9}(0) = 0°C$

71. $C = \dfrac{5}{9}(F - 32) = \dfrac{5}{9}(77 - 32) = \dfrac{5}{9}(45) = 25°C$

73. $(-4)^2 = (-4)(-4) = 16$

75. $4^{-2} = \dfrac{1}{4^2} = \dfrac{1}{16}$

77. $3^{-6} \cdot 3^4 = 3^{-6+4} = 3^{-2} = \dfrac{1}{3^2} = \dfrac{1}{9}$

79. $\left(3^{-2}\right)^{-1} = 3^{(-2)(-1)} = 3^2 = 9$

81. $\sqrt{25} = \sqrt{5^2} = 5$

83. $\sqrt{(-4)^2} = |-4| = 4$

85. $\left(8x^3\right)^2 = 8^2 \left(x^3\right)^2 = 64x^6$

87. $\left(x^2 y^{-1}\right)^2 = \left(x^2\right)^2 \cdot \left(y^{-1}\right)^2 = x^4 y^{-2} = \dfrac{x^4}{y^2}$

89. $\dfrac{x^2 y^3}{xy^4} = x^{2-1} y^{3-4} = x^1 y^{-1} = \dfrac{x}{y}$

91. $\dfrac{(-2)^3 x^4 (yz)^2}{3^2 x y^3 z} = \dfrac{-8x^4 y^2 z^2}{9x y^3 z}$

    $= \dfrac{-8}{9} x^{4-1} y^{2-3} z^{2-1}$

    $= \dfrac{-8}{9} x^3 y^{-1} z^1$

    $= -\dfrac{8x^3 z}{9y}$

93. $\left(\dfrac{3x^{-1}}{4y^{-1}}\right)^{-2} = \left(\dfrac{3y}{4x}\right)^{-2} = \left(\dfrac{4x}{3y}\right)^2 = \dfrac{4^2 x^2}{3^2 y^2} = \dfrac{16x^2}{9y^2}$

95. $2xy^{-1} = \dfrac{2x}{y} = \dfrac{2(2)}{(-1)} = -4$

97. $x^2 + y^2 = (2)^2 + (-1)^2 = 4 + 1 = 5$

99. $(xy)^2 = (2 \cdot (-1))^2 = (-2)^2 = 4$

101. $\sqrt{x^2} = |x| = |2| = 2$

103. $\sqrt{x^2 + y^2} = \sqrt{(2)^2 + (-1)^2} = \sqrt{4+1} = \sqrt{5}$

105. $x^y = 2^{-1} = \dfrac{1}{2}$

107. If $x = 2$,
$$2x^3 - 3x^2 + 5x - 4 = 2 \cdot 2^3 - 3 \cdot 2^2 + 5 \cdot 2 - 4$$
$$= 16 - 12 + 10 - 4$$
$$= 10$$

If $x = 1$,
$$2x^3 - 3x^2 + 5x - 4 = 2 \cdot 1^3 - 3 \cdot 1^2 + 5 \cdot 1 - 4$$
$$= 2 - 3 + 5 - 4$$
$$= 0$$

109. $\dfrac{(666)^4}{(222)^4} = \left(\dfrac{666}{222}\right)^4 = 3^4 = 81$

111. $(8.2)^6 \approx 304,006.671$

113. $(6.1)^{-3} \approx 0.004$

115. $(-2.8)^6 \approx 481.890$

117. $(-8.11)^{-4} \approx 0.000$

119. $454.2 = 4.542 \times 10^2$

121. $0.013 = 1.3 \times 10^{-2}$

123. $32,155 = 3.2155 \times 10^4$

125. $0.000423 = 4.23 \times 10^{-4}$

127. $6.15 \times 10^4 = 61,500$

129. $1.214 \times 10^{-3} = 0.001214$

131. $1.1 \times 10^8 = 110,000,000$

133. $8.1 \times 10^{-2} = 0.081$

135. $A = l \cdot w$

137. $C = \pi \cdot d$

139. $A = \dfrac{\sqrt{3}}{4} \cdot x^2$

141. $V = \dfrac{4}{3}\pi \cdot r^3$

143. $V = x^3$

145. a. If $x = 1000$,
$$C = 4000 + 2x$$
$$= 4000 + 2(1000)$$
$$= 4000 + 2000$$
$$= \$6000$$
The cost of producing 1000 watches is $6000.

b. If $x = 2000$,
$$C = 4000 + 2x$$
$$= 4000 + 2(2000)$$
$$= 4000 + 4000$$
$$= \$8000$$
The cost of producing 2000 watches is $8000.

147. a. $|x - 115| = |113 - 115| = |-2| = 2 \leq 5$
113 volts is acceptable.

b. $|x - 115| = |109 - 115| = |-6| = 6 \nleq 5$
109 volts is *not* acceptable.

149. a. $|x - 3| = |2.999 - 3|$
$$= |-0.001|$$
$$= 0.001 \leq 0.01$$
A radius of 2.999 centimeters is acceptable.

b. $|x - 3| = |2.89 - 3|$
$$= |-0.11|$$
$$= 0.11 \nleq 0.01$$
A radius of 2.89 centimeters is *not* acceptable.

151. The distance from Earth to the Moon is about $4 \times 10^8 = 400,000,000$ meters.

153. The wavelength of visible light is about $5 \times 10^{-7} = 0.0000005$ meters.

155. The smallest commercial copper wire has a diameter of about $0.0005 = 5 \times 10^{-4}$ inches.

157. $186,000 \cdot 60 \cdot 60 \cdot 24 \cdot 365$
$$= (1.86 \times 10^5)(6 \times 10^1)^2 (2.4 \times 10^1)(3.65 \times 10^2)$$
$$= 586.5696 \times 10^{10} = 5.865696 \times 10^{12}$$
There are about 5.9×10^{12} miles in one light-year.

159. $\frac{1}{3} = 0.333333... > 0.333$

$\frac{1}{3}$ is larger by approximately $0.0003333...$

161. No. For any positive number a, the value $\frac{a}{2}$ is smaller and therefore closer to 0.

163. Answers will vary.

Section R.3

1. right; hypotenuse

3. $C = 2\pi r$

5. True. $6^2 + 8^2 = 36 + 64 = 100 = 10^2$

7. $a = 5,\ b = 12,$
$c^2 = a^2 + b^2$
$= 5^2 + 12^2$
$= 25 + 144$
$= 169 \Rightarrow c = 13$

9. $a = 10,\ b = 24,$
$c^2 = a^2 + b^2$
$= 10^2 + 24^2$
$= 100 + 576$
$= 676 \Rightarrow c = 26$

11. $a = 7,\ b = 24,$
$c^2 = a^2 + b^2$
$= 7^2 + 24^2$
$= 49 + 576$
$= 625 \Rightarrow c = 25$

13. $5^2 = 3^2 + 4^2$
$25 = 9 + 16$
$25 = 25$
The given triangle is a right triangle. The hypotenuse is 5.

15. $6^2 = 4^2 + 5^2$
$36 = 16 + 25$
$36 = 41$ false
The given triangle is not a right triangle.

17. $25^2 = 7^2 + 24^2$
$625 = 49 + 576$
$625 = 625$
The given triangle is a right triangle. The hypotenuse is 25.

19. $6^2 = 3^2 + 4^2$
$36 = 9 + 16$
$36 = 25$ false
The given triangle is not a right triangle.

21. $A = l \cdot w = 4 \cdot 2 = 8 \text{ in}^2$

23. $A = \frac{1}{2} b \cdot h = \frac{1}{2}(2)(4) = 4 \text{ in}^2$

25. $A = \pi r^2 = \pi(5)^2 = 25\pi \text{ m}^2$
$C = 2\pi r = 2\pi(5) = 10\pi \text{ m}$

27. $V = lwh = 8 \cdot 4 \cdot 7 = 224 \text{ ft}^3$
$S = 2lw + 2lh + 2wh$
$= 2(8)(4) + 2(8)(7) + 2(4)(7)$
$= 64 + 112 + 56$
$= 232 \text{ ft}^2$

29. $V = \frac{4}{3}\pi r^3 = \frac{4}{3}\pi \cdot 4^3 = \frac{256}{3}\pi \text{ cm}^3$
$S = 4\pi r^2 = 4\pi \cdot 4^2 = 64\pi \text{ cm}^2$

31. $V = \pi r^2 h = \pi(9)^2(8) = 648\pi \text{ in}^3$
$S = 2\pi r^2 + 2\pi r h$
$= 2\pi(9)^2 + 2\pi(9)(8)$
$= 162\pi + 144\pi$
$= 306\pi \approx 961.33 \text{ in}^2$

33. The diameter of the circle is 2, so its radius is 1.
$A = \pi r^2 = \pi(1)^2 = \pi$ square units

35. The diameter of the circle is the length of the diagonal of the square.
$$d^2 = 2^2 + 2^2$$
$$= 4 + 4$$
$$= 8$$
$$d = \sqrt{8} = 2\sqrt{2}$$
$$r = \frac{d}{2} = \frac{2\sqrt{2}}{2} = \sqrt{2}$$
The area of the circle is:
$$A = \pi r^2 = \pi\left(\sqrt{2}\right)^2 = 2\pi \text{ square units}$$

37. The total distance traveled is 4 times the circumference of the wheel.
Total Distance $= 4C = 4(\pi d) = 4\pi \cdot 16$
$$= 64\pi \approx 201.1 \text{ inches} \approx 16.8 \text{ feet}$$

39. Area of the border = area of EFGH – area of ABCD $= 10^2 - 6^2 = 100 - 36 = 64 \text{ ft}^2$

41. Area of the window = area of the rectangle + area of the semicircle.
$$A = (6)(4) + \frac{1}{2} \cdot \pi \cdot 2^2 = 24 + 2\pi \approx 30.28 \text{ ft}^2$$
Perimeter of the window = 2 heights + width + one-half the circumference.
$$P = 2(6) + 4 + \frac{1}{2} \cdot \pi(4) = 12 + 4 + 2\pi$$
$$= 16 + 2\pi \approx 22.28 \text{ feet}$$

43. Convert 20 feet to miles, and solve the Pythagorean Theorem to find the distance:
$$20 \text{ feet} = 20 \text{ feet} \cdot \frac{1 \text{ mile}}{5280 \text{ feet}} = 0.003788 \text{ miles}$$
$$d^2 = (3960 + 0.003788)^2 - 3960^2 = 30 \text{ sq. miles}$$
$$d \approx 5.477 \text{ miles}$$

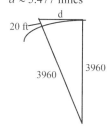

45. Convert 100 feet to miles, and solve the Pythagorean Theorem to find the distance:
$$100 \text{ feet} = 100 \text{ feet} \cdot \frac{1 \text{ mile}}{5280 \text{ feet}}$$
$$= 0.018939 \text{ miles}$$
$$d^2 = (3960 + 0.018939)^2 - 3960^2 \approx 150 \text{ sq. miles}$$
$$d \approx 12.247 \text{ miles}$$
Convert 150 feet to miles, and solve the Pythagorean Theorem to find the distance:
$$150 \text{ feet} = 150 \text{ feet} \cdot \frac{1 \text{ mile}}{5280 \text{ feet}}$$
$$= 0.028409 \text{ miles}$$
$$d^2 = (3960 + 0.028409)^2 - 3960^2 \approx 225 \text{ sq. miles}$$
$$d \approx 15 \text{ miles}$$

47. Let l = length of the rectangle and w = width of the rectangle. Notice that
$$(l+w)^2 - (l-w)^2$$
$$= [(l+w) + (l-w)][(l+w) - (l-w)]$$
$$= (2l)(2w) = 4lw = 4A$$
So $A = \frac{1}{4}[(l+w)^2 - (l-w)^2]$

Since $(l-w)^2 \geq 0$, the largest area will occur when $l - w = 0$ or $l = w$; that is, when the rectangle is a square. But
$$1000 = 2l + 2w = 2(l+w)$$
$$500 = l + w = 2l$$
$$250 = l = w$$
The largest possible area is $250^2 = 62500$ sq ft.
A circular pool with circumference = 1000 feet yields the equation: $2\pi r = 1000 \Rightarrow r = \frac{500}{\pi}$
The area enclosed by the circular pool is:
$$A = \pi r^2 = \pi\left(\frac{500}{\pi}\right)^2$$
$$= \frac{500^2}{\pi} \approx 79577.47 \text{ square feet}$$
Thus, a circular pool will enclose the most area.

Section R.4

1. 4; 3

3. $x^3 - 8$

5. True

7. $2x^3$ Monomial; Variable: x; Coefficient: 2; Degree: 3

9. $\dfrac{8}{x}$ Not a monomial.

11. $-2xy^2$ Monomial; Variable: x, y; Coefficient: -2; Degree: 3

13. $\dfrac{8x}{y}$ Not a monomial

15. $x^2 + y^2$ Not a monomial.

17. $3x^2 - 5$ Polynomial; Degree: 2

19. 5 Polynomial; Degree: 0

21. $3x^2 - \dfrac{5}{x}$ Not a polynomial.

23. $2y^3 - \sqrt{2}$ Polynomial; Degree: 3

25. $\dfrac{x^2 + 5}{x^3 - 1}$ Not a polynomial.

27. $(x^2 + 4x + 5) + (3x - 3)$
$= x^2 + (4x + 3x) + (5 - 3)$
$= x^2 + 7x + 2$

29. $(x^3 - 2x^2 + 5x + 10) - (2x^2 - 4x + 3)$
$= x^3 - 2x^2 + 5x + 10 - 2x^2 + 4x - 3$
$= x^3 + (-2x^2 - 2x^2) + (5x + 4x) + (10 - 3)$
$= x^3 - 4x^2 + 9x + 7$

31. $(6x^5 + x^3 + x) + (5x^4 - x^3 + 3x^2)$
$= 6x^5 + 5x^4 + 3x^2 + x$

33. $(x^2 - 3x + 1) + 2(3x^2 + x - 4)$
$= x^2 - 3x + 1 + 6x^2 + 2x - 8$
$= 7x^2 - x - 7$

35. $6(x^3 + x^2 - 3) - 4(2x^3 - 3x^2)$
$= 6x^3 + 6x^2 - 18 - 8x^3 + 12x^2$
$= -2x^3 + 18x^2 - 18$

37. $(x^2 - x + 2) + (2x^2 - 3x + 5) - (x^2 + 1)$
$= x^2 - x + 2 + 2x^2 - 3x + 5 - x^2 - 1$
$= 2x^2 - 4x + 6$

39. $9(y^2 - 3y + 4) - 6(1 - y^2)$
$= 9y^2 - 27y + 36 - 6 + 6y^2$
$= 15y^2 - 27y + 30$

41. $x(x^2 + x - 4) = x^3 + x^2 - 4x$

43. $-2x^2(4x^3 + 5) = -8x^5 - 10x^2$

45. $(x+1)(x^2 + 2x - 4)$
$= x(x^2 + 2x - 4) + 1(x^2 + 2x - 4)$
$= x^3 + 2x^2 - 4x + x^2 + 2x - 4$
$= x^3 + 3x^2 - 2x - 4$

47. $(x+2)(x+4) = x^2 + 4x + 2x + 8$
$= x^2 + 6x + 8$

49. $(2x+5)(x+2) = 2x^2 + 4x + 5x + 10$
$= 2x^2 + 9x + 10$

51. $(x-4)(x+2) = x^2 + 2x - 4x - 8$
$= x^2 - 2x - 8$

53. $(x-3)(x-2) = x^2 - 2x - 3x + 6$
$= x^2 - 5x + 6$

55. $(2x+3)(x-2) = 2x^2 - 4x + 3x - 6$
$= 2x^2 - x - 6$

57. $(-2x+3)(x-4) = -2x^2 + 8x + 3x - 12$
$= -2x^2 + 11x - 12$

59. $(-x-2)(-2x-4) = 2x^2 + 4x + 4x + 8$
$= 2x^2 + 8x + 8$

61. $(x-2y)(x+y) = x^2 + xy - 2xy - 2y^2$
$= x^2 - xy - 2y^2$

63. $(-2x-3y)(3x+2y) = -6x^2 - 4xy - 9xy - 6y^2$
$= -6x^2 - 13xy - 6y^2$

65. $(x-7)(x+7) = x^2 - 7^2 = x^2 - 49$

67. $(2x+3)(2x-3) = (2x)^2 - 3^2 = 4x^2 - 9$

69. $(x+4)^2 = x^2 + 2 \cdot x \cdot 4 + 4^2 = x^2 + 8x + 16$

71. $(x-4)^2 = x^2 - 2 \cdot x \cdot 4 + 4^2 = x^2 - 8x + 16$

73. $(3x+4)(3x-4) = (3x)^2 - 4^2 = 9x^2 - 16$

75. $(2x-3)^2 = (2x)^2 - 2(2x)(3) + 3^2$
$= 4x^2 - 12x + 9$

77. $(x+y)(x-y) = (x)^2 - (y)^2 = x^2 - y^2$

79. $(3x+y)(3x-y) = (3x)^2 - (y)^2 = 9x^2 - y^2$

81. $(x+y)^2 = x^2 + 2xy + y^2$

83. $(x-2y)^2 = x^2 + 2(x \cdot (-2y)) + (2y)^2$
$= x^2 - 4xy + 4y^2$

85. $(x-2)^3 = x^3 - 3 \cdot x^2 \cdot 2 + 3 \cdot x \cdot 2^2 - 2^3$
$= x^3 - 6x^2 + 12x - 8$

87. $(2x+1)^3 = (2x)^3 + 3(2x)^2(1) + 3(2x) \cdot 1^2 + 1^3$
$= 8x^3 + 12x^2 + 6x + 1$

89.
$$\begin{array}{r} 4x^2 - 11x + 23 \\ x+2 \overline{) 4x^3 - 3x^2 + x + 1} \\ \underline{4x^3 + 8x^2} \\ -11x^2 + x \\ \underline{-11x^2 - 22x} \\ 23x + 1 \\ \underline{23x + 46} \\ -45 \end{array}$$

Check:
$(x+2)(4x^2 - 11x + 23) + (-45)$
$= 4x^3 - 11x^2 + 23x + 8x^2 - 22x + 46 - 45$
$= 4x^3 - 3x^2 + x + 1$
Quotient: $4x^2 - 11x + 23$; remainder: -45.

91.
$$\begin{array}{r} 4x - 3 \\ x^2 \overline{) 4x^3 - 3x^2 + x + 1} \\ \underline{4x^3} \\ -3x^2 + x + 1 \\ \underline{-3x^2} \\ x + 1 \end{array}$$

Check:
$(x^2)(4x-3) + (x+1) = 4x^3 - 3x^2 + x + 1$
The quotient is $4x - 3$; the remainder is $x + 1$.

93.
$$\begin{array}{r} 5x^2 - 13 \\ x^2 + 2 \overline{) 5x^4 + 0x^3 - 3x^2 + x + 1} \\ \underline{5x^4 + 10x^2} \\ -13x^2 + x + 1 \\ \underline{-13x^2 - 26} \\ x + 27 \end{array}$$

Check:
$(x^2 + 2)(5x^2 - 13) + (x + 27)$
$= 5x^4 + 10x^2 - 13x^2 - 26 + x + 27$
$= 5x^4 - 3x^2 + x + 1$
The quotient is $5x^2 - 13$; the remainder is $x + 27$.

95.
$$\begin{array}{r} 2x^2 \\ 2x^3-1 \overline{) 4x^5+0x^4+0x^3-3x^2+x+1} \\ \underline{4x^5 -2x^2 } \\ -x^2+x+1 \end{array}$$

Check:
$(2x^3-1)(2x^2)+(-x^2+x+1)$
$=4x^5-2x^2-x^2+x+1=4x^5-3x^2+x+1$

The quotient is $2x^2$; the remainder is $-x^2+x+1$.

97.
$$\begin{array}{r} x^2-2x+\frac{1}{2} \\ 2x^2+x+1 \overline{) 2x^4-3x^3+0x^2+x+1} \\ \underline{2x^4+x^3+x^2 } \\ -4x^3-x^2+x \\ \underline{-4x^3-2x^2-2x } \\ x^2+3x+1 \\ \underline{x^2+\frac{1}{2}x+\frac{1}{2}} \\ \frac{5}{2}x+\frac{1}{2} \end{array}$$

Check:
$\left(2x^2+x+1\right)\left(x^2-2x+\frac{1}{2}\right)+\frac{5}{2}x+\frac{1}{2}$
$=2x^4-4x^3+x^2+x^3-2x^2+\frac{1}{2}x$
$+x^2-2x+\frac{1}{2}+\frac{5}{2}x+\frac{1}{2}$
$=2x^4-3x^3+x+1$

The quotient is $x^2-2x+\frac{1}{2}$; the remainder is $\frac{5}{2}x+\frac{1}{2}$.

99.
$$\begin{array}{r} -4x^2-3x-3 \\ x-1 \overline{) -4x^3+x^2+0x-4} \\ \underline{-4x^3+4x^2 } \\ -3x^2 \\ \underline{-3x^2+3x } \\ -3x-4 \\ \underline{-3x+3} \\ -7 \end{array}$$

Check:
$(x-1)(-4x^2-3x-3)+(-7)$
$=-4x^3-3x^2-3x+4x^2+3x+3-7$
$=-4x^3+x^2-4$

The quotient is $-4x^2-3x-3$; the remainder is -7.

101.
$$\begin{array}{r} x^2-x-1 \\ x^2+x+1 \overline{) x^4+0x^3-x^2+0x+1} \\ \underline{x^4+x^3+x^2 } \\ -x^3-2x^2 \\ \underline{-x^3-x^2-x } \\ -x^2+x+1 \\ \underline{-x^2-x-1} \\ 2x+2 \end{array}$$

Check:
$(x^2+x+1)(x^2-x-1)+2x+2$
$=x^4+x^3+x^2-x^3-x^2-x-x^2-x$
$-1+2x+2$
$=x^4-x^2+1$

The quotient is x^2-x-1; the remainder is $2x+2$.

SSM: College Algebra EGU   Chapter R: Review

103.
$$\require{enclose}
\begin{array}{r}
x^2 + ax + a^2 \\
x-a \enclose{longdiv}{x^3 + 0x^2 + 0x - a^3} \\
\underline{x^3 - ax^2} \\
ax^2 \\
\underline{ax^2 - a^2x} \\
a^2x - a^3 \\
\underline{a^2x - a^3} \\
0
\end{array}$$

Check:
$(x-a)(x^2 + ax + a^2) + 0$
$= x^3 + ax^2 + a^2x - ax^2 - a^2x - a^3$
$= x^3 - a^3$

The quotient is $x^2 + ax + a^2$; the remainder is 0.

105. When we multiply polynomials $p_1(x)$ and $p_2(x)$, each term of $p_1(x)$ will be multiplied by each term of $p_2(x)$. So when the highest-powered term of $p_1(x)$ multiplies by the highest powered term of $p_2(x)$, the exponents on the variables in those terms will add according to the basic rules of exponents. Therefore, the highest powered term of the product polynomial will have degree equal to the sum of the degrees of $p_1(x)$ and $p_2(x)$.

107. When we add two polynomials $p_1(x)$ and $p_2(x)$, where the degree of $p_1(x)$ = the degree of $p_2(x)$, the new polynomial will have degree ≤ the degree of $p_1(x)$ and $p_2(x)$.

109. Answers will vary.

Section R.5

1. $3x(x-2)(x+2)$

3. True; $x^2 + 4$ is prime over the set of real numbers.

5. $3x + 6 = 3(x + 2)$

7. $ax^2 + a = a(x^2 + 1)$

9. $x^3 + x^2 + x = x(x^2 + x + 1)$

11. $2x^2 - 2x = 2x(x - 1)$

13. $3x^2y - 6xy^2 + 12xy = 3xy(x - 2y + 4)$

15. $x^2 - 1 = x^2 - 1^2 = (x-1)(x+1)$

17. $4x^2 - 1 = (2x)^2 - 1^2 = (2x-1)(2x+1)$

19. $x^2 - 16 = x^2 - 4^2 = (x-4)(x+4)$

21. $25x^2 - 4 = (5x-2)(5x+2)$

23. $x^2 + 2x + 1 = (x+1)^2$

25. $x^2 + 4x + 4 = (x+2)^2$

27. $x^2 - 10x + 25 = (x-5)^2$

29. $4x^2 + 4x + 1 = (2x+1)^2$

31. $16x^2 + 8x + 1 = (4x+1)^2$

33. $x^3 - 27 = x^3 - 3^3 = (x-3)(x^2 + 3x + 9)$

35. $x^3 + 27 = x^3 + 3^3 = (x+3)(x^2 - 3x + 9)$

37. $8x^3 + 27 = (2x)^3 + 3^3$
$= (2x+3)(4x^2 - 6x + 9)$

39. $x^2 + 5x + 6 = (x+2)(x+3)$

41. $x^2 + 7x + 6 = (x+6)(x+1)$

43. $x^2 + 7x + 10 = (x+2)(x+5)$

45. $x^2 - 10x + 16 = (x-2)(x-8)$

47. $x^2 - 7x - 8 = (x+1)(x-8)$

49. $x^2 + 7x - 8 = (x+8)(x-1)$

Chapter R: *Review*   *SSM:* College Algebra EGU

51. $2x^2 + 4x + 3x + 6 = 2x(x+2) + 3(x+2)$
$= (x+2)(2x+3)$

53. $2x^2 - 4x + x - 2 = 2x(x-2) + 1(x-2)$
$= (x-2)(2x+1)$

55. $6x^2 + 9x + 4x + 6 = 3x(2x+3) + 2(2x+3)$
$= (2x+3)(3x+2)$

57. $3x^2 + 4x + 1 = (3x+1)(x+1)$

59. $2z^2 + 5z + 3 = (2z+3)(z+1)$

61. $3x^2 + 2x - 8 = (3x-4)(x+2)$

63. $3x^2 - 2x - 8 = (3x+4)(x-2)$

65. $3x^2 + 14x + 8 = (3x+2)(x+4)$

67. $3x^2 + 10x - 8 = (3x-2)(x+4)$

69. $x^2 - 36 = (x-6)(x+6)$

71. $2 - 8x^2 = 2(1-4x^2) = 2(1-2x)(1+2x)$

73. $x^2 + 7x + 10 = (x+2)(x+5)$

75. $x^2 - 10x + 21 = (x-7)(x-3)$

77. $4x^2 - 8x + 32 = 4(x^2 - 2x + 8)$

79. $x^2 + 4x + 16$ is prime over the reals because there are no factors of 16 whose sum is 4.

81. $15 + 2x - x^2 = -(x^2 - 2x - 15) = -(x-5)(x+3)$

83. $3x^2 - 12x - 36 = 3(x^2 - 4x - 12) = 3(x-6)(x+2)$

85. $y^4 + 11y^3 + 30y^2 = y^2(y^2 + 11y + 30)$
$= y^2(y+5)(y+6)$

87. $4x^2 + 12x + 9 = (2x+3)^2$

89. $6x^2 + 8x + 2 = 2(3x^2 + 4x + 1) = 2(3x+1)(x+1)$

91. $x^4 - 81 = (x^2)^2 - 9^2 = (x^2 - 9)(x^2 + 9)$
$= (x-3)(x+3)(x^2+9)$

93. $x^6 - 2x^3 + 1 = (x^3 - 1)^2$
$= \left[(x-1)(x^2+x+1)\right]^2$
$= (x-1)^2(x^2+x+1)^2$

95. $x^7 - x^5 = x^5(x^2 - 1) = x^5(x-1)(x+1)$

97. $16x^2 + 24x + 9 = (4x+3)^2$

99. $5 + 16x - 16x^2 = -(16x^2 - 16x - 5)$
$= -(4x-5)(4x+1)$

101. $4y^2 - 16y + 15 = (2y-5)(2y-3)$

103. $1 - 8x^2 - 9x^4 = -(9x^4 + 8x^2 - 1)$
$= -(9x^2 - 1)(x^2 + 1)$
$= -(3x-1)(3x+1)(x^2+1)$

105. $x(x+3) - 6(x+3) = (x+3)(x-6)$

107. $(x+2)^2 - 5(x+2) = (x+2)[(x+2) - 5]$
$= (x+2)(x-3)$

109. $(3x-2)^3 - 27$
$= (3x-2)^3 - 3^3$
$= [(3x-2) - 3][(3x-2)^2 + 3(3x-2) + 9]$
$= (3x-5)(9x^2 - 12x + 4 + 9x - 6 + 9)$
$= (3x-5)(9x^2 - 3x + 7)$

111. $3(x^2 + 10x + 25) - 4(x+5)$
$= 3(x+5)^2 - 4(x+5)$
$= (x+5)[3(x+5) - 4]$
$= (x+5)(3x + 15 - 4)$
$= (x+5)(3x + 11)$

113. $x^3 + 2x^2 - x - 2 = x^2(x+2) - 1(x+2)$
$= (x+2)(x^2 - 1)$
$= (x+2)(x-1)(x+1)$

115. $x^4 - x^3 + x - 1 = x^3(x-1) + 1(x-1)$
$= (x-1)(x^3 + 1)$
$= (x-1)(x+1)(x^2 - x + 1)$

117. $2(3x+4)^2 + (2x+3) \cdot 2(3x+4) \cdot 3$
$= 2(3x+4)\big((3x+4) + (2x+3) \cdot 3\big)$
$= 2(3x+4)(3x+4+6x+9)$
$= 2(3x+4)(9x+13)$

119. $2x(2x+5) + x^2 \cdot 2 = 2x\big((2x+5) + x\big)$
$= 2x(2x+5+x)$
$= 2x(3x+5)$

121. $2(x+3)(x-2)^3 + (x+3)^2 \cdot 3(x-2)^2$
$= (x+3)(x-2)^2\big(2(x-2) + (x+3) \cdot 3\big)$
$= (x+3)(x-2)^2(2x-4+3x+9)$
$= (x+3)(x-2)^2(5x+5)$
$= 5(x+3)(x-2)^2(x+1)$

123. $(4x-3)^2 + x \cdot 2(4x-3) \cdot 4$
$= (4x-3)\big((4x-3) + 8x\big)$
$= (4x-3)(4x-3+8x)$
$= (4x-3)(12x-3)$
$= 3(4x-3)(4x-1)$

125. $2(3x-5) \cdot 3(2x+1)^3 + (3x-5)^2 \cdot 3(2x+1)^2 \cdot 2$
$= 6(3x-5)(2x+1)^2\big((2x+1) + (3x-5)\big)$
$= 6(3x-5)(2x+1)^2(2x+1+3x-5)$
$= 6(3x-5)(2x+1)^2(5x-4)$

127. Factors of 4: 1, 4   2, 2   −1, −4   −2, −2
Sum: 5   4   −5   −4
None of the sums of the factors is 0, so $x^2 + 4$ is prime.

129. Answers will vary.

Section R.6

1. quotient; divisor; remainder

3. True

5. $2 \overline{) \; 1 \;\; -1 \;\;\; 2 \;\;\; 4}$
$\phantom{2 \overline{)} \;\;\;\;\;\;\;} 2 \;\;\; 2 \;\;\; 8$
$\phantom{2 \overline{)}} \;\; 1 \;\;\;\; 1 \;\;\; 4 \;\; 12$

Quotient: $x^2 + x + 4$
Remainder: 12

7. $3 \overline{) \; 3 \;\;\; 2 \;\; -1 \;\;\; 3}$
$\phantom{3 \overline{)} \;\;\;\;\;\;\;} 9 \;\; 33 \;\; 96$
$\phantom{3 \overline{)}} \;\; 3 \;\; 11 \;\; 32 \;\; 99$

Quotient: $3x^2 + 11x + 32$
Remainder: 99

9. $-3 \overline{) \; 1 \;\;\; 0 \;\; -4 \;\;\; 0 \;\;\; 1 \;\;\; 0}$
$\phantom{-3 \overline{)} \;\;\;\;\;\;\;} -3 \;\;\; 9 \;\; -15 \;\; 45 \;\; -138$
$\phantom{-3 \overline{)}} \;\; 1 \;\; -3 \;\;\; 5 \;\; -15 \;\; 46 \;\; -138$

Quotient: $x^4 - 3x^3 + 5x^2 - 15x + 46$
Remainder: -138

11. $1 \overline{) \; 4 \;\;\; 0 \;\; -3 \;\;\; 0 \;\;\; 1 \;\;\; 0 \;\;\; 5}$
$\phantom{1 \overline{)} \;\;\;\;\;\;\;} 4 \;\;\; 4 \;\;\; 1 \;\;\; 1 \;\;\; 2 \;\;\; 2$
$\phantom{1 \overline{)}} \;\; 4 \;\;\; 4 \;\;\; 1 \;\;\; 1 \;\;\; 2 \;\;\; 2 \;\;\; 7$

Quotient: $4x^5 + 4x^4 + x^3 + x^2 + 2x + 2$
Remainder: 7

13. $-1.1 \overline{) \; 0.1 \;\;\; 0 \;\;\; 0.2 \;\;\; 0}$
$\phantom{-1.1 \overline{)} \;\;\;\;\;\;\;} -0.11 \;\; 0.121 \;\; -0.3531$
$\phantom{-1.1 \overline{)}} \;\; 0.1 \;\; -0.11 \;\; 0.321 \;\; -0.3531$

Quotient: $0.1x^2 - 0.11x + 0.321$
Remainder: -0.3531

15. $1 \overline{) \; 1 \;\;\; 0 \;\;\; 0 \;\;\; 0 \;\;\; 0 \;\; -1}$
$\phantom{1 \overline{)} \;\;\;\;\;\;\;} 1 \;\;\; 1 \;\;\; 1 \;\;\; 1 \;\;\; 1$
$\phantom{1 \overline{)}} \;\; 1 \;\;\; 1 \;\;\; 1 \;\;\; 1 \;\;\; 1 \;\;\; 0$

Quotient: $x^4 + x^3 + x^2 + x + 1$
Remainder: 0

17.
$$\begin{array}{r|rrrr} 2) & 4 & -3 & -8 & 4 \\ & & 8 & 10 & 4 \\ \hline & 4 & 5 & 2 & 8 \end{array}$$

Remainder $= 8 \neq 0$; therefore $x - 2$ is not a factor of $4x^3 - 3x^2 - 8x + 4$.

19.
$$\begin{array}{r|rrrrr} 2) & 3 & -6 & 0 & -5 & 10 \\ & & 6 & 0 & 0 & -10 \\ \hline & 3 & 0 & 0 & -5 & 0 \end{array}$$

Remainder $= 0$; therefore $x - 2$ is a factor of $3x^4 - 6x^3 - 5x + 10$.

21.
$$\begin{array}{r|rrrrrrr} -3) & 3 & 0 & 0 & 82 & 0 & 0 & 27 \\ & & -9 & 27 & -81 & -3 & 9 & -27 \\ \hline & 3 & -9 & 27 & 1 & -3 & 9 & 0 \end{array}$$

Remainder $= 0$; therefore $x + 3$ is a factor of $3x^6 + 82x^3 + 27$.

23.
$$\begin{array}{r|rrrrrrr} -4) & 4 & 0 & -64 & 0 & 1 & 0 & -15 \\ & & -16 & 64 & 0 & 0 & -4 & 16 \\ \hline & 4 & -16 & 0 & 0 & 1 & -4 & 1 \end{array}$$

Remainder $= 1 \neq 0$; therefore $x + 3$ is not a factor of $4x^6 - 64x^4 + x^2 - 15$.

25.
$$\begin{array}{r|rrrrr} \frac{1}{2}) & 2 & -1 & 0 & 2 & -1 \\ & & 1 & 0 & 0 & 1 \\ \hline & 2 & 0 & 0 & 2 & 0 \end{array}$$

Remainder $= 0$; therefore $x - \frac{1}{2}$ is a factor of $2x^4 - x^3 + 2x - 1$.

27.
$$\begin{array}{r|rrrr} -2) & 1 & -2 & 3 & 5 \\ & & -2 & 8 & -22 \\ \hline & 1 & -4 & 11 & -17 \end{array}$$

$\frac{x^3 - 2x^2 + 3x + 5}{x + 2} = x^2 - 4x + 11 + \frac{-17}{x + 2}$

$a + b + c + d = 1 - 4 + 11 - 17 = -9$

Section R.7

1. lowest terms

3. True; $\frac{2x^3 - 4x}{x - 2} = \frac{2x(x^2 - 2)}{x - 2}$

5. $\frac{3x + 9}{x^2 - 9} = \frac{3(x + 3)}{(x - 3)(x + 3)} = \frac{3}{x - 3}$

7. $\frac{x^2 - 2x}{3x - 6} = \frac{x(x - 2)}{3(x - 2)} = \frac{x}{3}$

9. $\frac{24x^2}{12x^2 - 6x} = \frac{24x^2}{6x(2x - 1)} = \frac{4x}{2x - 1}$

11. $\frac{y^2 - 25}{2y^2 - 8y - 10} = \frac{(y + 5)(y - 5)}{2(y^2 - 4y - 5)}$

$= \frac{(y + 5)(y - 5)}{2(y - 5)(y + 1)}$

$= \frac{y + 5}{2(y + 1)}$

13. $\frac{x^2 + 4x - 5}{x^2 - 2x + 1} = \frac{(x + 5)(x - 1)}{(x - 1)(x - 1)} = \frac{x + 5}{x - 1}$

15. $\frac{x^2 + 5x - 14}{2 - x} = \frac{(x + 7)(x - 2)}{2 - x}$

$= \frac{(x + 7)(x - 2)}{(-1)(-2 + x)}$

$= \frac{(x + 7)(x - 2)}{(-1)(x - 2)}$

$= -(x + 7)$

17. $\frac{3x + 6}{5x^2} \cdot \frac{x}{x^2 - 4} = \frac{3(x + 2)}{5x^2} \cdot \frac{x}{(x - 2)(x + 2)}$

$= \frac{3}{5x(x - 2)}$

19. $\frac{4x^2}{x^2 - 16} \cdot \frac{x - 4}{2x} = \frac{4x^2}{(x - 4)(x + 4)} \cdot \frac{x - 4}{2x} = \frac{2x}{x + 4}$

21. $\dfrac{4x-8}{-3x} \cdot \dfrac{12}{12-6x} = \dfrac{4(x-2)}{-3x} \cdot \dfrac{12}{6(2-x)}$

$\phantom{\dfrac{4x-8}{-3x} \cdot \dfrac{12}{12-6x}} = \dfrac{4(x-2)}{-3x} \cdot \dfrac{2}{(-1)(x-2)}$

$\phantom{\dfrac{4x-8}{-3x} \cdot \dfrac{12}{12-6x}} = \dfrac{8}{3x}$

23. $\dfrac{x^2-3x-10}{x^2+2x-35} \cdot \dfrac{x^2+4x-21}{x^2+9x+14}$

$= \dfrac{(x-5)(x+2)}{(x+7)(x-5)} \cdot \dfrac{(x+7)(x-3)}{(x+7)(x+2)}$

$= \dfrac{x-3}{x+7}$

25. $\dfrac{\left(\dfrac{6x}{x^2-4}\right)}{\left(\dfrac{3x-9}{2x+4}\right)} = \dfrac{6x}{x^2-4} \cdot \dfrac{2x+4}{3x-9}$

$= \dfrac{6x}{(x-2)(x+2)} \cdot \dfrac{2(x+2)}{3(x-3)}$

$= \dfrac{4x}{(x-2)(x-3)}$

27. $\dfrac{\left(\dfrac{8x}{x^2-1}\right)}{\left(\dfrac{10x}{x+1}\right)} = \dfrac{8x}{x^2-1} \cdot \dfrac{x+1}{10x}$

$= \dfrac{8x}{(x-1)(x+1)} \cdot \dfrac{x+1}{10x}$

$= \dfrac{4}{5(x-1)}$

29. $\dfrac{\left(\dfrac{4-x}{4+x}\right)}{\left(\dfrac{4x}{x^2-16}\right)} = \dfrac{4-x}{4+x} \cdot \dfrac{x^2-16}{4x}$

$= \dfrac{4-x}{4+x} \cdot \dfrac{(x+4)(x-4)}{4x}$

$= \dfrac{(4-x)(x-4)}{4x}$

$= -\dfrac{(x-4)^2}{4x}$

31. $\dfrac{\left(\dfrac{x^2+7x+12}{x^2-7x+12}\right)}{\left(\dfrac{x^2+x-12}{x^2-x-12}\right)} = \dfrac{x^2+7x+12}{x^2-7x+12} \cdot \dfrac{x^2-x-12}{x^2+x-12}$

$= \dfrac{(x+3)(x+4)}{(x-3)(x-4)} \cdot \dfrac{(x-4)(x+3)}{(x+4)(x-3)}$

$= \dfrac{(x+3)^2}{(x-3)^2}$

33. $\dfrac{\left(\dfrac{2x^2-x-28}{3x^2-x-2}\right)}{\left(\dfrac{4x^2+16x+7}{3x^2+11x+6}\right)}$

$= \dfrac{2x^2-x-28}{3x^2-x-2} \cdot \dfrac{3x^2+11x+6}{4x^2+16x+7}$

$= \dfrac{(2x+7)(x-4)}{(3x+2)(x-1)} \cdot \dfrac{(3x+2)(x+3)}{(2x+7)(2x+1)}$

$= \dfrac{(x-4)(x+3)}{(x-1)(2x+1)}$

35. $\dfrac{x}{2} + \dfrac{5}{2} = \dfrac{x+5}{2}$

37. $\dfrac{x^2}{2x-3} - \dfrac{4}{2x-3} = \dfrac{x^2-4}{2x-3} = \dfrac{(x+2)(x-2)}{2x-3}$

39. $\dfrac{x+1}{x-3} + \dfrac{2x-3}{x-3} = \dfrac{x+1+2x-3}{x-3} = \dfrac{3x-2}{x-3}$

41. $\dfrac{3x+5}{2x-1} - \dfrac{2x-4}{2x-1} = \dfrac{(3x+5)-(2x-4)}{2x-1}$

$= \dfrac{3x+5-2x+4}{2x-1}$

$= \dfrac{x+9}{2x-1}$

43. $\dfrac{4}{x-2} + \dfrac{x}{2-x} = \dfrac{4}{x-2} - \dfrac{x}{x-2} = \dfrac{4-x}{x-2}$

45. $\dfrac{4}{x-1} - \dfrac{2}{x+2} = \dfrac{4(x+2)}{(x-1)(x+2)} - \dfrac{2(x-1)}{(x+2)(x-1)}$

$= \dfrac{4x+8-2x+2}{(x+2)(x-1)}$

$= \dfrac{2x+10}{(x+2)(x-1)}$

$= \dfrac{2(x+5)}{(x+2)(x-1)}$

47. $\dfrac{x}{x+1} + \dfrac{2x-3}{x-1} = \dfrac{x(x-1)}{(x+1)(x-1)} + \dfrac{(2x-3)(x+1)}{(x-1)(x+1)}$

$= \dfrac{x^2 - x + 2x^2 - x - 3}{(x-1)(x+1)}$

$= \dfrac{3x^2 - 2x - 3}{(x-1)(x+1)}$

49. $\dfrac{x-3}{x+2} - \dfrac{x+4}{x-2} = \dfrac{(x-3)(x-2)}{(x+2)(x-2)} - \dfrac{(x+4)(x+2)}{(x-2)(x+2)}$

$= \dfrac{x^2 - 5x + 6 - (x^2 + 6x + 8)}{(x+2)(x-2)}$

$= \dfrac{x^2 - 5x + 6 - x^2 - 6x - 8}{(x+2)(x-2)}$

$= \dfrac{-11x - 2}{(x+2)(x-2)}$

51. $\dfrac{x}{x^2-4} + \dfrac{1}{x} = \dfrac{x^2 + x^2 - 4}{(x)(x^2-4)}$

$= \dfrac{2x^2 - 4}{(x)(x^2-4)}$

$= \dfrac{2(x^2 - 2)}{(x)(x-2)(x+2)}$

53. $x^2 - 4 = (x+2)(x-2)$

$x^2 - x - 2 = (x+1)(x-2)$

$\therefore$ LCM is $(x+2)(x-2)(x+1)$

55. $x^3 - x = x(x^2 - 1) = x(x+1)(x-1)$

$x^2 - x = x(x-1)$

$\therefore$ LCM is $x(x+1)(x-1)$

57. $4x^3 - 4x^2 + x = x(4x^2 - 4x + 1)$

$= x(2x-1)(2x-1)$

$2x^3 - x^2 = x^2(2x-1)$

x^3

$\therefore$ LCM is $x^3(2x-1)^2$

59. $x^3 - x = x(x^2 - 1) = x(x+1)(x-1)$

$x^3 - 2x^2 + x = x(x^2 - 2x + 1) = x(x-1)^2$

$x^3 - 1 = (x-1)(x^2 + x + 1)$

$\therefore$ LCM is $x(x+1)(x-1)^2(x^2+x+1)$

61. $\dfrac{x}{x^2 - 7x + 6} - \dfrac{x}{x^2 - 2x - 24}$

$= \dfrac{x}{(x-6)(x-1)} - \dfrac{x}{(x-6)(x+4)}$

$= \dfrac{x(x+4)}{(x-6)(x-1)(x+4)} - \dfrac{x(x-1)}{(x-6)(x+4)(x-1)}$

$= \dfrac{x^2 + 4x - x^2 + x}{(x-6)(x+4)(x-1)} = \dfrac{5x}{(x-6)(x+4)(x-1)}$

63. $\dfrac{4x}{x^2 - 4} - \dfrac{2}{x^2 + x - 6}$

$= \dfrac{4x}{(x-2)(x+2)} - \dfrac{2}{(x+3)(x-2)}$

$= \dfrac{4x(x+3)}{(x-2)(x+2)(x+3)} - \dfrac{2(x+2)}{(x+3)(x-2)(x+2)}$

$= \dfrac{4x^2 + 12x - 2x - 4}{(x-2)(x+2)(x+3)}$

$= \dfrac{4x^2 + 10x - 4}{(x-2)(x+2)(x+3)}$

$= \dfrac{2(2x^2 + 5x - 2)}{(x-2)(x+2)(x+3)}$

65. $\dfrac{3}{(x-1)^2(x+1)} + \dfrac{2}{(x-1)(x+1)^2}$

$= \dfrac{3(x+1) + 2(x-1)}{(x-1)^2(x+1)^2}$

$= \dfrac{3x + 3 + 2x - 2}{(x-1)^2(x+1)^2}$

$= \dfrac{5x + 1}{(x-1)^2(x+1)^2}$

67. $\dfrac{x+4}{x^2-x-2} - \dfrac{2x+3}{x^2+2x-8}$

$= \dfrac{x+4}{(x-2)(x+1)} - \dfrac{2x+3}{(x+4)(x-2)}$

$= \dfrac{(x+4)(x+4)}{(x-2)(x+1)(x+4)} - \dfrac{(2x+3)(x+1)}{(x+4)(x-2)(x+1)}$

$= \dfrac{x^2+8x+16-(2x^2+5x+3)}{(x-2)(x+1)(x+4)}$

$= \dfrac{-x^2+3x+13}{(x-2)(x+1)(x+4)}$

69. $\dfrac{1}{x} - \dfrac{2}{x^2+x} + \dfrac{3}{x^3-x^2}$

$= \dfrac{1}{x} - \dfrac{2}{x(x+1)} + \dfrac{3}{x^2(x-1)}$

$= \dfrac{x(x+1)(x-1) - 2x(x-1) + 3(x+1)}{x^2(x+1)(x-1)}$

$= \dfrac{x(x^2-1) - 2x^2 + 2x + 3x + 3}{x^2(x+1)(x-1)}$

$= \dfrac{x^3 - x - 2x^2 + 5x + 3}{x^2(x+1)(x-1)}$

$= \dfrac{x^3 - 2x^2 + 4x + 3}{x^2(x+1)(x-1)}$

71. $\dfrac{1}{h}\left(\dfrac{1}{x+h} - \dfrac{1}{x}\right) = \dfrac{1}{h}\left(\dfrac{1 \cdot x}{(x+h)x} - \dfrac{1(x+h)}{x(x+h)}\right)$

$= \dfrac{1}{h}\left(\dfrac{x-x-h}{x(x+h)}\right)$

$= \dfrac{-h}{hx(x+h)}$

$= \dfrac{-1}{x(x+h)}$

73. $\dfrac{1+\dfrac{1}{x}}{1-\dfrac{1}{x}} = \dfrac{\left(\dfrac{x}{x}+\dfrac{1}{x}\right)}{\left(\dfrac{x}{x}-\dfrac{1}{x}\right)} = \dfrac{\left(\dfrac{x+1}{x}\right)}{\left(\dfrac{x-1}{x}\right)} = \dfrac{x+1}{x} \cdot \dfrac{x}{x-1} = \dfrac{x+1}{x-1}$

75. $\dfrac{2-\dfrac{x+1}{x}}{3+\dfrac{x-1}{x+1}} = \dfrac{\dfrac{2x}{x}-\dfrac{x+1}{x}}{\dfrac{3(x+1)}{x+1}+\dfrac{x-1}{x+1}} = \dfrac{\dfrac{2x-x-1}{x}}{\dfrac{3x+3+x-1}{x+1}}$

$= \dfrac{\dfrac{x-1}{x}}{\dfrac{4x+2}{x+1}} = \dfrac{x-1}{x} \cdot \dfrac{x+1}{2(2x+1)}$

$= \dfrac{(x-1)(x+1)}{2x(2x+1)}$

77. $\dfrac{\left(\dfrac{x+4}{x-2} - \dfrac{x-3}{x+1}\right)}{x+1}$

$= \dfrac{\left(\dfrac{(x+4)(x+1)}{(x-2)(x+1)} - \dfrac{(x-3)(x-2)}{(x+1)(x-2)}\right)}{x+1}$

$= \dfrac{\left(\dfrac{x^2+5x+4-(x^2-5x+6)}{(x-2)(x+1)}\right)}{x+1}$

$= \dfrac{10x-2}{(x-2)(x+1)} \cdot \dfrac{1}{x+1}$

$= \dfrac{2(5x-1)}{(x-2)(x+1)^2}$

79. $\dfrac{\left(\dfrac{x-2}{x+2}+\dfrac{x-1}{x+1}\right)}{\left(\dfrac{x}{x+1}-\dfrac{2x-3}{x}\right)}$

$= \dfrac{\left(\dfrac{(x-2)(x+1)}{(x+2)(x+1)}+\dfrac{(x-1)(x+2)}{(x+1)(x+2)}\right)}{\left(\dfrac{x^2}{(x+1)(x)}-\dfrac{(2x-3)(x+1)}{x(x+1)}\right)}$

$= \dfrac{\left(\dfrac{x^2-x-2+x^2+x-2}{(x+2)(x+1)}\right)}{\left(\dfrac{x^2-(2x^2-x-3)}{x(x+1)}\right)}$

$= \dfrac{\left(\dfrac{2x^2-4}{(x+2)(x+1)}\right)}{\left(\dfrac{-x^2+x+3}{x(x+1)}\right)}$

$= \dfrac{2(x^2-2)}{(x+2)(x+1)} \cdot \dfrac{x(x+1)}{-(x^2-x-3)}$

$= \dfrac{2x(x^2-2)}{-(x+2)(x^2-x-3)}$

$= \dfrac{-2x(x^2-2)}{(x+2)(x^2-x-3)}$

81. $1-\dfrac{1}{\left(1-\dfrac{1}{x}\right)} = 1-\dfrac{1}{\left(\dfrac{x-1}{x}\right)}$

$= 1-\dfrac{x}{x-1}$

$= \dfrac{x-1-x}{x-1}$

$= \dfrac{-1}{x-1}$

83. $\dfrac{(2x+3)\cdot 3-(3x-5)\cdot 2}{(3x-5)^2} = \dfrac{6x+9-6x+10}{(3x-5)^2}$

$= \dfrac{19}{(3x-5)^2}$

85. $\dfrac{x\cdot 2x-(x^2+1)\cdot 1}{(x^2+1)^2} = \dfrac{2x^2-x^2-1}{(x^2+1)^2} = \dfrac{x^2-1}{(x^2+1)^2}$

$= \dfrac{(x-1)(x+1)}{(x^2+1)^2}$

87. $\dfrac{(3x+1)\cdot 2x-x^2\cdot 3}{(3x+1)^2} = \dfrac{6x^2+2x-3x^2}{(3x+1)^2} = \dfrac{3x^2+2x}{(3x+1)^2}$

$= \dfrac{x(3x+2)}{(3x+1)^2}$

89. $\dfrac{(x^2+1)\cdot 3-(3x+4)\cdot 2x}{(x^2+1)^2} = \dfrac{3x^2+3-6x^2-8x}{(x^2+1)^2}$

$= \dfrac{-3x^2-8x+3}{(x^2+1)^2}$

$= \dfrac{-(3x^2+8x-3)}{(x^2+1)^2}$

$= -\dfrac{(3x-1)(x+3)}{(x^2+1)^2}$

91. $\dfrac{1}{f} = (n-1)\left(\dfrac{1}{R_1}+\dfrac{1}{R_2}\right)$

$\dfrac{1}{f} = (n-1)\left(\dfrac{R_2+R_1}{R_1\cdot R_2}\right)$

$\dfrac{R_1\cdot R_2}{f} = (n-1)(R_2+R_1)$

$\dfrac{f}{R_1\cdot R_2} = \dfrac{1}{(n-1)(R_2+R_1)}$

$f = \dfrac{R_1\cdot R_2}{(n-1)(R_2+R_1)}$

$f = \dfrac{0.1(0.2)}{(1.5-1)(0.2+0.1)}$

$= \dfrac{0.02}{0.5(0.3)} = \dfrac{0.02}{0.15} = \dfrac{2}{15}$ meters

93. $1 + \dfrac{1}{x} = \dfrac{x+1}{x} \Rightarrow a = 1, b = 1, c = 0$

$1 + \dfrac{1}{1 + \dfrac{1}{x}} = 1 + \dfrac{1}{\left(\dfrac{x+1}{x}\right)} = 1 + \dfrac{x}{x+1}$

$= \dfrac{x+1+x}{x+1} = \dfrac{2x+1}{x+1}$

$\Rightarrow a = 2, b = 1, c = 1$

$1 + \dfrac{1}{1 + \dfrac{1}{1 + \dfrac{1}{x}}} = 1 + \dfrac{1}{\left(\dfrac{2x+1}{x+1}\right)} = 1 + \dfrac{x+1}{2x+1}$

$= \dfrac{2x+1+x+1}{2x+1} = \dfrac{3x+2}{2x+1}$

$\Rightarrow a = 3, b = 2, c = 1$

$1 + \dfrac{1}{1 + \dfrac{1}{1 + \dfrac{1}{1 + \dfrac{1}{x}}}} = 1 + \dfrac{1}{\left(\dfrac{3x+2}{2x+1}\right)} = 1 + \dfrac{2x+1}{3x+2}$

$= \dfrac{3x+2+2x+1}{3x+2} = \dfrac{5x+3}{3x+2}$

$\Rightarrow a = 5, b = 3, c = 2$

If we continue this process, the values of a, b and c produce the following sequences:
$a: 1, 2, 3, 5, 8, 13, 21, \ldots$
$b: 1, 1, 2, 3, 5, 8, 13, 21, \ldots$
$c: 0, 1, 1, 2, 3, 5, 8, 13, 21, \ldots$

In each case we have a *Fibonacci Sequence*, where the next value in the list is obtained from the sum of the previous 2 values in the list.

95. Answers will vary.

Section R.8

1. 9; -9

3. index

5. cube root

7. $\sqrt[3]{27} = \sqrt[3]{3^3} = 3$

9. $\sqrt[3]{-8} = \sqrt[3]{(-2)^3} = -2$

11. $\sqrt{8} = \sqrt{4 \cdot 2} = 2\sqrt{2}$

13. $\sqrt[3]{-8x^4} = \sqrt[3]{-8x^3 \cdot x} = -2x\sqrt[3]{x}$

15. $\sqrt[4]{x^{12} y^8} = \sqrt[4]{(x^3)^4 (y^2)^4} = x^3 y^2$

17. $\sqrt[4]{\dfrac{x^9 y^7}{xy^3}} = \sqrt[4]{x^8 y^4} = x^2 y$

19. $\sqrt{36x} = 6\sqrt{x}$

21. $\sqrt{3x^2}\sqrt{12x} = \sqrt{36x^2 \cdot x} = 6x\sqrt{x}$

23. $\left(\sqrt{5}\sqrt[3]{9}\right)^2 = \left(\sqrt{5}\right)^2 \left(\sqrt[3]{9}\right)^2$
$= 5 \cdot \sqrt[3]{9^2}$
$= 5\sqrt[3]{81}$
$= 5 \cdot 3\sqrt[3]{3}$
$= 15\sqrt[3]{3}$

25. $(3\sqrt{6})(2\sqrt{2}) = 6\sqrt{12} = 6\sqrt{4 \cdot 3} = 12\sqrt{3}$

27. $3\sqrt{2} + 4\sqrt{2} = (3+4)\sqrt{2} = 7\sqrt{2}$

29. $-\sqrt{18} + 2\sqrt{8} = -\sqrt{9 \cdot 2} + 2\sqrt{4 \cdot 2}$
$= -3\sqrt{2} + 4\sqrt{2}$
$= (-3+4)\sqrt{2}$
$= \sqrt{2}$

31. $(\sqrt{3} + 3)(\sqrt{3} - 1) = (\sqrt{3})^2 + 3\sqrt{3} - \sqrt{3} - 3$
$= 3 + 2\sqrt{3} - 3$
$= 2\sqrt{3}$

33. $5\sqrt[3]{2} - 2\sqrt[3]{54} = 5\sqrt[3]{2} - 2 \cdot 3\sqrt[3]{2}$
$= 5\sqrt[3]{2} - 6\sqrt[3]{2}$
$= (5-6)\sqrt[3]{2}$
$= -\sqrt[3]{2}$

35. $(\sqrt{x} - 1)^2 = (\sqrt{x})^2 - 2\sqrt{x} + 1$
$= x - 2\sqrt{x} + 1$

Chapter R: *Review*   **SSM:** College Algebra EGU

37. $\sqrt[3]{16x^4} - \sqrt[3]{2x} = \sqrt[3]{8x^3 \cdot 2x} - \sqrt[3]{2x}$
$= 2x\sqrt[3]{2x} - \sqrt[3]{2x}$
$= (2x-1)\sqrt[3]{2x}$

39. $\sqrt{8x^3} - 3\sqrt{50x} = \sqrt{4x^2 \cdot 2x} - 3\sqrt{25 \cdot 2x}$
$= 2x\sqrt{2x} - 15\sqrt{2x}$
$= (2x-15)\sqrt{2x}$

41. $\sqrt[3]{16x^4 y} - 3x\sqrt[3]{2xy} + 5\sqrt[3]{-2xy^4}$
$= \sqrt[3]{8x^3 \cdot 2xy} - 3x\sqrt[3]{2xy} + 5\sqrt[3]{-y^3 \cdot 2xy}$
$= 2x\sqrt[3]{2xy} - 3x\sqrt[3]{2xy} - 5y\sqrt[3]{2xy}$
$= (2x - 3x - 5y)\sqrt[3]{2xy}$
$= (-x - 5y)\sqrt[3]{2xy}$

43. $\dfrac{1}{\sqrt{2}} = \dfrac{1}{\sqrt{2}} \cdot \dfrac{\sqrt{2}}{\sqrt{2}} = \dfrac{\sqrt{2}}{2}$

45. $\dfrac{-\sqrt{3}}{\sqrt{5}} = \dfrac{-\sqrt{3}}{\sqrt{5}} \cdot \dfrac{\sqrt{5}}{\sqrt{5}} = \dfrac{-\sqrt{15}}{5}$

47. $\dfrac{\sqrt{3}}{5-\sqrt{2}} = \dfrac{\sqrt{3}}{5-\sqrt{2}} \cdot \dfrac{5+\sqrt{2}}{5+\sqrt{2}}$
$= \dfrac{\sqrt{3}(5+\sqrt{2})}{25-2}$
$= \dfrac{\sqrt{3}(5+\sqrt{2})}{23}$

49. $\dfrac{2-\sqrt{5}}{2+3\sqrt{5}} = \dfrac{2-\sqrt{5}}{2+3\sqrt{5}} \cdot \dfrac{2-3\sqrt{5}}{2-3\sqrt{5}}$
$= \dfrac{4 - 2\sqrt{5} - 6\sqrt{5} + 15}{4 - 45}$
$= \dfrac{19 - 8\sqrt{5}}{-41}$
$= \dfrac{8\sqrt{5} - 19}{41}$

51. $\dfrac{5}{\sqrt[3]{2}} = \dfrac{5}{\sqrt[3]{2}} \cdot \dfrac{\sqrt[3]{4}}{\sqrt[3]{4}} = \dfrac{5\sqrt[3]{4}}{2}$

53. $\dfrac{\sqrt{x+h} - \sqrt{x}}{\sqrt{x+h} + \sqrt{x}} = \dfrac{\sqrt{x+h} - \sqrt{x}}{\sqrt{x+h} + \sqrt{x}} \cdot \dfrac{\sqrt{x+h} - \sqrt{x}}{\sqrt{x+h} - \sqrt{x}}$
$= \dfrac{(x+h) - 2\sqrt{x(x+h)} + x}{(x+h) - x}$
$= \dfrac{x + h - 2\sqrt{x^2 + xh} + x}{x + h - x}$
$= \dfrac{2x + h - 2\sqrt{x^2 + xh}}{h}$

55. $8^{2/3} = \left(\sqrt[3]{8}\right)^2 = 2^2 = 4$

57. $(-27)^{1/3} = \sqrt[3]{-27} = -3$

59. $16^{3/2} = \left(\sqrt{16}\right)^3 = 4^3 = 64$

61. $9^{-3/2} = \dfrac{1}{9^{3/2}} = \dfrac{1}{\left(\sqrt{9}\right)^3} = \dfrac{1}{3^3} = \dfrac{1}{27}$

63. $\left(\dfrac{9}{8}\right)^{3/2} = \left(\sqrt{\dfrac{9}{8}}\right)^3 = \left(\dfrac{3}{2\sqrt{2}}\right)^3 = \dfrac{3^3}{2^3\left(\sqrt{2}\right)^3}$
$= \dfrac{27}{8 \cdot 2\sqrt{2}} = \dfrac{27}{16\sqrt{2}} = \dfrac{27}{16\sqrt{2}} \cdot \dfrac{\sqrt{2}}{\sqrt{2}}$
$= \dfrac{27\sqrt{2}}{32}$

65. $\left(\dfrac{8}{9}\right)^{-3/2} = \left(\dfrac{9}{8}\right)^{3/2} = \left(\sqrt{\dfrac{9}{8}}\right)^3 = \left(\dfrac{3}{2\sqrt{2}}\right)^3$
$= \dfrac{3^3}{2^3\left(\sqrt{2}\right)^3} = \dfrac{27}{8 \cdot 2\sqrt{2}} = \dfrac{27}{16\sqrt{2}}$
$= \dfrac{27}{16\sqrt{2}} \cdot \dfrac{\sqrt{2}}{\sqrt{2}} = \dfrac{27\sqrt{2}}{32}$

67. $x^{3/4} x^{1/3} x^{-1/2} = x^{3/4 + 1/3 - 1/2} = x^{7/12}$

69. $\left(x^3 y^6\right)^{1/3} = \left(x^3\right)^{1/3} \left(y^6\right)^{1/3} = xy^2$

71. $\left(x^2 y\right)^{1/3} \left(xy^2\right)^{2/3} = \left(x^2\right)^{1/3} (y)^{1/3} \left(x\right)^{2/3} \left(y^2\right)^{2/3}$
$= x^{2/3} y^{1/3} x^{2/3} y^{4/3}$
$= x^{2/3 + 2/3} y^{1/3 + 4/3}$
$= x^{4/3} y^{5/3}$

73. $\left(16x^2 y^{-1/3}\right)^{3/4} = 16^{3/4}\left(x^2\right)^{3/4}\left(y^{-1/3}\right)^{3/4}$

$= \left(\sqrt[4]{16}\right)^3 x^{3/2} y^{-1/4}$

$= \dfrac{2^3 x^{3/2}}{y^{1/4}}$

$= \dfrac{8x^{3/2}}{y^{1/4}}$

75. $\dfrac{x}{(1+x)^{1/2}} + 2(1+x)^{1/2} = \dfrac{x + 2(1+x)^{1/2}(1+x)^{1/2}}{(1+x)^{1/2}}$

$= \dfrac{x + 2(1+x)}{(1+x)^{1/2}}$

$= \dfrac{x + 2 + 2x}{(1+x)^{1/2}}$

$= \dfrac{3x + 2}{(1+x)^{1/2}}$

77. $2x\left(x^2+1\right)^{1/2} + x^2 \cdot \dfrac{1}{2}\left(x^2+1\right)^{-1/2} \cdot 2x$

$= 2x\left(x^2+1\right)^{1/2} + \dfrac{x^3}{\left(x^2+1\right)^{1/2}}$

$= \dfrac{2x\left(x^2+1\right)^{1/2} \cdot \left(x^2+1\right)^{1/2} + x^3}{\left(x^2+1\right)^{1/2}}$

$= \dfrac{2x\left(x^2+1\right)^{1/2+1/2} + x^3}{\left(x^2+1\right)^{1/2}} = \dfrac{2x\left(x^2+1\right)^1 + x^3}{\left(x^2+1\right)^{1/2}}$

$= \dfrac{2x^3 + 2x + x^3}{\left(x^2+1\right)^{1/2}} = \dfrac{3x^3 + 2x}{\left(x^2+1\right)^{1/2}}$

$= \dfrac{x\left(3x^2+2\right)}{\left(x^2+1\right)^{1/2}}$

79. $\sqrt{4x+3} \cdot \dfrac{1}{2\sqrt{x-5}} + \sqrt{x-5} \cdot \dfrac{1}{5\sqrt{4x+3}},\ x > 5$

$= \dfrac{\sqrt{4x+3}}{2\sqrt{x-5}} + \dfrac{\sqrt{x-5}}{5\sqrt{4x+3}}$

$= \dfrac{\sqrt{4x+3} \cdot 5 \cdot \sqrt{4x+3} + \sqrt{x-5} \cdot 2 \cdot \sqrt{x-5}}{10\sqrt{x-5}\sqrt{4x+3}}$

$= \dfrac{5(4x+3) + 2(x-5)}{10\sqrt{(x-5)(4x+3)}} = \dfrac{20x + 15 + 2x - 10}{10\sqrt{(x-5)(4x+3)}}$

$= \dfrac{22x + 5}{10\sqrt{(x-5)(4x+3)}}$

81. $\dfrac{\left(\sqrt{1+x} - x \cdot \dfrac{1}{2\sqrt{1+x}}\right)}{1+x} = \dfrac{\left(\sqrt{1+x} - \dfrac{x}{2\sqrt{1+x}}\right)}{1+x}$

$= \dfrac{\left(\dfrac{2\sqrt{1+x}\sqrt{1+x} - x}{2\sqrt{1+x}}\right)}{1+x}$

$= \dfrac{2(1+x) - x}{2(1+x)^{1/2}} \cdot \dfrac{1}{1+x}$

$= \dfrac{2+x}{2(1+x)^{3/2}}$

83. $\dfrac{(x+4)^{1/2} - 2x(x+4)^{-1/2}}{x+4}$

$= \dfrac{\left((x+4)^{1/2} - \dfrac{2x}{(x+4)^{1/2}}\right)}{x+4}$

$= \dfrac{\left((x+4)^{1/2} \cdot \dfrac{(x+4)^{1/2}}{(x+4)^{1/2}} - \dfrac{2x}{(x+4)^{1/2}}\right)}{x+4}$

$= \dfrac{\left(\dfrac{x+4-2x}{(x+4)^{1/2}}\right)}{x+4}$

$= \dfrac{-x+4}{(x+4)^{1/2}} \cdot \dfrac{1}{x+4}$

$= \dfrac{-x+4}{(x+4)^{3/2}}$

$= \dfrac{4-x}{(x+4)^{3/2}}$

85. $\dfrac{\left(\dfrac{x^2}{(x^2-1)^{1/2}}-(x^2-1)^{1/2}\right)}{x^2}, x<-1 \text{ or } x>1$

$= \dfrac{\left(\dfrac{x^2-(x^2-1)^{1/2}\cdot(x^2-1)^{1/2}}{(x^2-1)^{1/2}}\right)}{x^2}$

$= \dfrac{x^2-(x^2-1)^{1/2}\cdot(x^2-1)^{1/2}}{(x^2-1)^{1/2}}\cdot\dfrac{1}{x^2}$

$= \dfrac{x^2-(x^2-1)}{(x^2-1)^{1/2}}\cdot\dfrac{1}{x^2}$

$= \dfrac{x^2-x^2+1}{(x^2-1)^{1/2}}\cdot\dfrac{1}{x^2}$

$= \dfrac{1}{x^2(x^2-1)^{1/2}}$

87. $\dfrac{\dfrac{1+x^2}{2\sqrt{x}}-2x\sqrt{x}}{(1+x^2)^2}, x>0$

$= \dfrac{\left(\dfrac{1+x^2-(2\sqrt{x})(2x\sqrt{x})}{2\sqrt{x}}\right)}{(1+x^2)^2}$

$= \dfrac{1+x^2-(2\sqrt{x})(2x\sqrt{x})}{2\sqrt{x}}\cdot\dfrac{1}{(1+x^2)^2}$

$= \dfrac{1+x^2-4x^2}{2\sqrt{x}}\cdot\dfrac{1}{(1+x^2)^2}$

$= \dfrac{1-3x^2}{2\sqrt{x}(1+x^2)^2}$

89. $(x+1)^{3/2}+x\cdot\dfrac{3}{2}(x+1)^{1/2} = (x+1)^{1/2}\left(x+1+\dfrac{3}{2}x\right)$

$= (x+1)^{1/2}\left(\dfrac{5}{2}x+1\right)$

$= \dfrac{1}{2}(x+1)^{1/2}(5x+2)$

91. $6x^{1/2}(x^2+x)-8x^{3/2}-8x^{1/2}$

$= 2x^{1/2}(3(x^2+x)-4x-4)$

$= 2x^{1/2}(3x^2-x-4)$

$= 2x^{1/2}(3x-4)(x+1)$

93. $3(x^2+4)^{4/3}+x\cdot 4(x^2+4)^{1/3}\cdot 2x$

$= (x^2+4)^{1/3}\left[3(x^2+4)+8x^2\right]$

$= (x^2+4)^{1/3}\left[3x^2+12+8x^2\right]$

$= (x^2+4)^{1/3}(11x^2+12)$

95. $4(3x+5)^{1/3}(2x+3)^{3/2}+3(3x+5)^{4/3}(2x+3)^{1/2}$

$= (3x+5)^{1/3}(2x+3)^{1/2}\left[4(2x+3)+3(3x+5)\right]$

$= (3x+5)^{1/3}(2x+3)^{1/2}(8x+12+9x+15)$

$= (3x+5)^{1/3}(2x+3)^{1/2}(17x+27)$

where $x\geq -\dfrac{3}{2}$.

97. $3x^{-1/2}+\dfrac{3}{2}x^{1/2}, x>0$

$= \dfrac{3}{x^{1/2}}+\dfrac{3}{2}x^{1/2}$

$= \dfrac{3\cdot 2+3x^{1/2}\cdot x^{1/2}}{2x^{1/2}}$

$= \dfrac{6+3x}{2x^{1/2}}$

$= \dfrac{3(2+x)}{2x^{1/2}}$

99. $\sqrt{2}\approx 1.41$

```
√(2)
         1.414213562
```

101. $\sqrt[3]{4}\approx 1.59$

```
³√(4)
         1.587401052
```

103. $\dfrac{2+\sqrt{3}}{3-\sqrt{5}} \approx 4.89$

```
(2+√(3))/(3-√(5))
          4.885317931
```

105. $\dfrac{3\sqrt[3]{5}-\sqrt{2}}{\sqrt{3}} \approx 2.15$

```
(3*³√(5)-√(2))/√(3)
          2.145268638
```

107. a. $V = 40(12)^2 \sqrt{\dfrac{96}{12} - 0.608}$
    ≈ 15660 gallons

  b. $V = 40(1)^2 \sqrt{\dfrac{96}{1} - 0.608} \approx 391$ gallons

109. $T = 2\pi \sqrt{\dfrac{64}{32}} = 2\pi\sqrt{2} \approx 8.89$ seconds

111. 8 inches = 8/12 = 2/3 feet

  $T = 2\pi \sqrt{\dfrac{\left(\dfrac{2}{3}\right)}{32}} = 2\pi \sqrt{\dfrac{1}{48}} = 2\pi \left(\dfrac{1}{4\sqrt{3}}\right)$

  $= \dfrac{\pi}{2\sqrt{3}} = \dfrac{\pi\sqrt{3}}{6} \approx 0.91$ seconds

113. If $a = -5$, then $\sqrt{a^2} = \sqrt{(-5)^2} = \sqrt{25} = 5 \neq a$.
  Since we use the principal square root, which is always non-negative,
  $\sqrt{a^2} = \begin{cases} a & \text{if } a \geq 0 \\ -a & \text{if } a < 0 \end{cases}$
  which is the definition of $|a|$, so $\sqrt{a^2} = |a|$.

Chapter R Review

1. a. none

  b. $\{-10\}$

  c. $\left\{-10, 0.65, 1.343434\ldots, \dfrac{1}{9}\right\}$

  d. $\{\sqrt{7}\}$

  e. $\left\{-10, 0.65, 1.343434\ldots, \sqrt{7}, \dfrac{1}{9}\right\}$

3. a. 782.57

  b. 782.56

5. a. 9.62

  b. 9.62

7. $-6 + 4 \cdot (8-3) = -6 + 4 \cdot (5) = -6 + 20 = 14$

9. $\dfrac{4}{3} \cdot \dfrac{9}{16} = \dfrac{4 \cdot 3 \cdot 3}{3 \cdot 4 \cdot 4} = \dfrac{\cancel{4} \cdot \cancel{3} \cdot 3}{\cancel{3} \cdot \cancel{4} \cdot 4} = \dfrac{3}{4}$

11. $\sqrt{(-3)^2} = |-3| = 3$

13. $4(x-3) = 4x - 12$

15. $x > 3$

17. $d(P,Q) = d(-2,3) = |3-(-2)| = |3+2| = |5| = 5$

19. $\dfrac{4x}{x+y} = \dfrac{4(-5)}{-5+7} = \dfrac{-20}{2} = -10$

```
-5→X:7→Y:4X/(X+Y)
                 -10
```

23

21. $5x^{-1}y^2 = 5(-5)^{-1}(7)^2 = 5 \cdot \dfrac{1}{-5} \cdot 49 = -49$

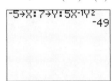

23. $\sqrt{x^2} = \sqrt{(-5)^2} = |-5| = 5$

25. $\dfrac{3}{x-6}$ Domain $= \{x \mid x \neq 6\}$

27. $\sqrt{32} = \sqrt{16 \cdot 2} = 4\sqrt{2}$

29. $\sqrt[3]{-16} = \sqrt[3]{-8 \cdot 2} = -2\sqrt[3]{2}$

31. $5\sqrt{8} - 2\sqrt{32} = 5\sqrt{4 \cdot 2} - 2\sqrt{16 \cdot 2}$
$= 5 \cdot 2\sqrt{2} - 2 \cdot 4\sqrt{2}$
$= 10\sqrt{2} - 8\sqrt{2}$
$= (10-8)\sqrt{2}$
$= 2\sqrt{2}$

33. $\dfrac{(x^2 y)^{-4}}{(xy)^{-3}} = \dfrac{x^{-8} y^{-4}}{x^{-3} y^{-3}} = \dfrac{x^3 y^3}{x^8 y^4} = \dfrac{1}{x^5 y}$

35. $\left(25x^{-4/3} y^{-2/3}\right)^{3/2} = \left(5^2 x^{-4/3} y^{-2/3}\right)^{3/2}$
$= 5^3 x^{-2} y^{-1}$
$= \dfrac{125}{x^2 y}$

37. $(1.5)^4 = 5.0625$

39. $3x^5 + 4x^4 - 2x^3 + 5x - 12$, the coefficients are 3, 4, –2, 5, and –12 and the degree = 5

41. $\left(2x^4 - 8x^3 + 5x - 1\right) + \left(6x^3 + x^2 + 4\right)$
$= 2x^4 - 8x^3 + 5x - 1 + 6x^3 + x^2 + 4$
$= 2x^4 - 2x^3 + x^2 + 5x + 3$

43. $(2x+1)(3x-5) = 6x^2 - 10x + 3x - 5$
$= 6x^2 - 7x - 5$

45. $(4x+1)(4x-1) = 16x^2 - 1$

47. $(x+1)(x+2)(x-3)$
$= \left(x^2 + 2x + x + 2\right)(x-3)$
$= \left(x^2 + 3x + 2\right)(x-3)$
$= x^3 - 3x^2 + 3x^2 - 9x + 2x - 6$
$= x^3 - 7x - 6$

49. $(3x+y)^2 = (3x+y)(3x+y)$
$= 9x^2 + 3xy + 3xy + y^2$
$= 9x^2 + 6xy + y^2$

51. $(x+2y)^3$
$= (x+2y)(x+2y)(x+2y)$
$= \left(x^2 + 2xy + 2xy + 4y^2\right)(x+2y)$
$= \left(x^2 + 4xy + 4y^2\right)(x+2y)$
$= x^3 + 4x^2 y + 4xy^2 + 2x^2 y + 8xy^2 + 8y^3$
$= x^3 + 6x^2 y + 12xy^2 + 8y^3$

53.
$$\begin{array}{r}
3x^2 + 8x + 25 \\
x-3 \overline{\smash{)}3x^3 - x^2 + x + 4} \\
\underline{3x^3 - 9x^2} \\
8x^2 + x + 4 \\
\underline{8x^2 - 24x} \\
25x + 4 \\
\underline{25x - 75} \\
79
\end{array}$$

Check:
$(x-3)(3x^2 + 8x + 25) + (79)$
$= 3x^3 + 8x^2 + 25x - 9x^2 - 24x - 75 + 79$
$= 3x^3 - x^2 + x + 4$

The quotient is $3x^2 + 8x + 25$; the remainder is 79.

55.
$$\begin{array}{r} -3x^2 + 4 \\ x^2+1{\overline{\smash{\big)}\,-3x^4 + 0\cdot x^3 + x^2 + 0\cdot x + 2}} \\ \underline{-3x^4 \qquad\quad -3x^2} \\ 4x^2 \qquad\quad +2 \\ \underline{4x^2 \qquad\quad +4} \\ -2 \end{array}$$

Check:
$(x^2+1)(-3x^2+4)+(-2)$
$= -3x^4 + 4x^2 - 3x^2 + 4 - 2$
$= -3x^4 + x^2 + 2$

The quotient is $-3x^2 + 4$; the remainder is -2.

57.
$$\begin{array}{r} x^4 - x^3 + x^2 - x + 1 \\ x+1{\overline{\smash{\big)}\,x^5 + 0\cdot x^4 + 0\cdot x^3 + 0\cdot x^2 + 0\cdot x + 1}} \\ \underline{x^5 + x^4} \\ -x^4 \qquad\qquad\qquad +1 \\ \underline{-x^4 - x^3} \\ x^3 \qquad\qquad +1 \\ \underline{x^3 + x^2} \\ -x^2 \qquad +1 \\ \underline{-x^2 - x \quad +1} \\ x \quad +1 \\ \underline{x \quad +1} \\ 0 \end{array}$$

Check:
$(x+1)(x^4 - x^3 + x^2 - x + 1) + (0)$
$= x^5 - x^4 + x^3 - x^2 + x + x^4 - x^3 + x^2 - x + 1$
$= x^5 + 1$

The quotient is $x^4 - x^3 + x^2 - x + 1$; the remainder is 0.

59. $x^2 + 5x - 14 = (x+7)(x-2)$

61. $6x^2 - 5x - 6 = (3x+2)(2x-3)$

63. $3x^2 - 15x - 42 = 3(x^2 - 5x - 14)$
$= 3(x+2)(x-7)$

65. $8x^3 + 1 = (2x+1)\left((2x)^2 - (2x)(1) + 1^2\right)$
$= (2x+1)(4x^2 - 2x + 1)$

67. $2x^3 + 3x^2 - 2x - 3 = x^2(2x+3) - (2x+3)$
$= (2x+3)(x^2-1)$
$= (2x+3)(x-1)(x+1)$

69. $25x^2 - 4 = (5x+2)(5x-2)$

71. $9x^2 + 1$; a sum of perfect squares is always prime over the set of real numbers

73. $x^2 + 8x + 16 = (x+4)(x+4) = (x+4)^2$

75. $\dfrac{2x^2 + 11x + 14}{x^2 - 4} = \dfrac{(2x+7)(x+2)}{(x+2)(x-2)} = \dfrac{2x+7}{x-2}$

77. $\dfrac{9x^2-1}{x^2-9} \cdot \dfrac{3x-9}{9x^2+6x+1} = \dfrac{(3x+1)(3x-1)}{(x+3)(x-3)} \cdot \dfrac{3(x-3)}{(3x+1)^2}$
$= \dfrac{3(3x-1)}{(x+3)(3x+1)}$

79. $\dfrac{x+1}{x-1} - \dfrac{x-1}{x+1} = \dfrac{(x+1)(x+1) - (x-1)(x-1)}{(x-1)(x+1)}$
$= \dfrac{(x^2+2x+1) - (x^2-2x+1)}{(x-1)(x+1)}$
$= \dfrac{x^2+2x+1-x^2+2x-1}{(x-1)(x+1)}$
$= \dfrac{4x}{(x-1)(x+1)}$

81. $\dfrac{3x+4}{x^2-4} - \dfrac{2x-3}{x^2+4x+4}$

$= \dfrac{3x+4}{(x+2)(x-2)} - \dfrac{2x-3}{(x+2)^2}$

$= \dfrac{(3x+4)(x+2) - (2x-3)(x-2)}{(x+2)^2(x-2)}$

$= \dfrac{3x^2+6x+4x+8 - (2x^2-4x-3x+6)}{(x+2)^2(x-2)}$

$= \dfrac{3x^2+10x+8 - (2x^2-7x+6)}{(x+2)^2(x-2)}$

$= \dfrac{3x^2+10x+8-2x^2+7x-6}{(x+2)^2(x-2)}$

$= \dfrac{x^2+17x+2}{(x+2)^2(x-2)}$

83. $\dfrac{\left(\dfrac{x^2-1}{x^2-5x+6}\right)}{\left(\dfrac{x+1}{x-2}\right)} = \left(\dfrac{x^2-1}{x^2-5x+6}\right)\left(\dfrac{x-2}{x+1}\right)$

$= \left(\dfrac{(x+1)(x-1)}{(x-3)(x-2)}\right)\left(\dfrac{x-2}{x+1}\right)$

$= \dfrac{x-1}{x-3}$

85. $\sqrt{\dfrac{9x^2}{25y^4}} = \dfrac{\sqrt{9x^2}}{\sqrt{25y^4}} = \dfrac{3x}{5y^2}$

87. $\sqrt[3]{27x^4y^{12}} = \left(\sqrt[3]{27}\right)\left(\sqrt[3]{x^4}\right)\left(\sqrt[3]{y^{12}}\right) = 3xy^4\sqrt[3]{x}$

89. $\dfrac{4}{\sqrt{5}} \cdot \dfrac{\sqrt{5}}{\sqrt{5}} = \dfrac{4\sqrt{5}}{5}$

91. $\dfrac{2}{1-\sqrt{2}} \cdot \dfrac{1+\sqrt{2}}{1+\sqrt{2}} = \dfrac{2(1+\sqrt{2})}{1-(\sqrt{2})^2} = \dfrac{2(1+\sqrt{2})}{1-2}$

$= \dfrac{2(1+\sqrt{2})}{-1} = -2(1+\sqrt{2})$

93. $\dfrac{1+\sqrt{5}}{1-\sqrt{5}} \cdot \dfrac{1+\sqrt{5}}{1+\sqrt{5}} = \dfrac{1+2\sqrt{5}+(\sqrt{5})^2}{1-(\sqrt{5})^2} = \dfrac{1+2\sqrt{5}+5}{1-5}$

$= \dfrac{6+2\sqrt{5}}{-4} = \dfrac{3+\sqrt{5}}{-2}$

$= -\dfrac{3+\sqrt{5}}{2}$

95. $(2+x^2)^{1/2} + x \cdot \dfrac{1}{2}(2+x^2)^{-1/2} \cdot 2x$

$= (2+x^2)^{1/2} + \dfrac{2x^2}{2(2+x^2)^{1/2}}$

$= (2+x^2)^{1/2} + \dfrac{x^2}{(2+x^2)^{1/2}}$

$= \dfrac{(2+x^2)^{1/2}(2+x^2)^{1/2} + x^2}{(2+x^2)^{1/2}}$

$= \dfrac{2+x^2+x^2}{(2+x^2)^{1/2}}$

$= \dfrac{2+2x^2}{(2+x^2)^{1/2}}$

$= \dfrac{2(1+x^2)}{(2+x^2)^{1/2}}$

97. $\dfrac{(x+4)^{1/2} \cdot 2x - x^2 \cdot \frac{1}{2}(x+4)^{-1/2}}{x+4}$

$= \dfrac{(x+4)^{1/2} \cdot 2x - \dfrac{x^2}{2(x+4)^{1/2}}}{x+4}$

$= \dfrac{\left(\dfrac{2(x+4)^{1/2}(x+4)^{1/2} \cdot 2x - x^2}{2(x+4)^{1/2}}\right)}{x+4}$

$= \dfrac{\left(\dfrac{2(x+4) \cdot 2x - x^2}{2(x+4)^{1/2}}\right)}{x+4}$

$= \dfrac{\left(\dfrac{4x^2 + 16x - x^2}{2(x+4)^{1/2}}\right)}{x+4}$

$= \left(\dfrac{3x^2 + 16x}{2(x+4)^{1/2}}\right)\left(\dfrac{1}{x+4}\right)$

$= \dfrac{3x^2 + 16x}{2(x+4)^{3/2}}$

$= \dfrac{x(3x+16)}{2(x+4)^{3/2}}$

99. $\dfrac{\dfrac{x^2}{\sqrt{x^2-1}} - \sqrt{x^2-1}}{x^2} = \dfrac{\dfrac{x^2}{\sqrt{x^2-1}} - \dfrac{\sqrt{x^2-1}}{1} \cdot \dfrac{\sqrt{x^2-1}}{\sqrt{x^2-1}}}{x^2}$

$= \dfrac{\dfrac{x^2 - (x^2-1)}{\sqrt{x^2-1}}}{x^2} = \dfrac{1}{\sqrt{x^2-1}} \cdot \dfrac{1}{x^2}$

$= \dfrac{1}{x^2\sqrt{x^2-1}} \cdot \dfrac{\sqrt{x^2-1}}{\sqrt{x^2-1}}$

$= \dfrac{\sqrt{x^2-1}}{x^2(x^2-1)}$

101. $3(x^2+4)^{4/3} + x \cdot 4(x^2+4)^{1/3} \cdot 2x$

$= (x^2+4)^{1/3}\left(3(x^2+4) + 8x^2\right)$

$= (x^2+4)^{1/3}\left(3x^2 + 12 + 8x^2\right)$

$= (x^2+4)^{1/3}\left(11x^2 + 12\right)$

103. $281{,}421{,}906 = 2.81421906 \times 10^8$

105. $12^2 + 16^2 = 144 + 256 = 400 = 20^2$, therefore we have a right triangle by the converse of the Pythagorean Theorem.

107. Total annual earnings per share = (1st quarter earnings) + (2nd quarter earnings) + (3rd quarter earnings) + (4th quarter earnings)
= 1.2 − 0.75 − 0.30 + 0.20 = $0.35 per share.

109. Deck Area = Total Area − Pool Area
$= (20+6)(10+6) - (20)(10)$
$= 26 \cdot 16 - 200$
$= 216$ square feet

Deck Perimeter $= 2(26+16) = 84$ feet

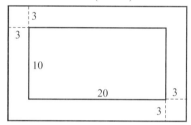

111. Recall that 1 mile = 5280 feet. Consider the diagram (not drawn to scale), using the radius of Earth as 3960 miles.

Let B refer to the furthest point the pilot can see (the horizon). Note also that

$35{,}000$ feet $= \dfrac{35{,}000}{5280}$ miles.

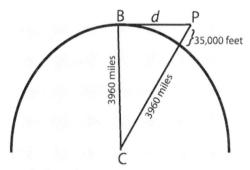

Apply the Pythagorean Theorem to $\triangle CBP$.

Chapter R: Review   SSM: College Algebra EGU

$$(3960)^2 + (d)^2 = \left(3960 + \frac{35000}{5280}\right)^2$$

$$(d)^2 = \left(3960 + \frac{35000}{5280}\right)^2 - (3960)^2$$

$$d = \sqrt{\left(3960 + \frac{35000}{5280}\right)^2 - (3960)^2}$$

$$\approx 229.22 \text{ miles.}$$

Therefore, the pilot can see 229.22 miles. Since the bridge is only 139 miles away, the pilot can easily see the bridge.

113. Answers will vary.

115. Answers will vary.

Chapter R Test

1. a. 6 is the only natural number.
   b. -5, 0, and 6 are integers.
   c. -5, 0, $\frac{4}{3}$, 1.5, $2.333... = 2.\overline{3}$, and 6 are rational numbers.
   d. $\frac{\pi}{4}$ and $\sqrt{3}$ are irrational numbers.
   e. All the numbers listed are real numbers.

2. For -2.537, the final digit is 3, since it is two decimal places from the decimal point.
   a. The digit following the final digit 3 is the digit 7. Since 7 is 5 or more, we add 1 to the final digit and truncate. Therefore, -2.537 rounded to two decimal places is -2.54.
   b. To truncate, we remove all digits following the final digit 3. Therefore, -2.537 truncated to two decimal places is -2.53.

3. For 3.14159, the final digit is 4, since it is two decimal places from the decimal point.
   a. The digit following the final digit 4 is the digit 1. Since 1 is 4 or less, we truncate without changing the final digit. Therefore, 3.14159 rounded to two decimal places is 3.14.
   b. To truncate, we remove all digits following the final digit 4. Therefore, 3.14159 truncated to two decimal places is 3.14.

4. $2 - 3(5 - 7) = 2 - 3(-2) = 2 + 6 = 8$

5. $\frac{2}{5} + \frac{1}{5} \cdot \frac{4}{3} = \frac{2}{5} + \frac{4}{15}$
   $= \frac{2 \cdot 3}{5 \cdot 3} + \frac{4}{15} = \frac{6}{15} + \frac{4}{15}$
   $= \frac{6 + 4}{15} = \frac{10}{15}$
   $= \frac{2}{3}$

6. $\sqrt{(-2)^2 + 5} = \sqrt{4 + 5} = \sqrt{9} = 3$

7. $4 \cdot 81^{-3/4} = 4 \cdot \frac{1}{81^{3/4}} = \frac{4}{\left(\sqrt[4]{81}\right)^3} = \frac{4}{3^3} = \frac{4}{27}$

8. $(x - 3)(4x + 1)(x - 2)$
   $= (x - 3)(4x^2 - 8x + x - 2)$
   $= (x - 3)(4x^2 - 7x - 2)$
   $= x(4x^2 - 7x - 2) - 3(4x^2 - 7x - 2)$
   $= 4x^3 - 7x^2 - 2x - 12x^2 + 21x + 6$
   $= 4x^3 - 19x^2 + 19x + 6$

9. $(2a - 1)(2a + 1) = (2a)^2 - 1^2 = 4a^2 - 1$

10. $(3y + 2z)(y - 5z)$
    $= 3y \cdot y - 3y \cdot 5z + 2z \cdot y - 2z \cdot 5z$
    $= 3y^2 - 15yz + 2yz - 10z^2$
    $= 3y^2 - 13yz - 10z^2$

11. $\frac{4x}{x^2 - 5x - 14} - \frac{1}{x + 2}$
    $= \frac{4x}{(x - 7)(x + 2)} - \frac{1}{x + 2}$
    $= \frac{4x}{(x - 7)(x + 2)} - \frac{(x - 7) \cdot 1}{(x - 7)(x + 2)}$
    $= \frac{4x - x + 7}{(x - 7)(x + 2)}$
    $= \frac{3x + 7}{(x - 7)(x + 2)}$

SSM: College Algebra EGU Chapter R: Review

12. $(2x^3 - 7x + 9) - (x^2 - 2x - 3)$
$= 2x^3 - 7x + 9 - x^2 + 2x + 3$
$= 2x^3 - x^2 - 5x + 12$

13. $(3m + 7n)^2 = (3m)^2 + 2(3m)(7n) + (7n)^2$
$= 9m^2 + 42mn + 49n^2$

14. $2x + 4 + \dfrac{4}{x+3} = \dfrac{2x(x+3)}{x+3} + \dfrac{4(x+3)}{x+3} + \dfrac{4}{x+3}$
$= \dfrac{2x^2 + 6x + 4x + 12 + 4}{x+3}$
$= \dfrac{2x^2 + 10x + 16}{x+3}$
$= \dfrac{2(x^2 + 5x + 8)}{x+3}$

15. $(2b+1)(4b^2 - 2b + 1)$
$= 2b(4b^2 - 2b + 1) + 1(4b^2 - 2b + 1)$
$= 8b^3 - 4b^2 + 2b + 4b^2 - 2b + 1$
$= 8b^3 + 1$

16. $\dfrac{\frac{x+2}{4}}{\frac{3}{x} + \frac{3}{2}} = \dfrac{\frac{x+2}{4}}{\frac{6}{2x} + \frac{3x}{2x}} = \dfrac{x+2}{4} \cdot \dfrac{2x}{3x+6}$
$= \dfrac{2x(x+2)}{4(3x+6)} = \dfrac{2x(x+2)}{4 \cdot 3(x+2)}$
$= \dfrac{x}{6}$

17. $\dfrac{x^2 - 9}{x+2} \cdot \dfrac{x^2 - 2x - 8}{x+3}$
$= \dfrac{(x-3)(x+3)(x-4)(x+2)}{(x+2)(x+3)}$
$= \dfrac{(x-3)\cancel{(x+3)}(x-4)\cancel{(x+2)}}{\cancel{(x+2)}\cancel{(x+3)}}$
$= (x-3)(x-4)$
$= x^2 - 7x + 12$, $x \neq -3, -2$

18. $\sqrt{8x^3 y} - 2\sqrt{32xy^3} + 2x\sqrt{2xy}$
$= \sqrt{4x^2 \cdot 2xy} - 2\sqrt{16y^2 \cdot 2xy} + 2x\sqrt{2xy}$
$= \sqrt{4x^2}\sqrt{2xy} - 2 \cdot \sqrt{16y^2}\sqrt{2xy} + 2x\sqrt{2xy}$
$= 2x\sqrt{2xy} - 8y\sqrt{2xy} + 2x\sqrt{2xy}$
$= (2x - 8y + 2x)\sqrt{2xy}$
$= (4x - 8y)\sqrt{2xy}$

19. $\dfrac{15a^2 b}{3b^2} \cdot \dfrac{ab^4 c^{-1}}{-10ac} = \dfrac{15(a^2 \cdot a)(b \cdot b^4)c^{-1}}{-30ab^2 c}$
$= \dfrac{a^3 b^5 c^{-1}}{-2ab^2 c} = -\dfrac{1}{2} a^{3-1} b^{5-2} c^{-1-1}$
$= -\dfrac{1}{2} a^2 b^3 c^{-2} = -\dfrac{a^2 b^3}{2c^2}$

20. $\left(\dfrac{x^2 yz^3}{xy^2 z}\right)^{-2} = \left(\dfrac{xy^2 z}{x^2 yz^3}\right)^2 = \left(x^{1-2} y^{2-1} z^{1-3}\right)^2$
$= \left(x^{-1} y^1 z^{-2}\right)^2 = \left(\dfrac{y}{xz^2}\right)^2$
$= \dfrac{y^2}{(x)^2 (z^2)^2} = \dfrac{y^2}{x^2 z^4}$

21. $\dfrac{x + \frac{2}{x+1}}{x - 1} = \dfrac{\frac{x(x+1)}{x+1} + \frac{2}{x+1}}{x-1}$
$= \dfrac{x^2 + x + 2}{x+1} \cdot \dfrac{1}{x-1}$
$= \dfrac{x^2 + x + 2}{(x+1)(x-1)}$

22. $\left(x^{1/2} y^{3/4}\right)^{2/3} \left(xy^{-2}\right)^{1/2}$
$= \left(x^{1/2}\right)^{2/3} \left(y^{3/4}\right)^{2/3} x^{1/2} \left(y^{-2}\right)^{1/2}$
$= x^{1/3} \cdot y^{1/2} \cdot x^{1/2} \cdot y^{-1}$
$= \left(x^{1/3} \cdot x^{1/2}\right)\left(y^{1/2} \cdot y^{-1}\right)$
$= x^{1/3 + 1/2} \cdot y^{1/2 + (-1)}$
$= x^{5/6} y^{-1/2}$
$= \dfrac{x^{5/6}}{y^{1/2}}$

23. $\dfrac{5x+1}{x^3-2x^2}$

Let $x = -2$:
$$\dfrac{5x+1}{x^3-2x^2} = \dfrac{5(-2)+1}{(-2)^3-2(-2)^2} = \dfrac{-10+1}{-8-2(4)}$$
$$= \dfrac{-9}{-8-8} = \dfrac{-9}{-16}$$
$$= \dfrac{9}{16}$$

$x = -2$ is in the domain of the variable.

Let $x = 3$:
$$\dfrac{5x+1}{x^3-2x^2} = \dfrac{5(3)+1}{(3)^3-2(3)^2} = \dfrac{15+1}{27-2(9)}$$
$$= \dfrac{16}{27-18} = \dfrac{16}{9}$$

$x = 3$ is in the domain of the variable.

$x = 0$ must be excluded from the domain of the variable because this value yields division by 0 which is undefined.
$$\dfrac{5x+1}{x^3-2x^2} = \dfrac{5(0)+1}{0^3-2(0)^2} = \dfrac{0+1}{0-2(0)} = \dfrac{1}{0}$$

This result is undefined so $x = 0$ must be excluded from the domain.

24. $\dfrac{3}{\sqrt{2}-1} = \dfrac{3}{\sqrt{2}-1} \cdot \dfrac{\sqrt{2}+1}{\sqrt{2}+1} = \dfrac{3(\sqrt{2}+1)}{(\sqrt{2})^2-1^2}$
$$= \dfrac{3\sqrt{2}+3}{2-1} = 3\sqrt{2}+3$$

25. $z^2 - 4z - 21$

We are looking for two factors of $C = -21$ whose sum is $B = -4$. Since C is negative, the factors will have opposite signs. Since B is also negative, the factor with the larger absolute value will be negative.

factors	$-21, 1$	$-7, 3$
sum	-20	-4

The two factors are -7 and 3. Thus, $z^2 - 4z - 21 = (z-7)(z+3)$.

26. $2x^2 - 6x + 7$
$AC = 2 \cdot 7 = 14$

We are looking for two factors of 14 whose sum is $B = -6$. Since the product is positive, the factors have the same sign. Since the sum is negative, the factors must both be negative.

factors	$-1, -14$	$-2, -7$
sum	-15	-9

Since none of the sums is -6 as needed, we conclude that $2x^2 - 6x + 7$ is prime.

27. $64w^3 + 125$

Since $64w^3 = (4w)^3$ and $125 = 5^3$, we have the sum of two cubes.
$$64w^3 + 125 = (4w)^3 + 5^3$$
$$= (4w+5)\left((4w)^2 - 5(4w) + 5^2\right)$$
$$= (4w+5)(16w^2 - 20w + 25)$$

28. $16c^4 - 9c^2$

We start by pulling out the greatest common factor, c^2.
$$16c^4 - 9c^2 = c^2(16c^2 - 9)$$

Since $16c^2 = (4c)^2$ and $9 = 3^2$, the expression in parentheses is the difference of two squares.
$$16c^4 - 9c^2 = c^2(16c^2 - 9)$$
$$= c^2(4c-3)(4c+3)$$

29. $5a^3 - 30a^2b + 45ab^2$

We start by pulling out the greatest common factor, $5a$.
$$5a^3 - 30a^2b + 45ab^2 = 5a(a^2 - 6ab + 9b^2)$$

The first term of the expression in parentheses, a^2, and the third term, $9b^2 = (3b)^2$, are perfect squares. The middle term, $-6ab$, is -2 times the product of a and $3b$ so the expression is a perfect square.
$$5a^3 - 30a^2b + 45ab^2 = 5a(a^2 - 6ab + 9b^2)$$
$$= 5a(a-3b)^2$$

30. $8x^2 + 2x - 15$
    $AC = 8 \cdot (-15) = -120$

    We need two factors of -120 whose sum is 2. Since the product is negative, the factors will have opposite signs. Since the sum is positive, the factor with the larger absolute value must be positive. The sum is small, so the absolute value of the factors will be close to each other.

factors	$-8, 15$	$-10, 12$
sum	7	2

    The factors -10 and 12 yield the required sum.
    $8x^2 + 2x - 15 = 8x^2 - 10x + 12x - 15$
    $= 2x(4x - 5) + 3(4x - 5)$
    $= (4x - 5)(2x + 3)$

31. $3y^3 - 12y^2 + 2y - 8 = (3y^3 - 12y^2) + (2y - 8)$
    $= 3y^2(y - 4) + 2(y - 4)$
    $= (y - 4)(3y^2 + 2)$

32. Using long division:

    $$\begin{array}{r} 3x^2 - 8x + 3 \\ x+2 \overline{\smash{)}3x^3 - 2x^2 - 13x + 8} \\ \underline{-(3x^3 + 6x^2)} \\ -8x^2 - 13x \\ \underline{-(-8x^2 - 16x)} \\ 3x + 8 \\ \underline{-(3x + 6)} \\ 2 \end{array}$$

    Therefore, the quotient is $3x^2 - 8x + 3$ and the remainder is 2.
    $\dfrac{3x^3 - 2x^2 - 13x + 8}{x + 2} = 3x^2 - 8x + 3 + \dfrac{2}{x + 2}$

    Using synthetic division:
    $x + 2 = x - (-2)$ so $c = -2$.

    $$\begin{array}{r|rrrr} -2 & 3 & -2 & -13 & 8 \\ & & -6 & 16 & -6 \\ \hline & 3 & -8 & 3 & \underline{|2} \end{array}$$

    Therefore, the quotient is $3x^2 - 8x + 3$ and the remainder is 2.
    $\dfrac{3x^3 - 2x^2 - 13x + 8}{x + 2} = 3x^2 - 8x + 3 + \dfrac{2}{x + 2}$

33. Let x = the width of the room in feet. Then the length of the room is $x + 12$.

    a. $A_{\text{not covered}} = A_{\text{room}} - A_{\text{mats}}$
       $= x(x + 12) - 640$
       $= x^2 + 12x - 640$
       $= (x - 20)(x + 32)$

    b. $x = 40$
       $A = (x - 20)(x + 32)$
       $= (40 - 20)(40 + 32)$
       $= 20 \cdot 72$
       $= 1,440$

       If the room is 40 feet wide, then 1440 square feet of the room will not be covered by mats.

34. Each guyline is the hypotenuse of a right triangle whose base is the ground and whose height is the height of the tower. We can use the Pythagorean Theorem to find the height of the tower.
    $a^2 + b^2 = c^2$
    $15^2 + b^2 = 25^2$
    $225 + b^2 = 625$
    $b^2 = 400$
    $b = 20$
    The tower was 20 feet tall.

35. The water in the pool forms a right circular cylinder. Therefore, we need the formula for the volume of a cylinder, $V = \pi r^2 h$.
    The diameter is $d = 2$ meters so the radius is
    $r = \dfrac{d}{2} = \dfrac{2}{2} = 1$ meter. The height is
    $h = 45 \text{ cm} \cdot \dfrac{1 \text{ m}}{100 \text{ cm}} = 0.45 \text{ m}$.
    $V = \pi r^2 h = \pi (1)^2 (0.45) = 0.45\pi \approx 1.41 \text{ m}^3$

    There is about 1.41 m^3 of water in the pool.

Chapter 1
Graphs, Equations, and Inequalities

Section 1.1

1. 0

3. $\sqrt{3^2+4^2}=\sqrt{25}=5$

5. x-coordinate, or abscissa; y-coordinate, or ordinate.

7. midpoint

9. False; points that lie in Quadrant IV will have a positive x-coordinate and a negative y-coordinate. The point $(-1,4)$ lies in Quadrant II.

11. (a) Quadrant II
    (b) Positive x-axis
    (c) Quadrant III
    (d) Quadrant I
    (e) Negative y-axis
    (f) Quadrant IV

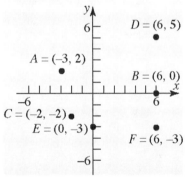

13. The points will be on a vertical line that is two units to the right of the y-axis.

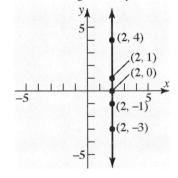

15. $(-1,4)$; Quadrant II

17. $(3, 1)$; Quadrant I

19. $X\min = -11$
    $X\max = 5$
    $X\operatorname{scl} = 1$
    $Y\min = -3$
    $Y\max = 6$
    $Y\operatorname{scl} = 1$

21. $X\min = -30$
    $X\max = 50$
    $X\operatorname{scl} = 10$
    $Y\min = -90$
    $Y\max = 50$
    $Y\operatorname{scl} = 10$

23. $X\min = -10$
    $X\max = 110$
    $X\operatorname{scl} = 10$
    $Y\min = -10$
    $Y\max = 160$
    $Y\operatorname{scl} = 10$

25. $X\min = -6$
    $X\max = 6$
    $X\operatorname{scl} = 2$
    $Y\min = -4$
    $Y\max = 4$
    $Y\operatorname{scl} = 2$

27. $X\min = -6$
    $X\max = 6$
    $X\operatorname{scl} = 2$
    $Y\min = -1$
    $Y\max = 3$
    $Y\operatorname{scl} = 1$

29. $X \min = 3$
    $X \max = 9$
    $X \text{ scl} = 1$
    $Y \min = 2$
    $Y \max = 10$
    $Y \text{ scl} = 2$

31. $d(P_1, P_2) = \sqrt{(2-0)^2 + (1-0)^2} = \sqrt{4+1} = \sqrt{5}$

33. $d(P_1, P_2) = \sqrt{(-2-1)^2 + (2-1)^2} = \sqrt{9+1} = \sqrt{10}$

35. $d(P_1, P_2) = \sqrt{(5-3)^2 + (4-(-4))^2} = \sqrt{2^2 + (8)^2}$
    $= \sqrt{4+64} = \sqrt{68} = 2\sqrt{17}$

37. $d(P_1, P_2) = \sqrt{(6-(-3))^2 + (0-2)^2}$
    $= \sqrt{9^2 + (-2)^2}$
    $= \sqrt{81+4}$
    $= \sqrt{85}$

39. $d(P_1, P_2) = \sqrt{(6-4)^2 + (4-(-3))^2} = \sqrt{2^2 + 7^2}$
    $= \sqrt{4+49} = \sqrt{53}$

41. $d(P_1, P_2) = \sqrt{(2.3-(-0.2))^2 + (1.1-0.3)^2}$
    $= \sqrt{(2.5)^2 + (0.8)^2} = \sqrt{6.25 + 0.64}$
    $= \sqrt{6.89} \approx 2.625$

43. $d(P_1, P_2) = \sqrt{(0-a)^2 + (0-b)^2} = \sqrt{a^2 + b^2}$

45. $P_1 = (1, 3); P_2 = (5, 15)$
    $d(P_1, P_2) = \sqrt{(5-1)^2 + (15-3)^2}$
    $= \sqrt{(4)^2 + (12)^2}$
    $= \sqrt{16 + 144}$
    $= \sqrt{160} = 4\sqrt{10}$

47. $P_1 = (-4, 6); P_2 = (4, -8)$
    $d(P_1, P_2) = \sqrt{(4-(-4))^2 + (-8-6)^2}$
    $= \sqrt{(8)^2 + (-14)^2}$
    $= \sqrt{64 + 196}$
    $= \sqrt{260} = 2\sqrt{65}$

49. $A = (-2, 5)$, $B = (1, 3)$, $C = (-1, 0)$
    $d(A, B) = \sqrt{(1-(-2))^2 + (3-5)^2}$
    $= \sqrt{3^2 + (-2)^2} = \sqrt{9+4}$
    $= \sqrt{13}$
    $d(B, C) = \sqrt{(-1-1)^2 + (0-3)^2}$
    $= \sqrt{(-2)^2 + (-3)^2} = \sqrt{4+9}$
    $= \sqrt{13}$
    $d(A, C) = \sqrt{(-1-(-2))^2 + (0-5)^2}$
    $= \sqrt{1^2 + (-5)^2} = \sqrt{1+25}$
    $= \sqrt{26}$

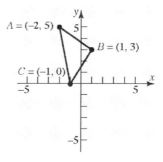

Verifying that $\triangle ABC$ is a right triangle by the Pythagorean Theorem:
$[d(A,B)]^2 + [d(B,C)]^2 = [d(A,C)]^2$
$(\sqrt{13})^2 + (\sqrt{13})^2 = (\sqrt{26})^2$
$13 + 13 = 26$
$26 = 26$

The area of a triangle is $A = \frac{1}{2} \cdot bh$. In this

Chapter 1: *Graphs, Equations, and Inequalities*   *SSM:* College Algebra EGU

problem,
$$A = \frac{1}{2} \cdot [d(A,B)] \cdot [d(B,C)]$$
$$= \frac{1}{2} \cdot \sqrt{13} \cdot \sqrt{13}$$
$$= \frac{1}{2} \cdot 13$$
$$= \frac{13}{2} \text{ square units}$$

51. $A = (-5,3), \ B = (6,0), \ C = (5,5)$
$$d(A,B) = \sqrt{(6-(-5))^2 + (0-3)^2}$$
$$= \sqrt{11^2 + (-3)^2} = \sqrt{121+9}$$
$$= \sqrt{130}$$
$$d(B,C) = \sqrt{(5-6)^2 + (5-0)^2}$$
$$= \sqrt{(-1)^2 + 5^2} = \sqrt{1+25}$$
$$= \sqrt{26}$$
$$d(A,C) = \sqrt{(5-(-5))^2 + (5-3)^2}$$
$$= \sqrt{10^2 + 2^2} = \sqrt{100+4}$$
$$= \sqrt{104}$$
$$= 2\sqrt{26}$$

[graph showing triangle with $A = (-5, 3)$, $B = (6, 0)$, $C = (5, 5)$]

Verifying that Δ ABC is a right triangle by the Pythagorean Theorem:
$$[d(A,C)]^2 + [d(B,C)]^2 = [d(A,B)]^2$$
$$\left(\sqrt{104}\right)^2 + \left(\sqrt{26}\right)^2 = \left(\sqrt{130}\right)^2$$
$$104 + 26 = 130$$
$$130 = 130$$

The area of a triangle is $A = \frac{1}{2}bh$. In this

problem,
$$A = \frac{1}{2} \cdot [d(A,C)] \cdot [d(B,C)]$$
$$= \frac{1}{2} \cdot \sqrt{104} \cdot \sqrt{26}$$
$$= \frac{1}{2} \cdot 2\sqrt{26} \cdot \sqrt{26}$$
$$= \frac{1}{2} \cdot 2 \cdot 26$$
$$= 26 \text{ square units}$$

53. $A = (4,-3), \ B = (0,-3), \ C = (4,2)$
$$d(A,B) = \sqrt{(0-4)^2 + (-3-(-3))^2}$$
$$= \sqrt{(-4)^2 + 0^2} = \sqrt{16+0}$$
$$= \sqrt{16}$$
$$= 4$$
$$d(B,C) = \sqrt{(4-0)^2 + (2-(-3))^2}$$
$$= \sqrt{4^2 + 5^2} = \sqrt{16+25}$$
$$= \sqrt{41}$$
$$d(A,C) = \sqrt{(4-4)^2 + (2-(-3))^2}$$
$$= \sqrt{0^2 + 5^2} = \sqrt{0+25}$$
$$= \sqrt{25}$$
$$= 5$$

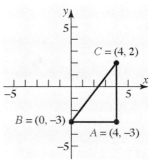

Verifying that Δ ABC is a right triangle by the Pythagorean Theorem:
$$[d(A,B)]^2 + [d(A,C)]^2 = [d(B,C)]^2$$
$$4^2 + 5^2 = \left(\sqrt{41}\right)^2$$
$$16 + 25 = 41$$
$$41 = 41$$

The area of a triangle is $A = \frac{1}{2}bh$. In this

problem,
$$A = \frac{1}{2} \cdot [d(A,B)] \cdot [d(A,C)]$$
$$= \frac{1}{2} \cdot 4 \cdot 5$$
$$= 10 \text{ square units}$$

55. The coordinates of the midpoint are:
$$(x, y) = \left(\frac{x_1 + x_2}{2}, \frac{y_1 + y_2}{2}\right)$$
$$= \left(\frac{3+5}{2}, \frac{-4+4}{2}\right)$$
$$= \left(\frac{8}{2}, \frac{0}{2}\right)$$
$$= (4, 0)$$

57. The coordinates of the midpoint are:
$$(x, y) = \left(\frac{x_1 + x_2}{2}, \frac{y_1 + y_2}{2}\right)$$
$$= \left(\frac{-3+6}{2}, \frac{2+0}{2}\right)$$
$$= \left(\frac{3}{2}, \frac{2}{2}\right)$$
$$= \left(\frac{3}{2}, 1\right)$$

59. The coordinates of the midpoint are:
$$(x, y) = \left(\frac{x_1 + x_2}{2}, \frac{y_1 + y_2}{2}\right)$$
$$= \left(\frac{4+6}{2}, \frac{-3+1}{2}\right)$$
$$= \left(\frac{10}{2}, \frac{-2}{2}\right)$$
$$= (5, -1)$$

61. The coordinates of the midpoint are:
$$(x, y) = \left(\frac{x_1 + x_2}{2}, \frac{y_1 + y_2}{2}\right)$$
$$= \left(\frac{-0.2+2.3}{2}, \frac{0.3+1.1}{2}\right)$$
$$= \left(\frac{2.1}{2}, \frac{1.4}{2}\right)$$
$$= (1.05, 0.7)$$

63. The coordinates of the midpoint are:
$$(x, y) = \left(\frac{x_1 + x_2}{2}, \frac{y_1 + y_2}{2}\right)$$
$$= \left(\frac{a+0}{2}, \frac{b+0}{2}\right)$$
$$= \left(\frac{a}{2}, \frac{b}{2}\right)$$

65. $y = x^4 - \sqrt{x}$

$0 = 0^4 - \sqrt{0}$ $1 = 1^4 - \sqrt{1}$ $0 = (-1)^4 - \sqrt{-1}$
$0 = 0$ $1 \neq 0$ $0 \neq 1 - \sqrt{-1}$

(0, 0) is on the graph of the equation.

67. $y^2 = x^2 + 9$

$3^2 = 0^2 + 9$ $0^2 = 3^2 + 9$ $0^2 = (-3)^2 + 9$
$9 = 9$ $0 \neq 18$ $0 \neq 18$

(0, 3) is on the graph of the equation.

69. $x^2 + y^2 = 4$

$0^2 + 2^2 = 4$ $(-2)^2 + 2^2 = 4$ $\sqrt{2}^2 + \sqrt{2}^2 = 4$
$4 = 4$ $8 \neq 4$ $4 = 4$

(0, 2) and $(\sqrt{2}, \sqrt{2})$ are on the graph of the equation.

71. $(-1, 0), (1, 0)$

73. $\left(-\frac{\pi}{2}, 0\right), \left(\frac{\pi}{2}, 0\right), (0, 1)$

75. $(0, 0)$

77. $(-4, 0), (-1, 0), (4, 0), (0, -3)$

79. $y = x + 2$

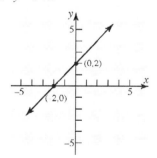

81. $y = 2x + 8$

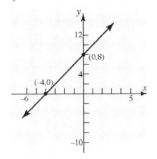

83. $y = x^2 - 1$

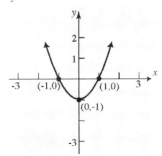

85. $y = -x^2 + 4$

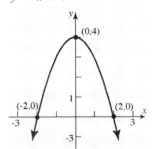

87. $2x + 3y = 6$

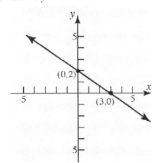

89. $9x^2 + 4y = 36$

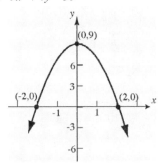

91. $y = 2x - 13$

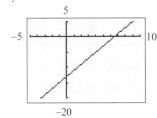

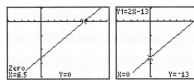

The x-intercept is $x = 6.5$ and the y-intercept is $y = -13$.

93. $y = 2x^2 - 15$

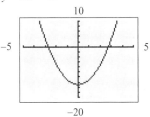

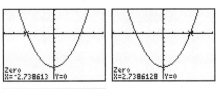

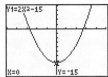

The x-intercepts are $x = -2.74$ and $x = 2.74$. The y-intercept is $y = -15$.

95. $3x - 2y = 43$ or $y = \dfrac{3x-43}{2}$

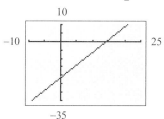

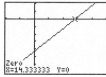 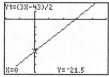

The x-intercept is $x = 14.33$ and the y-intercept is $y = -21.5$.

97. $5x^2 + 3y = 37$ or $y = \dfrac{-5x^2 + 37}{3}$

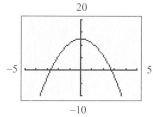

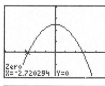 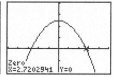

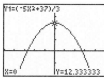

The x-intercepts are $x = -2.72$ and $x = 2.72$. The y-intercept is $y = 12.33$.

99. The midpoint of AB is: $D = \left(\dfrac{0+6}{2}, \dfrac{0+0}{2}\right)$
$= (3, 0)$

The midpoint of AC is: $E = \left(\dfrac{0+4}{2}, \dfrac{0+4}{2}\right)$
$= (2, 2)$

The midpoint of BC is: $F = \left(\dfrac{6+4}{2}, \dfrac{0+4}{2}\right)$
$= (5, 2)$

$d(C, D) = \sqrt{(0-4)^2 + (3-4)^2}$
$= \sqrt{(-4)^2 + (-1)^2} = \sqrt{16+1}$
$= \sqrt{17}$

$d(B, E) = \sqrt{(2-6)^2 + (2-0)^2}$
$= \sqrt{(-4)^2 + 2^2} = \sqrt{16+4}$
$= \sqrt{20} = 2\sqrt{5}$

$d(A, F) = \sqrt{(2-0)^2 + (5-0)^2}$
$= \sqrt{2^2 + 5^2}$
$= \sqrt{4+25}$
$= \sqrt{29}$

101. $d(P_1, P_2) = \sqrt{(-4-2)^2 + (1-1)^2}$
$= \sqrt{(-6)^2 + 0^2}$
$= \sqrt{36}$
$= 6$

$d(P_2, P_3) = \sqrt{(-4-(-4))^2 + (-3-1)^2}$
$= \sqrt{0^2 + (-4)^2}$
$= \sqrt{16}$
$= 4$

$d(P_1, P_3) = \sqrt{(-4-2)^2 + (-3-1)^2}$
$= \sqrt{(-6)^2 + (-4)^2}$
$= \sqrt{36+16}$
$= \sqrt{52}$
$= 2\sqrt{13}$

Since $[d(P_1, P_2)]^2 + [d(P_2, P_3)]^2 = [d(P_1, P_3)]^2$, the triangle is a right triangle.

103. $d(P_1,P_2) = \sqrt{(0-(-2))^2+(7-(-1))^2}$
$= \sqrt{2^2+8^2} = \sqrt{4+64} = \sqrt{68}$
$= 2\sqrt{17}$

$d(P_2,P_3) = \sqrt{(3-0)^2+(2-7)^2}$
$= \sqrt{3^2+(-5)^2} = \sqrt{9+25}$
$= \sqrt{34}$

$d(P_1,P_3) = \sqrt{(3-(-2))^2+(2-(-1))^2}$
$= \sqrt{5^2+3^2} = \sqrt{25+9}$
$= \sqrt{34}$

Since $d(P_2,P_3) = d(P_1,P_3)$, the triangle is isosceles.

Since $[d(P_1,P_3)]^2 + [d(P_2,P_3)]^2 = [d(P_1,P_2)]^2$, the triangle is also a right triangle.

Therefore, the triangle is an isosceles right triangle.

105. Using the Pythagorean Theorem:
$90^2 + 90^2 = d^2$
$8100 + 8100 = d^2$
$16200 = d^2$
$d = \sqrt{16200} = 90\sqrt{2} \approx 127.28$ feet

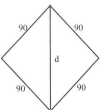

107. a. First: (90, 0), Second: (90, 90)
Third: (0, 90)

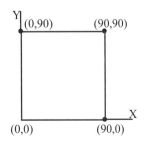

b. Using the distance formula:
$d = \sqrt{(310-90)^2+(15-90)^2}$
$= \sqrt{220^2+(-75)^2}$
$= \sqrt{54025} \approx 232.4$ feet

c. Using the distance formula:
$d = \sqrt{(300-0)^2+(300-90)^2}$
$= \sqrt{300^2+210^2}$
$= \sqrt{134100} \approx 366.2$ feet

109. The Neon heading east moves a distance $30t$ after t hours. The truck heading south moves a distance $40t$ after t hours. Their distance apart after t hours is:

$d = \sqrt{(30t)^2+(40t)^2}$
$= \sqrt{900t^2+1600t^2}$
$= \sqrt{2500t^2}$
$= 50t$

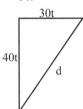

111. a.

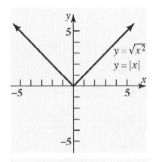

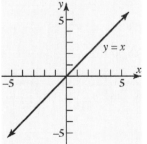

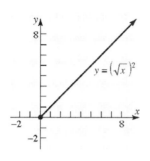

b. Since $\sqrt{x^2} = |x|$ for all x, the graphs of $y = \sqrt{x^2}$ and $y = |x|$ are the same.

c. For $y = \left(\sqrt{x}\right)^2$, the domain of the variable x is $x \geq 0$; for $y = x$, the domain of the variable x is all real numbers. Thus, $\left(\sqrt{x}\right)^2 = x$ only for $x \geq 0$.

d. For $y = \sqrt{x^2}$, the range of the variable y is $y \geq 0$; for $y = x$, the range of the variable y is all real numbers. Also, $\sqrt{x^2} = x$ only if $x \geq 0$.

113. Answers will vary

Section 1.2

1. No, $x = 4$ is not in the domain of $\dfrac{3}{x-4}$ because it makes the denominator equal 0.

3. equivalent equations

5. linear; first-degree

7. True; equations that are contradictions have no solution.

9. Divide both sides by 7 to get $x = 3$.

11. Subtract 15 from both sides, then divide both sides by 3 to get $x = -5$.

13. Add 3 to both sides, then divide both sides by 2 to get $x = \dfrac{3}{2}$.

15. Multiply both sides by 3 to get $x = \dfrac{5}{4}$.

17. $x^3 - 4x + 2 = 0$; Use ZERO (or ROOT) on the graph of $y_1 = x^3 - 4x + 2$.

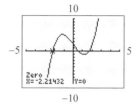 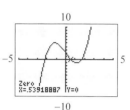

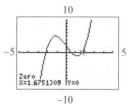

The solution set is $\{-2.21, 0.54, 1.68\}$.

19. $-2x^4 + 5 = 3x - 2$; Use INTERSECT on the graphs of $y_1 = -2x^4 + 5$ and $y_2 = 3x - 2$.

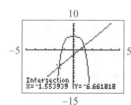 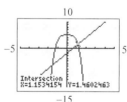

The solution set is $\{-1.55, 1.15\}$.

21. $x^4 - 2x^3 + 3x - 1 = 0$; Use ZERO (or ROOT) on the graph of $y_1 = x^4 - 2x^3 + 3x - 1$.

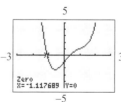 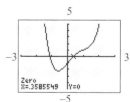

The solution set is $\{-1.12, 0.36\}$.

23. $-x^3 - \dfrac{5}{3}x^2 + \dfrac{7}{2}x + 2 = 0$;

Use ZERO (or ROOT) on the graph of
$y_1 = -x^3 - (5/3)x^2 + (7/2)x + 2$.

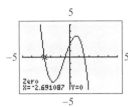
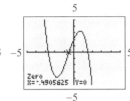
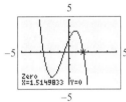

The solution set is $\{-2.69, -0.49, 1.51\}$.

25. $-\dfrac{2}{3}x^4 - 2x^3 + \dfrac{5}{2}x = -\dfrac{2}{3}x^2 + \dfrac{1}{2}$

Use INTERSECT on the graphs of
$y_1 = -(2/3)x^4 - 2x^3 + (5/2)x$ and
$y_2 = -(2/3)x^2 + 1/2$.

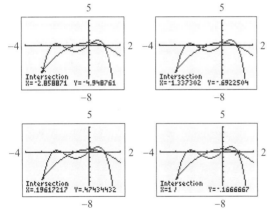

The solution set is $\{-2.86, -1.34, 0.20, 1.00\}$.

27. $x^4 - 5x^2 + 2x + 11 = 0$

Use ZERO (or ROOT) on the graph of
$y_1 = x^4 - 5x^2 + 2x + 11$.

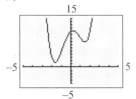

There are no real solutions.

29. $\quad 3x + 4 = x$

$3x + 4 - 4 = x - 4$

$3x = x - 4$

$3x - x = x - 4 - x$

$2x = -4$

$\dfrac{2x}{2} = \dfrac{-4}{2}$

$x = -2$

The solution set is $\{-2\}$.

31. $\quad 2t - 6 = 3 - t$

$2t - 6 + 6 = 3 - t + 6$

$2t = 9 - t$

$2t + t = 9 - t + t$

$3t = 9$

$\dfrac{3t}{3} = \dfrac{9}{3}$

$t = 3$

The solution set is $\{3\}$.

33. $\quad 6 - x = 2x + 9$

$6 - x - 6 = 2x + 9 - 6$

$-x = 2x + 3$

$-x - 2x = 2x + 3 - 2x$

$-3x = 3$

$\dfrac{-3x}{-3} = \dfrac{3}{-3}$

$x = -1$

The solution set is $\{-1\}$.

SSM: College Algebra EGU **Chapter 1:** Graphs, Equations, and Inequalities

35. $3 + 2n = 4n + 7$
$3 + 2n - 3 = 4n + 7 - 3$
$2n = 4n + 4$
$2n - 4n = 4n + 4 - 4n$
$-2n = 4$
$\dfrac{-2n}{-2} = \dfrac{4}{-2}$
$n = -2$
The solution set is $\{-2\}$.

37. $2(3 + 2x) = 3(x - 4)$
$6 + 4x = 3x - 12$
$6 + 4x - 6 = 3x - 12 - 6$
$4x = 3x - 18$
$4x - 3x = 3x - 18 - 3x$
$x = -18$
The solution set is $\{-18\}$.

39. $8x - (3x + 2) = 3x - 10$
$8x - 3x - 2 = 3x - 10$
$5x - 2 = 3x - 10$
$5x - 2 + 2 = 3x - 10 + 2$
$5x = 3x - 8$
$5x - 3x = 3x - 8 - 3x$
$2x = -8$
$\dfrac{2x}{2} = \dfrac{-8}{2}$
$x = -4$
The solution set is $\{-4\}$.

41. $\dfrac{3}{2}x + 2 = \dfrac{1}{2} - \dfrac{1}{2}x$
$2\left(\dfrac{3}{2}x + 2\right) = 2\left(\dfrac{1}{2} - \dfrac{1}{2}x\right)$
$3x + 4 = 1 - x$
$3x + 4 - 4 = 1 - x - 4$
$3x = -x - 3$
$3x + x = -x - 3 + x$
$4x = -3$
$\dfrac{4x}{4} = \dfrac{-3}{4}$
$x = -\dfrac{3}{4}$
The solution set is $\left\{-\dfrac{3}{4}\right\}$.

43. $\dfrac{1}{2}x - 5 = \dfrac{3}{4}x$
$4\left(\dfrac{1}{2}x - 5\right) = 4\left(\dfrac{3}{4}x\right)$
$2x - 20 = 3x$
$2x - 20 + 20 = 3x + 20$
$2x = 3x + 20$
$2x - 3x = 3x + 20 - 3x$
$-x = 20$
$-1(-x) = -1(20)$
$x = -20$
The solution set is $\{-20\}$.

45. $\dfrac{2}{3}p = \dfrac{1}{2}p - \dfrac{1}{3}$
$6\left(\dfrac{2}{3}p\right) = 6\left(\dfrac{1}{2}p - \dfrac{1}{3}\right)$
$4p = 3p - 2$
$4p - 3p = 3p - 2 - 3p$
$p = -2$
The solution set is $\{-2\}$.

47. $0.9t = 0.4 + 0.1t$
$0.9t - 0.1t = 0.4 + 0.1t - 0.1t$
$0.8t = 0.4$
$\dfrac{0.8t}{0.8} = \dfrac{0.4}{0.8}$
$t = 0.5$
The solution set is $\{0.5\}$.

49. $\dfrac{x+1}{3} + \dfrac{x+2}{7} = 2$

$21\left(\dfrac{x+1}{3}\right) + 21\left(\dfrac{x+2}{7}\right) = 21(2)$

$7x + 7 + 3x + 6 = 42$

$10x + 13 = 42$

$10x + 13 - 13 = 42 - 13$

$10x = 29$

$\dfrac{10x}{10} = \dfrac{29}{10}$

$x = 2.9$

The solution set is $\{2.9\}$.

51. $\dfrac{2}{y} + \dfrac{4}{y} = 3$

$\dfrac{6}{y} = 3$

$\left(\dfrac{6}{y}\right)^{-1} = 3^{-1}$

$\dfrac{y}{6} = \dfrac{1}{3}$

$6 \cdot \dfrac{y}{6} = 6 \cdot \dfrac{1}{3}$

$y = 2$

The solution set is $\{2\}$.

53. $\dfrac{1}{2} + \dfrac{2}{x} = \dfrac{3}{4}$

$4x\left(\dfrac{1}{2} + \dfrac{2}{x}\right) = 4x\left(\dfrac{3}{4}\right)$

$2x + 8 = 3x$

$2x + 8 - 2x = 3x - 2x$

$8 = x$ or $x = 8$

The solution set is $\{8\}$.

55. $(x+7)(x-1) = (x+1)^2$

$x^2 - x + 7x - 7 = x^2 + 2x + 1$

$x^2 + 6x - 7 = x^2 + 2x + 1$

$x^2 + 6x - 7 - x^2 = x^2 + 2x + 1 - x^2$

$6x - 7 = 2x + 1$

$6x - 7 - 2x = 2x + 1 - 2x$

$4x - 7 = 1$

$4x - 7 + 7 = 1 + 7$

$4x = 8$

$\dfrac{4x}{4} = \dfrac{8}{4} \Rightarrow x = 2$

The solution set is $\{2\}$.

57. $x(2x-3) = (2x+1)(x-4)$

$2x^2 - 3x = 2x^2 - 7x - 4$

$2x^2 - 3x - 2x^2 = 2x^2 - 7x - 4 - 2x^2$

$-3x = -7x - 4$

$-3x + 7x = -7x - 4 + 7x$

$4x = -4$

$\dfrac{4x}{4} = \dfrac{-4}{4} \Rightarrow x = -1$

The solution set is $\{-1\}$.

59. $z(z^2 + 1) = 3 + z^3$

$z^3 + z = 3 + z^3$

$z^3 + z - z^3 = 3 + z^3 - z^3$

$z = 3$

The solution set is $\{3\}$.

61. $\dfrac{x}{x-2} + 3 = \dfrac{2}{x-2}$

$(x-2)\left(\dfrac{x}{x-2} + 3\right) = (x-2)\left(\dfrac{2}{x-2}\right)$

$x + 3(x-2) = 2$

$x + 3x - 6 = 2$

$4x - 6 = 2$

$4x = 8$

$x = 2$

This solution is not in the domain of the variable so it must be discarded. The equation has no solution. The solution set is $\{\ \}$ or $\varnothing$.

63.
$$\frac{2x}{x^2-4} = \frac{4}{x^2-4} - \frac{3}{x+2}$$
$$(x^2-4)\left(\frac{2x}{x^2-4}\right) = (x^2-4)\left(\frac{4}{x^2-4} - \frac{3}{x+2}\right)$$
$$2x = 4 - 3(x-2)$$
$$2x = 4 - 3x + 6$$
$$2x = 10 - 3x$$
$$5x = 10$$
$$x = 2$$
This solution is not in the domain of the variable so it must be discarded. The equation has no solution. The solution set is $\{\ \}$ or $\varnothing$.

65.
$$\frac{x}{x+2} = \frac{3}{2}$$
$$2x = 3(x+2)$$
$$2x = 3x + 6$$
$$-x = 6$$
$$x = -6$$
The solution set is $\{-6\}$.

67.
$$\frac{5}{2x-3} = \frac{3}{x+5}$$
$$5(x+5) = 3(2x-3)$$
$$5x + 25 = 6x - 9$$
$$5x = 6x - 34$$
$$-x = -34$$
$$x = 34$$
The solution set is $\{34\}$.

69.
$$\frac{6t+7}{4t-1} = \frac{3t+8}{2t-4}$$
$$(6t+7)(2t-4) = (3t+8)(4t-1)$$
$$12t^2 + 14t - 24t - 28 = 12t^2 + 32t - 3t - 8$$
$$12t^2 - 10t - 28 = 12t^2 + 29t - 8$$
$$-10t - 28 = 29t - 8$$
$$-10t = 29t + 20$$
$$-39t = 20$$
$$t = \frac{20}{-39}$$
$$t = -\frac{20}{39}$$
The solution set is $\left\{-\frac{20}{39}\right\}$.

71.
$$\frac{4}{x-2} = \frac{-3}{x+5} + \frac{7}{(x+5)(x-2)}$$
LCD $= (x-2)(x+5)$
$$\frac{4(x+5)}{(x-2)(x+5)} = \frac{-3(x-2)}{(x-2)(x+5)} + \frac{7}{(x+5)(x-2)}$$
$$4(x+5) = -3(x-2) + 7$$
$$4x + 20 = -3x + 6 + 7$$
$$4x + 20 = -3x + 13$$
$$4x = -3x - 7$$
$$7x = -7$$
$$x = -1$$
The solution set is $\{-1\}$.

73.
$$\frac{2}{y+3} + \frac{3}{y-4} = \frac{5}{y+6}$$
LCD $= (y+3)(y-4)(y+6)$
$$(y+3)(y-4)(y+6)\left(\frac{2}{y+3} + \frac{3}{y-4} = \frac{5}{y+6}\right)$$
$$2(y-4)(y+6) + 3(y+3)(y+6) = 5(y+3)(y-4)$$
$$2(y^2+2y-24) + 3(y^2+9y+18) = 5(y^2-y-12)$$
$$2y^2 + 4y - 48 + 3y^2 + 27y + 54 = 5y^2 - 5y - 60$$
$$5y^2 + 31y + 6 = 5y^2 - 5y - 60$$
$$31y + 6 = -5y - 60$$
$$31y = -5y - 66$$
$$36y = -66$$
$$y = -\frac{66}{36}$$
$$y = -\frac{11}{6}$$
The solution set is $\left\{-\frac{11}{6}\right\}$.

Chapter 1: *Graphs, Equations, and Inequalities* SSM: College Algebra EGU

75. $\dfrac{x}{x^2-1} - \dfrac{x+3}{x^2-x} = \dfrac{-3}{x^2+x}$

    LCD $= x(x+1)(x-1)$

    $x(x+1)(x-1)\left(\dfrac{x}{x^2-1} - \dfrac{x+3}{x^2-x} = \dfrac{-3}{x^2+x}\right)$

    $x^2 - (x+1)(x+3) = -3(x-1)$

    $x^2 - (x^2+4x+3) = -3x+3$

    $x^2 - x^2 - 4x - 3 = -3x + 3$

    $-4x - 3 = -3x + 3$

    $-4x = -3x + 6$

    $-x = 6$

    $x = -6$

    The solution set is $\{-6\}$.

77. $y = 5x + 4$

    $2 = 5a + 4$

    $-2 = 5a$

    $a = -\dfrac{2}{5}$

79. $2x + 3y = 6$

    $2a + 3b = 6$

    $3b = -2a + 6$

    $b = -\dfrac{2}{3}a + 2$

81. $ax - b = c, \ a \ne 0$

    $ax - b + b = c + b$

    $ax = c + b$

    $\dfrac{ax}{a} = \dfrac{c+b}{a}$

    $x = \dfrac{c+b}{a}$

83. $\dfrac{x}{a} + \dfrac{x}{b} = c, \ a \ne 0, b \ne 0, a \ne -b$

    $ab\left(\dfrac{x}{a} + \dfrac{x}{b}\right) = ab \cdot c$

    $bx + ax = abc$

    $x(a+b) = abc$

    $\dfrac{x(a+b)}{a+b} = \dfrac{abc}{a+b}$

    $x = \dfrac{abc}{a+b}$

85. $\dfrac{1}{x-a} + \dfrac{1}{x+a} = \dfrac{2}{x-1}$

    Multiply both sides by the LCD, $(x-a)(x+a)(x-1)$, to get:

    $(x+a)(x-1) + (x-a)(x-1) = 2(x-a)(x+a)$

    $x^2 + ax - x - a + x^2 - ax - x + a = 2x^2 - 2a^2$

    $2x^2 - 2x = 2x^2 - 2a^2$

    $2x^2 - 2x - 2x^2 = 2x^2 - 2a^2 - 2x^2$

    $-2x = -2a^2$

    $\dfrac{-2x}{-2} = \dfrac{-2a^2}{-2}$

    $x = a^2$

    where $a \ne -1, 0, 1$.

87. $x + 2a = 16 + ax - 6a$

    when $x = 4$:

    $4 + 2a = 16 + a(4) - 6a$

    $4 + 2a = 16 + 4a - 6a$

    $4 + 2a = 16 - 2a$

    $4a = 12$

    $a = 3$

89. Solving for R:

    $\dfrac{1}{R} = \dfrac{1}{R_1} + \dfrac{1}{R_2}$

    $RR_1R_2\left(\dfrac{1}{R}\right) = RR_1R_2\left(\dfrac{1}{R_1} + \dfrac{1}{R_2}\right)$

    $R_1R_2 = RR_2 + RR_1$

    $R_1R_2 = R(R_2 + R_1)$

    $\dfrac{R_1R_2}{R_2+R_1} = \dfrac{R(R_2+R_1)}{R_2+R_1}$

    $\dfrac{R_1R_2}{R_2+R_1} = R$

91. Solving for R:

    $F = \dfrac{mv^2}{R}$

    $RF = R\left(\dfrac{mv^2}{R}\right)$

    $RF = mv^2$

    $\dfrac{RF}{F} = \dfrac{mv^2}{F} \Rightarrow R = \dfrac{mv^2}{F}$

93. Solving for r:
$$S = \frac{a}{1-r}$$
$$(1-r) \cdot S = (1-r) \cdot \frac{a}{1-r}$$
$$(1-r)S = a$$
$$\frac{(1-r)S}{S} = \frac{a}{S}$$
$$1 - r = \frac{a}{S}$$
$$1 - r - 1 = \frac{a}{S} - 1$$
$$-r = \frac{a}{S} - 1$$
$$r = 1 - \frac{a}{S} \quad \text{or} \quad r = \frac{S-a}{S}$$

95. Let x = amount invested in bonds.
Invested in CDs: $x - 3000$
$$x + (x - 3{,}000) = 20{,}000$$
$$2x - 3{,}000 = 20{,}000$$
$$2x = 23{,}000$$
$$x = 11{,}500$$
$11,500 will be invested in bonds and $8,500 will be invested in CD's.

97. Let x = amount that Scott gets.
Amount for Alice: $\frac{3}{4}x$
Amount for Tricia: $\frac{1}{2}x$
$$x + \left(\frac{3}{4}x\right) + \left(\frac{1}{2}x\right) = 900{,}000$$
$$\frac{9}{4}x = 900{,}000$$
$$x = \frac{4}{9}(900{,}000)$$
$$x = 400{,}000$$
Scott receives $400,000; Alice receives $300,000; Tricia receives $200,000.

99. Let x Sandra's regular hourly wage.
$$40x + (1.5x)(8) = 442$$
$$40x + 12x = 442$$
$$52x = 442$$
$$x = \frac{442}{52} = 8.50$$
Sandra's regular hourly wage is $8.50.

101. Let x = final exam score.
Compute the final average and set equal to 80.
$$\left(\frac{1}{7}\right)(80 + 83 + 71 + 61 + 95 + x + x) = 80$$
Now solve for x:
$$\left(\frac{1}{7}\right)(390 + 2x) = 80$$
$$390 + 2x = 560$$
$$2x = 170$$
$$x = 85$$
Brooke needs to score an 85 on the final exam to get an average of 80 in the course.

103. Let x represent the original price of the house. Then $0.15x$ represents the reduction in the price of the house.
original price − reduction = new price
$$x - 0.15x = 125{,}000$$
$$0.85x = 125{,}000$$
$$x = 147{,}058.82$$
The original price of the house was $147,058.82. The amount of the savings is $0.15(\$147{,}058.82) = \$22{,}058.82$.

105. Let x represent the price the bookstore pays for the book (publisher price). Then $0.35x$ represents the mark up on the book. The selling price of the book is $94.50.
publisher price + mark up = selling price
$$x + 0.35x = 94.50$$
$$1.35x = 94.50$$
$$x = \frac{94.50}{1.35} = 70$$
The bookstore pays $70.00 for the book.

107. Let a = number of adult patrons.
Child patrons: $5200 - a$
$$8.50a + 6.00(5200 - a) = 32{,}200$$
$$8.50a + 31{,}200 - 6.00a = 32{,}200$$
$$2.50a + 31{,}200 = 32{,}200$$
$$2.50a = 1{,}000$$
$$\frac{2.50a}{2.50} = \frac{1{,}000}{2.50}$$
$$a = 400$$
The theater had 400 adult patrons.

109. Let w = width.
Length: $l = w + 8$
Perimeter: $P = 2l + 2w$
$$2l + 2w = 60$$
$$2(w+8) + 2w = 60$$
$$2w + 16 + 2w = 60$$
$$4w + 16 = 60$$
$$4w = 44$$
$$w = 11$$
The rectangle has a width of 11 feet and a length of 19 feet.

111. Step 7 is only allowed if $x \neq 2$, otherwise we are dividing by 0. But step 1 states that $x = 2$, so we have a contradiction.

113. Answers will vary. One example is $3x + 1 = 3x + 6$.

Section 1.3

1. $x^2 - 5x - 6 = (x-6)(x+1)$

3. $\left\{-\dfrac{5}{3}, 3\right\}$

5. repeated; multiplicity two

7. second

9. False; a quadratic equation may have no real solutions.

11. $x^2 - 9x = 0$
$x(x-9) = 0$
$x = 0$
or $x - 9 = 0 \Rightarrow x = 9$
The solution set is $\{0, 9\}$.

13. $x^2 - 25 = 0$
$(x+5)(x-5) = 0$
$x + 5 = 0 \Rightarrow x = -5$
or $x - 5 = 0 \Rightarrow x = 5$
The solution set is $\{-5, 5\}$.

15. $z^2 + z - 6 = 0$
$(z+3)(z-2) = 0$
$z + 3 = 0 \Rightarrow z = -3$
or $z - 2 = 0 \Rightarrow z = 2$
The solution set is $\{-3, 2\}$.

17. $2x^2 - 5x - 3 = 0$
$(2x+1)(x-3) = 0$
$2x + 1 = 0 \Rightarrow x = -\dfrac{1}{2}$
or $x - 3 = 0 \Rightarrow x = 3$
The solution set is $\left\{-\dfrac{1}{2}, 3\right\}$

19. $3t^2 - 48 = 0$
$3(t^2 - 16) = 0$
$3(t+4)(t-4) = 0$
$t + 4 = 0 \Rightarrow t = -4$
or $t - 4 = 0 \Rightarrow t = 4$
The solution set is $\{-4, 4\}$.

21. $x(x+8) + 12 = 0$
$x^2 + 8x + 12 = 0$
$(x+6)(x+2) = 0 \Rightarrow x = -6, x = -2$
The solution set is $\{-2, -6\}$.

23. $4x^2 + 9 = 12x$
$4x^2 - 12x + 9 = 0$
$(2x-3)^2 = 0 \Rightarrow x = \dfrac{3}{2}$
The solution set is $\left\{\dfrac{3}{2}\right\}$.

25. $2x^2 - x = 15$
$2x^2 - x - 15 = 0$
$(2x+5)(x-3) = 0$
$2x + 5 = 0 \Rightarrow x = -\dfrac{5}{2}$
or $x - 3 = 0 \Rightarrow x = 3$
The solution set is $\left\{-\dfrac{5}{2}, 3\right\}$.

27. $\dfrac{4(x-2)}{x-3} + \dfrac{3}{x} = \dfrac{-3}{x(x-3)}$

$x(x-3)\left(\dfrac{4(x-2)}{x-3} + \dfrac{3}{x}\right) = \left(\dfrac{-3}{x(x-3)}\right)x(x-3)$

$x(x-3)\left(\dfrac{4(x-2)}{x-3}\right) + x(x-3)\left(\dfrac{3}{x}\right) = -3$

$x(4(x-2)) + (x-3)(3) = -3$

$4x^2 - 8x + 3x - 9 = -3$

$4x^2 - 5x - 6 = 0$

$(4x+3)(x-2) = 0$

$4x + 3 = 0 \Rightarrow x = -\dfrac{3}{4}$

or $x - 2 \Rightarrow x = 2$

Since neither of these values causes a denominator to equal zero, the solution set is $\left\{-\dfrac{3}{4}, 2\right\}$.

29. $x^2 = 25 \Rightarrow x = \pm\sqrt{25} \Rightarrow x = \pm 5$
The solution set is $\{-5, 5\}$.

31. $(x-1)^2 = 4$

$x - 1 = \pm\sqrt{4}$

$x - 1 = \pm 2$

$x - 1 = 2$ or $x - 1 = -2$

$\Rightarrow x = 3$ or $x = -1$

The solution set is $\{-1, 3\}$.

33. $(2x+3)^2 = 9$

$2x + 3 = \pm\sqrt{9}$

$2x + 3 = \pm 3$

$2x + 3 = 3$ or $2x + 3 = -3$

$\Rightarrow x = 0$ or $x = -3$

The solution set is $\{-3, 0\}$.

35. $\left(\dfrac{8}{2}\right)^2 = 4^2 = 16$

37. $\left(\dfrac{\left(-\dfrac{1}{2}\right)}{2}\right)^2 = \left(-\dfrac{1}{4}\right)^2 = \dfrac{1}{16}$

39. $x^2 + 4x = 21$

$x^2 + 4x + 4 = 21 + 4$

$(x+2)^2 = 25$

$x + 2 = \pm\sqrt{25} \Rightarrow x + 2 = \pm 5$

$x = -2 \pm 5 \Rightarrow x = 3$ or $x = -7$

The solution set is $\{-7, 3\}$.

41. $x^2 - \dfrac{1}{2}x - \dfrac{3}{16} = 0$

$x^2 - \dfrac{1}{2}x = \dfrac{3}{16}$

$x^2 - \dfrac{1}{2}x + \dfrac{1}{16} = \dfrac{3}{16} + \dfrac{1}{16}$

$\left(x - \dfrac{1}{4}\right)^2 = \dfrac{1}{4}$

$x - \dfrac{1}{4} = \pm\sqrt{\dfrac{1}{4}} = \pm\dfrac{1}{2}$

$x = \dfrac{1}{4} \pm \dfrac{1}{2} \Rightarrow x = \dfrac{3}{4}$ or $x = -\dfrac{1}{4}$

The solution set is $\left\{-\dfrac{1}{4}, \dfrac{3}{4}\right\}$.

43. $3x^2 + x - \dfrac{1}{2} = 0$

$x^2 + \dfrac{1}{3}x - \dfrac{1}{6} = 0$

$x^2 + \dfrac{1}{3}x = \dfrac{1}{6}$

$x^2 + \dfrac{1}{3}x + \dfrac{1}{36} = \dfrac{1}{6} + \dfrac{1}{36}$

$\left(x + \dfrac{1}{6}\right)^2 = \dfrac{7}{36}$

$x + \dfrac{1}{6} = \pm\sqrt{\dfrac{7}{36}}$

$x + \dfrac{1}{6} = \pm\dfrac{\sqrt{7}}{6}$

$x = \dfrac{-1 \pm \sqrt{7}}{6}$

The solution set is $\left\{\dfrac{-1-\sqrt{7}}{6}, \dfrac{-1+\sqrt{7}}{6}\right\}$.

45. $x^2 - 4x + 2 = 0$
$a = 1, \ b = -4, \ c = 2$
$x = \dfrac{-(-4) \pm \sqrt{(-4)^2 - 4(1)(2)}}{2(1)}$
$= \dfrac{4 \pm \sqrt{16 - 8}}{2} = \dfrac{4 \pm \sqrt{8}}{2}$
$= \dfrac{4 \pm 2\sqrt{2}}{2} = 2 \pm \sqrt{2}$
The solution set is $\{2 - \sqrt{2}, \ 2 + \sqrt{2}\}$.

47. $x^2 - 4x - 1 = 0$
$a = 1, \ b = -4, \ c = -1$
$x = \dfrac{-(-4) \pm \sqrt{(-4)^2 - 4(1)(-1)}}{2(1)}$
$= \dfrac{4 \pm \sqrt{16 + 4}}{2} = \dfrac{4 \pm \sqrt{20}}{2}$
$= \dfrac{4 \pm 2\sqrt{5}}{2} = 2 \pm \sqrt{5}$
The solution set is $\{2 - \sqrt{5}, \ 2 + \sqrt{5}\}$.

49. $2x^2 - 5x + 3 = 0$
$a = 2, \ b = -5, \ c = 3$
$x = \dfrac{-(-5) \pm \sqrt{(-5)^2 - 4(2)(3)}}{2(2)}$
$= \dfrac{5 \pm \sqrt{25 - 24}}{4} = \dfrac{5 \pm 1}{4}$
The solution set is $\left\{1, \ \dfrac{3}{2}\right\}$.

51. $4y^2 - y + 2 = 0$
$a = 4, \ b = -1, \ c = 2$
$y = \dfrac{-(-1) \pm \sqrt{(-1)^2 - 4(4)(2)}}{2(4)}$
$= \dfrac{1 \pm \sqrt{1 - 32}}{8} = \dfrac{1 \pm \sqrt{-31}}{8}$
No real solution.

53. $4x^2 = 1 - 2x$
$4x^2 + 2x - 1 = 0$
$a = 4, \ b = 2, \ c = -1$
$x = \dfrac{-2 \pm \sqrt{2^2 - 4(4)(-1)}}{2(4)}$
$= \dfrac{-2 \pm \sqrt{4 + 16}}{8} = \dfrac{-2 \pm \sqrt{20}}{8}$
$= \dfrac{-2 \pm 2\sqrt{5}}{8} = \dfrac{-1 \pm \sqrt{5}}{4}$
The solution set is $\left\{\dfrac{-1 - \sqrt{5}}{4}, \ \dfrac{-1 + \sqrt{5}}{4}\right\}$.

55. $4x^2 = 9x + 2$
$4x^2 - 9x - 2 = 0$
$a = 4, \ b = -9, \ c = -2$
$x = \dfrac{-(-9) \pm \sqrt{(-9)^2 - 4(4)(-2)}}{2(4)}$
$= \dfrac{9 \pm \sqrt{81 + 32}}{8} = \dfrac{9 \pm \sqrt{113}}{8}$
The solution set is $\left\{\dfrac{9 - \sqrt{113}}{8}, \ \dfrac{9 + \sqrt{113}}{8}\right\}$.

57. $9t^2 - 6t + 1 = 0$
$a = 9, \ b = -6, \ c = 1$
$t = \dfrac{-(-6) \pm \sqrt{(-6)^2 - 4(9)(1)}}{2(9)}$
$= \dfrac{6 \pm \sqrt{36 - 36}}{18} = \dfrac{6 \pm 0}{18} = \dfrac{1}{3}$
The solution set is $\left\{\dfrac{1}{3}\right\}$.

59. $\dfrac{3}{4}x^2 - \dfrac{1}{4}x - \dfrac{1}{2} = 0$

$4\left(\dfrac{3}{4}x^2 - \dfrac{1}{4}x - \dfrac{1}{2}\right) = (0)(4)$

$3x^2 - x - 2 = 0$

$a = 3, \quad b = -1, \quad c = -2$

$x = \dfrac{-(-1) \pm \sqrt{(-1)^2 - 4(3)(-2)}}{2(3)}$

$= \dfrac{1 \pm \sqrt{1+24}}{6} = \dfrac{1 \pm \sqrt{25}}{6} = \dfrac{1 \pm 5}{6}$

$\Rightarrow x = \dfrac{1+5}{6}$ or $x = \dfrac{1-5}{6}$

$x = \dfrac{6}{6} = 1$ or $x = \dfrac{-4}{6} = -\dfrac{2}{3}$

The solution set is $\left\{-\dfrac{2}{3}, 1\right\}$.

61. $4 - \dfrac{1}{x} - \dfrac{2}{x^2} = 0$

$(x^2)\left(4 - \dfrac{1}{x} - \dfrac{2}{x^2}\right) = (0)(x^2)$

$4x^2 - x - 2 = 0$

$a = 4, \quad b = -1, \quad c = -2$

$x = \dfrac{-(-1) \pm \sqrt{(-1)^2 - 4(4)(-2)}}{2(4)}$

$= \dfrac{1 \pm \sqrt{1+32}}{8} = \dfrac{1 \pm \sqrt{33}}{8}$

Since neither of these values causes a denominator to equal zero, the solution set is $\left\{\dfrac{1+\sqrt{33}}{8}, \dfrac{1-\sqrt{33}}{8}\right\}$.

63. $x^2 - 5 = 0$

$x^2 = 5 \Rightarrow x = \pm\sqrt{5}$

The solution set is $\{-\sqrt{5}, \sqrt{5}\}$.

65. $16x^2 - 8x + 1 = 0$

$(4x-1)(4x-1) = 0$

$4x - 1 = 0 \Rightarrow x = \dfrac{1}{4}$

The solution set is $\left\{\dfrac{1}{4}\right\}$.

67. $10x^2 - 19x - 15 = 0$

$(5x+3)(2x-5) = 0$

$5x + 3 = 0$ or $2x - 5 = 0$

$\Rightarrow x = -\dfrac{3}{5}$ or $x = \dfrac{5}{2}$

The solution set is $\left\{-\dfrac{3}{5}, \dfrac{5}{2}\right\}$.

69. $2 + z = 6z^2$

$0 = 6z^2 - z - 2$

$0 = (3z-2)(2z+1)$

$3z - 2 = 0$ or $2z + 1 = 0 \Rightarrow z = \dfrac{2}{3}$ or $z = -\dfrac{1}{2}$

The solution set is $\left\{-\dfrac{1}{2}, \dfrac{2}{3}\right\}$.

71. $x^2 + \sqrt{2}x = \dfrac{1}{2}$

$x^2 + \sqrt{2}x - \dfrac{1}{2} = 0$

$2\left(x^2 + \sqrt{2}x - \dfrac{1}{2}\right) = (0)(2)$

$2x^2 + 2\sqrt{2}x - 1 = 0$

$a = 2, \quad b = 2\sqrt{2}, \quad c = -1$

$x = \dfrac{-(2\sqrt{2}) \pm \sqrt{(2\sqrt{2})^2 - 4(2)(-1)}}{2(2)}$

$= \dfrac{-2\sqrt{2} \pm \sqrt{8+8}}{4} = \dfrac{-2\sqrt{2} \pm \sqrt{16}}{4}$

$= \dfrac{-2\sqrt{2} \pm 4}{4} = \dfrac{-\sqrt{2} \pm 2}{2}$

The solution set is $\left\{\dfrac{-\sqrt{2}+2}{2}, \dfrac{-\sqrt{2}-2}{2}\right\}$

73. $x^2 + x = 4$

$x^2 + x - 4 = 0$

$a = 1, \quad b = 1, \quad c = -4$

$x = \dfrac{-(1) \pm \sqrt{(1)^2 - 4(1)(-4)}}{2(1)}$

$= \dfrac{-1 \pm \sqrt{1+16}}{2} = \dfrac{-1 \pm \sqrt{17}}{2}$

The solution set is $\left\{\dfrac{-1+\sqrt{17}}{2}, \dfrac{-1-\sqrt{17}}{2}\right\}$

Chapter 1: *Graphs, Equations, and Inequalities*   *SSM*: College Algebra EGU

75. $2x^2 - 6x + 7 = 0$
$a = 2, \ b = -6, \ c = 7$
$b^2 - 4ac = (-6)^2 - 4(2)(7)$
$= 36 - 56 = -20$
Since the discriminant < 0, we have no real solutions.

77. $9x^2 - 30x + 25 = 0$
$a = 9, \ b = -30, \ c = 25$
$b^2 - 4ac = (-30)^2 - 4(9)(25)$
$= 900 - 900 = 0$
Since the discriminant = 0, we have one repeated real solution.

79. $3x^2 + 5x - 8 = 0$
$a = 3, \ b = 5, \ c = -8$
$b^2 - 4ac = (5)^2 - 4(3)(-8)$
$= 25 + 96 = 121$
Since the discriminant > 0, we have two unequal real solutions.

81. Let w represent the width of window.
Then $l = w + 2$ represents the length of the window.
Since the area is 143 square feet, we have:
$w(w + 2) = 143$
$w^2 + 2w - 143 = 0$
$(w + 13)(w - 11) = 0$
$\cancel{w = -13}$ or $w = 11$
Discard the negative solution since width cannot be negative. The width of the rectangular window is 11 feet and the length is 13 feet.

83. Let l represent the length of the rectangle.
Let w represent the width of the rectangle.
The perimeter is 26 meters and the area is 40 square meters.
$2l + 2w = 26$
$l + w = 13 \quad$ so $\quad w = 13 - l$
$lw = 40$
$l(13 - l) = 40$
$13l - l^2 = 40$
$l^2 - 13l + 40 = 0$
$(l - 8)(l - 5) = 0$
$l = 8$ or $l = 5$
$w = 5 \quad\quad w = 8$
The dimensions are 5 meters by 8 meters.

85. Let x = length of side of original sheet in feet.
Length of box: $x - 2$ feet
Width of box: $x - 2$ feet
Height of box: 1 foot
$V = l \cdot w \cdot h$
$4 = (x - 2)(x - 2)(1)$
$4 = x^2 - 4x + 4$
$0 = x^2 - 4x$
$0 = x(x - 4)$
$x = 0$ or $x = 4$
Discard $x = 0$ since that is not a feasible length for the original sheet. Therefore, the original sheet should measure 4 feet on each side.

87. a. When the ball strikes the ground, the distance from the ground will be 0. Therefore, we solve
$96 + 80t - 16t^2 = 0$
$-16t^2 + 80t + 96 = 0$
$t^2 - 5t - 6 = 0$
$(t - 6)(t + 1) = 0$
$t = 6$ or $t = -1$
Discard the negative solution since the time of flight must be positive. The ball will strike the ground after 6 seconds.

b. When the ball passes the top of the building, it will be 96 feet from the ground. Therefore, we solve
$96 + 80t - 16t^2 = 96$
$-16t^2 + 80t = 0$
$t^2 - 5t = 0$
$t(t - 5) = 0$
$t = 0$ or $t = 5$
The ball is at the top of the building at time $t = 0$ when it is thrown. It will pass the top of the building on the way down after 5 seconds.

89. Let x represent the number of centimeters the length and width should be reduced.
$12-x$ = the new length, $7-x$ = the new width.
The new volume is 90% of the old volume.
$$(12-x)(7-x)(3) = 0.9(12)(7)(3)$$
$$3x^2 - 57x + 252 = 226.8$$
$$3x^2 - 57x + 25.2 = 0$$
$$x^2 - 19x + 8.4 = 0$$
$$x = \frac{-(-19) \pm \sqrt{(-19)^2 - 4(1)(8.4)}}{2(1)}$$
$$= \frac{19 \pm \sqrt{327.4}}{2}$$
$$= \frac{19 \pm 18.1}{2} \rightarrow x = 0.45 \text{ or } 18.55$$

Since 18.55 exceeds the dimensions, it is discarded. The dimensions of the new chocolate bar are: 11.55 cm by 6.55 cm by 3 cm.

91. Let x represent the width of the border measured in feet.
The radius of the pool is 5 feet.
Then $x+5$ represents the radius of the circle, including both the pool and the border.
The total area of the pool and border is
$$A_T = \pi(x+5)^2.$$
The area of the pool is $A_P = \pi(5)^2 = 25\pi$.
The area of the border is
$$A_B = A_T - A_P = \pi(x+5)^2 - 25\pi.$$
Since the concrete is 3 inches or 0.25 feet thick, the volume of the concrete in the border is
$$0.25 A_B = 0.25\left(\pi(x+5)^2 - 25\pi\right)$$
Solving the volume equation:
$$0.25\left(\pi(x+5)^2 - 25\pi\right) = 27$$
$$\pi\left(x^2 + 10x + 25 - 25\right) = 108$$
$$\pi x^2 + 10\pi x - 108 = 0$$
$$x = \frac{-10\pi \pm \sqrt{(10\pi)^2 - 4(\pi)(-108)}}{2(\pi)}$$
$$= \frac{-31.42 \pm \sqrt{2344.1285}}{6.28}$$
$$= \frac{-31.42 \pm 48.42}{6.28}$$
$$= 2.71 \text{ or } -12.71$$

The width of the border is approximately 2.71 feet.

93. Let x represent the width of the border measured in feet.
The total area is $A_T = (6+2x)(10+2x)$.
The area of the garden is $A_G = 6 \cdot 10 = 60$.
The area of the border is
$A_B = A_T - A_G = (6+2x)(10+2x) - 60$.
Since the concrete is 3 inches or 0.25 feet thick, the volume of the concrete in the border is
$0.25 A_B = 0.25\left((6+2x)(10+2x) - 60\right)$
Solving the volume equation:
$$0.25\left((6+2x)(10+2x) - 60\right) = 27$$
$$60 + 32x + 4x^2 - 60 = 108$$
$$4x^2 + 32x - 108 = 0$$
$$x^2 + 8x - 27 = 0$$
$$x = \frac{-8 \pm \sqrt{8^2 - 4(1)(-27)}}{2(1)}$$
$$= \frac{-8 \pm \sqrt{172}}{2}$$
$$= \frac{-8 \pm 13.11}{2}$$
$$= 2.56 \text{ or } -10.56$$
Discard the negative solution. The width of the border is approximately 2.56 feet.

95. $$\frac{1}{2}n(n+1) = 666$$
$$n(n+1) = 1332$$
$$n^2 + n - 1332 = 0$$
$$(n-36)(n+37) = 0$$
$$n = 36 \text{ or } n = -37$$
Since the number of consecutive integers cannot be negative, we discard the negative value. We must add 36 consecutive integers, beginning at 1, in order to get a sum of 666.

97. The roots of a quadratic equation are
$$x_1 = \frac{-b - \sqrt{b^2 - 4ac}}{2a} \text{ and } x_2 = \frac{-b + \sqrt{b^2 - 4ac}}{2a}$$
$$x_1 + x_2 = \frac{-b - \sqrt{b^2 - 4ac}}{2a} + \frac{-b + \sqrt{b^2 - 4ac}}{2a}$$
$$= \frac{-b - \sqrt{b^2 - 4ac} - b + \sqrt{b^2 - 4ac}}{2a}$$
$$= \frac{-2b}{2a}$$
$$= -\frac{b}{a}$$

99. In order to have one repeated solution, we need the discriminant to be 0.
$$b^2 - 4ac = 0$$
$$1^2 - 4(k)(k) = 0$$
$$1 - 4k^2 = 0$$
$$4k^2 = 1$$
$$k^2 = \frac{1}{4}$$
$$k = \pm\sqrt{\frac{1}{4}}$$
$$k = \frac{1}{2} \quad \text{or} \quad k = -\frac{1}{2}$$

101. For $ax^2 + bx + c = 0$:
$$x_1 = \frac{-b - \sqrt{b^2 - 4ac}}{2a} \quad \text{and} \quad x_2 = \frac{-b + \sqrt{b^2 - 4ac}}{2a}$$
For $ax^2 - bx + c = 0$:
$$x_1^* = \frac{-(-b) - \sqrt{(-b)^2 - 4ac}}{2a}$$
$$= \frac{b - \sqrt{b^2 - 4ac}}{2a}$$
$$= -\left(\frac{-b + \sqrt{b^2 - 4ac}}{2a}\right) = -x_2$$
and
$$x_2^* = \frac{-(-b) + \sqrt{(-b)^2 - 4ac}}{2a}$$
$$= \frac{b + \sqrt{b^2 - 4ac}}{2a}$$
$$= -\left(\frac{-b - \sqrt{b^2 - 4ac}}{2a}\right) = -x_1$$

103. a. $x^2 = 9$ and $x = 3$ are not equivalent because they do not have the same solution set. In the first equation we can also have $x = -3$.

   b. $x = \sqrt{9}$ and $x = 3$ are equivalent because $\sqrt{9} = 3$.

   c. $(x-1)(x-2) = (x-1)^2$ and $x - 2 = x - 1$ are not equivalent because they do not have the same solution set.
   The first equation has the solution set $\{1\}$ while the second equation has no solutions.

105. Answers will vary. Knowing the discriminant allows us to know how many real solutions the equation will have.

107. Answers will vary.

Section 1.4

1. Integers: $\{-3, 0\}$
   Rationals: $\left\{-3, 0, \frac{6}{5}\right\}$

3. $\dfrac{3}{2+\sqrt{3}} = \dfrac{3}{2+\sqrt{3}} \cdot \dfrac{2-\sqrt{3}}{2-\sqrt{3}}$
   $= \dfrac{3(2-\sqrt{3})}{2^2 - (\sqrt{3})^2}$
   $= \dfrac{3(2-\sqrt{3})}{4-3}$
   $= 3(2-\sqrt{3})$

5. $\{-2i, 2i\}$

7. True; the set of real numbers is a subset of the complex numbers.

9. $(2-3i) + (6+8i) = (2+6) + (-3+8)i = 8 + 5i$

11. $(-3+2i) - (4-4i) = (-3-4) + (2-(-4))i$
    $= -7 + 6i$

13. $(2-5i) - (8+6i) = (2-8) + (-5-6)i$
    $= -6 - 11i$

15. $3(2-6i) = 6 - 18i$

17. $2i(2-3i) = 4i - 6i^2 = 4i - 6(-1) = 6 + 4i$

19. $(3-4i)(2+i) = 6 + 3i - 8i - 4i^2$
    $= 6 - 5i - 4(-1)$
    $= 10 - 5i$

21. $(-6+i)(-6-i) = 36 + 6i - 6i - i^2$
    $= 36 - (-1)$
    $= 37$

23. $\dfrac{10}{3-4i} = \dfrac{10}{3-4i} \cdot \dfrac{3+4i}{3+4i} = \dfrac{30+40i}{9+12i-12i-16i^2}$
    $= \dfrac{30+40i}{9-16(-1)} = \dfrac{30+40i}{25}$
    $= \dfrac{30}{25} + \dfrac{40}{25}i$
    $= \dfrac{6}{5} + \dfrac{8}{5}i$

25. $\dfrac{2+i}{i} = \dfrac{2+i}{i} \cdot \dfrac{-i}{-i} = \dfrac{-2i - i^2}{-i^2}$
    $= \dfrac{-2i - (-1)}{-(-1)} = \dfrac{1-2i}{1}$
    $= 1 - 2i$

27. $\dfrac{6-i}{1+i} = \dfrac{6-i}{1+i} \cdot \dfrac{1-i}{1-i} = \dfrac{6 - 6i - i + i^2}{1 - i + i - i^2}$
    $= \dfrac{6 - 7i + (-1)}{1 - (-1)} = \dfrac{5 - 7i}{2}$
    $= \dfrac{5}{2} - \dfrac{7}{2}i$

29. $\left(\dfrac{1}{2} + \dfrac{\sqrt{3}}{2}i\right)^2 = \dfrac{1}{4} + 2\left(\dfrac{1}{2}\right)\left(\dfrac{\sqrt{3}}{2}i\right) + \dfrac{3}{4}i^2$
    $= \dfrac{1}{4} + \dfrac{\sqrt{3}}{2}i + \dfrac{3}{4}(-1)$
    $= -\dfrac{1}{2} + \dfrac{\sqrt{3}}{2}i$

31. $(1+i)^2 = 1 + 2i + i^2 = 1 + 2i + (-1) = 2i$

33. $i^{23} = i^{22+1} = i^{22} \cdot i = \left(i^2\right)^{11} \cdot i = (-1)^{11} i = -i$

Chapter 1: Graphs, Equations, and Inequalities   **SSM:** College Algebra EGU

35. $i^{-15} = \dfrac{1}{i^{15}} = \dfrac{1}{i^{14+1}} = \dfrac{1}{i^{14} \cdot i} = \dfrac{1}{(i^2)^7 \cdot i}$

    $= \dfrac{1}{(-1)^7 i} = \dfrac{1}{-i} = \dfrac{1}{-i} \cdot \dfrac{i}{i} = \dfrac{i}{-i^2} = \dfrac{i}{-(-1)}$

    $= i$

37. $i^6 - 5 = (i^2)^3 - 5 = (-1)^3 - 5 = -1 - 5 = -6$

39. $6i^3 - 4i^5 = i^3(6 - 4i^2)$

    $= i^2 \cdot i(6 - 4(-1))$

    $= -1 \cdot i(10)$

    $= -10i$

41. $(1+i)^3 = (1+i)(1+i)(1+i) = (1+2i+i^2)(1+i)$

    $= (1+2i-1)(1+i) = 2i(1+i)$

    $= 2i + 2i^2 = 2i + 2(-1)$

    $= -2 + 2i$

43. $i^7(1+i^2) = i^7(1+(-1)) = i^7(0) = 0$

45. $i^6 + i^4 + i^2 + 1 = (i^2)^3 + (i^2)^2 + i^2 + 1$

    $= (-1)^3 + (-1)^2 + (-1) + 1$

    $= -1 + 1 - 1 + 1$

    $= 0$

47. $\sqrt{-4} = 2i$

49. $\sqrt{-25} = 5i$

51. $\sqrt{(3+4i)(4i-3)} = \sqrt{12i - 9 + 16i^2 - 12i}$

    $= \sqrt{-9 + 16(-1)}$

    $= \sqrt{-25}$

    $= 5i$

53. $x^2 + 4 = 0$

    $x^2 = -4$

    $x = \pm\sqrt{-4}$

    $x = \pm 2i$

    The solution set is $\{\pm 2i\}$.

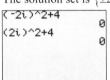

55. $x^2 - 16 = 0$

    $(x+4)(x-4) = 0 \Rightarrow x = -4, x = 4$

    The solution set is $\{\pm 4\}$.

57. $x^2 - 6x + 13 = 0$

    $a = 1, b = -6, c = 13,$

    $b^2 - 4ac = (-6)^2 - 4(1)(13) = 36 - 52 = -16$

    $x = \dfrac{-(-6) \pm \sqrt{-16}}{2(1)} = \dfrac{6 \pm 4i}{2} = 3 \pm 2i$

    The solution set is $\{3 - 2i, 3 + 2i\}$.

59. $x^2 - 6x + 10 = 0$
$a = 1, b = -6, c = 10$
$b^2 - 4ac = (-6)^2 - 4(1)(10) = 36 - 40 = -4$
$x = \dfrac{-(-6) \pm \sqrt{-4}}{2(1)} = \dfrac{6 \pm 2i}{2} = 3 \pm i$
The solution set is $\{3 - i, 3 + i\}$.

61. $8x^2 - 4x + 1 = 0$
$a = 8, b = -4, c = 1$
$b^2 - 4ac = (-4)^2 - 4(8)(1) = 16 - 32 = -16$
$x = \dfrac{-(-4) \pm \sqrt{-16}}{2(8)} = \dfrac{4 \pm 4i}{16} = \dfrac{1}{4} \pm \dfrac{1}{4}i$
The solution set is $\left\{\dfrac{1}{4} - \dfrac{1}{4}i, \dfrac{1}{4} + \dfrac{1}{4}i\right\}$.

63. $5x^2 + 2x + 1 = 0$
$a = 5, b = 2, c = 1$
$b^2 - 4ac = (2)^2 - 4(5)(1) = 4 - 20 = -16$
$x = \dfrac{-2 \pm \sqrt{-16}}{2(5)} = \dfrac{-2 \pm 4i}{10} = -\dfrac{1}{5} \pm \dfrac{2}{5}i$
The solution set is $\left\{-\dfrac{1}{5} - \dfrac{2}{5}i, -\dfrac{1}{5} + \dfrac{2}{5}i\right\}$.

65. $x^2 + x + 1 = 0$
$a = 1, b = 1, c = 1,$
$b^2 - 4ac = 1^2 - 4(1)(1) = 1 - 4 = -3$
$x = \dfrac{-1 \pm \sqrt{-3}}{2(1)} = \dfrac{-1 \pm \sqrt{3}i}{2} = -\dfrac{1}{2} \pm \dfrac{\sqrt{3}}{2}i$
The solution set is $\left\{-\dfrac{1}{2} - \dfrac{\sqrt{3}}{2}i, -\dfrac{1}{2} + \dfrac{\sqrt{3}}{2}i\right\}$.

67. $x^3 - 8 = 0$
$(x - 2)(x^2 + 2x + 4) = 0$
$x - 2 = 0 \Rightarrow x = 2$
$x^2 + 2x + 4 = 0$
$a = 1, b = 2, c = 4$
$b^2 - 4ac = 2^2 - 4(1)(4) = 4 - 16 = -12$
$x = \dfrac{-2 \pm \sqrt{-12}}{2(1)} = \dfrac{-2 \pm 2\sqrt{3}i}{2} = -1 \pm \sqrt{3}i$
The solution set is $\left\{2, -1 - \sqrt{3}i, -1 + \sqrt{3}i\right\}$.

69. $x^4 - 16 = 0$
$(x^2 - 4)(x^2 + 4) = 0 \Rightarrow (x - 2)(x + 2)(x^2 + 4) = 0$
$x - 2 = 0 \Rightarrow x = 2$
$x + 2 = 0 \Rightarrow x = -2$
$x^2 + 4 = 0 \Rightarrow x = \pm 2i$
The solution set is $\{-2, 2, -2i, 2i\}$.

71. $x^4 + 13x^2 + 36 = 0$
$(x^2 + 9)(x^2 + 4) = 0$
$x^2 + 9 = 0 \Rightarrow x = \pm 3i$
$x^2 + 4 = 0 \Rightarrow x = \pm 2i$
The solution set is $\{-3i, 3i, -2i, 2i\}$.

73. $3x^2 - 3x + 4 = 0$
$a = 3, b = -3, c = 4$
$b^2 - 4ac = (-3)^2 - 4(3)(4) = 9 - 48 = -39$
The equation has two complex conjugate solutions.

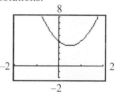

75. $2x^2 + 3x - 4 = 0$
$a = 2, b = 3, c = -4$
$b^2 - 4ac = 3^2 - 4(2)(-4) = 9 + 32 = 41$
The equation has two unequal real solutions.

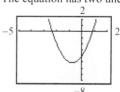

77. $9x^2 - 12x + 4 = 0$
$a = 9, b = -12, c = 4$
$b^2 - 4ac = (-12)^2 - 4(9)(4) = 144 - 144 = 0$
The equation has a repeated real solution.

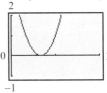

79. The other solution is the conjugate of $2 + 3i$, or $2 - 3i$.

81. $z + \bar{z} = 3 - 4i + \overline{3 - 4i} = 3 - 4i + 3 + 4i = 6$

83. $z \cdot \bar{z} = (3 - 4i)\overline{(3 - 4i)} = (3 - 4i)(3 + 4i)$
$= 9 + 12i - 12i - 16i^2 = 9 - 16(-1)$
$= 25$

85. $z + \bar{z} = a + bi + \overline{a + bi}$
$= a + bi + a - bi$
$= 2a$
$z - \bar{z} = a + bi - \overline{(a + bi)}$
$= a + bi - (a - bi)$
$= a + bi - a + bi$
$= 2bi$

87. $\overline{z + w} = \overline{(a + bi) + (c + di)}$
$= \overline{(a + c) + (b + d)i}$
$= (a + c) - (b + d)i$
$= (a - bi) + (c - di)$
$= \overline{a + bi} + \overline{c + di}$
$= \bar{z} + \bar{w}$

89. Answers will vary.

91. Answers will vary.

Section 1.5

1. True

3. $6x^3 - 2x^2 = 2x^2(3x - 1)$

5. $-a$

7. quadratic in form

9. $\sqrt{x} = 3$
$(\sqrt{x})^2 = 3^2$
$x = 9$
Check: $\sqrt{9} = 3$
The solution is $x = 9$.

11. $\sqrt{y + 3} = 5$
$(\sqrt{y + 3})^2 = 5^2$
$y + 3 = 25 \to y = 22$
Check: $\sqrt{22 + 3} = \sqrt{25} = 5$
The solution is $y = 22$.

13. $\sqrt{2t - 1} = 1$
$(\sqrt{2t - 1})^2 = 1^2$
$2t - 1 = 1 \to 2t = 2 \to t = 1$
Check: $\sqrt{2(1) - 1} = \sqrt{1} = 1$
The solution is $t = 1$.

15. $\sqrt{3t + 4} = -6$
Since the principal square root is never negative, the equation has no real solution.

17. $\sqrt[3]{1 - 2x} - 3 = 0$
$\sqrt[3]{1 - 2x} = 3$
$(\sqrt[3]{1 - 2x})^3 = 3^3$
$1 - 2x = 27 \to -2x = 26 \to x = -13$
Check: $\sqrt[3]{1 - 2(-13)} - 3 = \sqrt[3]{27} - 3 = 0$
The solution is $x = -13$.

SSM: College Algebra EGU ***Chapter 1:** Graphs, Equations, and Inequalities*

19. $\sqrt[4]{5x-4} = 2$
$\left(\sqrt[4]{5x-4}\right)^4 = 2^4$
$5x - 4 = 16$
$5x = 20$
$x = 4$
Check:
$\sqrt[4]{5(4)-4} = \sqrt[4]{16} = 2 \checkmark$
The solution is $x = 4$.

21. $\sqrt[5]{x^2 + 2x} = -1$
$\left(\sqrt[5]{x^2 + 2x}\right)^5 = (-1)^5$
$x^2 + 2x = -1$
$x^2 + 2x + 1 = 0$
$(x+1)^2 = 0$
$x + 1 = 0$
$x = -1$
Check:
$\sqrt[5]{(-1)^2 + 2(-1)} = \sqrt[5]{1-2} = \sqrt[5]{-1} = -1 \checkmark$
The solution is $x = -1$.

23. $x = 8\sqrt{x}$
$(x)^2 = \left(8\sqrt{x}\right)^2$
$x^2 = 64x$
$x^2 - 64x = 0$
$x(x - 64) = 0$
$x = 0$ or $x = 64$
Check:
$0 = 8\sqrt{0}$ $\quad 64 = 8\sqrt{64}$
$0 = 0 \checkmark$ $\quad 64 = 8 \cdot 8$
$\quad\quad\quad\quad\quad 64 = 64 \checkmark$
The solution set is $\{0, 64\}$.

25. $\sqrt{15 - 2x} = x$
$\left(\sqrt{15 - 2x}\right)^2 = x^2$
$15 - 2x = x^2 \to x^2 + 2x - 15 = 0$
$(x+5)(x-3) = 0 \to x = -5$ or $x = 3$
Check -5: $\sqrt{15 - 2(-5)} = \sqrt{25}$
$= 5 \neq -5$
Check 3: $\sqrt{15 - 2(3)} = \sqrt{9} = 3 = 3$
The solution is $x = 3$.

27. $x = 2\sqrt{x-1}$
$x^2 = \left(2\sqrt{x-1}\right)^2$
$x^2 = 4(x-1) \to x^2 = 4x - 4$
$x^2 - 4x + 4 = 0 \to (x-2)^2 = 0 \to x = 2$
Check: $2 = 2\sqrt{2-1} \to 2 = 2$
The solution is $x = 2$.

29. $\sqrt{x^2 - x - 4} = x + 2$
$\left(\sqrt{x^2 - x - 4}\right)^2 = (x+2)^2$
$x^2 - x - 4 = x^2 + 4x + 4$
$-8 = 5x \to -\dfrac{8}{5} = x$
Check:
$\sqrt{\left(-\dfrac{8}{5}\right)^2 - \left(-\dfrac{8}{5}\right) - 4} = \left(-\dfrac{8}{5}\right) + 2$
$\sqrt{\dfrac{64}{25} + \dfrac{8}{5} - 4} = \dfrac{2}{5}$
$\sqrt{\dfrac{4}{25}} = \dfrac{2}{5}$
$\dfrac{2}{5} = \dfrac{2}{5} \checkmark$
The solution is $x = -\dfrac{8}{5}$.

31. $3+\sqrt{3x+1}=x$
$\sqrt{3x+1}=x-3$
$\left(\sqrt{3x+1}\right)^2=(x-3)^2$
$3x+1=x^2-6x+9$
$0=x^2-9x+8$
$0=(x-1)(x-8)$
$x=1 \text{ or } x=8$
Check 1: $3+\sqrt{3(1)+1}=3+\sqrt{4}=5 \neq 1$
Check 8: $3+\sqrt{3(8)+1}=3+\sqrt{25}=8=8$
Discard $x=1$ as extraneous.
The solution is $x=8$.

33. $\sqrt{2x+3}-\sqrt{x+1}=1$
$\sqrt{2x+3}=1+\sqrt{x+1}$
$\left(\sqrt{2x+3}\right)^2=\left(1+\sqrt{x+1}\right)^2$
$2x+3=1+2\sqrt{x+1}+x+1$
$x+1=2\sqrt{x+1}$
$(x+1)^2=\left(2\sqrt{x+1}\right)^2$
$x^2+2x+1=4(x+1)$
$x^2+2x+1=4x+4$
$x^2-2x-3=0$
$(x+1)(x-3)=0 \rightarrow x=-1 \text{ or } x=3$
Check -1: $\sqrt{2(-1)+3}-\sqrt{-1+1}$
$\quad =\sqrt{1}-\sqrt{0}=1-0=1=1$
Check 3: $\sqrt{2(3)+3}-\sqrt{3+1}$
$\quad =\sqrt{9}-\sqrt{4}=3-2=1=1$
The solution set is $\{-1,3\}$.

35. $\sqrt{3x+1}-\sqrt{x-1}=2$
$\sqrt{3x+1}=2+\sqrt{x-1}$
$\left(\sqrt{3x+1}\right)^2=\left(2+\sqrt{x-1}\right)^2$
$3x+1=4+4\sqrt{x-1}+x-1$
$2x-2=4\sqrt{x-1}$
$(2x-2)^2=\left(4\sqrt{x-1}\right)^2$
$4x^2-8x+4=16(x-1)$
$x^2-2x+1=4x-4$
$x^2-6x+5=0$
$(x-1)(x-5)=0 \rightarrow x=1 \text{ or } x=5$
Check 1: $\sqrt{3(1)+1}-\sqrt{1-1}$
$\quad =\sqrt{4}-\sqrt{0}=2-0=2=2$
Check 5: $\sqrt{3(5)+1}-\sqrt{5-1}$
$\quad =\sqrt{16}-\sqrt{4}=4-2=2=2$
The solution set is $\{1,5\}$.

37. $\sqrt{3-2\sqrt{x}}=\sqrt{x}$
$\left(\sqrt{3-2\sqrt{x}}\right)^2=\left(\sqrt{x}\right)^2$
$3-2\sqrt{x}=x$
$-2\sqrt{x}=x-3$
$\left(-2\sqrt{x}\right)^2=(x-3)^2$
$4x=x^2-6x+9$
$0=x^2-10x+9$
$0=(x-9)(x-1)$
$x=1 \text{ or } x=9$
Check:
$\sqrt{3-2\sqrt{1}}=\sqrt{1}$ $\sqrt{3-2\sqrt{9}}=\sqrt{9}$
$\sqrt{3-2}=1$ $\sqrt{3-2\cdot 3}=3$
$\sqrt{1}=1$ $\sqrt{-3} \neq 3$
$1=1 \checkmark$
Discard $x=9$ as extraneous. The only solution is $x=1$

39. $(3x+1)^{1/2} = 4$

$\left((3x+1)^{1/2}\right)^2 = (4)^2$

$3x+1 = 16 \rightarrow 3x = 15 \rightarrow x = 5$

Check:

$(3(5)+1)^{1/2} = 4$

$16^{1/2} = 4$

$4 = 4$ ✓

The solution is $x = 5$.

41. $(5x-2)^{1/3} = 2$

$\left((5x-2)^{1/3}\right)^3 = (2)^3$

$5x-2 = 8 \rightarrow 5x = 10 \rightarrow x = 2$

Check:

$(5(2)-2)^{1/3} = 2$

$8^{1/3} = 2$

$2 = 2$

The solution is $x = 2$.

43. $(x^2+9)^{1/2} = 5$

$\left((x^2+9)^{1/2}\right)^2 = (5)^2$

$x^2+9 = 25 \rightarrow x^2 = 16$

$x = -4$ or $x = 4$

Check:

$\left((-4)^2+9\right)^{1/2} = 5 \qquad \left((4)^2+9\right)^{1/2} = 5$

$25^{1/2} = 5 \qquad\qquad\qquad 25^{1/2} = 5$

$5 = 5 \qquad\qquad\qquad\qquad 5 = 5$

The solution set is $\{-4, 4\}$.

45. $x^{3/2} - 3x^{1/2} = 0$

$x^{1/2}(x-3) = 0$

$x = 0$ or $x = 3$

Check:

$0^{3/2} - 3 \cdot 0^{1/2} = 0 - 0 = 0$ ✓

$3^{3/2} - 3 \cdot 3^{1/2} = 3\sqrt{3} - 3\sqrt{3} = 0$ ✓

The solution set is $\{0, 3\}$.

47. $t^4 - 16 = 0$

$(t^2+4)(t^2-4) = 0$

$(t^2+4)(t+2)(t-2) = 0$

$t^2 + 4 = 0$ has no real solution, so we only need to consider

$t + 2 = 0$ or $t - 2 = 0$

$t = -2 \qquad\qquad t = 2$

49. $x^4 - 5x^2 + 4 = 0$

$(x^2-4)(x^2-1) = 0$

$x^2 - 4 = 0$ or $x^2 - 1 = 0$

$x = \pm 2$ or $x = \pm 1$

The solution set is $\{-2, -1, 1, 2\}$.

51. $3x^4 - 2x^2 - 1 = 0$

$(3x^2+1)(x^2-1) = 0$

$3x^2 + 1 = 0$ or $x^2 - 1 = 0$

$3x^2 = -1$, which is impossible

or $x = \pm 1$

The solution set is $\{-1, 1\}$.

53. $x^6 + 7x^3 - 8 = 0$

$(x^3+8)(x^3-1) = 0$

$x^3 + 8 = 0$ or $x^3 - 1 = 0$

$x^3 = -8 \rightarrow x = -2$

or $x^3 = 1 \rightarrow x = 1$

The solution set is $\{-2, 1\}$.

55. $(x+2)^2 + 7(x+2) + 12 = 0$

let $p = x+2 \rightarrow p^2 = (x+2)^2$

$p^2 + 7p + 12 = 0$

$(p+3)(p+4) = 0$

$p + 3 = 0$ or $p + 4 = 0$

$p = -3 \rightarrow x + 2 = -3 \rightarrow x = -5$

or $p = -4 \rightarrow x + 2 = -4 \rightarrow x = -6$

The solution set is $\{-6, -5\}$.

Chapter 1: Graphs, Equations, and Inequalities

57. $(3x+4)^2 - 6(3x+4) + 9 = 0$
let $p = 3x+4 \rightarrow p^2 = (3x+4)^2$
$p^2 - 6p + 9 = 0$
$(p-3)(p-3) = 0$
$p - 3 = 0$
$p = 3 \rightarrow 3x + 4 = 3 \rightarrow x = -\frac{1}{3}$
The solution set is $\left\{-\frac{1}{3}\right\}$.

59. $2(s+1)^2 - 5(s+1) = 3$
let $p = s+1 \rightarrow p^2 = (s+1)^2$
$2p^2 - 5p = 3$
$2p^2 - 5p - 3 = 0$
$(2p+1)(p-3) = 0$
$2p + 1 = 0$ or $p - 3 = 0$
$p = -\frac{1}{2} \rightarrow s + 1 = -\frac{1}{2} \rightarrow s = -\frac{3}{2}$
or $p = 3 \rightarrow s + 1 = 3 \rightarrow s = 2$
The solution set is $\left\{-\frac{3}{2}, 2\right\}$.

61. $x - 4\sqrt{x} = 0$
$x = 4\sqrt{x}$
$(x)^2 = (4\sqrt{x})^2$
$x^2 = 16x \rightarrow 0 = 16x - x^2$
$0 = x(16-x) \rightarrow x = 0$
or $16 - x = 0 \rightarrow x = 16$
Check:
$x = 0$: $0 - 4\sqrt{0} = 0$
$0 = 0$
$x = 16$: $(16) - 4\sqrt{16} = 0$
$16 - 16 = 0$
$0 = 0$
The solution set is $\{0, 16\}$.

63. $x + \sqrt{x} = 20$
let $p = \sqrt{x} \rightarrow p^2 = x$
$p^2 + p = 20$
$p^2 + p - 20 = 0$
$(p+5)(p-4) = 0$
$p + 5 = 0$ or $p - 4 = 0$
$p = -5 \rightarrow \sqrt{x} = -5$ non-real
or $p = 4 \rightarrow \sqrt{x} = 4 \rightarrow x = 16$
Check:
$x = 16$: $16 + \sqrt{16} = 20$
$16 + 4 = 20$
The solution set is $\{16\}$.

65. $t^{1/2} - 2t^{1/4} + 1 = 0$
let $p = t^{1/4} \rightarrow p^2 = t^{1/2}$
$p^2 - 2p + 1 = 0$
$(p-1)(p-1) = 0$
$p - 1 = 0$
$p = 1 \rightarrow t^{1/4} = 1 \rightarrow t = 1$
Check:
$t = 1$: $1^{1/2} - 2(1)^{1/4} + 1 = 0$
$1 - 2 + 1 = 0 \rightarrow 0 = 0$
The solution set is $\{1\}$.

67. $4x^{1/2} - 9x^{1/4} + 4 = 0$
let $p = x^{1/4} \rightarrow p^2 = x^{1/2}$
$4p^2 - 9p + 4 = 0$
$p = \frac{9 \pm \sqrt{81 - 64}}{8} = \frac{9 \pm \sqrt{17}}{8}$
$x^{1/4} = \frac{9 \pm \sqrt{17}}{8} \rightarrow x = \left(\frac{9 \pm \sqrt{17}}{8}\right)^4$

Check:

$$4\left(\left(\frac{9+\sqrt{17}}{8}\right)^4\right)^{1/2} - 9\left(\left(\frac{9+\sqrt{17}}{8}\right)^4\right)^{1/4} + 4 = 0$$

$$4\left(\frac{9+\sqrt{17}}{8}\right)^2 - 9\left(\frac{9+\sqrt{17}}{8}\right) + 4 = 0$$

$$4\frac{(9+\sqrt{17})^2}{64} - 9\left(\frac{9+\sqrt{17}}{8}\right) + 4 = 0$$

$$64\left(4\frac{(9+\sqrt{17})^2}{64} - 9\left(\frac{9+\sqrt{17}}{8}\right) + 4\right) = (0)(64)$$

$$4(9+\sqrt{17})^2 - 72(9+\sqrt{17}) + 256 = 0$$

$$4(81+18\sqrt{17}+17) - 72(9+\sqrt{17}) + 256 = 0$$

$$324 + 72\sqrt{17} + 68 - 648 - 72\sqrt{17} + 256 = 0$$

$$0 = 0$$

$$4\left(\left(\frac{9-\sqrt{17}}{8}\right)^4\right)^{1/2} - 9\left(\left(\frac{9-\sqrt{17}}{8}\right)^4\right)^{1/4} + 4 = 0$$

$$4\left(\frac{9-\sqrt{17}}{8}\right)^2 - 9\left(\frac{9-\sqrt{17}}{8}\right) + 4 = 0$$

$$4(81-18\sqrt{17}+17) - 72(9-\sqrt{17}) + 256 = 0$$

$$324 - 72\sqrt{17} + 68 - 648 + 72\sqrt{17} + 256 = 0$$

$$0 = 0$$

The solution set is $\left\{\left(\frac{9-\sqrt{17}}{8}\right)^4, \left(\frac{9+\sqrt{17}}{8}\right)^4\right\}$.

69. $\sqrt[4]{5x^2-6} = x$

$\left(\sqrt[4]{5x^2-6}\right)^4 = x^4$

$5x^2 - 6 = x^4$

$0 = x^4 - 5x^2 + 6$

let $p = x^2 \to p^2 = x^4$

$0 = p^2 - 5p + 6 \to (p-3)(p-2) = 0$

$p = 3 \to x^2 = 3 \to x = \pm\sqrt{3}$

or $p = 2 \to x^2 = 2 \to x = \pm\sqrt{2}$

Check:

$x = -\sqrt{3}: \sqrt[4]{5(-\sqrt{3})^2 - 6} = -\sqrt{3}$

$\sqrt[4]{15-6} = -\sqrt{3} \to \sqrt[4]{9} \neq -\sqrt{3}$

$x = \sqrt{3}: \sqrt[4]{5(\sqrt{3})^2 - 6} = \sqrt{3}$

$\sqrt[4]{15-6} = \sqrt{3}$

$\sqrt[4]{9} = \sqrt{3} \to \sqrt{3} = \sqrt{3}$

$x = -\sqrt{2}: \sqrt[4]{5(-\sqrt{2})^2 - 6} = -\sqrt{2}$

$\sqrt[4]{10-6} = -\sqrt{2} \to \sqrt[4]{4} \neq -\sqrt{2}$

$x = \sqrt{2}: \sqrt[4]{5(\sqrt{2})^2 - 6} = \sqrt{2}$

$\sqrt[4]{10-6} = \sqrt{2}$

$\sqrt[4]{4} = \sqrt{2} \to \sqrt{2} = \sqrt{2}$

The solution set is $\{\sqrt{2}, \sqrt{3}\}$.

71. $\dfrac{1}{(x+1)^2} = \dfrac{1}{x+1} + 2$

let $p = \dfrac{1}{x+1} \to p^2 = \left(\dfrac{1}{x+1}\right)^2$

$p^2 = p + 2 \to p^2 - p - 2 = 0$

$(p+1)(p-2) = 0 \to p = -1$ or $p = 2$

$p = -1 \to \dfrac{1}{x+1} = -1 \to 1 = -x-1 \to x = -2$

or

$p = 2 \to \dfrac{1}{x+1} = 2 \to 1 = 2x+2 \to x = -\dfrac{1}{2}$

Check:

$x = -2: \dfrac{1}{(-2+1)^2} = \dfrac{1}{-2+1} + 2$

$1 = -1 + 2$

$1 = 1$

$x = -\dfrac{1}{2}: \dfrac{1}{\left(-\frac{1}{2}+1\right)^2} = \dfrac{1}{\left(-\frac{1}{2}+1\right)} + 2$

$4 = 2 + 2$

$4 = 4$

The solution set is $\left\{-2, -\dfrac{1}{2}\right\}$.

73. $3x^{-2} - 7x^{-1} - 6 = 0$

let $p = x^{-1} \to p^2 = x^{-2}$

$3p^2 - 7p - 6 = 0$

$(3p+2)(p-3) = 0$

$p = -\dfrac{2}{3}$ or $p = 3$

$p = -\dfrac{2}{3} \to x^{-1} = -\dfrac{2}{3} \to \left(x^{-1}\right)^{-1} = \left(-\dfrac{2}{3}\right)^{-1}$

$\to x = -\dfrac{3}{2}$

$p = 3 \to x^{-1} = 3 \to \left(x^{-1}\right)^{-1} = (3)^{-1} \to x = \dfrac{1}{3}$

Check:

$x = -\dfrac{3}{2}:\ 3\left(-\dfrac{3}{2}\right)^{-2} - 7\left(-\dfrac{3}{2}\right)^{-1} - 6 = 0$

$3\left(\dfrac{4}{9}\right) - 7\left(-\dfrac{2}{3}\right) - 6 = 0$

$\dfrac{4}{3} + \dfrac{14}{3} - 6 = 0$

$0 = 0$

$x = \dfrac{1}{3}:\ 3\left(\dfrac{1}{3}\right)^{-2} - 7\left(\dfrac{1}{3}\right)^{-1} - 6 = 0$

$3(9) - 7(3) - 6 = 0$

$27 - 21 - 6 = 0$

$0 = 0$

The solution set is $\left\{-\dfrac{3}{2}, \dfrac{1}{3}\right\}$.

75. $2x^{2/3} - 5x^{1/3} - 3 = 0$

let $p = x^{1/3} \to p^2 = x^{2/3}$

$2p^2 - 5p - 3 = 0 \to (2p+1)(p-3) = 0$

$p = -\dfrac{1}{2}$ or $p = 3$

$p = -\dfrac{1}{2} \to x^{1/3} = -\dfrac{1}{2}$

$\to \left(x^{1/3}\right)^3 = \left(-\dfrac{1}{2}\right)^3 \to x = -\dfrac{1}{8}$

or

$p = 3 \to x^{1/3} = 3 \to \left(x^{1/3}\right)^3 = (3)^3$

$\to x = 27$

Check:

$x = -\dfrac{1}{8}:\ 2\left(-\dfrac{1}{8}\right)^{2/3} - 5\left(-\dfrac{1}{8}\right)^{1/3} - 3 = 0$

$2\left(\dfrac{1}{4}\right) - 5\left(-\dfrac{1}{2}\right) - 3 = 0$

$\dfrac{1}{2} + \dfrac{5}{2} - 3 = 0$

$3 - 3 = 0 \to 0 = 0$

$x = 27:\ 2(27)^{2/3} - 5(27)^{1/3} - 3 = 0$

$2(9) - 5(3) - 3 = 0$

$18 - 15 - 3 = 0$

$3 - 3 = 0 \to 0 = 0$

The solution set is $\left\{-\dfrac{1}{8}, 27\right\}$.

77. $\left(\dfrac{v}{v+1}\right)^2 + \dfrac{2v}{v+1} = 8$

$\left(\dfrac{v}{v+1}\right)^2 + 2\left(\dfrac{v}{v+1}\right) = 8$

let $p = \dfrac{v}{v+1} \to p^2 = \left(\dfrac{v}{v+1}\right)^2$

$p^2 + 2p = 8$

$p^2 + 2p - 8 = 0$

$(p+4)(p-2) = 0$

$p = -4$ or $p = 2$

$p = -4$ $\qquad$ or $\qquad$ $p = 2$

$\dfrac{v}{v+1} = -4$ $\qquad$ $\dfrac{v}{v+1} = 2$

$v = -4v - 4$ $\qquad$ $v = 2v + 2$

$v = -\dfrac{4}{5}$ $\qquad$ $v = -2$

Check:

$\left(\dfrac{-\dfrac{4}{5}}{-\dfrac{4}{5}+1}\right)^2 + \dfrac{2\left(-\dfrac{4}{5}\right)}{\left(-\dfrac{4}{5}\right)+1} = 8$

$\dfrac{\left(\dfrac{16}{25}\right)}{\left(\dfrac{1}{25}\right)} + \dfrac{\left(-\dfrac{8}{5}\right)}{\left(\dfrac{1}{5}\right)} = 8$

$16 - 8 = 8$

$8 = 8$

$$\left(\frac{-2}{-2+1}\right)^2 + \frac{2(-2)}{(-2)+1} = 8$$
$$4+4=8$$
$$8=8$$
The solution set is $\left\{-\frac{4}{5}, -2\right\}$.

79. $|x| = 6$
    $x = 6$ or $x = -6$
    The solution set is $\{-6, 6\}$.

81. $|2x+3| = 5$
    $2x+3 = 5$ or $2x+3 = -5$
    $2x = 2$ or $2x = -8$
    $x = 1$ or $x = -4$
    The solution set is $\{-4, 1\}$.

83. $|1-4t| + 8 = 13 \Rightarrow |1-4t| = 5$
    $1-4t = 5$ or $1-4t = -5$
    $-4t = 4$ or $-4t = -6$
    $t = -1$ or $t = \frac{3}{2}$
    The solution set is $\left\{-1, \frac{3}{2}\right\}$.

85. $|-2x| = 8$
    $-2x = 8$ or $-2x = -8$
    $x = -4$ or $x = 4$
    The solution set is $\{-4, 4\}$.

87. $4 - |2x| = 3 \Rightarrow |2x| = 1$
    $2x = 1$ or $2x = -1$
    $x = \frac{1}{2}$ or $x = -\frac{1}{2}$
    The solution set is $\left\{-\frac{1}{2}, \frac{1}{2}\right\}$.

89. $\frac{2}{3}|x| = 9$
    $|x| = \frac{27}{2} \rightarrow x = \frac{27}{2}$ or $x = -\frac{27}{2}$
    The solution set is $\left\{-\frac{27}{2}, \frac{27}{2}\right\}$.

91. $\left|\frac{x}{3} + \frac{2}{5}\right| = 2$
    $\frac{x}{3} + \frac{2}{5} = 2$ or $\frac{x}{3} + \frac{2}{5} = -2$
    $5x + 6 = 30$ or $5x + 6 = -30$
    $5x = 24$ or $5x = -36$
    $x = \frac{24}{5}$ or $x = -\frac{36}{5}$
    The solution set is $\left\{-\frac{36}{5}, \frac{24}{5}\right\}$.

93. $|u-2| = -\frac{1}{2}$
    impossible, since absolute value always yields a non-negative number.

95. $|x^2 - 9| = 0$
    $x^2 - 9 = 0$
    $x^2 = 9$
    $x = \pm 3$
    The solution set is $\{-3, 3\}$.

97. $|x^2 - 2x| = 3$
    $x^2 - 2x = 3$ or $x^2 - 2x = -3$
    $x^2 - 2x - 3 = 0$ or $x^2 - 2x + 3 = 0$
    $(x-3)(x+1) = 0$ or $x = \frac{2 \pm \sqrt{4-12}}{2}$
    $= \frac{2 \pm \sqrt{-8}}{2}$ no real sol.
    $x = 3$ or $x = -1$
    The solution set is $\{-1, 3\}$.

99. $|x^2 + x - 1| = 1$
    $x^2 + x - 1 = 1$ or $x^2 + x - 1 = -1$
    $x^2 + x - 2 = 0$ or $x^2 + x = 0$
    $(x-1)(x+2) = 0$ or $x(x+1) = 0$
    $x = 1, x = -2$ or $x = 0, x = -1$
    The solution set is $\{-2, -1, 0, 1\}$.

Chapter 1: *Graphs, Equations, and Inequalities*

101.
$$x^3 - 9x = 0$$
$$x(x^2 - 9) = 0$$
$$x(x-3)(x+3) = 0$$
$x = 0$ or $x - 3 = 0$ $x + 3 = 0$
$\quad\quad\quad\quad\quad x = 3 \quad\quad x = -3$
The solution set is $\{-3, 0, 3\}$.

103.
$$4x^3 = 3x^2$$
$$4x^3 - 3x^2 = 0$$
$$x^2(4x - 3) = 0$$
$x^2 = 0$ or $4x - 3 = 0$
$x = 0 \quad\quad\quad 4x = 3$
$\quad\quad\quad\quad\quad x = \dfrac{3}{4}$
The solution set is $\left\{0, \dfrac{3}{4}\right\}$.

105.
$$x^3 + x^2 - 20x = 0$$
$$x(x^2 + x - 20) = 0$$
$$x(x + 5)(x - 4) = 0$$
$x = 0$ or $x + 5 = 0$ or $x - 4 = 0$
$\quad\quad\quad\quad x = -5 \quad\quad\quad x = 4$
The solution set is $\{-5, 0, 4\}$.

107.
$$x^3 + x^2 - x - 1 = 0$$
$$x^2(x + 1) - 1(x + 1) = 0$$
$$(x + 1)(x^2 - 1) = 0$$
$$(x + 1)(x - 1)(x + 1) = 0$$
$x + 1 = 0$ or $x - 1 = 0$
$x = -1 \quad\quad\quad x = 1$
The solution set is $\{-1, 1\}$.

109.
$$x^3 - 3x^2 - 4x + 12 = 0$$
$$x^2(x - 3) - 4(x - 3) = 0$$
$$(x - 3)(x^2 - 4) = 0$$
$$(x - 3)(x - 2)(x + 2) = 0$$
$x - 3 = 0$ or $x - 2 = 0$ or $x + 2 = 0$
$x = 3 \quad\quad\quad x = 2 \quad\quad\quad x = -2$
The solution set is $\{-2, 2, 3\}$.

111.
$$2x^3 + 4 = x^2 + 8x$$
$$2x^3 - x^2 - 8x + 4 = 0$$
$$x^2(2x - 1) - 4(2x - 1) = 0$$
$$(2x - 1)(x^2 - 4) = 0$$
$$(2x - 1)(x - 2)(x + 2) = 0$$
$2x - 1 = 0$ or $x - 2 = 0$ or $x + 2 = 0$
$2x = 1 \quad\quad\quad x = 2 \quad\quad\quad x = -2$
$x = \dfrac{1}{2}$
The solution set is $\left\{-2, \dfrac{1}{2}, 2\right\}$.

113.
$$5x^3 + 45x = 2x^2 + 18$$
$$5x^3 - 2x^2 + 45x - 18 = 0$$
$$x^2(5x - 2) + 9(5x - 2) = 0$$
$$(5x - 2)(x^2 + 9) = 0$$
$5x - 2 = 0$ or $x^2 + 9 = 0$
$5x = 2 \quad\quad\quad x^2 = -9$ no real solutions
$x = \dfrac{2}{5}$
The solution set is $\left\{\dfrac{2}{5}\right\}$.

115. $x - 4x^{1/2} + 2 = 0$; Use ZERO (or ROOT) on the graph of $y_1 = x - 4\sqrt{x} + 2$.

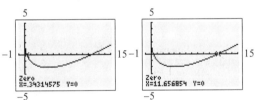

The solution set is $\{0.34, 11.66\}$.

117. $x^4 + \sqrt{3}x^2 - 3 = 0$; Use ZERO (or ROOT) on the graph of $y_1 = x^4 + \sqrt{3}x^2 - 3$.

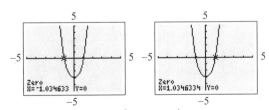

The solution set is $\{-1.03, 1.03\}$.

119. $\pi(1+t)^2 = \pi + 1 + t$; Use INTERSECT on the graphs of $y_1 = \pi(1+x)^2$ and $y_2 = \pi + 1 + x$.

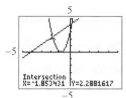 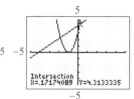

The solution set is $\{-1.85, 0.17\}$.

121. $k^2 - k = 12 \Rightarrow k^2 - k - 12 = 0$
$(k-4)(k+3) = 0$
$k = 4$ or $k = -3$

$\dfrac{x+3}{x-3} = 4$ \qquad $\dfrac{x+3}{x-3} = -3$

$x + 3 = 4x - 12$ \qquad $x + 3 = -3x + 9$

$3x = 15$ \qquad $x = \dfrac{3}{2}$

$x = 5$

And since neither of these x values causes a denominator to equal zero, the solution set is $\left\{\dfrac{3}{2}, 5\right\}$.

123. All points having an x-coordinate of 2 are of the form $(2, y)$. Those which are 5 units from $(-2, -1)$ are:

$\sqrt{(2-(-2))^2 + (y-(-1))^2} = 5$

$\sqrt{4^2 + (y+1)^2} = 5$

Squaring both sides:

$4^2 + (y+1)^2 = 25$

$16 + (y+1)^2 = 25$

$(y+1)^2 = 9$

$y + 1 = \pm 3$

$y = -1 \pm 3$

$y = -4$ or $y = 2$

Therefore, the points are $(2, -4)$ and $(2, 2)$.

125. All points on the x-axis are of the form $(x, 0)$. Those which are 5 units from $(4, -3)$ are:

$\sqrt{(x-4)^2 + (0-(-3))^2} = 5$

$\sqrt{(x-4)^2 + 3^2} = 5$

Squaring both sides:

$(x-4)^2 + 9 = 25$

$x^2 - 8x + 16 + 9 = 25$

$x^2 - 8x = 0$

$x(x-8) = 0$

$x = 0$ or $x = 8$

Therefore, the points are $(0, 0)$ and $(8, 0)$.

127. Graph the equations $y_1 = \sqrt{x}/4 + x/1100$ and $y_2 = 4$; then use INTERSECT to find the x-coordinate of the points of intersection:

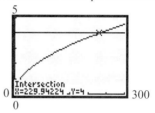

The distance to the water's surface is approximately 229.94 feet.

129. Answers will vary, one example is $x - \sqrt{x} - 2 = 0$

131. Answers will vary. Quadratic in form means that with an appropriate substitution, the equation can be written in the form $a \cdot u^2 + b \cdot u + c = 0$, where u is some expression involving the original variable.

Chapter 1: Graphs, Equations, and Inequalities

Section 1.6

1. mathematical modeling

3. uniform motion

5. True; this is the uniform motion formula.

7. Let A represent the area of the circle and r the radius. The area of a circle is the product of π times the square of the radius: $A = \pi r^2$

9. Let A represent the area of the square and s the length of a side. The area of the square is the square of the length of a side: $A = s^2$

11. Let F represent the force, m the mass, and a the acceleration. Force equals the product of the mass times the acceleration: $F = ma$

13. Let W represent the work, F the force, and d the distance. Work equals force times distance: $W = Fd$

15. C = total variable cost in dollars, x = number of dishwashers manufactured: $C = 150x$

17. Let x represent the amount of money invested in bonds. Then $50,000 - x$ represents the amount of money invested in CD's. Since the total interest is to be $6,000, we have:
$$0.15x + 0.07(50,000 - x) = 6,000$$
$$(100)(0.15x + 0.07(50,000 - x)) = (6,000)(100)$$
$$15x + 7(50,000 - x) = 600,000$$
$$15x + 350,000 - 7x = 600,000$$
$$8x + 350,000 = 600,000$$
$$8x = 250,000$$
$$x = 31,250$$
$31,250 should be invested in bonds at 15% and $18,750 should be invested in CD's at 7%.

19. Let x represent the amount of money loaned at 8%. Then $12,000 - x$ represents the amount of money loaned at 18%. Since the total interest is to be $1,000, we have:
$$0.08x + 0.18(12,000 - x) = 1,000$$
$$(100)(0.08x + 0.18(12,000 - x)) = (1,000)(100)$$
$$8x + 18(12,000 - x) = 100,000$$
$$8x + 216,000 - 18x = 100,000$$
$$-10x + 216,000 = 100,000$$
$$-10x = -116,000$$
$$x = 11,600$$
$11,600 is loaned at 8% and $400 is at 18%.

21. Let x represent the number of pounds of Earl Gray tea. Then $100 - x$ represents the number of pounds of Orange Pekoe tea.
$$5x + 3(100 - x) = 4.50(100)$$
$$5x + 300 - 3x = 450$$
$$2x + 300 = 450$$
$$2x = 150$$
$$x = 75$$
75 pounds of Earl Gray tea must be blended with 25 pounds of Orange Pekoe.

23. Let x represent the number of pounds of cashews. Then $x + 60$ represents the number of pounds in the mixture.
$$4x + 1.50(60) = 2.50(x + 60)$$
$$4x + 90 = 2.50x + 150$$
$$1.5x = 60$$
$$x = 40$$
40 pounds of cashews must be added to the 60 pounds of peanuts.

25. Let r represent the speed of the current.

	Rate	Time	Distance
Upstream	$16-r$	$\frac{20}{60} = \frac{1}{3}$	$\frac{16-r}{3}$
Downstream	$16+r$	$\frac{15}{60} = \frac{1}{4}$	$\frac{16+r}{4}$

Since the distance is the same in each direction:
$$\frac{16-r}{3} = \frac{16+r}{4}$$
$$4(16-r) = 3(16+r)$$
$$64 - 4r = 48 + 3r \rightarrow 16 = 7r \rightarrow r = \frac{16}{7} \approx 2.286$$

The speed of the current is approximately 2.286 miles per hour.

27. Let r represent the speed of the current.

	Rate	Time	Distance
Upstream	$15-r$	$\dfrac{10}{15-r}$	10
Downstream	$15+r$	$\dfrac{10}{15+r}$	10

Since the total time is 1.5 hours, we have:
$$\frac{10}{15-r}+\frac{10}{15+r}=1.5$$
$$10(15+r)+10(15-r)=1.5(15-r)(15+r)$$
$$150+10r+150-10r=1.5(225-r^2)$$
$$300=1.5(225-r^2)$$
$$200=225-r^2$$
$$r^2-25=0$$
$$(r-5)(r+5)=0$$
$$r=5 \text{ or } r=-5$$
The speed of the current is 5 miles per hour.

29. Let t represent the time it takes to do the job together.

	Time to do job	Part of job done in one minute
Trent	30	$\dfrac{1}{30}$
Lois	20	$\dfrac{1}{20}$
Together	t	$\dfrac{1}{t}$

$$\frac{1}{30}+\frac{1}{20}=\frac{1}{t}$$
$$2t+3t=60$$
$$5t=60$$
$$t=12$$
Working together, the job can be done in 12 minutes.

31. $l=$ length of the garden
    $w=$ width of the garden

  a. The length of the garden is to be twice its width. Thus, $l=2w$.
     The dimensions of the fence are $l+4$ and $w+4$.
     The perimeter is 46 feet, so:
     $$2(l+4)+2(w+4)=46$$
     $$2(2w+4)+2(w+4)=46$$
     $$4w+8+2w+8=46$$
     $$6w+16=46$$
     $$6w=30$$
     $$w=5$$
     The dimensions of the garden are 5 feet by 10 feet.

  b. Area $=l\cdot w=5\cdot 10=50$ square feet

  c. If the dimensions of the garden are the same, then the length and width of the fence are also the same $(l+4)$. The perimeter is 46 feet, so:
     $$2(l+4)+2(l+4)=46$$
     $$2l+8+2l+8=46$$
     $$4l+16=46$$
     $$4l=30$$
     $$l=7.5$$
     The dimensions of the garden are 7.5 feet by 7.5 feet.

  d. Area $=l\cdot w=7.5(7.5)=56.25$ square feet.

33. Let t represent the time it takes for the defensive back to catch the tight end.

	Time to run 100 yards	Time	Rate	Distance
Tight End	12 sec	t	$\dfrac{100}{12}=\dfrac{25}{3}$	$\dfrac{25}{3}t$
Def. Back	10 sec	t	$\dfrac{100}{10}=10$	$10t$

Since the defensive back has to run 5 yards farther, we have:
$$\frac{25}{3}t+5=10t$$
$$25t+15=30t$$
$$15=5t$$
$$t=3 \quad \rightarrow \quad 10t=30$$
The defensive back will catch the tight end at the 45 yard line $(15+30=45)$.

Chapter 1: Graphs, Equations, and Inequalities SSM: College Algebra EGU

35. Let x represent the number of gallons of pure water. Then $x+1$ represents the number of gallons in the 60% solution.
$$(\%)(\text{gallons}) + (\%)(\text{gallons}) = (\%)(\text{gallons})$$
$$0(x) + 1(1) = 0.60(x+1)$$
$$1 = 0.6x + 0.6$$
$$0.4 = 0.6x$$
$$x = \frac{4}{6} = \frac{2}{3}$$
$\frac{2}{3}$ gallon of pure water should be added.

37. Let x represent the number of ounces of water to be evaporated; the amount of salt remains the same. Therefore, we get
$$0.04(32) = 0.06(32 - x)$$
$$1.28 = 1.92 - 0.06x$$
$$0.06x = 0.64$$
$$x = \frac{0.64}{0.06} = \frac{64}{6} = \frac{32}{3} = 10\frac{2}{3}$$
$10\frac{2}{3}$ ounces of water need to be evaporated.

39. Let x represent the number of grams of pure gold. Then $60 - x$ represents the number of grams of 12 karat gold to be used.
$$x + \frac{1}{2}(60 - x) = \frac{2}{3}(60)$$
$$x + 30 - 0.5x = 40$$
$$0.5x = 10$$
$$x = 20$$
20 grams of pure gold should be mixed with 40 grams of 12 karat gold.

41. Let t represent the time it takes for Mike to catch up with Dan. Since the distances are the same, we have:
$$\frac{1}{6}t = \frac{1}{9}(t+1)$$
$$3t = 2t + 2$$
$$t = 2$$
Mike will pass Dan after 2 minutes, which is a distance of $\frac{1}{3}$ mile.

43. Let t represent the time the auxiliary pump needs to run. Since the two pumps are emptying one tanker, we have:
$$\frac{3}{4} + \frac{t}{9} = 1$$
$$27 + 4t = 36$$
$$4t = 9$$
$$t = \frac{9}{4} = 2.25$$
The auxiliary pump must run for 2.25 hours. It must be started at 9:45 a.m.

45. Let t represent the time for the tub to fill with the faucets on and the stopper removed. Since one tub is being filled, we have:
$$\frac{t}{15} + \left(-\frac{t}{20}\right) = 1$$
$$4t - 3t = 60$$
$$t = 60$$
60 minutes is required to fill the tub.

47. Let x represent the number of boxes ordered in excess of 150. The price per box is then $200 - x$ and the quantity ordered is $150 + x$.
$$\text{Charge} = (\text{price}) \cdot (\text{quanity ordered})$$
$$30,625 = (200 - x)(150 + x)$$
$$30,625 = 30,000 + 50x - x^2$$
$$0 = -625 + 50x - x^2$$
$$0 = x^2 - 50x + 625$$
$$0 = (x - 25)^2$$
$$0 = x - 25$$
$$25 = x$$
The customer ordered $150 + 25 = 175$ boxes.

49. Burke's rate is $\frac{100}{12}$ meters/sec.
In 9.99 seconds, Burke will run
$\frac{100}{12}(9.99) = 83.25$ meters.
Lewis would win by 16.75 meters.

51. Answers will vary.

53. Let t_1 and t_2 represent the times for the two segments of the trip. Since Atlanta is halfway between Chicago and Miami, the distances are equal.

$$45t_1 = 55t_2$$
$$t_1 = \frac{55}{45}t_2$$
$$t_1 = \frac{11}{9}t_2$$

Computing the average speed:

$$\text{Avg Speed} = \frac{\text{Distance}}{\text{Time}} = \frac{45t_1 + 55t_2}{t_1 + t_2}$$

$$= \frac{45\left(\frac{11}{9}t_2\right) + 55t_2}{\frac{11}{9}t_2 + t_2} = \frac{55t_2 + 55t_2}{\left(\frac{11t_2 + 9t_2}{9}\right)}$$

$$= \frac{110t_2}{\left(\frac{20t_2}{9}\right)} = \frac{990t_2}{20t_2}$$

$$= \frac{99}{2} = 49.5 \text{ miles per hour}$$

The average speed for the trip from Chicago to Miami is 49.5 miles per hour.

Section 1.7

1. $x \geq -2$

3. negative

5. multiplication property for inequalities

7. True; this follows from the addition property for inequalities.

9. False; since both sides of the inequality are being divided by a negative number, the sense, or direction, of the inequality must be reversed. That is, $\frac{a}{c} > \frac{b}{c}$.

11. Interval: $[0, 2]$
    Inequality: $0 \leq x \leq 2$

13. Interval: $(-1, 2)$
    Inequality: $-1 < x < 2$

15. Interval: $[0, 3)$
    Inequality: $0 \leq x < 3$

17. a. $3 < 5$
       $3 + 3 < 5 + 3$
       $6 < 8$

    b. $3 < 5$
       $3 - 5 < 5 - 5$
       $-2 < 0$

    c. $3 < 5$
       $3(3) < 3(5)$
       $9 < 15$

    d. $3 < 5$
       $-2(3) > -2(5)$
       $-6 > -10$

19. a. $4 > -3$
       $4 + 3 > -3 + 3$
       $7 > 0$

    b. $4 > -3$
       $4 - 5 > -3 - 5$
       $-1 > -8$

    c. $4 > -3$
       $3(4) > 3(-3)$
       $12 > -9$

    d. $4 > -3$
       $-2(4) < -2(-3)$
       $-8 < 6$

21. a. $2x + 1 < 2$
       $2x + 1 + 3 < 2 + 3$
       $2x + 4 < 5$

    b. $2x + 1 < 2$
       $2x + 1 - 5 < 2 - 5$
       $2x - 4 < -3$

    c. $2x + 1 < 2$
       $3(2x + 1) < 3(2)$
       $6x + 3 < 6$

d. $2x+1<2$
$-2(2x+1)>-2(2)$
$-4x-2>-4$

23. $[0, 4]$

25. $[4, 6)$

27. $[4, \infty)$

29. $(-\infty, -4)$

31. $2 \leq x \leq 5$

33. $-3 < x < -2$

35. $x \geq 4$

37. $x < -3$

39. If $x < 5$, then $x - 5 < 0$.

41. If $x > -4$, then $x + 4 > 0$.

43. If $x \geq -4$, then $3x \geq -12$.

45. If $x > 6$, then $-2x < -12$.

47. If $x \geq 5$, then $-4x \leq -20$.

49. If $2x < 6$, then $x < 3$.

51. If $-\frac{1}{2}x \leq 3$, then $x \geq -6$.

53. $x+1<5$
$x+1-1<5-1$
$x<4$
$\{x \mid x < 4\}$ or $(-\infty, 4)$

55. $1-2x \leq 3$
$-2x \leq 2$
$x \geq -1$
$\{x \mid x \geq -1\}$ or $[-1, \infty)$

57. $3x - 7 > 2$
$3x > 9$
$x > 3$
$\{x \mid x > 3\}$ or $(3, \infty)$

59. $3x - 1 \geq 3 + x$
$2x \geq 4$
$x \geq 2$
$\{x \mid x \geq 2\}$ or $[2, \infty)$

61. $-2(x+3) < 8$
$-2x - 6 < 8$
$-2x < 14$
$x > -7$
$\{x \mid x > -7\}$ or $(-7, \infty)$

63. $4 - 3(1-x) \leq 3$
$4 - 3 + 3x \leq 3$
$3x + 1 \leq 3$
$3x \leq 2$
$x \leq \frac{2}{3}$
$\{x \mid x \leq \frac{2}{3}\}$ or $(-\infty, \frac{2}{3}]$

65. $\frac{1}{2}(x-4) > x+8$

$\frac{1}{2}x - 2 > x + 8$

$-\frac{1}{2}x > 10$

$x < -20$

$\{x \mid x < -20\}$ or $(-\infty, -20)$

67. $\frac{x}{2} \geq 1 - \frac{x}{4}$

$2x \geq 4 - x$

$3x \geq 4$

$x \geq \frac{4}{3}$

$\{x \mid x \geq \frac{4}{3}\}$ or $[\frac{4}{3}, \infty)$

69. $0 \leq 2x - 6 \leq 4$

$6 \leq 2x \leq 10$

$3 \leq x \leq 5$

$\{x \mid 3 \leq x \leq 5\}$ or $[3, 5]$

71. $-5 \leq 4 - 3x \leq 2$

$-9 \leq -3x \leq -2$

$3 \geq x \geq \frac{2}{3}$

$\{x \mid \frac{2}{3} \leq x \leq 3\}$ or $[\frac{2}{3}, 3]$

73. $-3 < \frac{2x-1}{4} < 0$

$-12 < 2x - 1 < 0$

$-11 < 2x < 1$

$-\frac{11}{2} < x < \frac{1}{2}$

$\{x \mid -\frac{11}{2} < x < \frac{1}{2}\}$ or $(-\frac{11}{2}, \frac{1}{2})$

75. $1 < 1 - \frac{1}{2}x < 4$

$0 < -\frac{1}{2}x < 3$

$0 > x > -6$ or $-6 < x < 0$

$\{x \mid -6 < x < 0\}$ or $(-6, 0)$

77. $(x+2)(x-3) > (x-1)(x+1)$

$x^2 - x - 6 > x^2 - 1$

$-x - 6 > -1$

$-x > 5$

$x < -5$

$\{x \mid x < -5\}$ or $(-\infty, -5)$

79. $x(4x+3) \leq (2x+1)^2$

$4x^2 + 3x \leq 4x^2 + 4x + 1$

$3x \leq 4x + 1$

$-x \leq 1$

$x \geq -1$

$\{x \mid x \geq -1\}$ or $[-1, \infty)$

81. $\dfrac{1}{2} \le \dfrac{x+1}{3} < \dfrac{3}{4}$

$6 \le 4x+4 < 9$

$2 \le 4x < 5$

$\dfrac{1}{2} \le x < \dfrac{5}{4}$

$\left\{ x \mid \dfrac{1}{2} \le x < \dfrac{5}{4} \right\}$ or $\left[\dfrac{1}{2}, \dfrac{5}{4} \right)$

83. $|x| < 6$

$-6 < x < 6$

$\{x \mid -6 < x < 6\}$ or $(-6, 6)$

85. $|x| > 4$

$x < -4$ or $x > 4$

$\{x \mid x < -4 \text{ or } x > 4\}$ or $(-\infty, -4) \cup (4, \infty)$

87. $|2x| < 8$

$-8 < 2x < 8$

$-4 < x < 4$

$\{x \mid -4 < x < 4\}$ or $(-4, 4)$

89. $|3x| > 12$

$3x < -12$ or $3x > 12$

$x < -4$ or $x > 4$

$\{x \mid x < -4 \text{ or } x > 4\}$ or $(-\infty, -4) \cup (4, \infty)$

91. $|x-2| + 2 < 3$

$|x-2| < 1$

$-1 < x-2 < 1$

$1 < x < 3$

$\{x \mid 1 < x < 3\}$ or $(1, 3)$

93. $|3t-2| \le 4$

$-4 \le 3t-2 \le 4$

$-2 \le 3t \le 6$

$-\dfrac{2}{3} \le t \le 2$

$\left\{ t \mid -\dfrac{2}{3} \le t \le 2 \right\}$ or $\left[-\dfrac{2}{3}, 2 \right]$

95. $|x-3| \ge 2$

$x-3 \le -2$ or $x-3 \ge 2$

$x \le 1$ or $x \ge 5$

$\{x \mid x \le 1 \text{ or } x \ge 5\}$ or $(-\infty, 1] \cup [5, \infty)$

97. $|1-4x| - 7 < -2$

$|1-4x| < 5$

$-5 < 1-4x < 5$

$-6 < -4x < 4$

$\dfrac{-6}{-4} > x > \dfrac{4}{-4}$

$\dfrac{3}{2} > x > -1$ or $-1 < x < \dfrac{3}{2}$

$\left\{ x \mid -1 < x < \dfrac{3}{2} \right\}$ or $\left(-1, \dfrac{3}{2} \right)$

99. $|1-2x| > |-3|$

$|1-2x| > 3$

$1-2x < -3$ or $1-2x > 3$

$-2x < -4$ or $-2x > 2$

$x > 2$ or $x < -1$

$\{x \mid x < -1 \text{ or } x > 2\}$ or $(-\infty, -1) \cup (2, \infty)$

101. $|2x+1| < -1$

No solution since absolute value is always non-negative.

103. $-3 < x+5 < 2x$
$-3 < x+5$ and $x+5 < 2x$
$-8 < x$ $\qquad$ $5 < x$
$x > -8$ $\qquad$ $x > 5$
We need both $x > -8$ and $x > 5$. Therefore, $x > 5$ is sufficient to satisfy both inequalities.
$\{x \mid x > 5\}$ or $(5, \infty)$

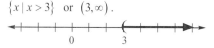

105. $x+2 < 2x-1 < 5x$
$x+2 < 2x-1$ and $2x-1 < 5x$
$2 < x-1$ $\qquad$ $-3x-1 < 0$
$3 < x$ $\qquad$ $-3x < 1$
$x > 3$ $\qquad$ $x > -\dfrac{1}{3}$
We need both $x > -\dfrac{1}{3}$ and $x > 3$. Therefore, $x > 3$ is sufficient to satisfy both inequalities.
$\{x \mid x > 3\}$ or $(3, \infty)$.

107. $|x-2| < 0.5$
$-0.5 < x-2 < 0.5$
$-0.5+2 < x < 0.5+2$
$1.5 < x < 2.5$
Solution set: $\{x \mid 1.5 < x < 2.5\}$

109. $|x-(-3)| > 2$
$x-(-3) < -2$ or $x-(-3) > 2$
$x+3 < -2$ or $x+3 > 2$
$x < -5$ or $x > -1$
Solution set: $\{x \mid x < -5 \text{ or } x > -1\}$

111. $21 <$ young adult's age < 30

113. A temperature x that differs from $98.6° F$ by at least $1.5°F$.
$|x-98.6°| \geq 1.5°$
$x-98.6° \leq -1.5°$ or $x-98.6° \geq 1.5°$
$x \leq 97.1°$ or $x \geq 100.1°$
The temperatures that are considered unhealthy are those that are less than $97.1°F$ or greater than $100.1°F$, inclusive.

115. a. Let x = age at death.
$x - 25 \geq 50.6$
$x \geq 75.6$
Therefore, the average life expectancy for a 25-year-old male will be greater than or equal to 75.6 years.

b. Let x = age at death.
$x - 25 \geq 55.4$
$x \geq 80.4$
Therefore, the average life expectancy for a 25-year-old female will be greater than or equal to 80.4 years.

c. By the given information, a female can expect to live $80.4 - 75.6 = 4.8$ years longer.

117. Let P represent the selling price and C represent the commission.
Calculating the commission:
$C = 45,000 + 0.25(P - 900,000)$
$= 45,000 + 0.25P - 225,000$
$= 0.25P - 180,000$
Calculate the commission range, given the price range:
$900,000 \leq P \leq 1,100,000$
$0.25(900,000) \leq 0.25P \leq 0.25(1,100,000)$
$225,000 \leq 0.25P \leq 275,000$
$225,000 - 180,000 \leq 0.25P - 180,000 \leq 275,000 - 180,000$
$45,000 \leq C \leq 95,000$
The agent's commission ranges from $45,000 to $95,000, inclusive.
$\dfrac{45,000}{900,000} = 0.05 = 5\%$ to
$\dfrac{95,000}{1,100,000} = 0.086 = 8.6\%$, inclusive.
As a percent of selling price, the commission ranges from 5% to 8.6%.

Chapter 1: Graphs, Equations, and Inequalities

119. Let W = weekly wages and T = tax withheld.
Calculating the withholding tax range, given the range of weekly wages:
$$600 \leq W \leq 700$$
$$600 - 592 \leq W - 592 \leq 700 - 592$$
$$8 \leq W - 592 \leq 108$$
$$0.25(8) \leq 0.25(W - 592) \leq 0.25(108)$$
$$2 \leq 0.25(W - 592) \leq 27$$
$$2 + 74.357 \leq 0.25(W - 592) + 74.357 \leq 27 + 74.357$$
$$76.357 \leq T \leq 101.357$$
The amount of withholding tax ranges from $76.36 to $101.36, inclusive.

121. Let K represent the monthly usage in kilowatt-hours and let C represent the monthly customer bill.
Calculating the bill:
$C = 0.08275K + 10.07$

Calculating the range of kilowatt-hours, given the range of bills:
$$65.96 \leq C \leq 217.02$$
$$65.96 \leq 0.08275K + 10.07 \leq 217.02$$
$$55.89 \leq 0.08275K \leq 206.95$$
$$675.41 \leq K \leq 2500.91$$
The range of usage in kilowatt-hours varied from 675.41 to 2500.91.

123. Let C represent the dealer's cost and M represent the markup over dealer's cost.
If the price is $8800, then
$8800 = C + MC = C(1 + M)$
Solving for C yields: $C = \dfrac{8800}{1+M}$
Calculating the range of dealer costs, given the range of markups:
$$0.12 \leq M \leq 0.18$$
$$1.12 \leq 1 + M \leq 1.18$$
$$\frac{1}{1.12} \geq \frac{1}{1+M} \geq \frac{1}{1.18}$$
$$\frac{8800}{1.12} \geq \frac{8800}{1+M} \geq \frac{8800}{1.18}$$
$$7857.14 \geq C \geq 7457.63$$
The dealer's cost ranged from $7457.63 to $7857.14, inclusive.

125. Let T represent the score on the last test and G represent the course grade.
Calculating the course grade and solving for the last test:
$$G = \frac{68 + 82 + 87 + 89 + T}{5}$$
$$= \frac{326 + T}{5}$$
$$5G = 326 + T$$
$$T = 5G - 326$$
Calculating the range of scores on the last test, given the grade range:
$$80 \leq G < 90$$
$$400 \leq 5G < 450$$
$$74 \leq 5G - 326 < 124$$
$$74 \leq T < 124$$
The fifth test must be greater than or equal to 74.

127. Let g represent the number of gallons of gasoline in the gas tank. Since the car averages 25 miles per gallon, a trip of at least 300 miles will require at least $\dfrac{300}{25} = 12$ gallons of gas.
Therefore the range of the amount of gasoline is $12 \leq g \leq 20$.

129. Since $a < b$
$$\frac{a}{2} < \frac{b}{2} \qquad\qquad \frac{a}{2} < \frac{b}{2}$$
$$\frac{a}{2} + \frac{a}{2} < \frac{a}{2} + \frac{b}{2} \qquad \frac{a}{2} + \frac{b}{2} < \frac{b}{2} + \frac{b}{2}$$
$$a < \frac{a+b}{2} \qquad\qquad \frac{a+b}{2} < b$$
Thus, $a < \dfrac{a+b}{2} < b$.

131. If $0 < a < b$, then
$$ab > a^2 > 0 \qquad\qquad b^2 > ab > 0$$
$$\left(\sqrt{ab}\right)^2 > a^2 \qquad\qquad b^2 > \left(\sqrt{ab}\right)^2$$
$$\sqrt{ab} > a \qquad\qquad b > \sqrt{ab}$$
Thus, $a < \sqrt{ab} < b$

133. For $0 < a < b$, $\dfrac{1}{h} = \dfrac{1}{2}\left(\dfrac{1}{a} + \dfrac{1}{b}\right)$

$$h \cdot \dfrac{1}{h} = \dfrac{1}{2}\left(\dfrac{b+a}{ab}\right) \cdot h$$

$$1 = \dfrac{1}{2}\left(\dfrac{b+a}{ab}\right) \cdot h$$

$$\dfrac{2ab}{a+b} = h$$

$$h - a = \dfrac{2ab}{a+b} - a = \dfrac{2ab - a(a+b)}{a+b}$$

$$= \dfrac{2ab - a^2 - ab}{a+b} = \dfrac{ab - a^2}{a+b}$$

$$= \dfrac{a(b-a)}{a+b} > 0$$

Therefore, $h > a$.

$$b - h = b - \dfrac{2ab}{a+b} = \dfrac{b(a+b) - 2ab}{a+b}$$

$$= \dfrac{ab + b^2 - 2ab}{a+b} = \dfrac{b^2 - ab}{a+b}$$

$$= \dfrac{b(b-a)}{a+b} > 0$$

Therefore, $h < b$ and we get $a < h < b$.

135. Answers will vary. One possibility:
No solution: $4x + 6 \le 2(x-5) + 2x$
One solution: $3x + 5 \le 2(x+3) + 1 \le 3(x+2) - 1$

137. Since $x^2 \ge 0$, we have
$x^2 + 1 \ge 0 + 1$
$x^2 + 1 \ge 1$
Therefore, the expression $x^2 + 1$ can never be less than -5.

Section 1.8

1. undefined; 0

3. $y = b$; y-intercept

5. False; the slope is $\dfrac{3}{2}$.
$2y = 3x + 5$
$y = \dfrac{3}{2}x + \dfrac{5}{2}$

7. **a.** Slope $= \dfrac{1-0}{2-0} = \dfrac{1}{2}$

   b. If x increases by 2 units, y will increase by 1 unit.

9. **a.** Slope $= \dfrac{1-2}{1-(-2)} = -\dfrac{1}{3}$

   b. If x increases by 3 units, y will decrease by 1 unit.

11. Slope $= \dfrac{y_2 - y_1}{x_2 - x_1} = \dfrac{0-3}{4-2} = \dfrac{-3}{2}$

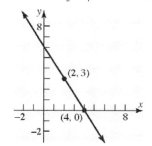

13. Slope $= \dfrac{y_2 - y_1}{x_2 - x_1} = \dfrac{1-3}{2-(-2)} = \dfrac{-2}{4} = -\dfrac{1}{2}$

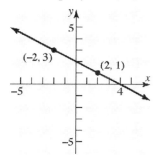

15. Slope $= \dfrac{y_2 - y_1}{x_2 - x_1} = \dfrac{-1-(-1)}{2-(-3)} = \dfrac{0}{5} = 0$

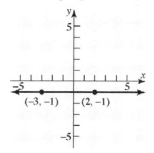

17. Slope $= \dfrac{y_2 - y_1}{x_2 - x_1} = \dfrac{-2-2}{-1-(-1)} = \dfrac{-4}{0}$ undefined.

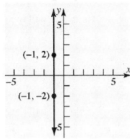

19. $P = (1,2); m = 3$

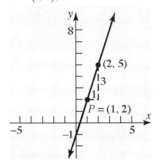

21. $P = (2,4); m = -\dfrac{3}{4}$

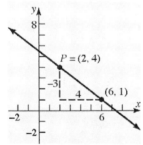

23. $P = (-1,3); m = 0$

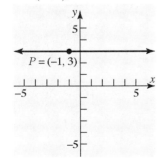

25. $P = (0,3)$; slope undefined

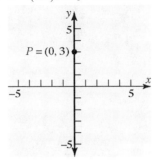

27. Slope $= 4 = \dfrac{4}{1}$

If x increases by 1 unit, then y increases by 4 units.
Answers will vary. Three possible points are:
$x = 1+1 = 2$ and $y = 2+4 = 6$
$(2,6)$
$x = 2+1 = 3$ and $y = 6+4 = 10$
$(3,10)$
$x = 3+1 = 4$ and $y = 10+4 = 14$
$(4,14)$

29. Slope $= -\dfrac{3}{2} = \dfrac{-3}{2}$

If x increases by 2 units, then y decreases by 3 units.
Answers will vary. Three possible points are:
$x = 2+2 = 4$ and $y = -4-3 = -7$
$(4,-7)$
$x = 4+2 = 6$ and $y = -7-3 = -10$
$(6,-10)$
$x = 6+2 = 8$ and $y = -10-3 = -13$
$(8,-13)$

31. Slope $= -2 = \dfrac{-2}{1}$

If x increases by 1 unit, then y decreases by 2 units.
Answers will vary. Three possible points are:
$x = -2 + 1 = -1$ and $y = -3 - 2 = -5$
$(-1, -5)$
$x = -1 + 1 = 0$ and $y = -5 - 2 = -7$
$(0, -7)$
$x = 0 + 1 = 1$ and $y = -7 - 2 = -9$
$(1, -9)$

33. $(0, 0)$ and $(2, 1)$ are points on the line.
Slope $= \dfrac{1-0}{2-0} = \dfrac{1}{2}$
y-intercept is 0; using $y = mx + b$:
$y = \dfrac{1}{2}x + 0$
$2y = x$
$0 = x - 2y$
$x - 2y = 0$ or $y = \dfrac{1}{2}x$

35. $(-1, 3)$ and $(1, 1)$ are points on the line.
Slope $= \dfrac{1-3}{1-(-1)} = \dfrac{-2}{2} = -1$
Using $y - y_1 = m(x - x_1)$
$y - 1 = -1(x - 1)$
$y - 1 = -x + 1$
$y = -x + 2$
$x + y = 2$ or $y = -x + 2$

37. $y - y_1 = m(x - x_1)$, $m = 2$
$y - 3 = 2(x - 3)$
$y - 3 = 2x - 6$
$y = 2x - 3$
$2x - y = 3$ or $y = 2x - 3$

39. $y - y_1 = m(x - x_1)$, $m = -\dfrac{1}{2}$
$y - 2 = -\dfrac{1}{2}(x - 1)$
$y - 2 = -\dfrac{1}{2}x + \dfrac{1}{2}$
$y = -\dfrac{1}{2}x + \dfrac{5}{2}$
$x + 2y = 5$ or $y = -\dfrac{1}{2}x + \dfrac{5}{2}$

41. Slope = 3; containing $(-2, 3)$
$y - y_1 = m(x - x_1)$
$y - 3 = 3(x - (-2))$
$y - 3 = 3x + 6$
$y = 3x + 9$
$3x - y = -9$ or $y = 3x + 9$

43. Slope $= -\dfrac{2}{3}$; containing $(1, -1)$
$y - y_1 = m(x - x_1)$
$y - (-1) = -\dfrac{2}{3}(x - 1)$
$y + 1 = -\dfrac{2}{3}x + \dfrac{2}{3}$
$y = -\dfrac{2}{3}x - \dfrac{1}{3}$
$2x + 3y = -1$ or $y = -\dfrac{2}{3}x - \dfrac{1}{3}$

45. Containing $(1, 3)$ and $(-1, 2)$
$m = \dfrac{2-3}{-1-1} = \dfrac{-1}{-2} = \dfrac{1}{2}$
$y - y_1 = m(x - x_1)$
$y - 3 = \dfrac{1}{2}(x - 1)$
$y - 3 = \dfrac{1}{2}x - \dfrac{1}{2}$
$y = \dfrac{1}{2}x + \dfrac{5}{2}$
$x - 2y = -5$ or $y = \dfrac{1}{2}x + \dfrac{5}{2}$

47. Slope = –3; y-intercept = 3
$$y = mx + b$$
$$y = -3x + 3$$
$$3x + y = 3 \text{ or } y = -3x + 3$$

49. x-intercept = 2; y-intercept = –1
Points are (2,0) and (0,–1)
$$m = \frac{-1-0}{0-2} = \frac{-1}{-2} = \frac{1}{2}$$
$$y = mx + b$$
$$y = \frac{1}{2}x - 1$$
$$x - 2y = 2 \text{ or } y = \frac{1}{2}x - 1$$

51. Slope undefined; passing through (2, 4)
This is a vertical line.
$x = 2$ No slope intercept form.

53. Parallel to $y = 2x$; Slope = 2
Containing (–1, 2)
$$y - y_1 = m(x - x_1)$$
$$y - 2 = 2(x - (-1))$$
$$y - 2 = 2x + 2 \rightarrow y = 2x + 4$$
$$2x - y = -4 \text{ or } y = 2x + 4$$

55. Parallel to $2x - y = -2$; Slope = 2
Containing the point (0, 0)
$$y - y_1 = m(x - x_1)$$
$$y - 0 = 2(x - 0)$$
$$y = 2x$$
$$2x - y = 0 \text{ or } y = 2x$$

57. Parallel to $x = 5$; Containing (4,2)
This is a vertical line.
$x = 4$ No slope intercept form.

59. Perpendicular to $y = \frac{1}{2}x + 4$; Containing (1, –2)
Slope of perpendicular = –2
$$y - y_1 = m(x - x_1)$$
$$y - (-2) = -2(x - 1)$$
$$y + 2 = -2x + 2 \rightarrow y = -2x$$
$$2x + y = 0 \text{ or } y = -2x$$

61. Perpendicular to $2x + y = 2$; Containing the point (–3, 0)
Slope of perpendicular = $\frac{1}{2}$
$$y - y_1 = m(x - x_1)$$
$$y - 0 = \frac{1}{2}(x - (-3)) \rightarrow y = \frac{1}{2}x + \frac{3}{2}$$
$$x - 2y = -3 \text{ or } y = \frac{1}{2}x + \frac{3}{2}$$

63. Perpendicular to $x = 8$; Containing (3, 4)
Slope of perpendicular = 0 (horizontal line)
$y = 4$

65. $y = 2x + 3$; Slope = 2; y-intercept = 3

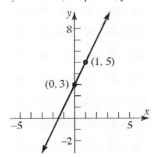

67. $\frac{1}{2}y = x - 1$; $y = 2x - 2$
Slope = 2; y-intercept = –2

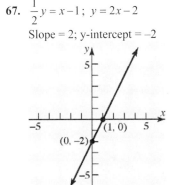

69. $y = \frac{1}{2}x + 2$; Slope = $\frac{1}{2}$; y-intercept = 2

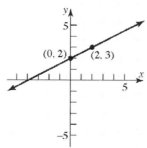

71. $x + 2y = 4$; $2y = -x + 4 \rightarrow y = -\frac{1}{2}x + 2$

Slope = $-\frac{1}{2}$; y-intercept = 2

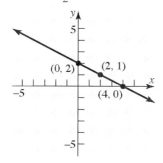

73. $2x - 3y = 6$; $-3y = -2x + 6 \rightarrow y = \frac{2}{3}x - 2$

Slope = $\frac{2}{3}$; y-intercept = -2

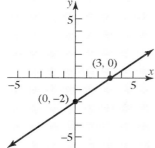

75. $x + y = 1$; $y = -x + 1$

Slope = -1; y-intercept = 1

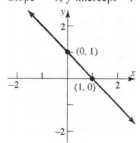

77. $x = -4$; Slope is undefined

y-intercept - none

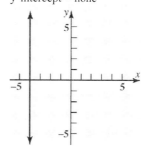

79. $y = 5$; Slope = 0; y-intercept = 5

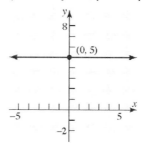

81. $y - x = 0$; $y = x$

Slope = 1; y-intercept = 0

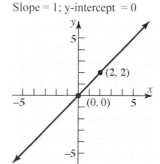

83. $2y - 3x = 0$; $2y = 3x \rightarrow y = \dfrac{3}{2}x$

Slope $= \dfrac{3}{2}$; y-intercept $= 0$

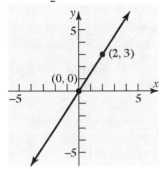

85. a. x-intercept: $2x + 3(0) = 6$
$$2x = 6$$
$$x = 3$$
The point $(3, 0)$ is on the graph.

y-intercept: $2(0) + 3y = 6$
$$3y = 6$$
$$y = 2$$
The point $(0, 2)$ is on the graph.

b.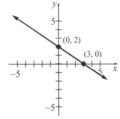

87. a. x-intercept: $-4x + 5(0) = 40$
$$-4x = 40$$
$$x = -10$$
The point $(-10, 0)$ is on the graph.

y-intercept: $-4(0) + 5y = 40$
$$5y = 40$$
$$y = 8$$
The point $(0, 8)$ is on the graph.

b.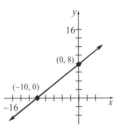

89. a. x-intercept: $7x + 2(0) = 21$
$$7x = 21$$
$$x = 3$$
The point $(3, 0)$ is on the graph.

y-intercept: $7(0) + 2y = 21$
$$2y = 21$$
$$y = \dfrac{21}{2}$$
The point $\left(0, \dfrac{21}{2}\right)$ is on the graph.

b.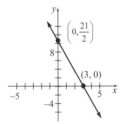

91. a. x-intercept: $\dfrac{1}{2}x + \dfrac{1}{3}(0) = 1$
$$\dfrac{1}{2}x = 1$$
$$x = 2$$
The point $(2, 0)$ is on the graph.

y-intercept: $\dfrac{1}{2}(0) + \dfrac{1}{3}y = 1$
$$\dfrac{1}{3}y = 1$$
$$y = 3$$
The point $(0, 3)$ is on the graph.

b.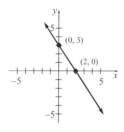

93. a. x-intercept: $0.2x - 0.5(0) = 1$
$$0.2x = 1$$
$$x = 5$$
The point $(5, 0)$ is on the graph.

y-intercept: $0.2(0) - 0.5y = 1$
$$-0.5y = 1$$
$$y = -2$$
The point $(0, -2)$ is on the graph.

b.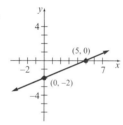

95. The equation of the x-axis is $y = 0$. (The slope is 0 and the y-intercept is 0.)

97. (b)

99. (d)

101. Intercepts: $(0, 2)$ and $(-2, 0)$. Thus, slope = 1.
$y = x + 2$ or $x - y = -2$

103. Intercepts: $(3, 0)$ and $(0, 1)$. Thus, slope $= -\frac{1}{3}$.
$y = -\frac{1}{3}x + 1$ or $x + 3y = 3$

105. Let x = number of miles driven, and let C = cost in dollars.
Total cost = (cost per mile)(number of miles) + fixed cost
$C = 0.07x + 29$
When $x = 110$, $C = (0.07)(110) + 29 = \$36.70$.
When $x = 230$, $C = (0.07)(230) + 29 = \$45.10$.

107. Let x = number newspapers delivered, and let C = cost in dollars.
Total cost = (delivery cost per paper)(number of papers delivered) + fixed cost
$C = 0.53x + 1,070,000$

109. a. $C = 0.08275x + 7.58$

b.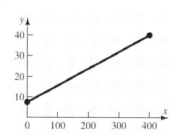

c. For 100 kWh,
$C = 0.08275(100) + 7.58 = \$15.86$

d. For 300 kWh,
$C = 0.08725(300) + 7.58 = \$32.41$

e. For each usage increase of 1 kWh, the monthly charge increases by 8.275 cents

111. $(°C, °F) = (0, 32); \quad (°C, °F) = (100, 212)$
$$\text{slope } = \frac{212 - 32}{100 - 0} = \frac{180}{100} = \frac{9}{5}$$
$$°F - 32 = \frac{9}{5}(°C - 0)$$
$$°F - 32 = \frac{9}{5}(°C)$$
$$°C = \frac{5}{9}(°F - 32)$$
If $°F = 70$, then
$$°C = \frac{5}{9}(70 - 32) = \frac{5}{9}(38)$$
$$°C \approx 21°$$

113. a. Let x = number of boxes sold, and A = money, in dollars, spent on advertising. We have the points
$(x_1, A_1) = (100,000, 40,000)$;
$(x_2, A_2) = (200,000, 60,000)$

$$\text{slope} = \frac{60,000 - 40,000}{200,000 - 100,000}$$
$$= \frac{20,000}{100,000} = \frac{1}{5}$$

$$A - 40,000 = \frac{1}{5}(x - 100,000)$$
$$A - 40,000 = \frac{1}{5}x - 20,000$$
$$A = \frac{1}{5}x + 20,000$$

b. If $x = 300,000$, then
$$A = \frac{1}{5}(300,000) + 20,000 = \$80,000$$

c. Each additional box sold requires an additional $0.20 in advertising.

115. $2x - y = C$
Graph the lines:
$2x - y = -4$
$2x - y = 0$
$2x - y = 2$
All the lines have the same slope, 2. The lines are parallel.

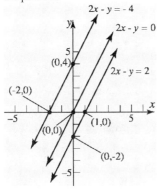

117. (b), (c), (e) and (g)

119. (c)

121. Answers will vary.

123. Answers will vary.

125. No, a line does not need to have both an x-intercept and a y-intercept. Vertical and horizontal lines have only one intercept (unless they are a coordinate axis). Every line must have at least one intercept.

127. Two lines that have the same x-intercept and y-intercept (assuming the x-intercept is not 0) are the same line since a line is uniquely defined by two distinct points.

129. Yes; two lines with the same y-intercept, but different slopes, can have the same x-intercept if the x-intercept is $x = 0$.

Section 1.9

1. False; for example, $x^2 + y^2 + 2x + 2y + 8 = 0$ is not a circle.

3. True; $r^2 = 9 \to r = 3$

5. Center = (2, 1)
Radius = distance from (0,1) to (2,1)
$$= \sqrt{(2-0)^2 + (1-1)^2} = \sqrt{4} = 2$$
$$(x-2)^2 + (y-1)^2 = 4$$

7. Center = midpoint of (1, 2) and (4, 2)
$$= \left(\frac{1+4}{2}, \frac{2+2}{2}\right) = \left(\frac{5}{2}, 2\right)$$
Radius = distance from $\left(\frac{5}{2}, 2\right)$ to (4,2)
$$= \sqrt{\left(4 - \frac{5}{2}\right)^2 + (2-2)^2} = \sqrt{\frac{9}{4}} = \frac{3}{2}$$
$$\left(x - \frac{5}{2}\right)^2 + (y-2)^2 = \frac{9}{4}$$

9. $(x-h)^2 + (y-k)^2 = r^2$
    $(x-0)^2 + (y-0)^2 = 2^2$
    $x^2 + y^2 = 4$
    General form: $x^2 + y^2 - 4 = 0$

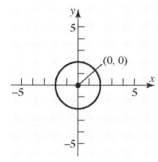

11. $(x-h)^2 + (y-k)^2 = r^2$
    $(x-0)^2 + (y-2)^2 = 2^2$
    $x^2 + (y-2)^2 = 4$
    General form: $x^2 + y^2 - 4y + 4 = 4$
    $x^2 + y^2 - 4y = 0$

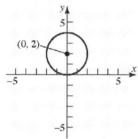

13. $(x-h)^2 + (y-k)^2 = r^2$
    $(x-4)^2 + (y-(-3))^2 = 5^2$
    $(x-4)^2 + (y+3)^2 = 25$
    General form:
    $x^2 - 8x + 16 + y^2 + 6y + 9 = 25$
    $x^2 + y^2 - 8x + 6y = 0$

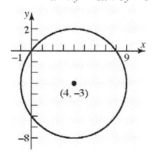

15. $(x-h)^2 + (y-k)^2 = r^2$
    $(x-(-2))^2 + (y-1)^2 = 4^2$
    $(x+2)^2 + (y-1)^2 = 16$
    General form: $x^2 + 4x + 4 + y^2 - 2y + 1 = 16$
    $x^2 + y^2 + 4x - 2y - 11 = 0$

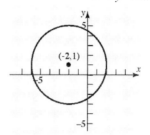

17. $(x-h)^2 + (y-k)^2 = r^2$
    $\left(x-\dfrac{1}{2}\right)^2 + (y-0)^2 = \left(\dfrac{1}{2}\right)^2$
    $\left(x-\dfrac{1}{2}\right)^2 + y^2 = \dfrac{1}{4}$
    General form: $x^2 - x + \dfrac{1}{4} + y^2 = \dfrac{1}{4}$
    $x^2 + y^2 - x = 0$

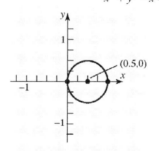

83

19. $x^2 + y^2 = 4$

   $x^2 + y^2 = 2^2$

   a. Center: (0,0); Radius = 2

   b.
   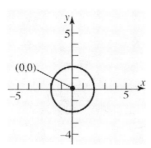

   c. x-intercepts: $y = 0$
   $$x^2 + 0 = 4$$
   $$x = \pm 2 \rightarrow (-2,0),(2,0)$$

   y-intercepts: $x = 0$
   $$0 + y^2 = 4$$
   $$y = \pm 2 \rightarrow (0,-2),(0,2)$$

21. $2(x-3)^2 + 2y^2 = 8$

   $(x-3)^2 + y^2 = 4$

   a. Center: (3, 0); Radius = 2

   b.
   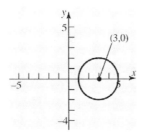

   c. x-intercepts: $y = 0$
   $$(x-3)^2 + 0 = 4$$
   $$(x-3)^2 = 4$$
   $$x - 3 = \pm 2$$
   $$x = 5, x = 1 \rightarrow (1,0),(5,0)$$

   y-intercepts: $x = 0$
   $$9 + y^2 = 4$$
   $$y^2 = -5$$
   no real solution $\Rightarrow$ no y-intercepts

23. $x^2 + y^2 - 2x + 4y - 4 = 0$
   $$x^2 - 2x + y^2 + 4y = 4$$
   $$(x^2 - 2x + 1) + (y^2 + 4y + 4) = 4 + 1 + 4$$
   $$(x-1)^2 + (y+2)^2 = 3^2$$

   a. Center: (1,–2); Radius = 3

   b.
   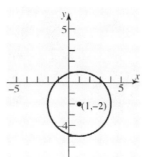

   c. x-intercepts: $y = 0$
   $$(x-1)^2 + 4 = 9$$
   $$(x-1)^2 = 5$$
   $$x - 1 = \pm\sqrt{5}$$
   $$x = 1 \pm \sqrt{5}$$
   $$\left(-\sqrt{5}+1,0\right),\left(\sqrt{5}+1,0\right)$$

   y-intercepts: $x = 0$
   $$1 + (y+2)^2 = 9$$
   $$(y+2)^2 = 8$$
   $$y + 2 = \pm\sqrt{8}$$
   $$y = -2 \pm \sqrt{8}$$
   $$\left(0,-\sqrt{8}-2\right),\left(0,\sqrt{8}-2\right)$$

25. $x^2 + y^2 + 4x - 4y - 1 = 0$
    $$x^2 + 4x + y^2 - 4y = 1$$
    $$(x^2 + 4x + 4) + (y^2 - 4y + 4) = 1 + 4 + 4$$
    $$(x+2)^2 + (y-2)^2 = 3^2$$

   a. Center: (–2, 2); Radius = 3

b.

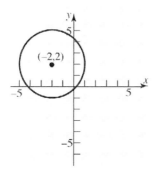

c. x-intercepts: $y = 0$
$(x+2)^2 + 4 = 9$
$(x+2)^2 = 5$
$x+2 = \pm\sqrt{5}$
$x = \sqrt{5} - 2, x = -\sqrt{5} - 2$
$\left(-\sqrt{5} - 2, 0\right), \left(\sqrt{5} - 2, 0\right)$

y-intercepts: $x = 0$
$4 + (y-2)^2 = 9$
$(y-2)^2 = 5$
$y - 2 = \pm\sqrt{5}$
$y = \sqrt{5} + 2, y = -\sqrt{5} + 2$
$\left(0, -\sqrt{5} + 2\right), \left(0, \sqrt{5} + 2\right)$

27. $x^2 + y^2 - x + 2y + 1 = 0$
$x^2 - x + y^2 + 2y = -1$
$\left(x^2 - x + \frac{1}{4}\right) + (y^2 + 2y + 1) = -1 + \frac{1}{4} + 1$
$\left(x - \frac{1}{2}\right)^2 + (y+1)^2 = \left(\frac{1}{2}\right)^2$

a. Center: $\left(\frac{1}{2}, -1\right)$; Radius $= \frac{1}{2}$

b.

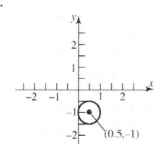

c. x-intercepts: $y = 0$
$\left(x - \frac{1}{2}\right)^2 + 1 = \frac{1}{4}$
$\left(x - \frac{1}{2}\right)^2 = -\frac{3}{4}$

no real solution $\Rightarrow$ no x-intercepts

y-intercepts: $x = 0$
$\frac{1}{4} + (y+1)^2 = \frac{1}{4}$
$(y+1)^2 = 0$
$y + 1 = 0$
$y = -1 \rightarrow (0, -1)$

29. $2x^2 + 2y^2 - 12x + 8y - 24 = 0$
$x^2 + y^2 - 6x + 4y = 12$
$x^2 - 6x + y^2 + 4y = 12$
$(x^2 - 6x + 9) + (y^2 + 4y + 4) = 12 + 9 + 4$
$(x-3)^2 + (y+2)^2 = 5^2$

a. Center: $(3, -2)$; Radius $= 5$

b.

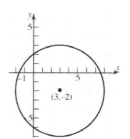

c. x-intercepts: $y = 0$
$(x-3)^2 + 4 = 25$
$(x-3)^2 = 21$
$x - 3 = \pm\sqrt{21}$
$x = \sqrt{21} + 3, x = -\sqrt{21} + 3$
$\left(-\sqrt{21} + 3, 0\right), \left(\sqrt{21} + 3, 0\right)$

y-intercepts: $x = 0$
$9 + (y+2)^2 = 25$
$(y+2)^2 = 16$
$y + 2 = \pm 4$
$y = 2, y = -6$
$(0, -6), (0, 2)$

31. Center at (0, 0); containing point (–2, 3).
$r = \sqrt{(-2-0)^2 + (3-0)^2} = \sqrt{4+9} = \sqrt{13}$
Equation: $(x-0)^2 + (y-0)^2 = (\sqrt{13})^2$
$$x^2 + y^2 = 13$$
$$x^2 + y^2 - 13 = 0$$

33. Center at (2, 3); tangent to the x-axis.
$r = 3$
Equation: $(x-2)^2 + (y-3)^2 = 3^2$
$$x^2 - 4x + 4 + y^2 - 6y + 9 = 9$$
$$x^2 + y^2 - 4x - 6y + 4 = 0$$

35. Endpoints of a diameter are (1, 4) and (–3, 2).
The center is at the midpoint of that diameter:
Center: $\left(\dfrac{1+(-3)}{2}, \dfrac{4+2}{2}\right) = (-1, 3)$
Radius: $r = \sqrt{(1-(-1))^2 + (4-3)^2}$
$= \sqrt{4+1}$
$= \sqrt{5}$
Equation: $(x-(-1))^2 + (y-3)^2 = (\sqrt{5})^2$
$$x^2 + 2x + 1 + y^2 - 6y + 9 = 5$$
$$x^2 + y^2 + 2x - 6y + 5 = 0$$

37. (c)

39. (b)

41. $(x+3)^2 + (y-1)^2 = 16$

43. $(x-2)^2 + (y-2)^2 = 9$

45. $x^2 + y^2 + 2x + 4y - 4091 = 0$
$x^2 + 2x + y^2 + 4y - 4091 = 0$
$x^2 + 2x + 1 + y^2 + 4y + 4 = 4091 + 5$
$(x+1)^2 + (y+2)^2 = 4096$
The circle representing Earth has center $(-1, -2)$ and radius $= \sqrt{4096} = 64$.
So the radius of the satellite's orbit is $64 + 0.6 = 64.6$ units.
The equation of the orbit is
$x^2 + y^2 + 2x + 4y - 4168.16 = 0$
$(x+1)^2 + (y+2)^2 = (64.6)^2$

47. $x^2 + y^2 = 9$
Center: (0, 0)
Slope from center to $(1, 2\sqrt{2})$ is
$\dfrac{2\sqrt{2}-0}{1-0} = \dfrac{2\sqrt{2}}{1} = 2\sqrt{2}$.
Slope of the tangent line is $\dfrac{-1}{2\sqrt{2}} = -\dfrac{\sqrt{2}}{4}$.
Equation of the tangent line is:
$y - 2\sqrt{2} = -\dfrac{\sqrt{2}}{4}(x-1)$
$y - 2\sqrt{2} = -\dfrac{\sqrt{2}}{4}x + \dfrac{\sqrt{2}}{4}$
$4y - 8\sqrt{2} = -\sqrt{2}\,x + \sqrt{2}$
$\sqrt{2}\,x + 4y = 9\sqrt{2}$

49. Let (h, k) be the center of the circle.
$x - 2y + 4 = 0 \to 2y = x + 4 \to y = \dfrac{1}{2}x + 2$
The slope of the tangent line is $\dfrac{1}{2}$. The slope from (h, k) to $(0, 2)$ is -2.
$\dfrac{2-k}{0-h} = -2 \to 2 - k = 2h$
The other tangent line is $y = 2x - 7$ and it has slope 2.
The slope from (h, k) to $(3, -1)$ is $-\dfrac{1}{2}$.
$\dfrac{-1-k}{3-h} = -\dfrac{1}{2}$
$2 + 2k = 3 - h$
$2k = 1 - h$
$h = 1 - 2k$
Solve the two equations in h and k:
$2 - k = 2(1 - 2k)$
$2 - k = 2 - 4k$
$3k = 0$
$k = 0$
$h = 1 - 2(0) = 1$
The center of the circle is (1, 0).

51. Consider the following diagram:

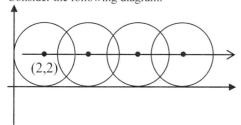

Therefore, the path of the center of the circle has the equation $y = 2$.

53. (b), (e) and (g)

55. Answers will vary.

Chapter 1 Review

1. $2 - \dfrac{x}{3} = 8$
$6 - x = 24$
$x = -18$

3. $-2(5 - 3x) + 8 = 4 + 5x$
$-10 + 6x + 8 = 4 + 5x$
$6x - 2 = 4 + 5x$
$x = 6$

5. $\dfrac{3x}{4} - \dfrac{x}{3} = \dfrac{1}{12}$
$9x - 4x = 1$
$5x = 1$
$x = \dfrac{1}{5}$

7. $\dfrac{x}{x-1} = \dfrac{6}{5}$
$5x = 6x - 6$
$6 = x$
Since $x = 6$ does not cause a denominator to equal zero, the solution set is $\{6\}$.

9. $x(1 - x) = 6$
$x - x^2 = 6$
$0 = x^2 - x + 6$
$b^2 - 4ac = (-1)^2 - 4(1)(6)$
$= 1 - 24 = -23$
Therefore, there are no real solutions.

11. $\dfrac{1}{2}\left(x - \dfrac{1}{3}\right) = \dfrac{3}{4} - \dfrac{x}{6}$
$(12)\left(\dfrac{1}{2}\right)\left(x - \dfrac{1}{3}\right) = \left(\dfrac{3}{4} - \dfrac{x}{6}\right)(12)$
$6x - 2 = 9 - 2x$
$8x = 11$
$x = \dfrac{11}{8}$
The solution set is $\left\{\dfrac{11}{8}\right\}$.

13. $(x - 1)(2x + 3) = 3$
$2x^2 + x - 3 = 3$
$2x^2 + x - 6 = 0$
$(2x - 3)(x + 2) = 0 \Rightarrow x = \dfrac{3}{2}$ or $x = -2$
The solution set is $\left\{-2, \dfrac{3}{2}\right\}$.

15. $2x + 3 = 4x^2$
$0 = 4x^2 - 2x - 3$
$x = \dfrac{2 \pm \sqrt{4 + 48}}{8} = \dfrac{2 \pm \sqrt{52}}{8}$
$= \dfrac{2 \pm 2\sqrt{13}}{8} = \dfrac{1 \pm \sqrt{13}}{4}$
The solution set is $\left\{\dfrac{1 - \sqrt{13}}{4}, \dfrac{1 + \sqrt{13}}{4}\right\}$.

17. $\sqrt[3]{x^2 - 1} = 2$
$\left(\sqrt[3]{x^2 - 1}\right)^3 = (2)^3$
$x^2 - 1 = 8$
$x^2 = 9 \Rightarrow x = \pm 3$

Check: $x = -3$ Check: $x = 3$
$\sqrt[3]{(-3)^2 - 1} = 2$ $\sqrt[3]{(3)^2 - 1} = 2$
$\sqrt[3]{9 - 1} = 2$ $\sqrt[3]{9 - 1} = 2$
$\sqrt[3]{8} = 2$ $\sqrt[3]{8} = 2$
$2 = 2$ $2 = 2$
The solution set is $\{-3, 3\}$.

19. $x(x+1) + 2 = 0$

$x^2 + x + 2 = 0$

$x = \dfrac{-1 \pm \sqrt{1-8}}{2} = \dfrac{-1 \pm \sqrt{-7}}{2}$

No real solutions.

21. $x^4 - 5x^2 + 4 = 0$

$(x^2 - 4)(x^2 - 1) = 0$

$x^2 - 4 = 0$ or $x^2 - 1 = 0$

$x = \pm 2$ or $x = \pm 1$

The solution set is $\{-2, -1, 1, 2\}$.

23. $\sqrt{2x-3} + x = 3$

$\sqrt{2x-3} = 3 - x$

$2x - 3 = 9 - 6x + x^2$

$x^2 - 8x + 12 = 0 \Rightarrow (x-2)(x-6) = 0$

$x = 2$ or $x = 6$

Check $x = 2$:

$\sqrt{2(2)-3} + 2 = \sqrt{1} + 2 = 3$

Check $x = 6$:

$\sqrt{2(6)-3} + 6 = \sqrt{9} + 6 = 9 \neq 3$

The solution set is $\{2\}$.

25. $\sqrt[4]{2x+3} = 2$

$\left(\sqrt[4]{2x+3}\right)^4 = 2^4$

$2x + 3 = 16$

$2x = 13$

$x = \dfrac{13}{2}$

Check $x = \dfrac{13}{2}$:

$\sqrt[4]{2\left(\dfrac{13}{2}\right) + 3} = \sqrt[4]{13+3} = \sqrt[4]{16} = 2$

The solution set is $\left\{\dfrac{13}{2}\right\}$.

27. $\sqrt{x+1} + \sqrt{x-1} = \sqrt{2x+1}$

$\left(\sqrt{x+1} + \sqrt{x-1}\right)^2 = \left(\sqrt{2x+1}\right)^2$

$x + 1 + 2\sqrt{x+1}\sqrt{x-1} + x - 1 = 2x + 1$

$2x + 2\sqrt{x+1}\sqrt{x-1} = 2x + 1$

$2\sqrt{x+1}\sqrt{x-1} = 1$

$\left(2\sqrt{x+1}\sqrt{x-1}\right)^2 = (1)^2$

$4(x+1)(x-1) = 1$

$4x^2 - 4 = 1$

$4x^2 = 5$

$x^2 = \dfrac{5}{4}$

$x = \pm\dfrac{\sqrt{5}}{2}$

Check $x = \dfrac{\sqrt{5}}{2}$:

$\sqrt{\dfrac{\sqrt{5}}{2}+1} + \sqrt{\dfrac{\sqrt{5}}{2}-1} = \sqrt{2\left(\dfrac{\sqrt{5}}{2}\right)+1}$

$1.79890743995 = 1.79890743995$

Check $x = -\dfrac{\sqrt{5}}{2}$:

$\sqrt{-\dfrac{\sqrt{5}}{2}+1} + \sqrt{-\dfrac{\sqrt{5}}{2}-1} = \sqrt{2\left(-\dfrac{\sqrt{5}}{2}\right)+1}$,

The second solution is not possible because it makes the radicand negative.

The solution set is $\left\{\dfrac{\sqrt{5}}{2}\right\}$.

29. $2x^{1/2} - 3 = 0$

$2x^{1/2} = 3$

$\left(2x^{1/2}\right)^2 = 3^2$

$4x = 9$

$x = \dfrac{9}{4}$

Check $x = \dfrac{9}{4}$:

$2\left(\dfrac{9}{4}\right)^{1/2} - 3 = 2\left(\dfrac{3}{2}\right) - 3 = 3 - 3 = 0$

The solution set is $\left\{\dfrac{9}{4}\right\}$.

31. $x^{-6} - 7x^{-3} - 8 = 0$

let $p = x^{-3} \Rightarrow p^2 = x^{-6}$

$p^2 - 7p - 8 = 0$
$(p-8)(p+1) = 0$
$\qquad p = 8 \text{ or } p = -1$
$\qquad p = 8 \Rightarrow x^{-3} = 8$
$\left(x^{-3}\right)^{-1/3} = (8)^{-1/3} \Rightarrow x = \dfrac{1}{2}$
$\qquad p = -1 \Rightarrow x^{-3} = -1$
$\left(x^{-3}\right)^{-1/3} = (-1)^{-1/3} \Rightarrow x = -1$

Check $x = \dfrac{1}{2}$:

$\left(\dfrac{1}{2}\right)^{-6} - 7\left(\dfrac{1}{2}\right)^{-3} - 8 = 64 - 56 - 8 = 0$

Check $x = -1$:

$(-1)^{-6} - 7(-1)^{-3} - 8 = 1 + 7 - 8 = 0$

The solution set is $\left\{-1, \dfrac{1}{2}\right\}$.

33.
$x^2 + m^2 = 2mx + (nx)^2$
$x^2 + m^2 = 2mx + n^2 x^2$
$x^2 - n^2 x^2 - 2mx + m^2 = 0$
$(1-n^2)x^2 - 2mx + m^2 = 0$

$x = \dfrac{2m \pm \sqrt{4m^2 - 4m^2(1-n^2)}}{2(1-n^2)}$

$= \dfrac{2m \pm \sqrt{4m^2(1-(1-n^2))}}{2(1-n^2)}$

$= \dfrac{2m \pm 2m\sqrt{1-(1-n^2)}}{2(1-n^2)}$

$= \dfrac{m \pm m\sqrt{n^2}}{1-n^2} = \dfrac{m \pm mn}{1-n^2} = \dfrac{m(1 \pm n)}{1-n^2}$

$x = \dfrac{m(1+n)}{1-n^2} = \dfrac{m(1+n)}{(1+n)(1-n)} = \dfrac{m}{1-n}$

or

$x = \dfrac{m(1-n)}{1-n^2} = \dfrac{m(1-n)}{(1+n)(1-n)} = \dfrac{m}{1+n}$

The solution set is $\left\{\dfrac{m}{1-n}, \dfrac{m}{1+n}\right\}$, $n \neq 1$, $n \neq -1$.

35. $10a^2 x^2 - 2abx - 36b^2 = 0$
$5a^2 x^2 - abx - 18b^2 = 0$
$(5ax + 9b)(ax - 2b) = 0$
$5ax + 9b = 0 \quad \text{or} \quad ax - 2b = 0$
$\qquad 5ax = -9b \qquad\qquad ax = 2b$
$\qquad x = -\dfrac{9b}{5a} \qquad\qquad x = \dfrac{2b}{a}$

The solution set is $\left\{-\dfrac{9b}{5a}, \dfrac{2b}{a}\right\}$, $a \neq 0$.

37.
$\sqrt{x^2 + 3x + 7} - \sqrt{x^2 - 3x + 9} + 2 = 0$
$\sqrt{x^2 + 3x + 7} = \sqrt{x^2 - 3x + 9} - 2$
$\left(\sqrt{x^2 + 3x + 7}\right)^2 = \left(\sqrt{x^2 - 3x + 9} - 2\right)^2$
$x^2 + 3x + 7 = x^2 - 3x + 9 - 4\sqrt{x^2 - 3x + 9} + 4$
$6x - 6 = -4\sqrt{x^2 - 3x + 9}$
$\left(6(x-1)\right)^2 = \left(-4\sqrt{x^2 - 3x + 9}\right)^2$
$36(x^2 - 2x + 1) = 16(x^2 - 3x + 9)$
$36x^2 - 72x + 36 = 16x^2 - 48x + 144$
$20x^2 - 24x - 108 = 0$
$5x^2 - 6x - 27 = 0$
$(5x + 9)(x - 3) = 0 \Rightarrow x = -\dfrac{9}{5} \text{ or } x = 3$

Check $x = -\dfrac{9}{5}$:

$\sqrt{\left(-\dfrac{9}{5}\right)^2 + 3\left(-\dfrac{9}{5}\right) + 7} - \sqrt{\left(-\dfrac{9}{5}\right)^2 - 3\left(-\dfrac{9}{5}\right) + 9} + 2$

$= \sqrt{\dfrac{81}{25} - \dfrac{27}{5} + 7} - \sqrt{\dfrac{81}{25} + \dfrac{27}{5} + 9} + 2$

$= \sqrt{\dfrac{81 - 135 + 175}{25}} - \sqrt{\dfrac{81 + 135 + 225}{25}} + 2$

$= \sqrt{\dfrac{121}{25}} - \sqrt{\dfrac{441}{25}} + 2 = \dfrac{11}{5} - \dfrac{21}{5} + 2 = 0$

Check $x = 3$:

$\sqrt{(3)^2 + 3(3) + 7} - \sqrt{(3)^2 - 3(3) + 9} + 2$

$= \sqrt{9 + 9 + 7} - \sqrt{9 - 9 + 9} + 2$

$= \sqrt{25} - \sqrt{9} + 2 = 2 + 2$

$= 4 \ne 0$

The solution set is $\left\{-\dfrac{9}{5}\right\}$.

39. $|2x + 3| = 7$
$2x + 3 = 7$ or $2x + 3 = -7$
$2x = 4$ or $2x = -10$
$x = 2$ or $x = -5$
The solution set is $\{-5, 2\}$.

41. $|2 - 3x| + 2 = 9$
$|2 - 3x| = 7$
$2 - 3x = 7$ or $2 - 3x = -7$
$-3x = 5$ $\qquad -3x = -9$
$x = -\dfrac{5}{3}$ $\qquad x = 3$
The solution set is $\left\{-\dfrac{5}{3}, 3\right\}$.

43. $2x^3 = 3x^2$
$2x^3 - 3x^2 = 0$
$x^2(2x - 3) = 0$
$x^2 = 0 \Rightarrow x = 0$
$2x - 3 = 0 \Rightarrow x = \dfrac{3}{2}$
The solution set is $\left\{0, \dfrac{3}{2}\right\}$.

45. $2x^3 + 5x^2 - 8x - 20 = 0$
$x^2(2x + 5) - 4(2x + 5) = 0$
$(2x + 5)(x^2 - 4) = 0$
$2x + 5 = 0$ or $x^2 - 4 = 0$
$x = -\dfrac{5}{2}$ or $x = \pm 2$
The solution set is $\left\{-\dfrac{5}{2}, -2, 2\right\}$.

47. $\dfrac{1}{x-1} + \dfrac{3}{x+2} = \dfrac{11}{x^2 + x - 2}$

$\dfrac{(x+2) + 3(x-1)}{(x-1)(x+2)} = \dfrac{11}{(x-1)(x+2)}$

$(x+2) + 3(x-1) = 11$
$x + 2 + 3x - 3 = 11$
$4x - 1 = 11$
$4x = 12$
$x = 3$

Since 3 does not make any denominator equal to 0, the solution set is $\{3\}$.

49. $(x - 2)^2 = 9$
$x - 2 = \pm\sqrt{9}$
$x - 2 = \pm 3$
$x = 2 \pm 3$
$x = 5$ or $x = -1$
The solution set is $\{-1, 5\}$.

51. $x^3 - 5x + 3 = 0$
Use the Zero option from the CALC menu.

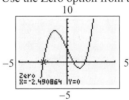

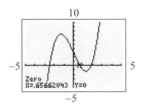

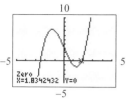

The solution set is $\{-2.49, 0.66, 1.83\}$.

53. $x^4 - 3 = 2x + 1$

Use the Intersect option on the CALC menu.

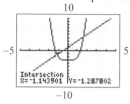

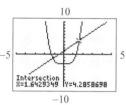

The solution set is $\{-1.14, 1.64\}$.

55. $\dfrac{2x-3}{5} + 2 \leq \dfrac{x}{2}$

$2(2x-3) + 10(2) \leq 5x$

$4x - 6 + 20 \leq 5x$

$14 \leq x$

$x \geq 14$

$\{x \mid x \geq 14\}$ or $[14, \infty)$

57. $-9 \leq \dfrac{2x+3}{-4} \leq 7$

$36 \geq 2x + 3 \geq -28$

$33 \geq 2x \geq -31$

$\dfrac{33}{2} \geq x \geq -\dfrac{31}{2}$

$-\dfrac{31}{2} \leq x \leq \dfrac{33}{2}$

$\left\{x \mid -\dfrac{31}{2} \leq x \leq \dfrac{33}{2}\right\}$ or $\left[-\dfrac{31}{2}, \dfrac{33}{2}\right]$

59. $2 < \dfrac{3-3x}{12} < 6$

$24 < 3 - 3x < 72$

$21 < -3x < 69$

$-7 > x > -23$

$\{x \mid -23 < x < -7\}$ or $(-23, -7)$

61. $|3x + 4| < \dfrac{1}{2}$

$-\dfrac{1}{2} < 3x + 4 < \dfrac{1}{2}$

$-\dfrac{9}{2} < 3x < -\dfrac{7}{2}$

$-\dfrac{3}{2} < x < -\dfrac{7}{6}$

$\left\{x \mid -\dfrac{3}{2} < x < -\dfrac{7}{6}\right\}$ or $\left(-\dfrac{3}{2}, -\dfrac{7}{6}\right)$

63. $|2x-5| \geq 9$
$2x-5 \leq -9$ or $2x-5 \geq 9$
$2x \leq -4$ or $2x \geq 14$
$x \leq -2$ or $x \geq 7$
$\{x | x \leq -2 \text{ or } x \geq 7\}$ or
$(-\infty, -2]$ or $[7, \infty)$

65. $2 + |2-3x| \leq 4$
$|2-3x| \leq 2$
$-2 \leq 2-3x \leq 2$
$-4 \leq -3x \leq 0$
$\frac{4}{3} \geq x \geq 0$
$0 \leq x \leq \frac{4}{3}$
$\left\{x \mid 0 \leq x \leq \frac{4}{3}\right\}$ or $\left[0, \frac{4}{3}\right]$

67. $1 - |2-3x| < -4$
$-|2-3x| < -5$
$|2-3x| > 5$
$2-3x < -5$ or $2-3x > 5$
$7 < 3x$ or $-3 > 3x$
$\frac{7}{3} < x$ or $-1 > x$
$x < -1$ or $x > \frac{7}{3}$
$\left\{x \mid x < -1 \text{ or } x > \frac{7}{3}\right\}$ or
$(-\infty, -1)$ or $\left(\frac{7}{3}, \infty\right)$

69. $\left(\frac{6}{2}\right)^2 = 9$

71. $\left(\frac{(-4/3)}{2}\right)^2 = \frac{4}{9}$

73. $(6+3i)-(2-4i) = (6-2)+(3-(-4))i = 4+7i$

75. $4(3-i)+3(-5+2i) = 12-4i-15+6i = -3+2i$

77. $\frac{3}{3+i} = \frac{3}{3+i} \cdot \frac{3-i}{3-i} = \frac{9-3i}{9-3i+3i-i^2} = \frac{9-3i}{10}$
$= \frac{9}{10} - \frac{3}{10}i$

79. $i^{50} = i^{48} \cdot i^2 = \left(i^4\right)^{12} \cdot i^2 = 1^{12}(-1) = -1$

81. $(2+3i)^3 = (2+3i)^2(2+3i)$
$= (4+12i+9i^2)(2+3i)$
$= (-5+12i)(2+3i)$
$= -10-15i+24i+36i^2$
$= -46+9i$

83. $x^2+x+1 = 0$
$a=1, b=1, c=1,$
$b^2-4ac = 1^2-4(1)(1) = 1-4 = -3$
$x = \frac{-1 \pm \sqrt{-3}}{2(1)} = \frac{-1 \pm \sqrt{3}i}{2} = -\frac{1}{2} \pm \frac{\sqrt{3}}{2}i$
The solution set is $\left\{-\frac{1}{2}-\frac{\sqrt{3}}{2}i, -\frac{1}{2}+\frac{\sqrt{3}}{2}i\right\}$.

85. $2x^2+x-2 = 0$
$a=2, b=1, c=-2,$
$b^2-4ac = 1^2-4(2)(-2) = 1+16 = 17$
$x = \frac{-1 \pm \sqrt{17}}{2(2)} = \frac{-1 \pm \sqrt{17}}{4}$
The solution set is $\left\{\frac{-1-\sqrt{17}}{4}, \frac{-1+\sqrt{17}}{4}\right\}$.

87. $x^2 + 3 = x$
$x^2 - x + 3 = 0$
$a = 1, b = -1, c = 3,$
$b^2 - 4ac = (-1)^2 - 4(1)(3) = 1 - 12 = -11$
$x = \dfrac{-(-1) \pm \sqrt{-11}}{2(1)} = \dfrac{1 \pm \sqrt{11}\,i}{2} = \dfrac{1}{2} \pm \dfrac{\sqrt{11}}{2}i$

The solution set is $\left\{ \dfrac{1}{2} - \dfrac{\sqrt{11}}{2}i,\ \dfrac{1}{2} + \dfrac{\sqrt{11}}{2}i \right\}$.

89. $x(1-x) = 6$
$-x^2 + x - 6 = 0$
$a = -1, b = 1, c = -6,$
$b^2 - 4ac = 1^2 - 4(-1)(-6) = 1 - 24 = -23$
$x = \dfrac{-1 \pm \sqrt{-23}}{2(-1)} = \dfrac{-1 \pm \sqrt{23}\,i}{-2} = \dfrac{1}{2} \pm \dfrac{\sqrt{23}}{2}i$

The solution set is $\left\{ \dfrac{1}{2} - \dfrac{\sqrt{23}}{2}i,\ \dfrac{1}{2} + \dfrac{\sqrt{23}}{2}i \right\}$.

91. $(0,0), (4,2)$

   a. distance $= \sqrt{(4-0)^2 + (2-0)^2}$
   $= \sqrt{16+4} = \sqrt{20}$
   $= 2\sqrt{5}$

   b. midpoint $= \left(\dfrac{0+4}{2}, \dfrac{0+2}{2} \right) = \left(\dfrac{4}{2}, \dfrac{2}{2} \right) = (2,1)$

   c. slope $= \dfrac{\Delta y}{\Delta x} = \dfrac{2-0}{4-0} = \dfrac{2}{4} = \dfrac{1}{2}$

   d. When x increases by 2 units, y increases by 1 unit

93. $(1,-1), (-2,3)$

   a. distance $= \sqrt{(-2-1)^2 + (3-(-1))^2}$
   $= \sqrt{9+16} = \sqrt{25} = 5$

   b. midpoint $= \left(\dfrac{1+(-2)}{2}, \dfrac{-1+3}{2} \right)$
   $= \left(\dfrac{-1}{2}, \dfrac{2}{2} \right) = \left(-\dfrac{1}{2}, 1 \right)$

   c. slope $= \dfrac{\Delta y}{\Delta x} = \dfrac{3-(-1)}{-2-1} = \dfrac{4}{-3} = -\dfrac{4}{3}$

   d. When x increases by 3 units, y decreases by 4 units.

95. $(4,-4), (4,8)$

   a. distance $= \sqrt{(4-4)^2 + (8-(-4))^2}$
   $= \sqrt{0+144} = \sqrt{144} = 12$

   b. midpoint $= \left(\dfrac{4+4}{2}, \dfrac{-4+8}{2} \right)$
   $= \left(\dfrac{8}{2}, \dfrac{4}{2} \right) = (4,2)$

   c. slope $= \dfrac{\Delta y}{\Delta x} = \dfrac{8-(-4)}{4-4} = \dfrac{12}{0}$, undefined

   d. Undefined slope means the points lie on a vertical line.

97. $y = -x^2 + 15$

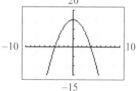

99. Slope $= -2$; containing $(3,-1)$
$y - y_1 = m(x - x_1)$
$y - (-1) = -2(x - 3)$
$y + 1 = -2x + 6$
$y = -2x + 5$
$2x + y = 5$ or $y = -2x + 5$

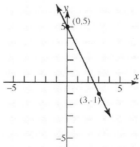

101. Slope undefined; containing (–3, 4).
This is a vertical line.
$x = -3$
No slope-intercept form.

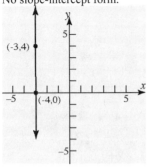

103. y-intercept = –2; containing (5,–3)
Points are (5,–3) and (0,–2)
$$m = \frac{-2-(-3)}{0-5} = \frac{1}{-5} = -\frac{1}{5}$$
$$y = mx + b$$
$$y = -\frac{1}{5}x - 2$$
$x + 5y = -10$ or $y = -\frac{1}{5}x - 2$

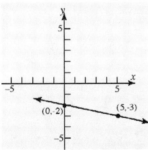

105. Parallel to $2x - 3y = -4$;
Slope $= \frac{2}{3}$; containing (–5,3)
$$y - y_1 = m(x - x_1)$$
$$y - 3 = \frac{2}{3}(x - (-5)) \rightarrow y - 3 = \frac{2}{3}x + \frac{10}{3}$$
$$y = \frac{2}{3}x + \frac{19}{3}$$
$2x - 3y = -19$ or $y = \frac{2}{3}x + \frac{19}{3}$

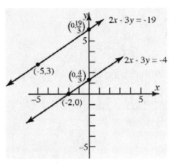

107. Perpendicular to $x + y = 2$;
Containing (4,–3)
Slope of perpendicular = 1
$$y - y_1 = m(x - x_1)$$
$$y - (-3) = 1(x - 4)$$
$$y + 3 = x - 4$$
$$y = x - 7$$
$x - y = 7$ or $y = x - 7$

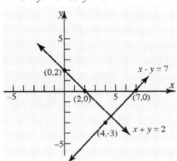

109. $4x + 6y = 36$
x-intercept: $4x + 6(0) = 36$
$$4x = 36$$
$$x = 9 \rightarrow (9, 0)$$
y-intercept: $4(0) + 6y = 36$
$$6y = 36$$
$$y = 6 \rightarrow (0, 6)$$

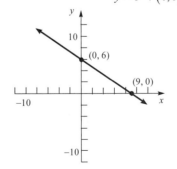

111. $\frac{1}{2}x + \frac{5}{2}y = 10$

x-intercept:
$$\frac{1}{2}x + \frac{5}{2}(0) = 10$$
$$\frac{1}{2}x = 10$$
$$x = 20 \rightarrow (20, 0)$$

y-intercept:
$$\frac{1}{2}(0) + \frac{5}{2}y = 10$$
$$\frac{5}{2}y = 10$$
$$y = 4 \rightarrow (0, 4)$$

113. $2x - 3y = 6$
$$-3y = -2x + 6$$
$$y = \frac{2}{3}x - 2$$

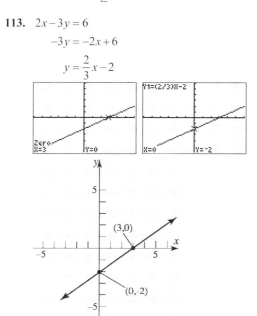

115. $y = x^2 - 9$

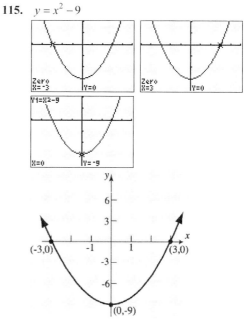

117. $x^2 + 2y = 16$
$$2y = -x^2 + 16$$
$$y = -\frac{1}{2}x^2 + 8$$

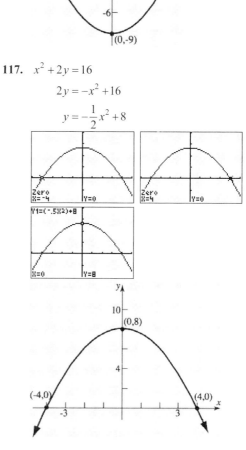

Chapter 1: Graphs, Equations, and Inequalities   SSM: College Algebra EGU

119. $(x-h)^2 + (y-k)^2 = r^2$
$(x-(-2))^2 + (y-3)^2 = 4^2$
$(x+2)^2 + (y-3)^2 = 16$

121. $(x-h)^2 + (y-k)^2 = r^2$
$(x-(-1))^2 + (y-(-2))^2 = 1^2$
$(x+1)^2 + (y+2)^2 = 1$

123. $x^2 + y^2 - 2x + 4y - 4 = 0$
$x^2 - 2x + y^2 + 4y = 4$
$(x^2 - 2x + 1) + (y^2 + 4y + 4) = 4 + 1 + 4$
$(x-1)^2 + (y+2)^2 = 9$
$(x-1)^2 + (y+2)^2 = 3^2$
Center: (1,–2)   Radius = 3

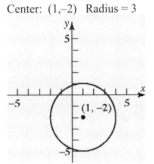

$x^2 + 0^2 - 2x + 4(0) - 4 = 0$
$x^2 - 2x - 4 = 0$
$x = \dfrac{-(-2) \pm \sqrt{(-2)^2 - 4(1)(-4)}}{2(1)} = \dfrac{2 \pm \sqrt{20}}{2}$
$= \dfrac{2 \pm 2\sqrt{5}}{2} = 1 \pm \sqrt{5}$

$0^2 + y^2 - 2(0) + 4y - 4 = 0$
$y^2 + 4y - 4 = 0$
$y = \dfrac{-4 \pm \sqrt{4^2 - 4(1)(-4)}}{2(1)} = \dfrac{-4 \pm \sqrt{32}}{2}$
$= \dfrac{-4 \pm 4\sqrt{2}}{2} = -2 \pm 2\sqrt{2}$

Intercepts:
$(1-\sqrt{5}, 0)$, $(1+\sqrt{5}, 0)$, $(0, -2-2\sqrt{2})$, and $(0, -2+2\sqrt{2})$

125. $3x^2 + 3y^2 - 6x + 12y = 0$
$x^2 + y^2 - 2x + 4y = 0$
$x^2 - 2x + y^2 + 4y = 0$
$(x^2 - 2x + 1) + (y^2 + 4y + 4) = 1 + 4$
$(x-1)^2 + (y+2)^2 = 5$
$(x-1)^2 + (y+2)^2 = (\sqrt{5})^2$
Center: (1,–2)   Radius = $\sqrt{5}$

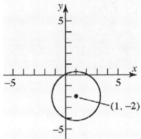

$3x^2 + 3(0)^2 - 6x + 12(0) = 0$
$3x^2 - 6x = 0$
$3x(x-2) = 0$
$x = 0$ or $x = 2$

$3(0)^2 + 3y^2 - 6(0) + 12y = 0$
$3y^2 + 12y = 0$
$3y(y+4) = 0$
$y = 0$ or $y = -4$
Intercepts:
$(0,0)$, $(2,0)$, and $(0,-4)$

127. Find the distance between each pair of points.
$d(A,B) = \sqrt{(1-3)^2 + (1-4)^2}$
$= \sqrt{4+9} = \sqrt{13}$
$d(B,C) = \sqrt{(-2-1)^2 + (3-1)^2}$
$= \sqrt{9+4} = \sqrt{13}$
$d(A,C) = \sqrt{(-2-3)^2 + (3-4)^2}$
$= \sqrt{25+1} = \sqrt{26}$
Since $AB = BC$, triangle ABC is isosceles.

129. slope of $\overline{AB} = \dfrac{1-5}{6-2} = -1$;

   slope of $\overline{AC} = \dfrac{-1-5}{8-2} = -1$;

   slope of $\overline{BC} = \dfrac{-1-1}{8-6} = -1$

   Therefore the points are collinear.

131. Endpoints of the diameter are (–3, 2) and (5,–6). The center is at the midpoint of the diameter:

   Center: $\left(\dfrac{-3+5}{2}, \dfrac{2+(-6)}{2}\right) = (1, -2)$

   Radius: $r = \sqrt{(1-(-3))^2 + (-2-2)^2} = \sqrt{16+16}$
   $= \sqrt{32} = 4\sqrt{2}$

   Equation: $(x-1)^2 + (y+2)^2 = \left(4\sqrt{2}\right)^2$
   $x^2 - 2x + 1 + y^2 + 4y + 4 = 32$
   $x^2 + y^2 - 2x + 4y - 27 = 0$

133. $p = 2l + 2w$

135. $I = P \cdot r \cdot t \Rightarrow I = (9000)(0.07)(1) = \$630$

137. Using $s = vt$, we have $t = 3$ and $v = 1100$.
   Finding the distance s in feet:
   $s = 1100(3) = 3300$
   The storm is 3300 feet away.

139. Let s represent the distance the plane can travel.

	With wind	Against wind
Rate	$250 + 30 = 280$	$250 - 30 = 220$
Time	$\dfrac{(s/2)}{280}$	$\dfrac{(s/2)}{220}$
Dist.	$\dfrac{s}{2}$	$\dfrac{s}{2}$

   Since the total time is at most 5 hours, we have:
   $\dfrac{(s/2)}{280} + \dfrac{(s/2)}{220} \le 5$

   $\dfrac{s}{560} + \dfrac{s}{440} \le 5$

   $11s + 14s \le 5(6160)$

   $25s \le 30{,}800$

   $s \le 1232$

   The plane can travel at most 1232 miles or 616 miles one way and return 616 miles.

141. Let t represent the time it takes the helicopter to reach the raft.

	Raft	Helicopter
Rate	5	90
Time	t	t
Dist.	$5t$	$90t$

   Since the total distance is 150 miles, we have:
   $5t + 90t = 150$
   $95t = 150$
   $t \approx 1.58$ hours $\approx$ 1 hour and 35 minutes

   The helicopter will reach the raft in about 1 hour and 35 minutes.

143. Let r represent the rate of the Metra train in miles per hour.

	Metra Train	Amtrak Train
Rate	r	$r + 50$
Time	3	1
Dist.	$3r$	$r + 50$

   The Amtrak Train has traveled 10 fewer miles than the Metra Train.
   $r + 50 = 3r - 10$
   $60 = 2r$
   $r = 30$

   The Metra Train is traveling 30 mph, and the Amtrak Train is traveling $30 + 50 = 80$ mph.

145. Let t represent the time it takes Clarissa to complete the job by herself.

	Clarissa	Shawna
Time to do job alone	t	$t + 5$
Part of job done in 1 day	$\dfrac{1}{t}$	$\dfrac{1}{t+5}$
Time on job (days)	6	6
Part of job done by each person	$\dfrac{6}{t}$	$\dfrac{6}{t+5}$

   Since the two people paint one house, we have:
   $\dfrac{6}{t} + \dfrac{6}{t+5} = 1$
   $6(t+5) + 6t = t(t+5)$
   $6t + 30 + 6t = t^2 + 5t$
   $t^2 - 7t - 30 = 0$
   $(t-10)(t+3) = 0 \Rightarrow t = 10$ or $t = -3$

   It takes Clarissa 10 days to paint the house when working by herself.

147. Let x represent the amount of the 15% solution added.

% acid	tot. amt.	amt. of acid
40%	60	$(0.40)(60)$
15%	x	$(0.15)(x)$
25%	$60+x$	$(0.25)(60+x)$

$$(0.40)(60)+(0.15)(x)=(0.25)(60+x)$$
$$24+0.15x=15+0.25x$$
$$9=0.1x$$
$$x=90$$

90 cubic centimeters of the 15% solution must be added, producing 150 cubic centimeters of the 25% solution.

149. Let x represent the amount of water added.

% salt	Tot. amt.	amt. of salt
10%	64	$(0.10)(64)$
0%	x	$(0.00)(x)$
2%	$64+x$	$(0.02)(64+x)$

$$(0.10)(64)+(0.00)(x)=(0.02)(64+x)$$
$$6.4=1.28+0.02x$$
$$5.12=0.02x$$
$$x=256$$

256 ounces of water must be added.

151. Let the length of leg 1 = x.
Then the length of leg 2 = $17-x$.
By the Pythagorean Theorem we have
$$x^2+(17-x)^2=(13)^2$$
$$x^2+x^2-34x+289=169$$
$$2x^2-34x+120=0$$
$$x^2-17x+60=0$$
$$(x-12)(x-5)=0 \Rightarrow x=12 \text{ or } x=5$$

the legs are 5 inches and 12 inches long.

153. The effective speed of the train (i.e., relative to the man) is $30-4=26$ miles per hour. The time is 5 sec $=\dfrac{5}{60}$ min $=\dfrac{5}{3600}$ hr $=\dfrac{1}{720}$ hr.

$$s=vt$$
$$=26\left(\dfrac{1}{720}\right)$$
$$=\dfrac{26}{720} \text{ miles}$$
$$=\dfrac{26}{720}\cdot 5280 \approx 190.67 \text{ feet}$$

The freight train is about 190.67 feet long.

155. Let t represent the time it takes the smaller pump to finish filling the tank.

	3hp Pump	8hp Pump
Time to do job alone	12	8
Part of job done in 1 hr	$\dfrac{1}{12}$	$\dfrac{1}{8}$
Time on job (hrs)	$t+4$	4
Part of job done by each pump	$\dfrac{t+4}{12}$	$\dfrac{4}{8}$

Since the two pumps fill one tank, we have:
$$\dfrac{t+4}{12}+\dfrac{4}{8}=1$$
$$\dfrac{t+4}{12}=\dfrac{1}{2}$$
$$t+4=6$$
$$t=2$$

It takes the small pump a total of 2 more hours to fill the tank.

157. Let x represent the number of passengers over 20. Then $20+x$ represents the total number of passengers, and $15-0.1x$ represents the fare for each passenger. Solving the equation for total cost, $482.40, we have:
$$(20+x)(15-0.1x)=482.40$$
$$300+13x-0.1x^2=482.40$$
$$-0.1x^2+13x-182.40=0$$
$$x^2-130x+1824=0$$
$$(x-114)(x-16)=0 \Rightarrow x=114 \text{ or } x=16$$

Since the capacity of the bus is 44, we discard the 114. Therefore, $20+16=36$ people went on the trip; each person paid $15-0.1(16)=\$13.40$.

159. Let r_S represent Scott's rate and let r_T represent Todd's rate. The time for Scott to run 95 meters is the same as for Todd to run 100 meters.
$$\frac{95}{r_S} = \frac{100}{r_T}$$
$$r_S = 0.95 r_T$$
$$d_S = t \cdot r_s = t(0.95 r_T) = 0.95 d_T$$
If Todd starts from 5 meters behind the start:
$$d_T = 105$$
$$d_S = 0.95 d_T = 0.95(105) = 99.75$$

a. The race does not end in a tie.
b. Todd wins the race.
c. Todd wins by 0.25 meters.
d. To end in a tie:
$$100 = 0.95(100 + x)$$
$$100 = 95 + 0.95x$$
$$5 = 0.95x$$
$$x = 5.263 \text{ meters}$$
e. $95 = 0.95(100)$ Therefore, the race ends in a tie.

Chapter 1 Test

1. a. $d = \sqrt{(x_2 - x_1)^2 + (y_2 - y_1)^2}$
$= \sqrt{(4-(-2))^2 + (5-(-3))^2}$
$= \sqrt{6^2 + 8^2}$
$= \sqrt{36 + 64}$
$= \sqrt{100}$
$= 10$

b. $M = \left(\dfrac{x_1 + x_2}{2}, \dfrac{y_1 + y_2}{2}\right)$
$= \left(\dfrac{4+(-2)}{2}, \dfrac{5+(-3)}{2}\right)$
$= \left(\dfrac{2}{2}, \dfrac{2}{2}\right)$
$= (1, 1)$

c. The distance between the points is the length of the diameter. The radius is half the diameter so we have $r = \dfrac{10}{2} = 5$.

The midpoint of the line segment connecting the two points is the center of the circle. Thus, $(h, k) = (1, 1)$.
$(x-h)^2 + (y-k)^2 = r^2$
$(x-1)^2 + (y-1)^2 = 5^2$
$(x-1)^2 + (y-1)^2 = 25$

d. $(x-1)^2 + (y-1)^2 = 25$

To graph the equation by hand, we plot the center and then use the radius to get points above, below, left, and right of the center.

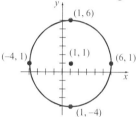

2. $2x^2 + 6x = x - 3$
$2x^2 + 5x + 3 = 0$
$(2x+3)(x+1) = 0$
$2x + 3 = 0 \quad \text{or} \quad x + 1 = 0$
$2x = -3 \qquad\qquad x = -1$
$x = -\dfrac{3}{2}$

The solution set is $\{-\tfrac{3}{2}, -1\}$.

3. $x + 1 = \sqrt{x + 7}$
$(x+1)^2 = \left(\sqrt{x+7}\right)^2$
$x^2 + 2x + 1 = x + 7$
$x^2 + x - 6 = 0$
$(x+3)(x-2) = 0$
$x + 3 = 0 \quad \text{or} \quad x - 2 = 0$
$x = -3 \qquad\qquad x = 2$

Check:
$$-3+1\stackrel{?}{=}\sqrt{-3+7} \qquad 2+1\stackrel{?}{=}\sqrt{2+7}$$
$$-2\stackrel{?}{=}\sqrt{4} \qquad 3\stackrel{?}{=}\sqrt{9}$$
$$-2\neq 2 \qquad 3\stackrel{?}{=}3\;\checkmark$$

The solution set is $\{2\}$.

4. $2-\dfrac{3}{m}=\dfrac{2}{m+2}$

   LCD: $m(m+2)$

   Restricted values: $m=-2, m=0$

   $$\dfrac{2m(m+2)}{m(m+2)}-\dfrac{3(m+2)}{m(m+2)}=\dfrac{2m}{m(m+2)}$$
   $$2m(m+2)-3(m+2)=2m$$
   $$2m^2+4m-3m-6=2m$$
   $$2m^2-m-6=0$$
   $$(2m+3)(m-2)=0$$
   $$2m+3=0 \quad\text{or}\quad m-2=0$$
   $$2m=-3 \qquad\qquad m=2$$
   $$m=-\dfrac{3}{2}$$

   Since neither solution is a restricted value, the solution set is $\left\{-\dfrac{3}{2},2\right\}$.

5. $5x-8=-4(x-1)+6$
   $$5x-8=-4x+4+6$$
   $$5x-8=-4x+10$$
   $$9x=18$$
   $$x=2$$
   The solution set is $\{2\}$.

6. $5|3-2b|-7=8$
   $$5|3-2b|=15$$
   $$|3-2b|=3$$
   $$3-2b=3 \quad\text{or}\quad 3-2b=-3$$
   $$-2b=0 \qquad\qquad -2b=-6$$
   $$b=0 \qquad\qquad\;\; b=3$$
   The solution set is $\{0,3\}$.

7. $x^4+x^2=3x^2+8$
   $$x^4-2x^2-8=0$$
   Let $u=x^2$. Then $u^2=\left(x^2\right)^2=x^4$, and we have

$u^2-2u-8=0$
$(u-4)(u+2)=0$
$u=4$ or $u=-2$
Since we are solving for x, we get
$x^2=4$ or $x^2=-2$
$x=\pm 2 \qquad x=\pm\sqrt{-2}$
$\qquad\qquad\qquad =\pm i\sqrt{2}$

The solution set is $\left\{-2,2,-i\sqrt{2},i\sqrt{2}\right\}$.

8. $x^2-4x+7=0$
   $a=1, b=-4, c=7$
   $$x=\dfrac{-b\pm\sqrt{b^2-4ac}}{2a}$$
   $$=\dfrac{-(-4)\pm\sqrt{(-4)^2-4(1)(7)}}{2(1)}$$
   $$=\dfrac{4\pm\sqrt{16-28}}{2}$$
   $$=\dfrac{4\pm\sqrt{-12}}{2}$$
   $$=\dfrac{4\pm 2i\sqrt{3}}{2}$$
   $$=2\pm i\sqrt{3}$$
   The solution set is $\left\{2-i\sqrt{3},2+i\sqrt{3}\right\}$.

9. $2x^2+x-1=x(x+7)+2$
   $$2x^2+x-1=x^2+7x+2$$
   $$x^2-6x-3=0$$
   $a=1, b=-6, c=-3$
   $$x=\dfrac{-b\pm\sqrt{b^2-4ac}}{2a}$$
   $$=\dfrac{-(-6)\pm\sqrt{(-6)^2-4(1)(-3)}}{2(1)}$$
   $$=\dfrac{6\pm\sqrt{36+12}}{2}$$
   $$=\dfrac{6\pm\sqrt{48}}{2}$$
   $$=\dfrac{6\pm 4\sqrt{3}}{2}$$
   $$=3\pm 2\sqrt{3}$$
   The solution set is $\left\{3-2\sqrt{3},3+2\sqrt{3}\right\}$.

10. $y - 14 = (x+1)^2$

$y = (x+1)^2 + 14$

x	$y = (x+1)^2 + 14$	(x, y)
-5	$y = (-5+1)^2 + 14 = 30$	$(-5, 30)$
-3	$y = (-3+1)^2 + 14 = 18$	$(-3, 18)$
-1	$y = (-1+1)^2 + 14 = 14$	$(-1, 14)$
1	$y = (1+1)^2 + 14 = 18$	$(1, 18)$
3	$y = (3+1)^2 + 14 = 30$	$(3, 30)$

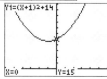

The only intercept is $(0, 15)$.

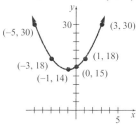

11. $2x - 7y = 21$

$-7y = -2x + 21$

$y = \frac{2}{7}x - 3$

x	$y = \frac{2}{7}x - 3$	(x, y)
-7	$y = \frac{2}{7}(-7) - 3 = -5$	$(-7, -5)$
0	$y = \frac{2}{7}(0) - 3 = -3$	$(0, -3)$
7	$y = \frac{2}{7}(7) - 3 = -1$	$(7, -1)$

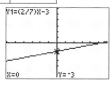

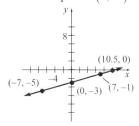

The intercepts are $(0, -3)$ and $(10.5, 0)$.

12. $y = x^3 - 3x + 2$

x	$y = x^3 - 3x + 2$	(x, y)
-3	$y = (-3)^3 - 3(-3) + 2 = -16$	$(-3, -16)$
-1	$y = (-1)^3 - 3(-1) + 2 = 4$	$(-1, 4)$
0	$y = (0)^3 - 3(0) + 2 = 2$	$(0, 2)$
1	$y = (1)^3 - 3(1) + 2 = 0$	$(1, 0)$
3	$y = (3)^3 - 3(3) + 2 = 20$	$(3, 20)$

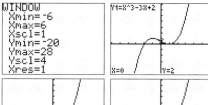
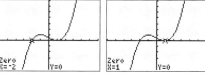

The intercepts are $(0, 2)$, $(-2, 0)$, and $(1, 0)$.

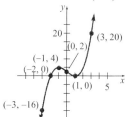

13. a. Start by putting the equation into slope-intercept form.
$$10x - 6y = 20$$
$$-6y = -10x + 20$$
$$y = \frac{-10x + 20}{-6}$$
$$y = \frac{5}{3}x - \frac{10}{3}$$
In this form, the slope is the coefficient of *x*. Therefore, the slope is $m = \frac{5}{3}$.

b. From part (a), we can see that the y-intercept is $-\frac{10}{3}$. To find the x-intercept, we let $y = 0$ in the original equation.
$$10x - 6(0) = 20$$
$$10x = 20$$
$$x = 2$$
The x-intercept is 2.

c. Plot the points $(2, 0)$ and $\left(0, -\frac{10}{3}\right)$, and then draw a line through the points.

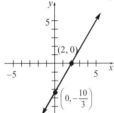

d. The slope of the given line is $\frac{5}{3}$. Since perpendicular lines have opposite-reciprocal slopes, the slope of our line will be $m = -\frac{3}{5}$.
Using the point-slope form of the equation of a line, we get
$$y - y_1 = m(x - x_1)$$
$$y - 5 = -\frac{3}{5}(x - (-1))$$
$$y - 5 = -\frac{3}{5}(x + 1)$$
$$y = -\frac{3}{5}x - \frac{3}{5} + 5$$
$$y = -\frac{3}{5}x + \frac{22}{5}$$

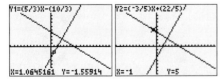

e. The slope of the given line is $\frac{5}{3}$. Since parallel lines have the same slope, the slope of our line will be $m = \frac{5}{3}$. Using the point-slope form of the equation of a line, we get
$$y - y_1 = m(x - x_1)$$
$$y - 5 = \frac{5}{3}(x - (-1))$$
$$y - 5 = \frac{5}{3}(x + 1)$$
$$y = \frac{5}{3}x + \frac{5}{3} + 5$$
$$y = \frac{5}{3}x + \frac{20}{3}$$

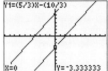 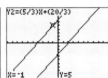

14. $2x^3 - x^2 - 2x + 1 = 0$
Since this equation has 0 on one side, we will use the Zero option from the CALC menu. It will be important to carefully select the window settings so as not to miss any solutions.

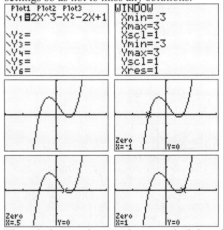

The solutions to the equation are -1, 0.5, and 1.

15. $x^4 - 5x^2 - 8 = 0$

Since this equation has 0 on one side, we will use the Zero option from the CALC menu.

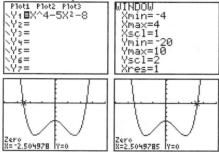

The solutions, rounded to two decimal places, are -2.50 and 2.50.

16. $-x^3 + 7x - 2 = x^2 + 3x - 3$

Since there are nonzero expressions on both sides of the equation, we will use the Intersect option from the CALC menu. We enter the left side of the equation in Y1 and the right side in Y2.

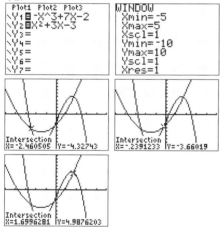

The solutions, rounded to two decimal places, are -2.46, -0.24, and 1.70.

17. $\dfrac{2x+3}{4} < -2$

$4 \cdot \dfrac{2x+3}{4} < 4 \cdot (-2)$

$2x + 3 < -8$

$2x + 3 - 3 < -8 - 3$

$2x < -11$

$\dfrac{2x}{2} < \dfrac{-11}{2}$

$x < -\dfrac{11}{2}$

Solution set: $\left\{ x \mid x < -\tfrac{11}{2} \right\}$

Interval: $\left(-\infty, -\tfrac{11}{2}\right)$

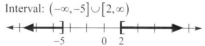

18. $|2x+3| - 4 \geq 3$

$|2x+3| - 4 + 4 \geq 3 + 4$

$|2x+3| \geq 7$

$2x + 3 \leq -7$ or $2x + 3 \geq 7$

$2x + 3 - 3 \leq -7 - 3$ $2x + 3 - 3 \geq 7 - 3$

$2x \leq -10$ $2x \geq 4$

$\dfrac{2x}{2} \leq \dfrac{-10}{2}$ $\dfrac{2x}{2} \geq \dfrac{4}{2}$

$x \leq -5$ $x \geq 2$

Solution set: $\{ x \mid x \leq -5 \text{ or } x \geq 2 \}$

Interval: $(-\infty, -5] \cup [2, \infty)$

19. $-7 < 3 - 5x \leq 8$

$-7 - 3 < 3 - 5x - 3 \leq 8 - 3$

$-10 < -5x \leq 5$

$\dfrac{-10}{-5} > \dfrac{-5x}{-5} \geq \dfrac{5}{-5}$

$2 > x \geq -1$

$-1 \leq x < 2$

Solution set: $\{ x \mid -1 \leq x < 2 \}$

Interval: $[-1, 2)$

Chapter 2: *Functions and Their Graphs*   SSM: College Algebra EGU

17. Intercepts: $(-1,0)$ and $(1,0)$
    Symmetric with respect to the x-axis, y-axis, and the origin.

19. Intercepts: $\left(-\frac{\pi}{2},0\right)$, $(0,1)$, and $\left(\frac{\pi}{2},0\right)$
    Symmetric with respect to the y-axis.

21. Intercepts: $(0,0)$
    Symmetric with respect to the x-axis.

23. Intercepts: $(-2,0)$, $(0,0)$, and $(2,0)$
    Symmetric with respect to the origin.

25.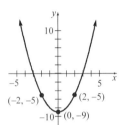

27.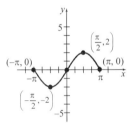

29. $y = x^3$

    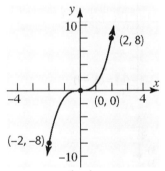

    x-intercept: $(0,0)$;   y-intercept: $(0,0)$

31. $y = \sqrt{x}$

    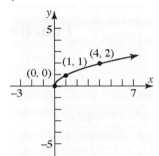

    x-intercept: $(0,0)$;   y-intercept: $(0,0)$

33. a. $0 = 3x$
    $0 = x$
    The x-intercept is $x = 0$.

    $y = 3(0) = 0$
    The y-intercept is $y = 0$.
    The lone intercept is $(0,0)$.

    b. x-axis (replace y by $-y$):
    $-y = 3x$
    $y = -3x$ different

    y-axis (replace x by $-x$):
    $y = 3(-x) = -3x$ different

    origin (replace x by $-x$ and y by $-y$):
    $-y = 3(-x)$
    $y = 3x$ same

    The equation has origin symmetry.

35. a. $0 = x^2 - 9$
    $x^2 = 9$
    $x = \pm 3$
    The x-intercepts are $x = -3$ and $x = 3$.

    $y = (0)^2 - 9 = -9$
    The y-intercept is $y = -9$.

    The intercepts are $(-3,0)$, $(3,0)$, and $(0,-9)$.

b. x-axis (replace y by $-y$):
$$-y = x^2 - 9$$
$$y = 9 - x^2 \text{ different}$$
y-axis (replace x by $-x$):
$$y = (-x)^2 - 9 = x^2 - 9 \text{ same}$$
origin (replace x by $-x$ and y by $-y$):
$$-y = (-x)^2 - 9$$
$$y = 9 - x^2 \text{ different}$$
The equation has y-axis symmetry.

37. a. $x = (0)^2 - 4$
$$x = -4$$
The x-intercept is $x = -4$.
$$0 = y^2 - 4$$
$$y^2 = 4 \rightarrow y = \pm 2$$
The y-intercepts are $y = -2$ and $y = 2$.
The intercepts are $(-4, 0)$, $(0, 2)$, and $(0, -2)$.

b. x-axis (replace y by $-y$):
$$x = (-y)^2 - 4$$
$$x = y^2 - 4 \text{ same}$$
y-axis (replace x by $-x$):
$$-x = y^2 - 4$$
$$x = 4 - y^2 \text{ different}$$
origin (replace x by $-x$ and y by $-y$):
$$-x = (-y)^2 - 4$$
$$x = 4 - y^2 \text{ different}$$
The equation has x-axis symmetry.

39. a. $0 = x^3 - 8x$
$$0 = x(x^2 - 8)$$
$$x = 0 \text{ or } x^2 - 8 = 0$$
$$x^2 = 8$$
$$x = \pm\sqrt{8} = \pm 2\sqrt{2}$$
The x-intercepts are $x = 0$, $x = -2\sqrt{2}$, and $x = 2\sqrt{2}$.

$$y = 0^3 - 8(0) = 0$$
The y-intercept is $y = 0$.
The intercepts are $(0, 0)$, $(-2\sqrt{2}, 0)$, and $(2\sqrt{2}, 0)$.

b. x-axis (replace y by $-y$):
$$-y = x^3 - 8x$$
$$y = 8x - x^3 \text{ different}$$
y-axis (replace x by $-x$):
$$y = (-x)^3 - 8(-x)$$
$$y = -x^3 + 8x \text{ different}$$
origin (replace x by $-x$ and y by $-y$):
$$-y = (-x)^3 - 8(-x)$$
$$-y = -x^3 + 8x$$
$$y = x^3 - 8x \text{ same}$$
The equation has origin symmetry.

41. a. $x^2 + (0)^2 = 9$
$$x^2 = 9$$
$$x = \pm 3$$
The x-intercepts are $x = -3$ and $x = 3$.
$$(0)^2 + y^2 = 9$$
$$y^2 = 9$$
$$y = \pm 3$$
The y-intercepts are $y = -3$ and $y = 3$.
The intercepts are $(-3, 0)$, $(3, 0)$, $(0, -3)$, and $(0, 3)$.

b. x-axis (replace y by $-y$):
$$x^2 + (-y)^2 = 9$$
$$x^2 + y^2 = 9 \text{ same}$$
y-axis (replace x by $-x$):
$$(-x)^2 + y^2 = 9$$
$$x^2 + y^2 = 9 \text{ same}$$
origin (replace x by $-x$ and y by $-y$):
$$(-x)^2 + (-y)^2 = 9$$
$$x^2 + y^2 = 9 \text{ same}$$
The equation has x-axis symmetry, y-axis symmetry, and origin symmetry.

43. a. $0 + 4 = x^2 - 3x$
$0 = x^2 - 3x - 4$
$0 = (x-4)(x+1)$
The x-intercepts are $x = 4$ and $x = -1$.
$y + 4 = 0^2 - 3(0)$
$y + 4 = 0$
$y = -4$
The y-intercept is $y = -4$. The intercepts are $(4, 0)$, $(-1, 0)$, and $(0, -4)$.

b. x-axis (replace y by $-y$):
$-y + 4 = x^2 - 3x$ different
y-axis (replace x by $-x$):
$y + 4 = (-x)^2 - 3(-x)$
$y + 4 = x^2 + 3x$ different
origin (replace x by $-x$ and y by $-y$):
$-y + 4 = (-x)^2 - 3(-x)$
$-y + 4 = x^2 + 3x$ different

The equation is not symmetric with respect to the x-axis, y-axis, or origin.

45. a. $0 + 4 = |x|$
$4 = |x|$
The x-intercepts are $x = -4$ and $x = 4$.
$y + 4 = |0|$
$y + 4 = 0$
$y = -4$
The y-intercept is $y = -4$.

The intercepts are $(-4, 0)$, $(4, 0)$, and $(0, -4)$.

b. x-axis (replace y by $-y$):
$-y + 4 = |x|$ different
y-axis (replace x by $-x$):
$y + 4 = |-x|$
$y + 4 = |x|$ same
origin (replace x by $-x$ and y by $-y$):
$-y + 4 = |-x|$
$-y + 4 = |x|$ different

The equation has y-axis symmetry.

47. a. $0 = \dfrac{x^2 - 9}{x} \rightarrow x^2 - 9 = 0$
$x^2 = 9$
$x = \pm 3$
The x-intercepts are $x = -3$ and $x = 3$.
There are no y-intercepts because $x = 0$ is not in the domain of the equation. The intercepts are $(-3, 0)$ and $(3, 0)$.

b. x-axis (replace y by $-y$):
$-y = \dfrac{x^2 - 9}{x}$
$y = \dfrac{9 - x^2}{x}$ different
y-axis (replace x by $-x$):
$y = \dfrac{(-x)^2 - 9}{(-x)} = \dfrac{x^2 - 9}{-x} = \dfrac{9 - x^2}{x}$ different
origin (replace x by $-x$ and y by $-y$):
$-y = \dfrac{(-x)^2 - 9}{(-x)}$
$-y = \dfrac{x^2 - 9}{-x}$
$y = \dfrac{x^2 - 9}{x}$ same

The equation has origin symmetry.

49. For a graph that is symmetric about the origin, if the point (x, y) is on the graph then the point $(-x, -y)$ is also on the graph. Since the point $(1, 2)$ is on the graph, the point $(-1, -2)$ must also be on the graph.

51. For a graph that is symmetric about the origin, if the point (x, y) is on the graph then the point $(-x, -y)$ is also on the graph. We know that an x-intercept is $x = -4$. This means that the point $(-4, 0)$ is on the graph. From the symmetry, the point $(4, 0)$ must also be on the graph. Therefore, another x-intercept is $x = 4$.

53. The graph of an equation could have symmetry with respect to the x-axis, y-axis, or the origin. There are other possible symmetries (such as about the line $x = 2$) but these three are special because they represent symmetry about a coordinate axis or the origin.

Section 2.2

1. $(-1, 3)$

3. We must not allow the denominator to be 0. $x + 4 \neq 0 \Rightarrow x \neq -4$; Domain: $\{x | x \neq -4\}$.

5. independent; dependent

7. $[0, 5]$
   We need the intersection of the intervals $[0, 7]$ and $[-2, 5]$.

9. $g(x) - f(x)$, or $(g - f)(x)$

11. True

13. False; if the domain is not specified, we assume it is the largest set of real numbers for which the value of f is a real number.

15. Function
    Domain: {Elvis, Colleen, Kaleigh, Marissa}
    Range: {Jan. 8, Mar. 15, Sept. 17}

17. Not a function

19. Not a function

21. Function
    Domain: {1, 2, 3, 4}
    Range: {3}

23. Not a function

25. Function
    Domain: {−2, −1, 0, 1}
    Range: {0, 1, 4}

27. Graph $y = x^2$. The graph passes the vertical line test. Thus, the equation represents a function.

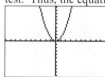

29. Graph $y = \dfrac{1}{x}$. The graph passes the vertical line test. Thus, the equation represents a function.

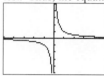

31. $y^2 = 4 - x^2$
    Solve for y: $y = \pm\sqrt{4 - x^2}$
    For $x = 0$, $y = \pm 2$. Thus, (0, 2) and (0, −2) are on the graph. This is not a function, since a distinct x corresponds to two different y's.

33. $x = y^2$
    Solve for y: $y = \pm\sqrt{x}$
    For $x = 1$, $y = \pm 1$. Thus, (1, 1) and (1, −1) are on the graph. This is not a function, since a distinct x corresponds to two different y's.

35. Graph $y = 2x^2 - 3x + 4$. The graph passes the vertical line test. Thus, the equation represents a function.

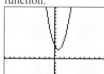

Chapter 2: *Functions and Their Graphs*   *SSM:* College Algebra EGU

37. $2x^2 + 3y^2 = 1$
    Solve for y: $2x^2 + 3y^2 = 1$
    $$3y^2 = 1 - 2x^2$$
    $$y^2 = \frac{1-2x^2}{3}$$
    $$y = \pm\sqrt{\frac{1-2x^2}{3}}$$
    For $x = 0$, $y = \pm\sqrt{\frac{1}{3}}$. Thus, $\left(0, \sqrt{\frac{1}{3}}\right)$ and $\left(0, -\sqrt{\frac{1}{3}}\right)$ are on the graph. This is not a function, since a distinct x corresponds to two different y's.

39. $f(x) = 3x^2 + 2x - 4$

    a. $f(0) = 3(0)^2 + 2(0) - 4 = -4$
    b. $f(1) = 3(1)^2 + 2(1) - 4 = 3 + 2 - 4 = 1$
    c. $f(-1) = 3(-1)^2 + 2(-1) - 4 = 3 - 2 - 4 = -3$
    d. $f(-x) = 3(-x)^2 + 2(-x) - 4 = 3x^2 - 2x - 4$
    e. $-f(x) = -(3x^2 + 2x - 4) = -3x^2 - 2x + 4$
    f. $f(x+1) = 3(x+1)^2 + 2(x+1) - 4$
       $= 3(x^2 + 2x + 1) + 2x + 2 - 4$
       $= 3x^2 + 6x + 3 + 2x + 2 - 4$
       $= 3x^2 + 8x + 1$
    g. $f(2x) = 3(2x)^2 + 2(2x) - 4 = 12x^2 + 4x - 4$
    h. $f(x+h) = 3(x+h)^2 + 2(x+h) - 4$
       $= 3(x^2 + 2xh + h^2) + 2x + 2h - 4$
       $= 3x^2 + 6xh + 3h^2 + 2x + 2h - 4$

41. $f(x) = \frac{x}{x^2 + 1}$

    a. $f(0) = \frac{0}{0^2 + 1} = \frac{0}{1} = 0$
    b. $f(1) = \frac{1}{1^2 + 1} = \frac{1}{2}$
    c. $f(-1) = \frac{-1}{(-1)^2 + 1} = \frac{-1}{1+1} = -\frac{1}{2}$
    d. $f(-x) = \frac{-x}{(-x)^2 + 1} = \frac{-x}{x^2 + 1}$
    e. $-f(x) = -\left(\frac{x}{x^2+1}\right) = \frac{-x}{x^2+1}$
    f. $f(x+1) = \frac{x+1}{(x+1)^2 + 1}$
       $= \frac{x+1}{x^2 + 2x + 1 + 1}$
       $= \frac{x+1}{x^2 + 2x + 2}$
    g. $f(2x) = \frac{2x}{(2x)^2 + 1} = \frac{2x}{4x^2 + 1}$
    h. $f(x+h) = \frac{x+h}{(x+h)^2 + 1} = \frac{x+h}{x^2 + 2xh + h^2 + 1}$

43. $f(x) = |x| + 4$

    a. $f(0) = |0| + 4 = 0 + 4 = 4$
    b. $f(1) = |1| + 4 = 1 + 4 = 5$
    c. $f(-1) = |-1| + 4 = 1 + 4 = 5$
    d. $f(-x) = |-x| + 4 = |x| + 4$
    e. $-f(x) = -(|x| + 4) = -|x| - 4$
    f. $f(x+1) = |x+1| + 4$
    g. $f(2x) = |2x| + 4 = 2|x| + 4$
    h. $f(x+h) = |x+h| + 4$

45. $f(x) = \frac{2x+1}{3x-5}$

    a. $f(0) = \frac{2(0)+1}{3(0)-5} = \frac{0+1}{0-5} = -\frac{1}{5}$
    b. $f(1) = \frac{2(1)+1}{3(1)-5} = \frac{2+1}{3-5} = \frac{3}{-2} = -\frac{3}{2}$
    c. $f(-1) = \frac{2(-1)+1}{3(-1)-5} = \frac{-2+1}{-3-5} = \frac{-1}{-8} = \frac{1}{8}$

d. $f(-x) = \dfrac{2(-x)+1}{3(-x)-5} = \dfrac{-2x+1}{-3x-5} = \dfrac{2x-1}{3x+5}$

e. $-f(x) = -\left(\dfrac{2x+1}{3x-5}\right) = \dfrac{-2x-1}{3x-5}$

f. $f(x+1) = \dfrac{2(x+1)+1}{3(x+1)-5} = \dfrac{2x+2+1}{3x+3-5} = \dfrac{2x+3}{3x-2}$

g. $f(2x) = \dfrac{2(2x)+1}{3(2x)-5} = \dfrac{4x+1}{6x-5}$

h. $f(x+h) = \dfrac{2(x+h)+1}{3(x+h)-5} = \dfrac{2x+2h+1}{3x+3h-5}$

47. $f(x) = -5x+4$
Domain: $\{x \mid x \text{ is any real number}\}$

49. $f(x) = \dfrac{x}{x^2+1}$
Domain: $\{x \mid x \text{ is any real number}\}$

51. $g(x) = \dfrac{x}{x^2-16}$
$x^2-16 \ne 0$
$x^2 \ne 16 \Rightarrow x \ne \pm 4$
Domain: $\{x \mid x \ne -4, x \ne 4\}$

53. $F(x) = \dfrac{x-2}{x^3+x}$
$x^3+x \ne 0$
$x(x^2+1) \ne 0$
$x \ne 0, \ x^2 \ne -1$
Domain: $\{x \mid x \ne 0\}$

55. $h(x) = \sqrt{3x-12}$
$3x-12 \ge 0$
$3x \ge 12$
$x \ge 4$
Domain: $\{x \mid x \ge 4\}$

57. $f(x) = \dfrac{4}{\sqrt{x-9}}$
$x-9 > 0$
$x > 9$
Domain: $\{x \mid x > 9\}$

59. $p(x) = \sqrt{\dfrac{2}{x-1}} = \dfrac{\sqrt{2}}{\sqrt{x-1}}$
$x-1 > 0$
$x > 1$
Domain: $\{x \mid x > 1\}$

61. $f(x) = 3x+4 \qquad g(x) = 2x-3$

a. $(f+g)(x) = 3x+4+2x-3 = 5x+1$
The domain is $\{x \mid x \text{ is any real number}\}$.

b. $(f-g)(x) = (3x+4)-(2x-3)$
$= 3x+4-2x+3$
$= x+7$
The domain is $\{x \mid x \text{ is any real number}\}$.

c. $(f \cdot g)(x) = (3x+4)(2x-3)$
$= 6x^2-9x+8x-12$
$= 6x^2-x-12$
The domain is $\{x \mid x \text{ is any real number}\}$.

d. $\left(\dfrac{f}{g}\right)(x) = \dfrac{3x+4}{2x-3}$
$2x-3 \ne 0 \Rightarrow 2x \ne 3 \Rightarrow x \ne \dfrac{3}{2}$
The domain is $\left\{x \mid x \ne \dfrac{3}{2}\right\}$.

63. $f(x) = x-1 \qquad g(x) = 2x^2$

a. $(f+g)(x) = x-1+2x^2 = 2x^2+x-1$
The domain is $\{x \mid x \text{ is any real number}\}$.

b. $(f-g)(x) = (x-1)-(2x^2)$
$= x-1-2x^2$
$= -2x^2+x-1$
The domain is $\{x \mid x \text{ is any real number}\}$.

Chapter 2: Functions and Their Graphs

    c. $(f \cdot g)(x) = (x-1)(2x^2) = 2x^3 - 2x^2$
       The domain is $\{x \mid x \text{ is any real number}\}$.

    d. $\left(\dfrac{f}{g}\right)(x) = \dfrac{x-1}{2x^2}$
       The domain is $\{x \mid x \neq 0\}$.

65. $f(x) = \sqrt{x} \qquad g(x) = 3x - 5$

    a. $(f+g)(x) = \sqrt{x} + 3x - 5$
       The domain is $\{x \mid x \geq 0\}$.

    b. $(f-g)(x) = \sqrt{x} - (3x - 5) = \sqrt{x} - 3x + 5$
       The domain is $\{x \mid x \geq 0\}$.

    c. $(f \cdot g)(x) = \sqrt{x}(3x - 5) = 3x\sqrt{x} - 5\sqrt{x}$
       The domain is $\{x \mid x \geq 0\}$.

    d. $\left(\dfrac{f}{g}\right)(x) = \dfrac{\sqrt{x}}{3x - 5}$
       $x \geq 0$ and $3x - 5 \neq 0$
       $3x \neq 5 \Rightarrow x \neq \dfrac{5}{3}$
       The domain is $\left\{x \mid x \geq 0 \text{ and } x \neq \dfrac{5}{3}\right\}$.

67. $f(x) = 1 + \dfrac{1}{x} \qquad g(x) = \dfrac{1}{x}$

    a. $(f+g)(x) = 1 + \dfrac{1}{x} + \dfrac{1}{x} = 1 + \dfrac{2}{x}$
       The domain is $\{x \mid x \neq 0\}$.

    b. $(f-g)(x) = 1 + \dfrac{1}{x} - \dfrac{1}{x} = 1$
       The domain is $\{x \mid x \neq 0\}$.

    c. $(f \cdot g)(x) = \left(1 + \dfrac{1}{x}\right)\dfrac{1}{x} = \dfrac{1}{x} + \dfrac{1}{x^2}$
       The domain is $\{x \mid x \neq 0\}$.

    d. $\left(\dfrac{f}{g}\right)(x) = \dfrac{1 + \dfrac{1}{x}}{\dfrac{1}{x}} = \dfrac{\dfrac{x+1}{x}}{\dfrac{1}{x}} = \dfrac{x+1}{x} \cdot \dfrac{x}{1} = x + 1$
       The domain is $\{x \mid x \neq 0\}$.

69. $f(x) = \dfrac{2x+3}{3x-2} \qquad g(x) = \dfrac{4x}{3x-2}$

    a. $(f+g)(x) = \dfrac{2x+3}{3x-2} + \dfrac{4x}{3x-2}$
       $= \dfrac{2x+3+4x}{3x-2}$
       $= \dfrac{6x+3}{3x-2}$
       $3x - 2 \neq 0$
       $3x \neq 2 \Rightarrow x \neq \dfrac{2}{3}$
       The domain is $\left\{x \mid x \neq \dfrac{2}{3}\right\}$.

    b. $(f-g)(x) = \dfrac{2x+3}{3x-2} - \dfrac{4x}{3x-2}$
       $= \dfrac{2x+3-4x}{3x-2}$
       $= \dfrac{-2x+3}{3x-2}$
       $3x - 2 \neq 0$
       $3x \neq 2 \Rightarrow x \neq \dfrac{2}{3}$
       The domain is $\left\{x \mid x \neq \dfrac{2}{3}\right\}$.

    c. $(f \cdot g)(x) = \left(\dfrac{2x+3}{3x-2}\right)\left(\dfrac{4x}{3x-2}\right) = \dfrac{8x^2 + 12x}{(3x-2)^2}$
       $3x - 2 \neq 0$
       $3x \neq 2 \Rightarrow x \neq \dfrac{2}{3}$
       The domain is $\left\{x \mid x \neq \dfrac{2}{3}\right\}$.

    d. $\left(\dfrac{f}{g}\right)(x) = \dfrac{\dfrac{2x+3}{3x-2}}{\dfrac{4x}{3x-2}} = \dfrac{2x+3}{3x-2} \cdot \dfrac{3x-2}{4x} = \dfrac{2x+3}{4x}$
       $3x - 2 \neq 0$ and $x \neq 0$
       $3x \neq 2$
       $x \neq \dfrac{2}{3}$
       The domain is $\left\{x \mid x \neq \dfrac{2}{3} \text{ and } x \neq 0\right\}$.

71. $f(x) = 3x+1 \quad (f+g)(x) = 6 - \frac{1}{2}x$

$6 - \frac{1}{2}x = 3x + 1 + g(x)$

$5 - \frac{7}{2}x = g(x)$

$g(x) = 5 - \frac{7}{2}x$

73. $f(x) = 4x + 3$

$\frac{f(x+h) - f(x)}{h} = \frac{4(x+h) + 3 - 4x - 3}{h}$

$= \frac{4x + 4h + 3 - 4x - 3}{h}$

$= \frac{4h}{h} = 4$

75. $f(x) = x^2 - x + 4$

$\frac{f(x+h) - f(x)}{h}$

$= \frac{(x+h)^2 - (x+h) + 4 - (x^2 - x + 4)}{h}$

$= \frac{x^2 + 2xh + h^2 - x - h + 4 - x^2 + x - 4}{h}$

$= \frac{2xh + h^2 - h}{h}$

$= 2x + h - 1$

77. $f(x) = x^3 - 2$

$\frac{f(x+h) - f(x)}{h}$

$= \frac{(x+h)^3 - 2 - (x^3 - 2)}{h}$

$= \frac{x^3 + 3x^2h + 3xh^2 + h^3 - 2 - x^3 + 2}{h}$

$= \frac{3x^2h + 3xh^2 + h^3}{h}$

$= 3x^2 + 3xh + h^2$

79. $f(x) = 2x^3 + Ax^2 + 4x - 5$ and $f(2) = 5$

$f(2) = 2(2)^3 + A(2)^2 + 4(2) - 5$

$5 = 16 + 4A + 8 - 5$

$5 = 4A + 19$

$-14 = 4A$

$A = -\frac{7}{2}$

81. $f(x) = \frac{3x+8}{2x-A}$ and $f(0) = 2$

$f(0) = \frac{3(0) + 8}{2(0) - A}$

$2 = \frac{8}{-A}$

$-2A = 8$

$A = -4$

83. $f(x) = \frac{2x - A}{x - 3}$ and $f(4) = 0$

$f(4) = \frac{2(4) - A}{4 - 3}$

$0 = \frac{8 - A}{1}$

$0 = 8 - A$

$A = 8$

f is undefined when $x = 3$.

85. Let x represent the length of the rectangle. Then, $\frac{x}{2}$ represents the width of the rectangle since the length is twice the width. The function for the area is:

$A(x) = x \cdot \frac{x}{2} = \frac{x^2}{2} = \frac{1}{2}x^2$

87. Let x represent the number of hours worked. The function for the gross salary is: $G(x) = 10x$

Chapter 2: *Functions and Their Graphs*

89. a. $H(1) = 20 - 4.9(1)^2$
$= 20 - 4.9$
$= 15.1$ meters
$H(1.1) = 20 - 4.9(1.1)^2 = 20 - 4.9(1.21)$
$= 20 - 5.929$
$= 14.071$ meters
$H(1.2) = 20 - 4.9(1.2)^2$
$= 20 - 4.9(1.44)$
$= 20 - 7.056$
$= 12.944$ meters
$H(1.3) = 20 - 4.9(1.3)^2$
$= 20 - 4.9(1.69)$
$= 20 - 8.281$
$= 11.719$ meters

b. $H(x) = 15$:
$15 = 20 - 4.9x^2$
$-5 = -4.9x^2$
$x^2 \approx 1.0204$
$x \approx 1.01$ seconds

$H(x) = 10$:
$10 = 20 - 4.9x^2$
$-10 = -4.9x^2$
$x^2 \approx 2.0408$
$x \approx 1.43$ seconds

$H(x) = 5$:
$5 = 20 - 4.9x^2$
$-15 = -4.9x^2$
$x^2 \approx 3.0612$
$x \approx 1.75$ seconds

c. $H(x) = 0$
$0 = 20 - 4.9x^2$
$-20 = -4.9x^2$
$x^2 \approx 4.0816$
$x \approx 2.02$ seconds

91. $C(x) = 100 + \dfrac{x}{10} + \dfrac{36,000}{x}$

a. $C(500) = 100 + \dfrac{500}{10} + \dfrac{36,000}{500}$
$= 100 + 50 + 72$
$= \$222$

b. $C(450) = 100 + \dfrac{450}{10} + \dfrac{36,000}{450}$
$= 100 + 45 + 80$
$= \$225$

c. $C(600) = 100 + \dfrac{600}{10} + \dfrac{36,000}{600}$
$= 100 + 60 + 60$
$= \$220$

d. $C(400) = 100 + \dfrac{400}{10} + \dfrac{36,000}{400}$
$= 100 + 40 + 90$
$= \$230$

93. $R(x) = \left(\dfrac{L}{P}\right)(x) = \dfrac{L(x)}{P(x)}$

95. $H(x) = (P \cdot I)(x) = P(x) \cdot I(x)$

97. a. $h(x) = 2x$
$h(a+b) = 2(a+b) = 2a + 2b$
$= h(a) + h(b)$
$h(x) = 2x$ has the property.

b. $g(x) = x^2$
$g(a+b) = (a+b)^2 = a^2 + 2ab + b^2$
Since
$a^2 + 2ab + b^2 \neq a^2 + b^2 = g(a) + g(b)$,
$g(x) = x^2$ does not have the property.

c. $F(x) = 5x - 2$
$F(a+b) = 5(a+b) - 2 = 5a + 5b - 2$
Since
$5a + 5b - 2 \neq 5a - 2 + 5b - 2 = F(a) + F(b)$,
$F(x) = 5x - 2$ does not have the property.

SSM: College Algebra EGU Chapter 2: Functions and Their Graphs

d. $G(x) = \dfrac{1}{x}$

$G(a+b) = \dfrac{1}{a+b} \ne \dfrac{1}{a} + \dfrac{1}{b} = G(a) + G(b)$

$G(x) = \dfrac{1}{x}$ does not have the property.

99. Answers will vary.

Section 2.3

1. $x^2 + 4y^2 = 16$

   x-intercepts:
   $x^2 + 4(0)^2 = 16$
   $x^2 = 16$
   $x = \pm 4 \Rightarrow (-4, 0), (4, 0)$

   y-intercepts:
   $(0)^2 + 4y^2 = 16$
   $4y^2 = 16$
   $y^2 = 4$
   $y = \pm 2 \Rightarrow (0, -2), (0, 2)$

3. vertical

5. $f(x) = ax^2 + 4$
   $a(-1)^2 + 4 = 2 \Rightarrow a = -2$

7. False; e.g. $y = \dfrac{1}{x}$.

9. a. $f(0) = 3$ since $(0, 3)$ is on the graph.
      $f(-6) = -3$ since $(-6, -3)$ is on the graph.

   b. $f(6) = 0$ since $(6, 0)$ is on the graph.
      $f(11) = 1$ since $(11, 1)$ is on the graph.

   c. $f(3)$ is positive since $f(3) \approx 3.7$.

   d. $f(-4)$ is negative since $f(-4) \approx -1$.

   e. $f(x) = 0$ when $x = -3$, $x = 6$, and $x = 10$.

   f. $f(x) > 0$ when $-3 < x < 6$, and $10 < x \le 11$.

   g. The domain of f is
      $\{x \mid -6 \le x \le 11\}$ or $[-6, 11]$.

   h. The range of f is
      $\{y \mid -3 \le y \le 4\}$ or $[-3, 4]$.

   i. The x-intercepts are $(-3, 0)$, $(6, 0)$, and $(10, 0)$.

   j. The y-intercept is $(0, 3)$.

   k. The line $y = \dfrac{1}{2}$ intersects the graph 3 times.

   l. The line $x = 5$ intersects the graph 1 time.

   m. $f(x) = 3$ when $x = 0$ and $x = 4$.

   n. $f(x) = -2$ when $x = -5$ and $x = 8$.

11. Not a function since vertical lines will intersect the graph in more than one point.

13. Function

    a. Domain: $\{x \mid -\pi \le x \le \pi\}$;
       Range: $\{y \mid -1 \le y \le 1\}$

    b. Intercepts: $\left(-\dfrac{\pi}{2}, 0\right)$, $\left(\dfrac{\pi}{2}, 0\right)$, $(0, 1)$

    c. Symmetry about y-axis.

15. Not a function since vertical lines will intersect the graph in more than one point.

17. Function

    a. Domain: $\{x \mid x > 0\}$;
       Range: $\{y \mid y$ is any real number$\}$

    b. Intercepts: $(1, 0)$

    c. None

19. Function

    a. Domain: $\{x \mid x$ is any real number$\}$;
       Range: $\{y \mid y \le 2\}$

    b. Intercepts: $(-3, 0)$, $(3, 0)$, $(0, 2)$

    c. Symmetry about y-axis.

21. Function

    a. Domain: $\{x \mid x$ is any real number$\}$;
       Range: $\{y \mid y \ge -3\}$

    b. Intercepts: $(1, 0)$, $(3, 0)$, $(0, 9)$

    c. None

Chapter 2: Functions and Their Graphs

23. $f(x) = 2x^2 - x - 1$

    a. $f(-1) = 2(-1)^2 - (-1) - 1 = 2$
    The point $(-1, 2)$ is on the graph of f.

    b. $f(-2) = 2(-2)^2 - (-2) - 1 = 9$
    The point $(-2, 9)$ is on the graph of f.

    c. Solve for x:
    $-1 = 2x^2 - x - 1$
    $0 = 2x^2 - x$
    $0 = x(2x - 1) \Rightarrow x = 0, x = \frac{1}{2}$
    $(0, -1)$ and $\left(\frac{1}{2}, -1\right)$ are on the graph of f.

    d. The domain of f is: $\{x \mid x \text{ is any real number}\}$.

    e. x-intercepts:
    $f(x) = 0 \Rightarrow 2x^2 - x - 1 = 0$
    $(2x + 1)(x - 1) = 0 \Rightarrow x = -\frac{1}{2}, x = 1$
    $\left(-\frac{1}{2}, 0\right)$ and $(1, 0)$

    f. y-intercept:
    $f(0) = 2(0)^2 - 0 - 1 = -1 \Rightarrow (0, -1)$

25. $f(x) = \frac{x+2}{x-6}$

    a. $f(3) = \frac{3+2}{3-6} = -\frac{5}{3} \neq 14$
    The point $(3, 14)$ is not on the graph of f.

    b. $f(4) = \frac{4+2}{4-6} = \frac{6}{-2} = -3$
    The point $(4, -3)$ is on the graph of f.

    c. Solve for x:
    $2 = \frac{x+2}{x-6}$
    $2x - 12 = x + 2$
    $x = 14$
    $(14, 2)$ is a point on the graph of f.

    d. The domain of f is $\{x \mid x \neq 6\}$.

    e. x-intercepts:
    $f(x) = 0 \Rightarrow \frac{x+2}{x-6} = 0$
    $x + 2 = 0 \Rightarrow x = -2 \Rightarrow (-2, 0)$

    f. y-intercept: $f(0) = \frac{0+2}{0-6} = -\frac{1}{3} \Rightarrow \left(0, -\frac{1}{3}\right)$

27. $f(x) = \frac{2x^2}{x^4 + 1}$

    a. $f(-1) = \frac{2(-1)^2}{(-1)^4 + 1} = \frac{2}{2} = 1$
    The point $(-1, 1)$ is on the graph of f.

    b. $f(2) = \frac{2(2)^2}{(2)^4 + 1} = \frac{8}{17}$
    The point $\left(2, \frac{8}{17}\right)$ is on the graph of f.

    c. Solve for x:
    $1 = \frac{2x^2}{x^4 + 1}$
    $x^4 + 1 = 2x^2$
    $x^4 - 2x^2 + 1 = 0$
    $(x^2 - 1)^2 = 0$
    $x^2 - 1 = 0 \Rightarrow x = \pm 1$
    $(1, 1)$ and $(-1, 1)$ are on the graph of f.

    d. The domain of f is $\{x \mid x \text{ is any real number}\}$.

    e. x-intercept:
    $f(x) = 0 \Rightarrow \frac{2x^2}{x^4 + 1} = 0$
    $2x^2 = 0 \Rightarrow x = 0 \Rightarrow (0, 0)$

    f. y-intercept:
    $f(0) = \frac{2(0)^2}{0^4 + 1} = \frac{0}{0+1} = 0 \Rightarrow (0, 0)$

29. $h(x) = \dfrac{-32x^2}{130^2} + x$

a. $h(100) = \dfrac{-32(100)^2}{130^2} + 100$

$= \dfrac{-320,000}{16,900} + 100 \approx 81.07$ feet

b. $h(300) = \dfrac{-32(300)^2}{130^2} + 300$

$= \dfrac{-2,880,000}{16,900} + 300 \approx 129.59$ feet

c. $h(500) = \dfrac{-32(500)^2}{130^2} + 500$

$= \dfrac{-8,000,000}{16,900} + 500 \approx 26.63$ feet

d. Solving $h(x) = \dfrac{-32x^2}{130^2} + x = 0$

$\dfrac{-32x^2}{130^2} + x = 0$

$x\left(\dfrac{-32x}{130^2} + 1\right) = 0$

$x = 0$ or $\dfrac{-32x}{130^2} + 1 = 0$

$1 = \dfrac{32x}{130^2}$

$130^2 = 32x$

$x = \dfrac{130^2}{32} = 528.125$ feet

Therefore, the golf ball travels 528.125 feet.

e. $y_1 = \dfrac{-32x^2}{130^2} + x$

f. Use INTERSECT on the graphs of $y_1 = \dfrac{-32x^2}{130^2} + x$ and $y_2 = 90$.

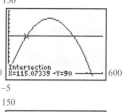

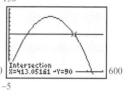

The ball reaches a height of 90 feet twice. The first time is when the ball has traveled approximately 115 feet, and the second time is when the ball has traveled approximately 413 feet.

g. The ball travels approximately 275 feet before it reaches its maximum height of approximately 131.8 feet.

h. The ball travels approximately 264 feet before it reaches its maximum height of approximately 132.03 feet.

31. $C(x) = 100 + \dfrac{x}{10} + \dfrac{36000}{x}$

   a. Graphing:

   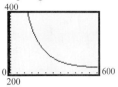

   b. TblStart = 0; ΔTbl = 50

X	Y1
0	ERROR
50	825
100	470
150	355
200	300
250	269
300	250

   Y1 = 100+X/10+360...

   c. The cost per passenger is minimized to about $220 when the ground speed is roughly 600 miles per hour.

X	Y1
450	225
500	222
550	220.45
600	220
650	220.38
700	221.43
750	223

   X=600

33. Answers will vary. From a graph, the domain can be found by visually locating the x-values for which the graph is defined. The range can be found in a similar fashion by visually locating the y-values for which the function is defined.

   If an equation is given, the domain can be found by locating any restricted values and removing them from the set of real numbers. The range can be found by using known properties of the graph of the equation, or estimated by means of a table of values.

35. The graph of a function can have at most one y-intercept.

37. (a) III; (b) IV; (c) I; (d) V; (e) II

39.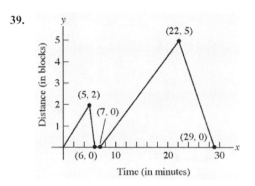

41. a. 2 hours elapsed; Kevin was between 0 and 3 miles from home.

   b. 0.5 hours elapsed; Kevin was 3 miles from home.

   c. 0.3 hours elapsed; Kevin was between 0 and 3 miles from home.

   d. 0.2 hours elapsed; Kevin was at home.

   e. 0.9 hours elapsed; Kevin was between 0 and 2.8 miles from home.

   f. 0.3 hours elapsed; Kevin was 2.8 miles from home.

   g. 1.1 hours elapsed; Kevin was between 0 and 2.8 miles from home.

   h. The farthest distance Kevin is from home is 3 miles.

   i. Kevin returned home 2 times.

43. Answers (graphs) will vary. Points of the form $(5, y)$ and of the form $(x, 0)$ cannot be on the graph of the function.

Section 2.4

1. $2 < x < 5$

3. The inequality is strict and has no upper bound. Therefore, we use parentheses and use infinity for our upper bound.
   $(3, \infty)$

5. $y = x^2 - 9$
   x-intercepts:
   $0 = x^2 - 9$
   $x^2 = 9 \rightarrow x = \pm 3$
   y-intercept:
   $y = (0)^2 - 9 = -9$
   The intercepts are $(-3, 0)$, $(3, 0)$, and $(0, -9)$.

7. even; odd

9. True

11. Yes

13. No, it only increases on (5, 10).

15. f is increasing on the intervals $(-8, -2)$, $(0, 2)$, $(5, \infty)$.

17. Yes. The local maximum at $x = 2$ is 10.

19. f has local maxima at $x = -2$ and $x = 2$. The local maxima are 6 and 10, respectively.

21. **a.** Intercepts: (–2, 0), (2, 0), and (0, 3).
    b. Domain: $\{x \mid -4 \leq x \leq 4\}$;
    Range: $\{y \mid 0 \leq y \leq 3\}$.
    c. Increasing: (–2, 0) and (2, 4);
    Decreasing: (–4, –2) and (0, 2).
    d. Since the graph is symmetric with respect to the y-axis, the function is <u>even</u>.

23. **a.** Intercepts: (0, 1).
    b. Domain: $\{x \mid x \text{ is any real number}\}$;
    Range: $\{y \mid y > 0\}$.
    c. Increasing: $(-\infty, \infty)$; Decreasing: never.

    d. Since the graph is not symmetric with respect to the y-axis or the origin, the function is <u>neither</u> even nor odd.

25. **a.** Intercepts: $(-\pi, 0), (\pi, 0)$, and $(0, 0)$.
    b. Domain: $\{x \mid -\pi \leq x \leq \pi\}$;
    Range: $\{y \mid -1 \leq y \leq 1\}$.
    c. Increasing: $\left(-\frac{\pi}{2}, \frac{\pi}{2}\right)$;
    Decreasing: $\left(-\pi, -\frac{\pi}{2}\right)$ and $\left(\frac{\pi}{2}, \pi\right)$.
    d. Since the graph is symmetric with respect to the origin, the function is <u>odd</u>.

27. **a.** Intercepts: $\left(\frac{1}{2}, 0\right), \left(\frac{5}{2}, 0\right)$, and $\left(0, \frac{1}{2}\right)$.
    b. Domain: $\{x \mid -3 \leq x \leq 3\}$;
    Range: $\{y \mid -1 \leq y \leq 2\}$.
    c. Increasing: $(2, 3)$; Decreasing: $(-1, 1)$;
    Constant: $(-3, -1)$ and $(1, 2)$
    d. Since the graph is not symmetric with respect to the y-axis or the origin, the function is <u>neither</u> even nor odd.

29. **a.** f has a local maximum of 3 at $x = 0$.
    b. f has a local minimum of 0 at both $x = -2$ and $x = 2$.

31. **a.** f has a local maximum of 1 at $x = \frac{\pi}{2}$.
    b. f has a local minimum of –1 at $x = -\frac{\pi}{2}$.

33. $f(x) = 4x^3$
    $f(-x) = 4(-x)^3 = -4x^3 = -f(x)$
    Therefore, f is odd.

35. $g(x) = -3x^2 - 5$
    $g(-x) = -3(-x)^2 - 5 = -3x^2 - 5 = g(x)$
    Therefore, g is even.

37. $F(x) = \sqrt[3]{x}$
$F(-x) = \sqrt[3]{-x} = -\sqrt[3]{x} = -F(x)$
Therefore, F is odd.

39. $f(x) = x + |x|$
$f(-x) = -x + |-x| = -x + |x|$
f is neither even nor odd.

41. $g(x) = \dfrac{1}{x^2}$
$g(-x) = \dfrac{1}{(-x)^2} = \dfrac{1}{x^2} = g(x)$
Therefore, g is even.

43. $h(x) = \dfrac{-x^3}{3x^2 - 9}$
$h(-x) = \dfrac{-(-x)^3}{3(-x)^2 - 9} = \dfrac{x^3}{3x^2 - 9} = -h(x)$
Therefore, h is odd.

45. $f(x) = x^3 - 3x + 2$ on the interval $(-2, 2)$
Use MAXIMUM and MINIMUM on the graph of $y_1 = x^3 - 3x + 2$.

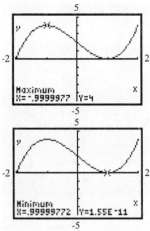

local maximum at: $(-1, 4)$;
local minimum at: $(1, 0)$
f is increasing on: $(-2, -1)$ and $(1, 2)$;
f is decreasing on: $(-1, 1)$

47. $f(x) = x^5 - x^3$ on the interval $(-2, 2)$
Use MAXIMUM and MINIMUM on the graph of $y_1 = x^5 - x^3$.

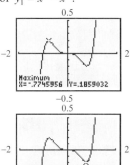

local maximum at: $(-0.77, 0.19)$;
local minimum at: $(0.77, -0.19)$;
f is increasing on: $(-2, -0.77)$ and $(0.77, 2)$;
f is decreasing on: $(-0.77, 0.77)$

49. $f(x) = -0.2x^3 - 0.6x^2 + 4x - 6$ on the interval $(-6, 4)$
Use MAXIMUM and MINIMUM on the graph of $y_1 = -0.2x^3 - 0.6x^2 + 4x - 6$.

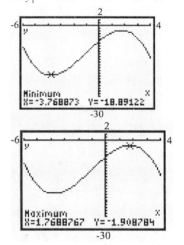

local maximum at: $(1.77, -1.91)$;
local minimum at: $(-3.77, -18.89)$
f is increasing on: $(-3.77, 1.77)$;
f is decreasing on: $(-6, -3.77)$ and $(1.77, 4)$

51. $f(x) = 0.25x^4 + 0.3x^3 - 0.9x^2 + 3$ on the interval $(-3, 2)$

Use MAXIMUM and MINIMUM on the graph of $y_1 = 0.25x^4 + 0.3x^3 - 0.9x^2 + 3$.

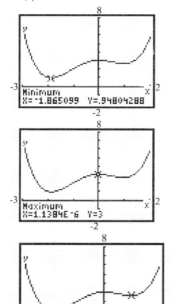

local maximum at: $(0, 3)$;

local minimum at: $(-1.87, 0.95)$, $(0.97, 2.65)$

f is increasing on: $(-1.87, 0)$ and $(0.97, 2)$;

f is decreasing on: $(-3, -1.87)$ and $(0, 0.97)$

53. $f(x) = -2x^2 + 4$

a. Average rate of change of f from $x = 0$ to $x = 2$

$$\frac{f(2) - f(0)}{2 - 0} = \frac{\left(-2(2)^2 + 4\right) - \left(-2(0)^2 + 4\right)}{2}$$

$$= \frac{(-4) - (4)}{2} = \frac{-8}{2} = -4$$

b. Average rate of change of f from $x = 1$ to $x = 3$:

$$\frac{f(3) - f(1)}{3 - 1} = \frac{\left(-2(3)^2 + 4\right) - \left(-2(1)^2 + 4\right)}{2}$$

$$= \frac{(-14) - (2)}{2} = \frac{-16}{2} = -8$$

c. Average rate of change of f from $x = 1$ to $x = 4$:

$$\frac{f(4) - f(1)}{4 - 1} = \frac{\left(-2(4)^2 + 4\right) - \left(-2(1)^2 + 4\right)}{3}$$

$$= \frac{(-28) - (2)}{3} = \frac{-30}{3} = -10$$

55. $g(x) = x^3 - 2x + 1$

a. Average rate of change of g from $x = -3$ to $x = -2$:

$$\frac{g(-2) - g(-3)}{-2 - (-3)}$$

$$= \frac{\left[(-2)^3 - 2(-2) + 1\right] - \left[(-3)^3 - 2(-3) + 1\right]}{1}$$

$$= \frac{(-3) - (-20)}{1} = \frac{17}{1}$$

$$= 17$$

b. Average rate of change of g from $x = -1$ to $x = 1$:

$$\frac{g(1) - g(-1)}{1 - (-1)}$$

$$= \frac{\left[(1)^3 - 2(1) + 1\right] - \left[(-1)^3 - 2(-1) + 1\right]}{2}$$

$$= \frac{(0) - (2)}{2} = \frac{-2}{2}$$

$$= -1$$

c. Average rate of change of g from $x = 1$ to $x = 3$:

$$\frac{g(3) - g(1)}{3 - 1}$$

$$= \frac{\left[(3)^3 - 2(3) + 1\right] - \left[(1)^3 - 2(1) + 1\right]}{2}$$

$$= \frac{(22) - (0)}{2} = \frac{22}{2}$$

$$= 11$$

57. $f(x) = 5x - 2$

a. Average rate of change of f from 1 to x:
$$\frac{f(x) - f(1)}{x - 1} = \frac{(5x-2) - (5(1)-2)}{x-1}$$
$$= \frac{5x - 2 - 3}{x-1} = \frac{5x-5}{x-1}$$
$$= \frac{5(x-1)}{x-1}$$
$$= 5$$

b. The average rate of change of f from 1 to x is a constant 5. Therefore, the average rate of change of f from 1 to 3 is 5. The slope of the secant line joining $(1, f(1))$ and $(3, f(3))$ is 5.

c. We use the point-slope form to find the equation of the secant line:
$$y - y_1 = m_{\sec}(x - x_1)$$
$$y - 3 = 5(x - 1)$$
$$y - 3 = 5x - 5$$
$$y = 5x - 2$$

d. The secant line coincides with the function so the graph only shows one line.

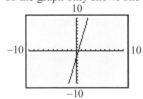

59. $g(x) = x^2 - 2$

a. Average rate of change of g from -2 to x:
$$\frac{g(x) - g(-2)}{x - (-2)} = \frac{[x^2 - 2] - [(-2)^2 - 2]}{x + 2}$$
$$= \frac{(x^2 - 2) - (2)}{x+2} = \frac{x^2 - 4}{x+2}$$
$$= \frac{(x+2)(x-2)}{x+2} = x - 2$$

b. The average rate of change of g from -2 to x is given by $x - 2$. Therefore, the average rate of change of g from -2 to 1 is $1 - 2 = -1$. The slope of the secant line joining $(-2, g(-2))$ and $(1, g(1))$ is -1.

c. We use the point-slope form to find the equation of the secant line:
$$y - y_1 = m_{\sec}(x - x_1)$$
$$y - 2 = -1(x - (-2))$$
$$y - 2 = -x - 2$$
$$y = -x$$

d. The graph below shows the graph of g along with the secant line $y = -x$.

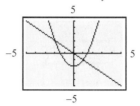

61. $h(x) = x^2 - 2x$

a. Average rate of change of h from 2 to x:
$$\frac{h(x) - h(2)}{x - 2} = \frac{[x^2 - 2x] - [(2)^2 - 2(2)]}{x - 2}$$
$$= \frac{(x^2 - 2x) - (0)}{x - 2} = \frac{x^2 - 2x}{x - 2}$$
$$= \frac{x(x-2)}{x-2} = x$$

b. The average rate of change of h from 2 to x is given by x. Therefore, the average rate of change of h from 2 to 4 is 4. The slope of the secant line joining $(2, h(2))$ and $(4, h(4))$ is 4.

c. We use the point-slope form to find the equation of the secant line:
$$y - y_1 = m_{\sec}(x - x_1)$$
$$y - 0 = 4(x - 2)$$
$$y = 4x - 8$$

d. The graph below shows the graph of h along with the secant line $y = 4x - 8$.

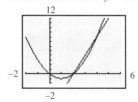

63. a. length = $24 - 2x$; width = $24 - 2x$; height = x
$V(x) = x(24 - 2x)(24 - 2x) = x(24 - 2x)^2$

b. $V(3) = 3(24 - 2(3))^2 = 3(18)^2$
$= 3(324) = 972$ cu.in.

c. $V(10) = 10(24 - 2(10))^2 = 10(4)^2$
$= 10(16) = 160$ cu.in.

d. $y_1 = x(24 - 2x)^2$

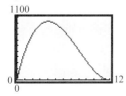

Use MAXIMUM.

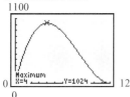

The volume is largest when $x = 4$ inches.

65. a. $y_1 = -16x^2 + 80x + 6$

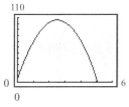

b. Use MAXIMUM. The maximum height occurs when $t = 2.5$ seconds.

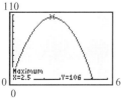

c. From the graph, the maximum height is 106 feet.

67. $\overline{C}(x) = 0.3x^2 + 21x - 251 + \dfrac{2500}{x}$

a. $y_1 = 0.3x^2 + 21x - 251 + \dfrac{2500}{x}$

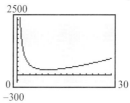

b. Use MINIMUM. The average cost is minimized when approximately 9.66 lawnmowers are produced per hour.

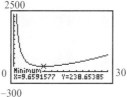

c. The minimum average cost is approximately $238.65.

69. (a), (b), (e)

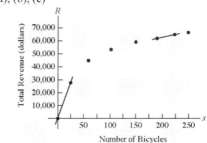

c. Average rate of change $= \dfrac{28000 - 0}{25 - 0} = \dfrac{28000}{25}$
$= 1120$ bicycles per hour

d. For each additional bicycle sold between 0 and 25, the total revenue increases by (an average of) $1120.

f. Average rate of change $= \dfrac{64835 - 62360}{223 - 190} = \dfrac{2475}{33}$
$= 75$ dollars per bicycle

g. For each additional bicycle sold between 190 and 223, the total revenue increases by (an average of) $75.

h. The average rate of change of revenue is decreasing as the number of bicycles increases.

71. $f(x) = x^2$

   a. Average rate of change of f from $x = 0$ to $x = 1$:

$$\frac{f(1) - f(0)}{1 - 0} = \frac{1^2 - 0^2}{1} = \frac{1}{1} = 1$$

   b. Average rate of change of f from $x = 0$ to $x = 0.5$:

$$\frac{f(0.5) - f(0)}{0.5 - 0} = \frac{(0.5)^2 - 0^2}{0.5} = \frac{0.25}{0.5} = 0.5$$

   c. Average rate of change of f from $x = 0$ to $x = 0.1$:

$$\frac{f(0.1) - f(0)}{0.1 - 0} = \frac{(0.1)^2 - 0^2}{0.1} = \frac{0.01}{0.1} = 0.1$$

   d. Average rate of change of f from $x = 0$ to $x = 0.01$:

$$\frac{f(0.01) - f(0)}{0.01 - 0} = \frac{(0.01)^2 - 0^2}{0.01}$$
$$= \frac{0.0001}{0.01} = 0.01$$

   e. Average rate of change of f from $x = 0$ to $x = 0.001$:

$$\frac{f(0.001) - f(0)}{0.001 - 0} = \frac{(0.001)^2 - 0^2}{0.001}$$
$$= \frac{0.000001}{0.001} = 0.001$$

   f. Graphing the secant lines:

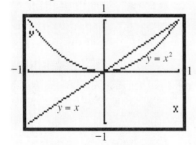

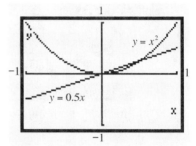

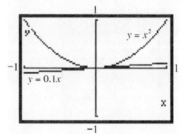

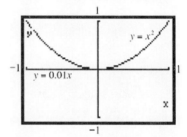

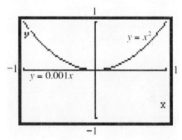

   g. The secant lines are beginning to look more and more like the tangent line to the graph of f at the point where $x = 0$.

   h. The slopes of the secant lines are getting smaller and smaller. They seem to be approaching the number zero.

73. $f(x) = 2x+5$

    a. $m_{sec} = \dfrac{f(x+h)-f(x)}{h}$

        $= \dfrac{2(x+h)+5-2x-5}{h}$

        $= \dfrac{2h}{h} = 2$

    b. When $x = 1$:
$h = 0.5 \Rightarrow m_{sec} = 2$
$h = 0.1 \Rightarrow m_{sec} = 2$
$h = 0.01 \Rightarrow m_{sec} = 2$
as $h \to 0$, $m_{sec} \to 2$

    c. Using the point $(1, f(1)) = (1, 7)$ and slope, $m = 2$, we get the secant line:
$y - 7 = 2(x-1)$
$y - 7 = 2x - 2$
$y = 2x + 5$

    d. Graphing:

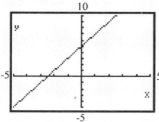

The graph and the secant line coincide.

75. $f(x) = x^2 + 2x$

    a. $m_{sec} = \dfrac{f(x+h)-f(x)}{h}$

        $= \dfrac{(x+h)^2 + 2(x+h) - (x^2 + 2x)}{h}$

        $= \dfrac{x^2 + 2xh + h^2 + 2x + 2h - x^2 - 2x}{h}$

        $= \dfrac{2xh + h^2 + 2h}{h}$

        $= 2x + h + 2$

    b. When $x = 1$,
$h = 0.5 \Rightarrow m_{sec} = 2 \cdot 1 + 0.5 + 2 = 4.5$
$h = 0.1 \Rightarrow m_{sec} = 2 \cdot 1 + 0.1 + 2 = 4.1$
$h = 0.01 \Rightarrow m_{sec} = 2 \cdot 1 + 0.01 + 2 = 4.01$
as $h \to 0$, $m_{sec} \to 2 \cdot 1 + 0 + 2 = 4$

    c. Using point $(1, f(1)) = (1, 3)$ and slope $= 4.01$, we get the secant line:
$y - 3 = 4.01(x-1)$
$y - 3 = 4.01x - 4.01$
$y = 4.01x - 1.01$

    d. Graphing:

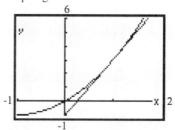

77. $f(x) = 2x^2 - 3x + 1$

    a. $m_{sec} = \dfrac{f(x+h)-f(x)}{h}$

        $= \dfrac{2(x+h)^2 - 3(x+h) + 1 - (2x^2 - 3x + 1)}{h}$

        $= \dfrac{2(x^2 + 2xh + h^2) - 3x - 3h + 1 - 2x^2 + 3x - 1}{h}$

        $= \dfrac{2x^2 + 4xh + 2h^2 - 3x - 3h + 1 - 2x^2 + 3x - 1}{h}$

        $= \dfrac{4xh + 2h^2 - 3h}{h}$

        $= 4x + 2h - 3$

    b. When $x = 1$,
$h = 0.5 \Rightarrow m_{sec} = 4 \cdot 1 + 2(0.5) - 3 = 2$
$h = 0.1 \Rightarrow m_{sec} = 4 \cdot 1 + 2(0.1) - 3 = 1.2$
$h = 0.01 \Rightarrow m_{sec} = 4 \cdot 1 + 2(0.01) - 3 = 1.02$
as $h \to 0$, $m_{sec} \to 4 \cdot 1 + 2(0) - 3 = 1$

    c. Using point $(1, f(1)) = (1, 0)$ and slope $= 1.02$, we get the secant line:
$y - 0 = 1.02(x-1)$
$y = 1.02x - 1.02$

d. Graphing:

   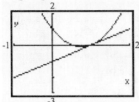

79. $f(x) = \dfrac{1}{x}$

   a. $m_{sec} = \dfrac{f(x+h) - f(x)}{h} = \dfrac{\left(\dfrac{1}{x+h} - \dfrac{1}{x}\right)}{h}$

   $= \dfrac{\left(\dfrac{x-(x+h)}{(x+h)x}\right)}{h} = \left(\dfrac{x-x-h}{(x+h)x}\right)\left(\dfrac{1}{h}\right)$

   $= \left(\dfrac{-h}{(x+h)x}\right)\left(\dfrac{1}{h}\right)$

   $= -\dfrac{1}{(x+h)x}$

   b. When $x = 1$,

   $h = 0.5 \Rightarrow m_{sec} = -\dfrac{1}{(1+0.5)(1)}$

   $= -\dfrac{1}{1.5} \approx -0.667$

   $h = 0.1 \Rightarrow m_{sec} = -\dfrac{1}{(1+0.1)(1)}$

   $= -\dfrac{1}{1.1} \approx -0.909$

   $h = 0.01 \Rightarrow m_{sec} = -\dfrac{1}{(1+0.01)(1)}$

   $= -\dfrac{1}{1.01} \approx -0.990$

   as $h \to 0$, $m_{sec} \to -\dfrac{1}{(1+0)(1)} = -\dfrac{1}{1} = -1$

   c. Using point $(1, f(1)) = (1,1)$ and slope $= -0.990$, we get the secant line:

   $y - 1 = -0.99(x - 1)$
   $y - 1 = -0.99x + 0.99$
   $y = -0.99x + 1.99$

   d. Graphing:

   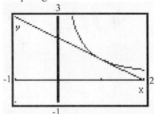

81. Answers will vary. One possibility follows:

   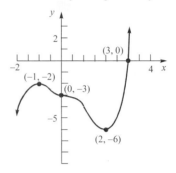

83. A function that is increasing on an interval can have at most one x-intercept on the interval. The graph of f could not "turn" and cross it again or it would start to decrease.

85. To be an even function we need $f(-x) = f(x)$ and to be an odd function we need $f(-x) = -f(x)$. In order for a function be both even and odd, we would need $f(x) = -f(x)$. This is only possible if $f(x) = 0$.

Section 2.5

1. From the equation $y = 2x - 3$, we see that the y-intercept is -3. Thus, the point $(0, -3)$ is on the graph. We can obtain a second point by choosing a value for x and finding the corresponding value for y.
   Let $x = 2$, then $y = 2(2) - 3 = 1$. Thus, the point $(2, 1)$ is also on the graph. Plotting the two points and connecting with a line yields the graph below.

   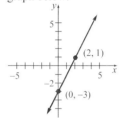

3. We can use the point-slope form of a line to obtain the equation.
   $y - y_1 = m(x - x_1)$
   $y - 5 = -3(x - (-1))$
   $y - 5 = -3(x + 1)$
   $y - 5 = -3x - 3$
   $y = -3x + 2$

5. slope; y-intercept

7. $y = kx$

9. True

11. $f(x) = 2x + 3$
    Slope = average rate of change = 2;
    y-intercept = 3

    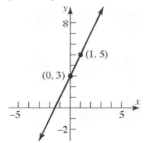

13. $h(x) = -3x + 4$
    Slope = average rate of change = -3;
    y-intercept = 4

    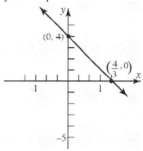

15. $f(x) = \dfrac{1}{4}x - 3$
    Slope = average rate of change = $\dfrac{1}{4}$;
    y-intercept = -3

    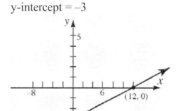

17. $F(x) = 4$
    Slope = average rate of change = 0;
    y-intercept = 4

    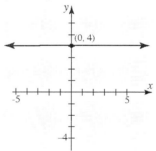

127

19. Linear, $m > 0$

21. Linear, $m < 0$

23. Nonlinear

25. **a.**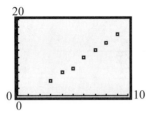

   b. Answers will vary. We select (3, 4) and (9, 16). The slope of the line containing these points is:
   $$m = \frac{16-4}{9-3} = \frac{12}{6} = 2$$
   The equation of the line is:
   $$y - y_1 = m(x - x_1)$$
   $$y - 4 = 2(x - 3)$$
   $$y - 4 = 2x - 6$$
   $$y = 2x - 2$$

   c.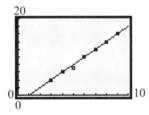

   d. Using the LINear REGresssion program, the line of best fit is:
   $$y = 2.0357x - 2.3571$$

   e.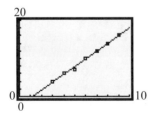

27. **a.**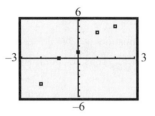

   b. Answers will vary. We select (–2,–4) and (1, 4). The slope of the line containing these points is:
   $$m = \frac{4-(-4)}{1-(-2)} = \frac{8}{3}$$
   The equation of the line is:
   $$y - y_1 = m(x - x_1)$$
   $$y - (-4) = \frac{8}{3}(x - (-2))$$
   $$y + 4 = \frac{8}{3}x + \frac{16}{3}$$
   $$y = \frac{8}{3}x + \frac{4}{3}$$

   c.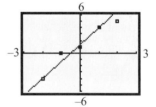

   d. Using the LINear REGresssion program, the line of best fit is:
   $$y = 2.2x + 1.2$$

   e.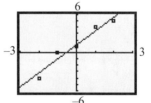

29. **a.**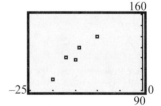

b. Answers will vary. We select $(-20, 100)$ and $(-15, 118)$. The slope of the line containing these points is:
$$m = \frac{118-100}{-15-(-20)} = \frac{18}{5} = 3.6$$
The equation of the line is:
$$y - y_1 = m(x - x_1)$$
$$y - 100 = 3.6(x - (-20))$$
$$y - 100 = 3.6x + 72$$
$$y = 3.6x + 172$$

c.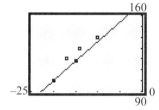

d. Using the LINear REGresssion program, the line of best fit is:
$$y = 3.8613x + 180.2920$$

e.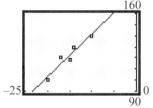

31. a. $C(x) = 0.25x + 35$
$C(40) = 0.25(40) + 35 = 45$
The moving truck will cost $45.00 if you drive 40 miles.

b. $80 = 0.25x + 35$
$45 = 0.25x$
$180 = x$
If the cost of the truck is $80.00, you drove for 180 miles.

c. $100 = 0.25x + 35$
$65 = 0.25x$
$260 = x$
To keep the cost below $100, you must drive less than 260 miles.

33. a. $B(t) = 19.25t + 585.72$
$B(10) = 19.25(10) + 585.72 = 778.22$
The average monthly benefit in 2000 was $778.22.

b. $893.72 = 19.25t + 585.72$
$308 = 19.25t$
$16 = t$
The average monthly benefit will be $893.72 in 2006.

c. $1000 = 19.25t + 585.72$
$414.28 = 19.25t$
$21.52 \approx t$
The average monthly benefit will exceed $1000 in 2012.

35. a. $S(p) = D(p)$
$-200 + 50p = 1000 - 25p$
$75p = 1200$
$p = 16$
The equilibrium price is $16.
$S(16) = -200 + 50(16) = 600$
The equilibrium quantity is 600 T-shirts.

b. $D(p) > S(p)$
$1000 - 25p > -200 + 50p$
$-75p > -1200$
$p < 16$
The quantity demanded will exceed the quantity supplied if $0 < p < \$16$.

c. If demand is higher than supply, generally the price will increase. The price will continue to increase towards the equilibrium point.

37. a. $R(x) = C(x)$
$8x = 4.5x + 17{,}500$
$3.5x = 17{,}500$
$x = 5000$
The company must sell 5000 units to break even.

b. To make a profit, the company must sell more than 5000 units.

39. a. Consider the data points (x, y), where x = the age in years of the computer and y = the value in dollars of the computer. So we have the points $(0, 3000)$ and $(3, 0)$. The slope formula yields:

$$\text{slope} = \frac{\Delta y}{\Delta x} = \frac{0 - 3000}{3 - 0}$$

$$= \frac{-3000}{3} = -1000 = m$$

$(0, 3000)$ is the y-intercept, so $b = 3000$
Therefore, the linear function is
$V(x) = mx + b = -1000x + 3000$.

b. The graph of $V(x) = -1000x + 3000$

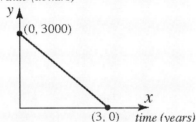

c. The computer's value after 2 years is given by
$V(2) = -1000(2) + 3000$
$= -2000 + 3000 = \$1000$

d. To find when the computer will be worth $2000, we solve the following:
$2000 = -1000x + 3000$
$-1000 = -1000x$
$1 = x$
The computer will be worth $2000 after 1 year.

41. a. Let x = the number of bicycles manufactured. We can use the cost function $C(x) = mx + b$, with $m = 90$ and $b = 1800$. Therefore $C(x) = 90x + 1800$

b. The graph of $C(x) = 90x + 1800$

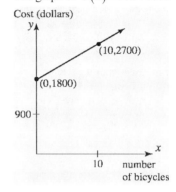

c. The cost of manufacturing 14 bicycles is given by $C(14) = 90(14) + 1800 = \$3060$.

d. To determine the number of bicycles, we solve the following:
$3780 = 90x + 1800$
$1980 = 90x$
$22 = x$
The company can manufacture 22 bicycles for $3780.

43. a. Let x = number of miles driven and C = cost in dollars to rent a truck for one day. Since the fixed daily charge is $29 and the variable mileage charge is $0.07 per mile, we have $C(x) = 0.07x + 29$.

b. $C(110) = 0.07(110) + 29 = 36.70$
$C(230) = 0.07(230) + 29 = 45.10$
It will cost $36.70 for one day if the truck is driven 110 miles, and it will cost $45.10 if the truck is driven 230 miles.

45. Let p = the monthly payment and B = the amount borrowed. Consider the ordered pair (B, p). We can use the points $(0, 0)$ and $(1000, 6.49)$.
Now compute the slope:

$$\text{slope} = \frac{\Delta y}{\Delta x} = \frac{6.49 - 0}{1000 - 0} = \frac{6.49}{1000} = 0.00649$$

Therefore we have the linear function
$p(B) = 0.00649B + 0 = 0.00649B$.
If $B = 145000$, then
$p = (0.00649)(145000) = \$941.05$.

47. Let R = the revenue and g = the number of gallons of gasoline sold. Consider the ordered pair (g, R). We can use the points $(0,0)$ and $(12, 23.40)$. Now compute the slope:

$$\text{slope} = \frac{\Delta y}{\Delta x} = \frac{23.40 - 0}{12 - 0} = \frac{23.40}{12} = 1.95$$

Therefore we have the linear function $R(g) = 1.95g + 0 = 1.95g$.

If $g = 10.5$, then $R = (1.95)(10.5) = \$20.48$.

49. $W = kS$
$1.875 = k(15)$
$0.125 = k$
For 40 gallons of sand:
$W = 0.125(40) = 5$ gallons of water.

51. a.

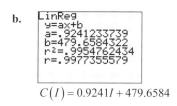

b.

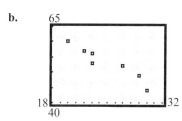

$C(I) = 0.9241I + 479.6584$

c. The slope indicates that for each \$1 increase in per capita disposable income, there is an increase of \$0.92 in per capita consumption.

d. $C(28,750) = 0.9241(28,750) + 479.6584$
$= 27,047.53$
When disposable income is \$28,750, the per capita consumption is about \$27,048.

e. $26,900 = 0.9241I + 479.6584$
$26,420.3416 = 0.9241I$
$28,590.35 = I$
When comsumption is \$26,500, the per capita disposable income is about \$28,590.

53. a.

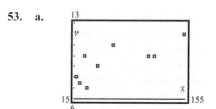

b. Using the LINear REGression program, the line of best fit is:
$L(G) = 0.0261G + 7.8738$

c. For each 1 day increase in Gestation period, the life expectancy increases by 0.0261 years (about 9.5 days).

d. $L(89) = 0.0261(89) + 7.8738 \approx 10.2$ years

55. a. The relation is not a function because 23 is paired with both 56 and 53.

b.

c. Using the LINear REGression program, the line of best fit is:
$D = -1.3355p + 86.1974$

d. As the price of the jeans increases by \$1, the demand for the jeans decreases by 1.3355 pairs per day.

e. $D(p) = -1.3355p + 86.1974$

f. Domain: $\{p \mid 0 < p < 64\}$
Note that the p-intercept is roughly 64.54 and that the number of pairs cannot be negative.

g. $D(28) = -1.3355(28) + 86.1974$
≈ 48.8034
Demand is about 49 pairs.

Chapter 2: *Functions and Their Graphs* **SSM:** *College Algebra EGU*

57. The data do not follow a linear pattern so it would not make sense to find the line of best fit.

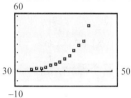

59. A linear function is odd if the y-intercept is 0. That is, if the line passes through the origin. A linear function can be even if the slope is 0.

61. A correlation coefficient of 0 implies that there is no linear relationship between the data.

Section 2.6

1. $y = \sqrt{x}$

3. $y = x^3 - 8$

y-intercept:
Let $x = 0$, then $y = (0)^3 - 8 = -8$.

x-intercept:
Let $y = 0$, then $0 = x^3 - 8$
$$x^3 = 8$$
$$x = 2$$

The intercepts are $(0, -8)$ and $(2, 0)$.

5. piecewise defined

7. False; the cube root function is odd and increasing on the interval $(-\infty, \infty)$.

9. C

11. E

13. B

15. F

17. $f(x) = x$

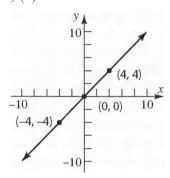

19. $f(x) = x^3$

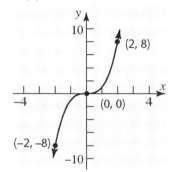

21. $f(x) = \dfrac{1}{x}$

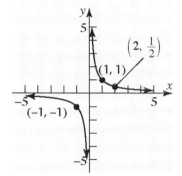

23. $f(x) = \sqrt[3]{x}$

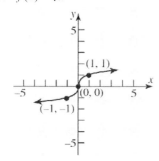

25. a. $f(-2) = (-2)^2 = 4$
    b. $f(0) = 2$
    c. $f(2) = 2(2) + 1 = 5$

27. a. $f(1.2) = \text{int}(2(1.2)) = \text{int}(2.4) = 2$
    b. $f(1.6) = \text{int}(2(1.6)) = \text{int}(3.2) = 3$
    c. $f(-1.8) = \text{int}(2(-1.8)) = \text{int}(-3.6) = -4$

29. $f(x) = \begin{cases} 2x & \text{if } x \neq 0 \\ 1 & \text{if } x = 0 \end{cases}$

    a. Domain: $\{x \mid x \text{ is any real number}\}$
    b. x-intercept: none
       y-intercept: $(0, 1)$
    c. Graph:

    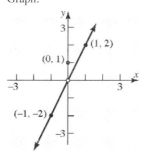

    d. Range: $\{y \mid y \neq 0\}$

31. $f(x) = \begin{cases} -2x + 3 & \text{if } x < 1 \\ 3x - 2 & \text{if } x \geq 1 \end{cases}$

    a. Domain: $\{x \mid x \text{ is any real number}\}$
    b. x-intercept: none
       y-intercept: $(0, 3)$

    c. Graph:

    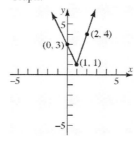

    d. Range: $\{y \mid y \geq 1\}$

33. $f(x) = \begin{cases} x + 3 & \text{if } -2 \leq x < 1 \\ 5 & \text{if } x = 1 \\ -x + 2 & \text{if } x > 1 \end{cases}$

    a. Domain: $\{x \mid x \geq -2\}$
    b. x-intercept: $(2, 0)$
       y-intercept: $(0, 3)$
    c. Graph:

    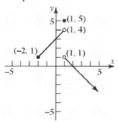

    d. Range: $\{y \mid y < 4 \text{ and } y = 5\}$

35. $f(x) = \begin{cases} 1 + x & \text{if } x < 0 \\ x^2 & \text{if } x \geq 0 \end{cases}$

    a. Domain: $\{x \mid x \text{ is any real number}\}$
    b. x-intercepts: $(-1, 0), (0, 0)$
       y-intercept: $(0, 0)$
    c. Graph:

    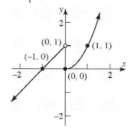

Chapter 2: Functions and Their Graphs

   d. Range: $\{y \mid y \text{ is any real number}\}$

37. $f(x) = \begin{cases} |x| & \text{if } -2 \leq x < 0 \\ 1 & \text{if } x = 0 \\ x^3 & \text{if } x > 0 \end{cases}$

   a. Domain: $\{x \mid x \geq -2\}$

   b. x-intercept: none
      y-intercept: (0, 1)

   c. Graph:

   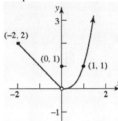

   d. Range: $\{y \mid y > 0\}$

39. $f(x) = 2\,\text{int}(x)$

   a. Domain: $\{x \mid x \text{ is any real number}\}$

   b. x-intercepts: all ordered pairs
      $(x, 0)$ when $0 \leq x < 1$.
      y-intercept: $(0,0)$

   c. Graph:

   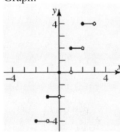

   d. Range: $\{y \mid y \text{ is an even integer}\}$

41. $f(x) = \begin{cases} -x & \text{if } -1 \leq x \leq 0 \\ \dfrac{1}{2}x & \text{if } 0 < x \leq 2 \end{cases}$

43. $f(x) = \begin{cases} -x & \text{if } x \leq 0 \\ -x+2 & \text{if } 0 < x \leq 2 \end{cases}$

45. $C = \begin{cases} 35 & \text{if } 0 < x \leq 300 \\ 0.40x - 85 & \text{if } x > 300 \end{cases}$

   a. $C(200) = \$35.00$

   b. $C(365) = 0.40(365) - 85 = \$61.00$

   c. $C(301) = 0.40(301) - 85 = \$35.40$

47. a. Charge for 50 therms:
      $C = 9.45 + 0.6338(50) + 0.36375(50)$
      $= \$59.33$

   b. Charge for 500 therms:
      $C = 9.45 + 0.36375(50) + 0.11445(450)$
      $\quad + 0.6338(500)$
      $= \$396.04$

   c. For $0 \leq x \leq 50$:
      $C = 9.45 + 0.36375x + 0.6338x$
      $= 9.45 + 0.99755x$

      For $x > 50$:
      $C = 9.45 + 0.36375(50) + 0.11445(x - 50)$
      $\quad + 0.6338x$
      $= 9.45 + 18.1875 + 0.11445x - 5.7225$
      $\quad + 0.6338x$
      $= 21.915 + 0.74825x$

      The monthly charge function:
      $C = \begin{cases} 9.45 + 0.99755x & \text{for } 0 \leq x \leq 50 \\ 21.915 + 0.74825x & \text{for } x > 50 \end{cases}$

   d. Graphing:

49. For schedule X:

$$f(x) = \begin{cases} 0.10x & \text{if } 0 < x \le 7150 \\ 715 + 0.15(x - 7150) & \text{if } 7150 < x \le 29,050 \\ 4000 + 0.25(x - 29,050) & \text{if } 29,050 < x \le 70,350 \\ 14,325 + 0.28(x - 70,350) & \text{if } 70,350 < x \le 146,750 \\ 35,717 + 0.33(x - 146,750) & \text{if } 146,750 < x \le 319,100 \\ 92,592.50 + 0.35(x - 319,100) & \text{if } x > 319,100 \end{cases}$$

51. a. Let x represent the number of miles and C be the cost of transportation.

$$C(x) = \begin{cases} 0.50x & \text{if } 0 \le x \le 100 \\ 0.50(100) + 0.40(x - 100) & \text{if } 100 < x \le 400 \\ 0.50(100) + 0.40(300) + 0.25(x - 400) & \text{if } 400 < x \le 800 \\ 0.50(100) + 0.40(300) + 0.25(400) + 0(x - 800) & \text{if } 800 < x \le 960 \end{cases}$$

$$C(x) = \begin{cases} 0.50x & \text{if } 0 \le x \le 100 \\ 10 + 0.40x & \text{if } 100 < x \le 400 \\ 70 + 0.25x & \text{if } 400 < x \le 800 \\ 270 & \text{if } 800 < x \le 960 \end{cases}$$

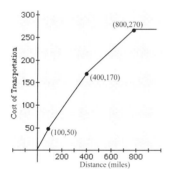

b. For hauls between 100 and 400 miles the cost is: $C(x) = 10 + 0.40x$.

c. For hauls between 400 and 800 miles the cost is: $C(x) = 70 + 0.25x$.

53. Let x = the amount of the bill in dollars. The minimum payment due is given by

$$f(x) = \begin{cases} x & \text{if } x < 10 \\ 10 & \text{if } 10 \leq x < 500 \\ 30 & \text{if } 500 \leq x < 1000 \\ 50 & \text{if } 1000 \leq x < 1500 \\ 70 & \text{if } 1500 \leq x \end{cases}$$

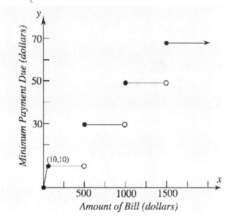

55. a. $W = 10°C$

   b. $W = 33 - \dfrac{(10.45 + 10\sqrt{5} - 5)(33 - 10)}{22.04}$
   $\approx 3.98°C$

   c. $W = 33 - \dfrac{(10.45 + 10\sqrt{15} - 15)(33 - 10)}{22.04}$
   $\approx -2.67°C$

   d. $W = 33 - 1.5958(33 - 10) = -3.7°C$

   e. When $0 \leq v < 1.79$, the wind speed is so small that there is no effect on the temperature.

   f. For each drop of 1° in temperature, the wind chill factor drops approximately 1.6°C. When the wind speed exceeds 20, there is a constant drop in temperature. That is, the windchill depends only on the temperature.

57. Each graph is that of $y = x^2$, but shifted vertically.

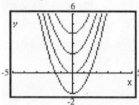

If $y = x^2 + k$, $k > 0$, the shift is up k units; if $y = x^2 + k$, $k < 0$, the shift is down $|k|$ units. The graph of $y = x^2 - 4$ is the same as the graph of $y = x^2$, but shifted down 4 units. The graph of $y = x^2 + 5$ is the graph of $y = x^2$, but shifted up 5 units.

59. Each graph is that of $y = |x|$, but either compressed or stretched vertically.

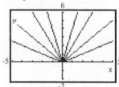

If $y = k|x|$ and $k > 1$, the graph is stretched; if $y = k|x|$ and $0 < k < 1$, the graph is compressed. The graph of $y = \dfrac{1}{4}|x|$ is the same as the graph of $y = |x|$, but compressed. The graph of $y = 5|x|$ is the same as the graph of $y = |x|$, but stretched.

61. The graph of $y = \sqrt{-x}$ is the reflection about the y-axis of the graph of $y = \sqrt{x}$.

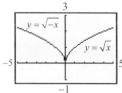

The same type of reflection occurs when graphing $y = 2x + 1$ and $y = 2(-x) + 1$.

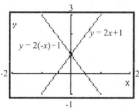

The graph of $y = f(-x)$ is the reflection about the y-axis of the graph of $y = f(x)$.

63. For the graph of $y = x^n$, n a positive even integer, as n increases, the graph of the function is narrower for $|x| > 1$ and flatter for $|x| < 1$.

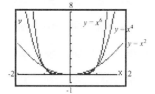

65. Yes, it is a function.
$$f(x) = \begin{cases} 1 & \text{if } x \text{ is rational} \\ 0 & \text{if } x \text{ is irrational} \end{cases}$$
$\{x \mid x \text{ is any real number}\}$ Range = $\{0, 1\}$

y-intercept: $x = 0 \Rightarrow x$ is rational $\Rightarrow y = 1$ So the y-intercept is (0, 1).

x-intercept: $y = 0 \Rightarrow x$ is irrational So the graph has infinitely many x-intercepts, namely, there is an x-intercept at each irrational value of x.

$f(-x) = 1 = f(x)$ when x is rational;
$f(-x) = 0 = f(x)$ when x is irrational, So f is even.

The graph of f consists of 2 infinite clusters of distinct points, extending horizontally in both directions. One cluster is located 1 unit above the x-axis, and the other is located along the x-axis.

Section 2.7

1. horizontal; right

3. -5, -2, and 2

5. False; to obtain the graph of $y = f(x+2) - 3$ you shift the graph of $y = f(x)$ to the left 2 units and down 3 units.

7. B

9. H

11. I

13. L

15. F

17. G

19. $y = (x-4)^3$

21. $y = x^3 + 4$

23. $y = (-x)^3 = -x^3$

25. $y = 4x^3$

27. (1) $y = \sqrt{x} + 2$
    (2) $y = -(\sqrt{x} + 2)$
    (3) $y = -(\sqrt{-x} + 2) = -\sqrt{-x} - 2$

29. (1) $y = -\sqrt{x}$
    (2) $y = -\sqrt{x} + 2$
    (3) $y = -\sqrt{x+3} + 2$

31. (c); To go from $y = f(x)$ to $y = -f(x)$ we reflect about the x-axis. This means we change the sign of the y-coordinate for each point on the graph of $y = f(x)$. Thus, the point $(3, 0)$ would remain the same.

33. (c); To go from $y = f(x)$ to $y = 2f(x)$, we multiply the y-coordinate of each point on the graph of $y = f(x)$ by 2. Thus, the point $(0, 3)$ would become $(0, 6)$.

35. $f(x) = x^2 - 1$

Using the graph of $y = x^2$, vertically shift downward 1 unit.

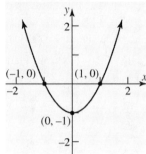

37. $g(x) = x^3 + 1$

Using the graph of $y = x^3$, vertically shift upward 1 unit.

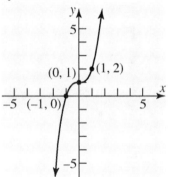

39. $h(x) = \sqrt{x-2}$

Using the graph of $y = \sqrt{x}$, horizontally shift to the right 2 units.

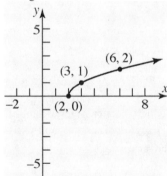

41. $f(x) = (x-1)^3 + 2$

Using the graph of $y = x^3$, horizontally shift to the right 1 unit, then vertically shift up 2 units.

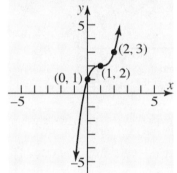

43. $g(x) = 4\sqrt{x}$

Using the graph of $y = \sqrt{x}$, vertically stretch by a factor of 4.

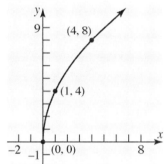

45. $h(x) = \dfrac{1}{2x} = \left(\dfrac{1}{2}\right)\left(\dfrac{1}{x}\right)$

Using the graph of $y = \dfrac{1}{x}$, vertically compress by a factor of $\dfrac{1}{2}$.

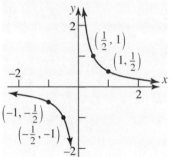

47. $f(x) = -\sqrt[3]{x}$

Reflect the graph of $y = \sqrt[3]{x}$, about the x-axis.

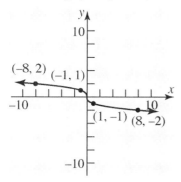

49. $g(x) = |-x|$

Reflect the graph of $y = |x|$ about the y-axis.

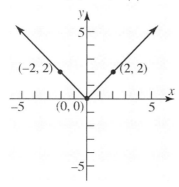

51. $h(x) = -x^3 + 2$

Reflect the graph of $y = x^3$ about the x-axis, then shift vertically upward 2 units.

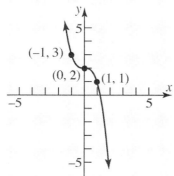

53. $f(x) = 2(x+1)^2 - 3$

Using the graph of $y = x^2$, horizontally shift to the left 1 unit, vertically stretch by a factor of 2, and vertically shift downward 3 units.

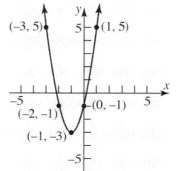

55. $g(x) = \sqrt{x-2} + 1$

Using the graph of $y = \sqrt{x}$, horizontally shift to the right 2 units and vertically shift upward 1 unit.

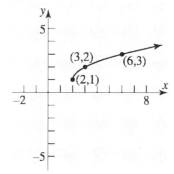

57. $h(x) = \sqrt{-x} - 2$

Reflect the graph of $y = \sqrt{x}$ about the y-axis and vertically shift downward 2 units.

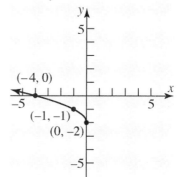

59. $f(x) = -(x+1)^3 - 1$

Using the graph of $y = x^3$, horizontally shift to the left 1 unit, reflect the graph about the x-axis, and vertically shift downward 1 unit.

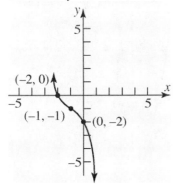

61. $g(x) = 2|1-x| = 2|-(-1+x)| = 2|x-1|$

Using the graph of $y = |x|$, horizontally shift to the right 1 unit, and vertically stretch by a factor or 2.

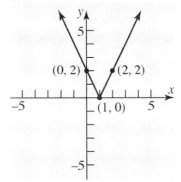

63. $h(x) = 2\operatorname{int}(x-1)$

Using the graph of $y = \operatorname{int}(x)$, horizontally shift to the right 1 unit, and vertically stretch by a factor of 2.

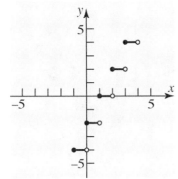

65. a. $F(x) = f(x) + 3$
Shift up 3 units.

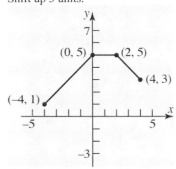

b. $G(x) = f(x+2)$
Shift left 2 units.

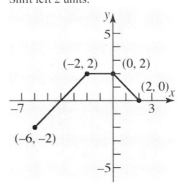

c. $P(x) = -f(x)$
Reflect about the x-axis.

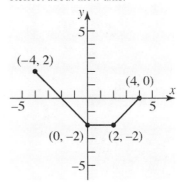

d. $H(x) = f(x+1) - 2$
Shift left 1 unit and shift down 2 units.

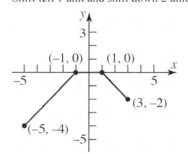

e. $Q(x) = \dfrac{1}{2} f(x)$

Compress vertically by a factor of $\dfrac{1}{2}$.

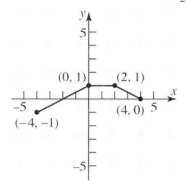

f. $g(x) = f(-x)$
Reflect about the y-axis.

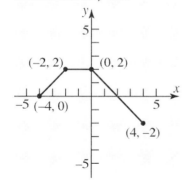

g. $h(x) = f(2x)$
Compress horizontally by a factor of $\dfrac{1}{2}$.

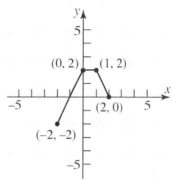

67. a. $F(x) = f(x) + 3$
Shift up 3 units.

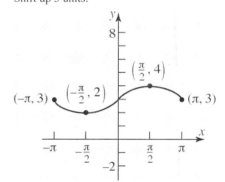

b. $G(x) = f(x+2)$
Shift left 2 units.

c. $P(x) = -f(x)$
Reflect about the x-axis.

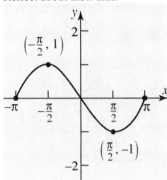

d. $H(x) = f(x+1) - 2$
Shift left 1 unit and shift down 2 units.

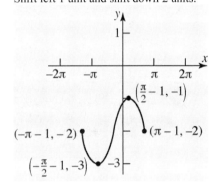

e. $Q(x) = \dfrac{1}{2} f(x)$
Compress vertically by a factor of $\dfrac{1}{2}$.

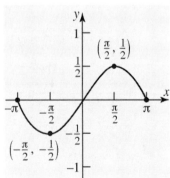

f. $g(x) = f(-x)$
Reflect about the y-axis.

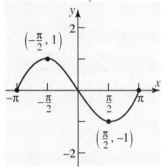

g. $h(x) = f(2x)$
Compress horizontally by a factor of $\dfrac{1}{2}$.

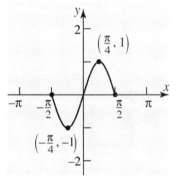

69. a. The graph of $y = f(x+2)$ is the same as the graph of $y = f(x)$, but shifted 2 units to the left. Therefore, the x-intercepts are -7 and 1.

b. The graph of $y = f(x-2)$ is the same as the graph of $y = f(x)$, but shifted 2 units to the right. Therefore, the x-intercepts are -3 and 5.

c. The graph of $y = 4f(x)$ is the same as the graph of $y = f(x)$, but stretched vertically by a factor of 4. Therefore, the x-intercepts are still -5 and 3 since the y-coordinate of each is 0.

d. The graph of $y = f(-x)$ is the same as the graph of $y = f(x)$, but reflected about the y-axis. Therefore, the x-intercepts are 5 and -3.

71. a. The graph of $y = f(x+2)$ is the same as the graph of $y = f(x)$, but shifted 2 units to the left. Therefore, the graph of $f(x+2)$ is increasing on the interval $(-3, 3)$.

   b. The graph of $y = f(x-5)$ is the same as the graph of $y = f(x)$, but shifted 5 units to the right. Therefore, the graph of $f(x-5)$ is increasing on the interval $(4, 10)$.

   c. The graph of $y = -f(x)$ is the same as the graph of $y = f(x)$, but reflected about the x-axis. Therefore, we can say that the graph of $y = -f(x)$ must be *decreasing* on the interval $(-1, 5)$.

   d. The graph of $y = f(-x)$ is the same as the graph of $y = f(x)$, but reflected about the y-axis. Therefore, we can say that the graph of $y = f(-x)$ must be *decreasing* on the interval $(-5, 1)$.

73. a. $y = |f(x)|$

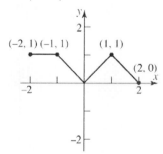

   b. $y = f(|x|)$

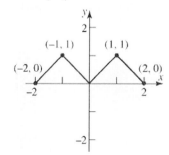

75. $f(x) = x^2 + 2x$
$f(x) = (x^2 + 2x + 1) - 1$
$f(x) = (x+1)^2 - 1$
Using $f(x) = x^2$, shift left 1 unit and shift down 1 unit.

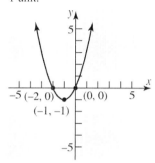

77. $f(x) = x^2 - 8x + 1$
$f(x) = (x^2 - 8x + 16) + 1 - 16$
$f(x) = (x-4)^2 - 15$
Using $f(x) = x^2$, shift right 4 units and shift down 15 units.

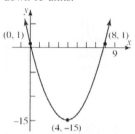

79. $f(x) = x^2 + x + 1$

$f(x) = \left(x^2 + x + \frac{1}{4}\right) + 1 - \frac{1}{4}$

$f(x) = \left(x + \frac{1}{2}\right)^2 + \frac{3}{4}$

Using $f(x) = x^2$, shift left $\frac{1}{2}$ unit and shift up $\frac{3}{4}$ unit.

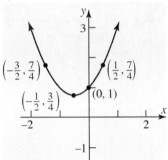

81. $f(x) = 2x^2 - 12x + 19$

$= 2(x^2 - 6x) + 19$

$= 2(x^2 - 6x + 9) + 19 - 18$

$= 2(x - 3)^2 + 1$

Using $f(x) = x^2$, shift right 3 units, vertically stretch by a factor of 2, and then shift up 1 unit.

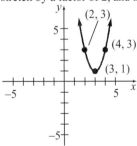

83. $f(x) = -3x^2 - 12x - 17$

$= -3(x^2 + 4x) - 17$

$= -3(x^2 + 4x + 4) - 17 + 12$

$= -3(x + 2)^2 - 5$

Using $f(x) = x^2$, shift left 2 units, stretch vertically by a factor of 3, reflect about the x-axis, and shift down 5 units.

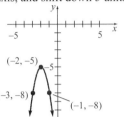

85. $y = (x - c)^2$

If $c = 0$, $y = x^2$.

If $c = 3$, $y = (x - 3)^2$; shift right 3 units.

If $c = -2$, $y = (x + 2)^2$; shift left 2 units.

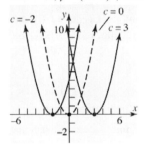

87. $F = \frac{9}{5}C + 32$

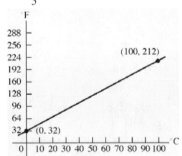

$F = \frac{9}{5}(K - 273) + 32$

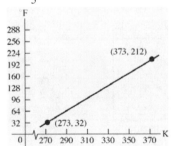

Shift the graph 273 units to the right.

89. a. $p(x) = -0.05x^2 + 100x - 2000$

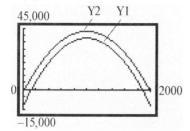

b. Select the 10% tax since the profits are higher.

c. The graph of Y1 is obtained by shifting the graph of $p(x)$ vertically down 10,000 units. The graph of Y2 is obtained by multiplying the y-coordinate of the graph of $p(x)$ by 0.9. Thus, Y2 is the graph of $p(x)$ vertically compressed by a factor of 0.9.

d. Select the 10% tax since the graph of $Y1 = 0.9p(x) \geq Y2 = -0.05x^2 + 100x - 6800$ for all x in the domain.

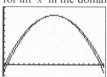

91. a. $y_1 = x+1$; $y_2 = |x+1|$

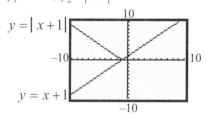

b. $y_1 = 4-x^2$; $y_2 = |4-x^2|$

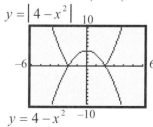

c. $y_1 = x^3 + x$; $y_2 = |x^3 + x|$

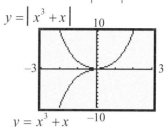

d. Any part of the graph of $y = f(x)$ that lies below the x-axis is reflected about the x-axis to obtain the graph of $y = |f(x)|$.

Section 2.8

1. $V = \pi r^2 h, \ h = 2r \Rightarrow V(r) = \pi r^2 \cdot (2r) = 2\pi r^3$

3. a. $R(x) = x\left(-\frac{1}{6}x + 100\right) = -\frac{1}{6}x^2 + 100x$

b. $R(200) = -\frac{1}{6}(200)^2 + 100(200)$

$= \frac{-20,000}{3} + 20,000$

$= \frac{40,000}{3} \approx \$13,333.33$

c.

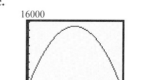

d. $x = 300$ maximizes revenue

$R(300) = -\frac{1}{6}(300)^2 + 100(300)$

$= -15,000 + 30,000$

$= \$15,000$

The maximum revenue is $15,000.

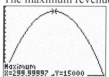

e. $p = -\frac{1}{6}(300) + 100 = -50 + 100 = \$50$ maximizes revenue

5. a. If $x = -5p + 100$, then $p = \dfrac{100-x}{5}$.

$R(x) = x\left(\dfrac{100-x}{5}\right) = -\dfrac{1}{5}x^2 + 20x$

b. $R(15) = -\dfrac{1}{5}(15)^2 + 20(15)$
$= -45 + 300 = \$255$

c.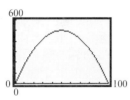

d. $x = 50$ maximizes revenue
$R(50) = -\dfrac{1}{5}(50)^2 + 20(50)$
$= -500 + 1000 = \$500$
The maximum revenue is $500.

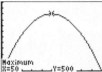

e. $p = \dfrac{100 - 50}{5} = \dfrac{50}{5} = \$10$
maximizes revenue.

7. a. Let $x =$ width and $y =$ length of the rectangular area.
$P = 2x + 2y = 400$
$y = \dfrac{400 - 2x}{2} = 200 - x$
Then
$A(x) = (200 - x)x$
$= 200x - x^2$
$= -x^2 + 200x$

b. We need
$x > 0$ and $y > 0 \Rightarrow 200 - x > 0 \Rightarrow 200 > x$
So the domain of A is $\{x \mid 0 < x < 200\}$

c. $x = 100$ yards maximizes area

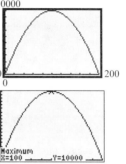

9. a. The distance d from P to the origin is $d = \sqrt{x^2 + y^2}$. Since P is a point on the graph of $y = x^2 - 8$, we have:
$d(x) = \sqrt{x^2 + (x^2 - 8)^2} = \sqrt{x^4 - 15x^2 + 64}$

b. $d(0) = \sqrt{0^4 - 15(0)^2 + 64} = \sqrt{64} = 8$

c. $d(1) = \sqrt{(1)^4 - 15(1)^2 + 64} = \sqrt{1 - 15 + 64}$
$= \sqrt{50} = 5\sqrt{2} \approx 7.07$

d.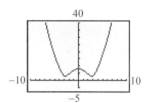

e. d is smallest when $x \approx -2.74$ and when $x \approx 2.74$.

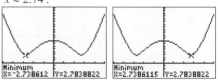

11. a. The distance d from P to the point $(1, 0)$ is $d = \sqrt{(x-1)^2 + y^2}$. Since P is a point on the graph of $y = \sqrt{x}$, we have:
$d(x) = \sqrt{(x-1)^2 + \left(\sqrt{x}\right)^2} = \sqrt{x^2 - x + 1}$
where $x \geq 0$.

b.

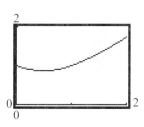

c. d is smallest when x is 0.50.

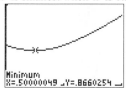

13. By definition, a triangle has area $A = \frac{1}{2}bh$, b = base, h = height. From the figure, we know that $b = x$ and $h = y$. Expressing the area of the triangle as a function of x, we have:
$$A(x) = \frac{1}{2}xy = \frac{1}{2}x(x^3) = \frac{1}{2}x^4.$$

15. a. $A(x) = xy = x(16 - x^2) = -x^3 + 16x$

    b. Domain: $\{x \mid 0 < x < 4\}$

    c. The area is largest when x is approximately 2.31.

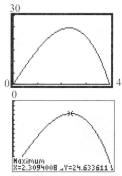

17. a. In Quadrant I, $x^2 + y^2 = 4 \rightarrow y = \sqrt{4 - x^2}$
$$A(x) = (2x)(2y) = 4x\sqrt{4 - x^2}$$

    b. $p(x) = 2(2x) + 2(2y) = 4x + 4\sqrt{4 - x^2}$

c. Graphing the area equation:

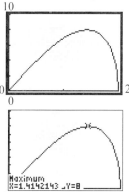

The area is largest when x is roughly 1.41.

d. Graphing the perimeter equation:

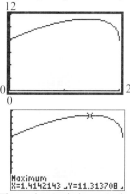

The perimeter is largest when x is approximately 1.41.

19. a. C = circumference, TA = total area, r = radius, x = side of square
$$C = 2\pi r = 10 - 4x \implies r = \frac{5 - 2x}{\pi}$$
Total Area = area$_{\text{square}}$ + area$_{\text{circle}}$
$$= x^2 + \pi r^2$$
$$TA(x) = x^2 + \pi\left(\frac{5 - 2x}{\pi}\right)^2$$
$$= x^2 + \frac{25 - 20x + 4x^2}{\pi}$$

b. Since the lengths must be positive, we have:
$10 - 4x > 0$ and $x > 0$
$-4x > -10$ and $x > 0$
$x < 2.5$ and $x > 0$
Domain: $\{x \mid 0 < x < 2.5\}$

c. The total area is smallest when x is approximately 1.40 meters.

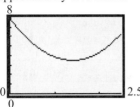

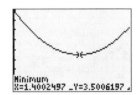

21. a. Since the wire of length x is bent into a circle, the circumference is x. Therefore, $C(x) = x$.

  b. Since $C = x = 2\pi r$, $r = \dfrac{x}{2\pi}$.

  $A(x) = \pi r^2 = \pi\left(\dfrac{x}{2\pi}\right)^2 = \dfrac{x^2}{4\pi}$.

23. a. A = area, r = radius; diameter = $2r$

  $A(r) = (2r)(r) = 2r^2$

  b. p = perimeter

  $p(r) = 2(2r) + 2r = 6r$

25. Area of the equilateral triangle

  $A = \dfrac{1}{2} x \cdot \dfrac{\sqrt{3}}{2} x = \dfrac{\sqrt{3}}{4} x^2$

  From problem 24, we have $r^2 = \dfrac{x^2}{3}$.

  Area inside the circle, but outside the triangle:

  $A(x) = \pi r^2 - \dfrac{\sqrt{3}}{4} x^2$

  $= \pi \dfrac{x^2}{3} - \dfrac{\sqrt{3}}{4} x^2$

  $= \left(\dfrac{\pi}{3} - \dfrac{\sqrt{3}}{4}\right) x^2$

27. a. $d^2 = d_1^2 + d_2^2$

  $d^2 = (2-30t)^2 + (3-40t)^2$

  $d(t) = \sqrt{(2-30t)^2 + (3-40t)^2}$

  $= \sqrt{4 - 120t + 900t^2 + 9 - 240t + 1600t^2}$

  $= \sqrt{2500t^2 - 360t + 13}$

  b. The distance is smallest at $t \approx 0.072$ hours.

29. r = radius of cylinder, h = height of cylinder, V = volume of cylinder

  By similar triangles: $\dfrac{H}{R} = \dfrac{H-h}{r}$

  $Hr = R(H-h)$

  $Hr = RH - Rh$

  $Rh = RH - Hr$

  $h = \dfrac{RH - Hr}{R} = H - \dfrac{Hr}{R}$

  $V = \pi r^2 h = \pi r^2\left(H - \dfrac{Hr}{R}\right) = H\pi r^2\left(1 - \dfrac{r}{R}\right)$

31. a. The time on the boat is given by $\dfrac{d_1}{3}$. The time on land is given by $\dfrac{12-x}{5}$.

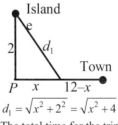

  $d_1 = \sqrt{x^2 + 2^2} = \sqrt{x^2 + 4}$

  The total time for the trip is:

  $T(x) = \dfrac{12-x}{5} + \dfrac{d_1}{3} = \dfrac{12-x}{5} + \dfrac{\sqrt{x^2+4}}{3}$

b. Domain: $\{x \mid 0 \le x \le 12\}$

c. $T(4) = \dfrac{12-4}{5} + \dfrac{\sqrt{4^2+4}}{3}$

$= \dfrac{8}{5} + \dfrac{\sqrt{20}}{3} \approx 3.09$ hours

d. $T(8) = \dfrac{12-8}{5} + \dfrac{\sqrt{8^2+4}}{3}$

$= \dfrac{4}{5} + \dfrac{\sqrt{68}}{3} \approx 3.55$ hours

Chapter 2 Review

1. $2x = 3y^2$

    x-axis: $2x = 3(-y)^2$
    $2x = 3y^2$ same

    y-axis: $2(-x) = 3y^2$
    $-2x = 3y^2$ different

    origin: $2(-x) = 3(-y)^2$
    $-2x = 3y^2$ different

    Therefore, the graph of the equation has x-axis symmetry.

3. $x^2 + 4y^2 = 16$

    x-axis: $x^2 + 4(-y)^2 = 16$
    $x^2 + 4y^2 = 16$ same

    y-axis: $(-x)^2 + 4y^2 = 16$
    $x^2 + 4y^2 = 16$ same

    origin: $(-x)^2 + 4(-y)^2 = 16$
    $x^2 + 4y^2 = 16$ same

    Therefore, the graph of the equation has x-axis, y-axis, and origin symmetry.

5. $y = x^4 + 2x^2 + 1$

    x-axis: $(-y) = x^4 + 2x^2 + 1$
    $-y = x^4 + 2x^2 + 1$ different

    y-axis: $y = (-x)^4 + 2(-x)^2 + 1$
    $y = x^4 + 2x^2 + 1$ same

    origin: $(-y) = (-x)^4 + 2(-x)^2 + 1$
    $-y = x^4 + 2x^2 + 1$ different

    Therefore, the equation of the graph has y-axis symmetry.

7. $x^2 + x + y^2 + 2y = 0$

    x-axis: $x^2 + x + (-y)^2 + 2(-y) = 0$
    $x^2 + x + y^2 - 2y = 0$ different

    y-axis: $(-x)^2 + (-x) + y^2 + 2y = 0$
    $x^2 - x + y^2 + 2y = 0$ different

    origin:
    $(-x)^2 + (-x) + (-y)^2 + 2(-y) = 0$
    $x^2 - x + y^2 - 2y = 0$ different

    Therefore, the graph of the equation has none of the three symmetries.

9.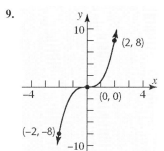

11. This relation represents a function.
    Domain = $\{-1, 2, 4\}$; Range = $\{0, 3\}$.

13. $f(x) = \dfrac{3x}{x^2 - 1}$

    a. $f(2) = \dfrac{3(2)}{(2)^2 - 1} = \dfrac{6}{4-1} = \dfrac{6}{3} = 2$

    b. $f(-2) = \dfrac{3(-2)}{(-2)^2 - 1} = \dfrac{-6}{4-1} = \dfrac{-6}{3} = -2$

c. $f(-x) = \dfrac{3(-x)}{(-x)^2 - 1} = \dfrac{-3x}{x^2 - 1}$

d. $-f(x) = -\left(\dfrac{3x}{x^2 - 1}\right) = \dfrac{-3x}{x^2 - 1}$

e. $f(x-2) = \dfrac{3(x-2)}{(x-2)^2 - 1}$
$= \dfrac{3x - 6}{x^2 - 4x + 4 - 1}$
$= \dfrac{3(x-2)}{x^2 - 4x + 3}$

f. $f(2x) = \dfrac{3(2x)}{(2x)^2 - 1} = \dfrac{6x}{4x^2 - 1}$

15. $f(x) = \sqrt{x^2 - 4}$

   a. $f(2) = \sqrt{2^2 - 4} = \sqrt{4 - 4} = \sqrt{0} = 0$

   b. $f(-2) = \sqrt{(-2)^2 - 4} = \sqrt{4 - 4} = \sqrt{0} = 0$

   c. $f(-x) = \sqrt{(-x)^2 - 4} = \sqrt{x^2 - 4}$

   d. $-f(x) = -\sqrt{x^2 - 4}$

   e. $f(x-2) = \sqrt{(x-2)^2 - 4}$
   $= \sqrt{x^2 - 4x + 4 - 4}$
   $= \sqrt{x^2 - 4x}$

   f. $f(2x) = \sqrt{(2x)^2 - 4} = \sqrt{4x^2 - 4}$
   $= \sqrt{4(x^2 - 1)} = 2\sqrt{x^2 - 1}$

17. $f(x) = \dfrac{x^2 - 4}{x^2}$

   a. $f(2) = \dfrac{2^2 - 4}{2^2} = \dfrac{4 - 4}{4} = \dfrac{0}{4} = 0$

   b. $f(-2) = \dfrac{(-2)^2 - 4}{(-2)^2} = \dfrac{4 - 4}{4} = \dfrac{0}{4} = 0$

   c. $f(-x) = \dfrac{(-x)^2 - 4}{(-x)^2} = \dfrac{x^2 - 4}{x^2}$

d. $-f(x) = -\left(\dfrac{x^2 - 4}{x^2}\right) = \dfrac{4 - x^2}{x^2} = -\dfrac{x^2 - 4}{x^2}$

e. $f(x-2) = \dfrac{(x-2)^2 - 4}{(x-2)^2} = \dfrac{x^2 - 4x + 4 - 4}{(x-2)^2}$
$= \dfrac{x^2 - 4x}{(x-2)^2} = \dfrac{x(x-4)}{(x-2)^2}$

f. $f(2x) = \dfrac{(2x)^2 - 4}{(2x)^2} = \dfrac{4x^2 - 4}{4x^2}$
$= \dfrac{4(x^2 - 1)}{4x^2} = \dfrac{x^2 - 1}{x^2}$

19. $f(x) = \dfrac{x}{x^2 - 9}$
The denominator cannot be zero:
$x^2 - 9 \neq 0$
$(x+3)(x-3) \neq 0$
$x \neq -3$ or 3
Domain: $\{x \mid x \neq -3, x \neq 3\}$

21. $f(x) = \sqrt{2 - x}$
The radicand must be non-negative:
$2 - x \geq 0$
$x \leq 2$
Domain: $\{x \mid x \leq 2\}$ or $(-\infty, 2]$

23. $f(x) = \dfrac{\sqrt{x}}{|x|}$
The radicand must be non-negative and the denominator cannot be zero: $x > 0$
Domain: $\{x \mid x > 0\}$ or $(0, \infty)$

25. $f(x) = \dfrac{x}{x^2 + 2x - 3}$
The denominator cannot be zero:
$x^2 + 2x - 3 \neq 0$
$(x+3)(x-1) \neq 0$
$x \neq -3$ or 1
Domain: $\{x \mid x \neq -3, x \neq 1\}$

27. $f(x) = 2-x \quad g(x) = 3x+1$

$(f+g)(x) = f(x) + g(x)$
$= 2-x+3x+1 = 2x+3$
Domain: $\{x \mid x \text{ is any real number}\}$

$(f-g)(x) = f(x) - g(x)$
$= 2-x-(3x+1)$
$= 2-x-3x-1$
$= -4x+1$
Domain: $\{x \mid x \text{ is any real number}\}$

$(f \cdot g)(x) = f(x) \cdot g(x)$
$= (2-x)(3x+1)$
$= 6x+2-3x^2-x$
$= -3x^2+5x+2$
Domain: $\{x \mid x \text{ is any real number}\}$

$\left(\dfrac{f}{g}\right)(x) = \dfrac{f(x)}{g(x)} = \dfrac{2-x}{3x+1}$

$3x+1 \neq 0$

$3x \neq -1 \Rightarrow x \neq -\dfrac{1}{3}$

Domain: $\left\{x \mid x \neq -\dfrac{1}{3}\right\}$

29. $f(x) = 3x^2+x+1 \quad g(x) = 3x$

$(f+g)(x) = f(x) + g(x)$
$= 3x^2+x+1+3x$
$= 3x^2+4x+1$
Domain: $\{x \mid x \text{ is any real number}\}$

$(f-g)(x) = f(x) - g(x)$
$= 3x^2+x+1-3x$
$= 3x^2-2x+1$
Domain: $\{x \mid x \text{ is any real number}\}$

$(f \cdot g)(x) = f(x) \cdot g(x)$
$= (3x^2+x+1)(3x)$
$= 9x^3+3x^2+3x$
Domain: $\{x \mid x \text{ is any real number}\}$

$\left(\dfrac{f}{g}\right)(x) = \dfrac{f(x)}{g(x)} = \dfrac{3x^2+x+1}{3x}$

$3x \neq 0 \Rightarrow x \neq 0$
Domain: $\{x \mid x \neq 0\}$

31. $f(x) = \dfrac{x+1}{x-1} \quad g(x) = \dfrac{1}{x}$

$(f+g)(x) = f(x) + g(x)$
$= \dfrac{x+1}{x-1} + \dfrac{1}{x}$
$= \dfrac{x(x+1)+1(x-1)}{x(x-1)}$
$= \dfrac{x^2+x+x-1}{x(x-1)}$
$= \dfrac{x^2+2x-1}{x(x-1)}$
Domain: $\{x \mid x \neq 0, x \neq 1\}$

$(f-g)(x) = f(x) - g(x)$
$= \dfrac{x+1}{x-1} - \dfrac{1}{x}$
$= \dfrac{x(x+1)-1(x-1)}{x(x-1)}$
$= \dfrac{x^2+x-x+1}{x(x-1)}$
$= \dfrac{x^2+1}{x(x-1)}$
Domain: $\{x \mid x \neq 0, x \neq 1\}$

$(f \cdot g)(x) = f(x) \cdot g(x) = \left(\dfrac{x+1}{x-1}\right)\left(\dfrac{1}{x}\right) = \dfrac{x+1}{x(x-1)}$

Domain: $\{x \mid x \neq 0, x \neq 1\}$

$\left(\dfrac{f}{g}\right)(x) = \dfrac{f(x)}{g(x)} = \dfrac{\dfrac{x+1}{x-1}}{\dfrac{1}{x}} = \left(\dfrac{x+1}{x-1}\right)\left(\dfrac{x}{1}\right) = \dfrac{x(x+1)}{x-1}$

Domain: $\{x \mid x \neq 0, x \neq 1\}$

33. $f(x) = -2x^2 + x + 1$

$\dfrac{f(x+h) - f(x)}{h}$

$= \dfrac{-2(x+h)^2 + (x+h) + 1 - (-2x^2 + x + 1)}{h}$

$= \dfrac{-2(x^2 + 2xh + h^2) + x + h + 1 + 2x^2 - x - 1}{h}$

$= \dfrac{-2x^2 - 4xh - 2h^2 + x + h + 1 + 2x^2 - x - 1}{h}$

$= \dfrac{-4xh - 2h^2 + h}{h} = \dfrac{h(-4x - 2h + 1)}{h}$

$= -4x - 2h + 1$

35. a. Domain: $\{x \mid -4 \le x \le 3\}$
    Range: $\{y \mid -3 \le y \le 3\}$

  b. x-intercept: $(0, 0)$; y-intercept: $(0, 0)$

  c. $f(-2) = -1$

  d. $f(x) = -3$ when $x = -4$

  e. $f(x) > 0$ when $0 < x \le 3$

  f. To graph $y = f(x - 3)$, shift the graph of f horizontally 3 units to the right.

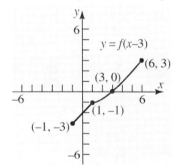

  g. To graph $y = f\left(\dfrac{1}{2}x\right)$, stretch the graph of f horizontally by a factor of 2.

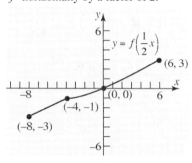

  h. To graph $y = -f(x)$, reflect the graph of f vertically about the y-axis.

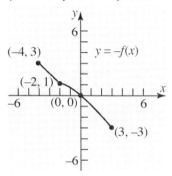

37. a. Domain: $\{x \mid -4 \le x \le 4\}$
    Range: $\{y \mid -3 \le y \le 1\}$

  b. Increasing: $(-4, -1)$ and $(3, 4)$;
     Decreasing: $(-1, 3)$

  c. Local minimum is -3 when $x = 3$;
     Local maximum is 1 when $x = -1$.
     Note that $x = 4$ and $x = -4$ do not yield local extrema because there is no open interval that contains either value.

  d. The graph is not symmetric with respect to the x-axis, the y-axis or the origin.

  e. The function is neither even nor odd.

  f. x-intercepts: $(-2, 0)$, $(0, 0)$, $(4, 0)$,
     y-intercept: $(0, 0)$

39. $f(x) = x^3 - 4x$

$f(-x) = (-x)^3 - 4(-x) = -x^3 + 4x$
$= -(x^3 - 4x) = -f(x)$

f is odd.

41. $h(x) = \dfrac{1}{x^4} + \dfrac{1}{x^2} + 1$

$h(-x) = \dfrac{1}{(-x)^4} + \dfrac{1}{(-x)^2} + 1 = \dfrac{1}{x^4} + \dfrac{1}{x^2} + 1 = h(x)$

h is even.

43. $G(x) = 1 - x + x^3$

$G(-x) = 1 - (-x) + (-x)^3$
$= 1 + x - x^3 \neq -G(x)$ or $G(x)$

G is neither even nor odd.

45. $f(x) = \dfrac{x}{1 + x^2}$

$f(-x) = \dfrac{-x}{1 + (-x)^2} = \dfrac{-x}{1 + x^2} = -f(x)$

f is odd.

47. $f(x) = 2x^3 - 5x + 1$ on the interval $(-3, 3)$
Use MAXIMUM and MINIMUM on the graph of $y_1 = 2x^3 - 5x + 1$.

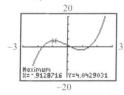

local maximum at: $(-0.91, 4.04)$;
local minimum at: $(0.91, -2.04)$
f is increasing on: $(-3, -0.91)$ and $(0.91, 3)$;
f is decreasing on: $(-0.91, 0.91)$

49. $f(x) = 2x^4 - 5x^3 + 2x + 1$ on the interval $(-2, 3)$
Use MAXIMUM and MINIMUM on the graph of $y_1 = 2x^4 - 5x^3 + 2x + 1$.

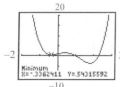 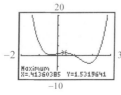

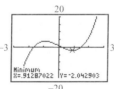

local maximum at: $(0.41, 1.53)$;
local minima at: $(-0.34, 0.54)$, $(1.80, -3..$
f is increasing on: $(-0.34, 0.41)$ and $(1.80, 3$
f is decreasing on: $(-2, -0.34)$ and $(0.41, 1.80$

51. $f(x) = 8x^2 - x$

a. $\dfrac{f(2) - f(1)}{2 - 1} = \dfrac{8(2)^2 - 2 - (8(1)^2 - 1)}{1}$
$= 32 - 2 - (7) = 23$

b. $\dfrac{f(1) - f(0)}{1 - 0} = \dfrac{8(1)^2 - 1 - (8(0)^2 - 0)}{1}$
$= 8 - 1 - (0) = 7$

c. $\dfrac{f(4) - f(2)}{4 - 2} = \dfrac{8(4)^2 - 4 - (8(2)^2 - 2)}{2}$
$= \dfrac{128 - 4 - (30)}{2} = \dfrac{94}{2} = 47$

53. $f(x) = 2 - 5x$

$\dfrac{f(x) - f(2)}{x - 2} = \dfrac{2 - 5x - (-8)}{x - 2} = \dfrac{-5x + 10}{x - 2}$
$= \dfrac{-5(x - 2)}{x - 2} = -5$

55. $f(x) = 3x - 4x^2$

$\dfrac{f(x) - f(2)}{x - 2} = \dfrac{3x - 4x^2 - (-10)}{x - 2}$
$= \dfrac{-4x^2 + 3x + 10}{x - 2}$
$= \dfrac{-(4x^2 - 3x - 10)}{x - 2}$
$= \dfrac{-(4x + 5)(x - 2)}{x - 2}$
$= -4x - 5$

57. (b) passes the Vertical Line Test and is therefore a function.

59. $f(x) = 2x - 5$

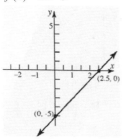

61. $h(x) = \dfrac{4}{5}x - 6$

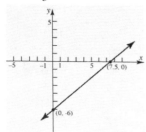

63. $f(x) = |x|$

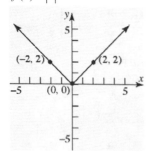

65. $F(x) = |x| - 4$

Using the graph of $y = |x|$, vertically shift the graph downward 4 units.

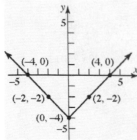

Intercepts: $(-4, 0), (4, 0), (0, -4)$
Domain: $\{x \mid x \text{ is any real number}\}$
Range: $\{y \mid y \geq -4\}$

67. $g(x) = -2|x|$

Reflect the graph of $y = |x|$ about the x-axis and vertically stretch the graph by a factor of 2.

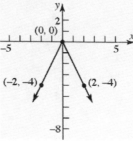

Intercepts: $(0, 0)$
Domain: $\{x \mid x \text{ is any real number}\}$
Range: $\{y \mid y \leq 0\}$

69. $h(x) = \sqrt{x - 1}$

Using the graph of $y = \sqrt{x}$, horizontally shift the graph to the right 1 unit.

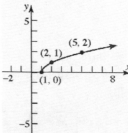

Intercept: $(1, 0)$
Domain: $\{x \mid x \geq 1\}$; Range: $\{y \mid y \geq 0\}$

71. $f(x) = \sqrt{1 - x} = \sqrt{-1(x - 1)}$

Reflect the graph of $y = \sqrt{x}$ about the y-axis and horizontally shift the graph to the right 1 unit.

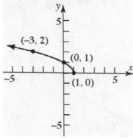

Intercepts: $(1, 0), (0, 1)$
Domain: $\{x \mid x \leq 1\}$
Range: $\{y \mid y \geq 0\}$

73. $h(x) = (x-1)^2 + 2$

Using the graph of $y = x^2$, horizontally shift the graph to the right 1 unit and vertically shift the graph up 2 units.

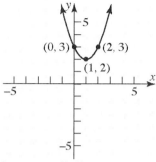

Intercepts: $(0, 3)$
Domain: $\{x \mid x \text{ is any real number}\}$
Range: $\{y \mid y \geq 2\}$

75. $g(x) = 3(x-1)^3 + 1$

Using the graph of $y = x^3$, horizontally shift the graph to the right 1 unit vertically stretch the graph by a factor of 3, and vertically shift the graph up 1 unit.

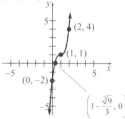

Intercepts: $(0, -2)$, $\left(1 - \dfrac{\sqrt[3]{9}}{3}, 0\right)$
Domain: $\{x \mid x \text{ is any real number}\}$
Range: $\{y \mid y \text{ is any real number}\}$

77. $f(x) = \begin{cases} 3x & \text{if } -2 < x \leq 1 \\ x+1 & \text{if } x > 1 \end{cases}$

a. Domain: $\{x \mid x > -2\}$

b. x-intercept: $(0,0)$
y-intercept: $(0,0)$

c. Graph:

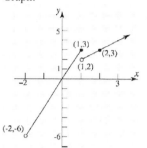

d. Range: $\{y \mid y > -6\}$

79. $f(x) = \begin{cases} x & \text{if } -4 \leq x < 0 \\ 1 & \text{if } x = 0 \\ 3x & \text{if } x > 0 \end{cases}$

a. Domain: $\{x \mid x \geq -4\}$

b. x-intercept: none
y-intercept: $(0, 1)$

c. Graph:

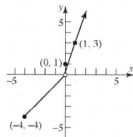

d. Range: $\{y \mid y \geq -4, y \neq 0\}$

81. $f(4) = -5$ gives the ordered pair $(4, -5)$
$f(0) = 3$ gives $f(0) = 3$ gives $(0, 3)$
Finding the slope: $m = \dfrac{3 - (-5)}{0 - 4} = \dfrac{8}{-4} = -2$
Using slope-intercept form: $f(x) = -2x + 3$

Chapter 2: Functions and Their Graphs

83. $f(x) = \dfrac{Ax+5}{6x-2}$ and $f(1) = 4$

$\dfrac{A(1)+5}{6(1)-2} = 4$

$\dfrac{A+5}{4} = 4$

$A+5 = 16$

$A = 11$

85. We have the points $(h_1, T_1) = (0, 30)$ and $(h_2, T_2) = (10000, 5)$.

slope $= \dfrac{\Delta T}{\Delta h} = \dfrac{5-30}{10000-0} = \dfrac{-25}{10000} = -0.0025$

Using the point-slope formula yields

$T - T_1 = m(h - h_1) \Rightarrow T - 30 = -0.0025(h - 0)$

$T - 30 = -0.0025h \Rightarrow T = -0.0025h + 30$

$T(h) = -0.0025h + 30,\ 0 \le x \le 10,000$

87. $S = 4\pi r^2 \Rightarrow r = \sqrt{\dfrac{S}{4\pi}}$

$V(S) = \dfrac{4}{3}\pi r^3 = \dfrac{4\pi}{3}\left(\sqrt{\dfrac{S}{4\pi}}\right)^3$

$= \dfrac{4\pi}{3} \cdot \dfrac{S}{4\pi}\sqrt{\dfrac{S}{4\pi}} = \dfrac{S}{6}\sqrt{\dfrac{S}{\pi}}$

$V(2S) = \dfrac{2S}{6}\sqrt{\dfrac{2S}{\pi}} = 2\sqrt{2}\left(\dfrac{S}{6}\sqrt{\dfrac{S}{\pi}}\right)$

The volume is $2\sqrt{2}$ times as large.

89. $S = kxd^3$, x = width; d = depth

In the diagram, depth = length of the rectangle. Therefore, we have

$\left(\dfrac{d}{2}\right)^2 + \left(\dfrac{x}{2}\right)^2 = 3^2$

$\dfrac{d^2}{4} + \dfrac{x^2}{4} = 9$

$d^2 + x^2 = 36$

$d = \sqrt{36 - x^2}$

$S(x) = kx\left(\sqrt{36-x^2}\right)^3 = kx\left(36-x^2\right)^{3/2}$

Domain: $\{x \mid 0 < x < 6\}$

91. **a.** The relation is a function. Each HS GPA value is paired with exactly one College GPA value.

b. Scatter diagram:

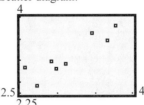

c. Using the LINear REGression program, the line of best fit is: $G = 0.964x + 0.072$

d. As the high school GPA increases by 0.1 point, the college GPA increases by 0.0964 point.

e. $G(x) = 0.964x + 0.072$

f. Domain: $\{x \mid 0 \le x \le 4\}$

g. $G(3.23) = (0.964)(3.23) + 0.072 \approx 3.19$
The college GPA is approximately 3.19.

93. Let R = the revenue in dollars, and g = the number of gallons of gasoline sold. Consider the ordered pair (g, R). We can use the points $(0, 0)$ and $(13.5, 28.89)$. Now compute the slope:

slope $= \dfrac{\Delta y}{\Delta x} = \dfrac{28.89 - 0}{13.5 - 0} = \dfrac{28.89}{13.5} \approx 2.14$

Therefore we have the linear function
$R(g) = 2.14g + 0 = 2.14g$.
If $g = 11.2$, then $R = (2.14)(11.2) = \$23.97$.

95. Let x represent the length and y represent the width of the rectangle.
$2x + 2y = 20 \rightarrow y = 10 - x$.
$x \cdot y = 16 \rightarrow x(10 - x) = 16$.
Solving the area equation:
$10x - x^2 = 16 \rightarrow x^2 - 10x + 16 = 0$
$(x-8)(x-2) = 0 \rightarrow x = 8$ or $x = 2$
The length and width of the rectangle are 8 feet by 2 feet.

97. $C(x) = 4.9x^2 - 617.40x + 19,600$;
$a = 4.9$, $b = -617.40$, $c = 19,600$.
Since $a = 4.9 > 0$, the graph opens up, so the vertex is a minimum point.

  a. The minimum marginal cost occurs at $x = 63$.

  b. The minimum marginal cost is
  $$C\left(\frac{-b}{2a}\right) = C(63)$$
  $$= 4.9(63)^2 - (617.40)(63) + 19600$$
  $$= \$151.90$$

99. Let $P = (4,1)$ and $Q = (x,y) = (x, x+1)$.
$$d(P,Q) = \sqrt{(x-4)^2 + (x+1-1)^2}$$
$$\rightarrow d^2(x) = (x-4)^2 + x^2$$
$$= x^2 - 8x + 16 + x^2$$
$$\therefore d^2(x) = 2x^2 - 8x + 16$$

Since $d^2(x) = 2x^2 - 8x + 16$ is a quadratic function with $a = 2 > 0$, the vertex corresponds to the minimum value for the function.

 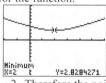

The vertex occurs at $x = 2$. Therefore the point Q on the line $y = x + 1$ will be closest to the point $P = (4,1)$ when $Q = (2,3)$.

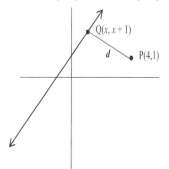

101. a. $x^2 h = 10 \Rightarrow h = \dfrac{10}{x^2}$
$$A(x) = 2x^2 + 4xh$$
$$= 2x^2 + 4x\left(\frac{10}{x^2}\right)$$
$$= 2x^2 + \frac{40}{x}$$

b. $A(1) = 2 \cdot 1^2 + \dfrac{40}{1} = 2 + 40 = 42$ ft^2

c. $A(2) = 2 \cdot 2^2 + \dfrac{40}{2} = 8 + 20 = 28$ ft^2

d. Graphing:

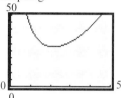

The area is smallest when $x \approx 2.15$ feet.

Chapter 2 Test

1. To check x-axis symmetry, we change y to $-y$ and see if an equivalent equation results.
$$5x - 2(-y)^2 = 7$$
$$5x - 2y^2 = 7$$
This equation is the same as the original equation, so there is x-axis symmetry.

   To check y-axis symmetry, we change x to $-x$ and see if an equivalent equation results.
$$5(-x) - 2y^2 = 7$$
$$-5x - 2y^2 = 7$$
This equation is not the same as the original equation, so there is no y-axis symmetry.

   To check origin symmetry, we change x to $-x$ and y to $-y$, and see if an equivalent equation results.
$$5(-x) - 2(-y)^2 = 7$$
$$-5x - 2y^2 = 7$$
This equation is not the same as the original equation, so there is no origin symmetry. (note: if an equation has any two of these symmetries, it must have all three. Since we saw that there was x-axis symmetry but not y-axis symmetry, we could have concluded that there would be no origin symmetry as well)

2. **a.** $\{(2,5),(4,6),(6,7),(8,8)\}$

   This relation is a function because there are no ordered pairs that have the same first element and different second elements.
   Domain: $\{2,4,6,8\}$
   Range: $\{5,6,7,8\}$

   b. $\{(1,3),(4,-2),(-3,5),(1,7)\}$

   This relation is not a function because there are two ordered pairs that have the same first element but different second elements.

   c. This relation is not a function because the graph fails the vertical line test.

   d. This relation is a function because it passes the vertical line test.
   Domain: $\mathbb{R}$ (all real numbers)
   Range: $\{y \mid y \geq 2\}$

3. $f(x) = \sqrt{4-5x}$

   The function tells us to take the square root of $4-5x$. Only nonnegative numbers have real square roots so we need $4-5x \geq 0$.
$$4 - 5x \geq 0$$
$$4 - 5x - 4 \geq 0 - 4$$
$$-5x \geq -4$$
$$\frac{-5x}{-5} \leq \frac{-4}{-5}$$
$$x \leq \frac{4}{5}$$
   Domain: $\left\{x \mid x \leq \frac{4}{5}\right\}$

$$f(-1) = \sqrt{4-5(-1)} = \sqrt{4+5} = \sqrt{9} = 3$$

4. $g(x) = \dfrac{x+2}{|x+2|}$

   The function tells us to divide $x+2$ by $|x+2|$. Division by 0 is undefined, so the denominator can never equal 0. This means that $x \neq -2$.
   Domain: $\{x \mid x \neq -2\}$

$$g(-1) = \frac{(-1)+2}{|(-1)+2|} = \frac{1}{|1|} = 1$$

5. $h(x) = \dfrac{x-4}{x^2+5x-36}$

   The function tells us to divide $x-4$ by $x^2+5x-36$. Since division by 0 is not defined, we need to exclude any values which make the denominator 0.
$$x^2 + 5x - 36 = 0$$
$$(x+9)(x-4) = 0$$
$$x = -9 \text{ or } x = 4$$
   Domain: $\{x \mid x \neq -9, x \neq 4\}$
   (note: there is a common factor of $x-4$ but we must determine the domain prior to simplifying)

$$h(-1) = \frac{(-1)-4}{(-1)^2+5(-1)-36} = \frac{-5}{-40} = \frac{1}{8}$$

6. a. To find the domain, note that all the points on the graph will have an x-coordinate between −5 and 5, inclusive. To find the range, note that all the points on the graph will have a y-coordinate between −3 and 3, inclusive.

   Domain: $\{x \mid -5 \leq x \leq 5\}$

   Range: $\{y \mid -3 \leq y \leq 3\}$

   b. The intercepts are $(0, 2)$, $(-2, 0)$, and $(2, 0)$.

   x-intercepts: 2, −2
   y-intercept: 2

   c. $f(1)$ is the value of the function when $x = 1$. According to the graph, $f(1) = 3$.

   d. Since $(-5, -3)$ and $(3, -3)$ are the only points on the graph for which $y = f(x) = -3$, we have $f(x) = -3$ when $x = -5$ and $x = 3$.

   e. To solve $f(x) < 0$, we want to find x-values such that the graph is below the x-axis. The graph is below the x-axis for values in the domain that are less than −2 and greater than 2. Therefore, the solution set is $\{x \mid -5 \leq x < -2 \text{ or } 2 < x \leq 5\}$. In interval notation we would write the solution set as $[-5, -2) \cup (2, 5]$.

7. $f(x) = -x^4 + 2x^3 + 4x^2 - 2$

   We set Xmin = −5 and Xmax = 5. The standard Ymin and Ymax will not be good enough to see the whole picture so some adjustment must be made.

   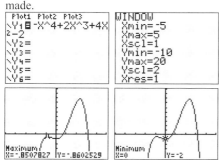

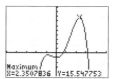

We see that the graph has a local maximum of −0.86 (rounded to two places) when $x = -0.85$ and another local maximum of 15.55 when $x = 2.35$. There is a local minimum of −2 when $x = 0$. Thus, we have

Local maxima: $f(-0.85) \approx -0.86$

$f(2.35) \approx 15.55$

Local minima: $f(0) = -2$

The function is increasing on the intervals $(-5, -0.85)$ and $(0, 2.35)$ and decreasing on the intervals $(-0.85, 0)$ and $(2.35, 5)$.

8. a. $f(x) = \begin{cases} 2x + 1 & x < -1 \\ x - 4 & x \geq -1 \end{cases}$

   To graph the function, we graph each "piece". First we graph the line $y = 2x + 1$ but only keep the part for which $x < -1$. Then we plot the line $y = x - 4$ but only keep the part for which $x \geq -1$.

   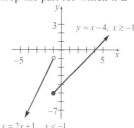

   b. To find the intercepts, notice that the only piece that hits either axis is $y = x - 4$.

$y = x - 4$	$y = x - 4$
$y = 0 - 4$	$0 = x - 4$
$y = -4$	$4 = x$

   The intercepts are $(0, -4)$ and $(4, 0)$.

   c. To find $g(-5)$ we first note that $x = -5$ so we must use the first "piece" because $-5 < -1$.

   $g(-5) = 2(-5) + 1 = -10 + 1 = -9$

d. To find $g(2)$ we first note that $x = 2$ so we must use the second "piece" because $2 \geq -1$.
$g(2) = 2 - 4 = -2$

9. The average rate of change from 3 to x is given by
$$\frac{\Delta y}{\Delta x} = \frac{f(x) - f(3)}{x - 3} \quad x \neq 3$$
$$= \frac{(3x^2 - 2x + 4) - (3(3)^2 - 2(3) + 4)}{x - 3}$$
$$= \frac{3x^2 - 2x + 4 - 25}{x - 3}$$
$$= \frac{3x^2 - 2x - 21}{x - 3}$$
$$= \frac{(x - 3)(3x + 7)}{x - 3}$$
$$= 3x + 7 \quad x \neq 3$$

10. a. $f - g = (2x^2 + 1) - (3x - 2)$
$= 2x^2 + 1 - 3x + 2$
$= 2x^2 - 3x + 3$

b. $f \cdot g = (2x^2 + 1)(3x - 2)$
$= 6x^3 - 4x^2 + 3x - 2$

c. $f(x + h) - f(x)$
$= (2(x + h)^2 + 1) - (2x^2 + 1)$
$= (2(x^2 + 2xh + h^2) + 1) - (2x^2 + 1)$
$= 2x^2 + 4xh + 2h^2 + 1 - 2x^2 - 1$
$= 4xh + 2h^2$

11. a. The basic function is $y = x^3$ so we start with the graph of this function.

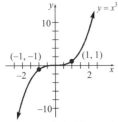

Next we shift this graph 1 unit to the left to obtain the graph of $y = (x + 1)^3$.

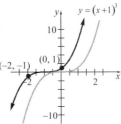

Next we reflect this graph about the x-axis to obtain the graph of $y = -(x + 1)^3$.

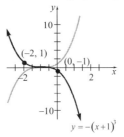

Next we stretch this graph vertically by a factor of 2 to obtain the graph of $y = -2(x + 1)^3$.

The last step is to shift this graph up 3 units to obtain the graph of $y = -2(x + 1)^3 + 3$.

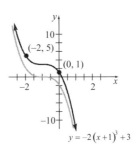

b. The basic function is $y = |x|$ so we start with the graph of this function.

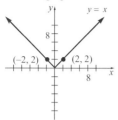

Next we shift this graph 4 units to the left to obtain the graph of $y = |x + 4|$.

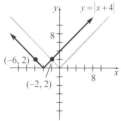

Next we shift this graph up 2 units to obtain the graph of $y = |x + 4| + 2$.

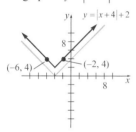

12. a. Graph of Set A:
Enter the x-values in L1 and the y-values in L2. Make a scatter diagram using STATPLOT and press [ZOOM] [9] to fit the data to your window.

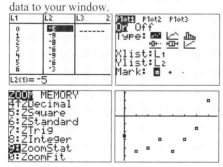

Graph of Set B:
Enter the x-values in L1 and the y-values in L2. Make a scatter diagram using STATPLOT and press [ZOOM] [9] to fit the data to your window.

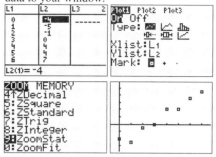

From the graphs, it appears that Set B is more linear. Set A has too much curvature.

b. Set B appeared to be the most linear so we will use that data set.
Press [STAT] [▷] [4] [ENTER] to get the equation of the line (assuming the data is entered already).

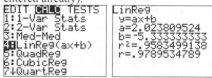

The line of best fit for Set B is roughly $y = 2.02x - 5.33$.

13. a. $r(x) = -0.115x^2 + 1.183x + 5.623$

For the years 1992 to 2004, we have values of x between 0 and 12. Therefore, we can let Xmin = 0 and Xmax = 12. Since r is the interest rate as a percent, we can try letting Ymin = 0 and Ymax = 10.

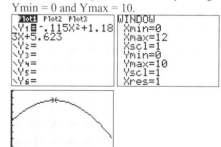

The highest rate during this period appears to be 8.67%, occurring in 1997 ($x \approx 5$).

Chapter 2: Functions and Their Graphs

b. For 2010, we have $x = 2010 - 1992 = 18$.
$$r(18) = -0.115(18)^2 + 1.183(18) + 5.623$$
$$= -10.343$$

```
Y1(18)
        -10.343
```

The model predicts that the interest rate will be -10.343%. This is not a reasonable value since it implies that the bank would be paying interest to the borrower.

14. a. Let x = width of the rink in feet. Then the length of the rectangular portion is given by $2x - 20$. The radius of the semicircular portions is half the width, or $r = \dfrac{x}{2}$.

To find the volume, we first find the area of the surface and multiply by the thickness of the ice. The two semicircles can be combined to form a complete circle, so the area is given by
$$A = l \cdot w + \pi r^2$$
$$= (2x - 20)(x) + \pi \left(\dfrac{x}{2}\right)^2$$
$$= 2x^2 - 20x + \dfrac{\pi x^2}{4}$$

We have expressed our measures in feet so we need to convert the thickness to feet as well.
$$0.75 \text{ in} \cdot \dfrac{1 \text{ ft}}{12 \text{ in}} = \dfrac{0.75}{12} \text{ ft} = \dfrac{1}{16} \text{ ft}$$

Now we multiply this by the area to obtain the volume. That is,
$$V(x) = \dfrac{1}{16}\left(2x^2 - 20x + \dfrac{\pi x^2}{4}\right)$$
$$V(x) = \dfrac{x^2}{8} - \dfrac{5x}{4} + \dfrac{\pi x^2}{64}$$

b. If the rink is 90 feet wide, then we have $x = 90$.
$$V(90) = \dfrac{90^2}{8} - \dfrac{5(90)}{4} + \dfrac{\pi(90)^2}{64} \approx 1297.61$$
The volume of ice is roughly 1297.61 ft^3.

Cumulative Review R-2

1. $-5x + 4 = 0$
$$-5x = -4$$
$$x = \dfrac{-4}{-5} = \dfrac{4}{5}$$
The solution set is $\left\{\dfrac{4}{5}\right\}$.

3. $3x^2 - 5x - 2 = 0$
$$(3x + 1)(x - 2) = 0 \Rightarrow x = -\dfrac{1}{3}, x = 2$$
The solution set is $\left\{-\dfrac{1}{3}, 2\right\}$.

5. $4x^2 - 2x + 4 = 0 \Rightarrow 2x^2 - x + 2 = 0$
$$x = \dfrac{-(-1) \pm \sqrt{(-1)^2 - 4(2)(2)}}{2(2)}$$
$$= \dfrac{1 \pm \sqrt{1 - 16}}{4} = \dfrac{1 \pm \sqrt{-15}}{4}$$
no real solution

7. $\sqrt[5]{1 - x} = 2$
$$\left(\sqrt[5]{1 - x}\right)^5 = 2^5$$
$$1 - x = 32$$
$$-x = 31$$
$$x = -31$$
The solution set is $\{-31\}$.

9. $4x^2 - 2x + 4 = 0 \Rightarrow 2x^2 - x + 2 = 0$
$$x = \dfrac{-(-1) \pm \sqrt{(-1)^2 - 4(2)(2)}}{2(2)}$$
$$= \dfrac{1 \pm \sqrt{1 - 16}}{4} = \dfrac{1 \pm \sqrt{-15}}{4} = \dfrac{1 \pm \sqrt{15}i}{4}$$
The solution set is $\left\{\dfrac{1 - \sqrt{15}i}{4}, \dfrac{1 + \sqrt{15}i}{4}\right\}$.

11. $-3x+4y=12 \Rightarrow 4y=3x+12$

$y=\dfrac{3}{4}x+3$

This is a line with slope $\dfrac{3}{4}$ and y-intercept $(0, 3)$.

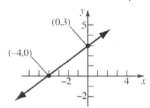

13. $x^2+y^2+2x-4y+4=0$

$x^2+2x+y^2-4y=-4$

$(x^2+2x+1)+(y^2-4y+4)=-4+1+4$

$(x+1)^2+(y-2)^2=1$

$(x+1)^2+(y-2)^2=1^2$

This is a circle with center $(-1,2)$ and radius 1.

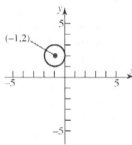

15. a. Domain: $\{x \mid -4 \leq x \leq 4\}$
   Range: $\{y \mid -1 \leq y \leq 3\}$

   b. Intercepts: $(-1,0)$, $(0,-1)$, $(1,0)$
   x-intercepts: $-1, 1$
   y-intercept: -1

   c. The graph is symmetric with respect to the y-axis.

   d. When $x=2$, the function takes on a value of 1. Therefore, $f(2)=1$.

   e. The function takes on the value 3 at $x=-4$ and $x=4$.

   f. $f(x)<0$ means that the graph lies below the x-axis. This happens for x values between -1 and 1. Thus, the solution set is $\{x \mid -1<x<1\}$.

   g. The graph of $y=f(x)+2$ is the graph of $y=f(x)$ but shifted up 2 units.

   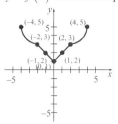

   h. The graph of $y=f(-x)$ is the graph of $y=f(x)$ but reflected about the y-axis.

   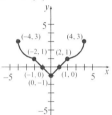

   i. The graph of $y=2f(x)$ is the graph of $y=f(x)$ but stretched vertically by a factor of 2. That is, the coordinate of each point is multiplied by 2.

   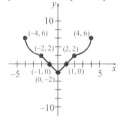

   j. Since the graph is symmetric about the y-axis, the function is even.

   k. The function is increasing on the open interval $(0,4)$.

   l. The function is decreasing on the open interval $(-4,0)$.

   m. There is a local minimum of -1 at $x=0$. There are no local maxima.

   n. $\dfrac{f(4)-f(1)}{4-1}=\dfrac{3-0}{3}=\dfrac{3}{3}=1$
   The average rate of change of the function from 1 to 4 is 1.

Chapter 2: Functions and Their Graphs

17. $y = x^3 - 3x + 1$

   (a) $(-2, -1)$
   $(-2)^3 - (3)(-2) + 1 = -8 + 6 + 1 = -1$
   $(-2, -1)$ is on the graph.

   (b) $(2, 3)$
   $(2)^3 - (3)(2) + 1 = 8 - 6 + 1 = 3$
   $(2, 3)$ is on the graph.

   (c) $(3, 1)$
   $(3)^3 - (3)(3) + 1 = 27 - 9 + 1 = 19 \neq 1$
   $(3, 1)$ is not on the graph.

19. Use ZERO (or ROOT) on the graph of $y_1 = x^4 - 3x^3 + 4x - 1$.

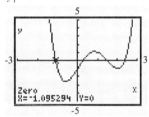

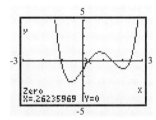

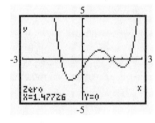

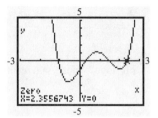

The solution set is $\{-1.10, 0.26, 1.48, 2.36\}$.

21. Yes, each x corresponds to exactly 1 y.

23. $h(z) = \dfrac{3z - 1}{z^2 - 6z - 7}$
    The denominator cannot be zero:
    $z^2 - 6z - 7 \neq 0$
    $(z + 1)(z - 7) \neq 0$
    $z \neq -1 \text{ or } 7$
    Domain: $\{z \mid z \neq -1, z \neq 7\}$

25. $f(x) = \dfrac{x}{x + 4}$

   a. $f(1) = \dfrac{1}{1 + 4} = \dfrac{1}{5} \neq \dfrac{1}{4}$
   $\left(1, \dfrac{1}{4}\right)$ is not on the graph of f

   b. $f(-2) = \dfrac{-2}{-2 + 4} = \dfrac{-2}{2} = -1$
   $(-2, -1)$ is on the graph of f

   c. Solve for x:
   $\dfrac{x}{x + 4} = 2$
   $x = 2(x + 4)$
   $x = 2x + 8$
   $-8 = x$
   $(-8, 2)$ is on the graph of f.

Chapter 3
Polynomial and Rational Functions

Section 3.1

1. $(0, -9), (-3, 0), (3, 0)$

3. $\dfrac{25}{4}$

5. parabola

7. $-\dfrac{b}{2a}$

9. True

11. C

13. F

15. G

17. H

19. $f(x) = \dfrac{1}{4}x^2$

   Using the graph of $y = x^2$, compress vertically by a factor of $\dfrac{1}{4}$.

   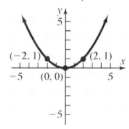

21. $f(x) = \dfrac{1}{4}x^2 - 2$

   Using the graph of $y = x^2$, compress vertically by a factor of 2, then shift down 2 units.

   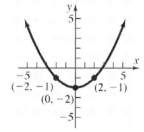

23. $f(x) = \dfrac{1}{4}x^2 + 2$

   Using the graph of $y = x^2$, compress vertically by a factor of $\dfrac{1}{4}$, then shift up 2 units.

   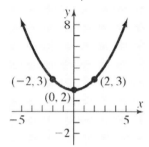

25. $f(x) = \dfrac{1}{4}x^2 + 1$

   Using the graph of $y = x^2$, compress vertically by a factor of $\dfrac{1}{4}$, then shift up 1 unit.

   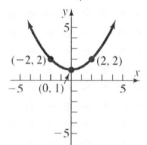

27. $f(x) = x^2 + 4x + 2$
   $= (x^2 + 4x + 4) + 2 - 4$
   $= (x+2)^2 - 2$

   Using the graph of $y = x^2$, shift left 2 units, then shift down 2 units.

   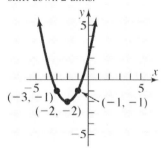

29. $f(x) = 2x^2 - 4x + 1$
$= 2(x^2 - 2x) + 1$
$= 2(x^2 - 2x + 1) + 1 - 2$
$= 2(x-1)^2 - 1$

Using the graph of $y = x^2$, shift right 1 unit, stretch vertically by a factor of 2, then shift down 1 unit.

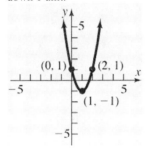

31. $f(x) = -x^2 - 2x$
$= -(x^2 + 2x)$
$= -(x^2 + 2x + 1) + 1$
$= -(x+1)^2 + 1$

Using the graph of $y = x^2$, shift left 1 unit, reflect across the x-axis, then shift up 1 unit.

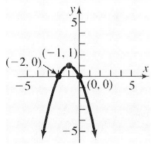

33. $f(x) = \frac{1}{2}x^2 + x - 1$
$= \frac{1}{2}(x^2 + 2x) - 1$
$= \frac{1}{2}(x^2 + 2x + 1) - 1 - \frac{1}{2}$
$= \frac{1}{2}(x+1)^2 - \frac{3}{2}$

Using the graph of $y = x^2$, shift left 1 unit, compress vertically by a factor of $\frac{1}{2}$, then shift down $\frac{3}{2}$ units.

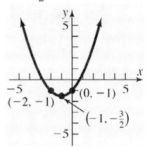

35. For $f(x) = x^2 + 2x$, $a = 1$, $b = 2$, $c = 0$.
Since $a = 1 > 0$, the graph opens up.
The x-coordinate of the vertex is
$x = \frac{-b}{2a} = \frac{-(2)}{2(1)} = \frac{-2}{2} = -1$.
The y-coordinate of the vertex is
$f\left(\frac{-b}{2a}\right) = f(-1) = (-1)^2 + 2(-1) = 1 - 2 = -1$.

Thus, the vertex is $(-1, -1)$.
The axis of symmetry is the line $x = -1$.
The discriminant is
$b^2 - 4ac = (2)^2 - 4(1)(0) = 4 > 0$, so the graph has two x-intercepts.
The x-intercepts are found by solving:
$x^2 + 2x = 0$
$x(x + 2) = 0$
$x = 0$ or $x = -2$
The x-intercepts are -2 and 0.
The y-intercept is $f(0) = 0$.

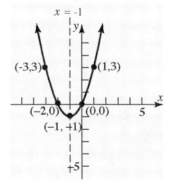

The domain is all real numbers and the range is $[-1, \infty)$; increasing on $(-1, \infty)$ and decreasing on $(-\infty, -1)$.

37. For $f(x)=-x^2-6x$, $a=-1$, $b=-6$, $c=0$.
Since $a=-1<0$, the graph opens down.
The x-coordinate of the vertex is
$$x=\frac{-b}{2a}=\frac{-(-6)}{2(-1)}=\frac{6}{-2}=-3.$$
The y-coordinate of the vertex is
$$f\left(\frac{-b}{2a}\right)=f(-3)=-(-3)^2-6(-3)$$
$$=-9+18=9.$$
Thus, the vertex is $(-3,9)$.
The axis of symmetry is the line $x=-3$.
The discriminant is:
$$b^2-4ac=(-6)^2-4(-1)(0)=36>0,$$
so the graph has two x-intercepts.
The x-intercepts are found by solving:
$$-x^2-6x=0$$
$$-x(x+6)=0$$
$$x=0 \text{ or } x=-6.$$
The x-intercepts are -6 and 0.
The y-intercepts are $f(0)=0$.

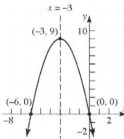

The domain is all real numbers and the range is $(-\infty, 9]$; increasing on $(-\infty, -3)$ and decreasing on $(-3, \infty)$.

39. For $f(x)=2x^2-8x$, $a=2$, $b=-8$, $c=0$.
Since $a=2>0$, the graph opens up.
The x-coordinate of the vertex is
$$x=\frac{-b}{2a}=\frac{-(-8)}{2(2)}=\frac{8}{4}=2.$$
The y-coordinate of the vertex is
$$f\left(\frac{-b}{2a}\right)=f(2)$$
$$=2(2)^2-8(2)=8-16=-8.$$
Thus, the vertex is $(2,-8)$.
The axis of symmetry is the line $x=2$.
The discriminant is:
$$b^2-4ac=(-8)^2-4(2)(0)=64>0,$$

so the graph has two x-intercepts.
The x-intercepts are found by solving:
$$2x^2-8x=0$$
$$2x(x-4)=0$$
$$x=0 \text{ or } x=4.$$
The x-intercepts are 0 and 4.
The y-intercepts is $f(0)=0$.

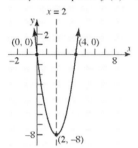

The domain is all real numbers and the range is $[-8, \infty)$; increasing on $(2, \infty)$ and decreasing on $(-\infty, 2)$.

41. For $f(x)=x^2+2x-8$, $a=1$, $b=2$, $c=-8$.
Since $a=1>0$, the graph opens up.
The x-coordinate of the vertex is
$$x=\frac{-b}{2a}=\frac{-2}{2(1)}=\frac{-2}{2}=-1.$$
The y-coordinate of the vertex is
$$f\left(\frac{-b}{2a}\right)=f(-1)$$
$$=(-1)^2+2(-1)-8=1-2-8=-9.$$
Thus, the vertex is $(-1,-9)$.
The axis of symmetry is the line $x=-1$.
The discriminant is:
$$b^2-4ac=2^2-4(1)(-8)=4+32=36>0,$$
so the graph has two x-intercepts.
The x-intercepts are found by solving:
$$x^2+2x-8=0$$
$$(x+4)(x-2)=0$$
$$x=-4 \text{ or } x=2.$$
The x-intercepts are -4 and 2.
The y-intercept is $f(0)=-8$.

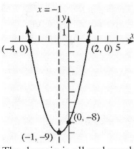

The domain is all real numbers and the range is $[-9, \infty)$; increasing on $(-1, \infty)$ and decreasing on $(-\infty, -1)$.

43. For $f(x) = x^2 + 2x + 1$, $a = 1$, $b = 2$, $c = 1$.
Since $a = 1 > 0$, the graph opens up.
The x-coordinate of the vertex is
$$x = \frac{-b}{2a} = \frac{-2}{2(1)} = \frac{-2}{2} = -1.$$
The y-coordinate of the vertex is
$$f\left(\frac{-b}{2a}\right) = f(-1)$$
$$= (-1)^2 + 2(-1) + 1$$
$$= 1 - 2 + 1$$
$$= 0.$$
Thus, the vertex is $(-1, 0)$.
The axis of symmetry is the line $x = -1$.
The discriminant is:
$$b^2 - 4ac = 2^2 - 4(1)(1) = 4 - 4 = 0,$$
so the graph has one x-intercept.
The x-intercept is found by solving:
$$x^2 + 2x + 1 = 0$$
$$(x+1)^2 = 0$$
$$x = -1.$$
The x-intercept is -1.
The y-intercept is $f(0) = 1$.

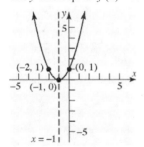

The domain is all real numbers and the range is $[0, \infty)$; increasing on $(-1, \infty)$ and decreasing on $(-\infty, -1)$.

45. For $f(x) = 2x^2 - x + 2$, $a = 2$, $b = -1$, $c = 2$.
Since $a = 2 > 0$, the graph opens up.
The x-coordinate of the vertex is
$$x = \frac{-b}{2a} = \frac{-(-1)}{2(2)} = \frac{1}{4}.$$
The y-coordinate of the vertex is
$$f\left(\frac{-b}{2a}\right) = f\left(\frac{1}{4}\right) = 2\left(\frac{1}{4}\right)^2 - \frac{1}{4} + 2$$
$$= \frac{1}{8} - \frac{1}{4} + 2 = \frac{15}{8}.$$
Thus, the vertex is $\left(\frac{1}{4}, \frac{15}{8}\right)$.
The axis of symmetry is the line $x = \frac{1}{4}$.
The discriminant is:
$$b^2 - 4ac = (-1)^2 - 4(2)(2) = 1 - 16 = -15,$$
so the graph has no x-intercepts.
The y-intercept is $f(0) = 2$.

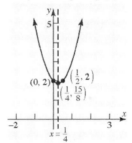

The domain is all real numbers and the range is $\left[\frac{15}{8}, \infty\right)$; increasing on $\left(\frac{1}{4}, \infty\right)$ and decreasing on $\left(-\infty, \frac{1}{4}\right)$.

47. For $f(x) = -2x^2 + 2x - 3$, $a = -2$, $b = 2$, $c = -3$. Since $a = -2 < 0$, the graph opens down.

The x-coordinate of the vertex is
$$x = \frac{-b}{2a} = \frac{-(2)}{2(-2)} = \frac{-2}{-4} = \frac{1}{2}.$$

The y-coordinate of the vertex is
$$f\left(\frac{-b}{2a}\right) = f\left(\frac{1}{2}\right) = -2\left(\frac{1}{2}\right)^2 + 2\left(\frac{1}{2}\right) - 3$$
$$= -\frac{1}{2} + 1 - 3 = -\frac{5}{2}.$$

Thus, the vertex is $\left(\frac{1}{2}, -\frac{5}{2}\right)$.

The axis of symmetry is the line $x = \frac{1}{2}$.

The discriminant is:
$$b^2 - 4ac = 2^2 - 4(-2)(-3) = 4 - 24 = -20,$$
so the graph has no x-intercepts.
The y-intercept is $f(0) = -3$.

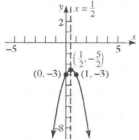

The domain is all real numbers and the range is $\left(-\infty, -\frac{5}{2}\right]$; increasing on $\left(-\infty, \frac{1}{2}\right)$ and decreasing on $\left(\frac{1}{2}, \infty\right)$.

49. For $f(x) = 3x^2 + 6x + 2$, $a = 3$, $b = 6$, $c = 2$. Since $a = 3 > 0$, the graph opens up.
The x-coordinate of the vertex is
$$x = \frac{-b}{2a} = \frac{-6}{2(3)} = \frac{-6}{6} = -1.$$

The y-coordinate of the vertex is
$$f\left(\frac{-b}{2a}\right) = f(-1) = 3(-1)^2 + 6(-1) + 2$$
$$= 3 - 6 + 2 = -1.$$
Thus, the vertex is $(-1, -1)$.
The axis of symmetry is the line $x = -1$.

The discriminant is:
$$b^2 - 4ac = 6^2 - 4(3)(2) = 36 - 24 = 12,$$
so the graph has two x-intercepts.
The x-intercepts are found by solving:
$$3x^2 + 6x + 2 = 0$$
$$x = \frac{-b \pm \sqrt{b^2 - 4ac}}{2a}$$
$$= \frac{-6 \pm \sqrt{12}}{6} = \frac{-6 \pm 2\sqrt{3}}{6}$$
$$= \frac{-3 \pm \sqrt{3}}{3}$$

The x-intercepts are $-1 - \frac{\sqrt{3}}{3}$ and $-1 + \frac{\sqrt{3}}{3}$.
The y-intercept is $f(0) = 2$.

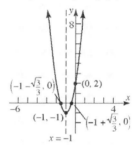

The domain is all real numbers and the range is $[-1, \infty)$; increasing on $(-1, \infty)$ and decreasing on $(-\infty, -1)$.

51. For $f(x) = -4x^2 - 6x + 2$, $a = -4$, $b = -6$, $c = 2$. Since $a = -4 < 0$, the graph opens down.
The x-coordinate of the vertex is
$$x = \frac{-b}{2a} = \frac{-(-6)}{2(-4)} = \frac{6}{-8} = -\frac{3}{4}.$$

The y-coordinate of the vertex is
$$f\left(\frac{-b}{2a}\right) = f\left(-\frac{3}{4}\right) = -4\left(-\frac{3}{4}\right)^2 - 6\left(-\frac{3}{4}\right) + 2$$
$$= -\frac{9}{4} + \frac{9}{2} + 2 = \frac{17}{4}.$$

Thus, the vertex is $\left(-\frac{3}{4}, \frac{17}{4}\right)$.

The axis of symmetry is the line $x = -\frac{3}{4}$.

The discriminant is:
$$b^2 - 4ac = (-6)^2 - 4(-4)(2) = 36 + 32 = 68,$$
so the graph has two x-intercepts.
The x-intercepts are found by solving:

$$-4x^2 - 6x + 2 = 0$$
$$x = \frac{-b \pm \sqrt{b^2 - 4ac}}{2a} = \frac{-(-6) \pm \sqrt{68}}{2(-4)}$$
$$= \frac{6 \pm \sqrt{68}}{-8} = \frac{6 \pm 2\sqrt{17}}{-8}$$
$$= \frac{3 \pm \sqrt{17}}{-4}$$

The x-intercepts are $\frac{-3 + \sqrt{17}}{4}$ and $\frac{-3 - \sqrt{17}}{4}$.

The y-intercept is $f(0) = 2$.

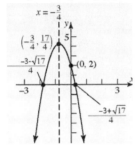

The domain is all real numbers and the range is $\left(-\infty, \frac{17}{4}\right]$; increasing on $\left(-\infty, -\frac{3}{4}\right)$ and decreasing on $\left(-\frac{3}{4}, \infty\right)$.

53. Since it is given that the graph passes through $(0, -1)$, then $f(0) = -1$.
So $f(x) = ax^2 + bx + c$
$$-1 = a(0)^2 + b(0) + c$$
$$-1 = 0 + 0 + c$$
$$c = -1.$$
Since it is given that the vertex is at $(-1, -2)$, then $f(-1) = -2$ and $\frac{-b}{2a} = -1 \Rightarrow b = 2a$.
So $f(x) = ax^2 + bx - 1$
$$-2 = a(-1)^2 + b(-1) - 1$$
$$-2 = a - b - 1$$
$$-1 = a - b$$
$$-1 = a - (2a)$$
$$-1 = -a$$
$$a = 1.$$
So, $b = 2a = 2(1) = 2$.
Therefore, $f(x) = ax^2 + bx + c = x^2 + 2x - 1$.

55. Since it is given that the graph passes through $(0, -4)$, then $f(0) = -4$.
So $f(x) = ax^2 + bx + c$
$$-4 = a(0)^2 + b(0) + c$$
$$-4 = 0 + 0 + c$$
$$c = -4.$$
Since it is given that the vertex is at $(-3, 5)$, then $f(-3) = 5$ and $\frac{-b}{2a} = -3 \Rightarrow b = 6a$.
So $f(x) = ax^2 + bx - 4$
$$5 = a(-3)^2 + b(-3) - 4$$
$$5 = 9a - 3b - 4$$
$$9 = 9a - 3b$$
$$9 = 9a - 3(6a)$$
$$9 = 9a - 18a$$
$$9 = -9a$$
$$a = -1.$$
So, $b = 6a = 6(-1) = -6$.
Therefore, $f(x) = ax^2 + bx + c = -x^2 - 6x - 4$.

57. Since it is given that the graph passes through $(3, 5)$, then $f(3) = 5$.
So $f(x) = ax^2 + bx + c$
$$5 = a(3)^2 + b(3) + c$$
$$5 = 9a + 3b + c$$
Since it is given that the vertex is at $(1, -3)$, then $f(1) = -3$ and $\frac{-b}{2a} = 1 \Rightarrow b = -2a$.
So $f(x) = ax^2 + bx + c$
$$-3 = a(1)^2 + b(1) + c$$
$$-3 = a + b + c$$
Since $5 = 9a + 3b + c$ and $-3 = a + b + c$, then
$$8 = 8a + 2b$$
$$8 = 8a + 2(-2a)$$
$$8 = 8a - 4a$$
$$8 = 4a$$
$$a = 2.$$
Then $b = -2a = -2(2) = -4$ and
$$-3 = a + b + c$$
$$-3 = 2 + (-4) + c$$
$$-3 = -2 + c$$
$$c = -1.$$
Therefore, $f(x) = ax^2 + bx + c = 2x^2 - 4x - 1$.

59. For $f(x) = 2x^2 + 12x$, $a = 2$, $b = 12$, $c = 0$. Since $a = 2 > 0$, the graph opens up, so the vertex is a minimum point. The minimum occurs at $x = \dfrac{-b}{2a} = \dfrac{-12}{2(2)} = \dfrac{-12}{4} = -3$.
The minimum value is
$$f\left(\dfrac{-b}{2a}\right) = f(-3) = 2(-3)^2 + 12(-3)$$
$$= 18 - 36$$
$$= -18$$

61. For $f(x) = 2x^2 + 12x - 3$, $a = 2$, $b = 12$, $c = -3$. Since $a = 2 > 0$, the graph opens up, so the vertex is a minimum point. The minimum occurs at $x = \dfrac{-b}{2a} = \dfrac{-12}{2(2)} = \dfrac{-12}{4} = -3$. The minimum value is
$$f\left(\dfrac{-b}{2a}\right) = f(-3) = 2(-3)^2 + 12(-3) - 3$$
$$= 18 - 36 - 3 = -21.$$

63. For $f(x) = -x^2 + 10x - 4$, $a = -1$, $b = 10$, $c = -4$. Since $a = -1 < 0$, the graph opens down, so the vertex is a maximum point. The maximum occurs at $x = \dfrac{-b}{2a} = \dfrac{-10}{2(-1)} = \dfrac{-10}{-2} = 5$.
The maximum value is
$$f\left(\dfrac{-b}{2a}\right) = f(5) = -(5)^2 + 10(5) - 4.$$
$$= -25 + 50 - 4$$
$$= 21.$$

65. For $f(x) = -3x^2 + 12x + 1$, $a = -3$, $b = 12$, $c = 1$. Since $a = -3 < 0$, the graph opens down, so the vertex is a maximum point. The maximum occurs at $x = \dfrac{-b}{2a} = \dfrac{-12}{2(-3)} = \dfrac{-12}{-6} = 2$.
The maximum value
is $f\left(\dfrac{-b}{2a}\right) = f(2) = -3(2)^2 + 12(2) + 1$.
$$= -12 + 24 + 1$$
$$= 13.$$

67. Use the form $f(x) = a(x-h)^2 + k$.
The vertex is $(0, 2)$, so $h = 0$ and $k = 2$.
$f(x) = a(x - 0)^2 + 2 = ax^2 + 2$.

Since the graph passes through $(1, 8)$, $f(1) = 8$.
$$f(x) = ax^2 + 2$$
$$8 = a(1)^2 + 2$$
$$8 = a + 2$$
$$6 = a$$
$f(x) = 6x^2 + 2$.
$a = 6, b = 0, c = 2$

69. a. For $a = 1$:
$$f(x) = a(x - r_1)(x - r_2)$$
$$= 1(x - (-3))(x - 1)$$
$$= (x + 3)(x - 1)$$
$$= x^2 + 2x - 3$$
For $a = 2$:
$$f(x) = 2(x - (-3))(x - 1)$$
$$= 2(x + 3)(x - 1)$$
$$= 2(x^2 + 2x - 3)$$
$$= 2x^2 + 4x - 6$$
For $a = -2$:
$$f(x) = -2(x - (-3))(x - 1)$$
$$= -2(x + 3)(x - 1)$$
$$= -2(x^2 + 2x - 3)$$
$$= -2x^2 - 4x + 6$$
For $a = 5$:
$$f(x) = 5(x - (-3))(x - 1)$$
$$= 5(x + 3)(x - 1)$$
$$= 5(x^2 + 2x - 3)$$
$$= 5x^2 + 10x - 15$$

b. The value of a multiplies the value of the y-intercept by the value of a. The values of the x-intercepts are not changed.

c. The axis of symmetry is unaffected by the value of a.

d. The y-coordinate of the vertex is multiplied by the value of a.

e. The x-coordinate of the vertex is the midpoint of the x-intercepts.

71. $R(p) = -4p^2 + 4000p$, $a = -4$, $b = 4000$, $c = 0$.
Since $a = -4 < 0$, the graph is a parabola that opens down, so the vertex is a maximum point. The maximum occurs at
$$p = \frac{-b}{2a} = \frac{-4000}{2(-4)} = 500.$$
Thus, the unit price should be $500 for maximum revenue. The maximum revenue is
$$R(500) = -4(500)^2 + 4000(500)$$
$$= -1000000 + 2000000$$
$$= \$1,000,000.$$

73. a. $R(x) = x\left(-\frac{1}{6}x + 100\right) = -\frac{1}{6}x^2 + 100x$

b. $R(200) = -\frac{1}{6}(200)^2 + 100(200)$
$$= \frac{-20000}{3} + 20000$$
$$= \frac{40000}{3} \approx \$13,333.33$$

c. $x = \frac{-b}{2a} = \frac{-100}{2\left(-\frac{1}{6}\right)} = \frac{-100}{\left(-\frac{1}{3}\right)} = \frac{300}{1} = 300$

The maximum revenue is
$$R(300) = -\frac{1}{6}(300)^2 + 100(300)$$
$$= -15000 + 30000$$
$$= \$15,000$$

d. $p = -\frac{1}{6}(300) + 100 = -50 + 100 = \$50$

75. a. If $x = -5p + 100$, then $p = \frac{100 - x}{5}$.
$$R(x) = x\left(\frac{100 - x}{5}\right) = -\frac{1}{5}x^2 + 20x$$

b. $R(15) = -\frac{1}{5}(15)^2 + 20(15)$
$$= -45 + 300$$
$$= \$255$$

c. $x = \frac{-b}{2a} = \frac{-20}{2\left(-\frac{1}{5}\right)} = \frac{-20}{\left(-\frac{2}{5}\right)} = \frac{100}{2} = 50$

The maximum revenue is
$$R(50) = -\frac{1}{5}(50)^2 + 20(50)$$
$$= -500 + 1000$$
$$= \$500$$

d. $p = \frac{100 - 50}{5} = \frac{50}{5} = \$10$

77. a. Let $x =$ width and $y =$ length of the rectangular area.
Solving $P = 2x + 2y = 400$ for y:
$$y = \frac{400 - 2x}{2} = 200 - x.$$
Then $A(x) = (200 - x)x = 200x - x^2$
$$= -x^2 + 200x.$$

b. $x = \frac{-b}{2a} = \frac{-200}{2(-1)} = \frac{-200}{-2} = 100$ yards

c. $A(100) = -100^2 + 200(100)$
$= -10000 + 20000$
$= 10,000$ sq yds.

79. Let $x =$ width and $y =$ length of the rectangular area. Solving $P = 2x + y = 4000$ for y:
$y = 4000 - 2x$. Then
$A(x) = (4000 - 2x)x = 4000x - 2x^2 = -2x^2 + 4000x$
$x = \frac{-b}{2a} = \frac{-4000}{2(-2)} = \frac{-4000}{-4} = 1000$ meters
maximizes area.
$A(1000) = -2(1000)^2 + 4000(1000)$
$= -2000000 + 4000000$
$= 2,000,000$
The largest area that can be enclosed is 2,000,000 square meters.

81. a. $a = -\frac{32}{2500}, b = 1, c = 200$. The maximum height occurs when
$x = \frac{-b}{2a} = \frac{-1}{2\left(-\frac{32}{2500}\right)} = \frac{2500}{64} = 39.0625$ ft.
from base of the cliff.

b. The maximum height is
$h(39.0625)$
$= \frac{-32(39.0625)^2}{2500} + 39.0625 + 200$
≈ 219.53 ft.

c. Solving when $h(x) = 0$:
$-\frac{32}{2500}x^2 + x + 200 = 0$

$x = \frac{-1 \pm \sqrt{1^2 - 4\left(-\frac{32}{2500}\right)(200)}}{2\left(-\frac{32}{2500}\right)} = \frac{-1 \pm \sqrt{11.24}}{-0.0256}$

So $x \approx -91.90$ or $x \approx 170.02$.
Since the distance cannot be negative, the projectile strikes the water 170.02 feet from the base of the cliff.

d. Graphing:

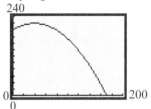

e. Solving when $h(x) = 100$:
Since $h(x) = -\frac{32}{2500}x^2 + x + 200 = 100$,
$-\frac{32}{2500}x^2 + x + 100 = 0$.

$x = \frac{-1 \pm \sqrt{1^2 - 4\left(-\frac{32}{2500}\right)(100)}}{2\left(-\frac{32}{2500}\right)} = \frac{-1 \pm \sqrt{6.12}}{-0.0256}$

$x \approx -57.57$ or $x \approx 135.70$
Since the distance cannot be negative, the projectile is 100 feet above the water 135.70 feet from the base of the cliff.

83. Locate the origin at the point where the cable touches the road. Then the equation of the parabola is of the form: $y = ax^2$, where $a > 0$.
Since the point (200, 75) is on the parabola, we can find the constant a:
Since $75 = a(200)^2$, then
$a = \frac{75}{200^2} = 0.001875$.
When $x = 100$, we have:
$y = 0.001875(100)^2 = 18.75$ meters.

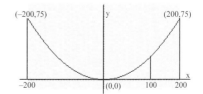

85. Let $x =$ the depth of the gutter and $y =$ the width of the gutter.
Then $A = xy$ is the cross-sectional area of the gutter. Since the aluminum sheets for the gutter are 12 inches wide, we have $2x + y = 12$.
Solving for y: $y = 12 - 2x$. The area is to be maximized,
so: $A = xy = x(12 - 2x) = -2x^2 + 12x$. This equation is a parabola opening down; thus, it has a maximum when $x = \dfrac{-b}{2a} = \dfrac{-12}{2(-2)} = \dfrac{-12}{-4} = 3$.

Thus, a depth of 3 inches produces a maximum cross-sectional area.

87. Let $x =$ the width of the rectangle or the diameter of the semicircle and let $y =$ the length of the rectangle.

The perimeter of each semicircle is $\dfrac{\pi x}{2}$.

The perimeter of the track is given by: $\dfrac{\pi x}{2} + \dfrac{\pi x}{2} + y + y = 1500$.

Solving for x:
$$\pi x + 2y = 1500$$
$$\pi x = 1500 - 2y$$
$$x = \dfrac{1500 - 2y}{\pi}$$

The area of the rectangle is:
$$A = xy = \left(\dfrac{1500 - 2y}{\pi}\right) y = \dfrac{-2}{\pi} y^2 + \dfrac{1500}{\pi} y.$$

This equation is a parabola opening down; thus, it has a maximum when
$$y = \dfrac{-b}{2a} = \dfrac{-\dfrac{1500}{\pi}}{2\left(\dfrac{-2}{\pi}\right)} = \dfrac{-1500}{-4} = 375..$$

Thus, $x = \dfrac{1500 - 2(375)}{\pi} = \dfrac{750}{\pi} \approx 238.73$

The dimensions for the rectangle with maximum area are $\dfrac{750}{\pi} \approx 238.73$ meters by 375 meters.

89. We are given: $V(x) = kx(a - x) = -kx^2 + akx$.
The reaction rate is a maximum when:
$$x = \dfrac{-b}{2a} = \dfrac{-ak}{2(-k)} = \dfrac{ak}{2k} = \dfrac{a}{2}.$$

91. $f(x) = -5x^2 + 8,\ h = 1$
$$\text{Area} = \dfrac{h}{3}\left(2ah^2 + 6c\right) = \dfrac{1}{3}\left(2(-5)(1)^2 + 6(8)\right)$$
$$= \dfrac{1}{3}(-10 + 48) = \dfrac{38}{3} \approx 12.67 \text{ sq. units}$$

93. $f(x) = x^2 + 3x + 5,\ h = 4$
$$\text{Area} = \dfrac{h}{3}\left(2ah^2 + 6c\right) = \dfrac{4}{3}\left(2(1)(4)^2 + 6(5)\right)$$
$$= \dfrac{4}{3}(32 + 30) = \dfrac{248}{3} \approx 82.67 \text{ sq. units.}$$

95. The area function is:
$$A(x) = x(10 - x) = -x^2 + 10x.$$
The maximum value occurs at the vertex:
$$x = \dfrac{-b}{2a} = \dfrac{-10}{2(-1)} = \dfrac{-10}{-2} = 5$$
The maximum area is:
$$A(5) = -(5)^2 + 10(5) = -25 + 50 = 25 \text{ sq. units.}$$

97. a. $a = -1.01,\ b = 114.3,\ c = 451.0$ The maximum number of hunters occurs when the income level is
$$x = -\dfrac{b}{2a} = -\dfrac{114.3}{2(-1.01)} = -\dfrac{114.3}{-2.02} \approx 56.584158$$
thousand dollars = \$56,584.16
The number of hunters earning this amount is:
$H(56.584158) =$
$-1.01(56.584158)^2 + 114.3(56.584158) + 451.0$
≈ 3685 hunters.

b. Graphing:

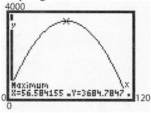

The function H is increasing on $(0, 56)$,
Therefore the number of hunters is increasing for individuals earning between \$20,000 and \$40,000.

99. a. $M(23) = .76(23)^2 - 107.00(23) + 3854.18$
   $= 1795.22$ victims.

   b. Solve for x:
   $M(x) = 0.76x^2 - 107.00x + 3854.18 = 1456$
   $0.46x^2 - 107.00x + 3854.18 = 1456$
   $0.76x^2 - 107.00x + 2398.18 = 0$
   $a = 0.76, b = -107.00, c = 2398.18$

   $x = \dfrac{-b \pm \sqrt{b^2 - 4ac}}{2a}$

   $= \dfrac{-(-107) \pm \sqrt{(-107)^2 - 4(0.76)(2398.18)}}{2(0.76)}$

   $= \dfrac{107 \pm \sqrt{4158.5328}}{1.52}$

   $\approx \dfrac{107 \pm 64.49}{1.52} \approx 112.82$ or 27.97

   Since the model is valid on the interval $20 \le x < 90$, the only solution is $x \approx 27.97$ years of age.

   c. Graphing:
   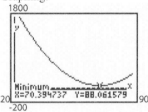

   d. As age increases between the ages of 20 and 70.39, the number of murder victims decreases. After age 70.39, the number of murder victims increases as age increases

101. a. Graphing: The data appear to be quadratic with $a < 0$.

   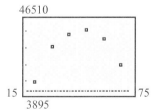

   b. Using the QUADratic REGression program

   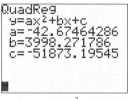

   $I(x) = -42.67x^2 + 3998.27x - 51873.2$

   c. $x = \dfrac{-b}{2a} = \dfrac{-3998.27}{2(-42.67)} \approx 46.85$

   An individual will earn the most income at an age of 46.85 years.

   d. The maximum income will be:
   $I(46.85)$
   $= -42.67(46.85)^2 + 3998.27(46.85) - 51873.2$
   $\approx \$41,788.41$

   e.
   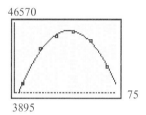

103. a. Graphing: The data appears to be quadratic with $a < 0$.

   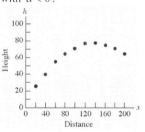

   b. Using the QUADratic REGression program, the quadratic function of best fit is:
   $h(x) = -0.0037x^2 + 1.0318x + 5.6667$

   c. $x = \dfrac{-b}{2a} = \dfrac{-1.0318}{2(-0.0037)} \approx 139.4$ feet

   The ball travels about 139.4 feet before reaching its maximum height.

Chapter 3: Polynomial and Rational Functions

d. The maximum height will be: (using the equation to many more decimal places)
$h(139) = -0.0037(139)^2 + 1.0318(139) + 5.6667$
≈ 77.6 feet

e. Graphing the quadratic function of best fit:

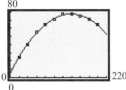

105. Answers will vary.

107. $y = x^2 - 4x + 1$; $y = x^2 + 1$; $y = x^2 + 4x + 1$

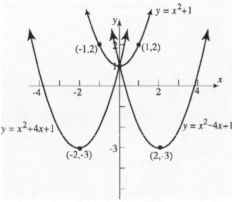

Each member of this family will be a parabola with the following characteristics:
(i) opens upwards since $a > 0$
(ii) y-intercept occurs at $(0, 1)$

109. By completing the square on the quadratic function $f(x) = ax^2 + bx + c$ we obtain the equation $y = a\left(x + \dfrac{b}{2a}\right)^2 + c - \dfrac{b^2}{4a}$. We can then draw the graph by applying transformations to the graph of the basic parabola $y = x^2$, which opens up. When $a > 0$, the basic parabola will either be stretched or compressed vertically. When $a < 0$, the basic parabola will either be stretched or compressed vertically as well as reflected across the x-axis. Therefore, when $a > 0$, the graph of $f(x) = ax^2 + bx + c$ will open up, and when $a < 0$, the graph of $f(x) = ax^2 + bx + c$ will open down.

Section 3.2

1. $(-2, 0), (2, 0), (0, 9)$

3. down; 4

5. smooth; continuous

7. touches

9. False

11. $f(x) = 4x + x^3$ is a polynomial function of degree 3.

13. $g(x) = \dfrac{1-x^2}{2} = \dfrac{1}{2} - \dfrac{1}{2}x^2$ is a polynomial function of degree 2.

15. $f(x) = 1 - \dfrac{1}{x} = 1 - x^{-1}$ is not a polynomial function because it contains a negative exponent.

17. $g(x) = x^{3/2} - x^2 + 2$ is not a polynomial function because it contains a fractional exponent.

19. $F(x) = 5x^4 - \pi x^3 + \dfrac{1}{2}$ is a polynomial function of degree 4.

21. $G(x) = 2(x-1)^2(x^2+1) = 2(x^2 - 2x + 1)(x^2 + 1)$
$= 2(x^4 + x^2 - 2x^3 - 2x + x^2 + 1)$
$= 2(x^4 - 2x^3 + 2x^2 - 2x + 1)$
$= 2x^4 - 4x^3 + 4x^2 - 4x + 2$
is a polynomial function of degree 4.

23. $f(x) = (x+1)^4$

Using the graph of $y = x^4$, shift the graph horizontally, 1 unit to the left.

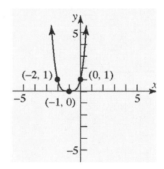

176

25. $f(x) = x^5 - 3$

Using the graph of $y = x^5$, shift the graph vertically, 3 units down.

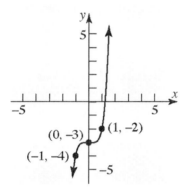

27. $f(x) = \frac{1}{2}x^4$

Using the graph of $y = x^4$, compress the graph vertically by a factor of $\frac{1}{2}$.

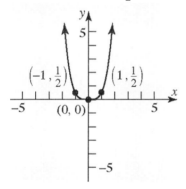

29. $f(x) = -x^5$

Using the graph of $y = x^5$, reflect the graph about the x-axis.

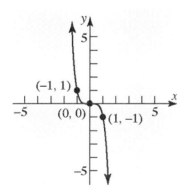

31. $f(x) = (x-1)^5 + 2$

Using the graph of $y = x^5$, shift the graph horizontally, 1 unit to the right, and shift vertically 2 units up.

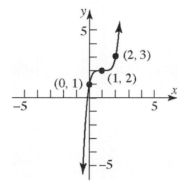

33. $f(x) = 2(x+1)^4 + 1$

Using the graph of $y = x^4$, shift the graph horizontally, 1 unit to the left, stretch vertically by a factor of 2, and shift vertically 1 unit up.

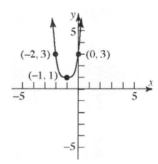

35. $f(x) = 4 - (x-2)^5 = -(x-2)^5 + 4$

Using the graph of $y = x^5$, shift the graph horizontally, 2 units to the right, reflect about the x-axis, and shift vertically 4 units up.

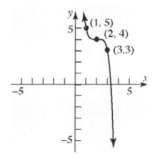

37. $f(x) = a(x-(-1))(x-1)(x-3)$
   For $a = 1$:
   $f(x) = (x+1)(x-1)(x-3) = (x^2-1)(x-3)$
   $= x^3 - 3x^2 - x + 3$

39. $f(x) = a(x-(-3))(x-0)(x-4)$
   For $a = 1$:
   $f(x) = (x+3)(x)(x-4) = (x^2+3x)(x-4)$
   $= x^3 - 4x^2 + 3x^2 - 12x$
   $= x^3 - x^2 - 12x$

41. $f(x) = a(x-(-4))(x-(-1))(x-2)(x-3)$
   For $a = 1$:
   $f(x) = (x+4)(x+1)(x-2)(x-3)$
   $= (x^2 + 5x + 4)(x^2 - 5x + 6)$
   $= x^4 - 5x^3 + 6x^2 + 5x^3 - 25x^2 + 30x + 4x^2 - 20x + 24$
   $= x^4 - 15x^2 + 10x + 24$

43. $f(x) = a(x-(-1))(x-3)^2$
   For $a = 1$:
   $f(x) = (x+1)(x-3)^2$
   $= (x+1)(x^2 - 6x + 9)$
   $= x^3 - 6x^2 + 9x + x^2 - 6x + 9$
   $= x^3 - 5x^2 + 3x + 9$

45. a. The real zeros of $f(x) = 3(x-7)(x+3)^2$ are: 7, with multiplicity one; and –3, with multiplicity two.
   b. The graph crosses the x-axis at 7 and touches it at –3.
   c. The function resembles $y = 3x^3$ for large values of $|x|$.

47. a. The real zeros of $f(x) = 4(x^2+1)(x-2)^3$ is: 2, with multiplicity three. $x^2 + 1 = 0$ has no real solution.
   b. The graph crosses the x-axis at 2.
   c. The function resembles $y = 4x^5$ for large values of $|x|$.

49. a. The real zero of $f(x) = -2\left(x+\frac{1}{2}\right)^2(x^2+4)^2$ is: $-\frac{1}{2}$, with multiplicity two. $x^2 + 4 = 0$ has no real solution.
   b. The graph touches the x-axis at $-\frac{1}{2}$.
   c. The function resembles $y = -2x^6$ for large values of $|x|$.

51. a. The real zeros of $f(x) = (x-5)^3(x+4)^2$ are: 5, with multiplicity three; and –4, with multiplicity two.
   b. The graph crosses the x-axis at 5 and touches it at –4.
   c. The function resembles $y = x^5$ for large values of $|x|$.

53. a. $f(x) = 3(x^2+8)(x^2+9)^2$ has no real zeros. $x^2 + 8 = 0$ and $x^2 + 9 = 0$ have no real solutions.
   b. The graph neither touches nor crosses the x-axis.
   c. The function resembles $y = 3x^6$ for large values of $|x|$.

55. a. The real zeros of $f(x) = -2x^2(x^2-2)$ are: $-\sqrt{2}$ and $\sqrt{2}$ with multiplicity one; and 0, with multiplicity two.
   b. The graph touches the x-axis at 0 and crosses the x-axis at $-\sqrt{2}$ and $\sqrt{2}$.
   c. The function resembles $y = -2x^4$ for large values of $|x|$.

57. $f(x) = (x-1)^2$
   a. Degree is 2; The function resembles $y = x^2$ for large values of $|x|$.
   b. y-intercept: $f(0) = (0-1)^2 = 1$
   x-intercept: solve $f(x) = 0$
   $(x-1)^2 = 0 \Rightarrow x = 1$

c. The graph touches the *x*-axis at $x = 1$, since this zero has multiplicity 2.

d. Graphing utility:

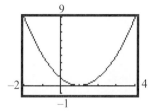

e. Local minimum at $x = 1$.

f. Graphing:

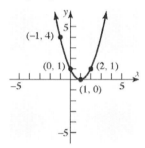

g. Domain: all real numbers
Range: $[0, \infty)$

h. Increasing on $(1, \infty)$; decreasing on $(-\infty, 1)$.

59. $f(x) = x^2(x-3)$

a. Degree is 3. The function resembles $y = x^3$ for large values of $|x|$.

b. *y*-intercept: $f(0) = 0^2(0-3) = 0$
x-intercepts: solve $f(x) = 0$
$0 = x^2(x-3)$
$x = 0, x = 3$

c. touches *x*-axis at $x = 0$; crosses *x*-axis at $x = 3$

d. Graphing utility:

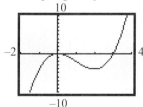

e. 2 turning points; local maximum: (0, 0); local minimum: (2, –4)

f. Graphing by hand:

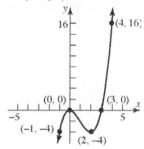

g. Domain: all real numbers
Range: all real numbers

h. Increasing on $(-\infty, 0)$ and $(2, \infty)$; decreasing on $(0, 2)$

61. $f(x) = (x+4)(x-2)^2$

a. Degree is 3. The function resembles $y = x^3$ for large values of $|x|$.

b. *y*-intercept: $f(0) = (0+4)(0-2)^2 = 16$
x-intercepts: solve $f(x) = 0$
$0 = (x+4)(x-2)^2$
$x = -4, 2$

c. crosses *x*-axis at $x = -4$; touches *x*-axis at $x = 2$

d. Graphing utility:

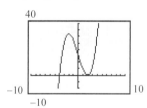

e. 2 turning points;
local maximum: $(-2, 32)$
local minimum: $(2, 0)$

179

f. Graphing by hand;

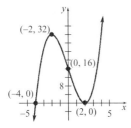

g. Domain: all real numbers
Range: all real numbers

h. Increasing on $(-\infty, -2)$ and $(2, \infty)$;
decreasing on $(-2, 2)$

63. $f(x) = -2(x+2)(x-2)^3$

a. Degree is 4. The function resembles $y = -2x^4$ for large values of $|x|$.

b. x-intercepts: $-2, 2$; y-intercept: 32

c. crosses x-axis at $x = -2$ and $x = 2$.

d. Graphing utility;

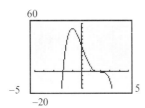

e. 1 turning point;
local maximum: $(-1, 54)$

f. Graphing by hand;

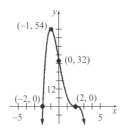

g. Domain: all real numbers
Range: all real numbers

h. Increasing on $(-\infty, -1)$;
decreasing on $(-1, \infty)$

65. $f(x) = (x+1)(x-2)(x+4)$

a. Degree is 3. The function resembles $y = x^3$ for large values of $|x|$.

b. y-intercept: $f(0) = (0+1)(0-2)(0+4)$
$= -8$
x-intercepts: solve $f(x) = 0$
$0 = (x+1)(x-2)(x+4)$
$x = -1, 2, -4$

c. crosses x-axis at $x = -1, 2, -4$

d. Graphing utility:

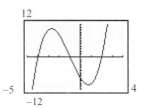

e. 2 turning points;
local maximum: $(-2.73, 10.39)$;
local minimum: $(0.73, -10.39)$

f. Graphing by hand;

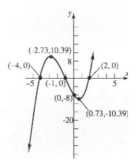

g. Domain: all real numbers
Range: all real numbers

h. Increasing on $(-\infty, -2.73)$ and $(0.73, \infty)$;
decreasing on $(-2.73, 0.73)$

67. $f(x) = 4x - x^3 = x(4-x^2) = x(2+x)(2-x)$

  a. Degree is 3. The function resembles $y = -x^3$ for large values of $|x|$.

  b. y-intercept: $f(0) = 4(0) - 0^3 = 0$
  x-intercepts: solve $f(x) = 0$
  $0 = x(2+x)(2-x)$
  $x = 0, -2, 2$

  c. crosses x-axis at $x = 0, x = -2, x = 2$

  d. Graphing utility:

  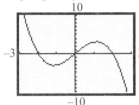

  e. 2 turning points;
  local maximum: (1.15, 3.08);
  local minimum: (−1.15, −3.08)

  f. Graphing by hand

  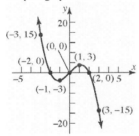

  g. Domain: all real numbers
  Range: all real numbers

  h. Increasing on $(-1.15, 1.15)$; decreasing on $(-\infty, -1.15)$ and $(1.15, \infty)$.

69. $f(x) = x^2(x-2)(x+2)$

  a. Degree is 4. The function resembles $y = x^4$ for large values of $|x|$.

  b. y-intercept: $f(0) = 0^2(0-2)(0+2) = 0$
  x-intercepts: solve $f(x) = 0$
  $0 = x^2(x-2)(x+2)$
  $x = 0, 2, -2$

  c. crosses x-axis at $x = 2, x = -2$; touches x-axis at $x = 0$

  d. Graphing utility:

  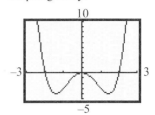

  e. 3 turning points;
  local maximum: (0, 0);
  local minima: (−1.41, −4), (1.41, −4)

  f. graphing by hand

  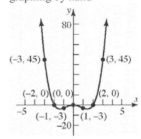

  g. Domain: all real numbers
  Range: $(-3, \infty)$

  h. Increasing on $(-1.41, 0)$ and $(1.41, \infty)$; decreasing on $(-\infty, -1.41)$ and $(0, 1.41)$.

71. $f(x) = (x+1)^2(x-2)^2$

  a. Degree is 4. The graph of the function resembles $y = x^4$ for large values of $|x|$.

  b. y-intercept: $f(0) = (0+1)^2(0-2)^2 = 4$
  x-intercepts: solve $f(x) = 0$
  $(x+1)^2(x-2)^2 = 0$
  $x = -1$ or $x = 2$

  c. The graph touches the x-axis at $x = -1$ and $x = 2$ since these zeroes have multiplicity is 2.

  d.

  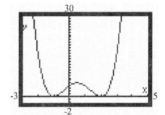

e. 3 turning points;
local maximum: (0.5, 5.06);
local minima: $(-1, 0), (2, 0)$

f. Graphing:

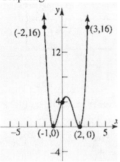

g. Domain: all real numbers
Range: $[0, \infty)$

h. Increasing on $(-1, 0.5)$ and $(2, \infty)$;
Decreasing on $(-\infty, -1)$ and $(0.5, 2)$.

73. $f(x) = x^2(x-3)(x+1)$

   a. Degree is 4. The graph of the function resembles $y = x^4$ for large values of $|x|$.

   b. y-intercept:
   $f(0) = (0)^2(0-3)(0+1) = -3$
   x-intercept: solve $f(x) = 0$
   $x^2(x-3)(x+1) = 0$
   $x = 0$ or $x = 3$ or $x = -1$

   c. The graph touches the x-axis at $x = 0$, since this zero has multiplicity 2. The graph crosses the x-axis at $x = 3$ and $x = -1$ since each zero has multiplicity 1.

   d. Graphing utility:

   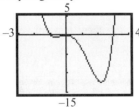

   e. 3 turning points; local maximum: (0, 0);
   local minima: (–0.69, –0.54), (2.19, –12.39)

f. Graphing by hand;

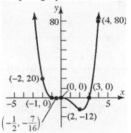

g. Domain: all real numbers
Range: $[-12.39, \infty)$

h. Increasing on $(-0.69, 0)$ and $(2.19, \infty)$;
Decreasing on $(-\infty, -0.69)$ and $(0, 2.19)$

75. $f(x) = (x+2)^2(x-4)^2$

   a. Degree is 4. The graph of the function resembles $y = x^4$ for large values of $|x|$.

   b. y-intercept: $f(0) = (0+2)^2(0-4)^2 = 64$
   x-intercept: solve $f(x) = 0$
   $(x+2)^2(x-4)^2 = 0 \Rightarrow x = -2$ or $x = 4$

   c. The graph touches the x-axis at $x = -2$ and $x = 4$ since each zero has multiplicity 2.

   d. Graphing utility:

   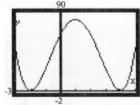

   e. 3 turning points; local maximum $(1, 81)$;
   local minima $(-2, 0)$ and $(4, 0)$

   f. Graphing by hand;

   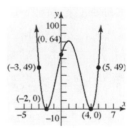

g. Domain: all real numbers
   Range: $[0, \infty)$

h. Increasing on $(-2, 1)$ and $(4, \infty)$
   Decreasing on $(-\infty, -2)$ and $(1, 4)$

77. $f(x) = x^2(x-2)(x^2+3)$

   a. Degree is 5. The graph of the function resembles $y = x^5$ for large values of $|x|$.

   b. y-intercept: $f(0) = 0^2(0-2)(0^2+3) = 0$
      x-intercept: solve $f(x) = 0$
      $x^2(x-2)(x^2+3) = 0 \Rightarrow x = 0$ or $x = 2$
      $x^2 + 3 = 0$ has no real solution

   c. The graph touches the x-axis at $x = 0$, since this zero has multiplicity 2. The graph crosses the x-axis at $x = 2$, since this zero has multiplicity 1.

   d. Graphing utility:
      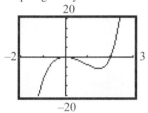

   e. 2 turning points; local maximum: $(0, 0)$; local minimum: $(1.48, -5.91)$

   f. Graphing by hand:
      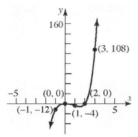

   g. Domain: all real numbers
      Range: all real numbers

   h. Increasing on $(-\infty, 0)$ and $(1.48, \infty)$;
      Decreasing on $(0, 1.48)$

79. $f(x) = -x^2(x^2-1)(x+1)$
    $= -x^2(x-1)(x+1)(x+1)$
    $= -x^2(x-1)(x+1)^2$

   a. Degree is 5. The graph of the function resembles $y = -x^5$ for large values of $|x|$.

   b. y-intercept: $f(0) = -0^2(0^2-1)(0+1) = 0$
      x-intercept: solve $f(x) = 0$
      $-x^2(x-1)(x+1)^2 = 0$
      $x = 0$ or $x = 1$ or $x = -1$

   c. The graph touches the x-axis at $x = 0$ and $x = -1$, since each zero has multiplicity 2. The graph crosses the x-axis at $x = 1$, since this zero has multiplicity 1.

   d. Graphing utility:
      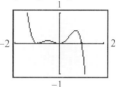

   e. 4 turning points;
      local maxima: $(-0.54, 0.10), (0.74, 0.43)$;
      local minima: $(-1, 0), (0, 0)$

   f. Graphing by hand:
      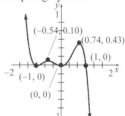

   g. Domain: all real numbers
      Range: all real numbers

   h. Increasing on $(-1, -0.54)$ and $(0, 0.74)$;
      Decreasing on $(-\infty, -1)$, $(-0.54, 0)$, and $(0.74, \infty)$

81. $f(x) = x^3 + 0.2x^2 - 1.5876x - 0.31752$

a. Degree = 3; The graph of the function resembles $y = x^3$ for large values of $|x|$.

b. Graphing utility

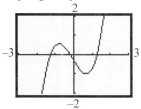

c. x-intercepts: $-1.26, -0.2, 1.26$; y-intercept: -0.31752

d.

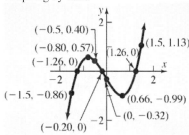

e. 2 turning points; local maximum: $(-0.80, 0.57)$; local minimum: $(0.66, -0.99)$

f. Graphing by hand

g. Domain: $\{x \mid x \text{ is any real number}\}$; Range: $\{y \mid y \text{ is any real number}\}$.

h. f is increasing on $(-\infty, -0.80)$ and $(0.66, \infty)$; f is decreasing on $(-0.80, 0.66)$

83. $f(x) = x^3 + 2.56x^2 - 3.31x + 0.89$

a. Degree = 3; The graph of the function resembles $y = x^3$ for large values of $|x|$.

b. Graphing utility

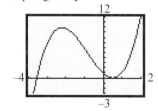

c. x-intercepts: $-3.56, 0.50$; y-intercept: 0.89

d.

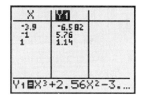

e. 2 turning points; local maximum: $(-2.21, 9.91)$; local minimum: $(0.50, 0)$

f. Graphing by hand

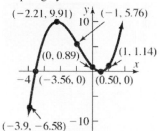

g. Domain: $\{x \mid x \text{ is any real number}\}$; Range: $\{y \mid y \text{ is any real number}\}$.

h. f is increasing on $(-\infty, -2.21)$ and $(0.50, \infty)$; f is decreasing on $(-2.21, 0.50)$

85. $f(x) = x^4 - 2.5x^2 + 0.5625$

  a. Degree = 4; The graph of the function r resembles $y = x^4$ for large values of $|x|$.

  b. Graphing utility
  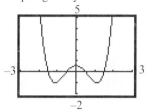

  c. x-intercepts: −1.50, −0.50, 0.50, 1.50; y-intercept: 0.5625

  d.
  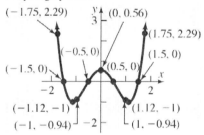

  e. 3 turning points:
  local maximum: (0, 0.5625);
  local minima: (−1.12, −1), (1.12, −1)

  f. Graphing by hand
  (−1.75, 2.29), (0, 0.56), (1.75, 2.29), (−0.5, 0), (1.5, 0), (−1.5, 0), (0.5, 0), (−1.12, −1), (1.12, −1), (−1, −0.94), (1, −0.94)

  g. Domain: $\{x \mid x \text{ is any real number}\}$;
  Range: $\{y \mid y \geq -1\}$.

  h. f is increasing on $(-1.12, 0)$ and $(1.12, \infty)$; f is decreasing on $(-\infty, -1.12)$ and $(0, 1.12)$

87. $f(x) = 2x^4 - \pi x^3 + \sqrt{5}x - 4$

  a. Degree = 4; The graph of the function resembles $y = 2x^4$ for large values of $|x|$.

  b. Graphing utility:
  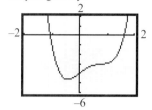

  c. x-intercepts: −1.07, 1.62; y-intercept: −4

  d.
  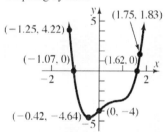

  e. 1 turning point;
  local minimum: (−0.42, −4.64)

  f. Graphing by hand
  (−1.25, 4.22), (1.75, 1.83), (−1.07, 0), (1.62, 0), (−0.42, −4.64), (0, −4)

  g. Domain: $\{x \mid x \text{ is any real number}\}$;
  Range: $\{y \mid y \geq -4.62\}$.

  h. f is increasing on $(-0.42, \infty)$; f is decreasing on $(-\infty, -0.42)$

89. $f(x) = -2x^5 - \sqrt{2}x^2 - x - \sqrt{2}$

  a. Degree = 5; The graph of the function resembles $y = -2x^5$ for large values of $|x|$.

  b. Graphing utility

  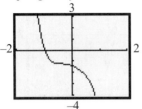

  c. x-intercept: -0.98; y-intercept: $-\sqrt{2} \approx -1.41$

  d.

  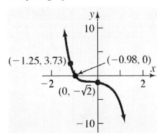

  e. No turning points; No local extrema

  f. Graphing by hand

  g. Domain: $\{x \mid x \text{ is any real number}\}$; Range: $\{y \mid y \text{ is any real number}\}$.

  h. f is decreasing on $(-\infty, \infty)$

91. Answers will vary. One possible answer is $f(x) = x(x-1)(x-2)$, since the graph crosses the x-axis at $x = 0, 1$ and 2.

93. Answers will vary. One possible answer is $f(x) = -\dfrac{1}{2}(x+1)(x-1)(x-2)$, since the graph crosses the x-axis at $x = -1, 1, 2$ and has a y-intercept at -1.

95. a. Graphing, we see that the graph may be a cubic relation.

  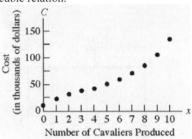

  b. Average rate of change $= \dfrac{50-43}{5-4} = \dfrac{7}{1} = 7$ thousand dollars per car

  c. Average rate of change $= \dfrac{105-85}{9-8} = \dfrac{20}{1} = 20$ thousand dollars per car

  d. $C(x) = 0.2x^3 - 2.3x^2 + 14.3x + 10.2$

  e. Graphing the cubic function of best fit:

  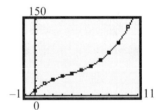

  f. $C(11) = 0.2(11)^3 - 2.3(11)^2 + 14.3(11) + 10.2$
  ≈ 155.4
  The cost of manufacturing 11 Cavaliers in 1 hour would be approximately $155,400.

  g. The y-intercept would indicate the fixed costs before any cars are made.

97. a. Graphing, we see that the graph may be a cubic relation..

  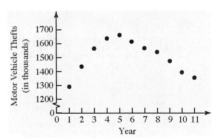

  b. $T(x) = 1.52x^3 - 39.81x^2 + 282.29x + 1035.5$

c. Graphing the cubic function of best fit:

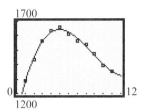

d. $T(12) = 1.52(12)^3 - 39.81(12)^2 + 282.29(12) + 1035.5$
$= 1316.9$
According to the function there would be approximately 1,316,900 motor vehicle thefts in 1998.

e. Answers will vary.

99. The graph of a polynomial function will always have a y-intercept since the domain of every polynomial function is the set of real numbers. Therefore $f(0)$ will always produce a y-coordinate on the graph. A polynomial function might have no x-intercepts. For example, $f(x) = x^2 + 1$ has no x-intercepts since the equation $x^2 + 1 = 0$ has no real solutions.

101. Answers will vary, one such polynomial is
$f(x) = x^2(x+1)(4-x)(x-2)^2$

103. $f(x) = \dfrac{1}{x}$ is smooth but not continuous;
$g(x) = |x|$ is continuous but not smooth.

Section 3.3

1. True

3.

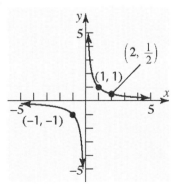

5. $y = 1$

7. proper

9. True

11. In $R(x) = \dfrac{4x}{x-3}$, the denominator, $q(x) = x - 3$, has a zero at 3. Thus, the domain of $R(x)$ is all real numbers except 3.

13. In $H(x) = \dfrac{-4x^2}{(x-2)(x+4)}$, the denominator, $q(x) = (x-2)(x+4)$, has zeros at 2 and –4. Thus, the domain of $H(x)$ is all real numbers except 2 and –4.

15. In $F(x) = \dfrac{3x(x-1)}{2x^2 - 5x - 3}$, the denominator, $q(x) = 2x^2 - 5x - 3 = (2x+1)(x-3)$, has zeros at $-\dfrac{1}{2}$ and 3. Thus, the domain of $F(x)$ is all real numbers except $-\dfrac{1}{2}$ and 3.

17. In $R(x) = \dfrac{x}{x^3 - 8}$, the denominator, $q(x) = x^3 - 8 = (x-2)(x^2+2x+4)$, has a zero at 2 ($x^2 + 2x + 4$ has no real zeros). Thus, the domain of $R(x)$ is all real numbers except 2.

19. In $H(x) = \dfrac{3x^2 + x}{x^2 + 4}$, the denominator, $q(x) = x^2 + 4$, has no real zeros. Thus, the domain of $H(x)$ is all real numbers.

21. In $R(x) = \dfrac{3(x^2 - x - 6)}{4(x^2 - 9)}$, the denominator, $q(x) = 4(x^2 - 9) = 4(x-3)(x+3)$, has zeros at 3 and –3. Thus, the domain of $R(x)$ is all real numbers except 3 and –3.

23. (a) Domain: $\{x \mid x \neq 2\}$; Range: $\{y \mid y \neq 1\}$

(b) Intercept: $(0, 0)$

(c) Horizontal Asymptote: $y = 1$

(d) Vertical Asymptote: $x = 2$

(e) Oblique Asymptote: none

25. (a) Domain: $\{x \mid x \neq 0\}$;
Range: all real numbers

(b) Intercepts: $(-1, 0)$ and $(1, 0)$

(c) Horizontal Asymptote: none

(d) Vertical Asymptote: $x = 0$

(e) Oblique Asymptote: $y = 2x$

27. (a) Domain: $\{x \mid x \neq -2, x \neq 2\}$;
Range: $\{y \mid y \leq 0 \text{ or } y > 1\}$

(b) Intercept: $(0, 0)$

(c) Horizontal Asymptote: $y = 1$

(d) Vertical Asymptotes: $x = -2, x = 2$

(e) Oblique Asymptote: none

29. $F(x) = 2 + \dfrac{1}{x}$; Using the function, $y = \dfrac{1}{x}$, shift the graph vertically 2 units up.

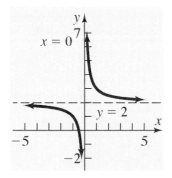

31. $R(x) = \dfrac{1}{(x-1)^2}$; Using the function, $y = \dfrac{1}{x^2}$, shift the graph horizontally 1 unit to the right.

33. $H(x) = \dfrac{-2}{x+1} = -2\left(\dfrac{1}{x+1}\right)$; Using the function $y = \dfrac{1}{x}$, shift the graph horizontally 1 unit to the left, reflect about the x-axis, and stretch vertically by a factor of 2.

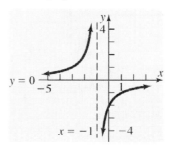

35. $R(x) = \dfrac{-1}{x^2 + 4x + 4} = -\dfrac{1}{(x+2)^2}$; Using the function $y = \dfrac{1}{x^2}$, shift the graph horizontally 2 units to the left, and reflect about the x-axis.

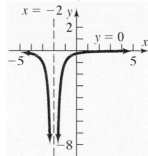

37. $G(x) = 1 + \dfrac{2}{(x-3)^2} = \dfrac{2}{(x-3)^2} + 1$;

$= 2\left(\dfrac{1}{(x-3)^2}\right) + 1$

Using the function $y = \dfrac{1}{x^2}$, shift the graph right 3 units, stretch vertically by a factor of 2, and shift vertically 1 unit up.

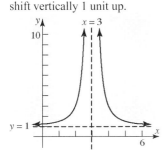

39. $R(x) = \dfrac{x^2 - 4}{x^2} = 1 - \dfrac{4}{x^2} = -4\left(\dfrac{1}{x^2}\right) + 1$; Using the function $y = \dfrac{1}{x^2}$, reflect about the x-axis, stretch vertically by a factor of 4, and shift vertically 1 unit up.

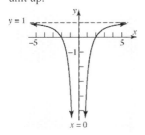

41. $R(x) = \dfrac{3x}{x+4}$; The degree of the numerator, $p(x) = 3x$, is $n = 1$. The degree of the denominator, $q(x) = x+4$, is $m = 1$. Since $n = m$, the line $y = \dfrac{3}{1} = 3$ is a horizontal asymptote. The denominator is zero at $x = -4$, so $x = -4$ is a vertical asymptote.

43. $H(x) = \dfrac{x^3 - 8}{x^2 - 5x + 6} = \dfrac{(x-2)(x^2 + 2x + 4)}{(x-2)(x-3)}$

$= \dfrac{x^2 + 2x + 4}{x - 3}$, where $x \ne 2, 3$

The degree of the numerator in the reduced expression is $n = 2$. The degree of the denominator in the reduced expression is $m = 1$. Since $n = m + 1$, there is an oblique asymptote. Dividing:

$$\begin{array}{r} x + 5 \\ x-3 \overline{\smash{\big)}\, x^2 + 2x + 4} \\ -(x^2 - 3x) \\ \hline 5x + 4 \\ -(5x - 15) \\ \hline 19 \end{array}$$

$H(x) = x + 5 + \dfrac{19}{x-3}$, $x \ne 2, 3$

Thus, the oblique asymptote is $y = x + 5$.
The denominator of the reduced expression is zero at $x = 3$ so $x = 3$ is a vertical asymptote.

45. $T(x) = \dfrac{x^3}{x^4 - 1}$; The degree of the numerator, $p(x) = x^3$, is $n = 3$. The degree of the denominator, $q(x) = x^4 - 1$ is $m = 4$. Since $n < m$, the line $y = 0$ is a horizontal asymptote. The denominator is zero at $x = -1$ and $x = 1$, so $x = -1$ and $x = 1$ are vertical asymptotes.

47. $Q(x) = \dfrac{5 - x^2}{3x^4}$; The degree of the numerator, $p(x) = 5 - x^2$, is $n = 2$. The degree of the denominator, $q(x) = 3x^4$ is $m = 4$. Since $n < m$, the line $y = 0$ is a horizontal asymptote. The denominator is zero at $x = 0$, so $x = 0$ is a vertical asymptote.

49. $R(x) = \dfrac{3x^4 + 4}{x^3 + 3x}$; The degree of the numerator, $p(x) = 3x^4 + 4$, is $n = 4$. The degree of the denominator, $q(x) = x^3 + 3x$ is $m = 3$. Since $n = m + 1$, there is an oblique asymptote. Dividing:

$$\begin{array}{r} 3x \\ x^3+3x\overline{\smash{)}3x^4+4} \\ \underline{3x^4+9x^2} \\ -9x^2+4 \end{array}$$

$R(x) = 3x + \dfrac{-9x^2+4}{x^3+3x}$

Thus, the oblique asymptote is $y = 3x$.

The denominator is zero at $x = 0$, so $x = 0$ is a vertical asymptote.

51. $G(x) = \dfrac{x^3-1}{x-x^2}$; The degree of the numerator, $p(x) = x^3 - 1$, is $n = 3$. The degree of the denominator, $q(x) = x - x^2$ is $m = 2$. Since $n = m+1$, there is an oblique asymptote. Dividing:

$$\begin{array}{r} -x-1 \\ -x^2+x\overline{\smash{)}x^3-1} \\ \underline{x^3-x^2} \\ x^2 \\ \underline{x^2-x} \\ x-1 \end{array}$$

$G(x) = -x-1 + \dfrac{x-1}{x-x^2} = -x-1-\dfrac{1}{x},\ x \neq 1$

Thus, the oblique asymptote is $y = -x-1$.

$G(x)$ must be in lowest terms to find the vertical asymptote:

$G(x) = \dfrac{x^3-1}{x-x^2} = \dfrac{(x-1)(x^2+x+1)}{-x(x-1)} = \dfrac{x^2+x+1}{-x}$

The denominator is zero at $x = 0$, so $x = 0$ is a vertical asymptote.

53. $g(h) = \dfrac{3.99 \times 10^{14}}{\left(6.374 \times 10^6 + h\right)^2}$

   (a) $g(0) = \dfrac{3.99 \times 10^{14}}{\left(6.374 \times 10^6 + 0\right)^2} \approx 9.8208$ m/s^2

   (b) $g(443) = \dfrac{3.99 \times 10^{14}}{\left(6.374 \times 10^6 + 443\right)^2}$
   ≈ 9.8195 m/s^2

   (c) $g(8848) = \dfrac{3.99 \times 10^{14}}{\left(6.374 \times 10^6 + 8848\right)^2}$
   ≈ 9.7936 m/s^2

   (d) $g(h) = \dfrac{3.99 \times 10^{14}}{\left(6.374 \times 10^6 + h\right)^2}$
   $\approx \dfrac{3.99 \times 10^{14}}{h^2} \to 0$ as $h \to \infty$
   Thus, $g = 0$ is the horizontal asymptote.

   (e) $g(h) = \dfrac{3.99 \times 10^{14}}{\left(6.374 \times 10^6 + h\right)^2} = 0$, to solve this equation would require that $3.99 \times 10^{14} = 0$, which is impossible. Therefore, there is no height above sea level at which $g = 0$. In other words, there is no point in the entire universe that is unaffected by the Earth's gravity!

55-57. Answers will vary.

Section 3.4

1. $(-10, 0), (0, 6)$

3. in lowest terms

5. False

In problems 7–43, we will use the terminology: $R(x) = \dfrac{p(x)}{q(x)}$, where the degree of $p(x) = n$ and the degree of $q(x) = m$.

7. $R(x) = \dfrac{x+1}{x(x+4)}$ $p(x) = x+1$; $q(x) = x(x+4) = x^2 + 4x$; $n = 1$; $m = 2$

   Step 1: Domain: $\{x \mid x \neq -4, x \neq 0\}$

   Step 2: (a) The x-intercept is the zero of $p(x)$: -1

   (b) There is no y-intercept; $R(0)$ is not defined, since $q(0) = 0$.

   Step 3: $R(-x) = \dfrac{-x+1}{-x(-x+4)} = \dfrac{-x+1}{x^2 - 4x}$; this is neither $R(x)$ nor $-R(x)$, so there is no symmetry.

   Step 4: $R(x) = \dfrac{x+1}{x(x+4)}$ is in lowest terms.

   The vertical asymptotes are the zeros of $q(x)$: $x = -4$ and $x = 0$.

   Step 5: Since $n < m$, the line $y = 0$ is the horizontal asymptote. Solve to find intersection points:
   $$\dfrac{x+1}{x(x+4)} = 0$$
   $$x + 1 = 0$$
   $$x = -1$$
   $R(x)$ intersects $y = 0$ at $(-1, 0)$.

   Step 6:

Interval	$(-\infty, -4)$	$(-4, -1)$	$(-1, 0)$	$(0, \infty)$
Number Chosen	-5	-2	$-\frac{1}{2}$	1
Value of R	$R(-5) = -\frac{4}{5}$	$R(-2) = \frac{1}{4}$	$R\left(-\frac{1}{2}\right) = -\frac{2}{7}$	$R(1) = \frac{2}{5}$
Location of Graph	Below x-axis	Above x-axis	Below x-axis	Above x-axis
Point on Graph	$\left(-5, -\frac{4}{5}\right)$	$\left(-2, \frac{1}{4}\right)$	$\left(-\frac{1}{2}, -\frac{2}{7}\right)$	$\left(1, \frac{2}{5}\right)$

   Step 7: Graphing

   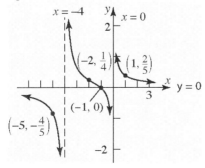

9. $R(x) = \dfrac{3x+3}{2x+4}$ $p(x) = 3x+3$; $q(x) = 2x+4$; $n=1$; $m=1$

Step 1: Domain: $\{x \mid x \neq -2\}$

Step 2: (a) The x-intercept is the zero of $p(x)$: -1

(b) The y-intercept is $R(0) = \dfrac{3(0)+3}{2(0)+4} = \dfrac{3}{4}$.

Step 3: $R(-x) = \dfrac{3(-x)+3}{2(-x)+4} = \dfrac{-3x+3}{-2x+4} = \dfrac{3x-3}{2x-4}$; this is neither $R(x)$ nor $-R(x)$, so there is no symmetry.

Step 4: $R(x) = \dfrac{3x+3}{2x+4}$ is in lowest terms.

The vertical asymptote is the zero of $q(x)$: $x = -2$

Step 5: Since $n = m$, the line $y = \dfrac{3}{2}$ is the horizontal asymptote.

Solve to find intersection points:
$$\dfrac{3x+3}{2x+4} = \dfrac{3}{2}$$
$$2(3x+3) = 3(2x+4)$$
$$6x+6 = 6x+4$$
$$0 \neq 2$$

$R(x)$ does not intersect $y = \dfrac{3}{2}$.

Step 6:

Interval	$(-\infty, -2)$	$(-2, -1)$	$(-1, \infty)$
Number Chosen	-3	$-\dfrac{3}{2}$	0
Value of R	$R(-3) = 3$	$R\left(-\dfrac{3}{2}\right) = -\dfrac{3}{2}$	$R(0) = \dfrac{3}{4}$
Location of Graph	Above x-axis	Below x-axis	Above x-axis
Point on Graph	$(-3, 3)$	$\left(-\dfrac{3}{2}, -\dfrac{3}{2}\right)$	$\left(0, \dfrac{3}{4}\right)$

Step 7: Graphing:

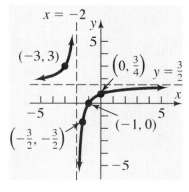

11. $R(x) = \dfrac{3}{x^2 - 4}$ $p(x) = 3$; $q(x) = x^2 - 4$; $n = 0$; $m = 2$

Step 1: Domain: $\{x \mid x \neq -2, x \neq 2\}$

Step 2: (a) There is no x-intercept.

(b) The y-intercept is $R(0) = \dfrac{3}{0^2 - 4} = \dfrac{3}{-4} = -\dfrac{3}{4}$.

Step 3: $R(-x) = \dfrac{3}{(-x)^2 - 4} = \dfrac{3}{x^2 - 4} = R(x)$; $R(x)$ is symmetric with respect to the y-axis.

Step 4: $R(x) = \dfrac{3}{x^2 - 4}$ is in lowest terms. The vertical asymptotes are the zeros of $q(x)$: $x = -2$ and $x = 2$

Step 5: Since $n < m$, the line $y = 0$ is the horizontal asymptote. Solve to find intersection points:

$$\dfrac{3}{x^2 - 4} = 0$$
$$3 = 0(x^2 - 4)$$
$$3 \neq 0$$

$R(x)$ does not intersect $y = 0$.

Step 6:

Interval	$(-\infty, -2)$	$(-2, 2)$	$(2, \infty)$
Number Chosen	-3	0	3
Value of R	$R(-3) = \dfrac{3}{5}$	$R(0) = -\dfrac{3}{4}$	$R(3) = \dfrac{3}{5}$
Location of Graph	Above x-axis	Below x-axis	Above x-axis
Point on Graph	$\left(-3, \dfrac{3}{5}\right)$	$\left(0, -\dfrac{3}{4}\right)$	$\left(3, \dfrac{3}{5}\right)$

Step 7: Graphing:

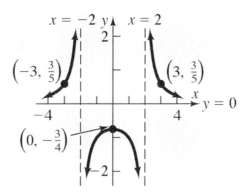

13. $P(x) = \dfrac{x^4 + x^2 + 1}{x^2 - 1}$ $p(x) = x^4 + x^2 + 1;\ q(x) = x^2 - 1;\ n = 4;\ m = 2$

Step 1: Domain: $\{x \mid x \neq -1,\ x \neq 1\}$

Step 2: (a) There is no x-intercept.

(b) The y-intercept is $P(0) = \dfrac{0^4 + 0^2 + 1}{0^2 - 1} = \dfrac{1}{-1} = -1$.

Step 3: $P(-x) = \dfrac{(-x)^4 + (-x)^2 + 1}{(-x)^2 - 1} = \dfrac{x^4 + x^2 + 1}{x^2 - 1} = P(x)$. $P(x)$ is symmetric with respect to the y-axis.

Step 4: $P(x) = \dfrac{x^4 + x^2 + 1}{x^2 - 1}$ is in lowest terms. The vertical asymptotes are the zeros of $q(x)$: $x = -1$ and $x = 1$

Step 5: Since $n > m + 1$, there is no horizontal or oblique asymptote.

Step 6:

Interval	$(-\infty, -1)$	$(-1, 1)$	$(1, \infty)$
Number Chosen	-2	0	2
Value of P	$P(-2) = 7$	$P(0) = -1$	$P(2) = 7$
Location of Graph	Above x-axis	Below x-axis	Above x-axis
Point on Graph	$(-2, 7)$	$(0, -1)$	$(2, 7)$

Step 7: Graphing:

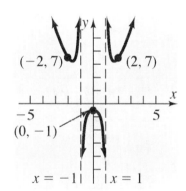

SSM: College Algebra EGU   Chapter 3: Polynomial and Rational Functions

15. $H(x) = \dfrac{x^3 - 1}{x^2 - 9} = \dfrac{(x-1)(x^2 + x + 1)}{(x+3)(x-3)}$ $p(x) = x^3 - 1;\ q(x) = x^2 - 9;\ n = 3;\ m = 2$

Step 1:   Domain: $\{x \mid x \neq -3,\ x \neq 3\}$

Step 2:   (a) The x-intercept is the zero of $p(x)$: 1.

(b) The y-intercept is $H(0) = \dfrac{0^3 - 1}{0^2 - 9} = \dfrac{-1}{-9} = \dfrac{1}{9}$.

Step 3: $H(-x) = \dfrac{(-x)^3 - 1}{(-x)^2 - 9} = \dfrac{-x^3 - 1}{x^2 - 9}$; this is neither $H(x)$ nor $-H(x)$, so there is no symmetry.

Step 4: $H(x) = \dfrac{x^3 - 1}{x^2 - 9}$ is in lowest terms. The vertical asymptotes are the zeros of $q(x)$: $x = -3$ and $x = 3$

Step 5:   Since $n = m + 1$, there is an oblique asymptote. Dividing:

$$x^2 - 9 \overline{\smash{\big)}\, x^3 + 0x^2 + 0x - 1} \quad\quad H(x) = x + \dfrac{9x - 1}{x^2 - 9}$$

with quotient x and working: $x^3 - 9x$, remainder $9x - 1$.

The oblique asymptote is $y = x$. Solve to find intersection points:

$$\dfrac{x^3 - 1}{x^2 - 9} = x$$
$$x^3 - 1 = x^3 - 9x$$
$$-1 = -9x$$
$$x = \dfrac{1}{9}$$

The oblique asymptote intersects $H(x)$ at $\left(\dfrac{1}{9}, \dfrac{1}{9}\right)$.

Step 6:

Interval	$(-\infty, -3)$	$(-3, 1)$	$(1, 3)$	$(3, \infty)$
Number Chosen	-4	0	2	4
Value of H	$H(-4) \approx -9.3$	$H(0) = \dfrac{1}{9}$	$H(2) = -1.4$	$H(4) = 9$
Location of Graph	Below x-axis	Above x-axis	Below x-axis	Above x-axis
Point on Graph	$(-4, -9.3)$	$\left(0, \dfrac{1}{9}\right)$	$(2, -1.4)$	$(4, 9)$

Step 7: Graphing:

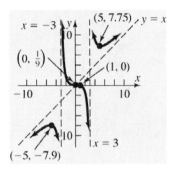

17. $R(x) = \dfrac{x^2}{x^2+x-6} = \dfrac{x^2}{(x+3)(x-2)}$ $p(x) = x^2$; $q(x) = x^2+x-6$; $n=2$; $m=2$

Step 1: Domain: $\{x \mid x \neq -3, x \neq 2\}$

Step 2: (a) The x-intercept is the zero of $p(x)$: 0

(b) The y-intercept is $R(0) = \dfrac{0^2}{0^2+0-6} = \dfrac{0}{-6} = 0$.

Step 3: $R(-x) = \dfrac{(-x)^2}{(-x)^2+(-x)-6} = \dfrac{x^2}{x^2-x-6}$; this is neither $R(x)$ nor $-R(x)$, so there is no symmetry.

Step 4: $R(x) = \dfrac{x^2}{x^2+x-6}$ is in lowest terms. The vertical asymptotes are the zeros of $q(x)$:

$x = -3$ and $x = 2$

Step 5: Since $n = m$, the line $y = 1$ is the horizontal asymptote. Solve to find intersection points:

$$\dfrac{x^2}{x^2+x-6} = 1$$
$$x^2 = x^2+x-6$$
$$0 = x-6$$
$$x = 6$$

$R(x)$ intersects $y = 1$ at $(6, 1)$.

Step 6:

Interval	$(-\infty, -3)$	$(-3, 0)$	$(0, 2)$	$(2, \infty)$
Number Chosen	-6	-1	1	3
Value of R	$R(-6) = 1.5$	$R(-1) = -\tfrac{1}{6}$	$R(1) = -0.25$	$R(3) = 1.5$
Location of Graph	Above x-axis	Below x-axis	Below x-axis	Above x-axis
Point on Graph	$(-6, 1.5)$	$\left(-1, -\tfrac{1}{6}\right)$	$(1, -0.25)$	$(3, 1.5)$

Step 7: Graphing:

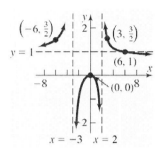

19. $G(x) = \dfrac{x}{x^2-4} = \dfrac{x}{(x+2)(x-2)}$ $p(x) = x$; $q(x) = x^2 - 4$; $n = 1$; $m = 2$

Step 1: Domain: $\{x \mid x \neq -2, x \neq 2\}$

Step 2: (a) The x-intercept is the zero of $p(x)$: 0

(b) The y-intercept is $G(0) = \dfrac{0}{0^2 - 4} = \dfrac{0}{-4} = 0$.

Step 3: $G(-x) = \dfrac{-x}{(-x)^2 - 4} = \dfrac{-x}{x^2 - 4} = -G(x)$; $G(x)$ is symmetric with respect to the origin.

Step 4: $G(x) = \dfrac{x}{x^2 - 4}$ is in lowest terms. The vertical asymptotes are the zeros of $q(x)$: $x = -2$ and $x = 2$

Step 5: Since $n < m$, the line $y = 0$ is the horizontal asymptote. Solve to find intersection points:

$\dfrac{x}{x^2 - 4} = 0$

$x = 0$

$G(x)$ intersects $y = 0$ at $(0, 0)$.

Step 6:

Interval	$(-\infty, -2)$	$(-2, 0)$	$(0, 2)$	$(2, \infty)$
Number Chosen	-3	-1	1	3
Value of G	$G(-3) = -\tfrac{3}{5}$	$G(-1) = \tfrac{1}{3}$	$G(1) = -\tfrac{1}{3}$	$G(3) = \tfrac{3}{5}$
Location of Graph	Below x-axis	Above x-axis	Below x-axis	Above x-axis
Point on Graph	$\left(-3, -\tfrac{3}{5}\right)$	$\left(-1, \tfrac{1}{3}\right)$	$\left(1, -\tfrac{1}{3}\right)$	$\left(3, \tfrac{3}{5}\right)$

Step 7: Graphing:

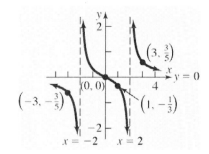

21. $R(x) = \dfrac{3}{(x-1)(x^2-4)} = \dfrac{3}{(x-1)(x+2)(x-2)}$ $p(x) = 3;\ q(x) = (x-1)(x^2-4);\ n = 0;\ m = 3$

Step 1: Domain: $\{x \mid x \neq -2,\ x \neq 1,\ x \neq 2\}$

Step 2: (a) There is no x-intercept.

(b) The y-intercept is $R(0) = \dfrac{3}{(0-1)(0^2-4)} = \dfrac{3}{4}$.

Step 3: $R(-x) = \dfrac{3}{(-x-1)\left((-x)^2-4\right)} = \dfrac{3}{(-x-1)(x^2-4)}$; this is neither $R(x)$ nor $-R(x)$, so there is no symmetry.

Step 4: $R(x) = \dfrac{3}{(x-1)(x^2-4)}$ is in lowest terms. The vertical asymptotes are the zeros of $q(x)$: $x = -2,\ x = 1,$ and $x = 2$.

Step 5: Since $n < m$, the line $y = 0$ is the horizontal asymptote. Solve to find intersection points:

$\dfrac{3}{(x-1)(x^2-4)} = 0$

$3 \neq 0$

$R(x)$ does not intersect $y = 0$.

Step 6:

Interval	$(-\infty, -2)$	$(-2, 1)$	$(1, 2)$	$(2, \infty)$
Number Chosen	-3	0	1.5	3
Value of R	$R(-3) = -\tfrac{3}{20}$	$R(0) = \tfrac{3}{4}$	$R(1.5) = -\tfrac{24}{7}$	$R(3) = \tfrac{3}{10}$
Location of Graph	Below x-axis	Above x-axis	Below x-axis	Above x-axis
Point on Graph	$\left(-3, -\tfrac{3}{20}\right)$	$\left(0, \tfrac{3}{4}\right)$	$\left(1.5, -\tfrac{24}{7}\right)$	$\left(3, \tfrac{3}{10}\right)$

Step 7: Graphing:

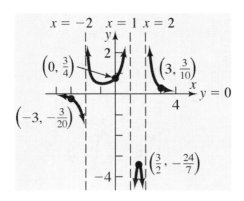

23. $H(x) = \dfrac{4(x^2-1)}{x^4-16} = \dfrac{4(x-1)(x+1)}{(x^2+4)(x+2)(x-2)}$ $p(x)=4(x^2-1);\ q(x)=x^4-16;\ n=2;\ m=4$

Step 1: Domain: $\{x \mid x \neq -2,\ x \neq 2\}$

Step 2: (a) The x-intercepts are the zeros of $p(x)$: -1 and 1

(b) The y-intercept is $H(0) = \dfrac{4(0^2-1)}{0^4-16} = \dfrac{-4}{-16} = \dfrac{1}{4}$.

Step 3: $H(-x) = \dfrac{4\left((-x)^2-1\right)}{(-x)^4-16} = \dfrac{4(x^2-1)}{x^4-16} = H(x)$; $H(x)$ is symmetric with respect to the y-axis.

Step 4: $H(x) = \dfrac{4(x^2-1)}{x^4-16}$ is in lowest terms. The vertical asymptotes are the zeros of $q(x)$: $x=-2$ and $x=2$

Step 5: Since $n < m$, the line $y = 0$ is the horizontal asymptote. Solve to find intersection points:

$\dfrac{4(x^2-1)}{x^4-16} = 0$

$4(x^2-1) = 0$

$x^2 - 1 = 0$

$x = \pm 1$

$H(x)$ intersects $y = 0$ at $(-1, 0)$ and $(1, 0)$.

Step 6:

Interval	$(-\infty, -2)$	$(-2, -1)$	$(-1, 1)$	$(1, 2)$	$(2, \infty)$
Number Chosen	-3	-1.5	0	1.5	3
Value of H	$H(-3) \approx 0.49$	$H(-1.5) \approx -0.46$	$H(0) = \tfrac{1}{4}$	$H(1.5) \approx -0.46$	$H(3) \approx 0.49$
Location of Graph	Above x-axis	Below x-axis	Above x-axis	Below x-axis	Above x-axis
Point on Graph	$(-3, 0.49)$	$(-1.5, -0.46)$	$(0, \tfrac{1}{4})$	$(1.5, -0.46)$	$(3, 0.49)$

Step 7: Graphing:

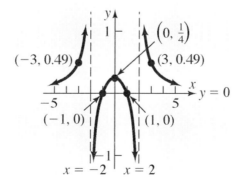

25. $F(x) = \dfrac{x^2 - 3x - 4}{x + 2} = \dfrac{(x+1)(x-4)}{x+2}$ $p(x) = x^2 - 3x - 4;\ q(x) = x + 2;\ n = 2;\ m = 1$

Step 1: Domain: $\{x \mid x \neq -2\}$

Step 2: (a) The x-intercepts are the zeros of $p(x)$: -1 and 4.

(b) The y-intercept is $F(0) = \dfrac{0^2 - 3(0) - 4}{0 + 2} = \dfrac{-4}{2} = -2$.

Step 3: $F(-x) = \dfrac{(-x)^2 - 3(-x) - 4}{-x + 2} = \dfrac{x^2 + 3x - 4}{-x + 2}$; this is neither $F(x)$ nor $-F(x)$, so there is no symmetry.

Step 4: $F(x) = \dfrac{x^2 - 3x - 4}{x + 2}$ is in lowest terms. The vertical asymptote is the zero of $q(x)$: $x = -2$

Step 5: Since $n = m + 1$, there is an oblique asymptote. Dividing:

$$\begin{array}{r} x - 5 \\ x+2\overline{)x^2 - 3x - 4} \\ \underline{x^2 + 2x} \\ -5x - 4 \\ \underline{-5x - 10} \\ 6 \end{array}$$

$F(x) = x - 5 + \dfrac{6}{x + 2}$

The oblique asymptote is $y = x - 5$. Solve to find intersection points:

$\dfrac{x^2 - 3x - 4}{x + 2} = x - 5$

$x^2 - 3x - 4 = x^2 - 3x - 10$

$-4 \neq -10$

The oblique asymptote does not intersect $F(x)$.

Step 6:

Interval	$(-\infty, -2)$	$(-2, -1)$	$(-1, 4)$	$(4, \infty)$
Number Chosen	-3	-1.5	0	5
Value of F	$F(-3) = -14$	$F(-1.5) = 5.5$	$F(0) = -2$	$F(5) \approx 0.86$
Location of Graph	Below x-axis	Above x-axis	Below x-axis	Above x-axis
Point on Graph	$(-3, -14)$	$(-1.5, 5.5)$	$(0, -2)$	$(5, 0.86)$

Step 7: Graphing:

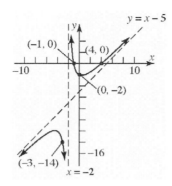

27. $R(x) = \dfrac{x^2 + x - 12}{x - 4} = \dfrac{(x+4)(x-3)}{x-4}$ $p(x) = x^2 + x - 12;\ q(x) = x - 4;\ n = 2;\ m = 1$

Step 1: Domain: $\{x \mid x \neq 4\}$

Step 2: (a) The x-intercepts are the zeros of $p(x)$: -4 and 3.

(b) The y-intercept is $R(0) = \dfrac{0^2 + 0 - 12}{0 - 4} = \dfrac{-12}{-4} = 3$.

Step 3: $R(-x) = \dfrac{(-x)^2 + (-x) - 12}{-x - 4} = \dfrac{x^2 - x - 12}{-x - 4}$; this is neither $R(x)$ nor $-R(x)$, so there is no symmetry.

Step 4: $R(x) = \dfrac{x^2 + x - 12}{x - 4}$ is in lowest terms. The vertical asymptote is the zero of $q(x)$: $x = 4$

Step 5: Since $n = m + 1$, there is an oblique asymptote. Dividing:

$$\begin{array}{r}
x + 5 \\
x - 4 \overline{)\, x^2 + x - 12\,} \\
\underline{x^2 - 4x} \\
5x - 12 \\
\underline{5x - 20} \\
8
\end{array}$$

$R(x) = x + 5 + \dfrac{8}{x - 4}$

The oblique asymptote is $y = x + 5$. Solve to find intersection points:

$\dfrac{x^2 + x - 12}{x - 4} = x + 5$

$x^2 + x - 12 = x^2 + x - 20$

$-12 \neq -20$

The oblique asymptote does not intersect $R(x)$.

Step 6:

Interval	$(-\infty, -4)$	$(-4, 3)$	$(3, 4)$	$(4, \infty)$
Number Chosen	-5	0	3.5	5
Value of R	$R(-5) = -\frac{8}{9}$	$R(0) = 3$	$R(3.5) = -7.5$	$R(5) = 18$
Location of Graph	Below x-axis	Above x-axis	Below x-axis	Above x-axis
Point on Graph	$\left(-5, -\frac{8}{9}\right)$	$(0, 3)$	$(3.5, -7.5)$	$(5, 18)$

Step 7: Graphing:

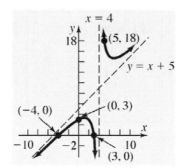

29. $F(x) = \dfrac{x^2 + x - 12}{x + 2} = \dfrac{(x+4)(x-3)}{x+2}$ $p(x) = x^2 + x - 12;\ q(x) = x + 2;\ n = 2;\ m = 1$

Step 1: Domain: $\{x \mid x \neq -2\}$

Step 2: (a) The x-intercepts are the zeros of $p(x)$: -4 and 3.

(b) The y-intercept is $F(0) = \dfrac{0^2 + 0 - 12}{0 + 2} = \dfrac{-12}{2} = -6$.

Step 3: $F(-x) = \dfrac{(-x)^2 + (-x) - 12}{-x + 2} = \dfrac{x^2 - x - 12}{-x + 2}$; this is neither $F(x)$ nor $-F(x)$, so there is no symmetry.

Step 4: $F(x) = \dfrac{x^2 + x - 12}{x + 2}$ is in lowest terms. The vertical asymptote is the zero of $q(x)$: $x = -2$

Step 5: Since $n = m + 1$, there is an oblique asymptote. Dividing:

$$\begin{array}{r} x - 1 \\ x+2 \overline{\smash{)}x^2 + x - 12} \\ \underline{x^2 + 2x } \\ -x - 12 \\ \underline{-x - 2} \\ -10 \end{array}$$

$F(x) = x - 1 + \dfrac{-10}{x + 2}$

The oblique asymptote is $y = x - 1$. Solve to find intersection points:

202

$$\frac{x^2+x-12}{x+2} = x-1$$
$$x^2+x-12 = x^2+x-2$$
$$-12 \neq -2$$

The oblique asymptote does not intersect $F(x)$.

Step 6:

Interval	$(-\infty, -4)$	$(-4, -2)$	$(-2, 3)$	$(3, \infty)$
Number Chosen	-5	-3	0	4
Value of F	$F(-5) = -\frac{8}{3}$	$F(-3) = 6$	$F(0) = -6$	$F(4) = \frac{4}{3}$
Location of Graph	Below x-axis	Above x-axis	Below x-axis	Above x-axis
Point on Graph	$\left(-5, -\frac{8}{3}\right)$	$(-3, 6)$	$(0, -6)$	$\left(4, \frac{4}{3}\right)$

Step 7: Graphing:

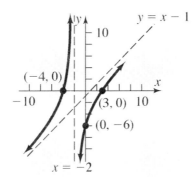

31. $R(x) = \frac{x(x-1)^2}{(x+3)^3}$ $p(x) = x(x-1)^2$; $q(x) = (x+3)^3$; $n=3$; $m=3$

    Step 1: Domain: $\{x \mid x \neq -3\}$

    Step 2: (a) The x-intercepts are the zeros of $p(x)$: 0 and 1

    (b) The y-intercept is $R(0) = \frac{0(0-1)^2}{(0+3)^3} = \frac{0}{27} = 0$.

    Step 3: $R(-x) = \frac{-x(-x-1)^2}{(-x+3)^3}$; this is neither $R(x)$ nor $-R(x)$, so there is no symmetry.

    Step 4: $R(x) = \frac{x(x-1)^2}{(x+3)^3}$ is in lowest terms. The vertical asymptote is the zero of $q(x)$: $x = -3$

Step 5: Since $n = m$, the line $y = 1$ is the horizontal asymptote. Solve to find intersection points:

$$\frac{x(x-1)^2}{(x+3)^3} = 1$$
$$x^3 - 2x^2 + x = x^3 + 9x^2 + 27x + 27$$
$$0 = 11x^2 + 26x + 27$$
$$b^2 - 4ac = 26^2 - 4(11)(27) = -512$$
no real solution

$R(x)$ does not intersect $y = 1$.

Step 6:

Interval	$(-\infty, -3)$	$(-3, 0)$	$(0, 1)$	$(1, \infty)$
Number Chosen	-4	-1	$\frac{1}{2}$	2
Value of R	$R(-4) = 100$	$R(-1) = -0.5$	$R\left(\frac{1}{2}\right) \approx 0.003$	$R(2) = 0.016$
Location of Graph	Above x-axis	Below x-axis	Above x-axis	Above x-axis
Point on Graph	$(-4, 100)$	$(-1, -0.5)$	$\left(\frac{1}{2}, 0.003\right)$	$(2, 0.016)$

Step 7: Graphing:

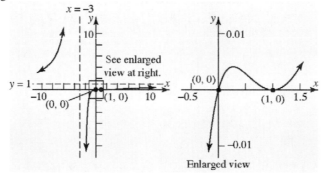

33. $R(x) = \dfrac{x^2 + x - 12}{x^2 - x - 6} = \dfrac{(x+4)(x-3)}{(x-3)(x+2)} = \dfrac{x+4}{x+2}$ $p(x) = x^2 + x - 12$; $q(x) = x^2 - x - 6$; $n = 2$; $m = 2$

Step 1: Domain: $\{x \mid x \neq -2, x \neq 3\}$

Step 2: (a) The x-intercept is the zero of $y = x + 4$: -4; Note: 3 is not a zero because reduced form must be used to find the zeros.

(b) The y-intercept is $R(0) = \dfrac{0^2 + 0 - 12}{0^2 - 0 - 6} = \dfrac{-12}{-6} = 2$.

Step 3: $R(-x) = \dfrac{(-x)^2 + (-x) - 12}{(-x)^2 - (-x) - 6} = \dfrac{x^2 - x - 12}{x^2 + x - 6}$; this is neither $R(x)$ nor $-R(x)$, so there is no symmetry.

Step 4: In lowest terms, $R(x) = \dfrac{x+4}{x+2}$, $x \neq 3$. The vertical asymptote is the zero of $f(x) = x+2$: $x = -2$;

Note: $x = 3$ is not a vertical asymptote because reduced form must be used to find the asymptotes. The graph has a hole at $\left(3, \dfrac{7}{5}\right)$.

Step 5: Since $n = m$, the line $y = 1$ is the horizontal asymptote. Solve to find intersection points:

$$\dfrac{x^2 + x - 12}{x^2 - x - 6} = 1$$

$$x^2 + x - 12 = x^2 - x - 6$$

$$2x = 6$$

$$x = 3$$

$R(x)$ does not intersect $y = 1$ because $R(x)$ is not defined at $x = 3$.

Step 6:

Interval	$(-\infty, -4)$	$(-4, -2)$	$(-2, 3)$	$(3, \infty)$
Number Chosen	-5	-3	0	4
Value of R	$R(-5) = \dfrac{1}{3}$	$R(-3) = -1$	$R(0) = 2$	$R(4) = \dfrac{4}{3}$
Location of Graph	Above x-axis	Below x-axis	Above x-axis	Above x-axis
Point on Graph	$\left(-5, \dfrac{1}{3}\right)$	$(-3, -1)$	$(0, 2)$	$\left(4, \dfrac{4}{3}\right)$

Step 7: Graphing:

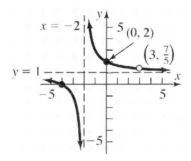

35. $R(x) = \dfrac{6x^2 - 7x - 3}{2x^2 - 7x + 6} = \dfrac{(3x+1)(2x-3)}{(2x-3)(x-2)} = \dfrac{3x+1}{x-2}$ $p(x) = 6x^2 - 7x - 3;$ $q(x) = 2x^2 - 7x + 6;$ $n = 2;$ $m = 2$

Step 1: Domain: $\left\{x \mid x \neq \dfrac{3}{2}, x \neq 2\right\}$

Step 2: (a) The x-intercept is the zero of $y = 3x+1$: $-\dfrac{1}{3}$; Note: $x = \dfrac{3}{2}$ is not a zero because reduced form must be used to find the zeros.

(b) The y-intercept is $R(0) = \dfrac{6(0)^2 - 7(0) - 3}{2(0)^2 - 7(0) + 6} = \dfrac{-3}{6} = -\dfrac{1}{2}$.

Step 3: $R(-x) = \dfrac{6(-x)^2 - 7(-x) - 3}{2(-x)^2 - 7(-x) + 6} = \dfrac{6x^2 + 7x - 3}{2x^2 + 7x + 6}$; this is neither $R(x)$ nor $-R(x)$, so there is no symmetry.

Step 4: In lowest terms, $R(x) = \dfrac{3x+1}{x-2}$, $x \neq \dfrac{3}{2}$. The vertical asymptote is the zero of $f(x) = x - 2$: $x = 2$;

Note: $x = \dfrac{3}{2}$ is not a vertical asymptote because reduced form must be used to find the asymptotes.

The graph has a hole at $\left(\dfrac{3}{2}, -11\right)$.

Step 5: Since $n = m$, the line $y = 3$ is the horizontal asymptote. Solve to find intersection points:

$$\dfrac{6x^2 - 7x - 3}{2x^2 - 7x + 6} = 3$$
$$6x^2 - 7x - 3 = 6x^2 - 21x + 18$$
$$14x = 21$$
$$x = \dfrac{3}{2}$$

$R(x)$ does not intersect $y = 3$ because $R(x)$ is not defined at $x = \dfrac{3}{2}$.

Step 6:

Interval	$\left(-\infty, -\dfrac{1}{3}\right)$	$\left(-\dfrac{1}{3}, \dfrac{3}{2}\right)$	$\left(\dfrac{3}{2}, 2\right)$	$(2, \infty)$
Number Chosen	-1	0	1.7	6
Value of R	$R(-1) = \dfrac{2}{3}$	$R(0) = -\dfrac{1}{2}$	$R(1.7) \approx -20.3$	$R(6) = 4.75$
Location of Graph	Above x-axis	Below x-axis	Below x-axis	Above x-axis
Point on Graph	$\left(-1, \dfrac{2}{3}\right)$	$\left(0, -\dfrac{1}{2}\right)$	$(1.7, -20.3)$	$(6, 4.75)$

Step 7: Graphing:

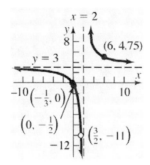

37. $R(x) = \dfrac{x^2+5x+6}{x+3} = \dfrac{(x+2)(x+3)}{x+3} = x+2$ $\quad p(x) = x^2+5x+6;\quad q(x) = x+3;\quad n=2;\ m=1$

Step 1: Domain: $\{x \mid x \neq -3\}$

Step 2: (a) The x-intercept is the zero of $y = x+2$: -2; Note: -3 is not a zero because reduced form must be used to find the zeros.

(b) The y-intercept is $R(0) = \dfrac{0^2 + 5(0) + 6}{0+3} = \dfrac{6}{3} = 2$.

Step 3: $R(-x) = \dfrac{(-x)^2 + 5(-x) + 6}{-x+3} = \dfrac{x^2 - 5x + 6}{-x+3}$; this is neither $R(x)$ nor $-R(x)$, so there is no symmetry.

Step 4: In lowest terms, $R(x) = x+2$, $x \neq -3$. There are no vertical asymptotes. Note: $x = -3$ is not a vertical asymptote because reduced form must be used to find the asymptotes. The graph has a hole at $(-3, -1)$.

Step 5: Since $n = m+1$ there is an oblique asymptote. The line $y = x+2$ is the oblique asymptote. Solve to find intersection points:
$$\dfrac{x^2+5x+6}{x+3} = x+2$$
$$x^2+5x+6 = (x+2)(x+3)$$
$$x^2+5x+6 = x^2+5x+6$$
$$0 = 0$$
The oblique asymptote intersects $R(x)$ at every point of the form $(x, x+2)$ except $(-3, -1)$.

Step 6:

Interval	$(-\infty, -3)$	$(-3, -2)$	$(-2, \infty)$
Number Chosen	-4	-2.5	0
Value of R	$R(-4) = -2$	$R(-2.5) = -\frac{1}{2}$	$R(0) = 2$
Location of Graph	Below x-axis	Below x-axis	Above x-axis
Point on Graph	$(-4, -2)$	$\left(-2.5, -\frac{1}{2}\right)$	$(0, 2)$

Step 7: Graphing:

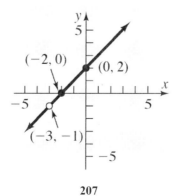

39. $f(x) = x + \dfrac{1}{x} = \dfrac{x^2+1}{x}$ $p(x) = x^2+1;$ $q(x) = x;$ $n = 2;$ $m = 1$

Step 1: Domain: $\{x \mid x \neq 0\}$

Step 2: (a) There are no x-intercepts since $x^2 + 1 = 0$ has no real solutions.

(b) There is no y-intercept because 0 is not in the domain.

Step 3: $f(-x) = \dfrac{(-x)^2+1}{-x} = \dfrac{x^2+1}{-x} = -f(x)$; The graph of $f(x)$ is symmetric with respect to the origin.

Step 4: $f(x) = \dfrac{x^2+1}{x}$ is in lowest terms. The vertical asymptote is the zero of $q(x)$: $x = 0$

Step 5: Since $n = m+1$, there is an oblique asymptote.
Dividing:

$$\begin{array}{r} x \\ x\overline{)x^2+1} \\ \underline{x^2} \\ 1 \end{array}$$ $f(x) = x + \dfrac{1}{x}$

The oblique asymptote is $y = x$.

Solve to find intersection points:
$$\dfrac{x^2+1}{x} = x$$
$$x^2 + 1 = x^2$$
$$1 \neq 0$$

The oblique asymptote does not intersect $f(x)$.

Step 6:

Interval	$(-\infty, 0)$	$(0, \infty)$
Number Chosen	-1	1
Value of f	$f(-1) = -2$	$f(1) = 2$
Location of Graph	Below x-axis	Above x-axis
Point on Graph	(-1, -2)	(1, 2)

Step 7: Graphing :

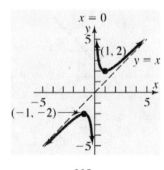

41. $f(x) = x^2 + \dfrac{1}{x} = \dfrac{x^3 + 1}{x}$ $p(x) = x^3 + 1;\ q(x) = x;\ n = 3;\ m = 1$

Step 1: Domain: $\{x \mid x \neq 0\}$

Step 2: (a) The x-intercept is the zero of $p(x)$: -1

(b) There is no y-intercept because 0 is not in the domain.

Step 3: $f(-x) = \dfrac{(-x)^3 + 1}{-x} = \dfrac{-x^3 + 1}{-x}$; this is neither $f(x)$ nor $-f(x)$, so there is no symmetry.

Step 4: $f(x) = \dfrac{x^3 + 1}{x}$ is in lowest terms. The vertical asymptote is the zero of $q(x)$: $x = 0$

Step 5: Since $n > m + 1$, there is no horizontal or oblique asymptote.

Step 6:

Interval	$(-\infty, -1)$	$(-1, 0)$	$(0, \infty)$
Number Chosen	-2	$-\frac{1}{2}$	1
Value of f	$f(-2) = 3.5$	$f(-\frac{1}{2}) = -1.75$	$f(1) = 2$
Location of Graph	Above x-axis	Below x-axis	Above x-axis
Point on Graph	$(-2, 3.5)$	$(-\frac{1}{2}, -1.75)$	$(1, 2)$

Step 7: Graphing:

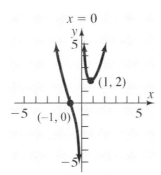

43. $f(x) = x + \dfrac{1}{x^3} = \dfrac{x^4 + 1}{x^3}$ $p(x) = x^4 + 1;\ q(x) = x^3;\ n = 4;\ m = 3$

Step 1: Domain: $\{x \mid x \neq 0\}$

Step 2: (a) There are no x-intercepts since $x^4 + 1 = 0$ has no real solutions.

(b) There is no y-intercept because 0 is not in the domain.

Step 3: $f(-x) = \dfrac{(-x)^4 + 1}{(-x)^3} = \dfrac{x^4 + 1}{-x^3} = -f(x)$; the graph of $f(x)$ is symmetric with respect to the origin.

Step 4: $f(x) = \dfrac{x^4 + 1}{x^3}$ is in lowest terms. The vertical asymptote is the zero of $q(x)$: $x = 0$

Step 5: Since $n = m + 1$, there is an oblique asymptote. Dividing:

$$x^3 \overline{)x^4 + 1} \qquad f(x) = x + \dfrac{1}{x^3}$$
$$\underline{x^4}$$
$$1$$

The oblique asymptote is $y = x$. Solve to find intersection points:

$$\dfrac{x^4 + 1}{x^3} = x$$
$$x^4 + 1 = x^4$$
$$1 \neq 0$$

The oblique asymptote does not intersect $f(x)$.

Step 6:

Interval	$(-\infty, 0)$	$(0, \infty)$
Number Chosen	-1	1
Value of f	$f(-1) = -2$	$f(1) = 2$
Location of Graph	Below x-axis	Above x-axis
Point on Graph	$(-1, -2)$	$(1, 2)$

Step 7: Graphing:

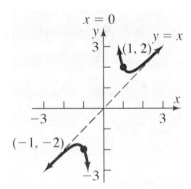

45. $f(x) = x + \dfrac{1}{x}, x > 0$

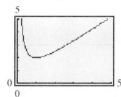

minimum value = 2.00 at $x = 1.00$

47. $f(x) = x^2 + \dfrac{1}{x}, x > 0$

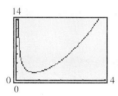

minimum value ≈ 1.89 at $x \approx 0.79$

49. $f(x) = x + \dfrac{1}{x^3}, x > 0$

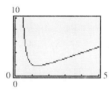

minimum value ≈ 1.75 at $x \approx 1.32$

51. One possibility: $f(x) = \dfrac{x^2}{x^2 - 4}$

53. One possibility: $f(x) = \dfrac{(x-1)(x-3)\left(x^2 + \dfrac{4}{3}\right)}{(x+1)^2 (x-2)^2}$

55. a. The degree of the numerator is 1 and the degree of the denominator is 2.

Thus, the horizontal asymptote is $C(t) = 0$.
The concentration of the drug decreases to 0 as time increases.

b. Graphing:

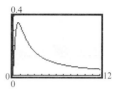

c. Using MAXIMUM, the concentration is highest when $t \approx 0.71$ hours.

57. a. The average cost function is:
$$\overline{C}(x) = \dfrac{0.2x^3 - 2.3x^2 + 14.3x + 10.2}{x}$$

b. $\overline{C}(6) = \dfrac{(.2)(6)^3 - (2.3)(6)^2 + (14.3)(6) + 10.2}{6}$

$= \dfrac{56.4}{6} = 9.4$

The average cost of producing 6 Cavaliers is $9400 per car.

c. $\overline{C}(9) = \dfrac{(.2)(9)^3 - (2.3)(9)^2 + 14.3(9) + 10.2}{9}$

$= \dfrac{98.4}{9} \approx 10.933$

The average cost of producing 9 Cavaliers is $10,933 per car.

d. Graphing:

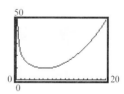

e. Using MINIMUM, the number of Cavaliers that should be produced to minimize cost is 6.38 cars (or about 6).

f. The minimum average cost is
$\overline{C}(6.38)$

$= \dfrac{0.2(6.38)^3 - 2.3(6.38)^2 + 14.3(6.38) + 10.2}{6.38}$

$\approx \$9366$ per car

59. a. The surface area is the sum of the areas of the six sides.

$S = xy + xy + xy + xy + x^2 + x^2 = 4xy + 2x^2$

The volume is

$x \cdot x \cdot y = x^2 y = 10{,}000 \Rightarrow y = \dfrac{10{,}000}{x^2}$

Thus, $S(x) = 4x\left(\dfrac{10{,}000}{x^2}\right) + 2x^2$

$= 2x^2 + \dfrac{40{,}000}{x}$

$= \dfrac{2x^3 + 40{,}000}{x}$

b. Graphing:

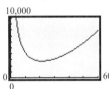

c. Using MINIMUM, the minimum surface area (amount of cardboard) is about 2785 square inches.

d. The surface area is a minimum when $x \approx 21.54$ inches.

$y = \dfrac{10{,}000}{(21.544)^2} \approx 21.54$ inches

The dimensions of the box are: 21.54 in. by 21.54 in. by 21.54 in.

61. a. $500 = \pi r^2 h \Rightarrow h = \dfrac{500}{\pi r^2}$

$C(r) = 6(2\pi r^2) + 4(2\pi r h)$

$= 12\pi r^2 + 8\pi r \left(\dfrac{500}{\pi r^2}\right)$

$= 12\pi r^2 + \dfrac{4000}{r}$

b. Graphing:

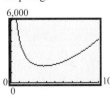

Using MINIMUM, the cost is least for $r \approx 3.76$ cm.

63. $D = \dfrac{k}{p}$

a. $D = 156$, $p = 2.75$;

$156 = \dfrac{k}{2.75}$

$k = 429$

So, $D = \dfrac{429}{p}$.

b. $D = \dfrac{429}{3} = 143$ bags of candy

65. $V = \dfrac{k}{P}$

a. $V = 600$, $P = 150$;

$600 = \dfrac{k}{150}$

$k = 90{,}000$

So, $V = \dfrac{90{,}000}{P}$

When $P = 200$;

$V = \dfrac{90{,}000}{200} = 450$ cubic centimeters

67. $W = \dfrac{k}{d^2}$

If $W = 125$, $d = 3960$ then

$125 = \dfrac{k}{3960^2}$ and $k = 1{,}960{,}200{,}000$

So, $W = \dfrac{1{,}960{,}200{,}000}{d}$

At the top of Mt. McKinley, we have $d = 3960 + 3.8 = 3963.8$

$W = \dfrac{1{,}960{,}200{,}000}{(3963.8)^2} = 124.76$ pounds

69. $V = \pi r^2 h$

71. $h = ksd^3$

$36 = k(75)(2)^3 \Rightarrow k = 0.06$

When $h = 45$ and $s = 125$,
$$45 = (0.06)(125)(d)^3$$
$$\Rightarrow d = \sqrt[3]{\frac{45}{7.5}} \approx 1.82 \text{ inches}$$

73. $K = kmv^2$

$400 = k(25)(100)^2 \Rightarrow k = 0.0016$

When $v = 150$,
$$K = (0.0016)(25)(150)^2 = 900 \text{ foot-lbs}$$

75. $S = \dfrac{kpd}{t}$

$100 = \dfrac{k(25)(5)}{(0.75)} \Rightarrow k = 0.6$

When $p = 40$, $d = 8$, and $t = 0.50$
$$S = \frac{(0.6)(40)(8)}{(0.50)} = 384 \text{ psi}$$

77. $y = \dfrac{x^2 - 1}{x - 1}$

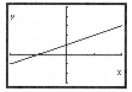

$y = \dfrac{x^3 - 1}{x - 1}$

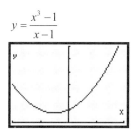

$y = \dfrac{x^4 - 1}{x - 1}$

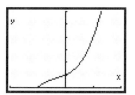

$y = \dfrac{x^5 - 1}{x - 1}$

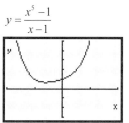

$x = 1$ is not a vertical asymptote because of the following behavior:

When $x \neq 1$:
$$y = \frac{x^2 - 1}{x - 1} = \frac{(x+1)(x-1)}{x-1} = x + 1$$
$$y = \frac{x^3 - 1}{x - 1} = \frac{(x-1)(x^2 + x + 1)}{x - 1} = x^2 + x + 1$$
$$y = \frac{x^4 - 1}{x - 1} = \frac{(x^2 + 1)(x^2 - 1)}{x - 1}$$
$$= \frac{(x^2 + 1)(x - 1)(x + 1)}{x - 1}$$
$$= x^3 + x^2 + x + 1$$
$$y = \frac{x^5 - 1}{x - 1}$$
$$= \frac{(x^4 + x^3 + x^2 + x + 1)(x - 1)}{x - 1}$$
$$= x^4 + x^3 + x^2 + x + 1$$

In general, the graph of
$$y = \frac{x^n - 1}{x - 1}, \ n \geq 1, \text{ an integer, will have a}$$
"hole" with coordinates $(1, n)$.

79. Answers will vary.

81. Answers will vary, one example is
$$R(x) = \frac{2(x - 3)(x + 2)^2}{(x - 1)^3}.$$

83-85. Answers will vary.

Section 3.5

1.

3. $(x-5)(x+2) < 0$

   $f(x) = (x-5)(x+2)$

   $x = 5, x = -2$ are the zeros of f.

Interval	$(-\infty, -2)$	$(-2, 5)$	$(5, \infty)$
Number Chosen	-3	0	6
Value of f	8	-10	8
Conclusion	Positive	Negative	Positive

   The solution set is $\{x \mid -2 < x < 5\}$.

5. $x^2 - 4x > 0$
   $x(x-4) > 0$

   $f(x) = x^2 - 4x$

   $x = 0, x = 4$ are the zeros of f.

Interval	$(-\infty, 0)$	$(0, 4)$	$(4, \infty)$
Number Chosen	-1	1	5
Value of f	5	-3	5
Conclusion	Positive	Negative	Positive

   The solution set is $\{x \mid x < 0 \text{ or } x > 4\}$.

7. $x^2 - 9 < 0$
   $(x-3)(x+3) < 0$

   $f(x) = x^2 - 9$

   $x = -3, x = 3$ are the zeros of f.

Interval	$(-\infty, -3)$	$(-3, 3)$	$(3, \infty)$
Number Chosen	-4	0	4
Value of f	7	-9	7
Conclusion	Positive	Negative	Positive

   The solution set is $\{x \mid -3 < x < 3\}$.

9. $x^2 + x \geq 12$
   $x^2 + x - 12 \geq 0$
   $(x+4)(x-3) \geq 0$

   $f(x) = x^2 + x - 12$

   $x = -4, x = 3$ are the zeros of f

Interval	$(-\infty, -4)$	$(-4, 3)$	$(3, \infty)$
Number Chosen	-5	0	4
Value of f	8	-12	8
Conclusion	Positive	Negative	Positive

   The solution set is $\{x \mid x \leq -4 \text{ or } x \geq 3\}$.

11. $2x^2 > 5x + 3$
    $2x^2 - 5x - 3 > 0$
    $(2x+1)(x-3) > 0$

    $f(x) = 2x^2 - 5x - 3$

    $x = -\dfrac{1}{2}, x = 3$ are the zeros of f

Interval	$\left(-\infty, -\dfrac{1}{2}\right)$	$\left(-\dfrac{1}{2}, 3\right)$	$(3, \infty)$
Number Chosen	-1	0	4
Value of f	4	-3	9
Conclusion	Positive	Negative	Positive

    The solution set is $\left\{x \mid x < -\dfrac{1}{2} \text{ or } x > 3\right\}$.

13. $x(x-7) < 8$
$x^2 - 7x < 8$
$x^2 - 7x - 8 < 0$
$(x+1)(x-8) < 0$

$f(x) = x^2 - 7x - 8$

$x = -1, x = 8$ are the zeros of f.

Interval	$(-\infty, -1)$	$(-1, 8)$	$(8, \infty)$
Number Chosen	-2	0	9
Value of f	10	-8	10
Conclusion	Positive	Negative	Positive

The solution set is $\{x \mid -1 < x < 8\}$.

15. $4x^2 + 9 < 6x$
$4x^2 - 6x + 9 < 0$

$f(x) = 4x^2 - 6x + 9$

$b^2 - 4ac = (-6)^2 - 4(4)(9)$
$= 36 - 144$
$= -108$

Since the discriminant is negative, f has no real zeros. Therefore, f is either always positive or always negative. Choose any value and test.
For $x = 0$,
$f(0) = 4(0)^2 - 6(0) + 9 = 9 > 0$
Thus, there is no real solution

17. $6(x^2 - 1) \geq 5x$
$6x^2 - 6 \geq 5x$
$6x^2 - 5x - 6 \geq 0$
$(3x+2)(2x-3) \geq 0$

$f(x) = 6x^2 - 5x - 6$

$x = -\dfrac{2}{3}, x = \dfrac{3}{2}$ are the zeros of f.

Interval	$\left(-\infty, -\dfrac{2}{3}\right)$	$\left(-\dfrac{2}{3}, \dfrac{3}{2}\right)$	$\left(\dfrac{3}{2}, \infty\right)$
Number Chosen	-1	0	2
Value of f	5	-6	8
Conclusion	Positive	Negative	Positive

The solution set is $\left\{x \mid x \leq -\dfrac{2}{3} \text{ or } x \geq \dfrac{3}{2}\right\}$.

19. $(x-1)(x^2 + x + 1) > 0$

$f(x) = (x-1)(x^2 + x + 1)$

$x = 1$ is the zero of f. $x^2 + x + 1 = 0$ has no real solution.

Interval	$(-\infty, 1)$	$(1, \infty)$
Number Chosen	0	2
Value of f	-1	7
Conclusion	Negative	Positive

The solution set is $\{x \mid x > 1\}$.

21. $(x-1)(x-2)(x-3) < 0$

$f(x) = (x-1)(x-2)(x-3)$

$x = 1, x = 2, x = 3$ are the zeros of f.

Interval	$(-\infty, 1)$	$(1, 2)$	$(2, 3)$	$(3, \infty)$
Number Chosen	0	1.5	2.5	4
Value of f	-6	0.375	-0.375	6
Conclusion	Negative	Positive	Negative	Positive

The solution set is $\{x \mid x < 1 \text{ or } 2 < x < 3\}$.

23. $x^3 - 2x^2 - 3x \geq 0$

$x(x^2 - 2x - 3) \geq 0$

$x(x+1)(x-3) \geq 0$

$f(x) = x^3 - 2x^2 - 3x$

$x = -1, x = 0, x = 3$ are the zeros of f.

Interval	$(-\infty, -1)$	$(-1, 0)$	$(0, 3)$	$(3, \infty)$
Number Chosen	-2	-0.5	1	4
Value of f	-10	0.875	-4	20
Conclusion	Negative	Positive	Negative	Positive

The solution set is $\{x \mid -1 \leq x \leq 0 \text{ or } x \geq 3\}$.

25. $x^4 > x^2$

$x^4 - x^2 > 0$

$x^2(x^2 - 1) > 0$

$x^2(x-1)(x+1) > 0$

$f(x) = x^2(x-1)(x+1)$

$x = -1, x = 0, x = 1$ are the zeros of f

Interval	$(-\infty, -1)$	$(-1, 0)$	$(0, 1)$	$(1, \infty)$
Number Chosen	-2	-0.5	0.5	2
Value of f	12	-0.1875	-0.1875	12
Conclusion	Positive	Negative	Negative	Positive

The solution set is $\{x \mid x < -1 \text{ or } x > 1\}$.

27. $x^3 > x^2$

$x^3 - x^2 > 0$

$x^2(x-1) > 0$

$f(x) = x^2(x-1)$

$x = 0, x = 1$ are the zeros of f

Interval	$(-\infty, 0)$	$(0, 1)$	$(1, \infty)$
Number Chosen	-1	0.5	2
Value of f	-2	-0.125	4
Conclusion	Negative	Negative	Positive

The solution set is $\{x \mid x > 1\}$.

29. $x^4 > 1$

$x^4 - 1 > 0$

$(x^2 - 1)(x^2 + 1) > 0$

$(x-1)(x+1)(x^2 + 1) > 0$

$f(x) = (x-1)(x+1)(x^2 + 1)$

$x = 1, x = -1$ are the zeros of f; $x^2 + 1$ has no real solution

Interval	$(-\infty, -1)$	$(-1, 1)$	$(1, \infty)$
Number Chosen	-2	0	2
Value of f	15	-1	15
Conclusion	Positive	Negative	Positive

The solution set is $\{x \mid x < -1 \text{ or } x > 1\}$.

31. $x^2 - 7x - 8 < 0$

$(x-8)(x+1) < 0$

$f(x) = (x-8)(x+1)$

$x = 8, x = -1$ are the zeros of f

Interval	$(-\infty, -1)$	$(-1, 8)$	$(8, \infty)$
Number Chosen	-2	0	9
Value of f	10	-8	10
Conclusion	Positive	Negative	Positive

The solution set is $\{x \mid -1 < x < 8\}$.

33. $$x^4 - 3x^2 - 4 > 0$$
$$(x^2 - 4)(x^2 + 1) > 0$$
$$(x - 2)(x + 2)(x^2 + 1) > 0$$

$$f(x) = (x - 2)(x + 2)(x^2 + 1)$$

$x = 2, x = -2$ are the zeros of f; $x^2 + 1$ has no real solution

Interval	$(-\infty, -2)$	$(-2, 2)$	$(2, \infty)$
Number Chosen	-3	0	3
Value of f	50	-4	50
Conclusion	Positive	Negative	Positive

The solution set is $\{x \mid x < -2 \text{ or } x > 2\}$

35. $$\frac{x+1}{x-1} > 0$$

$$f(x) = \frac{x+1}{x-1}$$

The zeros and values where f is undefined are $x = -1$ and $x = 1$.

Interval	$(-\infty, -1)$	$(-1, 1)$	$(1, \infty)$
Number Chosen	-2	0	2
Value of f	$\frac{1}{3}$	-1	3
Conclusion	Positive	Negative	Positive

The solution set is $\{x \mid x < -1 \text{ or } x > 1\}$.

37. $$\frac{(x-1)(x+1)}{x} < 0$$

$$f(x) = \frac{(x-1)(x+1)}{x}$$

The zeros and values where f is undefined are $x = -1, x = 0$ and $x = 1$.

Interval	$(-\infty, -1)$	$(-1, 0)$	$(0, 1)$	$(1, \infty)$
Number Chosen	-2	-0.5	0.5	2
Value of f	-1.5	1.5	-1.5	1.5
Conclusion	Negative	Positive	Negative	Positive

The solution set is $\{x \mid x < -1 \text{ or } 0 < x < 1\}$.

39. $$\frac{(x-2)^2}{x^2 - 1} \geq 0$$
$$\frac{(x-2)^2}{(x+1)(x-1)} \geq 0$$

$$f(x) = \frac{(x-2)^2}{x^2 - 1}$$

The zeros and values where f is undefined are $x = -1, x = 1$ and $x = 2$.

Interval	$(-\infty, -1)$	$(-1, 1)$	$(1, 2)$	$(2, \infty)$
Number Chosen	-2	0	1.5	3
Value of f	$\frac{16}{3}$	-4	0.2	0.125
Conclusion	Positive	Negative	Positive	Positive

The solution set is $\{x \mid x < -1 \text{ or } x > 1\}$.

41. $$6x - 5 < \frac{6}{x}$$
$$6x - 5 - \frac{6}{x} < 0$$
$$\frac{6x^2 - 5x - 6}{x} < 0$$
$$\frac{(2x - 3)(3x + 2)}{x} < 0$$

$$f(x) = 6x - 5 - \frac{6}{x}$$

The zeros and values where f is undefined are $x = -\frac{2}{3}, x = 0$ and $x = \frac{3}{2}$.

Interval	$\left(-\infty, -\dfrac{2}{3}\right)$	$\left(-\dfrac{2}{3}, 0\right)$	$\left(0, \dfrac{3}{2}\right)$	$\left(\dfrac{3}{2}, \infty\right)$
Number Chosen	-1	-0.5	1	2
Value of f	-5	4	-5	4
Conclusion	Negative	Positive	Negative	Positive

The solution set is $\left\{ x \mid x < -\dfrac{2}{3} \text{ or } 0 < x < \dfrac{3}{2} \right\}$.

43.
$$\dfrac{x+4}{x-2} \le 1$$
$$\dfrac{x+4}{x-2} - 1 \le 0$$
$$\dfrac{x+4-(x-2)}{x-2} \le 0$$
$$\dfrac{6}{x-2} \le 0$$

$$f(x) = \dfrac{x+4}{x-2} - 1$$

The value where f is undefined is $x = 2$.

Interval	$(-\infty, 2)$	$(2, \infty)$
Number Chosen	0	3
Value of f	-3	6
Conclusion	Negative	Positive

The solution set is $\{ x \mid x < 2 \}$.

45.
$$\dfrac{3x-5}{x+2} \le 2$$
$$\dfrac{3x-5}{x+2} - 2 \le 0$$
$$\dfrac{3x-5-2(x+2)}{x+2} \le 0$$
$$\dfrac{x-9}{x+2} \le 0$$

$$f(x) = \dfrac{3x-5}{x+2} - 2$$

The zeros and values where f is undefined are $x = -2$ and $x = 9$.

Interval	$(-\infty, -2)$	$(-2, 9)$	$(9, \infty)$
Number Chosen	-3	0	10
Value of f	12	-4.5	$\dfrac{1}{12}$
Conclusion	Positive	Negative	Positive

The solution set is $\{ x \mid -2 < x \le 9 \}$.

47.
$$\dfrac{1}{x-2} < \dfrac{2}{3x-9}$$
$$\dfrac{1}{x-2} - \dfrac{2}{3x-9} < 0$$
$$\dfrac{3x-9-2(x-2)}{(x-2)(3x-9)} < 0$$
$$\dfrac{x-5}{(x-2)(3x-9)} < 0$$

$$f(x) = \dfrac{1}{x-2} - \dfrac{2}{3x-9}$$

The zeros and values where f is undefined are $x = 2$, $x = 3$, and $x = 5$.

Interval	$(-\infty, 2)$	$(2, 3)$	$(3, 5)$	$(5, \infty)$
Number Chosen	0	2.5	4	6
Value of f	$-\dfrac{5}{18}$	$\dfrac{10}{3}$	$-\dfrac{1}{6}$	$\dfrac{1}{36}$
Conclusion	Negative	Positive	Negative	Positive

The solution set is $\{ x \mid x < 2 \text{ or } 3 < x < 5 \}$.

49.
$$\frac{2x+5}{x+1} > \frac{x+1}{x-1}$$
$$\frac{2x+5}{x+1} - \frac{x+1}{x-1} > 0$$
$$\frac{(2x+5)(x-1)-(x+1)(x+1)}{(x+1)(x-1)} > 0$$
$$\frac{2x^2+3x-5-\left(x^2+2x+1\right)}{(x+1)(x-1)} > 0$$
$$\frac{x^2+x-6}{(x+1)(x-1)} > 0$$
$$\frac{(x+3)(x-2)}{(x+1)(x-1)} > 0$$
$$f(x) = \frac{2x+5}{x+1} - \frac{x+1}{x-1}$$

The zeros and values where f is undefined are $x=-3, x=-1, x=1$ and $x=2$.

Interval	Number Chosen	Value of f	Conclusion
$(-\infty, -3)$	-4	0.4	Positive
$(-3, -1)$	-2	$-\dfrac{4}{3}$	Negative
$(-1, 1)$	0	6	Positive
$(1, 2)$	1.5	-1.8	Negative
$(2, \infty)$	3	0.75	Positive

The solution set is
$\{ x \mid x < -3 \text{ or } -1 < x < 1 \text{ or } x > 2 \}$.

51. $\dfrac{x^2(3+x)(x+4)}{(x+5)(x-1)} > 0$

$$f(x) = \frac{x^2(3+x)(x+4)}{(x+5)(x-1)}$$

The zeros and values where f is undefined are $x=-5, x=-4, x=-3, x=0$ and $x=1$.

Interval	Number Chosen	Value of f	Conclusion
$(-\infty, -5)$	-4	$\dfrac{216}{7}$	Positive
$(-5, -4)$	-4.5	$-\dfrac{243}{44}$	Negative
$(-4, -3)$	-3.5	$\dfrac{49}{108}$	Positive
$(-3, 0)$	-1	-0.75	Negative
$(0, 1)$	0.5	$-\dfrac{63}{44}$	Negative
$(1, \infty)$	2	$\dfrac{120}{7}$	Positive

The solution set is
$\{ x \mid x < -5 \text{ or } -4 < x < -3 \text{ or } x > 1 \}$.

53. $\dfrac{2x^2-x-1}{x-4} \leq 0$

$$\frac{(2x+1)(x-1)}{x-4} \leq 0$$

$$f(x) = \frac{2x^2-x-1}{x-4}$$

The zeros and values where f is undefined are $x=-\dfrac{1}{2}, x=1, x=4$.

Interval	$\left(-\infty, -\dfrac{1}{2}\right)$	$\left(-\dfrac{1}{2}, 1\right)$	$(1, 4)$	$(4, \infty)$
Number Chosen	-1	0	2	5
Value of f	-0.4	0.25	-2.5	44
Conclusion	Negative	Positive	Negative	Positive

The solution set is $\left\{ x \mid x \leq -\dfrac{1}{2} \text{ or } 1 \leq x < 4 \right\}$.

55. $\dfrac{x^2+3x-1}{x+3} > 0$

$f(x) = \dfrac{x^2+3x-1}{x+3}$

f is undefined at $x = -3$ and the zeros of f are:

$x = \dfrac{-b \pm \sqrt{b^2-4ac}}{2a} = \dfrac{-3 \pm \sqrt{3^2-4(1)(-1)}}{2(1)}$

$= \dfrac{-3 \pm \sqrt{9+4}}{2}$

$= \dfrac{-3 \pm \sqrt{13}}{2}$

Interval	Number Chosen	Value of f	Conclusion
$\left(-\infty, \dfrac{-3-\sqrt{13}}{2}\right)$	-4	-3	Negative
$\left(\dfrac{-3-\sqrt{13}}{2}, -3\right)$	-3.25	0.75	Positive
$\left(-3, \dfrac{-3+\sqrt{13}}{2}\right)$	0	-0.33	Negative
$\left(\dfrac{-3+\sqrt{13}}{2}, \infty\right)$	1	0.75	Positive

The solution set is

$\left\{ x \,\bigg|\, \dfrac{-3-\sqrt{13}}{2} < x < -3 \text{ or } x > \dfrac{-3+\sqrt{13}}{2} \right\}$.

57. Let x be the positive number. Then

$x^3 > 4x^2$

$x^3 - 4x^2 > 0$

$x^2(x-4) > 0$

$f(x) = x^3 - 4x^2$

$x = 0$ and $x = 4$ are the zeros of f.

Interval	$(-\infty, 0)$	$(0, 4)$	$(4, \infty)$
Number Chosen	-1	1	5
Value of f	-5	-3	25
Conclusion	Negative	Negative	Positive

The solution set is $\{x \mid x > 4\}$. Since x must be positive, all real numbers greater than 4 satisfy the condition.

59. The domain of the expression consists of all real numbers x for which

$x^2 - 16 \geq 0$

$(x+4)(x-4) \geq 0$

$p(x) = x^2 - 16$

$x = -4$ and $x = 4$ are the zeros of p.

Interval	$(-\infty, -4)$	$(-4, 4)$	$(4, \infty)$
Number Chosen	-5	0	5
Value of p	9	-16	9
Conclusion	Positive	Negative	Positive

The domain of f is $\{x \mid x \leq -4 \text{ or } x \geq 4\}$.

61. The domain of the expression includes all values for which

$\dfrac{x-2}{x+4} \geq 0$

The zeros and values where the expression is undefined are $x = -4$ and $x = 2$.

$f(x) = \dfrac{x-2}{x+4}$

Interval	$(-\infty, -4)$	$(-4, 2)$	$(2, \infty)$
Number Chosen	-5	0	3
Value of f	7	$-\dfrac{1}{2}$	$\dfrac{1}{7}$
Conclusion	Positive	Negative	Positive

The solution or domain is $\{x \mid x < -4 \text{ or } x \geq 2\}$.

63. a. Find the values of t for which

$$80t - 16t^2 > 96$$
$$-16t^2 + 80t - 96 > 0$$
$$16t^2 - 80t + 96 < 0$$
$$16(t^2 - 5t + 6) < 0$$
$$16(t-2)(t-3) < 0$$

The zeros are $t = 2$ and $t = 3$.

$s(t) = 16t^2 - 80t + 96$

Interval	$(-\infty, 2)$	$(2, 3)$	$(3, \infty)$
Number Chosen	1	2.5	4
$s(t)$	32	-4	32
Conclusion	Positive	Negative	Positive

The solution set is $\{t \mid 2 < t < 3\}$. The ball is more than 96 feet above the ground for times between 2 and 3 seconds.

b. Graphing: $s = 80t - 16t^2$

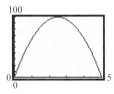

c. Using MAXIMUM, the maximum height is 100 feet.

d. The maximum height occurs at 2.5 seconds.

65. a. Profit = Revenue – Cost

$$x(40 - 0.2x) - 32x \geq 50$$
$$40x - 0.2x^2 - 32x \geq 50$$
$$-0.2x^2 + 8x - 50 \geq 0$$
$$2x^2 - 80x + 500 \leq 0$$
$$x^2 - 40x + 250 \leq 0$$

The zeros are approximately $x = 7.75$ and $x = 32.25$.

$f(x) = x^2 - 40x + 250$

Interval	$(0, 7.75)$	$(7.75, 32.25)$	$(32.25, \infty)$
Number Chosen	7	10	40
$f(x)$	19	-50	250
Conclusion	Positive	Negative	Positive

The profit is at least $50 when at least 8 and no more than 32 watches are sold.

b. Graphing the revenue function:
$(R(x) = x(40 - 0.2x))$

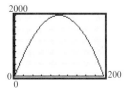

c. Using MAXIMUM, the maximum revenue is $2,000.

d. Using MAXIMUM, the company should sell 100 wristwatches to maximize revenue.

e. Graphing the profit function:
$P(x) = x(40 - 0.2x) - 32x$

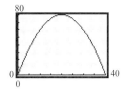

f. Using MAXIMUM, the maximum profit is $80.

g. Using MAXIMUM, the company should sell 20 watches for maximum profit.

h. Answers will vary.

67. The cost of manufacturing x Chevy Cavaliers in a day was found to be:

$$C(x) = 0.216x^3 - 2.347x^2 + 14.328x + 10.224$$

Since the budget constraints require that the cost must be less than or equal to $97,000, we need to solve: $C(x) \leq 97$ or

$$0.216x^3 - 2.347x^2 + 14.328x + 10.224 \leq 97$$

Graphing
$y_1 = .216x^3 - 2.347x^2 + 14.328x + 10.224$;
$y_2 = 97$

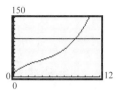

y_1 intersects y_2 at $x = 8.59$. $y_1 < y_2$ when $x < 8.59$. Chevy can produce at most eight Cavaliers in a day, assuming cars cannot be partially completed.

69. Prove that if a, b are real numbers and $a \geq 0, b \geq 0$, then $a \leq b$ is equivalent to $\sqrt{a} \leq \sqrt{b}$.

First note that if $a = 0$, then $a \leq b \Rightarrow b = 0$.

Similarly, if and $\sqrt{a} \leq \sqrt{b} \Rightarrow \sqrt{b} = 0 \Rightarrow b = 0$.

Therefore, $a \leq b$ is equivalent to $\sqrt{a} \leq \sqrt{b}$. Now suppose $a \neq 0$.

We must show that $a \leq b \Rightarrow \sqrt{a} \leq \sqrt{b}$ and that $\sqrt{a} \leq \sqrt{b} \Rightarrow a \leq b$.

Assume $a \leq b$, this means that $b - a \geq 0$. We can factor the left hand side by treating it as a difference of two perfect squares:
$(\sqrt{b} - \sqrt{a})(\sqrt{b} + \sqrt{a}) \geq 0$

Case 1: $\sqrt{b} - \sqrt{a} \geq 0$ and $\sqrt{b} + \sqrt{a} \geq 0$
$\Rightarrow \sqrt{b} \geq \sqrt{a}$ and $\sqrt{b} \geq -\sqrt{a}$
$\Rightarrow \sqrt{a} \leq \sqrt{b}$ and $-\sqrt{a} \leq \sqrt{b}$
or
Case 2: $\sqrt{b} \leq \sqrt{a}$ and $\sqrt{b} \leq -\sqrt{a}$
This is impossible since
$\sqrt{a} \geq 0$ and $\sqrt{b} \geq 0$ which means $\sqrt{b} \leq -\sqrt{a}$ is impossible since $a \neq 0$.

Since Case 2 is impossible, Case 1 must be true. Therefore, $a \leq b \Rightarrow \sqrt{a} \leq \sqrt{b}$.

Assume $\sqrt{a} \leq \sqrt{b}$. This means that $\sqrt{a} - \sqrt{b} \leq 0$. Since $\sqrt{a} \geq 0$ and $\sqrt{b} \geq 0$, we can conclude that $\sqrt{a} + \sqrt{b} \geq 0$. Therefore,

$(\sqrt{a} - \sqrt{b})(\sqrt{a} + \sqrt{b}) \leq 0(\sqrt{a} + \sqrt{b})$
$a - b \leq 0$
$a \leq b$

Thus $\sqrt{a} \leq \sqrt{b} \Rightarrow a \leq b$.

71. Rewriting the inequality:
$x^2 + 1 < -5$
$x^2 < -5 - 1$
$x^2 < -6$
x^2 is always positive and can never be less than -6. Thus, $x^2 + 1 < -5$ has no solution.

Section 3.6

1. Integers: $\{-2, 0\}$; Rational: $\{-2, 0, \frac{1}{2}, 4.5\}$

3. Quotient: $3x^2 + 4x^2 + 12x + 43$; Remainder: 125

5. Remainder; dividend

7. -4

9. False; the potential rational zeros are $\pm 1, \pm \frac{1}{2}$

11. $f(x) = 4x^3 - 3x^2 - 8x + 4$; $c = 3$
$f(3) = 4(3)^3 - 3(3)^2 - 8(3) + 4$
$= 108 - 27 - 24 + 4$
$= 61 \neq 0$
Thus, 3 is not a zero of f. So $x - 3$ is not a factor of f.

13. $f(x) = 3x^4 - 6x^3 - 5x + 10$; $c = 1$
$f(1) = 3(1)^4 - 6(1)^3 - 5(1) + 10$
$= 3 - 6 - 5 + 10$
$= 2 \neq 0$
Thus, 1 is not a zero of f. So $x - 1$ is not a factor of f.

15. $f(x) = 3x^6 + 2x^3 - 176$; $c = -2$

$f(-2) = 3(-2)^6 + 2(-2)^3 - 176$
$= 192 - 16 - 176$
$= 0$

Thus, -2 is a zero of f. So $x + 2$ is a factor of f.

Use synthetic division to find the factors.

$$\begin{array}{r|rrrrrrr} -2 & 3 & 0 & 0 & 2 & 0 & 0 & -176 \\ & & -6 & 12 & -24 & 44 & -88 & 176 \\ \hline & 3 & -6 & 12 & -22 & 44 & -88 & 0 \end{array}$$

The factored form is:
$f(x) = (x+2)(3x^5 - 6x^4 + 12x^3 - 22x^2 + 44x - 88)$.

17. $f(x) = 4x^6 - 64x^4 + x^2 - 16$; $c = 4$

$f(4) = 4(4)^6 - 64(4)^4 + (4)^2 - 16$
$= 16384 - 16384 + 16 - 16$
$= 0$

Thus, 4 is a zero of f. So $x - 4$ is a factor of f.

Use synthetic division to find the factors.

$$\begin{array}{r|rrrrrrr} 4 & 4 & 0 & -64 & 0 & 1 & 0 & -16 \\ & & 16 & 64 & 0 & 0 & 4 & 16 \\ \hline & 4 & 16 & 0 & 0 & 1 & 4 & 0 \end{array}$$

The factored form is:
$f(x) = (x-4)(4x^5 + 16x^4 + x + 4)$.

We can use grouping to factor

$4x^5 + 16x^4 + x + 4 = 4x^4(x+4) + x + 4$
$= (x+4)(4x^4 + 1)$

So $f(x) = (x-4)(x+4)(4x^4+1)$.

19. $f(x) = 2x^4 - x^3 + 2x - 1$; $c = -\frac{1}{2}$

$f\left(-\frac{1}{2}\right) = 2\left(-\frac{1}{2}\right)^4 - \left(-\frac{1}{2}\right)^3 + 2\left(-\frac{1}{2}\right) - 1$
$= \frac{1}{8} + \frac{1}{8} - 1 - 1$
$= -\frac{7}{4} \neq 0$

Thus, $-\frac{1}{2}$ is not a zero of f. So $x + \frac{1}{2}$ is not a factor of f.

21. $f(x) = 3x^4 - 3x^3 + x^2 - x + 1$

The maximum number of zeros is the degree of the polynomial which is 4.

p must be a factor of 1: $p = \pm 1$
q must be a factor of 3: $q = \pm 1, \pm 3$

The possible rational zeros are: $\frac{p}{q} = \pm 1, \pm \frac{1}{3}$

23. $f(x) = x^5 - 6x^2 + 9x - 3$

The maximum number of zeros is the degree of the polynomial which is 5.

p must be a factor of -3: $p = \pm 1, \pm 3$
q must be a factor of 1: $q = \pm 1$

The possible rational zeros are: $\frac{p}{q} = \pm 1, \pm 3$

25. $f(x) = -4x^3 - x^2 + x + 2$

The maximum number of zeros is the degree of the polynomial which is 3.

p must be a factor of 2: $p = \pm 1, \pm 2$
q must be a factor of -4: $q = \pm 1, \pm 2, \pm 4$

The possible rational zeros are:
$\frac{p}{q} = \pm 1, \pm 2, \pm \frac{1}{2}, \pm \frac{1}{4}$

Chapter 3: *Polynomial and Rational Functions*

27. $f(x) = 3x^4 - x^2 + 2$

    The maximum number of zeros is the degree of the polynomial which is 4.

    p must be a factor of 2: $p = \pm 1, \pm 2$
    q must be a factor of 3: $q = \pm 1, \pm 3$

    The possible rational zeros are:
    $$\frac{p}{q} = \pm 1, \pm \frac{1}{3}, \pm 2, \pm \frac{2}{3}$$

29. $f(x) = 2x^5 - x^3 + 2x^2 + 4$

    The maximum number of zeros is the degree of the polynomial which is 5.

    p must be a factor of 4: $p = \pm 1, \pm 2, \pm 4$
    q must be a factor of 2: $q = \pm 1, \pm 2$

    The possible rational zeros are:
    $$\frac{p}{q} = \pm 1, \pm \frac{1}{2}, \pm 2, \pm 4$$

31. $f(x) = 6x^4 + 2x^3 - x^2 + 2$

    The maximum number of zeros is the degree of the polynomial which is 4.

    p must be a factor of 2: $p = \pm 1, \pm 2$
    q must be a factor of 6: $q = \pm 1, \pm 2, \pm 3, \pm 6$

    The possible rational zeros are:
    $$\frac{p}{q} = \pm 1, \pm \frac{1}{2}, \pm \frac{1}{3}, \pm \frac{1}{6}, \pm 2, \pm \frac{2}{3}$$

33. $f(x) = 2x^3 + x^2 - 1 = 2\left(x^3 + \frac{1}{2}x^2 - \frac{1}{2}\right)$

    Note: The leading coefficient must be 1.

    $a_2 = \frac{1}{2}, a_1 = 0, a_0 = -\frac{1}{2}$

    $Max\left\{1, \left|-\frac{1}{2}\right| + |0| + \left|\frac{1}{2}\right|\right\} = Max\left\{1, \frac{1}{2} + 0 + \frac{1}{2}\right\}$
    $= Max\{1, 1\}$
    $= 1$

SSM: College Algebra EGU

$1 + Max\left\{\left|-\frac{1}{2}\right|, |0|, \left|\frac{1}{2}\right|\right\} = 1 + Max\left\{\frac{1}{2}, 0, \frac{1}{2}\right\}$
$= 1 + \frac{1}{2}$
$= 1.5$

The smaller of the two numbers is 1. Thus, every zero of f lies between -1 and 1.

Graphing using the bounds and ZOOM-FIT:

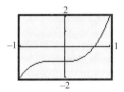

35. $f(x) = x^3 - 5x^2 - 11x + 11$

    $a_2 = -5, a_1 = -11, a_0 = 1$

    $Max\{1, |-5| + |-11| + |11|\} = Max\{1, 5 + 11 + 11\}$
    $= Max\{1, 27\}$
    $= 27$

    $1 + Max\{|-5|, |-11|, |11|\} = 1 + Max\{1, 5, 11, 11\}$
    $= 1 + 11$
    $= 12$

    The smaller of the two numbers is 12. Thus, every zero of f lies between -12 and 12.

    Graphing using the bounds and ZOOM-FIT: (Second graph has a better window.)

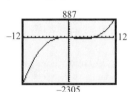

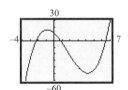

37. $f(x) = x^4 + 3x^3 - 5x^2 + 9$

$a_3 = 3, a_2 = -5, a_1 = 0, a_0 = 9$

$Max\{1, |9| + |0| + |-5| + |3|\} = Max\{1, 9+0+5+3\}$
$= Max\{1, 17\}$
$= 17$

$1 + Max\{|9|, |0|, |-5|, |3|\}$
$= 1 + Max\{9, 0, 5, 3\}$
$= 1 + 9$
$= 10$

The smaller of the two numbers is 10. Thus, every zero of f lies between -10 and 10.

Graphing using the bounds and ZOOM-FIT:
(Second graph has a better window.)

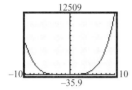

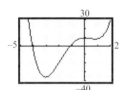

39. $f(x) = x^3 + 2x^2 - 5x - 6$

Step 1: $f(x)$ has at most 3 real zeros.

Step 2: Possible rational zeros:

$p = \pm 1, \pm 2, \pm 3, \pm 6; \quad q = \pm 1;$

$\dfrac{p}{q} = \pm 1, \pm 2, \pm 3, \pm 6$

Step 3: Using the Bounds on Zeros Theorem:

$a_2 = 2, \quad a_1 = -5, \quad a_0 = -6$

$Max\{1, |-6| + |-5| + |2|\} = Max\{1, 13\}$
$= 13$

$1 + Max\{|-6|, |-5|, |2|\} = 1 + 6 = 7$

The smaller of the two numbers is 7. Thus, every zero of f lies between -7 and 7.

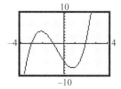

Step 4:
(a) From the graph it appears that there are x-intercepts at -3, -1, and 2.

(b) Using synthetic division:

$\begin{array}{r|rrrr} -3 & 1 & 2 & -5 & -6 \\ & & -3 & 3 & 6 \\ \hline & 1 & -1 & -2 & 0 \end{array}$

Since the remainder is 0, $x - (-3) = x + 3$ is a factor. The other factor is the quotient: $x^2 - x - 2$.

(c) Thus, $f(x) = (x+3)(x^2 - x - 2)$.
$= (x+3)(x+1)(x-2)$

The zeros are -3, -1, and 2.

41. $f(x) = 2x^3 - 13x^2 + 24x - 9$

Step 1: $f(x)$ has at most 3 real zeros.

Step 2: Possible rational zeros:

$p = \pm 1, \pm 3, \pm 9; \quad q = \pm 1, \pm 2;$

$\dfrac{p}{q} = \pm 1, \pm 3, \pm 9, \pm \dfrac{1}{2}, \pm \dfrac{3}{2}, \pm \dfrac{9}{2}$

Step 3: Using the Bounds on Zeros Theorem:

$f(x) = 2\left(x^3 - 6.5x^2 + 12x - 4.5\right)$

$a_2 = -6.5, \quad a_1 = 12, \quad a_0 = -4.5$

$Max\{1, |-4.5| + |12| + |-6.5|\} = Max\{1, 23\}$
$= 23$

$1 + Max\{|-4.5|, |12|, |-6.5|\} = 1 + 12$
$= 13$

The smaller of the two numbers is 13. Thus, every zero of f lies between -13 and 13.

Graphing using the bounds and ZOOM-FIT:
(Second graph has a better window.)

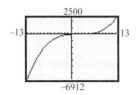

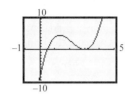

Step 4:
(a) From the graph it appears that there are x-intercepts at 0.5 and 3.

(b) Using synthetic division:

$$3\overline{)\begin{array}{cccc} 2 & -13 & 24 & -9 \\ & 6 & -21 & 9 \\ \hline 2 & -7 & 3 & 0 \end{array}}$$

Since the remainder is 0, $x-3$ is a factor. The other factor is the quotient: $2x^2 - 7x + 3$.

(c) Thus, $f(x) = (x-3)(2x^2 - 7x + 3)$.
$= (x-3)(2x-1)(x-3)$

The zeros are 0.5 and 3 (multiplicity 2).

43. $f(x) = 3x^3 + 4x^2 + 4x + 1$

Step 1: $f(x)$ has at most 3 real zeros..

Step 2: Possible rational zeros:

$p = \pm 1; \quad q = \pm 1, \pm 3; \quad \dfrac{p}{q} = \pm 1, \pm \dfrac{1}{3}$

Step 3: Using the Bounds on Zeros Theorem:

$f(x) = 3\left(x^3 + \dfrac{4}{3}x^2 + \dfrac{4}{3}x + \dfrac{1}{3}\right)$

$a_2 = \dfrac{4}{3}, \quad a_1 = \dfrac{4}{3}, \quad a_0 = \dfrac{1}{3}$

$\text{Max}\left\{1, \left|\dfrac{1}{3}\right| + \left|\dfrac{4}{3}\right| + \left|\dfrac{4}{3}\right|\right\} = \text{Max}\{1, 3\} = 3$

$1 + \text{Max}\left\{\left|\dfrac{1}{3}\right|, \left|\dfrac{4}{3}\right|, \left|\dfrac{4}{3}\right|\right\} = 1 + \dfrac{4}{3} = \dfrac{7}{3}$

The smaller of the two numbers is $\dfrac{7}{3}$. Thus, every zero of f lies between $-\dfrac{7}{3}$ and $\dfrac{7}{3}$.

Graphing using the bounds and ZOOM-FIT: (Second graph has a better window.)

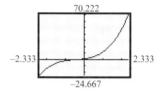

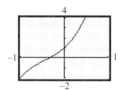

Step 4:
(a) From the graph it appears that there is an x-intercepts at $-\dfrac{1}{3}$.

(b) Using synthetic division:

$$-\dfrac{1}{3}\overline{)\begin{array}{cccc} 3 & 4 & 4 & 1 \\ & -1 & -1 & -1 \\ \hline 3 & 3 & 3 & 0 \end{array}}$$

Since the remainder is 0, $x - \left(-\dfrac{1}{3}\right) = x + \dfrac{1}{3}$ is a factor. The other factor is the quotient: $3x^2 + 3x + 3$.

(c) Thus, $f(x) = \left(x + \dfrac{1}{3}\right)(3x^2 + 3x + 3)$.

$= 3\left(x + \dfrac{1}{3}\right)(x^2 + x + 1)$

$= (3x + 1)(x^2 + x + 1)$

$x^2 + x + 1 = 0$ has no real solution.

The zero is $-\dfrac{1}{3}$.

45. $f(x) = x^3 - 8x^2 + 17x - 6$

Step 1: $f(x)$ has at most 3 real zeros.

Step 2: Possible rational zeros:

$p = \pm 1, \pm 2, \pm 3, \pm 6;\quad q = \pm 1;$

$\dfrac{p}{q} = \pm 1, \pm 2, \pm 3, \pm 6$

Step 3: Using the Bounds on Zeros Theorem:

$a_2 = -8,\ a_1 = 17,\ a_0 = -6$

Max $\{1, |-6| + |17| + |-8|\}$ = Max $\{1, 31\}$
$= 31$

$1 + $ Max $\{|-6|, |17|, |-8|\} = 1 + 17 = 18$

The smaller of the two numbers is 18. Thus, every zero of f lies between -18 and 18.

Graphing using the bounds and ZOOM-FIT: (Second graph has a better window.)

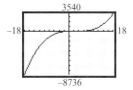

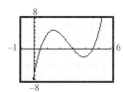

Step 4:
(a) From the graph it appears that there are x-intercepts at 0.5, 3, and 4.5.

(b) Using synthetic division:

$\begin{array}{r|rrrr} 3) & 1 & -8 & 17 & -6 \\ & & 3 & -15 & 6 \\ \hline & 1 & -5 & 2 & 0 \end{array}$

Since the remainder is 0, $x - 3$ is a factor. The other factor is the quotient: $x^2 - 5x + 2$.

(c) Thus, $f(x) = (x - 3)(x^2 - 5x + 2)$. Using the quadratic formula to find the solutions of the depressed equation $x^2 - 5x + 2 = 0$:

$x = \dfrac{-(-5) \pm \sqrt{(-5)^2 - 4(1)(2)}}{2(1)} = \dfrac{5 \pm \sqrt{17}}{2}$

Thus,

$f(x) = (x-3)\left(x - \left(\dfrac{5+\sqrt{17}}{2}\right)\right)\left(x - \left(\dfrac{5-\sqrt{17}}{2}\right)\right).$

The zeros are 3, $\dfrac{5+\sqrt{17}}{2}$, and $\dfrac{5-\sqrt{17}}{2}$ or 3, 4.56, and 0.44.

47. $f(x) = x^4 + x^3 - 3x^2 - x + 2$

Step 1: $f(x)$ has at most 4 real zeros.

Step 2: Possible rational zeros:

$p = \pm 1, \pm 2;\quad q = \pm 1;\quad \dfrac{p}{q} = \pm 1, \pm 2$

Step 3: Using the Bounds on Zeros Theorem:

$a_3 = 1,\ a_2 = -3,\ a_1 = -1,\ a_0 = 2$

Max $\{1, |2| + |-1| + |-3| + |1|\}$ = Max $\{1, 7\}$
$= 7$

$1 + $ Max $\{|2|, |-1|, |-3|, |1|\} = 1 + 3 = 4$

The smaller of the two numbers is 4. Thus, every zero of f lies between -4 and 4.

Graphing using the bounds and ZOOM-FIT: (Second graph has a better window.)

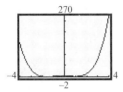

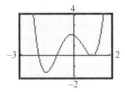

Step 4:
(a) From the graph it appears that there are x-intercepts at -2, -1, and 1.

(b) Using synthetic division:

$$\begin{array}{r|rrrr} -2 & 1 & -3 & -1 & 2 \\ & & -2 & 2 & 2 & -2 \\ \hline & 1 & -1 & -1 & 1 & 0 \end{array}$$

$$\begin{array}{r|rrr} -1 & 1 & -1 & -1 & 1 \\ & & -1 & 2 & -1 \\ \hline & 1 & -2 & 1 & 0 \end{array}$$

Since the remainder is 0, $x+2$ and $x+1$ are factors. The other factor is the quotient: $x^2 - 2x + 1$.

(c) Thus, $f(x) = (x+2)(x+1)(x-1)^2$.

The zeros are -2, -1, and 1 (multiplicity 2).

49. $f(x) = 2x^4 + 17x^3 + 35x^2 - 9x - 45$

Step 1: $f(x)$ has at most 4 real zeros.

Step 2: Possible rational zeros:

$p = \pm 1, \pm 3, \pm 5, \pm 9, \pm 15, \pm 45; \quad q = \pm 1, \pm 2;$

$\dfrac{p}{q} = \pm 1, \pm 3, \pm 5, \pm 9, \pm 15, \pm 45, \pm \dfrac{1}{2},$

$\pm \dfrac{3}{2}, \pm \dfrac{5}{2}, \pm \dfrac{9}{2}, \pm \dfrac{15}{2}, \pm \dfrac{45}{2}$

Step 3: Using the Bounds on Zeros Theorem:

$f(x) = 2\left(x^4 + 8.5x^3 + 17.5x^2 - 4.5x - 22.5\right)$

$a_3 = 8.5, \; a_2 = 17.5, \; a_1 = -4.5, \; a_0 = -22.5$

Max $\{1, |-22.5| + |-4.5| + |17.5| + |8.5|\}$

= Max $\{1, 53\}$

= 53

$1 + $ Max $\{|-22.5|, |-4.5|, |17.5|, |8.5|\}$

= 1 + 22.5

= 23.5

The smaller of the two numbers is 23.5. Thus, every zero of f lies between -23.5 and 23.5.

Graphing using the bounds and ZOOM-FIT:
(Second graph has a better window.)

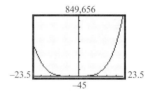

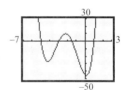

Step 4:
(a) From the graph it appears that there are x-intercepts at -5, -3, -1.5, and 1.

(b) Using synthetic division:

$$\begin{array}{r|rrrr} -5 & 2 & 17 & 35 & -9 & -45 \\ & & -10 & -35 & 0 & 45 \\ \hline & 2 & 7 & 0 & -9 & 0 \end{array}$$

$$\begin{array}{r|rrr} -3 & 2 & 7 & 0 & -9 \\ & & -6 & -3 & 9 \\ \hline & 2 & 1 & -3 & 0 \end{array}$$

Since the remainder is 0, $x+5$ and $x+3$ are factors. The other factor is the quotient: $2x^2 + x - 3$.

(c) Thus, $f(x) = (x+5)(x+3)(2x+3)(x-1)$.

The zeros are -5, -3, -1.5 and 1.

51. $f(x) = 2x^4 - 3x^3 - 21x^2 - 2x + 24$

Step 1: $f(x)$ has at most 4 real zeros.

Step 2: Possible rational zeros:

$p = \pm 1, \pm 2, \pm 3, \pm 4, \pm 6, \pm 8, \pm 12, \pm 24;$
$q = \pm 1, \pm 2;$
$\dfrac{p}{q} = \pm 1, \pm 2, \pm 3, \pm 4, \pm 6,$
$\pm 8, \pm 12, \pm 24, \pm \dfrac{1}{2}, \pm \dfrac{3}{2}$

Step 3: Using the Bounds on Zeros Theorem:

$f(x) = 2\left(x^4 - 1.5x^3 - 10.5x^2 - x + 12\right)$
$a_3 = -1.5,\ a_2 = -10.5,\ a_1 = -1,\ a_0 = 12$
Max $\{1, |12| + |-1| + |-10.5| + |-1.5|\}$
$= $ Max $\{1, 25\}$
$= 25$
$1 + $ Max $\{|12|, |-1|, |-10.5|, |-1.5|\}$
$= 1 + 12$
$= 13$

The smaller of the two numbers is 13. Thus, every zero of f lies between -13 and 13.

Graphing using the bounds and ZOOM-FIT:
(Second graph has a better window.)

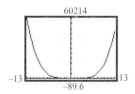

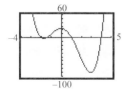

Step 4:
(a) From the graph it appears that there are x-intercepts at $-2, -1.5, 1,$ and 4.

(b) Using synthetic division:

$\begin{array}{r|rrrrr} -2) & 2 & -3 & -21 & -2 & 24 \\ & & -4 & 14 & 14 & -24 \\ \hline & 2 & -7 & -7 & 12 & 0 \end{array}$

$\begin{array}{r|rrrr} 4) & 2 & -7 & -7 & 12 \\ & & 8 & 4 & -12 \\ \hline & 2 & 1 & -3 & 0 \end{array}$

Since the remainder is 0, $x + 2$ and $x - 4$ are factors. The other factor is the quotient: $2x^2 + x - 3$.

(c) Thus, $f(x) = (x+2)(2x+3)(x-1)(x-4)$.

The zeros are $-2, -1.5, 1$ and 4.

53. $f(x) = 4x^4 + 7x^2 - 2$

Step 1: $f(x)$ has at most 4 real zeros.

Step 2: Possible rational zeros:

$p = \pm 1, \pm 2;\quad q = \pm 1, \pm 2, \pm 4;$
$\dfrac{p}{q} = \pm 1, \pm 2, \pm \dfrac{1}{2}, \pm \dfrac{1}{4}$

Step 3: Using the Bounds on Zeros Theorem:

$f(x) = 4\left(x^4 + 1.75x^2 - 0.5\right)$
$a_3 = 0,\ a_2 = 1.75,\ a_1 = 0,\ a_0 = -0.5$
Max $\{1, |-0.5| + |0| + |1.75| + |0|\}$
$= $ Max $\{1, 2.25\}$
$= 2.25$
$1 + $ Max $\{|-0.5|, |0|, |1.75|, |0|\}$
$= 1 + 1.75$
$= 2.75$

The smaller of the two numbers is 2.25. Thus, every zero of f lies between -2.25 and 2.25.

Graphing using the bounds and ZOOM-FIT:
(Second graph has a better window.)

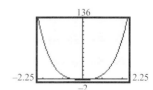

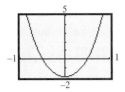

Step 4:
(a) From the graph it appears that there are x-intercepts at –0.5, and 0.5.

(b) Using synthetic division:

$$\begin{array}{r|rrrrr} -0.5 & 4 & 0 & 7 & 0 & -2 \\ & & -2 & 1 & -4 & 2 \\ \hline & 4 & -2 & 8 & -4 & 0 \end{array}$$

$$\begin{array}{r|rrrr} 0.5 & 4 & -2 & 8 & -4 \\ & & 2 & 0 & 4 \\ \hline & 4 & 0 & 8 & 0 \end{array}$$

Since the remainder is 0, $x+0.5$ and $x-0.5$ are factors. The other factor is the quotient: $4x^2 + 8$.

(c) Thus, $f(x) = 4(x+0.5)(x-0.5)(x^2+2)$.
$= (2x+1)(2x-1)(x^2+2)$

The depressed equation has no real zeros.

The zeros are –0.5, and 0.5.

55. $f(x) = 4x^5 - 8x^4 - x + 2$

Step 1: $f(x)$ has at most 5 real zeros.

Step 2: Possible rational zeros:
$p = \pm 1, \pm 2; \quad q = \pm 1, \pm 2, \pm 4;$
$\dfrac{p}{q} = \pm 1, \pm 2, \pm \dfrac{1}{2}, \pm \dfrac{1}{4}$

Step 3: Using the Bounds on Zeros Theorem:
$f(x) = 4\left(x^5 - 2x^4 - 0.25x + 0.5\right)$
$a_4 = -2,\ a_3 = 0,\ a_2 = 0,\ a_1 = -0.25,\ a_0 = 0.5$
Max $\{1, |0.5| + |-0.25| + |0| + |0| + |-2|\}$
$= $ Max $\{1, 2.75\}$
$= 2.75$

$1 + $ Max $\{|0.5|, |-0.25|, |0|, |0|, |-2|\}$
$= 1 + 2$
$= 3$

The smaller of the two numbers is 2.75. Thus, every zero of f lies between –2.75 and 2.75.

Graphing using the bounds and ZOOM-FIT:
(Second graph has a better window.)

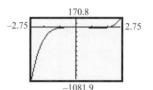

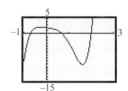

Step 4:
(a) From the graph it appears that there are x-intercepts at –0.7, 0.7 and 2.

(b) Using synthetic division:

$$\begin{array}{r|rrrrrr} 2 & 4 & -8 & 0 & 0 & -1 & 2 \\ & & 8 & 0 & 0 & 0 & -2 \\ \hline & 4 & 0 & 0 & 0 & -1 & 0 \end{array}$$

Since the remainder is 0, $x-2$ is a factor. The other factor is the quotient: $4x^4 - 1$.

(c) Factoring,
$f(x) = (x-2)(4x^4 - 1)$
$= (x-2)(2x^2 - 1)(2x^2 + 1)$
$= (x-2)(\sqrt{2}x - 1)(\sqrt{2}x + 1)(2x^2 + 1)$

The zeros are $-\dfrac{\sqrt{2}}{2}, \dfrac{\sqrt{2}}{2},$ and 2 or –0.71, –0.71, and 2.

57. $f(x) = x^3 + 3.2x^2 - 16.83x - 5.31$

$f(x)$ has at most 3 real zeros.

Solving by graphing (using ZERO):

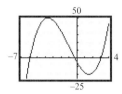

The zeros are approximately –5.9, –0.3, and 3.

59. $f(x) = x^4 - 1.4x^3 - 33.71x^2 + 23.94x + 292.41$

$f(x)$ has at most 4 real zeros.

Solving by graphing (using ZERO):

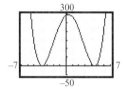

The zeros are approximately –3.80 and 4.50.

These zeros are each of multiplicity 2.

61. $f(x) = x^3 + 19.5x^2 - 1021x + 1000.5$

$f(x)$ has at most 3 real zeros.

Solving by graphing (using ZERO):

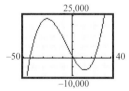

The zeros are approximately –43.5, 1, and 23.

63. $x^4 - x^3 + 2x^2 - 4x - 8 = 0$

The solutions of the equation are the zeros of $f(x) = x^4 - x^3 + 2x^2 - 4x - 8$.

Step 1: $f(x)$ has at most 4 real zeros.

Step 2: Possible rational zeros:

$p = \pm 1, \pm 2, \pm 4, \pm 8; \quad q = \pm 1;$

$\dfrac{p}{q} = \pm 1, \pm 2, \pm 4, \pm 8$

Step 3: Using the Bounds on Zeros Theorem:

$a_3 = -1, \; a_2 = 2, \; a_1 = -4, \; a_0 = -8$

Max $\{1, |-8| + |-4| + |2| + |-1|\}$

$= $ Max $\{1, 15\}$

$= 15$

$1 + $ Max $\{|-8|, |-4|, |2|, |-1|\}$

$= 1 + 8$

$= 9$

The smaller of the two numbers is 9. Thus, every zero of f lies between –9 and 9.

Graphing using the bounds and ZOOM-FIT:
(Second graph has a better window.)

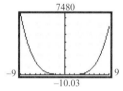

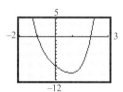

Step 4:
(a) From the graph it appears that there are x-intercepts at –1 and 2.

(b) Using synthetic division:

$$\begin{array}{r|rrrrr} -1) & 1 & -1 & 2 & -4 & -8 \\ & & -1 & 2 & -4 & 8 \\ \hline & 1 & -2 & 4 & -8 & 0 \end{array}$$

$$\begin{array}{r|rrr}
2) & 1 & -2 & 4 & -8 \\
 & & 2 & 0 & 8 \\
\hline
 & 1 & 0 & 4 & 0
\end{array}$$

Since the remainder is 0, $x+1$ and $x-2$ are factors. The other factor is the quotient: x^2+4.

(c) The zeros are -1 and 2. ($x^2+4=0$ has no real solutions.)

65. $3x^3+4x^2-7x+2=0$

The solutions of the equation are the zeros of $f(x)=3x^3+4x^2-7x+2$.

Step 1: $f(x)$ has at most 3 real zeros.

Step 2: Possible rational zeros:
$p=\pm 1, \pm 2;\quad q=\pm 1, \pm 3;$
$\dfrac{p}{q}=\pm 1, \pm 2, \pm\dfrac{1}{3}, \pm\dfrac{2}{3}$

Step 3: Using the Bounds on Zeros Theorem:

$f(x)=3\left(x^3+\dfrac{4}{3}x^2-\dfrac{7}{3}x+\dfrac{2}{3}\right)$

$a_2=\dfrac{4}{3},\ a_1=-\dfrac{7}{3},\ a_0=\dfrac{2}{3}$

$\text{Max}\left\{1,\left|\dfrac{2}{3}\right|+\left|-\dfrac{7}{3}\right|+\left|\dfrac{4}{3}\right|\right\}=\text{Max}\left\{1,\dfrac{13}{3}\right\}$

$=\dfrac{13}{3}\approx 4.333$

$1+\text{Max}\left\{\left|\dfrac{2}{3}\right|,\left|-\dfrac{7}{3}\right|,\left|\dfrac{4}{3}\right|\right\}=1+\dfrac{7}{3}$

$=\dfrac{10}{3}\approx 3.333$

The smaller of the two numbers is 3.33. Thus, every zero of f lies between -3.33 and 3.33.

Graphing using the bounds and ZOOM-FIT: (Second graph has a better window.)

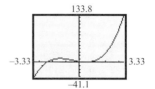

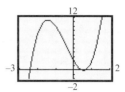

Step 4:
(a) From the graph it appears that there are x-intercepts at $\dfrac{1}{3}, \dfrac{2}{3}$, and -2.4.

(b) Using synthetic division:

$$\begin{array}{r|rrr}
\dfrac{2}{3}) & 3 & 4 & -7 & 2 \\
 & & 2 & 4 & -2 \\
\hline
 & 3 & 6 & -3 & 0
\end{array}$$

Since the remainder is 0, $x-\dfrac{2}{3}$ is a factor. The other factor is the quotient: $3x^2+6x-3$.

$f(x)=\left(x-\dfrac{2}{3}\right)(3x^2+6x-3)$

$=3\left(x-\dfrac{2}{3}\right)(x^2+2x-1)$

Using the quadratic formula to solve $x^2+2x-1=0$:

$x=\dfrac{-2\pm\sqrt{4-4(1)(-1)}}{2(1)}=\dfrac{-2\pm\sqrt{8}}{2}$

$=\dfrac{-2\pm 2\sqrt{2}}{2}=-1\pm\sqrt{2}$

(c) The zeros are $\dfrac{2}{3}, -1+\sqrt{2}$, and $-1-\sqrt{2}$ or $0.67, 0.41$, and -2.41.

67. $3x^3 - x^2 - 15x + 5 = 0$

Solving by factoring:
$$x^2(3x-1) - 5(3x-1) = 0$$
$$(3x-1)(x^2 - 5) = 0$$
$$(3x-1)(x-\sqrt{5})(x+\sqrt{5}) = 0$$

The solutions of the equation are $\frac{1}{3}, \sqrt{5},$ and $-\sqrt{5}$ or 0.33, 2.24, and –2.24.

69. $x^4 + 4x^3 + 2x^2 - x + 6 = 0$

The solutions of the equation are the zeros of $f(x) = x^4 + 4x^3 + 2x^2 - x + 6$.

Step 1: $f(x)$ has at most 4 real zeros.

Step 2: Possible rational zeros:
$$p = \pm 1, \pm 2, \pm 3, \pm 6; \quad q = \pm 1;$$
$$\frac{p}{q} = \pm 1, \pm 2, \pm 3, \pm 6$$

Step 3: Using the Bounds on Zeros Theorem:
$a_3 = 4, \; a_2 = 2, \; a_1 = -1, \; a_0 = 6$
$$\text{Max}\{1, |6| + |-1| + |2| + |4|\} = \text{Max}\{1, 13\}$$
$$= 13$$
$$1 + \text{Max}\{|6|, |-1|, |2|, |4|\} = 1 + 6 = 7$$

The smaller of the two numbers is 7. Thus, every zero of f lies between –7 and 7.

Graphing using the bounds and ZOOM-FIT:
(Second graph has a better window.)

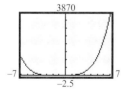

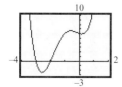

Step 4:
(a) From the graph it appears that there are x-intercepts at –3 and –2.

(b) Using synthetic division:

$$\begin{array}{r|rrrrr} -3) & 1 & 4 & 2 & -1 & 6 \\ & & -3 & -3 & 3 & -6 \\ \hline & 1 & 1 & -1 & 2 & 0 \end{array}$$

$$\begin{array}{r|rrrr} -2) & 1 & 1 & -1 & 2 \\ & & -2 & 2 & -2 \\ \hline & 1 & -1 & 1 & 0 \end{array}$$

Since the remainder is 0, $x+3$ and $x+2$ are factors. The other factor is the quotient: $x^2 - x + 1$.

(c) The zeros are –3 and –2. ($x^2 - x + 1 = 0$ has no real solutions.)

71. $x^3 - \frac{2}{3}x^2 + \frac{8}{3}x + 1 = 0$

The solutions of the equation are the zeros of $f(x) = x^3 - \frac{2}{3}x^2 + \frac{8}{3}x + 1 = 0$.

Step 1: $f(x)$ has at most 3 real zeros.

Step 2: Use the equivalent equation $3x^3 - 2x^2 + 8x + 3 = 0$ to find the possible rational zeros:
$$p = \pm 1, \pm 3; \quad q = \pm 1, \pm 3; \quad \frac{p}{q} = \pm 1, \pm 3, \pm \frac{1}{3}$$

Step 3: Using the Bounds on Zeros Theorem:
$$a_2 = -\frac{2}{3}, \; a_1 = \frac{8}{3}, \; a_0 = 1$$

$$\text{Max}\left\{1, |1| + \left|\frac{8}{3}\right| + \left|-\frac{2}{3}\right|\right\} = \text{Max}\left\{1, \frac{13}{3}\right\}$$
$$= \frac{13}{3}$$
$$\approx 4.333$$

$$1 + \text{Max}\left\{|1|, \left|\frac{8}{3}\right|, \left|-\frac{2}{3}\right|\right\} = 1 + \frac{8}{3} = \frac{11}{3} \approx 3.667$$

The smaller of the two numbers is 3.67. Thus, every zero of f lies between -3.67 and 3.67.

Graphing using the bounds and ZOOM-FIT: (Second graph has a better window.)

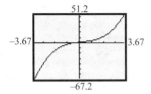

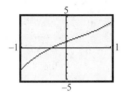

Step 4:
(a) From the graph it appears that there is an x-intercept at $-\dfrac{1}{3}$.

(b) Using synthetic division:

$$-\tfrac{1}{3} \overline{\big)\ \begin{array}{cccc} 1 & -\tfrac{2}{3} & \tfrac{8}{2} & 1 \\ & -\tfrac{1}{3} & \tfrac{1}{3} & -1 \\ \hline 1 & -1 & 3 & 0 \end{array}}$$

Since the remainder is 0, $x + \dfrac{1}{3}$ is a factor.

The other factor is the quotient: $x^2 - x + 3$.

(c) The real zero is $-\dfrac{1}{3}$. ($x^2 - x + 3 = 0$ has no real solutions.)

73. Using the TABLE feature to show that there is a zero in the interval:

$f(x) = 8x^4 - 2x^2 + 5x - 1; \quad [0, 1]$

X	Y1
-1	0
0	-1
1	10
2	129
3	644
4	2035
5	4974

X = -1

$f(0) = -1 < 0$ and $f(1) = 10 > 0$
Since one is positive and one is negative, there is a zero in the interval.

Using the TABLE feature to approximate the zero to two decimal places:

X	Y1
.213	-.0093
.214	-.0048
.215	-4E-4
.216	.0041
.217	.00856
.218	.01302
.219	.01748

X = .219

The zero is approximately 0.22.

75. Using the TABLE feature to show that there is a zero in the interval:

$f(x) = 2x^3 + 6x^2 - 8x + 2; \quad [-5, -4]$

X	Y1
-8	-574
-7	-334
-6	-166
-5	-58
-4	2
-3	26
-2	26

Y1 = 2X^3+6X²-8X+2

$f(-5) = -58 < 0$ and $f(-4) = 2 > 0$
Since one is positive and one is negative, there is a zero in the interval.

Using the TABLE feature to approximate the zero to two decimal places:

X	Y1
-4.052	-.129
-4.051	-.0871
-4.05	-.0453
-4.049	-.0035
-4.048	.03831
-4.047	.08003
-4.046	.12172

Y1 = 2X^3+6X²-8X+2

The zero is approximately -4.05.

77. Using the TABLE feature to show that there is a zero in the interval:

$f(x) = x^5 - x^4 + 7x^3 - 7x^2 - 18x + 18;\ [1.4, 1.5]$

$f(1.4) = -0.1754 < 0$ and $f(1.5) = 1.4063 > 0$
Since one is positive and one is negative, there is a zero in the interval.

Using the TABLE feature to approximate the zero to two decimal places:

The zero is approximately 1.41.

79. $C(x) = 3000$

$0.216x^3 - 2.347x^2 + 14.328x + 10.224 = 3000$
$0.216x^3 - 2.347x^2 + 14.328x - 2989.776 = 0$

There are at most 3 solutions.
Graphing and solving:

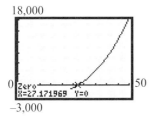

About 27 Cavaliers can be manufactured.

81. $x - 2$ is a factor of $f(x) = x^3 - kx^2 + kx + 2$ only if the remainder that results when $f(x)$ is divided by $x - 2$ is 0. Dividing, we have:

$$\begin{array}{r|rrrr} 2) & 1 & -k & k & 2 \\ & & 2 & -2k+4 & -2k+8 \\ \hline & 1 & -k+2 & -k+4 & -2k+10 \end{array}$$

Since we want the remainder to equal 0, set the remainder equal to zero and solve:

$-2k + 10 = 0 \Rightarrow -2k = -10 \Rightarrow k = 5$

83. By the Remainder Theorem we know that the remainder from synthetic division by c is equal to $f(c)$. Thus the easiest way to find the remainder is to evaluate:

$f(1) = 2(1)^{20} - 8(1)^{10} + 1 - 2 = 2 - 8 + 1 - 2 = -7$

The remainder is –7.

85. $x^3 - 8x^2 + 16x - 3 = 0$ has solution $x = 3$, so $x - 3$ is a factor of $f(x) = x^3 - 8x^2 + 16x - 3$.

Using synthetic division

$$\begin{array}{r|rrrr} 3) & 1 & -8 & 16 & -3 \\ & & 3 & -15 & 3 \\ \hline & 1 & -5 & 1 & 0 \end{array}$$

Thus, $f(x) = x^3 - 8x^2 + 16x - 3$
$ = (x-3)(x^2 - 5x + 1)$

Solving $x^2 - 5x + 1 = 0$,

$x = \dfrac{5 \pm \sqrt{25-4}}{2} = \dfrac{5 \pm \sqrt{21}}{2}$

The sum of these two roots is

$\dfrac{5+\sqrt{21}}{2} + \dfrac{5-\sqrt{21}}{2} = \dfrac{10}{2} = 5$.

87. Let x be the length of a side of the original cube.

    After removing the 1 inch slice, one dimension will be $x-1$.

    The volume of the new solid will be:
    $$(x-1)\cdot x \cdot x = 294$$
    $$x^3 - x^2 = 294$$
    $$x^3 - x^2 - 294 = 0$$

    The possible rational zeros are:
    $p = \pm 1, \pm 2, \pm 3, \pm 6, \pm 7, \pm 14, \pm 21,$
    $\quad \pm 42, \pm 49, \pm 98, \pm 147, \pm 294;$
    $q = \pm 1$

    The rational zeros are the same as the values for p.

    Using synthetic division:

    ```
    7)  1   -1    0   -294
             7   42    294
        ─────────────────────
        1    6   42     0
    ```

    7 is a zero, so the length of the original edge of the cube was 7 inches.

89. We want to prove that $x - c$ is a factor of $x^n - c^n$, for any positive integer n. By the Factor Theorem, $x - c$ will be a factor of $f(x)$ provided $f(c) = 0$. Here, $f(x) = x^n - c^n$, so that $f(c) = c^n - c^n = 0$. Therefore, $x - c$ is a factor of $x^n - c^n$.

91. $f(x) = x^n + a_{n-1}x^{n-1} + a_{n-2}x^{n-2} + \ldots + a_1 x + a_0$; where $a_{n-1}, a_{n-2}, \ldots a_1, a_0$ are integers

    If r is a real zero of f, then r is either rational or irrational. We know that the rational roots of f must be of the form $\dfrac{p}{q}$ where p is a divisor of a_0 and q is a divisor of 1. This means that $q = \pm 1$. So if r is rational, then $r = \dfrac{p}{q} = \pm p$.

    Therefore, r is an integer or r is irrational.

93. $f(x) = 2x^3 + 3x^2 - 6x + 7$

    If $x = \dfrac{1}{3}$ is zero of $f(x)$, then $f\left(\dfrac{1}{3}\right) = 0$:

    $$f\left(\dfrac{1}{3}\right) = 2\left(\dfrac{1}{3}\right)^3 + 3\left(\dfrac{1}{3}\right)^2 - 6\left(\dfrac{1}{3}\right) + 7$$
    $$= 2\left(\dfrac{1}{27}\right) + 3\left(\dfrac{1}{9}\right) - 2 + 7$$
    $$= \dfrac{2}{27} + \dfrac{1}{3} + 5$$
    $$= \dfrac{146}{27}$$
    $$\neq 0$$

    Thus, $x = \dfrac{1}{3}$ is not a zero of $f(x)$.

95. $f(x) = 2x^6 - 5x^4 + x^3 - x + 1$

    If $x = \dfrac{3}{5}$ is zero of $f(x)$, then $f\left(\dfrac{3}{5}\right) = 0$:

    $$f(x) = 2x^6 - 5x^4 + x^3 - x + 1$$
    $$f\left(\dfrac{3}{5}\right) = 2\left(\dfrac{3}{5}\right)^6 - 5\left(\dfrac{3}{5}\right)^4 + \left(\dfrac{3}{5}\right)^3 - \left(\dfrac{3}{5}\right) + 7$$
    $$= 0.061312$$
    $$\neq 0$$

    Thus, $x = \dfrac{3}{5}$ is not a zero of $f(x)$.

Section 3.7

1. Sum: $3i$; product: $1 + 21i$

3. one

5. True

7. Since complex zeros appear in conjugate pairs, $4 + i$, the conjugate of $4 - i$, is the remaining zero of f.

9. Since complex zeros appear in conjugate pairs, $-i$, the conjugate of i, and $1 - i$, the conjugate of $1 + i$, are the remaining zeros of f.

11. Since complex zeros appear in conjugate pairs, $-i$, the conjugate of i, and $-2i$, the conjugate of $2i$, are the remaining zeros of f.

13. Since complex zeros appear in conjugate pairs, $-i$, the conjugate of i, is the remaining zero.

15. Since complex zeros appear in conjugate pairs, $2-i$, the conjugate of $2+i$, and $-3+i$, the conjugate of $-3-i$, are the remaining zeros.

For 17–21, we will use $a = 1$ in the polynomial. Also note that
$$(x-(a+bi))(x-(a-bi)) = ((x-a)-bi)((x-a)+bi)$$
$$= (x-a)^2 - (bi)^2$$

17. Since $3+2i$ is a zero, its conjugate $3-2i$ is also a zero of f.

$$f(x) = (x-4)(x-4)(x-(3+2i))(x-(3-2i))$$
$$= (x^2 - 8x + 16)((x-3)-2i)((x-3)+2i)$$
$$= (x^2 - 8x + 16)(x^2 - 6x + 9 - 4i^2)$$
$$= (x^2 - 8x + 16)(x^2 - 6x + 13)$$
$$= x^4 - 6x^3 + 13x^2 - 8x^3 + 48x^2$$
$$\quad - 104x + 16x^2 - 96x + 208$$
$$= x^4 - 14x^3 + 77x^2 - 200x + 208$$

19. Since $-i$ is a zero, its conjugate i is also a zero, and since $1+i$ is a zero, its conjugate $1-i$ is also a zero of f.

$$f(x) = (x-2)(x+i)(x-i)(x-(1+i))(x-(1-i))$$
$$= (x-2)(x^2 - i^2)((x-1)-i)((x-1)+i)$$
$$= (x-2)(x^2 + 1)(x^2 - 2x + 1 - i^2)$$
$$= (x^3 - 2x^2 + x - 2)(x^2 - 2x + 2)$$
$$= x^5 - 2x^4 + 2x^3 - 2x^4 + 4x^3 - 4x^2$$
$$\quad + x^3 - 2x^2 + 2x - 2x^2 + 4x - 4$$
$$= x^5 - 4x^4 + 7x^3 - 8x^2 + 6x - 4$$

21. Since $-i$ is a zero, its conjugate i is also a zero.
$$f(x) = (x-3)(x-3)(x+i)(x-i)$$
$$= (x^2 - 6x + 9)(x^2 - i^2)$$
$$= (x^2 - 6x + 9)(x^2 + 1)$$
$$= x^4 + x^2 - 6x^3 - 6x + 9x^2 + 9$$
$$= x^4 - 6x^3 + 10x^2 - 6x + 9$$

23. Since $2i$ is a zero, its conjugate $-2i$ is also a zero of f. $x - 2i$ and $x + 2i$ are factors of f. Thus, $(x-2i)(x+2i) = x^2 + 4$ is a factor of f. Using division to find the other factor:

$$\begin{array}{r} x - 4 \\ x^2 + 4 \overline{\smash{)}\, x^3 - 4x^2 + 4x - 16} \\ \underline{x^3 + 4x} \\ -4x^2 - 16 \\ \underline{-4x^2 - 16} \end{array}$$

$x - 4$ is a factor, so the remaining zero is 4. The zeros of f are $4, 2i, -2i$.

25. Since $-2i$ is a zero, its conjugate $2i$ is also a zero of f. $x - 2i$ and $x + 2i$ are factors of f. Thus, $(x-2i)(x+2i) = x^2 + 4$ is a factor of f. Using division to find the other factor:

$$\begin{array}{r} 2x^2 + 5x - 3 \\ x^2 + 4 \overline{\smash{)}\, 2x^4 + 5x^3 + 5x^2 + 20x - 12} \\ \underline{2x^4 + 8x^2} \\ 5x^3 - 3x^2 + 20x \\ \underline{5x^3 + 20x} \\ -3x^2 - 12 \\ \underline{-3x^2 - 12} \end{array}$$

$2x^2 + 5x - 3 = (2x-1)(x+3)$

The remaining zeros are $\frac{1}{2}$ and -3.

The zeros of f are $2i, -2i, -3, \frac{1}{2}$.

27. Since $3-2i$ is a zero, its conjugate $3+2i$ is also a zero of h. $x-(3-2i)$ and $x-(3+2i)$ are factors of h.
Thus,
$(x-(3-2i))(x-(3+2i)) = ((x-3)+2i)((x-3)-2i)$
$= x^2 - 6x + 9 - 4i^2$
$= x^2 - 6x + 13$
is a factor of h.
Using division to find the other factor:

$$\begin{array}{r} x^2 - 3x - 10 \\ x^2 - 6x + 13 \overline{\smash{\big)}\, x^4 - 9x^3 + 21x^2 + 21x - 130} \\ \underline{x^4 - 6x^3 + 13x^2} \\ -3x^3 + 8x^2 + 21x \\ \underline{-3x^3 + 18x^2 - 39x} \\ -10x^2 + 60x - 130 \\ \underline{-10x^2 + 60x - 130} \\ \end{array}$$

$x^2 - 3x - 10 = (x+2)(x-5)$
The remaining zeros are –2 and 5.
The zeros of h are $3-2i, 3+2i, -2, 5$.

29. Since $-4i$ is a zero, its conjugate $4i$ is also a zero of h. $x-4i$ and $x+4i$ are factors of h.
Thus, $(x-4i)(x+4i) = x^2 + 16$ is a factor of h.
Using division to find the other factor:

$$\begin{array}{r} 3x^3 + 2x^2 - 33x - 22 \\ x^2 + 16 \overline{\smash{\big)}\, 3x^5 + 2x^4 + 15x^3 + 10x^2 - 528x - 352} \\ \underline{3x^5 + 48x^3} \\ 2x^4 - 33x^3 + 10x^2 \\ \underline{2x^4 + 32x^2} \\ -33x^3 - 22x^2 - 528x \\ \underline{-33x^3 - 528x} \\ -22x^2 - 352 \\ \underline{-22x^2 - 352} \\ 0 \end{array}$$

$3x^3 + 2x^2 - 33x - 22 = x^2(3x+2) - 11(3x+2)$
$= (3x+2)(x^2 - 11)$
$= (3x+2)(x - \sqrt{11})(x + \sqrt{11})$

The remaining zeros are $-\dfrac{2}{3}, \sqrt{11}$, and $-\sqrt{11}$.

The zeros of h are $4i, -4i, -\sqrt{11}, \sqrt{11}, -\dfrac{2}{3}$.

31. $f(x) = x^3 - 1 = (x-1)(x^2 + x + 1)$ The solutions of $x^2 + x + 1 = 0$ are:
$x = \dfrac{-1 \pm \sqrt{1^2 - 4(1)(1)}}{2(1)} = \dfrac{-1 \pm \sqrt{-3}}{2}$
$= -\dfrac{1}{2} + \dfrac{\sqrt{3}}{2}i$ and $-\dfrac{1}{2} - \dfrac{\sqrt{3}}{2}i$

The zeros are: $1, -\dfrac{1}{2} + \dfrac{\sqrt{3}}{2}i, -\dfrac{1}{2} - \dfrac{\sqrt{3}}{2}i$.

33. $f(x) = x^3 - 8x^2 + 25x - 26$

Step 1: $f(x)$ has 3 complex zeros.

Step 2: By Descartes Rule of Signs, there are three positive real zeros or there is one positive real zero.

$f(-x) = (-x)^3 - 8(-x)^2 + 25(-x) - 26$, thus,
$= -x^3 - 8x^2 - 25x - 26$
there are no negative real zeros.

Step 3: Possible rational zeros:
$p = \pm 1, \pm 2, \pm 13, \pm 26; \quad q = \pm 1;$
$\dfrac{p}{q} = \pm 1, \pm 2, \pm 13, \pm 26$

Step 4: Using synthetic division:
We try $x - 2$:

$$\begin{array}{r|rrrr} 2) & 1 & -8 & 25 & -26 \\ & & 2 & -12 & 26 \\ \hline & 1 & -6 & 13 & 0 \end{array}$$

$x - 2$ is a factor. The other factor is the quotient: $x^2 - 6x + 13$.

The solutions of $x^2 - 6x + 13 = 0$ are:

$$x = \frac{-(-6) \pm \sqrt{(-6)^2 - 4(1)(13)}}{2(1)}$$

$$= \frac{6 \pm \sqrt{-16}}{2}$$

$$= \frac{6 \pm 4i}{2} = 3 \pm 2i$$

The zeros are $2, 3-2i, 3+2i$.

35. $f(x) = x^4 + 5x^2 + 4 = (x^2 + 4)(x^2 + 1)$
    $= (x + 2i)(x - 2i)(x + i)(x - i)$

    The zeros are: $-2i, -i, i, 2i$.

37. $f(x) = x^4 + 2x^3 + 22x^2 + 50x - 75$

    Step 1: $f(x)$ has 4 complex zeros.

    Step 2: By Descartes Rule of Signs, there is 1 positive real zero.

    $f(-x) = (-x)^4 + 2(-x)^3 + 22(-x)^2 + 50(-x) - 75$
    $= x^4 - 2x^3 + 22x^2 - 50x - 75$

    Thus, there are three negative real zeros or there is one negative real zero.

    Step 3: Possible rational zeros:

    $p = \pm 1, \pm 3, \pm 5, \pm 15, \pm 25, \pm 75; \quad q = \pm 1;$

    $\frac{p}{q} = \pm 1, \pm 3, \pm 5, \pm 15, \pm 25, \pm 75$

    Step 4: Using synthetic division:

    We try $x + 3$:

    ```
    -3) 1   2   22   50  -75
           -3    3  -75   75
        ─────────────────────
        1  -1   25  -25    0
    ```

    $x + 3$ is a factor. The other factor is the quotient: $x^3 - x^2 + 25x - 25$.

    $x^3 - x^2 + 25x - 25 = x^2(x - 1) + 25(x - 1)$
    $= (x - 1)(x^2 + 25)$
    $= (x - 1)(x + 5i)(x - 5i)$

    The zeros are $-3, 1, -5i, 5i$.

39. $f(x) = 3x^4 - x^3 - 9x^2 + 159x - 52$

    Step 1: $f(x)$ has 4 complex zeros.

    Step 2: By Descartes Rule of Signs, there are three positive real zeros or there is one positive real zero.

    $f(-x) = 3(-x)^4 - (-x)^3 - 9(-x)^2 + 159(-x) - 52$
    $= 3x^4 + x^3 - 9x^2 - 159x - 52$

    Thus, there is 1 negative real zero.

    Step 3: Possible rational zeros:

    $p = \pm 1, \pm 2, \pm 4, \pm 13, \pm 26, \pm 52;$

    $q = \pm 1, \pm 3;$

    $\frac{p}{q} = \pm 1, \pm 2, \pm 4, \pm 13, \pm 26, \pm 52,$

    $\pm \frac{1}{3}, \pm \frac{2}{3}, \pm \frac{4}{3}, \pm \frac{13}{3}, \pm \frac{26}{3}, \pm \frac{52}{3}$

    Step 4: Using synthetic division:

    We try $x + 4$:

    ```
    -4) 3  -1   -9   159  -52
           -12   52  -172   52
        ──────────────────────
        3  -13   43   -13    0
    ```

    $x + 4$ is a factor and the quotient is $3x^3 - 13x^2 + 43x - 13$.

    We try $x - \frac{1}{3}$ on $3x^3 - 13x^2 + 43x - 13$:

    ```
    1/3) 3  -13   43  -13
              1   -4   13
         ───────────────────
         3  -12   39    0
    ```

    $x - \frac{1}{3}$ is a factor and the quotient is $3x^2 - 12x + 39$.

    $3x^2 - 12x + 39 = 3(x^2 - 4x + 13)$

The solutions of $x^2 - 4x + 13 = 0$ are:

$$x = \frac{-(-4) \pm \sqrt{(-4)^2 - 4(1)(13)}}{2(1)}$$

$$= \frac{4 \pm \sqrt{-36}}{2}$$

$$= \frac{4 \pm 6i}{2}$$

$$= 2 \pm 3i$$

The zeros are $-4, \frac{1}{3}, 2-3i, 2+3i$.

41. If the coefficients are real numbers and $2+i$ is a zero, then $2-i$ would also be a zero. This would then require a polynomial of degree 4.

43. If the coefficients are real numbers, then complex zeros must appear in conjugate pairs. We have a conjugate pair and one real zero. Thus, there is only one remaining zero, and it must be real because a complex zero would require a pair of complex conjugates.

Chapter Review

1. $f(x) = (x-2)^2 + 2$

   Using the graph of $y = x^2$, shift right 2 units, then shift up 2 units.

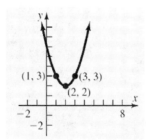

3. $f(x) = -(x-4)^2$

   Using the graph of $y = x^2$, shift the graph 4 units right, then reflect about the x-axis.

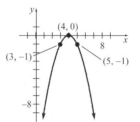

5. $f(x) = 2(x+1)^2 + 4$

   Using the graph of $y = x^2$, stretch vertically by a factor of 2, then shift 1 unit left, then shift 4 units up.

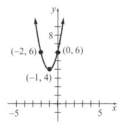

7. $f(x) = (x-2)^2 + 2$
   $= x^2 - 4x + 4 + 2$
   $= x^2 - 4x + 6$

   $a = 1, b = -4, c = 6$. Since $a = 1 > 0$, the graph opens up.

   The x-coordinate of the vertex is
   $$x = -\frac{b}{2a} = -\frac{-4}{2(1)} = \frac{4}{2} = 2.$$

   The y-coordinate of the vertex is
   $$f\left(-\frac{b}{2a}\right) = f(2) = (2)^2 - 4(2) + 6 = 2.$$

   Thus, the vertex is (2, 2).

   The axis of symmetry is the line $x = 2$.

   The discriminant is:
   $b^2 - 4ac = (-4)^2 - 4(1)(6) = -8 < 0$,
   so the graph has no x-intercepts.

The y-intercept is $f(0) = 6$.

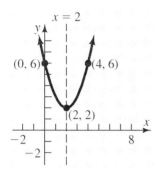

9. $f(x) = \dfrac{1}{4}x^2 - 16$,

   $a = \dfrac{1}{4}, b = 0, c = -16.$ Since $a = \dfrac{1}{4} > 0,$ the graph opens up.

   The x-coordinate of the vertex is
   $$x = -\dfrac{b}{2a} = -\dfrac{-0}{2\left(\dfrac{1}{4}\right)} = -\dfrac{0}{\dfrac{1}{2}} = 0.$$

   The y-coordinate of the vertex is
   $$f\left(-\dfrac{b}{2a}\right) = f(0) = \dfrac{1}{4}(0)^2 - 16 = -16.$$

   Thus, the vertex is $(0, -16)$.

   The axis of symmetry is the line $x = 0$.

   The discriminant is:
   $$b^2 - 4ac = (0)^2 - 4\left(\dfrac{1}{4}\right)(-16) = 16 > 0,$$
   so the graph has two x-intercepts.

   The x-intercepts are found by solving:
   $$\dfrac{1}{4}x^2 - 16 = 0$$
   $$x^2 - 64 = 0$$
   $$x^2 = 64$$
   $$x = 8 \text{ or } x = -8$$

   The x-intercepts are -8 and 8.

   The y-intercept is $f(0) = -16$.

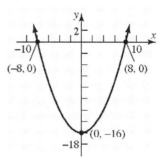

11. $f(x) = -4x^2 + 4x$,

    $a = -4, b = 4, c = 0.$ Since $a = -4 < 0,$ the graph opens down.

    The x-coordinate of the vertex is
    $$x = -\dfrac{b}{2a} = -\dfrac{4}{2(-4)} = -\dfrac{4}{-8} = \dfrac{1}{2}.$$

    The y-coordinate of the vertex is
    $$\begin{aligned}f\left(-\dfrac{b}{2a}\right) &= f\left(\dfrac{1}{2}\right) \\ &= -4\left(\dfrac{1}{2}\right)^2 + 4\left(\dfrac{1}{2}\right) \\ &= -1 + 2 \\ &= 1\end{aligned}$$

    Thus, the vertex is $\left(\dfrac{1}{2}, 1\right)$.

    The axis of symmetry is the line $x = \dfrac{1}{2}$.

    The discriminant is:
    $$b^2 - 4ac = 4^2 - 4(-4)(0) = 16 > 0,$$
    so the graph has two x-intercepts.

    The x-intercepts are found by solving:
    $$-4x^2 + 4x = 0$$
    $$-4x(x - 1) = 0$$
    $$x = 0 \text{ or } x = 1$$

    The x-intercepts are 0 and 1.

    The y-intercept is $f(0) = -4(0)^2 + 4(0) = 0$.

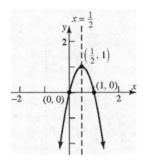

13. $f(x) = \dfrac{9}{2}x^2 + 3x + 1$

$a = \dfrac{9}{2}, b = 3, c = 1.$ Since $a = \dfrac{9}{2} > 0,$ the graph opens up.

The x-coordinate of the vertex is
$$x = -\dfrac{b}{2a} = -\dfrac{3}{2\left(\dfrac{9}{2}\right)} = -\dfrac{3}{9} = -\dfrac{1}{3}.$$

The y-coordinate of the vertex is
$$f\left(-\dfrac{b}{2a}\right) = f\left(-\dfrac{1}{3}\right)$$
$$= \dfrac{9}{2}\left(-\dfrac{1}{3}\right)^2 + 3\left(-\dfrac{1}{3}\right) + 1$$
$$= \dfrac{1}{2} - 1 + 1$$
$$= \dfrac{1}{2}$$

Thus, the vertex is $\left(-\dfrac{1}{3}, \dfrac{1}{2}\right).$

The axis of symmetry is the line $x = -\dfrac{1}{3}.$

The discriminant is:
$$b^2 - 4ac = 3^2 - 4\left(\dfrac{9}{2}\right)(1) = 9 - 18 = -9 < 0,$$
so the graph has no x-intercepts.

The y-intercept is $f(0) = \dfrac{9}{2}(0)^2 + 3(0) + 1 = 1.$

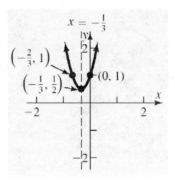

15. $f(x) = 3x^2 + 4x - 1,$

$a = 3, b = 4, c = -1.$ Since $a = 3 > 0,$ the graph opens up.

The x-coordinate of the vertex is
$$x = -\dfrac{b}{2a} = -\dfrac{4}{2(3)} = -\dfrac{4}{6} = -\dfrac{2}{3}.$$

The y-coordinate of the vertex is
$$f\left(-\dfrac{b}{2a}\right) = f\left(-\dfrac{2}{3}\right)$$
$$= 3\left(-\dfrac{2}{3}\right)^2 + 4\left(-\dfrac{2}{3}\right) - 1$$
$$= \dfrac{4}{3} - \dfrac{8}{3} - 1$$
$$= -\dfrac{7}{3}$$

Thus, the vertex is $\left(-\dfrac{2}{3}, -\dfrac{7}{3}\right).$

The axis of symmetry is the line $x = -\dfrac{2}{3}.$

The discriminant is:
$b^2 - 4ac = (4)^2 - 4(3)(-1) = 28 > 0,$ so the graph has two x-intercepts.

The x-intercepts are found by solving:
$3x^2 + 4x - 1 = 0$
$$x = \dfrac{-b \pm \sqrt{b^2 - 4ac}}{2a}$$
$$= \dfrac{-4 \pm \sqrt{28}}{2(3)}$$
$$= \dfrac{-4 \pm 2\sqrt{7}}{6} = \dfrac{-2 \pm \sqrt{7}}{3}$$

The x-intercepts are $\dfrac{-2-\sqrt{7}}{3}$ and $\dfrac{-2+\sqrt{7}}{3}$.

The y-intercept is $f(0) = 3(0)^2 + 4(0) - 1 = -1$.

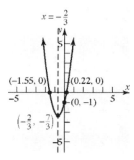

17. $f(x) = 3x^2 - 6x + 4$

    $a = 3, b = -6, c = 4$. Since $a = 3 > 0$, the graph opens up, so the vertex is a minimum point.

    The minimum occurs at
    $x = -\dfrac{b}{2a} = -\dfrac{-6}{2(3)} = \dfrac{6}{6} = 1$.

    The minimum value is
    $f\left(-\dfrac{b}{2a}\right) = f(1)$
    $= 3(1)^2 - 6(1) + 4$
    $= 3 - 6 + 4$
    $= 1$

19. $f(x) = -x^2 + 8x - 4$

    $a = -1, b = 8, c = -4$. Since $a = -1 < 0$, the graph opens down, so the vertex is a maximum point. The maximum occurs at
    $x = -\dfrac{b}{2a} = -\dfrac{8}{2(-1)} = -\dfrac{8}{-2} = 4$.

    The maximum value is
    $f\left(-\dfrac{b}{2a}\right) = f(4)$
    $= -(4)^2 + 8(4) - 4$
    $= -16 + 32 - 4$
    $= 12$

21. $f(x) = -3x^2 + 12x + 4$

    $a = -3, b = 12, c = 4$. Since $a = -3 < 0$, the graph opens down, so the vertex is a maximum point. The maximum occurs at
    $x = -\dfrac{b}{2a} = -\dfrac{12}{2(-3)} = -\dfrac{12}{-6} = 2$.

    The maximum value is
    $f\left(-\dfrac{b}{2a}\right) = f(2)$
    $= -3(2)^2 + 12(2) + 4$
    $= -12 + 24 + 4$
    $= 16$

23. $f(x) = 4x^5 - 3x^2 + 5x - 2$ is a polynomial of degree 5.

25. $f(x) = 3x^2 + 5x^{1/2} - 1$ is not a polynomial because the variable x is raised to the $\dfrac{1}{2}$ power, which is not a nonnegative integer.

27. $f(x) = (x+2)^3$

    Using the graph of $y = x^3$, shift left 2 units.

    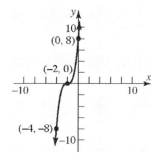

29. $f(x) = -(x-1)^4$

Using the graph of $y = x^4$, shift right 1 unit, then reflect about the x-axis.

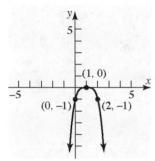

31. $f(x) = 2(x+1)^4 + 2$

Using the graph of $y = x^4$, stretch vertically by a factor of 2, then shift left 1 unit, then shift up 2 units.

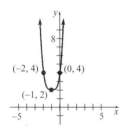

33. $f(x) = x(x+2)(x+4)$

   (a) y-intercept: $f(0) = (0)(0+2)(0+4) = 0$
   x-intercepts: solve $f(x) = 0$:
   $x(x+2)(x+4) = 0$
   $x = 0$ or $x = -2$ or $x = -4$

   (b) The graph crosses the x-axis at $x = -4$, $x = -2$ and $x = 0$ since each zero has multiplicity 1.

   (c) The function resembles $y = x^3$ for large values of $|x|$.

   (d) Graphing utility:

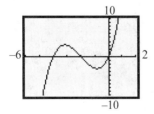

   (e) The graph has at most 2 turning points;
   Local maximum: $(-3.15, 3.08)$
   Local minimum: $(-0.85, -3.08)$

   (f) Graphing:

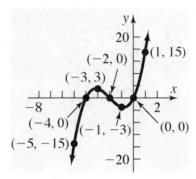

   (g) Domain: all real numbers
   Range: all real numbers

   (h) Increasing on $(-\infty, -3.15)$ and $(-0.85, \infty)$;
   decreasing on $(-3.15, -0.85)$

35. $f(x) = (x-2)^2(x+4)$

   (a) y-intercept: $f(0) = (0-2)^2(0+4) = 16$
   x-intercepts: solve $f(x) = 0$:
   $(x-2)^2(x+4) = 0 \Rightarrow x = 2$ or $x = -4$

   (b) The graph crosses the x-axis at $x = -4$ since this zero has multiplicity 1. The graph touches the x-axis at $x = 2$ since this zero has multiplicity 2.

   (c) The function resembles $y = x^3$ for large values of $|x|$.

   (d) Graphing utility:

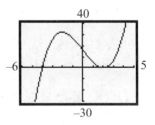

   (e) 2 turning points; Local maximum: $(-2, 32)$
   Local minimum: $(2, 0)$

(f) Graphing:

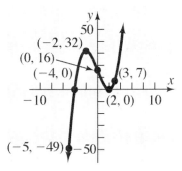

(g) Domain: all real numbers
Range: all real numbers

(h) Increasing on $(-\infty, -2)$ and $(2, \infty)$;
decreasing on $(-2, 2)$

37. $f(x) = x^3 - 4x^2 = x^2(x-4)$

(a) x-intercepts: 0, 4; y-intercept: 0

(b) crosses x axis at $x = 4$ and touches the x axis at $x = 0$

(c) The function resembles $y = x^3$ for large values of $|x|$.

(d) Graphing utility

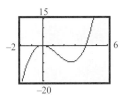

(e) 2 turning points; Local maximum: $(0, 0)$
Local minimum: $(2.67, -9.48)$

(f) Graphing by hand

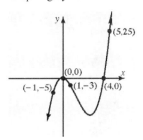

(g) Domain: all real numbers
Range: all real numbers

(h) Increasing on $(-\infty, 0)$ and $(2.67, \infty)$;
decreasing on $(0, 2.67)$

39. $f(x) = (x-1)^2 (x+3)(x+1)$

(a) y-intercept: $f(0) = (0-1)^2 (0+3)(0+1)$
$= 3$
x-intercepts: solve $f(x) = 0$:
$(x-1)^2 (x+3)(x+1) = 0$
$x = 1$ or $x = -3$ or $x = -1$

(b) The graph crosses the x-axis at $x = -3$ and $x = -1$ since each zero has multiplicity 1. The graph touches the x-axis at $x = 1$ since this zero has multiplicity 2.

(c) The function resembles $y = x^4$ for large values of $|x|$.

(d) Graphing utility:

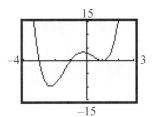

(e) 3 turning points;
Local maximum: $(-0.22, 3.23)$
Local minima: $(-2.28, -9.91), (1,0)$

(f) Graphing:

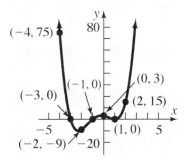

(g) Domain: all real numbers
Range: $[-9.91, \infty)$

(h) Increasing on $(-2.28, -0.22)$ and $(1, \infty)$;
decreasing on $(-\infty, -2.28)$ and d $(-0.22, 1)$

Chapter 3: Polynomial and Rational Functions

41. $R(x) = \dfrac{x+2}{x^2-9} = \dfrac{x+2}{(x+3)(x-3)}$ is in lowest terms. The denominator has zeros at -3 and 3. Thus, the domain is $\{x \mid x \neq -3, x \neq 3\}$. The degree of the numerator, $p(x) = x+2$, is $n = 1$. The degree of the denominator, $q(x) = x^2 - 9$, is $m = 2$. Since $n < m$, the line $y = 0$ is a horizontal asymptote. Since the denominator is zero at -3 and 3, $x = -3$ and $x = 3$ are vertical asymptotes.

43. $R(x) = \dfrac{x^2+3x+2}{(x+2)^2} = \dfrac{(x+2)(x+1)}{(x+2)^2} = \dfrac{x+1}{x+2}$ is in lowest terms. The denominator has a zero at -2. Thus, the domain is $\{x \mid x \neq -2\}$. The degree of the numerator, $p(x) = x^2 + 3x + 2$, is $n = 2$. The degree of the denominator, $q(x) = (x+2)^2 = x^2 + 4x + 4$, is $m = 2$. Since $n = m$, the line $y = \dfrac{1}{1} = 1$ is a horizontal asymptote. Since the denominator of $y = \dfrac{x+1}{x+2}$ is zero at -2, $x = -2$ is a vertical asymptote.

45. $R(x) = \dfrac{2x-6}{x}$ $p(x) = 2x-6;\ q(x) = x;\ n = 1;\ m = 1$

Step 1: Domain: $\{x \mid x \neq 0\}$

Step 2: (a) The x-intercept is the zero of $p(x)$: 3

(b) There is no y-intercept because 0 is not in the domain.

Step 3: $R(-x) = \dfrac{2(-x)-6}{-x} = \dfrac{-2x-6}{-x} = \dfrac{2x+6}{x}$; this is neither $R(x)$ nor $-R(x)$, so there is no symmetry.

Step 4: $R(x) = \dfrac{2x-6}{x}$ is in lowest terms. The vertical asymptote is the zero of $q(x)$: $x = 0$.

Step 5: Since $n = m$, the line $y = \dfrac{2}{1} = 2$ is the horizontal asymptote. Solve to find intersection points:

$$\dfrac{2x-6}{x} = 2$$
$$2x - 6 = 2x$$
$$-6 \neq 0$$

$R(x)$ does not intersect $y = 2$.

Step 6:

Interval	$(-\infty, 0)$	$(0, 3)$	$(3, \infty)$
Number Chosen	-2	1	4
Value of R	$R(-2) = 5$	$R(1) = -4$	$R(4) = \frac{1}{2}$
Location of Graph	Above x-axis	Below x-axis	Above x-axis
Point on Graph	$(-2, 5)$	$(1, -4)$	$(4, \frac{1}{2})$

Step 7: Graphing:

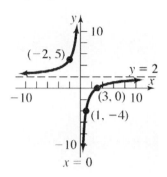

47. $H(x) = \dfrac{x+2}{x(x-2)}$ $p(x) = x+2$; $q(x) = x(x-2) = x^2 - 2x$; $n = 1$; $m = 2$

Step 1: Domain: $\{x \mid x \neq 0, x \neq 2\}$.

Step 2: (a) The x-intercept is the zero of $p(x)$: -2
(b) There is no y-intercept because 0 is not in the domain.

Step 3: $H(-x) = \dfrac{-x+2}{-x(-x-2)} = \dfrac{-x+2}{x^2+2x}$; this is neither $H(x)$ nor $-H(x)$, so there is no symmetry.

Step 4: $H(x) = \dfrac{x+2}{x(x-2)}$ is in lowest terms. The vertical asymptotes are the zeros of $q(x)$: $x = 0$ and $x = 2$.

Step 5: Since $n < m$, the line $y = 0$ is the horizontal asymptote. Solve to find intersection points:

$$\dfrac{x+2}{x(x-2)} = 0$$
$$x + 2 = 0$$
$$x = -2$$

$H(x)$ intersects $y = 0$ at $(-2, 0)$.

Step 6:

Interval	$(-\infty, -2)$	$(-2, 0)$	$(0, 2)$	$(2, \infty)$
Number Chosen	-3	-1	1	3
Value of H	$H(-3) = -\dfrac{1}{15}$	$H(-1) = \dfrac{1}{3}$	$H(1) = -3$	$H(3) = \dfrac{5}{3}$
Location of Graph	Below x-axis	Above x-axis	Below x-axis	Above x-axis
Point on Graph	$\left(-3, -\dfrac{1}{15}\right)$	$\left(-1, \dfrac{1}{3}\right)$	$(1, -3)$	$\left(3, \dfrac{5}{3}\right)$

Step 7: Graphing:

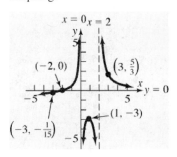

49. $R(x) = \dfrac{x^2+x-6}{x^2-x-6} = \dfrac{(x+3)(x-2)}{(x-3)(x+2)}$ $p(x) = x^2+x-6;\ q(x) = x^2-x-6;$

Step 1: Domain: $\{x \mid x \neq -2, x \neq 3\}$.

Step 2: (a) The x-intercepts are the zeros of $p(x)$: -3 and 2.

(b) The y-intercept is $R(0) = \dfrac{0^2+0-6}{0^2-0-6} = \dfrac{-6}{-6} = 1$.

Step 3: $R(-x) = \dfrac{(-x)^2+(-x)-6}{(-x)^2-(-x)-6} = \dfrac{x^2-x-6}{x^2+x-6}$; this is neither $R(x)$ nor $-R(x)$, so there is no symmetry.

Step 4: $R(x) = \dfrac{x^2+x-6}{x^2-x-6}$ is in lowest terms. The vertical asymptotes are the zeros of $q(x)$:
$x = -2$ and $x = 3$.

Step 5: Since $n = m$, the line $y = \dfrac{1}{1} = 1$ is the horizontal asymptote. Solve to find intersection points:

$$\dfrac{x^2+x-6}{x^2-x-6} = 1$$
$$x^2+x-6 = x^2-x-6$$
$$2x = 0$$
$$x = 0$$

$R(x)$ intersects $y = 1$ at $(0, 1)$.

Step 6:

Interval	$(-\infty, -3)$	$(-3, -2)$	$(-2, 2)$	$(2, 3)$	$(3, \infty)$
Number Chosen	-4	-2.5	0	2.5	4
Value of R	$R(-4) \approx 0.43$	$R(-2.5) \approx -0.82$	$R(0) = 1$	$R(2.5) \approx -1.22$	$R(4) = \tfrac{7}{3}$
Location of Graph	Above x-axis	Below x-axis	Above x-axis	Below x-axis	Above x-axis
Point on Graph	$(-4, 0.43)$	$(-2.5, -0.82)$	$(0, 1)$	$(2.5, -1.22)$	$\left(4, \tfrac{7}{3}\right)$

Step 7: Graphing:

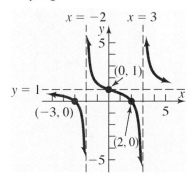

51. $F(x) = \dfrac{x^3}{x^2 - 4} = \dfrac{x^3}{(x+2)(x-2)}$ $\quad p(x) = x^3;\ q(x) = x^2 - 4;\ n = 3;\ m = 2$

Step 1: Domain: $\{x \mid x \neq -2, x \neq 2\}$.

Step 2: (a) The x-intercept is the zero of $p(x)$: 0.

(b) The y-intercept is $F(0) = \dfrac{0^3}{0^2 - 4} = \dfrac{0}{-4} = 0$.

Step 3: $F(-x) = \dfrac{(-x)^3}{(-x)^2 - 4} = \dfrac{-x^3}{x^2 - 4} = -F(x)$; $F(x)$ is symmetric with respect to the origin.

Step 4: $F(x) = \dfrac{x^3}{x^2 - 4}$ is in lowest terms. The vertical asymptotes are the zeros of $q(x)$: $x = -2$ and $x = 2$.

Step 5: Since $n = m + 1$, there is an oblique asymptote. Dividing:

$$\begin{array}{r} x \\ x^2 - 4 \overline{) x^3 } \\ \underline{x^3 - 4x} \\ 4x \end{array} \qquad \dfrac{x^3}{x^2 - 4} = x + \dfrac{4x}{x^2 - 4}$$

The oblique asymptote is $y = x$. Solve to find intersection points:

$$\dfrac{x^3}{x^2 - 4} = x$$
$$x^3 = x^3 - 4x$$
$$4x = 0$$
$$x = 0$$

$F(x)$ intersects $y = x$ at $(0, 0)$.

Step 6:

Interval	$(-\infty, -2)$	$(-2, 0)$	$(0, 2)$	$(2, \infty)$
Number Chosen	-3	-1	1	3
Value of F	$F(-3) = -\frac{27}{5}$	$F(-1) = \frac{1}{3}$	$F(1) = -\frac{1}{3}$	$F(3) = \frac{27}{5}$
Location of Graph	Below x-axis	Above x-axis	Below x-axis	Above x-axis
Point on Graph	$\left(-3, -\frac{27}{5}\right)$	$\left(-1, \frac{1}{3}\right)$	$\left(1, -\frac{1}{3}\right)$	$\left(3, \frac{27}{5}\right)$

Step 7: Graphing:

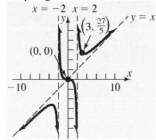

53. $R(x) = \dfrac{2x^4}{(x-1)^2}$ $p(x) = 2x^4$; $q(x) = (x-1)^2$; $n = 4$; $m = 2$

   Step 1: Domain: $\{x \mid x \neq 1\}$.

   Step 2: (a) The x-intercept is the zero of $p(x)$: 0.

   (b) The y-intercept is $R(0) = \dfrac{2(0)^4}{(0-1)^2} = \dfrac{0}{1} = 0$.

   Step 3: $R(-x) = \dfrac{2(-x)^4}{(-x-1)^2} = \dfrac{2x^4}{(x+1)^2}$; this is neither $R(x)$ nor $-R(x)$, so there is no symmetry.

   Step 4: $R(x) = \dfrac{2x^4}{(x-1)^2}$ is in lowest terms. The vertical asymptote is the zero of $q(x)$: $x = 1$.

   Step 5: Since $n > m + 1$, there is no horizontal asymptote and no oblique asymptote.

   Step 6:

Interval	$(-\infty, 0)$	$(0, 1)$	$(1, \infty)$
Number Chosen	-2	$\frac{1}{2}$	2
Value of R	$R(-2) \approx \frac{32}{9}$	$R\left(\frac{1}{2}\right) = \frac{1}{2}$	$R(2) = 32$
Location of Graph	Above x-axis	Above x-axis	Above x-axis
Point on Graph	$\left(-2, \frac{32}{9}\right)$	$\left(\frac{1}{2}, \frac{1}{2}\right)$	$(2, 32)$

Step 7: Graphing:

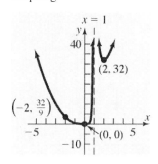

55. $G(x) = \dfrac{x^2-4}{x^2-x-2} = \dfrac{(x+2)(x-2)}{(x-2)(x+1)} = \dfrac{x+2}{x+1}$ $p(x) = x^2-4;\ q(x) = x^2-x-2;$

Step 1: Domain: $\{x \mid x \neq -1,\ x \neq 2\}$.

Step 2: (a) The x-intercept is the zero of $y = x+2$: -2; Note: 2 is not a zero because reduced form must be used to find the zeros.

(b) The y-intercept is $G(0) = \dfrac{0^2 - 4}{0^2 - 0 - 2} = \dfrac{-4}{-2} = 2$.

Step 3: $G(-x) = \dfrac{(-x)^2 - 4}{(-x)^2 - (-x) - 2} = \dfrac{x^2 - 4}{x^2 + x - 2}$; this is neither $G(x)$ nor $-G(x)$, so there is no symmetry.

Step 4: In lowest terms, $G(x) = \dfrac{x+2}{x+1}$, $x \neq 2$. The vertical asymptote is the zero of $f(x) = x+1$: $x = -1$; Note: $x = 2$ is not a vertical asymptote because reduced form must be used to find the asymptotes. The graph has a hole at $\left(2, \dfrac{4}{3}\right)$.

Step 5: Since $n = m$, the line $y = \dfrac{1}{1} = 1$ is the horizontal asymptote. Solve to find intersection points:

$$\dfrac{x^2-4}{x^2-x-2} = 1$$
$$x^2 - 4 = x^2 - x - 2$$
$$x = 2$$

$G(x)$ does not intersect $y = 1$ because $G(x)$ is not defined at $x = 2$.

Step 6:

Interval	$(-\infty, -2)$	$(-2, -1)$	$(-1, 2)$	$(2, \infty)$
Number Chosen	-3	-1.5	0	3
Value of G	$G(-3) = \dfrac{1}{2}$	$G(-1.5) = -1$	$G(0) = 2$	$G(3) = 1.25$
Location of Graph	Above x-axis	Below x-axis	Above x-axis	Above x-axis
Point on Graph	$\left(-3, \dfrac{1}{2}\right)$	$(-1.5, -1)$	$(0, 2)$	$(3, 1.25)$

Step 7: Graphing:

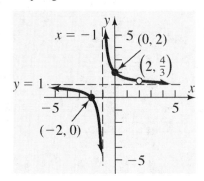

57. Solving algebraically:
$2x^2 + 5x - 12 < 0$
$f(x) = 2x^2 + 5x - 12$
$(x+4)(2x-3) < 0$
$x = -4, x = \frac{3}{2}$ are the zeros of f.

Interval	$(-\infty, -4)$	$\left(-4, \frac{3}{2}\right)$	$\left(\frac{3}{2}, \infty\right)$
Number Chosen	-5	0	2
Value of f	13	-12	6
Conclusion	Positive	Negative	Positive

The solution set is $\left\{x \mid -4 < x < \frac{3}{2}\right\}$.

Solving graphically:
Graph $f(x) = 2x^2 + 5x - 12$.

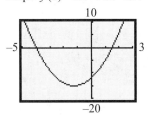

The x-intercepts are $x = -4$ and $x = 1.5$. The graph of f is below the x-axis for $-4 < x < 1.5$. Thus, the solution set is $\left\{x \mid -4 < x < 1.5\right\}$.

The solution set is $\left\{x \mid -4 < x < \frac{3}{2}\right\}$.

59. Solving algebraically:
$\dfrac{6}{x+3} \geq 1$

$f(x) = \dfrac{6}{x+3} - 1$

$\dfrac{6}{x+3} - 1 \geq 0 \Rightarrow \dfrac{6 - 1(x+3)}{x+3} \geq 0 \Rightarrow \dfrac{-x+3}{x+3} \geq 0$

Interval	$(-\infty, -3)$	$(-3, 3)$	$(3, \infty)$
Number Chosen	-4	0	4
Value of f	-7	1	$-\dfrac{1}{7}$
Conclusion	Negative	Positive	Negative

The solution set is $\left\{ x \mid -3 < x \leq 3 \right\}$.

Solving graphically:
Graph $y_1 = \dfrac{6}{x+3}$, $y_2 = 1$.

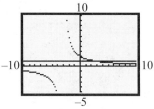

y_1 is undefined at $x = -3$.
y_1 intersects y_2 at $x = 3$. $y_1 > y_2$ for $-3 < x < 3$. Thus, the solution set is $\left\{ x \mid -3 < x \leq 3 \right\}$.

61. Solving algebraically:
$$\frac{2x-6}{1-x} < 2$$
$$f(x) = \frac{2x-6}{1-x} - 2$$
$$\frac{2x-6}{1-x} - 2 < 0$$
$$\frac{2x-6-2(1-x)}{1-x} < 0$$
$$\frac{4x-8}{1-x} < 0$$

The zeros and values where the expression is undefined are $x = 1$, and $x = 2$.

Interval	$(-\infty, 1)$	$(1, 2)$	$(2, \infty)$
Number Chosen	0	1.5	3
Value of f	-8	4	-2
Conclusion	Negative	Positive	Negative

The solution set is $\{x \mid x < 1 \text{ or } x > 2\}$.

Solving graphically:

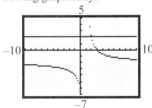

Graph $y_1 = \frac{2x-6}{1-x}$, $y_2 = 2$.

y_1 is undefined at $x = 1$. y_1 intersects y_2 at $x = 2$. $y_1 < y_2$ for $x < 1$ or $x > 2$. Thus, the solution set is $\{x \mid x < 1 \text{ or } x > 2\}$.

63. Solving algebraically:
$$\frac{(x-2)(x-1)}{x-3} > 0$$
$$f(x) = \frac{(x-2)(x-1)}{x-3}$$

The zeros and values where the expression is undefined are $x = 1$, $x = 2$, and $x = 3$.

Interval	$(-\infty, 1)$	$(1, 2)$	$(2, 3)$	$(3, \infty)$
Number Chosen	0	1.5	2.5	4
Value of f	$-\frac{2}{3}$	$\frac{1}{6}$	$-\frac{3}{2}$	6
Conclusion	Negative	Positive	Negative	Positive

The solution set is $\{x \mid 1 < x < 2 \text{ or } x > 3\}$.

Solving graphically:
Graph $f(x) = \frac{(x-2)(x-1)}{x-3}$.

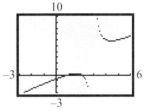

The x-intercepts are $x = 2$ and $x = 1$. The expression is undefined at $x = 3$. The graph of f is above the x-axis for $1 < x < 2$ or $x > 3$. Thus, the solution set is $\{x \mid 1 < x < 2 \text{ or } x > 3\}$.

65. Solving algebraically:
$$\frac{x^2 - 8x + 12}{x^2 - 16} > 0$$
$$f(x) = \frac{x^2 - 8x + 12}{x^2 - 16}$$
$$\frac{(x-2)(x-6)}{(x+4)(x-4)} > 0$$

The zeros and values where the expression is undefined are $x = -4$, $x = 2$, $x = 4$, and $x = 6$.

Interval	Number Chosen	Value of f	Conclusion
$(-\infty, -4)$	-5	$\frac{77}{9}$	Positive
$(-4, 2)$	0	$-\frac{3}{4}$	Negative
$(2, 4)$	3	$\frac{3}{7}$	Positive
$(4, 6)$	5	$-\frac{1}{3}$	Negative
$(6, \infty)$	7	$\frac{5}{33}$	Positive

The solution set is
$\{x \mid x < -4, 2 < x < 4, x > 6\}$.

Solving graphically:

Graph $f(x) = \dfrac{x^2 - 8x + 12}{x^2 - 16}$.

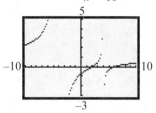

The x-intercepts are $x = 2$, and $x = 6$. The expression is undefined at $x = -4$ and $x = 4$. The graph of f is above the x-axis for $x < -4$, $2 < x < 4$, or $x > 6$. Thus, the solution set is $\{x \mid x < -4, 2 < x < 4, x > 6\}$.

67. $f(x) = 8x^3 - 3x^2 + x + 4$

Since $g(x) = x - 1$ then $c = 1$. From the Remainder Theorem, the remainder R when $f(x)$ is divided by $g(x)$ is $f(c)$:
$$f(1) = 8(1)^3 - 3(1)^2 + 1 + 4$$
$$= 8 - 3 + 1 + 4$$
$$= 10$$

So $R = 10$ and g is not a factor of f.

69. $f(x) = x^4 - 2x^3 + 15x - 2$

Since $g(x) = x + 2$ then $c = -2$. From the Remainder Theorem, the remainder R when $f(x)$ is divided by $g(x)$ is $f(c)$:
$$f(-2) = (-2)^4 - 2(-2)^3 + 15(-2) - 2$$
$$= 16 - 2(-8) - 30 - 2$$
$$= 16 + 16 - 30 - 2$$
$$= 0$$

So $R = 0$ and g is a factor of f.

71.
$$\begin{array}{r|rrrrrr} 4) & 12 & 0 & -8 & 0 & 0 & 0 & 1 \\ & & 48 & 192 & 736 & 2944 & 11{,}776 & 47{,}104 \\ \hline & 12 & 48 & 184 & 736 & 2944 & 11{,}776 & 47{,}105 \end{array}$$

$f(4) = 47{,}105$

73. $f(x) = 2x^8 - x^7 + 8x^4 - 2x^3 + x + 3$

The degree is 8 so the maximum number of zeros is 8.

p must be a factor of 3: $p = \pm 1, \pm 3$

q must be a factor of 2: $q = \pm 1, \pm 2$

The possible rational zeros are:
$$\dfrac{p}{q} = \pm 1, \pm 3, \pm \dfrac{1}{2}, \pm \dfrac{3}{2}$$

75. $f(x) = x^3 - 3x^2 - 6x + 8$

Step 1: $f(x)$ has at most 3 real zeros.

Step 2: By Descartes' Rule of Signs, there are two positive real zeros or no positive real zeros.

$f(-x) = (-x)^3 - 3(-x)^2 - 6(-x) + 8$, there is one
$= -x^3 - 3x^2 + 6x + 8$
negative real zero.

Step 3: Possible rational zeros:
$p = \pm 1, \pm 2, \pm 4, \pm 8;\ \ q = \pm 1;$
$\dfrac{p}{q} = \pm 1, \pm 2, \pm 4, \pm 8$

Step 4: Using the Bounds on Zeros Theorem:
$a_2 = -3,\ \ a_1 = -6,\ \ a_0 = 8$
Max $\{1, |8| + |-6| + |-3|\}$ = Max $\{1, 17\}$
$= 17$
$1 + $ Max $\{|8|, |-6|, |-3|\} = 1 + 8 = 9$

The smaller of the two numbers is 9. Thus, every real zero of f lies between -9 and 9.

Step 5: Using synthetic division:
We try $x + 2$:

$$\begin{array}{r|rrrr} -2) & 1 & -3 & -6 & 8 \\ & & -2 & 10 & -8 \\ \hline & 1 & -5 & 4 & 0 \end{array}$$

$x + 2$ is a factor. The other factor is the quotient: $x^2 - 5x + 4$.

Thus, $f(x) = (x+2)(x^2 - 5x + 4)$.
$= (x+2)(x-1)(x-4)$

The zeros are -2, 1, and 4, each of multiplicity 1.

77. $f(x) = 4x^3 + 4x^2 - 7x + 2$

Step 1: $f(x)$ has at most 3 real zeros.

Step 2: By Descartes' Rule of Signs, there are two positive real zeros or no positive real zeros.

$f(-x) = 4(-x)^3 + 4(-x)^2 - 7(-x) + 2$; thus, there
$= -4x^3 + 4x^2 + 7x + 2$

is one negative real zero.

Step 3: Possible rational zeros:
$p = \pm 1, \pm 2; \quad q = \pm 1, \pm 2, \pm 4;$
$\dfrac{p}{q} = \pm 1, \pm 2, \pm \dfrac{1}{2}, \pm \dfrac{1}{4}$

Step 4: Using the Bounds on Zeros Theorem:
$f(x) = 4\left(x^3 + x^2 - \dfrac{7}{4}x + \dfrac{1}{2}\right) \Rightarrow$
$a_2 = 1, \quad a_1 = -\dfrac{7}{4}, \quad a_0 = \dfrac{1}{2}$
$\text{Max}\left\{1, \left|\dfrac{1}{2}\right| + \left|-\dfrac{7}{4}\right| + |1|\right\} = \text{Max}\left\{1, \dfrac{13}{4}\right\}$
$= \dfrac{13}{4} = 3.25$
$1 + \text{Max}\left\{\left|\dfrac{1}{2}\right|, \left|-\dfrac{7}{4}\right|, |1|\right\} = 1 + \dfrac{7}{4} = \dfrac{11}{4} = 2.75$

The smaller of the two numbers is 2.75. Thus, every real zero of f lies between -2.75 and 2.75.

Step 5: Using synthetic division:
We try $x + 2$:

$$\begin{array}{r|rrrr} -2) & 4 & 4 & -7 & 2 \\ & & -8 & 8 & -2 \\ \hline & 4 & -4 & 1 & 0 \end{array}$$

$x + 2$ is a factor. The other factor is the quotient: $4x^2 - 4x + 1$.

Thus, $f(x) = (x+2)(4x^2 - 4x + 1)$
$= (x+2)(2x-1)(2x-1)$

The zeros are -2, of multiplicity 1 and $\dfrac{1}{2}$, of multiplicity 2.

79. $f(x) = x^4 - 4x^3 + 9x^2 - 20x + 20$

Step 1: $f(x)$ has at most 4 real zeros.

Step 2: By Descartes' Rule of Signs, there are four positive real zeros or two positive real zeros or no positive real zeros.

$f(-x) = (-x)^4 - 4(-x)^3 + 9(-x)^2 - 20(-x) + 20$
$= x^4 + 4x^3 + 9x^2 + 20x + 20;$

Thus, there are no negative real zeros.

Step 3: Possible rational zeros:
$p = \pm 1, \pm 2, \pm 4, \pm 5, \pm 10, \pm 20; \quad q = \pm 1;$
$\dfrac{p}{q} = \pm 1, \pm 2, \pm 4, \pm 5, \pm 10, \pm 20$

Step 4: Using the Bounds on Zeros Theorem:
$a_3 = -4, \ a_2 = 9, \ a_1 = -20, \ a_0 = 20$
$\text{Max}\{1, |20| + |-20| + |9| + |-4|\} = \text{Max}\{1, 53\}$
$= 53$
$1 + \text{Max}\{|20|, |-20|, |9|, |-4|\} = 1 + 20 = 21$

The smaller of the two numbers is 21. Thus, every real zero of f lies between -21 and 21.

Step 5: Using synthetic division:
We try $x - 2$:

$$\begin{array}{r|rrrrr} 2) & 1 & -4 & 9 & -20 & 20 \\ & & 2 & -4 & 10 & -20 \\ \hline & 1 & -2 & 5 & -10 & 0 \end{array}$$

$x - 2$ is a factor and the quotient is $x^3 - 2x^2 + 5x - 10$

We try $x - 2$ on $x^3 - 2x^2 + 5x - 10$

$$\begin{array}{r|rrrr} 2) & 1 & -2 & 5 & -10 \\ & & 2 & 0 & 10 \\ \hline & 1 & 0 & 5 & 0 \end{array}$$

$x - 2$ is a factor and the quotient is $x^2 + 5$.
$x - 2$ is a factor twice. The other factor is the quotient: $x^2 + 5$.

Thus, $f(x) = (x-2)(x-2)(x^2 + 5)$
$= (x-2)^2 (x^2 + 5)$

Since $x^2 + 5 = 0$ has no real solutions, the only zero is 2, of multiplicity 2.

81. $f(x) = 2x^3 - 11.84x^2 - 9.116x + 82.46$

$f(x)$ has at most 3 real zeros.

Solving by graphing (using ZERO):

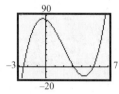

The zeros are approximately –2.5, 3.1, and 5.32.

83. $g(x) = 15x^4 - 21.5x^3 - 1718.3x^2 + 5308x + 3796.8$

$g(x)$ has at most 4 real zeros.

Solving by graphing (using ZERO):

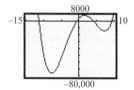

The zeros are approximately –11.3, –0.6, 4, and 9.33.

85. $2x^4 + 2x^3 - 11x^2 + x - 6 = 0$

The solutions of the equation are the zeros of $f(x) = 2x^4 + 2x^3 - 11x^2 + x - 6$.

Step 1: $f(x)$ has at most 4 real zeros.

Step 2: By Descartes' Rule of Signs, there are three positive real zeros or there is one positive real zero.

$f(-x) = 2(-x)^4 + 2(-x)^3 - 11(-x)^2 + (-x) - 6$;
$= 2x^4 - 2x^3 - 11x^2 - x - 6$

Thus, there is one negative real zero.

Step 3: Possible rational zeros:
$p = \pm 1, \pm 2, \pm 3, \pm 6; \quad q = \pm 1, \pm 2;$
$\dfrac{p}{q} = \pm 1, \pm 2, \pm 3, \pm 6, \pm \dfrac{1}{2}, \pm \dfrac{3}{2}$

Step 4: Using the Bounds on Zeros Theorem:

$f(x) = 2\left(x^4 + x^3 - \dfrac{11}{2}x^2 + \dfrac{1}{2}x - 3\right) \Rightarrow$

$a_3 = 1, \; a_2 = -\dfrac{11}{2}, \; a_1 = \dfrac{1}{2}, \; a_0 = -3$

$\text{Max}\left\{1, |-3| + \left|\dfrac{1}{2}\right| + \left|-\dfrac{11}{2}\right| + |1|\right\} = \text{Max}\{1, 10\}$
$= 10$

$1 + \text{Max}\left\{|-3|, \left|\dfrac{1}{2}\right|, \left|-\dfrac{11}{2}\right|, |1|\right\} = 1 + \dfrac{11}{2}$

$= \dfrac{13}{2} = 6.5$

The smaller of the two numbers is 6.5. Thus, every real zero of f lies between –6.5 and 6.5.

Step 5: Using synthetic division:

We try $x + 3$:

$$\begin{array}{r|rrrrr} -3) & 2 & 2 & -11 & 1 & -6 \\ & & -6 & 12 & -3 & 6 \\ \hline & 2 & -4 & 1 & -2 & 0 \end{array}$$

$x + 3$ is a factor and the quotient is $2x^3 - 4x^2 + x - 2$.

We try $x - 2$ on $2x^3 - 4x^2 + x - 2$:

$$\begin{array}{r|rrrr} 2) & 2 & -4 & 1 & -2 \\ & & 4 & 0 & 2 \\ \hline & 2 & 0 & 1 & 0 \end{array}$$

$x - 2$ is a factor and the quotient is $2x^2 + 1$.

$x + 3$ and $x - 2$ are factors. The other factor is the quotient: $2x^2 + 1$.

Thus, $f(x) = (x+3)(x-2)\left(2x^2 + 1\right)$.

Since $2x^2 + 1 = 0$ has no real solutions, the solution set is $\{-3, 2\}$.

87. $2x^4 + 7x^3 + x^2 - 7x - 3 = 0$

The solutions of the equation are the zeros of $f(x) = 2x^4 + 7x^3 + x^2 - 7x - 3$.

Step 1: $f(x)$ has at most 4 real zeros.

Step 2: By Descartes' Rule of Signs, there is one positive real zero.
$f(-x) = 2(-x)^4 + 7(-x)^3 + (-x)^2 - 7(-x) - 3$;
$= 2x^4 - 7x^3 + x^2 + 7x - 3$

Thus, there are three negative real zeros or there is one negative real zero.

Step 3: Possible rational zeros:
$p = \pm 1, \pm 3; \quad q = \pm 1, \pm 2;$
$\dfrac{p}{q} = \pm 1, \pm 3, \pm \dfrac{1}{2}, \pm \dfrac{3}{2}$

Step 4: Using the Bounds on Zeros Theorem:
$f(x) = 2\left(x^4 + \dfrac{7}{2}x^3 + \dfrac{1}{2}x^2 - \dfrac{7}{2}x - \dfrac{3}{2}\right)$
$a_3 = \dfrac{7}{2}, \; a_2 = \dfrac{1}{2}, \; a_1 = -\dfrac{7}{2}, \; a_0 = -\dfrac{3}{2}$

$\text{Max}\left\{1, \left|-\dfrac{3}{2}\right| + \left|-\dfrac{7}{2}\right| + \left|\dfrac{1}{2}\right| + \left|\dfrac{7}{2}\right|\right\} = \text{Max}\{1, 9\}$
$= 9$

$1 + \text{Max}\left\{\left|-\dfrac{3}{2}\right|, \left|-\dfrac{7}{2}\right|, \left|\dfrac{1}{2}\right|, \left|\dfrac{7}{2}\right|\right\} = 1 + \dfrac{7}{2} = \dfrac{9}{2}$
$= 4.5$

The smaller of the two numbers is 4.5. Thus, every real zero of f lies between -4.5 and 4.5.

Step 5: Using synthetic division:
We try $x + 3$:

```
-3) 2   7   1   -7   -3
       -6  -3    6    3
    ─────────────────────
    2   1   -2  -1    0
```

$x + 3$ is a factor and the quotient is $2x^3 + x^2 - 2x - 1$.

We try $x + 3$:

```
-3) 2   7   1   -7   -3
       -6  -3    6    3
    ─────────────────────
    2   1   -2  -1    0
```

$x + 3$ is a factor and the quotient is $2x^3 + x^2 - 2x - 1$.

We try $x + 1$ on $2x^3 + x^2 - 2x - 1$

```
-1) 2    1   -2   -1
        -2    1    1
    ─────────────────
    2   -1   -1    0
```

$x + 1$ is a factor and the quotient is $2x^2 - x - 1$.

$x + 3$ and $x + 1$ are factors. The other factor is the quotient: $2x^2 - x - 1$.

Thus, $f(x) = (x+3)(x+1)(2x^2 - x - 1)$.
$= (x+3)(x+1)(2x+1)(x-1)$

The solution set is $\left\{-3, \; -1, \; -\dfrac{1}{2}, \; 1\right\}$.

89. $f(x) = x^3 - x^2 - 4x + 2$

$a_2 = -1, \; a_1 = -4, \; a_0 = 2$

$\text{Max}\{1, |2| + |-4| + |-1|\} = \text{Max}\{1, 7\} = 7$

$1 + \text{Max}\{|2|, |-4|, |-1|\} = 1 + 4 = 5$

The smaller of the two numbers is 5, so every real zero of f lies between -5 and 5.

91. $f(x) = 2x^3 - 7x^2 - 10x + 35$
$= 2\left(x^3 - \dfrac{7}{2}x^2 - 5x + \dfrac{35}{2}\right)$

$a_2 = -\dfrac{7}{2}, \; a_1 = -5, \; a_0 = \dfrac{35}{2}$

$\text{Max}\left\{1, \left|\dfrac{35}{2}\right| + |-5| + \left|-\dfrac{7}{2}\right|\right\} = \text{Max}\{1, 26\}$
$= 26$

$1 + \text{Max}\left\{\left|\dfrac{35}{2}\right|, |-5|, \left|-\dfrac{7}{2}\right|\right\} = 1 + \dfrac{35}{2}$
$= \dfrac{37}{2} = 18.5$

The smaller of the two numbers is 18.5, so every real zero of f lies between -18.5 and 18.5.

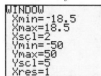 

93. $f(x) = 3x^3 - x - 1;\ [0, 1]$

    $f(0) = -1 < 0$ and $f(1) = 1 > 0$

    Since one value is positive and one is negative, there is a zero in the interval.

     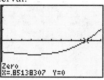

    The zero is roughly 0.85.

95. $f(x) = 8x^4 - 4x^3 - 2x - 1;\ [0, 1]$

    $f(0) = -1 < 0$ and $f(1) = 1 > 0$

    Since one value is positive and one is negative, there is a zero in the interval.

     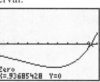

    The zero is roughly 0.94.

97. Since complex zeros appear in conjugate pairs, $4 - i$, the conjugate of $4 + i$, is the remaining zero of f.

    $f(x) = (x-6)(x-4-i)(x-4+i)$
    $= x^3 - 14x^2 + 65x - 102$

    $f(x) = (x-5)(x-3-4i)(x-3+4i)$
    $= x^3 - 11x^2 + 55x - 125$

99. Since complex zeros appear in conjugate pairs, $-i$, the conjugate of i, and $1 - i$, the conjugate of $1 + i$, are the remaining zeros of f.

    $f(x) = (x-i)(x+i)(x-1-i)(x-1+i)$
    $= x^4 - 2x^3 + 3x^2 - 2x + 2$

101. $x^2 + x + 1 = 0$
     $a = 1, b = 1, c = 1,$
     $b^2 - 4ac = 1^2 - 4(1)(1) = 1 - 4 = -3$
     $x = \dfrac{-1 \pm \sqrt{-3}}{2(1)} = \dfrac{-1 \pm \sqrt{3}\,i}{2} = -\dfrac{1}{2} \pm \dfrac{\sqrt{3}}{2}i$
     The solution set is $\left\{-\dfrac{1}{2} - \dfrac{\sqrt{3}}{2}i,\ -\dfrac{1}{2} + \dfrac{\sqrt{3}}{2}i\right\}$.

103. $2x^2 + x - 2 = 0$
     $a = 2, b = 1, c = -2,$
     $b^2 - 4ac = 1^2 - 4(2)(-2) = 1 + 16 = 17$
     $x = \dfrac{-1 \pm \sqrt{17}}{2(2)} = \dfrac{-1 \pm \sqrt{17}}{4}$
     The solution set is $\left\{\dfrac{-1 - \sqrt{17}}{4},\ \dfrac{-1 + \sqrt{17}}{4}\right\}$.

105. $x^2 + 3 = x$
     $x^2 - x + 3 = 0$
     $a = 1, b = -1, c = 3,$
     $b^2 - 4ac = (-1)^2 - 4(1)(3) = 1 - 12 = -11$
     $x = \dfrac{-(-1) \pm \sqrt{-11}}{2(1)} = \dfrac{1 \pm \sqrt{11}\,i}{2} = \dfrac{1}{2} \pm \dfrac{\sqrt{11}}{2}i$
     The solution set is $\left\{\dfrac{1}{2} - \dfrac{\sqrt{11}}{2}i,\ \dfrac{1}{2} + \dfrac{\sqrt{11}}{2}i\right\}$.

107. $x(1 - x) = 6$
     $-x^2 + x - 6 = 0$
     $a = -1, b = 1, c = -6,$
     $b^2 - 4ac = 1^2 - 4(-1)(-6) = 1 - 24 = -23$
     $x = \dfrac{-1 \pm \sqrt{-23}}{2(-1)} = \dfrac{-1 \pm \sqrt{23}\,i}{-2} = \dfrac{1}{2} \pm \dfrac{\sqrt{23}}{2}i$
     The solution set is $\left\{\dfrac{1}{2} - \dfrac{\sqrt{23}}{2}i,\ \dfrac{1}{2} + \dfrac{\sqrt{23}}{2}i\right\}$.

109. $x^4 + 2x^2 - 8 = 0$
     $(x^2 + 4)(x^2 - 2) = 0$
     $x^2 + 4 = 0$ or $x^2 - 2 = 0$
     $x^2 = -4$ or $x^2 = 2$
     $x = \pm 2i$ or $x = \pm\sqrt{2}$
     The solution set is $\left\{-2i,\ 2i,\ -\sqrt{2},\ \sqrt{2}\right\}$.

111. $x^3 - x^2 - 8x + 12 = 0$

The solutions of the equation are the zeros of the function $f(x) = x^3 - x^2 - 8x + 12$.

Step 1: $f(x)$ has 3 complex zeros.

Step 2: Possible rational zeros:
$p = \pm 1, \pm 2, \pm 3, \pm 4, \pm 6, \pm 12; \quad q = \pm 1;$
$\dfrac{p}{q} = \pm 1, \pm 2, \pm 3, \pm 4, \pm 6, \pm 12$

Step 3: Graphing the function:

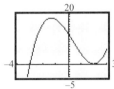

From the graph it appears that there are x-intercepts at -3 and 2.

Step 4: Using synthetic division:

$$\begin{array}{r|rrrr} 2 & 1 & -1 & -8 & 12 \\ & & 2 & 2 & -12 \\ \hline & 1 & 1 & -6 & 0 \end{array}$$

Since the remainder is 0, $x - 2$ is a factor. The other factor is the quotient:
$x^2 + x - 6 = (x+3)(x-2)$.

The complex zeros are -3, 2 (multiplicity 2).

113. $3x^4 - 4x^3 + 4x^2 - 4x + 1 = 0$

The solutions of the equation are the zeros of the function
$f(x) = 3x^4 - 4x^3 + 4x^2 - 4x + 1$

Step 1: $f(x)$ has 4 complex zeros.

Step 2: Possible rational zeros:
$p = \pm 1; \quad q = \pm 1, \pm 3; \quad \dfrac{p}{q} = \pm 1, \pm \dfrac{1}{3}$

Step 3: Graphing the function:

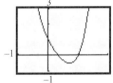

From the graph it appears that there are x-intercepts at $\dfrac{1}{3}$ and 1.

Step 4: Using synthetic division:

$$\begin{array}{r|rrrrr} 1 & 3 & -4 & 4 & -4 & 1 \\ & & 3 & -1 & 3 & -1 \\ \hline & 3 & -1 & 3 & -1 & 0 \end{array} \quad \begin{array}{r|rrrr} \frac{1}{3} & 3 & -1 & 3 & -1 \\ & & 1 & 0 & 1 \\ \hline & 3 & 0 & 3 & 0 \end{array}$$

Since the remainder is 0, $x - 1$ and $x - \dfrac{1}{3}$ are factors. The other factor is the quotient:
$3x^2 + 3 = 3(x^2 + 1)$.

Solving $x^2 + 1 = 0$:
$x^2 = -1$
$x = \pm i$

The complex zeros are 1, $\dfrac{1}{3}$, $-i$, i.

115. The distance between the point $P(x, y)$ and $Q(3, 1)$ is $d(P, Q) = \sqrt{(x-3)^2 + (y-1)^2}$.

If P is on the line $y = x$, then the distance is
$d(P, Q) = \sqrt{(x-3)^2 + (x-1)^2}$
$d^2(x) = (x-3)^2 + (x-1)^2$
$ = x^2 - 6x + 9 + x^2 - 2x + 1$
$ = 2x^2 - 8x + 10$

Since $d^2(x) = 2x^2 - 8x + 10$ is a quadratic function with $a = 2 > 0$, the vertex corresponds to the minimum value for the function.

The vertex occurs at $x = -\dfrac{b}{2a} = -\dfrac{-8}{2(2)} = 2$.

Therefore the point on the line $y = x$ closest to the point $(3, 1)$ is $(2, 2)$.

117. Consider the diagram

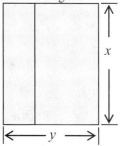

Total amount of fence $= 3x + 2y = 10{,}000$

$$y = \frac{10{,}000 - 3x}{2} = 5000 - \frac{3}{2}x$$

Total area enclosed $= (x)(y) = (x)\left(5000 - \frac{3}{2}x\right)$

$A(x) = 5000x - \frac{3}{2}x^2 = -\frac{3}{2}x^2 + 5000x$ is a

quadratic function with $a = -\frac{3}{2} < 0$.

So the vertex corresponds to the maximum value for this function.
The vertex occurs when

$$x = -\frac{b}{2a} = -\frac{5000}{2\left(-\frac{3}{2}\right)} = \frac{5000}{3}$$

The maximum area is:

$$A\left(\frac{5000}{3}\right) = -\frac{3}{2}\left(\frac{5000}{3}\right)^2 + 5000\left(\frac{5000}{3}\right)$$

$$= -\frac{3}{2}\left(\frac{25{,}000{,}000}{9}\right) + \frac{25{,}000{,}000}{3}$$

$$= -\frac{12{,}500{,}000}{3} + \frac{25{,}000{,}000}{3}$$

$$= \frac{12{,}500{,}000}{3}$$

$$\approx 4{,}166{,}666.67 \text{ square meters}$$

119. Consider the diagram

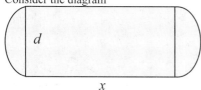

d = diameter of the semicircles
  = width of the rectangle

x = length of the rectangle

100 = outside dimension length
$100 = 2x + 2(\text{circumference of a semicircle})$
$100 = 2x + \text{circumference of a circle}$
$100 = 2x + \pi d$

$$\Rightarrow x = \frac{100 - \pi d}{2} = 50 - \frac{1}{2}\pi d$$

We need an expression for the area of a rectangle in terms of a single variable.

$A_{\text{rectangle}} = x \cdot d$

$$= \left(50 - \frac{1}{2}\pi d\right) \cdot d$$

$$= 50d - \frac{1}{2}\pi d^2$$

This is a quadratic function with $a = -\frac{1}{2}\pi < 0$.

Therefore, the x-coordinate of the vertex represents the value for d that maximizes the area of the rectangle and the y-coordinate of the vertex is the maximum area of the rectangle.
The vertex occurs at

$$d = -\frac{b}{2a} = -\frac{50}{2\left(-\frac{1}{2}\pi\right)}$$

$$= \frac{-50}{-\pi} = \frac{50}{\pi}$$

This gives us

$$x = 50 - \frac{1}{2}\pi d$$

$$= 50 - \frac{1}{2}\pi\left(\frac{50}{\pi}\right)$$

$$= 50 - 25 = 25$$

Therefore, the side of the rectangle with the semicircle should be $\frac{50}{\pi}$ feet and the other side should be 25 feet.
The maximum area is

$$(25)\left(\frac{50}{\pi}\right) = \frac{1250}{\pi} \approx 397.89 \text{ ft}^2.$$

121. $C(x) = 4.9x^2 - 617.40x + 19,600$;
$a = 4.9$, $b = -617.40$, $c = 19,600$. Since $a = 4.9 > 0$, the graph opens up, so the vertex is a minimum point.

(a) The minimum marginal cost occurs at
$$x = -\frac{b}{2a} = -\frac{-617.40}{2(4.9)} = \frac{617.40}{9.8} = 63.$$

Thus, 63 golf clubs should be manufactured in order to minimize the marginal cost.

(b) The minimum marginal cost is
$$C\left(-\frac{b}{2a}\right) = C(63)$$
$$= 4.9(63)^2 - (617.40)(63) + 19600$$
$$= \$151.90$$

123. $W = \dfrac{k}{d^2}$
$$200 = \frac{k}{3960^2} \Rightarrow k = 3,136,320,000$$
When $d = 3961$,
$$W = \frac{3,136,320,000}{3961^2} \approx 199.90 \text{ pounds}$$

125. (a) Graphing, the data appear to be quadratic.

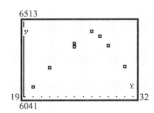

(b) Using the QUADratic REGression program, the quadratic function of best fit is:
$$R(A) = -7.76A^2 + 411.88A + 942.72$$

(c) The maximum revenue occurs at
$$A = \frac{-b}{2a} = \frac{-(411.88)}{2(-7.76)}$$
$$= \frac{-411.88}{-15.52}$$
$$\approx 26.53865979 \text{ thousand dollars}$$
$$\approx \$26,538.66$$

(d) The maximum revenue is
$$R\left(\frac{-b}{2a}\right)$$
$$= R(26.53866)$$
$$= -7.76(26.53866)^2 + (411.88)(26.53866) + 942.72$$
$$\approx 6408.091683 \text{ thousand dollars}$$
$$\approx \$6,408,091.68$$

(e) graphing:

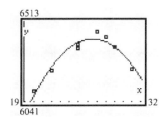

127. (a) $250 = \pi r^2 h \Rightarrow h = \dfrac{250}{\pi r^2}$;
$$A(r) = 2\pi r^2 + 2\pi rh$$
$$= 2\pi r^2 + 2\pi r\left(\frac{250}{\pi r^2}\right)$$
$$= 2\pi r^2 + \frac{500}{r}$$

(b) $A(3) = 2\pi \cdot 3^2 + \dfrac{500}{3}$
$$= 18\pi + \frac{500}{3} \approx 223.22 \text{ square cm}$$

(c) $A(5) = 2\pi \cdot 5^2 + \dfrac{500}{5}$
$$= 50\pi + 100 \approx 257.08 \text{ square cm}$$

(d) Use MINIMUM on the graph of
$$y_1 = 2\pi x^2 + \frac{500}{x}$$

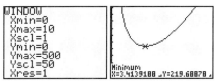

The area is smallest when the radius is approximately 3.41 cm.

129. Answers will vary.

Chapter 3 Test

1. $f(x) = (x-3)^4 - 2$

    Using the graph of $y = x^4$, shift right 3 units, then shift down 2 units.

    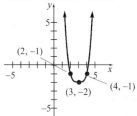

2. a. The leading coefficient (i.e. the coefficient on x^2) is positive so the graph will open up. Thus, the graph has a minimum.

    b. The x-coordinate of the vertex is given by
    $$x = -\frac{b}{2a} = -\frac{(-12)}{2(3)} = \frac{12}{6} = 2$$
    The y-coordinate of the vertex is given by
    $$f\left(-\frac{b}{2a}\right) = f(2) = 3(2)^2 - 12(2) + 4 = -8$$
    The vertex is $(2, -8)$.

    c. The axis of symmetry is $x = -\dfrac{b}{2a} = 2$.

    d. y-intercept:
    $$f(0) = 3(0)^2 - 12(0) + 4 = 4$$
    The y-intercept is 4.

    x-intercepts:
    $$f(x) = 0$$
    $$3x^2 - 12x + 4 = 0$$
    $$a = 3, b = -12, c = 4$$

    $$x = \frac{-(-12) \pm \sqrt{(-12)^2 - 4(3)(4)}}{2(3)}$$
    $$= \frac{12 \pm \sqrt{144-48}}{6}$$
    $$= \frac{12 \pm \sqrt{96}}{6}$$
    $$= \frac{12 \pm 4\sqrt{6}}{6}$$
    $$= \frac{6 \pm 2\sqrt{6}}{3}$$

    The x-intercepts are $\dfrac{6 \pm 2\sqrt{6}}{3}$.

    The intercepts are $(0, 4)$, $\left(\dfrac{6 - 2\sqrt{6}}{3}, 0\right)$, and $\left(\dfrac{6 + 2\sqrt{6}}{3}, 0\right)$.

    e. Graph:

    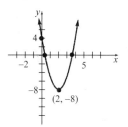

3. a. The maximum number of real zeros is the degree, $n = 3$.

    b. First we write the polynomial so that the leading coefficient is 1.
    $$g(x) = 2\left(x^3 + \frac{5}{2}x^2 - 14x - \frac{15}{2}\right)$$
    For the expression in parentheses, we have
    $a_2 = \dfrac{5}{2}$, $a_1 = -14$, and $a_0 = -\dfrac{15}{2}$.
    $$\max\{1, |a_0| + |a_1| + |a_2|\}$$
    $$= \max\left\{1, \left|-\tfrac{15}{2}\right| + |-14| + \left|\tfrac{5}{2}\right|\right\}$$
    $$= \max\{1, 24\}$$
    $$= 24$$

$$1 + \max\{|a_0|, |a_1|, |a_2|\}$$
$$= 1 + \max\left\{\left|-\tfrac{15}{2}\right|, |-14|, \left|\tfrac{5}{2}\right|\right\}$$
$$= 1 + 14$$
$$= 15$$

The smaller of the two numbers, 15, is the bound. Every zero of g lies between -15 and 15.

c. $g(x) = 2x^3 + 5x^2 - 28x - 15$

We list all integers p that are factors of $a_0 = -15$ and all the integers q that are factors of $a_3 = 2$.

$p: \pm 1, \pm 3, \pm 5, \pm 15$

$q: \pm 1, \pm 2$

Now we form all possible ratios $\dfrac{p}{q}$:

$\dfrac{p}{q}: \pm\tfrac{1}{2}, \pm 1, \pm\tfrac{3}{2}, \pm\tfrac{5}{2}, \pm 3, \pm 5, \pm\tfrac{15}{2}, \pm 15$

If g has a rational zero, it must be one of the 16 possibilities listed.

d. From part (b), we know that all the zeros must lie between -15 and 15. This allows us to set our Xmin and Xmax values before graphing.

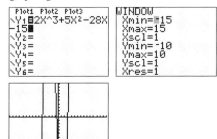

Note that we see three zeros, the maximum number possible.

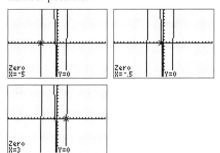

The zeros of the polynomial are $-5, -0.5,$ and 3.

$$(x+5)(x-3)\left(x+\tfrac{1}{2}\right) = (x^2 + 2x - 15)\left(x + \tfrac{1}{2}\right)$$
$$= x^3 + \tfrac{5}{2}x^2 - 14x - \tfrac{15}{2}$$

Therefore,
$$g(x) = 2(x+5)(x-3)\left(x+\tfrac{1}{2}\right)$$
$$= (x+5)(x-3)(2x+1)$$

4. $x^3 - 4x^2 + 25x - 100 = 0$
$$x^2(x-4) + 25(x-4) = 0$$
$$(x-4)(x^2 + 25) = 0$$
$$x - 4 = 0 \quad \text{or} \quad x^2 + 25 = 0$$
$$x = 4 \qquad\qquad x^2 = -25$$
$$x = \pm\sqrt{-25}$$
$$x = \pm 5i$$

The solution set is $\{4, -5i, 5i\}$.

5. $3x^3 + 2x - 1 = 8x^2 - 4$
$$3x^3 - 8x^2 + 2x + 3 = 0$$

If we let the left side of the equation be $f(x)$, then we are simply finding the zeros of f. We list all integers p that are factors of $a_0 = 3$ and all the integers q that are factors of $a_3 = 3$.

$p: \pm 1, \pm 3$

$q: \pm 1, \pm 3$

Now we form all possible ratios $\dfrac{p}{q}$:

$\dfrac{p}{q}: \pm\tfrac{1}{3}, \pm 1, \pm 3$

It appears that there is a zero near $x = 1$.
$$f(1) = 3(1)^3 - 8(1)^2 + 2(1) + 3 = 0$$

Therefore, $x=1$ is a zero and $(x-1)$ is a factor of $f(x)$. We can reduce the polynomial expression by using synthetic division.

$$\begin{array}{r} 1\overline{)3 \quad -8 \quad 2 \quad 3} \\ \underline{3 \quad -5 \quad -3} \\ 3 \quad -5 \quad -3 \quad 0 \end{array}$$

Thus, $f(x) = (x-1)(3x^2 - 5x - 3)$. We can find the remaining zeros by using the quadratic formula.
$3x^2 - 5x - 3 = 0$
$a = 3, b = -5, c = -3$
$x = \dfrac{-(-5) \pm \sqrt{(-5)^2 - 4(3)(-3)}}{2(3)}$
$= \dfrac{5 \pm \sqrt{25 + 36}}{6}$
$= \dfrac{5 \pm \sqrt{61}}{6}$

Thus, the solution set is
$\left\{ 1, \dfrac{5 - \sqrt{61}}{6} \approx -0.468, \dfrac{5 + \sqrt{61}}{6} \approx 2.135 \right\}$.

6. We start by factoring the numerator and denominator.
$g(x) = \dfrac{2x^2 - 14x + 24}{x^2 + 6x - 40} = \dfrac{2(x-3)(x-4)}{(x+10)(x-4)}$
The domain of f is $\{x \mid x \neq -10, x \neq 4\}$.

In lowest terms, $g(x) = \dfrac{2(x-3)}{x+10}$ with $x \neq 4$.

The graph has one vertical asymptote, $x = -10$, since $x + 10$ is the only factor of the denominator of g in lowest terms. The graph is still undefined at $x = 4$, but there is a hole in the graph there instead of an asymptote.

Since the degree of the numerator is the same as the degree of the denominator, the graph has a horizontal asymptote equal to the quotient of the leading coefficients. The leading coefficient in the numerator is 2 and the leading coefficient in the denominator is 1. Therefore, the graph has the horizontal asymptote $y = \tfrac{2}{1} = 2$.

7. $r(x) = \dfrac{x^2 + 2x - 3}{x + 1}$

Start by factoring the numerator.
$r(x) = \dfrac{(x+3)(x-1)}{x+1}$
The domain of the function is $\{x \mid x \neq -1\}$.

Asymptotes:
Since the function is in lowest terms, the graph has one vertical asymptote, $x = -1$.
The degree of the numerator is one more than the degree of the denominator so the graph will have an oblique asymptote. To find it, we need to use long division (note: we could also use synthetic division in this case because the dividend is linear).

$$\begin{array}{r} x + 1 \\ x+1\overline{)x^2 + 2x - 3} \\ \underline{-(x^2 + x)} \\ x - 3 \\ \underline{-(x+1)} \\ -4 \end{array}$$

The oblique asymptote is $y = x + 1$.

8. From problem 7 we know that the domain is $\{x \mid x \neq -1\}$ and that the graph has one vertical asymptote, $x = -1$, and one oblique asymptote, $y = x + 1$.

x-intercepts:
To find the x-intercepts, we need to set the numerator equal to 0 and solve the resulting equation.
$(x+3)(x-1) = 0$
$x + 3 = 0 \quad \text{or} \quad x - 1 = 0$
$x = -3 \qquad\qquad x = 1$
The x-intercepts are -3 and 1.

y-intercept:
$r(0) = \dfrac{0^2 + 2(0) - 3}{0 + 1} = -3$
The y-intercept is -3.

Test for symmetry:
$r(-x) = \dfrac{(-x)^2 + 2(-x) - 3}{(-x) + 1} = \dfrac{x^2 - 2x - 3}{-x + 1}$
Since $r(-x) \neq r(x)$, the graph is not symmetric with respect to the y-axis.

Since $r(-x) \neq -r(x)$, the graph is not symmetric with respect to the origin.

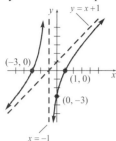

9. Since the polynomial has real coefficients, we can apply the Conjugate Pairs Theorem to find the remaining zero. If $3+i$ is a zero, then its conjugate, $3-i$, must also be a zero. Thus, the four zeros are -2, 0, $3-i$, and $3+i$. The Factor Theorem says that if $f(c)=0$, then $(x-c)$ is a factor of the polynomial. This allows us to write the following function:
$f(x) = a(x-(-2))(x-0)(x-(3-i))(x-(3+i))$
where a is any real number. Letting $a=1$ we get
$f(x) = (x+2)(x)(x-3+i)(x-3-i)$
$= (x^2+2x)(x-3+i)(x-3-i)$
$= (x^2+2x)(x^2-6x+10)$
$= x^4 - 6x^3 + 10x^2 + 2x^3 - 12x^2 + 20x$
$= x^4 - 4x^3 - 2x^2 + 20x$

10. Since the domain excludes 4 and 9, the denominator must contain the factors $(x-4)$ and $(x-9)$. However, because there is only one vertical asymptote, $x=4$, the numerator must also contain the factor $(x-9)$.

The horizontal asymptote, $y=2$, indicates that the degree of the numerator must be the same as the degree of the denominator and that the ratio of the leading coefficients needs to be 2. We can accomplish this by including another factor in the numerator, $(x-a)$, where $a \neq 4$, along with a factor of 2.

Therefore, we have $r(x) = \dfrac{2(x-9)(x-a)}{(x-4)(x-9)}$. If we let $a=1$, we get

$r(x) = \dfrac{2(x-9)(x-1)}{(x-4)(x-9)} = \dfrac{2x^2 - 20x + 18}{x^2 - 13x + 36}$.

11. Since we have a polynomial function and polynomials are continuous, we simply need to show that $f(a)$ and $f(b)$ have opposite signs (where a and b are the endpoints of the interval).
$f(0) = -2(0)^2 - 3(0) + 8 = 8$
$f(4) = -2(4)^2 - 3(4) + 8 = -36$
Since $f(0) = 8 > 0$ and $f(4) = -36 < 0$, the Intermediate Value Theorem guarantees that there is at least one real zero between 0 and 4.

12. Rearrange terms to get 0 on the right side.
$3x^2 - x - 4 \geq x^2 - 3x + 8$
$2x^2 + 2x - 12 \geq 0$
This is an equivalent inequality to the one we wish to solve.
Next we find the zeros of $f(x) = 2x^2 + 2x - 12$ by solving
$2x^2 + 2x - 12 = 0$
$2(x^2 + x - 6) = 0$
$x^2 + x - 6 = 0$
$(x+3)(x-2) = 0$
$x+3 = 0$ or $x-2 = 0$
$x = -3$ $\quad$ $x = 2$
We use the zeros to divide the real number line into three subintervals.

Interval	$(-\infty, -3)$	$(-3, 2)$	$(2, \infty)$
Num. chosen	-4	0	3
Value of f	12	-12	12
Conclusion	positive	negative	positive

Since we want to know where $f(x)$ is positive, we conclude that values of x for which $x < -3$ or $x > 2$ are solutions. Since the inequality is not strict, we need to include the values for x that make $f(x) = 0$. Thus, the solution set is $\{x \mid -\infty < x \leq -3 \text{ or } 2 \leq x < \infty\}$, or in interval notation we would write $(-\infty, -3] \cup [2, \infty)$.

13. $\dfrac{x+2}{x-3} < 2$

We note that the domain of the variable consists of all real numbers except 3.
Rearrange the terms so that the right side is 0.
$\dfrac{x+2}{x-3} - 2 < 0$

For $f(x) = \dfrac{x+2}{x-3} - 2$, we find the zeros of f and the values of x at which f is undefined. To do this, we need to write f as a single rational expression.

$f(x) = \dfrac{x+2}{x-3} - 2$

$= \dfrac{x+2}{x-3} - 2 \cdot \dfrac{x-3}{x-3}$

$= \dfrac{x+2-2x+6}{x-3}$

$= \dfrac{-x+8}{x-3}$

The zero of f is $x = 8$ and f is undefined at $x = 3$. We use these two values to divide the real number line into three subintervals.

Interval	$(-\infty, 3)$	$(3, 8)$	$(8, \infty)$
Num. chosen	0	4	9
Value of f	$-\dfrac{8}{3}$	4	$-\dfrac{1}{6}$
Conclusion	negative	positive	negative

Since we want to know where $f(x)$ is negative, we conclude that values of x for which $x < 3$ or $x > 8$ are solutions. The inequality is strict so the solution set is $\{x \mid x < 3 \text{ or } x > 8\}$. In interval notation we write $(-\infty, 3) \cup (8, \infty)$.

14. a. y-intercept:
$f(0) = -2(0-1)^2(0+2) = -4$
The y-intercept is -4.

x-intercepts:
The x-intercepts are also the zeros of the function. Since the polynomial is already in factored form so we can see that the x-intercepts are 1 and -2.

The intercepts are $(0,-4)$, $(-2,0)$, and $(1,0)$.

b. The x-intercept 1 is a zero of multiplicity 2, so the graph of f will touch the x-axis at 1. The x-intercept -2 is a zero of multiplicity 1, so the graph will cross the x-axis at -2.

c. If we multiplied the factors together, the leading term would be $-2x^3$. Thus, the graph of f resembles the power function $y = -2x^3$ for large values of $|x|$.

d. Because the power function has an odd degree, we expect the ends of the graph to go in opposite directions. The leading coefficient is negative, so the left side of the graph will go up and the right side will go down.

Reading the graph from left to right, we would expect to see it decreasing and cross the x-axis at $x = -2$. Somewhere between $x = -2$ and $x = 1$ the graph turns so that it is increasing, touches the x-axis at $x = 1$, turns around and decreases from that point on.

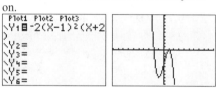

e. $f(x) = -2(x-1)^2(x+2)$

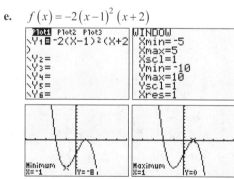

The turning points are $(-1,-8)$ and $(1,0)$.

f. Based on the above information, the graph of the function would be as follows:

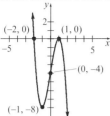

Notice that the graph crosses the x-axis at $x = -2$ and touches at $x = 1$.

15. Press the [STAT] button and select EDIT to enter the data. Enter the values for x in the first list and the values for y in the second list.

The best-fit quadratic model is given as $A(x) = -0.604x^2 + 6.038x + 25.350$.

b. For the year 2004, we have $x = 9$.
$$A(9) = -0.604(9)^2 + 6.038(9) + 25.350$$
$$= 30.768$$
The model predicts that the average 2004 home game attendance for the St. Louis Cardinals will be 30,768.

Cumulative Review

1. $P = (1, 3)$, $Q = (-4, 2)$
$$d_{P,Q} = \sqrt{(-4-1)^2 + (2-3)^2}$$
$$= \sqrt{(-5)^2 + (-1)^2}$$
$$= \sqrt{25+1}$$
$$= \sqrt{26}$$

3. $x^2 - 3x < 4$
$$x^2 - 3x - 4 < 0$$
$$(x-4)(x+1) < 0$$
$$f(x) = x^2 - 3x - 4$$
$x = -1$, $x = 4$ are the zeros of f.

Interval	$(-\infty, -1)$	$(-1, 4)$	$(4, \infty)$
Number Chosen	-2	0	5
Value of f	6	-4	6
Conclusion	Positive	Negative	Positive

The solution set is $\{x \mid -1 < x < 4\}$.

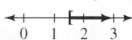

5. Parallel to $y = 2x + 1$; Slope 2, Containing the point (3, 5)
Using the point-slope formula yields:
$$y - y_1 = m(x - x_1)$$
$$y - 5 = 2(x - 3)$$
$$y - 5 = 2x - 6$$
$$y = 2x - 1$$

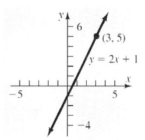

7. This relation is not a function because the ordered pairs (3, 6) and (3, 8) have the same first element, but different second elements.

9. $3x + 2 \leq 5x - 1$
$$3 \leq 2x$$
$$\frac{3}{2} \leq x$$
$$x \geq \frac{3}{2}$$

The solution set is $\left\{x \mid x \geq \frac{3}{2}\right\}$.

11. $y = x^3 - 9x$

x-intercepts: $0 = x^3 - 9x$
$0 = x(x^2 - 9)$
$0 = x(x+3)(x-3)$
$x = 0, -3, \text{ and } 3$
$(0,0), (-3,0), (3,0)$

y-intercepts: $y = 0^3 - 9(0) = 0 \Rightarrow (0,0)$

Test for symmetry:

x-axis: Replace y by $-y$: $-y = x^3 - 9x$, which is not equivalent to $y = x^3 - 9x$.

y-axis: Replace x by $-x$: $y = (-x)^3 - 9(-x)$
$= -x^3 + 9x$

which is not equivalent to $y = x^3 - 9x$.

Origin: Replace x by $-x$ and y by $-y$:
$-y = (-x)^3 - 9(-x)$
$y = -x^3 + 9x$

which is equivalent to $y = x^3 - 9x$. Therefore, the graph is symmetric with respect to origin.

13. Not a function, since the graph fails the Vertical Line Test, for example, when $x = 0$.

15. $f(x) = \dfrac{x+5}{x-1}$

a. Domain $\{x \mid x \neq 1\}$.

b. $f(2) = \dfrac{2+5}{2-1} = \dfrac{7}{1} = 7 \neq 6$;

$(2,6)$ is not on the graph of f.

c. $f(3) = \dfrac{3+5}{3-1} = \dfrac{8}{2} = 4$;

$(3,4)$ is on the graph of f.

d. Solve for x
$\dfrac{x+5}{x-1} = 9$
$x + 5 = 9(x-1)$
$x + 5 = 9x - 9$
$14 = 8x$
$x = \dfrac{14}{8} = \dfrac{7}{4}$

Therefore, $\left(\dfrac{7}{4}, 9\right)$ is on the graph of f.

17. $f(x) = 2x^2 - 4x + 1$

$a = 2$, $b = -4$, $c = 1$. Since $a = 2 > 0$, the graph opens up.

The x-coordinate of the vertex is
$x = -\dfrac{b}{2a} = -\dfrac{-4}{2(2)} = 1$.

The y-coordinate of the vertex is
$f\left(-\dfrac{b}{2a}\right) = f(1) = 2(1)^2 - 4(1) + 1 = -1$.

Thus, the vertex is $(1, -1)$.
The axis of symmetry is the line $x = 1$.

The discriminant is:
$b^2 - 4ac = (-4)^2 - 4(2)(1) = 8 > 0$, so the graph has two x-intercepts.

The x-intercepts are found by solving:
$2x^2 - 4x + 1 = 0$
$x = \dfrac{-(-4) \pm \sqrt{8}}{2(2)}$
$= \dfrac{4 \pm 2\sqrt{2}}{4} = \dfrac{2 \pm \sqrt{2}}{2}$

The x-intercepts are $\dfrac{2-\sqrt{2}}{2}$ and $\dfrac{2+\sqrt{2}}{2}$.

The y-intercept is $f(0) = 1$.

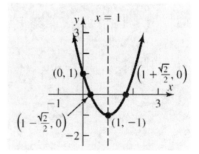

19. a. x-intercepts: $(-5,0); (-1,0); (5,0)$;
   y-intercept: $(0,-3)$

b. The graph is not symmetric with respect to the origin, x-axis or y-axis.

c. The function is neither even nor odd.

d. f is increasing on $(-\infty, -3)$ and $(2, \infty)$; f is decreasing on $(-3, 2)$;

e. f has a local maximum at $x = -3$, and the local maximum is $f(-3) = 5$.

f. f has a local minimum at $x = 2$, and the local minimum is $f(2) = -6$.

21. $f(x) = \begin{cases} 2x+1 & \text{if } -3 < x < 2 \\ -3x+4 & \text{if } x \geq 2 \end{cases}$

a. Domain: $\{x \mid x > -3\}$

b. x-intercept: $\left(-\dfrac{1}{2}, 0\right)$
   y-intercept: $(0,1)$

c.

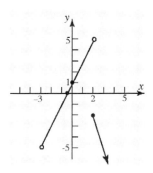

d. Range: $\{y \mid y < 5\}$

23. $f(x) = x^2 - 5x + 1 \qquad g(x) = -4x - 7$

a. $(f+g)(x) = x^2 - 5x + 1 + (-4x - 7)$
   $= x^2 - 9x - 6$
   The domain is: $\{x \mid x \text{ is a real number}\}$.

b. $\left(\dfrac{f}{g}\right)(x) = \dfrac{f(x)}{g(x)} = \dfrac{x^2 - 5x + 1}{-4x - 7}$
   The domain is: $\left\{x \mid x \neq -\dfrac{7}{4}\right\}$.

Chapter 4
Exponential and Logarithmic Functions

Section 4.1

1. $f(3) = -4(3)^2 + 5(3)$
    $= -4(9) + 15$
    $= -36 + 15$
    $= -21$

3. $f(x) = \dfrac{x^2 - 1}{x^2 - 4}$
    $x^2 - 4 \neq 0$
    $(x+2)(x-2) \neq 0$
    $x \neq -2, \quad x \neq 2$
    Domain: $\{x \mid x \neq -2, x \neq 2\}$

5. False

7. a. $(f \circ g)(1) = f(g(1)) = f(0) = -1$
    b. $(f \circ g)(-1) = f(g(-1)) = f(0) = -1$
    c. $(g \circ f)(-1) = g(f(-1)) = g(-3) = 8$
    d. $(g \circ f)(0) = g(f(0)) = g(-1) = 0$
    e. $(g \circ g)(-2) = g(g(-2)) = g(3) = 8$
    f. $(f \circ f)(-1) = f(f(-1)) = f(-3) = -7$

9. a. $g(f(-1)) = g(1) = 4$
    b. $g(f(0)) = g(0) = 5$
    c. $f(g(-1)) = f(3) = -1$
    d. $f(g(4)) = f(2) = -2$

11. $f(x) = 2x \qquad g(x) = 3x^2 + 1$
    a. $(f \circ g)(4) = f(g(4))$
    $= f(3(4)^2 + 1)$
    $= f(49)$
    $= 2(49)$
    $= 98$

    b. $(g \circ f)(2) = g(f(2))$
    $= g(2 \cdot 2)$
    $= g(4)$
    $= 3(4)^2 + 1$
    $= 48 + 1$
    $= 49$

    c. $(f \circ f)(1) = f(f(1))$
    $= f(2(1))$
    $= f(2)$
    $= 2(2)$
    $= 4$

    d. $(g \circ g)(0) = g(g(0))$
    $= g(3(0)^2 + 1)$
    $= g(1)$
    $= 3(1)^2 + 1$
    $= 4$

13. $f(x) = 4x^2 - 3 \qquad g(x) = 3 - \dfrac{1}{2}x^2$
    a. $(f \circ g)(4) = f(g(4))$
    $= f\left(3 - \dfrac{1}{2}(4)^2\right)$
    $= f(-5)$
    $= 4(-5)^2 - 3$
    $= 97$

    b. $(g \circ f)(2) = g(f(2))$
    $= g(4(2)^2 - 3)$
    $= g(13)$
    $= 3 - \dfrac{1}{2}(13)^2$
    $= 3 - \dfrac{169}{2}$
    $= -\dfrac{163}{2}$

c. $(f \circ f)(1) = f(f(1))$
$= f(4(1)^2 - 3)$
$= f(1)$
$= 4(1)^2 - 3$
$= 1$

d. $(g \circ g)(0) = g(g(0))$
$= g\left(3 - \frac{1}{2}(0)^2\right)$
$= g(3)$
$= 3 - \frac{1}{2}(3)^2$
$= 3 - \frac{9}{2}$
$= -\frac{3}{2}$

15. $f(x) = \sqrt{x} \qquad g(x) = 2x$

a. $(f \circ g)(4) = f(g(4))$
$= f(2(4))$
$= f(8)$
$= \sqrt{8}$
$= 2\sqrt{2}$

b. $(g \circ f)(2) = g(f(2))$
$= g\left(\sqrt{2}\right)$
$= 2\sqrt{2}$

c. $(f \circ f)(1) = f(f(1))$
$= f\left(\sqrt{1}\right)$
$= f(1)$
$= \sqrt{1}$
$= 1$

d. $(g \circ g)(0) = g(g(0))$
$= g(2(0))$
$= g(0)$
$= 2(0)$
$= 0$

17. $f(x) = |x| \qquad g(x) = \dfrac{1}{x^2 + 1}$

a. $(f \circ g)(4) = f(g(4))$
$= f\left(\dfrac{1}{4^2 + 1}\right)$
$= f\left(\dfrac{1}{17}\right)$
$= \left|\dfrac{1}{17}\right|$
$= \dfrac{1}{17}$

b. $(g \circ f)(2) = g(f(2))$
$= g(|2|)$
$= g(2)$
$= \dfrac{1}{2^2 + 1}$
$= \dfrac{1}{5}$

c. $(f \circ f)(1) = f(f(1))$
$= f(|1|)$
$= f(1)$
$= |1|$
$= 1$

d. $(g \circ g)(0) = g(g(0))$
$= g\left(\dfrac{1}{0^2 + 1}\right)$
$= g(1)$
$= \dfrac{1}{1^2 + 1}$
$= \dfrac{1}{2}$

Chapter 4: Exponential and Logarithmic Functions   SSM: College Algebra EGU

19. $f(x) = \dfrac{3}{x+1} \qquad g(x) = \sqrt[3]{x}$

   a. $(f \circ g)(4) = f(g(4))$
$= f\left(\sqrt[3]{4}\right)$
$= \dfrac{3}{\sqrt[3]{4}+1}$

   b. $(g \circ f)(2) = g(f(2))$
$= g\left(\dfrac{3}{2+1}\right)$
$= g\left(\dfrac{3}{3}\right)$
$= g(1)$
$= \sqrt[3]{1}$
$= 1$

   c. $(f \circ f)(1) = f(f(1))$
$= f\left(\dfrac{3}{1+1}\right)$
$= f\left(\dfrac{3}{2}\right)$
$= \dfrac{3}{\dfrac{3}{2}+1}$
$= \dfrac{3}{\dfrac{5}{2}}$
$= \dfrac{6}{5}$

   d. $(g \circ g)(0) = g(g(0))$
$= g\left(\sqrt[3]{0}\right)$
$= g(0)$
$= \sqrt[3]{0}$
$= 0$

21. The domain of g is $\{x \mid x \neq 0\}$. The domain of f is $\{x \mid x \neq 1\}$. Thus, $g(x) \neq 1$, so we solve:
$g(x) = 1$
$\dfrac{2}{x} = 1$
$x = 2$
Thus, $x \neq 2$; so the domain of $f \circ g$ is $\{x \mid x \neq 0, x \neq 2\}$.

23. The domain of g is $\{x \mid x \neq 0\}$. The domain of f is $\{x \mid x \neq 1\}$. Thus, $g(x) \neq 1$, so we solve:
$g(x) = 1$
$-\dfrac{4}{x} = 1$
$x = -4$
Thus, $x \neq -4$; so the domain of $f \circ g$ is $\{x \mid x \neq -4, x \neq 0\}$.

25. The domain of g is $\{x \mid x \text{ is any real number}\}$. The domain of f is $\{x \mid x \geq 0\}$. Thus, $g(x) \geq 0$, so we solve:
$2x + 3 \geq 0$
$x \geq -\dfrac{3}{2}$
Thus, the domain of $f \circ g$ is $\left\{x \mid x \geq -\dfrac{3}{2}\right\}$.

27. The domain of g is $\{x \mid x \geq 1\}$. The domain of f is $\{x \mid x \text{ is any real number}\}$. Thus, the domain of $f \circ g$ is $\{x \mid x \geq 1\}$.

29. $f(x) = 2x + 3 \qquad g(x) = 3x$
The domain of f is $\{x \mid x \text{ is any real number}\}$.
The domain of g is $\{x \mid x \text{ is any real number}\}$.

   a. $(f \circ g)(x) = f(g(x))$
$= f(3x)$
$= 2(3x) + 3$
$= 6x + 3$
Domain: $\{x \mid x \text{ is any real number}\}$.

b. $(g \circ f)(x) = g(f(x))$
$= g(2x+3)$
$= 3(2x+3)$
$= 6x+9$
Domain: $\{x \mid x \text{ is any real number}\}$.

c. $(f \circ f)(x) = f(f(x))$
$= f(2x+3)$
$= 2(2x+3)+3$
$= 4x+6+3$
$= 4x+9$
Domain: $\{x \mid x \text{ is any real number}\}$.

d. $(g \circ g)(x) = g(g(x))$
$= g(3x)$
$= 3(3x)$
$= 9x$
Domain: $\{x \mid x \text{ is any real number}\}$.

d. $(g \circ g)(x) = g(g(x))$
$= g(2x-4)$
$= 2(2x-4)-4$
$= 4x-8-4$
$= 4x-12$
Domain: $\{x \mid x \text{ is any real number}\}$.

31. $f(x) = 3x+1 \qquad g(x) = x^2$
The domain of f is $\{x \mid x \text{ is any real number}\}$.
The domain of g is $\{x \mid x \text{ is any real number}\}$.

a. $(f \circ g)(x) = f(g(x))$
$= f(x^2)$
$= 3x^2+1$
Domain: $\{x \mid x \text{ is any real number}\}$.

b. $(g \circ f)(x) = g(f(x))$
$= g(3x+1)$
$= (3x+1)^2$
$= 9x^2+6x+1$
Domain: $\{x \mid x \text{ is any real number}\}$.

c. $(f \circ f)(x) = f(f(x))$
$= f(3x+1)$
$= 3(3x+1)+1$
$= 9x+3+1$
$= 9x+4$
Domain: $\{x \mid x \text{ is any real number}\}$.

d. $(g \circ g)(x) = g(g(x))$
$= g(x^2)$
$= (x^2)^2$
$= x^4$
Domain: $\{x \mid x \text{ is any real number}\}$.

33. $f(x) = x^2 \qquad g(x) = x^2+4$
The domain of f is $\{x \mid x \text{ is any real number}\}$.
The domain of g is $\{x \mid x \text{ is any real number}\}$.

a. $(f \circ g)(x) = f(g(x))$
$= f(x^2+4)$
$= (x^2+4)^2$
$= x^4+8x^2+16$
Domain: $\{x \mid x \text{ is any real number}\}$.

b. $(g \circ f)(x) = g(f(x))$
$= g(x^2)$
$= (x^2)^2+4$
$= x^4+4$
Domain: $\{x \mid x \text{ is any real number}\}$.

c. $(f \circ f)(x) = f(f(x))$
$= f(x^2)$
$= (x^2)^2$
$= x^4$
Domain: $\{x \mid x \text{ is any real number}\}$.

d. $(g \circ g)(x) = g(g(x))$
$= g(x^2+4)$
$= (x^2+4)^2+4$
$= x^4+8x^2+16+4$
$= x^4+8x^2+20$
Domain: $\{x \mid x \text{ is any real number}\}$.

35. $f(x) = \dfrac{3}{x-1}$ $g(x) = \dfrac{2}{x}$

The domain of f is $\{x \mid x \neq 1\}$. The domain of g is $\{x \mid x \neq 0\}$.

a. $(f \circ g)(x) = f(g(x))$
$$= f\left(\dfrac{2}{x}\right)$$
$$= \dfrac{3}{\dfrac{2}{x} - 1}$$
$$= \dfrac{3}{\dfrac{2-x}{x}}$$
$$= \dfrac{3x}{2-x}$$
Domain $\{x \mid x \neq 0, x \neq 2\}$.

b. $(g \circ f)(x) = g(f(x))$
$$= g\left(\dfrac{3}{x-1}\right)$$
$$= \dfrac{2}{\dfrac{3}{x-1}}$$
$$= \dfrac{2(x-1)}{3}$$
Domain $\{x \mid x \neq 1\}$

c. $(f \circ f)(x) = f(f(x))$
$$= f\left(\dfrac{3}{x-1}\right)$$
$$= \dfrac{3}{\dfrac{3}{x-1} - 1}$$
$$= \dfrac{3}{\dfrac{3-(x-1)}{x-1}}$$
$$= \dfrac{3(x-1)}{4-x}$$
Domain $\{x \mid x \neq 1, x \neq 4\}$.

d. $(g \circ g)(x) = g(g(x)) = g\left(\dfrac{2}{x}\right) = \dfrac{2}{\dfrac{2}{x}} = \dfrac{2x}{2} = x$

Domain $\{x \mid x \neq 0\}$.

37. $f(x) = \dfrac{x}{x-1}$ $g(x) = -\dfrac{4}{x}$

The domain of f is $\{x \mid x \neq 1\}$. The domain of g is $\{x \mid x \neq 0\}$.

a. $(f \circ g)(x) = f(g(x))$
$$= f\left(-\dfrac{4}{x}\right)$$
$$= \dfrac{-\dfrac{4}{x}}{-\dfrac{4}{x} - 1}$$
$$= \dfrac{-\dfrac{4}{x}}{\dfrac{-4-x}{x}}$$
$$= \dfrac{-4}{-4-x}$$
$$= \dfrac{4}{4+x}$$
Domain $\{x \mid x \neq -4, x \neq 0\}$.

b. $(g \circ f)(x) = g(f(x))$
$$= g\left(\dfrac{x}{x-1}\right)$$
$$= -\dfrac{4}{\dfrac{x}{x-1}}$$
$$= \dfrac{-4(x-1)}{x}$$
Domain $\{x \mid x \neq 0, x \neq 1\}$.

c. $(f \circ f)(x) = f(f(x))$
$$= f\left(\frac{x}{x-1}\right)$$
$$= \frac{\frac{x}{x-1}}{\frac{x}{x-1}-1}$$
$$= \frac{\frac{x}{x-1}}{\frac{x-(x-1)}{x-1}}$$
$$= \frac{\frac{x}{x-1}}{\frac{1}{x-1}}$$
$$= x$$
Domain $\{x \mid x \neq 1\}$.

d. $(g \circ g)(x) = g(g(x))$
$$= g\left(\frac{-4}{x}\right)$$
$$= -\frac{4}{-\frac{4}{x}}$$
$$= \frac{-4x}{-4}$$
$$= x$$
Domain $\{x \mid x \neq 0\}$.

39. $f(x) = \sqrt{x}$ $g(x) = 2x+3$
The domain of f is $\{x \mid x \geq 0\}$. The domain of g is $\{x \mid x \text{ is any real number}\}$.

a. $(f \circ g)(x) = f(g(x)) = f(2x+3) = \sqrt{2x+3}$
Domain $\left\{x \mid x \geq -\frac{3}{2}\right\}$.

b. $(g \circ f)(x) = g(f(x)) = g(\sqrt{x}) = 2\sqrt{x}+3$
Domain $\{x \mid x \geq 0\}$.

c. $(f \circ f)(x) = f(f(x))$
$$= f(\sqrt{x})$$
$$= \sqrt{\sqrt{x}}$$
$$= \left(x^{1/2}\right)^{1/2}$$
$$= x^{1/4}$$
$$= \sqrt[4]{x}$$
Domain $\{x \mid x \geq 0\}$.

d. $(g \circ g)(x) = g(g(x))$
$$= g(2x+3)$$
$$= 2(2x+3)+3$$
$$= 4x+6+3$$
$$= 4x+9$$
Domain $\{x \mid x \text{ is any real number}\}$.

41. $f(x) = x^2+1$ $g(x) = \sqrt{x-1}$
The domain of f is $\{x \mid x \text{ is any real number}\}$.
The domain of g is $\{x \mid x \geq 1\}$.

a. $(f \circ g)(x) = f(g(x))$
$$= f(\sqrt{x-1})$$
$$= (\sqrt{x-1})^2 + 1$$
$$= x-1+1$$
$$= x$$
Domain $\{x \mid x \geq 1\}$.

b. $(g \circ f)(x) = g(f(x))$
$$= g(x^2+1)$$
$$= \sqrt{x^2+1-1}$$
$$= \sqrt{x^2}$$
$$= |x|$$
Domain $\{x \mid x \text{ is any real number}\}$.

c. $(f \circ f)(x) = f(f(x))$
$$= f(x^2+1)$$
$$= (x^2+1)^2 + 1$$
$$= x^4 + 2x^2 + 1 + 1$$
$$= x^4 + 2x^2 + 2$$
Domain $\{x \mid x \text{ is any real number}\}$.

d. $(g \circ g)(x) = g(g(x))$
$= g(\sqrt{x-1})$
$= \sqrt{\sqrt{x-1}-1}$

Now, $\sqrt{x-1}-1 \geq 0$
$\sqrt{x-1} \geq 1$
$x-1 \geq 1$
$x \geq 2$
Domain $\{x \mid x \geq 2\}$.

43. $f(x) = ax+b \quad g(x) = cx+d$
The domain of f is $\{x \mid x \text{ is any real number}\}$.
The domain of g is $\{x \mid x \text{ is any real number}\}$.

a. $(f \circ g)(x) = f(g(x))$
$= f(cx+d)$
$= a(cx+d)+b$
$= acx+ad+b$
Domain $\{x \mid x \text{ is any real number}\}$.

b. $(g \circ f)(x) = g(f(x))$
$= g(ax+b)$
$= c(ax+b)+d$
$= acx+bc+d$
Domain $\{x \mid x \text{ is any real number}\}$.

c. $(f \circ f)(x) = f(f(x))$
$= f(ax+b)$
$= a(ax+b)+b$
$= a^2x+ab+b$
Domain $\{x \mid x \text{ is any real number}\}$.

d. $(g \circ g)(x) = g(g(x))$
$= g(cx+d)$
$= c(cx+d)+d$
$= c^2x+cd+d$
Domain $\{x \mid x \text{ is any real number}\}$.

45. $(f \circ g)(x) = f(g(x)) = f\left(\frac{1}{2}x\right) = 2\left(\frac{1}{2}x\right) = x$
$(g \circ f)(x) = g(f(x)) = g(2x) = \frac{1}{2}(2x) = x$

47. $(f \circ g)(x) = f(g(x)) = f\left(\sqrt[3]{x}\right) = \left(\sqrt[3]{x}\right)^3 = x$
$(g \circ f)(x) = g(f(x)) = g\left(x^3\right) = \sqrt[3]{x^3} = x$

49. $(f \circ g)(x) = f(g(x))$
$= f\left(\frac{1}{2}(x+6)\right)$
$= 2\left(\frac{1}{2}(x+6)\right)-6$
$= x+6-6$
$= x$
$(g \circ f)(x) = g(f(x))$
$= g(2x-6)$
$= \frac{1}{2}((2x-6)+6)$
$= \frac{1}{2}(2x)$
$= x$

51. $(f \circ g)(x) = f(g(x))$
$= f\left(\frac{1}{a}(x-b)\right)$
$= a\left(\frac{1}{a}(x-b)\right)+b$
$= x-b+b$
$= x$
$(g \circ f)(x) = g(f(x))$
$= g(ax+b)$
$= \frac{1}{a}((ax+b)-b)$
$= \frac{1}{a}(ax)$
$= x$

53. $H(x) = (2x+3)^4$
Answers may vary. One possibility is
$f(x) = x^4, \quad g(x) = 2x+3$

SSM: College Algebra EGU   *Chapter 4: Exponential and Logarithmic Functions*

55. $H(x) = \sqrt{x^2 + 1}$
Answers may vary. One possibility is
$f(x) = \sqrt{x}, \quad g(x) = x^2 + 1$

57. $H(x) = |2x + 1|$
Answers may vary. One possibility is
$f(x) = |x|, \quad g(x) = 2x + 1$

59. $f(x) = 2x^3 - 3x^2 + 4x - 1 \quad g(x) = 2$
$(f \circ g)(x) = f(g(x))$
$\qquad = f(2)$
$\qquad = 2(2)^3 - 3(2)^2 + 4(2) - 1$
$\qquad = 16 - 12 + 8 - 1$
$\qquad = 11$
$(g \circ f)(x) = g(f(x)) = g(2x^3 - 3x^2 + 4x - 1) = 2$

61. $f(x) = 2x^2 + 5 \quad g(x) = 3x + a$
$(f \circ g)(x) = f(g(x)) = f(3x + a) = 2(3x + a)^2 + 5$
When $x = 0$, $(f \circ g)(0) = 23$.
Solving: $2(3 \cdot 0 + a)^2 + 5 = 23$
$\qquad\qquad 2a^2 + 5 = 23$
$\qquad\qquad 2a^2 - 18 = 0$
$\qquad\qquad 2(a + 3)(a - 3) = 0$
$\qquad\qquad a = -3 \ \text{ or } \ a = 3$

63. $S(r) = 4\pi r^2 \qquad r(t) = \dfrac{2}{3}t^3, \ t \geq 0$
$S(r(t)) = S\left(\dfrac{2}{3}t^3\right)$
$\qquad = 4\pi\left(\dfrac{2}{3}t^3\right)^2$
$\qquad = 4\pi\left(\dfrac{4}{9}t^6\right)$
$\qquad = \dfrac{16}{9}\pi t^6$

65. $N(t) = 100t - 5t^2, \ 0 \leq t \leq 10$
$C(N) = 15{,}000 + 8000N$
$C(N(t)) = C\left(100t - 5t^2\right)$
$\qquad = 15{,}000 + 8000\left(100t - 5t^2\right)$
$\qquad = 15{,}000 + 800{,}000t - 40{,}000t^2$

67. $p = -\dfrac{1}{4}x + 100, \quad 0 \leq x \leq 400$
$\dfrac{1}{4}x = 100 - p$
$x = 4(100 - p)$

$C = \dfrac{\sqrt{x}}{25} + 600$
$ = \dfrac{\sqrt{4(100 - p)}}{25} + 600$
$ = \dfrac{2\sqrt{100 - p}}{25} + 600, \quad 0 \leq p \leq 100$

69. $V = \pi r^2 h \qquad h = 2r$
$V(r) = \pi r^2 (2r) = 2\pi r^3$

71. $f(x)$ = the number of Euros bought for x dollars;
$g(x)$ = the number of yen bought for x Euros

   a. $f(x) = 0.857118x$

   b. $g(x) = 128.6054x$

   c. $g(f(x)) = g(0.857118x)$
$\qquad\qquad = 128.6054(0.857118x)$
$\qquad\qquad = 110.2300032372x$

   d. $g(f(1000)) = 110.2300032372(1000)$
$\qquad\qquad\quad = 110{,}230.0032372$ yen

73. Given that f is odd and g is even, we know that $f(-x) = -f(x)$ and $g(-x) = g(x)$ for all x in the domain of f and g, respectively. The composite function $(f \circ g)(x) = f(g(x))$ has the following property:
$(f \circ g)(-x) = f(g(-x))$
$\qquad\qquad = f(g(x)) \quad \text{since } g \text{ is even}$
$\qquad\qquad = (f \circ g)(x)$
Thus, $f \circ g$ is an even function.

The composite function $(g \circ f)(x) = g(f(x))$ has the following property:

$(g \circ f)(-x) = g(f(-x))$
$= g(-f(x))$ since f is odd
$= g(f(x))$ since g is even
$= (g \circ f)(x)$

Thus, $g \circ f$ is an even function.

Section 4.2

1. The set of ordered pairs is a function because there are no ordered pairs with the same first element and different second elements.

3. The function is not defined when $x^2 + 3x - 18 = 0$. Solve: $x^2 + 3x - 18 = 0$
$(x+6)(x-3) = 0$
$x = -6$ or $x = 3$
The domain is $\{x \mid x \neq -6, x \neq 3\}$.

5. $y = x$

7. False. If f and g are inverse functions, then the range of f is the domain of g and the domain of f is the range of g.

9. The function is one-to-one because there are no two distinct inputs that correspond to the same output.

11. The function is not one-to-one because there are two different inputs, 20 Hours and 50 Hours, that correspond to the same output, $200.

13. The function is one-to-one because there are no two distinct inputs that correspond to the same output.

15. The function is one-to-one because there are no two distinct inputs that correspond to the same output.

17. The function f is one-to-one because every horizontal line intersects the graph at exactly one point.

19. The function f is not one-to-one because there are horizontal lines (for example, $y = 1$) that intersect the graph at more than one point.

21. The function f is one-to-one because every horizontal line intersects the graph at exactly one point.

23. To find the inverse, interchange the elements in the domain with the elements in the range:

Domain: {460.00, 202.01, 196.46, 191.02, 182.87}

Range: {Mt Waialeale, Monrovia, Pago Pago, Moulmein, Lea}

25. To find the inverse, interchange the elements in the domain with the elements in the range:

Monthly Cost of Life Insurance	Age
$7.09	30
$8.40	40
$11.29	45

Domain: {$7.09, $8.40, $11.29}

Range: {30, 40, 45}

27. Interchange the entries in each ordered pair:
$\{(5,-3), (9,-2), (2,-1), (11,0), (-5,1)\}$

Domain: $\{5, 9, 2, 11, -5\}$

Range: $\{-3, -2, -1, 0, 1\}$

29. Interchange the entries in each ordered pair:
$\{(1,-2), (2,-3), (0,-10), (9,1), (4,2)\}$

Domain: $\{1, 2, 0, 9, 4\}$

Range: $\{-2, -3, -10, 1, 2\}$

31. Graphing the inverse:

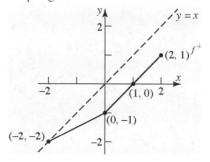

33. Graphing the inverse:

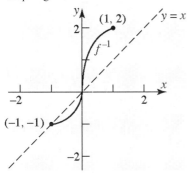

35. Graphing the inverse:

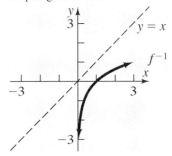

37. $f(x) = 3x+4;\qquad g(x) = \dfrac{1}{3}(x-4)$

$$f(g(x)) = f\left(\dfrac{1}{3}(x-4)\right)$$
$$= 3\left(\dfrac{1}{3}(x-4)\right)+4$$
$$= (x-4)+4$$
$$= x$$

$$g(f(x)) = g(3x+4)$$
$$= \dfrac{1}{3}((3x+4)-4)$$
$$= \dfrac{1}{3}(3x)$$
$$= x$$

Thus, f and g are inverses of each other.

39. $f(x) = 4x-8;\qquad g(x) = \dfrac{x}{4}+2$

$$f(g(x)) = f\left(\dfrac{x}{4}+2\right)$$
$$= 4\left(\dfrac{x}{4}+2\right)-8$$
$$= x+8-8$$
$$= x$$

$$g(f(x)) = g(4x-8)$$
$$= \dfrac{4x-8}{4}+2$$
$$= x-2+2$$
$$= x$$

Thus, f and g are inverses of each other.

41. $f(x) = x^3-8;\qquad g(x) = \sqrt[3]{x+8}$

$$f(g(x)) = f\left(\sqrt[3]{x+8}\right)$$
$$= \left(\sqrt[3]{x+8}\right)^3 - 8$$
$$= x+8-8$$
$$= x$$

$$g(f(x)) = g(x^3-8)$$
$$= \sqrt[3]{(x^3-8)+8}$$
$$= \sqrt[3]{x^3}$$
$$= x$$

Thus, f and g are inverses of each other.

43. $f(x) = \dfrac{1}{x};\qquad g(x) = \dfrac{1}{x}$

$$f(g(x)) = f\left(\dfrac{1}{x}\right) = \dfrac{1}{\dfrac{1}{x}} = 1\cdot\dfrac{x}{1} = x$$

$$g(f(x)) = g\left(\dfrac{1}{x}\right) = \dfrac{1}{\dfrac{1}{x}} = 1\cdot\dfrac{x}{1} = x$$

Thus, f and g are inverses of each other.

Chapter 4: Exponential and Logarithmic Functions

45. $f(x) = \dfrac{2x+3}{x+4}$; $g(x) = \dfrac{4x-3}{2-x}$

$f(g(x)) = f\left(\dfrac{4x-3}{2-x}\right) = \dfrac{2\left(\dfrac{4x-3}{2-x}\right)+3}{\dfrac{4x-3}{2-x}+4}$

$= \dfrac{\left(2\left(\dfrac{4x-3}{2-x}\right)+3\right)(2-x)}{\left(\dfrac{4x-3}{2-x}+4\right)(2-x)}$

$= \dfrac{2(4x-3)+3(2-x)}{4x-3+4(2-x)}$

$= \dfrac{8x-6+6-3x}{4x-3+8-4x}$

$= \dfrac{5x}{5}$

$= x$

$g(f(x)) = g\left(\dfrac{2x+3}{x+4}\right)$

$= \dfrac{4\left(\dfrac{2x+3}{x+4}\right)-3}{2-\dfrac{2x+3}{x+4}}$

$= \dfrac{\left(4\left(\dfrac{2x+3}{x+4}\right)-3\right)(x+4)}{\left(2-\dfrac{2x+3}{x+4}\right)(x+4)}$

$= \dfrac{4(2x+3)-3(x+4)}{2(x+4)-(2x+3)}$

$= \dfrac{8x+12-3x-12}{2x+8-2x-3}$

$= \dfrac{5x}{5}$

$= x$

Thus, f and g are inverses of each other.

47. $f(x) = 3x$
   $y = 3x$
   $x = 3y$ Inverse
   $y = \dfrac{x}{3}$
   $f^{-1}(x) = \dfrac{x}{3}$

Verifying: $f(f^{-1}(x)) = f\left(\dfrac{x}{3}\right) = 3\left(\dfrac{x}{3}\right) = x$

$f^{-1}(f(x)) = f^{-1}(3x) = \dfrac{3x}{3} = x$

Domain of f = Range of f^{-1} = All real numbers
Range of f = Domain of f^{-1} = All real numbers

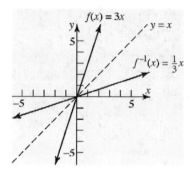

49. $f(x) = 4x+2$
   $y = 4x+2$
   $x = 4y+2$ Inverse
   $4y = x-2$
   $y = \dfrac{x-2}{4}$
   $f^{-1}(x) = \dfrac{x-2}{4}$

Verifying:

$f(f^{-1}(x)) = f\left(\dfrac{x-2}{4}\right) = 4\left(\dfrac{x-2}{4}\right)+2 = x-2+2 = x$

$f^{-1}(f(x)) = f^{-1}(4x+2) = \dfrac{(4x+2)-2}{4} = \dfrac{4x}{4} = x$

Domain of f = Range of f^{-1} = All real numbers
Range of f = Domain of f^{-1} = All real numbers

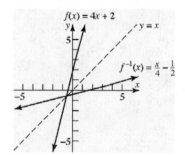

51. $f(x) = x^3 - 1$
$y = x^3 - 1$
$x = y^3 - 1$ Inverse
$y^3 = x + 1$
$y = \sqrt[3]{x+1}$
$f^{-1}(x) = \sqrt[3]{x+1}$

Verifying:
$f(f^{-1}(x)) = f(\sqrt[3]{x+1}) = (\sqrt[3]{x+1})^3 - 1 = x + 1 - 1 = x$
$f^{-1}(f(x)) = f^{-1}(x^3 - 1) = \sqrt[3]{(x^3-1)+1} = \sqrt[3]{x^3} = x$

Domain of f = Range of f^{-1} = All real numbers
Range of f = Domain of f^{-1} = All real numbers

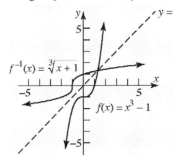

53. $f(x) = x^2 + 4$, $x \geq 0$
$y = x^2 + 4$ $x \geq 0$
$x = y^2 + 4$ $y \geq 0$ Inverse
$y^2 = x - 4$ $y \geq 0$
$y = \sqrt{x-4}$
$f^{-1}(x) = \sqrt{x-4}$

Verifying: $f(f^{-1}(x)) = f(\sqrt{x-4})$
$= (\sqrt{x-4})^2 + 4$
$= x - 4 + 4$
$= x$

$f^{-1}(f(x)) = f^{-1}(x^2 + 4)$
$= \sqrt{(x^2+4)-4}$
$= \sqrt{x^2}$
$= |x|$
$= x, x \geq 0$

Domain of f = Range of f^{-1} = $\{x \mid x \geq 0\}$ or $[0, \infty)$
Range of f = Domain of f^{-1} = $\{x \mid x \geq 4\}$ or $[4, \infty)$

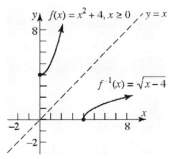

55. $f(x) = \dfrac{4}{x}$
$y = \dfrac{4}{x}$
$x = \dfrac{4}{y}$ Inverse
$xy = 4$
$y = \dfrac{4}{x}$
$f^{-1}(x) = \dfrac{4}{x}$

Verifying: $f(f^{-1}(x)) = f\left(\dfrac{4}{x}\right) = \dfrac{4}{\frac{4}{x}} = 4 \cdot \left(\dfrac{x}{4}\right) = x$

$f^{-1}(f(x)) = f^{-1}\left(\dfrac{4}{x}\right) = \dfrac{4}{\frac{4}{x}} = 4 \cdot \left(\dfrac{x}{4}\right) = x$

Domain of f = Range of f^{-1}
  = All real numbers except 0.
Range of f = Domain of f^{-1}
  = All real numbers except 0.

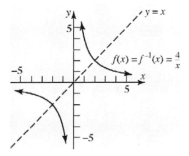

57. $f(x) = \dfrac{1}{x-2}$

$y = \dfrac{1}{x-2}$

$x = \dfrac{1}{y-2}$ Inverse

$xy - 2x = 1$

$xy = 2x + 1$

$y = \dfrac{2x+1}{x}$

$f^{-1}(x) = \dfrac{2x+1}{x}$

Verifying: $f\left(f^{-1}(x)\right) = f\left(\dfrac{2x+1}{x}\right)$

$= \dfrac{1}{\dfrac{2x+1}{x} - 2}$

$= \dfrac{1 \cdot x}{\left(\dfrac{2x+1}{x} - 2\right)x}$

$= \dfrac{x}{2x+1-2x}$

$= \dfrac{x}{1}$

$= x$

$f^{-1}(f(x)) = f^{-1}\left(\dfrac{1}{x-2}\right)$

$= \dfrac{2\left(\dfrac{1}{x-2}\right)+1}{\dfrac{1}{x-2}}$

$= \dfrac{\left(2\left(\dfrac{1}{x-2}\right)+1\right)(x-2)}{\left(\dfrac{1}{x-2}\right)(x-2)}$

$= \dfrac{2+(x-2)}{1}$

$= \dfrac{x}{1}$

$= x$

Domain of f = Range of f^{-1}
          = All real numbers except 0.
Range of f = Domain of f^{-1}
          = All real numbers except 2.

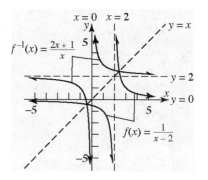

59. $f(x) = \dfrac{2}{3+x}$

$y = \dfrac{2}{3+x}$

$x = \dfrac{2}{3+y}$ Inverse

$x(3+y) = 2$

$3x + xy = 2$

$xy = 2 - 3x$

$y = \dfrac{2-3x}{x}$

$f^{-1}(x) = \dfrac{2-3x}{x}$

Verifying: $f\left(f^{-1}(x)\right) = f\left(\dfrac{2-3x}{x}\right)$

$= \dfrac{2}{3 + \dfrac{2-3x}{x}}$

$= \dfrac{2 \cdot x}{\left(3 + \dfrac{2-3x}{x}\right)x}$

$= \dfrac{2x}{3x + 2 - 3x}$

$= \dfrac{2x}{2}$

$= x$

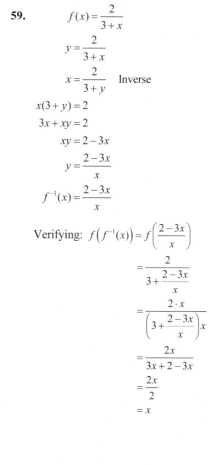

$$f^{-1}(f(x)) = f^{-1}\left(\frac{2}{3+x}\right)$$

$$= \frac{2 - 3\left(\frac{2}{3+x}\right)}{\frac{2}{3+x}}$$

$$= \frac{\left(2 - 3\left(\frac{2}{3+x}\right)\right)(3+x)}{\left(\frac{2}{3+x}\right)(3+x)}$$

$$= \frac{2(3+x) - 3(2)}{2}$$

$$= \frac{6 + 2x - 6}{2}$$

$$= \frac{2x}{2}$$

$$= x$$

Domain of f = Range of f^{-1}
    = All real numbers except -3.
Range of f = Domain of f^{-1}
    = All real numbers except 0.

61. $f(x) = \dfrac{3x}{x+2}$

$$y = \frac{3x}{x+2}$$

$$x = \frac{3y}{y+2} \quad \text{Inverse}$$

$$x(y+2) = 3y$$

$$xy + 2x = 3y$$

$$xy - 3y = -2x$$

$$y(x-3) = -2x$$

$$y = \frac{-2x}{x-3}$$

$$f^{-1}(x) = \frac{-2x}{x-3}$$

Verifying: $f(f^{-1}(x)) = f\left(\dfrac{-2x}{x-3}\right) = \dfrac{3\left(\dfrac{-2x}{x-3}\right)}{\dfrac{-2x}{x-3} + 2}$

$$= \frac{\left(3\left(\dfrac{-2x}{x-3}\right)\right)(x-3)}{\left(\dfrac{-2x}{x-3} + 2\right)(x-3)}$$

$$= \frac{-6x}{-2x + 2x - 6}$$

$$= \frac{-6x}{-6}$$

$$= x$$

$$f^{-1}(f(x)) = f^{-1}\left(\frac{3x}{x+2}\right) = \frac{-2\left(\dfrac{3x}{x+2}\right)}{\dfrac{3x}{x+2} - 3}$$

$$= \frac{\left(-2\left(\dfrac{3x}{x+2}\right)\right)(x+2)}{\left(\dfrac{3x}{x+2} - 3\right)(x+2)}$$

$$= \frac{-6x}{3x - 3x - 6}$$

$$= \frac{-6x}{-6}$$

$$= x$$

Domain of f = Range of f^{-1}
    = All real numbers except -2.
Range of f = Domain of f^{-1}
    = All real numbers except 3.

63. $f(x) = \dfrac{2x}{3x-1}$

$y = \dfrac{2x}{3x-1}$

$x = \dfrac{2y}{3y-1}$ Inverse

$3xy - x = 2y$

$3xy - 2y = x$

$y(3x-2) = x$

$y = \dfrac{x}{3x-2}$

$f^{-1}(x) = \dfrac{x}{3x-2}$

Verifying: $f(f^{-1}(x)) = f\left(\dfrac{x}{3x-2}\right) = \dfrac{2\left(\dfrac{x}{3x-2}\right)}{3\left(\dfrac{x}{3x-2}\right) - 1}$

$= \dfrac{\left(2\left(\dfrac{x}{3x-2}\right)\right)(3x-2)}{\left(3\left(\dfrac{x}{3x-2}\right) - 1\right)(3x-2)}$

$= \dfrac{2x}{3x - (3x-2)}$

$= \dfrac{2x}{2}$

$= x$

$f^{-1}(f(x)) = f\left(\dfrac{2x}{3x-1}\right) = \dfrac{\dfrac{2x}{3x-1}}{3\left(\dfrac{2x}{3x-1}\right) - 2}$

$= \dfrac{\left(\dfrac{2x}{3x-1}\right)(3x-1)}{\left(3\left(\dfrac{2x}{3x-1}\right) - 2\right)(3x-1)}$

$= \dfrac{2x}{3(2x) - 2(3x-1)}$

$= \dfrac{2x}{6x - 6x + 2}$

$= \dfrac{2x}{2}$

$= x$

Domain of f = Range of f^{-1}

     = All real numbers except $1/3$.

Range of f = Domain of f^{-1}

     = All real numbers except $2/3$.

65. $f(x) = \dfrac{3x+4}{2x-3}$

$y = \dfrac{3x+4}{2x-3}$

$x = \dfrac{3y+4}{2y-3}$ Inverse

$x(2y-3) = 3y+4$

$2xy - 3x = 3y + 4$

$2xy - 3y = 3x + 4$

$y(2x-3) = 3x+4$

$y = \dfrac{3x+4}{2x-3}$

$f^{-1}(x) = \dfrac{3x+4}{2x-3}$

Verifying: $f(f^{-1}(x)) = f\left(\dfrac{x}{x+2}\right) = \dfrac{3\left(\dfrac{3x+4}{2x-3}\right) + 4}{2\left(\dfrac{3x+4}{2x-3}\right) - 3}$

$= \dfrac{\left(3\left(\dfrac{3x+4}{2x-3}\right) + 4\right)(2x-3)}{\left(2\left(\dfrac{3x+4}{2x-3}\right) - 3\right)(2x-3)}$

$= \dfrac{3(3x+4) + 4(2x-3)}{2(3x+4) - 3(2x-3)}$

$= \dfrac{9x + 12 + 8x - 12}{6x + 8 - 6x + 9}$

$= \dfrac{17x}{17}$

$= x$

$f^{-1}(f(x)) = f^{-1}\left(\dfrac{3x+4}{2x-3}\right) = \dfrac{3\left(\dfrac{3x+4}{2x-3}\right) + 4}{2\left(\dfrac{3x+4}{2x-3}\right) - 3}$

$= \dfrac{\left(3\left(\dfrac{3x+4}{2x-3}\right) + 4\right)(2x-3)}{\left(2\left(\dfrac{3x+4}{2x-3}\right) - 3\right)(2x-3)}$

$= \dfrac{3(3x+4) + 4(2x-3)}{2(3x+4) - 3(2x-3)}$

$= \dfrac{9x + 12 + 8x - 12}{6x + 8 - 6x + 9}$

$= \dfrac{17x}{17}$

$= x$

Domain of f = Range of f^{-1}

     = All real numbers except $3/2$.

Range of f = Domain of f^{-1}

     = All real numbers except $3/2$.

67. $f(x) = \dfrac{2x+3}{x+2}$

$y = \dfrac{2x+3}{x+2}$

$x = \dfrac{2y+3}{y+2}$ Inverse

$xy + 2x = 2y + 3$

$xy - 2y = -2x + 3$

$y(x-2) = -2x + 3$

$y = \dfrac{-2x+3}{x-2}$

$f^{-1}(x) = \dfrac{-2x+3}{x-2}$

Verifying:

$f(f^{-1}(x)) = f\left(\dfrac{-2x+3}{x-2}\right) = \dfrac{2\left(\dfrac{-2x+3}{x-2}\right)+3}{\dfrac{-2x+3}{x-2}+2}$

$= \dfrac{\left(2\left(\dfrac{-2x+3}{x-2}\right)+3\right)(x-2)}{\left(\dfrac{-2x+3}{x-2}+2\right)(x-2)}$

$= \dfrac{2(-2x+3) + 3(x-2)}{-2x+3+2(x-2)}$

$= \dfrac{-4x+6+3x-6}{-2x+3+2x-4}$

$= \dfrac{-x}{-1}$

$= x$

$f^{-1}(f(x)) = f^{-1}\left(\dfrac{2x+3}{x+2}\right) = \dfrac{-2\left(\dfrac{2x+3}{x+2}\right)+3}{\dfrac{2x+3}{x+2}-2}$

$= \dfrac{\left(-2\left(\dfrac{2x+3}{x+2}\right)+3\right)(x+2)}{\left(\dfrac{2x+3}{x+2}-2\right)(x+2)}$

$= \dfrac{-2(2x+3)+3(x+2)}{2x+3-2(x+2)}$

$= \dfrac{-4x-6+3x+6}{2x+3-2x-4}$

$= \dfrac{-x}{-1}$

$= x$

Domain of f = Range of f^{-1}
 = All real numbers except -2.
Range of f = Domain of f^{-1}
 = All real numbers except 2.

69. $f(x) = \dfrac{x^2-4}{2x^2}, \; x > 0$

$y = \dfrac{x^2-4}{2x^2}, \; x > 0$

$x = \dfrac{y^2-4}{2y^2}, \; y > 0$ Inverse

$2xy^2 = y^2 - 4, \; x < \dfrac{1}{2}$

$2xy^2 - y^2 = -4, \; x < \dfrac{1}{2}$

$y^2(2x-1) = -4, \; x < \dfrac{1}{2}$

$y^2(1-2x) = 4, \; x < \dfrac{1}{2}$

$y^2 = \dfrac{4}{1-2x}, \; x < \dfrac{1}{2}$

$y = \sqrt{\dfrac{4}{1-2x}}, \; x < \dfrac{1}{2}$

$y = \dfrac{2}{\sqrt{1-2x}}, \; x < \dfrac{1}{2}$

$f^{-1}(x) = \dfrac{2}{\sqrt{1-2x}}, \; x < \dfrac{1}{2}$

Verifying: $f(f^{-1}(x)) = f\left(\dfrac{2}{\sqrt{1-2x}}\right)$

$= \dfrac{\left(\dfrac{2}{\sqrt{1-2x}}\right)^2 - 4}{2\left(\dfrac{2}{\sqrt{1-2x}}\right)^2}$

$= \dfrac{\dfrac{4}{1-2x} - 4}{2\left(\dfrac{4}{1-2x}\right)}$

$= \dfrac{\left(\dfrac{4}{1-2x} - 4\right)(1-2x)}{\left(2\left(\dfrac{4}{1-2x}\right)\right)(1-2x)}$

$= \dfrac{4 - 4(1-2x)}{2(4)}$

$= \dfrac{4 - 4 + 8x}{8}$

$= \dfrac{8x}{8}$

$= x$

Chapter 4: Exponential and Logarithmic Functions

$$f^{-1}(f(x)) = f^{-1}\left(\frac{x^2-4}{2x^2}\right)$$

$$= \frac{2}{\sqrt{1-2\left(\frac{x^2-4}{2x^2}\right)}}$$

$$= \frac{2}{\sqrt{1-\frac{x^2-4}{x^2}}}$$

$$= \frac{2}{\sqrt{1-1+\frac{4}{x^2}}}$$

$$= \frac{2}{\sqrt{\frac{4}{x^2}}}$$

$$= \frac{2}{\frac{2}{|x|}}$$

$$= 2 \cdot \frac{|x|}{2}$$

$$= |x|$$

$$= x, \quad x > 0$$

Domain of f = Range of f^{-1}
  $= \{x \mid x > 0\}$ or $(0, \infty)$

Range of f = Domain of f^{-1}
  $= \left\{x \mid x < \frac{1}{2}\right\}$ or $\left(-\infty, \frac{1}{2}\right)$

71. **a.** Because the ordered pair $(-1, 0)$ is on the graph, $f(-1) = 0$.

   b. Because the ordered pair $(1, 2)$ is on the graph, $f(1) = 2$.

   c. Because the ordered pair $(0, 1)$ is on the graph, $f^{-1}(1) = 0$.

   d. Because the ordered pair $(1, 2)$ is on the graph, $f^{-1}(2) = 1$.

73. $f(x) = mx + b, \quad m \neq 0$
  $y = mx + b$
  $x = my + b$ Inverse
  $x - b = my$
  $y = \frac{x-b}{m}$
  $f^{-1}(x) = \frac{x-b}{m}, \quad m \neq 0$

75. If (a, b) is on the graph of f, then (b, a) is on the graph of f^{-1}. Since the graph of f^{-1} lies in quadrant I, both coordinates of (a, b) are positive, which means that both coordinates of (b, a) are positive. Thus, the graph of f^{-1} must lie in quadrant I.

77. Answers may vary. One possibility follows:
  $f(x) = |x|, x \geq 0$ is one-to-one.
  Thus, $f(x) = x, x \geq 0$
  $y = x, x \geq 0$
  $f^{-1}(x) = x, x \geq 0$

79. **a.** $H(C) = 2.15C - 10.53$
  $H = 2.15C - 10.53$
  $H + 10.53 = 2.15C$
  $\frac{H + 10.53}{2.15} = C$
  $C(H) = \frac{H + 10.53}{2.15}$

   b. $C(26) = \frac{26 + 10.53}{2.15} \approx 16.99$ inches

81. $p(x) = 300 - 50x, \quad x \geq 0$
  $p = 300 - 50x$
  $50x = 300 - p$
  $x = \frac{300 - p}{50}$
  $x(p) = \frac{300 - p}{50}, \quad p \leq 300$

83. $f(x) = \dfrac{ax+b}{cx+d}$

$y = \dfrac{ax+b}{cx+d}$

$x = \dfrac{ay+b}{cy+d}$ Inverse

$x(cy+d) = ay+b$

$cxy + dx = ay + b$

$cxy - ay = b - dx$

$y(cx - a) = b - dx$

$y = \dfrac{b - dx}{cx - a}$

$f^{-1}(x) = \dfrac{-dx + b}{cx - a}$

Now, $f = f^{-1}$ provided that $\dfrac{ax+b}{cx+d} = \dfrac{-dx+b}{cx-a}$.
This is only true if $a = -d$.

85. Answers will vary.

87. The only way the function $y = f(x)$ can be both even and one-to-one is if the domain of $y = f(x)$ is $\{x \mid x \geq 0\}$. Otherwise, its graph will fail the Horizontal Line Test.

89. If the graph of a function and its inverse intersect, they must intersect at a point on the line $y = x$. They cannot intersect anywhere else. The graphs do not have to intersect.

Section 4.3

1. $4^3 = 64$; $8^{2/3} = \left(\sqrt[3]{8}\right)^2 = 2^2 = 4$; $3^{-2} = \dfrac{1}{3^2} = \dfrac{1}{9}$

3. False. To obtain the graph of $y = (x-2)^3$, we would shift the graph of $y = x^3$ to the *right* 2 units.

5. True.

7. 1

9. False.

11. a. $3^{2.2} \approx 11.212$
    b. $3^{2.23} \approx 11.587$
    c. $3^{2.236} \approx 11.664$
    d. $3^{\sqrt{5}} \approx 11.665$

13. a. $2^{3.14} \approx 8.815$
    b. $2^{3.141} \approx 8.821$
    c. $2^{3.1415} \approx 8.824$
    d. $2^{\pi} \approx 8.825$

15. a. $3.1^{2.7} \approx 21.217$
    b. $3.14^{2.71} \approx 22.217$
    c. $3.141^{2.718} \approx 22.440$
    d. $\pi^e \approx 22.459$

17. $e^{1.2} \approx 3.320$

19. $e^{-0.85} \approx 0.427$

21.

x	$y = f(x)$	$\dfrac{f(x+1)}{f(x)}$
-1	3	$\dfrac{6}{3} = 2$
0	6	$\dfrac{12}{6} = 2$
1	12	$\dfrac{18}{12} = \dfrac{3}{2}$
2	18	
3	30	

Not an exponential function since the ratio of consecutive terms is not constant.

23.

x	$y = H(x)$	$\dfrac{H(x+1)}{H(x)}$
-1	$\dfrac{1}{4}$	$\dfrac{1}{(1/4)} = 4$
0	1	$\dfrac{4}{1} = 4$
1	4	$\dfrac{16}{4} = 4$
2	16	$\dfrac{64}{16} = 4$
3	64	

Yes, an exponential function since the ratio of consecutive terms is constant with $a = 4$. So the base is 4.

25.

x	$y=f(x)$	$\dfrac{f(x+1)}{f(x)}$
-1	$\dfrac{3}{2}$	$\dfrac{3}{(3/2)} = 3 \cdot \dfrac{2}{3} = 2$
0	3	$\dfrac{6}{3} = 2$
1	6	$\dfrac{12}{6} = 2$
2	12	$\dfrac{24}{12} = 2$
3	24	

Yes, an exponential function since the ratio of consecutive terms is constant with $a = 2$. So the base is 2.

27.

x	$y=H(x)$	$\dfrac{H(x+1)}{H(x)}$
-1	2	$\dfrac{4}{2} = 2$
0	4	$\dfrac{6}{4} = \dfrac{3}{2}$
1	6	
2	8	
3	10	

Not an exponential function since the ratio of consecutive terms is not constant.

29. B

31. D

33. A

35. E

37. $f(x) = 2^x + 1$

Using the graph of $y = 2^x$, shift the graph up 1 unit.
Domain: All real numbers
Range: $\{y \mid y > 0\}$ or $(1, \infty)$
Horizontal Asymptote: $y = 1$

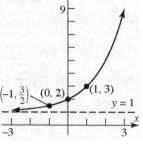

39. $f(x) = 3^{-x} - 2$

Using the graph of $y = 3^x$, reflect the graph about the y-axis, and shift down 2 units.
Domain: All real numbers
Range: $\{y \mid y > -2\}$ or $(-2, \infty)$
Horizontal Asymptote: $y = -2$

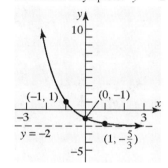

41. $f(x) = 2 + 3(4^x)$

Using the graph of $y = 4^x$, stretch the graph vertically by a factor of 3, and shift up 2 units.
Domain: All real numbers
Range: $\{y \mid y > 2\}$ or $(2, \infty)$
Horizontal Asymptote: $y = 2$

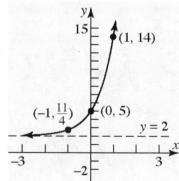

43. $f(x) = 2 + 3^{x/2}$

Using the graph of $y = 3^x$, stretch the graph horizontally by a factor of 2, and shift up 2 units.
Domain: All real number
Range: $\{y \mid y > 2\}$ or $(2, \infty)$
Horizontal Asymptote: $y = 2$

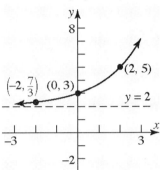

45. $f(x) = e^{-x}$

Using the graph of $y = e^x$, reflect the graph about the y-axis.
Domain: All real numbers
Range: $\{y \mid y > 0\}$ or $(0, \infty)$
Horizontal Asymptote: $y = 0$

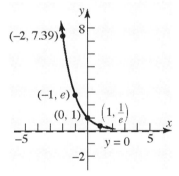

47. $f(x) = e^{x+2}$

Using the graph of $y = e^x$, shift the graph 2 units to the left.
Domain: All real numbers
Range: $\{y \mid y > 0\}$ or $(0, \infty)$
Horizontal Asymptote: $y = 0$

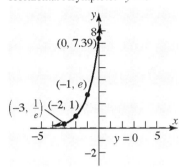

49. $f(x) = 5 - e^{-x}$

Using the graph of $y = e^x$, reflect the graph about the y-axis, reflect about the x-axis, and shift up 5 units.
Domain: All real numbers
Range: $\{y \mid y < 5\}$ or $(-\infty, 5)$
Horizontal Asymptote: $y = 5$

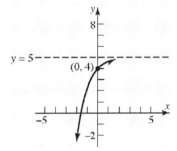

51. $f(x) = 2 - e^{-x/2}$

Using the graph of $y = e^x$, reflect the graph about the y-axis, stretch horizontally by a factor of 2, reflect about the x-axis, and shift up 2 units.
Domain: All real numbers
Range: $\{y \mid y < 2\}$ or $(-\infty, 2)$
Horizontal Asymptote: $y = 2$

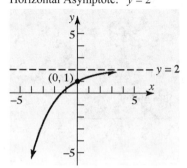

53. $2^{2x+1} = 4$
$2^{2x+1} = 2^2$
$2x + 1 = 2$
$2x = 1$
$x = \frac{1}{2}$

The solution set is $\left\{\frac{1}{2}\right\}$.

55. $3^{x^3} = 9^x$
$3^{x^3} = (3^2)^x$
$3^{x^3} = 3^{2x}$
$x^3 = 2x$
$x^3 - 2x = 0$
$x(x^2 - 2) = 0$
$x = 0$ or $x^2 - 2 = 0$
$x^2 = 2$
$x = \pm\sqrt{2}$

The solution set is $\{-\sqrt{2}, 0, \sqrt{2}\}$.

57. $8^{x^2 - 2x} = \frac{1}{2}$
$(2^3)^{x^2 - 2x} = 2^{-1}$
$2^{3x^2 - 6x} = 2^{-1}$
$3x^2 - 6x = -1$
$3x^2 - 6x + 1 = 0$
$x = \frac{-(-6) \pm \sqrt{(-6)^2 - 4(3)(1)}}{2(3)}$
$= \frac{6 \pm \sqrt{24}}{6}$
$= \frac{6 \pm 2\sqrt{6}}{6}$
$= \frac{3 \pm \sqrt{6}}{3}$

The solution set is $\left\{1 - \frac{\sqrt{6}}{3}, 1 + \frac{\sqrt{6}}{3}\right\}$.

59. $2^x \cdot 8^{-x} = 4^x$
$2^x \cdot (2^3)^{-x} = (2^2)^x$
$2^x \cdot 2^{-3x} = 2^{2x}$
$2^{-2x} = 2^{2x}$
$-2x = 2x$
$-4x = 0$
$x = 0$

The solution set is $\{0\}$.

61. $\left(\frac{1}{5}\right)^{2-x} = 25$
$(5^{-1})^{2-x} = 5^2$
$5^{x-2} = 5^2$
$x - 2 = 2$
$x = 4$

The solution set is $\{4\}$.

63.
$$4^x = 8$$
$$(2^2)^x = 2^3$$
$$2^{2x} = 2^3$$
$$2x = 3$$
$$x = \frac{3}{2}$$
The solution set is $\left\{\frac{3}{2}\right\}$.

65.
$$e^{x^2} = e^{3x} \cdot \frac{1}{e^2}$$
$$e^{x^2} = e^{3x-2}$$
$$x^2 = 3x - 2$$
$$x^2 - 3x + 2 = 0$$
$$(x-1)(x-2) = 0$$
$$x = 1 \text{ or } x = 2$$
The solution set is $\{1, 2\}$.

67. If $4^x = 7$, then $(4^x)^{-2} = 7^{-2}$
$$4^{-2x} = \frac{1}{7^2}$$
$$4^{-2x} = \frac{1}{49}$$

69. If $3^{-x} = 2$, then $(3^{-x})^{-2} = 2^{-2}$
$$3^{2x} = \frac{1}{2^2}$$
$$3^{2x} = \frac{1}{4}$$

71. We need a function of the form $f(x) = k \cdot a^{p \cdot x}$, with $a > 0$, $a \neq 1$. The graph contains the points $\left(-1, \frac{1}{3}\right)$, $(0, 1)$, $(1, 3)$, and $(2, 9)$. In other words, $f(-1) = \frac{1}{3}$, $f(0) = 1$, $f(1) = 3$, and $f(2) = 9$. Therefore, $f(0) = k \cdot a^{p \cdot (0)}$
$$1 = k \cdot a^0$$
$$1 = k \cdot 1$$
$$1 = k$$
and $f(1) = a^{p \cdot (1)}$
$$3 = a^p$$

Let's use $a = 3$, $p = 1$. Then $f(x) = 3^x$. Now we need to verify that this function yields the other known points on the graph.
$$f(-1) = 3^{-1} = \frac{1}{3};$$
$$f(2) = 3^2 = 9$$
So we have the function $f(x) = 3^x$.

73. We need a function of the form $f(x) = k \cdot a^{p \cdot x}$, with $a > 0$, $a \neq 1$. The graph contains the points $\left(-1, -\frac{1}{6}\right)$, $(0, -1)$, $(1, -6)$, and $(2, -36)$. In other words, $f(-1) = -\frac{1}{6}$, $f(0) = -1$, $f(1) = -6$, and $f(2) = -36$. Therefore, $f(0) = k \cdot a^{p \cdot (0)}$
$$-1 = k \cdot a^0$$
$$-1 = k \cdot 1$$
$$-1 = k$$
and $f(1) = -a^{p \cdot (1)}$
$$-6 = -a^p$$
$$6 = a^p$$
Let's use $a = 6$, $p = 1$. Then $f(x) = -6^x$. Now we need to verify that this function yields the other known points on the graph.
$$f(-1) = -6^{-1} = -\frac{1}{6}$$
$$f(2) = -6^2 = -36$$
So we have the function $f(x) = -6^x$.

75. $p(n) = 100(0.97)^n$

  a. $p(10) = 100(0.97)^{10} \approx 74\%$ of light

  b. $p(25) = 100(0.97)^{25} \approx 47\%$ of light

77. $p(x) = 16,630(0.90)^x$

  a. $p(3) = 16,630(0.90)^3 \approx \$12,123$

  b. $p(9) = 16,630(0.90)^9 \approx \$6,443$

79. $D(h) = 5e^{-0.4h}$

$D(1) = 5e^{-0.4(1)} = 5e^{-0.4} \approx 3.35$

After 1 hours, 3.35 milligrams will be present.

$D(6) = 5e^{-0.4(6)} = 5e^{-2.4} \approx 0.45$ milligrams

After 6 hours, 0.45 milligrams will be present.

81. $F(t) = 1 - e^{-0.1t}$

a. $F(10) = 1 - e^{-0.1(10)} = 1 - e^{-1} \approx 0.63$

The probability that a car will arrive within 10 minutes of 12:00 PM is 0.63.

b. $F(40) = 1 - e^{-0.1(40)} = 1 - e^{-4} \approx 0.98$

The probability that a car will arrive within 40 minutes of 12:00 PM is 0.98.

c. As $t \to \infty$, $F(t) = 1 - e^{-0.1t} \to 1 - 0 = 1$

d. Graphing the function:

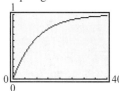

e. $F(7) \approx 0.50$, so about 7 minutes are needed for the probability to reach 50%.

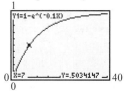

83. $P(x) = \dfrac{20^x e^{-20}}{x!}$

a. $P(15) = \dfrac{20^{15} e^{-20}}{15!} \approx 0.0516$ or 5.16%

The probability that 15 cars will arrive between 5:00 PM and 6:00 PM is 5.16%.

b. $P(20) = \dfrac{20^{20} e^{-20}}{20!} \approx 0.0888$ or 8.88%

The probability that 20 cars will arrive between 5:00 PM and 6:00 PM is 8.88%.

85. $R = 10^{\left(\frac{4221}{T+459.4} - \frac{4221}{D+459.4} + 2\right)}$

a. $R = 10^{\left(\frac{4221}{50+459.4} - \frac{4221}{41+459.4} + 2\right)} \approx 70.95\%$

b. $R = 10^{\left(\frac{4221}{68+459.4} - \frac{4221}{59+459.4} + 2\right)} \approx 72.62\%$

c. $R = 10^{\left(\frac{4221}{T+459.4} - \frac{4221}{T+459.4} + 2\right)} = 10^2 = 100\%$

87. $I = \dfrac{E}{R}\left[1 - e^{-\left(\frac{R}{L}\right)t}\right]$

a. $I_1 = \dfrac{120}{10}\left[1 - e^{-\left(\frac{10}{5}\right)0.3}\right] = 12\left[1 - e^{-0.6}\right] \approx 5.414$

amperes after 0.3 second

$I_1 = \dfrac{120}{10}\left[1 - e^{-\left(\frac{10}{5}\right)0.5}\right] = 12\left[1 - e^{-1}\right] \approx 7.585$

amperes after 0.5 second

$I_1 = \dfrac{120}{10}\left[1 - e^{-\left(\frac{10}{5}\right)1}\right] = 12\left[1 - e^{-2}\right] \approx 10.376$

amperes after 1 second

b. As $t \to \infty$, $e^{-\left(\frac{10}{5}\right)t} \to 0$. Therefore, as,

$t \to \infty$, $I_1 = \dfrac{120}{10}\left[1 - e^{-\left(\frac{10}{5}\right)t}\right] \to 12[1-0] = 12$,

which means the maximum current is 12 amperes.

c. See the graph at the bottom of the page.

d. $I_2 = \dfrac{120}{5}\left[1 - e^{-\left(\frac{5}{10}\right)0.3}\right]$

$= 24\left[1 - e^{-0.15}\right]$

≈ 3.343 amperes after 0.3 second

$I_2 = \dfrac{120}{5}\left[1 - e^{-\left(\frac{5}{10}\right)0.5}\right]$

$= 24\left[1 - e^{-0.25}\right]$

≈ 5.309 amperes after 0.5 second

$$I_2 = \frac{120}{5}\left[1-e^{-\left(\frac{5}{10}\right)1}\right]$$
$$= 24\left[1-e^{-0.5}\right]$$
≈ 9.443 amperes after 1 second

e. As $t \to \infty$, $e^{-\left(\frac{5}{10}\right)t} \to 0$. Therefore, as,

$t \to \infty$, $I_1 = \frac{120}{5}\left[1-e^{-\left(\frac{10}{5}\right)t}\right] \to 24[1-0] = 24$,

which means the maximum current is 24 amperes.

f. See the graph at the bottom of the page.

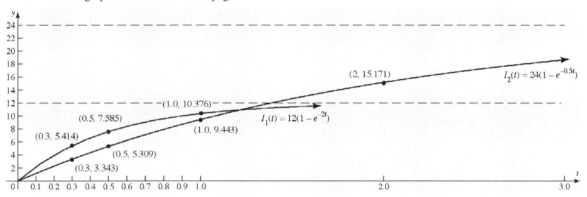

89. $2 + \frac{1}{2!} + \frac{1}{3!} + \frac{1}{4!} + \ldots + \frac{1}{n!}$

$n = 4$; $2 + \frac{1}{2!} + \frac{1}{3!} + \frac{1}{4!} \approx 2.7083$

$n = 6$; $2 + \frac{1}{2!} + \frac{1}{3!} + \frac{1}{4!} + \frac{1}{5!} + \frac{1}{6!} \approx 2.7181$

$n = 8$; $2 + \frac{1}{2!} + \frac{1}{3!} + \frac{1}{4!} + \frac{1}{5!} + \frac{1}{6!} + \frac{1}{7!} + \frac{1}{8!}$
≈ 2.7182788

$n = 10$; $2 + \frac{1}{2!} + \frac{1}{3!} + \frac{1}{4!} + \frac{1}{5!} + \frac{1}{6!} + \frac{1}{7!} + \frac{1}{8!} + \frac{1}{9!} + \frac{1}{10!}$
≈ 2.7182818

$e \approx 2.718281828$

91. $f(x) = a^x$

$$\frac{f(x+h)-f(x)}{h} = \frac{a^{x+h}-a^x}{h}$$
$$= \frac{a^x a^h - a^x}{h}$$
$$= \frac{a^x(a^h-1)}{h}$$
$$= a^x\left(\frac{a^h-1}{h}\right)$$

93. $f(x) = a^x$

$$f(-x) = a^{-x} = \frac{1}{a^x} = \frac{1}{f(x)}$$

95. $\sinh x = \frac{1}{2}(e^x - e^{-x})$

   a. $f(-x) = \sinh(-x)$
   $= \frac{1}{2}(e^{-x} - e^x)$
   $= -\frac{1}{2}(e^x - e^{-x})$
   $= -\sinh x$
   $= -f(x)$
   Therefore, $f(x) = \sinh x$ is an odd function.

   b. Let $Y_1 = \frac{1}{2}(e^x - e^{-x})$.

   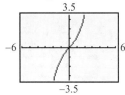

97. $f(x) = 2^{(2^x)} + 1$

   $f(1) = 2^{(2^1)} + 1 = 2^2 + 1 = 4 + 1 = 5$
   $f(2) = 2^{(2^2)} + 1 = 2^4 + 1 = 16 + 1 = 17$
   $f(3) = 2^{(2^3)} + 1 = 2^8 + 1 = 256 + 1 = 257$
   $f(4) = 2^{(2^4)} + 1 = 2^{16} + 1 = 65,536 + 1 = 65,537$
   $f(5) = 2^{(2^5)} + 1 = 2^{32} + 1 = 4,294,967,296 + 1$
   $= 4,294,967,297$
   $= 641 \times 6,700,417$

99. Answers will vary.

101. Given the function $f(x) = a^x$, with $a > 1$,
   If $x > 0$, the graph becomes steeper as a increases.
   If $x < 0$, the graph becomes less steep as a increases.

Section 4.4

1. $3x - 7 \leq 8 - 2x$
   $5x \leq 15$
   $x \leq 3$
   The solution set is $\{x \mid x \leq 3\}$.

3. $\dfrac{x-1}{x+4} > 0$

   $f(x) = \dfrac{x-1}{x+4}$

   f is zero or undefined when $x = 1$ or $x = -4$.

Interval	$(-\infty, -4)$	$(-4, 1)$	$(1, \infty)$
Test Value	-5	0	2
Value of f	6	$-\dfrac{1}{4}$	$\dfrac{1}{6}$
Conclusion	positive	negative	positive

   The solution set is $\{x \mid x < -4 \text{ or } x > 1\}$.

5. $(1, 0)$, $(a, 1)$, $\left(\dfrac{1}{a}, -1\right)$

7. False. If $y = \log_a x$, then $x = a^y$.

9. $9 = 3^2$ is equivalent to $2 = \log_3 9$.

11. $a^2 = 1.6$ is equivalent to $2 = \log_a 1.6$.

13. $1.1^2 = M$ is equivalent to $2 = \log_{1.1} M$.

15. $2^x = 7.2$ is equivalent to $x = \log_2 7.2$.

17. $x^{\sqrt{2}} = \pi$ is equivalent to $\sqrt{2} = \log_x \pi$.

19. $e^x = 8$ is equivalent to $x = \ln 8$.

21. $\log_2 8 = 3$ is equivalent to $2^3 = 8$.

23. $\log_a 3 = 6$ is equivalent to $a^6 = 3$.

25. $\log_3 2 = x$ is equivalent to $3^x = 2$.

27. $\log_2 M = 1.3$ is equivalent to $2^{1.3} = M$.

29. $\log_{\sqrt{2}} \pi = x$ is equivalent to $(\sqrt{2})^x = \pi$.

31. $\ln 4 = x$ is equivalent to $e^x = 4$.

33. $\log_2 1 = 0$ since $2^0 = 1$.

35. $\log_5 25 = 2$ since $5^2 = 25$.

37. $\log_{1/2} 16 = -4$ since $\left(\dfrac{1}{2}\right)^{-4} = 2^4 = 16$.

39. $\log_{10} \sqrt{10} = \dfrac{1}{2}$ since $10^{1/2} = \sqrt{10}$.

41. $\log_{\sqrt{2}} 4 = 4$ since $\left(\sqrt{2}\right)^4 = 4$.

43. $\ln \sqrt{e} = \dfrac{1}{2}$ since $e^{1/2} = \sqrt{e}$.

45. $f(x) = \ln(x-3)$ requires $x - 3 > 0$.
    $x - 3 > 0$
    $x > 3$
    The domain of f is $\{x \mid x > 3\}$.

47. $F(x) = \log_2 x^2$ requires $x^2 > 0$.
    $x^2 > 0$ for all $x \neq 0$.
    The domain of F is $\{x \mid x \neq 0\}$.

49. $f(x) = 3 - 2\log_4 \dfrac{x}{2}$ requires $\dfrac{x}{2} > 0$.
    $\dfrac{x}{2} > 0$
    $x > 0$
    The domain of f is $\{x \mid x > 0\}$.

51. $f(x) = \ln\left(\dfrac{1}{x+1}\right)$ requires $\dfrac{1}{x+1} > 0$.
    $p(x) = \dfrac{1}{x+1}$ is undefined when $x = -1$.

Interval	$(-\infty, -1)$	$(-1, \infty)$
Test Value	-2	0
Value of p	-1	1
Conclusion	negative	positive

    The domain of f is $\{x \mid x > -1\}$.

53. $g(x) = \log_5\left(\dfrac{x+1}{x}\right)$ requires $\dfrac{x+1}{x} > 0$.

    $p(x) = \dfrac{x+1}{x}$ is zero or undefined when $x = -1$ or $x = 0$.

Interval	$(-\infty, -1)$	$(-1, 0)$	$(0, \infty)$
Test Value	-2	$-\dfrac{1}{2}$	1
Value of p	$\dfrac{1}{2}$	-1	2
Conclusion	positive	negative	positive

    The domain of g is $\{x \mid x < -1 \text{ or } x > 0\}$.

55. $f(x) = \sqrt{\ln x}$ requires $\ln x \geq 0$ and $x > 0$
    $\ln x \geq 0$
    $x \geq e^0$
    $x \geq 1$
    The domain of h is $\{x \mid x \geq 1\}$.

57. $\ln\left(\dfrac{5}{3}\right) \approx 0.511$

59. $\dfrac{\ln \dfrac{10}{3}}{0.04} \approx 30.099$

61. If the graph of $f(x) = \log_a x$ contains the point $(2, 2)$, then $f(2) = \log_a 2 = 2$. Thus,
    $\log_a 2 = 2$
    $a^2 = 2$
    $a = \pm\sqrt{2}$
    Since the base a must be positive by definition, we have that $a = \sqrt{2}$.

63. $y = \log_3 x$

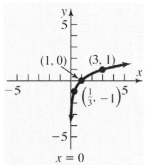

295

65. $y = \log_{1/4} x$

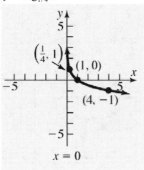

67. B

69. D

71. A

73. E

75. $f(x) = \ln(x+4)$
Using the graph of $y = \ln x$, shift the graph 4 units to the left.
Domain: $(-4, \infty)$
Range: $(-\infty, \infty)$
Vertical Asymptote: $x = -4$

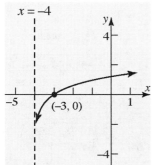

77. $f(x) = 2 + \ln x = \ln x + 2$
Using the graph of $y = \ln x$, shift up 2 units.
Domain: $(0, \infty)$
Range: $(-\infty, \infty)$
Vertical Asymptote: $x = 0$

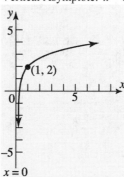

79. $g(x) = \ln(2x)$
Using the graph of $y = \ln x$, compress the graph horizontally by a factor of $\frac{1}{2}$.
Domain: $(0, \infty)$
Range: $(-\infty, \infty)$
Vertical Asymptote: $x = 0$

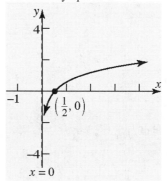

81. $f(x) = 3\ln x$

Using the graph of $y = \ln x$, stretch the graph vertically by a factor of 3.
Domain: $(0, \infty)$
Range: $(-\infty, \infty)$
Vertical Asymptote: $x = 0$

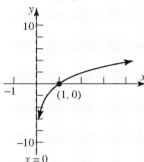

83. $f(x) = \log(x-4)$

Using the graph of $y = \log x$, shift 4 units to the right.
Domain: $(4, \infty)$
Range: $(-\infty, \infty)$
Vertical Asymptote: $x = 4$

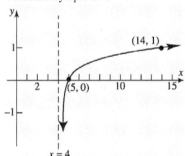

85. $h(x) = 4\log x$

Using the graph of $y = \log x$, stretch the graph vertically by a factor of 4.
Domain: $(0, \infty)$
Range: $(-\infty, \infty)$
Vertical Asymptote: $x = 0$

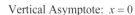
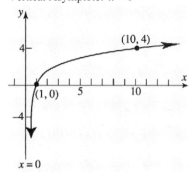

87. $F(x) = \log(2x)$

Using the graph of $y = \log x$, compress the graph horizontally by a factor of $\dfrac{1}{2}$.
Domain: $(0, \infty)$
Range: $(-\infty, \infty)$
Vertical Asymptote: $x = 0$

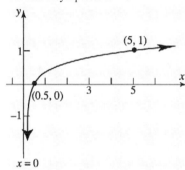

89. $f(x) = 3 + \log(x+2) = \log(x+2) + 3$

Using the graph of $y = \log x$, shift 2 units to the left, and shift up 3 units.
Domain: $(-2, \infty)$
Range: $(-\infty, \infty)$
Vertical Asymptote: $x = -2$

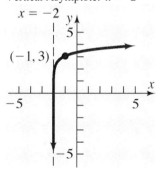

91. $\log_3 x = 2$
$x = 3^2$
$x = 9$
The solution set is $\{9\}$.

93. $\log_2(2x+1) = 3$
$2x+1 = 2^3$
$2x+1 = 8$
$2x = 7$
$x = \dfrac{7}{2}$
The solution set is $\left\{\dfrac{7}{2}\right\}$.

95. $\log_x 4 = 2$
$x^2 = 4$
$x = 2$ $(x \neq -2,$ base is positive$)$
The solution set is $\{2\}$.

97. $\ln e^x = 5$
$e^x = e^5$
$x = 5$
The solution set is $\{5\}$.

99. $\log_4 64 = x$
$4^x = 64$
$4^x = 4^3$
$x = 3$
The solution set is $\{3\}$.

101. $\log_3 243 = 2x+1$
$3^{2x+1} = 243$
$3^{2x+1} = 3^5$
$2x+1 = 5$
$2x = 4$
$x = 2$
The solution set is $\{2\}$.

103. $e^{3x} = 10$
$3x = \ln 10$
$x = \dfrac{\ln 10}{3}$
The solution set is $\left\{\dfrac{\ln 10}{3}\right\}$.

105. $e^{2x+5} = 8$
$2x+5 = \ln 8$
$2x = -5 + \ln 8$
$x = \dfrac{-5 + \ln 8}{2}$
The solution set is $\left\{\dfrac{-5+\ln 8}{2}\right\}$.

107. $\log_3(x^2+1) = 2$
$x^2 + 1 = 3^2$
$x^2 + 1 = 9$
$x^2 = 8$
$x = \pm\sqrt{8} = \pm 2\sqrt{2}$
The solution set is $\{-2\sqrt{2},\ 2\sqrt{2}\}$.

109. $\log_2 8^x = -3$
$8^x = 2^{-3}$
$(2^3)^x = 2^{-3}$
$2^{3x} = 2^{-3}$
$3x = -3$
$x = -1$
The solution set is $\{-1\}$.

111. a. Graphing $f(x) = 2^x$:

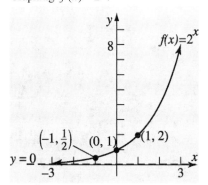

Domain: $(-\infty, \infty)$
Range: $(0, \infty)$
Horizontal asymptote: $y = 0$

b. Finding the inverse:
$$f(x) = 2^x$$
$$y = 2^x$$
$$x = 2^y \quad \text{Inverse}$$
$$y = \log_2 x$$
$$f^{-1}(x) = \log_2 x$$

c. Graphing the inverse:

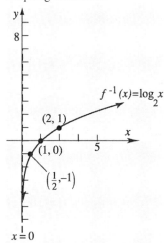

Domain: $(0, \infty)$
Range: $(-\infty, \infty)$
Vertical asymptote: $x = 0$

113. a. Graphing $f(x) = 2^{x+3}$:

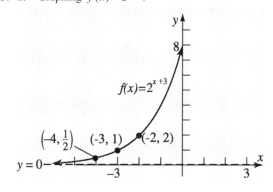

Domain: $(-\infty, \infty)$
Range: $(0, \infty)$
Horizontal asymptote: $y = 0$

b. Finding the inverse:
$$f(x) = 2^{x+3}$$
$$y = 2^{x+3}$$
$$x = 2^{y+3} \quad \text{Inverse}$$
$$y + 3 = \log_2 x$$
$$y = -3 + \log_2 x$$
$$f^{-1}(x) = -3 + \log_2 x$$

c. Graphing the inverse:

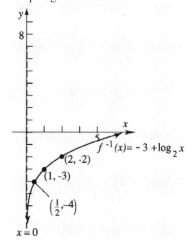

Domain: $(0, \infty)$
Range: $(-\infty, \infty)$
Vertical asymptote: $x = 0$

115. $\text{pH} = -\log_{10}\left[H^+\right]$

a. $\text{pH} = -\log_{10}[0.1] = -(-1) = 1$

b. $\text{pH} = -\log_{10}[0.01] = -(-2) = 2$

c. $\text{pH} = -\log_{10}[0.001] = -(-3) = 3$

d. As the H^+ decreases, the pH increases.

e. $3.5 = -\log_{10}\left[H^+\right]$
$-3.5 = \log_{10}\left[H^+\right]$
$\left[H^+\right] = 10^{-3.5}$
$\approx 3.16 \times 10^{-4}$
$= 0.000316$

f. $7.4 = -\log_{10}\left[H^+\right]$

$-7.4 = \log_{10}\left[H^+\right]$

$\left[H^+\right] = 10^{-7.4}$

$\approx 3.981 \times 10^{-8}$

$= 0.00000003981$

117. $p = 760e^{-0.145h}$

a. $320 = 760e^{-0.145h}$

$\dfrac{320}{760} = e^{-0.145h}$

$\ln\left(\dfrac{320}{760}\right) = -0.145h$

$h = \dfrac{\ln\left(\dfrac{320}{760}\right)}{-0.145} \approx 5.97$

Approximately 5.97 kilometers.

b. $667 = 760e^{-0.145h}$

$\dfrac{667}{760} = e^{-0.145h}$

$\ln\left(\dfrac{667}{760}\right) = -0.145h$

$h = \dfrac{\ln\left(\dfrac{667}{760}\right)}{-0.145} \approx 0.90$

Approximately 0.90 kilometers.

119. $F(t) = 1 - e^{-0.1t}$

a. $0.5 = 1 - e^{-0.1t}$

$-0.5 = -e^{-0.1t}$

$0.5 = e^{-0.1t}$

$\ln(0.5) = -0.1t$

$t = \dfrac{\ln(0.5)}{-0.1} \approx 6.93$

Approximately 6.93 minutes.

b. $0.8 = 1 - e^{-0.1t}$

$-0.2 = -e^{-0.1t}$

$0.2 = e^{-0.1t}$

$\ln(0.2) = -0.1t$

$t = \dfrac{\ln(0.2)}{-0.1} \approx 16.09$

Approximately 16.09 minutes.

c. It is impossible for the probability to reach 100% because $e^{-0.1t}$ will never equal zero; thus, $F(t) = 1 - e^{-0.1t}$ will never equal 1.

121. $D = 5e^{-0.4h}$

$2 = 5e^{-0.4h}$

$0.4 = e^{-0.4h}$

$\ln(0.4) = -0.4h$

$h = \dfrac{\ln(0.4)}{-0.4} \approx 2.29$

Approximately 2.29 hours, or 2 hours and 17 minutes.

123. $I = \dfrac{E}{R}\left[1 - e^{-(R/L)t}\right]$

Substituting $E = 12$, $R = 10$, $L = 5$, and $I = 0.5$, we obtain:

$0.5 = \dfrac{12}{10}\left[1 - e^{-(10/5)t}\right]$

$\dfrac{5}{12} = 1 - e^{-2t}$

$e^{-2t} = \dfrac{7}{12}$

$-2t = \ln(7/12)$

$t = \dfrac{\ln(7/12)}{-2} \approx 0.2695$

It takes approximately 0.2695 second to obtain a current of 0.5 ampere.

Substituting $E = 12$, $R = 10$, $L = 5$, and $I = 1.0$, we obtain:

$1.0 = \dfrac{12}{10}\left[1 - e^{-(10/5)t}\right]$

$\dfrac{10}{12} = 1 - e^{-2t}$

$e^{-2t} = \dfrac{1}{6}$

$-2t = \ln(1/6)$

$t = \dfrac{\ln(1/6)}{-2} \approx 0.8959$

It takes approximately 0.8959 second to obtain a current of 0.5 ampere.

Graphing:

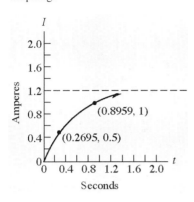

125. $L(10^{-7}) = 10 \log\left(\dfrac{10^{-7}}{10^{-12}}\right)$
$= 10 \log(10^5)$
$= 10 \cdot 5$
$= 50$ decibels

127. $L(10^{-1}) = 10 \log\left(\dfrac{10^{-1}}{10^{-12}}\right)$
$= 10 \log(10^{11})$
$= 10 \cdot 11$
$= 110$ decibels

129. $M(125{,}892) = \log\left(\dfrac{125{,}892}{10^{-3}}\right) \approx 8.1$

131. $R = 3e^{kx}$

   a. $10 = 3e^{k(0.06)}$
   $3.3333 = e^{0.06k}$
   $\ln(3.3333) = 0.06k$
   $k = \dfrac{\ln(3.3333)}{0.06}$
   $k \approx 20.07$

   b. $R = 3e^{20.066(0.17)} = 3e^{3.41122} \approx 91\%$

   c. $100 = 3e^{20.066x}$
   $33.3333 = e^{20.066x}$
   $\ln(33.3333) = 20.066x$
   $x = \dfrac{\ln(33.3333)}{20.07}$
   $x \approx 0.175$

   d. $15 = 3e^{20.066x}$
   $5 = e^{20.066x}$
   $\ln 5 = 20.066x$
   $x = \dfrac{\ln 5}{20.066}$
   $x \approx 0.08$

   e. Answers will vary.

133. If the base of a logarithmic function equals 1, we would have the following:
$f(x) = \log_1(x)$
$f^{-1}(x) = 1^x = 1$ for every real number x.
In other words, f^{-1} would be a constant function and, therefore, f^{-1} would not be one-to-one.

Section 4.5

1. sum

3. $r \log_a M$

5. False

7. $\log_3 3^{71} = 71$

9. $\ln e^{-4} = -4$

11. $2^{\log_2 7} = 7$

13. $\log_8 2 + \log_8 4 = \log_8(4 \cdot 2) = \log_8 8 = 1$

15. $\log_6 18 - \log_6 3 = \log_6 \dfrac{18}{3} = \log_6 6 = 1$

17. $\log_2 6 \cdot \log_6 4 = \log_6 4^{\log_2 6}$
$= \log_6 (2^2)^{\log_2 6}$
$= \log_6 2^{2\log_2 6}$
$= \log_6 2^{\log_2 6^2}$
$= \log_6 6^2$
$= 2$

19. $3^{\log_3 5 - \log_3 4} = 3^{\log_3 \frac{5}{4}} = \dfrac{5}{4}$

21. $e^{\log_{e^2} 16}$
Let $a = \log_{e^2} 16$, then $(e^2)^a = 16$.
$e^{2a} = 16$
$e^{2a} = 4^2$
$(e^{2a})^{1/2} = (4^2)^{1/2}$
$e^a = 4$
$a = \ln 4$
Thus, $e^{\log_{e^2} 16} = e^{\ln 4} = 4$.

23. $\ln 6 = \ln(2 \cdot 3) = \ln 2 + \ln 3 = a + b$

25. $\ln 1.5 = \ln \dfrac{3}{2} = \ln 3 - \ln 2 = b - a$

27. $\ln 8 = \ln 2^3 = 3 \cdot \ln 2 = 3a$

29. $\ln \sqrt[5]{6} = \ln 6^{1/5}$
$= \dfrac{1}{5} \ln 6$
$= \dfrac{1}{5} \ln(2 \cdot 3)$
$= \dfrac{1}{5}(\ln 2 + \ln 3)$
$= \dfrac{1}{5}(a + b)$

31. $\log_5(25x) = \log_5 25 + \log_5 x = 2 + \log_5 x$

33. $\log_2 z^3 = 3 \log_2 z$

35. $\ln(ex) = \ln e + \ln x = 1 + \ln x$

37. $\ln(xe^x) = \ln x + \ln e^x = \ln x + x$

39. $\log_a(u^2 v^3) = \log_a u^2 + \log_a v^3$
$= 2\log_a u + 3\log_a v$

41. $\ln(x^2 \sqrt{1-x}) = \ln x^2 + \ln \sqrt{1-x}$
$= \ln x^2 + \ln(1-x)^{1/2}$
$= 2\ln x + \dfrac{1}{2}\ln(1-x)$

43. $\log_2\left(\dfrac{x^3}{x-3}\right) = \log_2 x^3 - \log_2(x-3)$
$= 3\log_2 x - \log_2(x-3)$

45. $\log\left[\dfrac{x(x+2)}{(x+3)^2}\right] = \log[x(x+2)] - \log(x+3)^2$
$= \log x + \log(x+2) - 2\log(x+3)$

47. $\ln\left[\dfrac{x^2 - x - 2}{(x+4)^2}\right]^{1/3}$
$= \dfrac{1}{3}\ln\left[\dfrac{(x-2)(x+1)}{(x+4)^2}\right]$
$= \dfrac{1}{3}\left[\ln(x-2)(x+1) - \ln(x+4)^2\right]$
$= \dfrac{1}{3}[\ln(x-2) + \ln(x+1) - 2\ln(x+4)]$
$= \dfrac{1}{3}\ln(x-2) + \dfrac{1}{3}\ln(x+1) - \dfrac{2}{3}\ln(x+4)$

49. $\ln \dfrac{5x\sqrt{1+3x}}{(x-4)^3}$

$= \ln\left(5x\sqrt{1+3x}\right) - \ln(x-4)^3$

$= \ln 5 + \ln x + \ln\sqrt{1+3x} - 3\ln(x-4)$

$= \ln 5 + \ln x + \ln(1+3x)^{1/2} - 3\ln(x-4)$

$= \ln 5 + \ln x + \dfrac{1}{2}\ln(1+3x) - 3\ln(x-4)$

51. $3\log_5 u + 4\log_5 v = \log_5 u^3 + \log_5 v^4$

$\qquad = \log_5\left(u^3 v^4\right)$

53. $\log_3\sqrt{x} - \log_3 x^3 = \log_3\left(\dfrac{\sqrt{x}}{x^3}\right)$

$\qquad = \log_3\left(\dfrac{x^{1/2}}{x^3}\right)$

$\qquad = \log_3 x^{-5/2}$

$\qquad = -\dfrac{5}{2}\log_3 x$

55. $\log_4(x^2-1) - 5\log_4(x+1)$

$= \log_4(x^2-1) - \log_4(x+1)^5$

$= \log_4\left(\dfrac{x^2-1}{(x+1)^5}\right)$

$= \log_4\left(\dfrac{(x+1)(x-1)}{(x+1)^5}\right)$

$= \log_4\left(\dfrac{x-1}{(x+1)^4}\right)$

57. $\ln\left(\dfrac{x}{x-1}\right) + \ln\left(\dfrac{x+1}{x}\right) - \ln(x^2-1)$

$= \ln\left[\dfrac{x}{x-1}\cdot\dfrac{x+1}{x}\right] - \ln(x^2-1)$

$= \ln\left[\dfrac{x+1}{x-1} \div (x^2-1)\right]$

$= \ln\left[\dfrac{x+1}{(x-1)(x^2-1)}\right]$

$= \ln\left[\dfrac{x+1}{(x-1)(x-1)(x+1)}\right]$

$= \ln\left(\dfrac{1}{(x-1)^2}\right)$

$= \ln(x-1)^{-2}$

$= -2\ln(x-1)$

59. $8\log_2\sqrt{3x-2} - \log_2\left(\dfrac{4}{x}\right) + \log_2 4$

$= \log_2\left(\sqrt{3x-2}\right)^8 - (\log_2 4 - \log_2 x) + \log_2 4$

$= \log_2(3x-2)^4 - \log_2 4 + \log_2 x + \log_2 4$

$= \log_2(3x-2)^4 + \log_2 x$

$= \log_2\left[x(3x-2)^4\right]$

61. $2\log_a(5x^3) - \dfrac{1}{2}\log_a(2x+3)$

$= \log_a(5x^3)^2 - \log_a(2x-3)^{1/2}$

$= \log_a(25x^6) - \log_a\sqrt{2x-3}$

$= \log_a\left[\dfrac{25x^6}{\sqrt{2x-3}}\right]$

63. $2\log_2(x+1) - \log_2(x+3) - \log_2(x-1)$

$= \log_2(x+1)^2 - \log_2(x+3) - \log_2(x-1)$

$= \log_2\dfrac{(x+1)^2}{(x+3)} - \log_2(x-1)$

$= \log_2\left(\dfrac{(x+1)^2}{(x+3)(x-1)}\right)$

65. $\log_3 21 = \dfrac{\log 21}{\log 3} \approx 2.771$

67. $\log_{1/3} 71 = \dfrac{\log 71}{\log(1/3)} = \dfrac{\log 71}{-\log 3} \approx -3.880$

69. $\log_{\sqrt{2}} 7 = \dfrac{\log 7}{\log \sqrt{2}} \approx 5.615$

71. $\log_\pi e = \dfrac{\ln e}{\ln \pi} \approx 0.874$

73. $y = \log_4 x = \dfrac{\ln x}{\ln 4}$ or $y = \dfrac{\log x}{\log 4}$

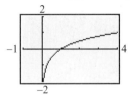

75. $y = \log_2(x+2) = \dfrac{\ln(x+2)}{\ln 2}$ or $y = \dfrac{\log(x+2)}{\log 2}$

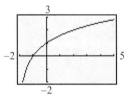

77. $y = \log_{x-1}(x+1) = \dfrac{\ln(x+1)}{\ln(x-1)}$ or $y = \dfrac{\log(x+1)}{\log(x-1)}$

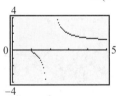

79. $\ln y = \ln x + \ln C$
$\ln y = \ln(xC)$
$y = Cx$

81. $\ln y = \ln x + \ln(x+1) + \ln C$
$\ln y = \ln(x(x+1)C)$
$y = Cx(x+1)$

83. $\ln y = 3x + \ln C$
$\ln y = \ln e^{3x} + \ln C$
$\ln y = \ln(Ce^{3x})$
$y = Ce^{3x}$

85. $\ln(y-3) = -4x + \ln C$
$\ln(y-3) = \ln e^{-4x} + \ln C$
$\ln(y-3) = \ln(Ce^{-4x})$
$y - 3 = Ce^{-4x}$
$y = Ce^{-4x} + 3$

87. $3\ln y = \dfrac{1}{2}\ln(2x+1) - \dfrac{1}{3}\ln(x+4) + \ln C$

$\ln y^3 = \ln(2x+1)^{1/2} - \ln(x+4)^{1/3} + \ln C$

$\ln y^3 = \ln\left[\dfrac{C(2x+1)^{1/2}}{(x+4)^{1/3}}\right]$

$y^3 = \dfrac{C(2x+1)^{1/2}}{(x+4)^{1/3}}$

$y = \left[\dfrac{C(2x+1)^{1/2}}{(x+4)^{1/3}}\right]^{1/3}$

$y = \dfrac{\sqrt[3]{C}(2x+1)^{1/6}}{(x+4)^{1/9}}$

89. $\log_2 3 \cdot \log_3 4 \cdot \log_4 5 \cdot \log_5 6 \cdot \log_6 7 \cdot \log_7 8$

$= \dfrac{\log 3}{\log 2} \cdot \dfrac{\log 4}{\log 3} \cdot \dfrac{\log 5}{\log 4} \cdot \dfrac{\log 6}{\log 5} \cdot \dfrac{\log 7}{\log 6} \cdot \dfrac{\log 8}{\log 7}$

$= \dfrac{\log 8}{\log 2} = \dfrac{\log 2^3}{\log 2}$

$= \dfrac{3\log 2}{\log 2}$

$= 3$

91. $\log_2 3 \cdot \log_3 4 \cdots \log_n(n+1) \cdot \log_{n+1} 2$

$= \dfrac{\log 3}{\log 2} \cdot \dfrac{\log 4}{\log 3} \cdots \dfrac{\log(n+1)}{\log n} \cdot \dfrac{\log 2}{\log(n+1)}$

$= \dfrac{\log 2}{\log 2}$

$= 1$

93. $\log_a\left(x+\sqrt{x^2-1}\right)+\log_a\left(x-\sqrt{x^2-1}\right)$:

$= \log_a\left[\left(x+\sqrt{x^2-1}\right)\left(x-\sqrt{x^2-1}\right)\right]$
$= \log_a\left[x^2-\left(x^2-1\right)\right]$
$= \log_a\left[x^2-x^2+1\right]$
$= \log_a 1$
$= 0$

95. $2x+\ln\left(1+e^{-2x}\right) = \ln e^{2x}+\ln\left(1+e^{-2x}\right)$
$= \ln\left(e^{2x}\left(1+e^{-2x}\right)\right)$
$= \ln\left(e^{2x}+e^0\right)$
$= \ln\left(e^{2x}+1\right)$

97. $f(x) = \log_a x$ means that $x = a^{f(x)}$.
Now, raising both sides to the -1 power, we
obtain $x^{-1} = \left(a^{f(x)}\right)^{-1} = \left(a^{-1}\right)^{f(x)} = \left(\dfrac{1}{a}\right)^{f(x)}$.

$x^{-1} = \left(\dfrac{1}{a}\right)^{f(x)}$ means that $\log_{1/a} x^{-1} = f(x)$.

Thus, $\log_{1/a} x^{-1} = f(x)$
$-\log_{1/a} x = f(x)$
$-f(x) = \log_{1/a} x$

99. $f(x) = \log_a x$

$f\left(\dfrac{1}{x}\right) = \log_a\left(\dfrac{1}{x}\right)$
$= \log_a 1 - \log_a x$
$= -\log_a x$
$= -f(x)$

101. If $A = \log_a M$ and $B = \log_a N$, then $a^A = M$ and $a^B = N$.

$\log_a\left(\dfrac{M}{N}\right) = \log_a\left(\dfrac{a^A}{a^B}\right)$
$= \log_a a^{A-B}$
$= A - B$
$= \log_a M - \log_a N$

103. $Y_1 = \log x^2$ $\qquad$ $Y_2 = 2\log x$

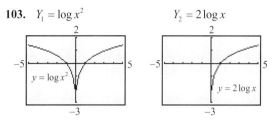

The domain of $Y_1 = \log_a x^2$ is $\{x \mid x \neq 0\}$. The domain of $Y_2 = 2\log_a x$ is $\{x \mid x > 0\}$. These two domains are different because the logarithm property $\log_a x^n = n \cdot \log_a x$ holds only when $\log_a x$ exists.

105. Answers may vary. One possibility follows:
Let $a = 2$, $x = 8$, and $r = 3$. Then
$\left(\log_a x\right)^r = \left(\log_2 8\right)^3 = 3^3 = 27$. But
$r\log_a x = 3\log_2 8 = 3 \cdot 3 = 9$. Thus,
$\left(\log_2 8\right)^3 \neq 3\log_2 8$ and, in general,
$\left(\log_a x\right)^r \neq r\log_a x$.

Section 4.6

1. $x^2 - 7x - 30 = 0$
$(x+3)(x-10) = 0$
$x+3 = 0 \quad \text{or} \quad x-10 = 0$
$x = -3 \quad \text{or} \quad x = 10$
The solution set is $\{-3, 10\}$.

3. $x^3 = x^2 - 5$
Using INTERSECT to solve:
$y_1 = x^3; \ y_2 = x^2 - 5$

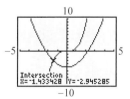

Thus, $x \approx -1.43$, so the solution set is $\{-1.43\}$.

5. $\log_4(x+2) = \log_4 8$
$x + 2 = 8$
$x = 6$
The solution set is $\{6\}$.

7. $\dfrac{1}{2}\log_3 x = 2\log_3 2$
$\log_3 x^{1/2} = \log_3 2^2$
$x^{1/2} = 4$
$x = 16$
The solution set is $\{16\}$.

9. $2\log_5 x = 3\log_5 4$
$\log_5 x^2 = \log_5 4^3$
$x^2 = 64$
$x = \pm 8$
Since $\log_5(-8)$ is undefined, the solution set is $\{8\}$.

11. $3\log_2(x-1) + \log_2 4 = 5$
$\log_2(x-1)^3 + \log_2 4 = 5$
$\log_2\left(4(x-1)^3\right) = 5$
$4(x-1)^3 = 2^5$
$(x-1)^3 = \dfrac{32}{4}$
$(x-1)^3 = 8$
$x - 1 = 2$
$x = 3$
The solution set is $\{3\}$.

13. $\log x + \log(x+15) = 2$
$\log\left(x(x+15)\right) = 2$
$x(x+15) = 10^2$
$x^2 + 15x - 100 = 0$
$(x+20)(x-5) = 0$
$x = -20$ or $x = 5$
Since $\log(-20)$ is undefined, the solution set is $\{5\}$.

15. $\ln x + \ln(x+2) = 4$
$\ln\left(x(x+2)\right) = 4$
$x(x+2) = e^4$
$x^2 + 2x - e^4 = 0$
$x = \dfrac{-2 \pm \sqrt{2^2 - 4(1)(-e^4)}}{2(1)}$
$= \dfrac{-2 \pm \sqrt{4 + 4e^4}}{2}$
$= \dfrac{-2 \pm 2\sqrt{1+e^4}}{2}$
$= -1 \pm \sqrt{1+e^4}$
$x = -1 - \sqrt{1+e^4}$ or $x = -1 + \sqrt{1+e^4}$
≈ -8.456 ≈ 6.456
Since $\ln(-8.456)$ is undefined, the solution set is $\left\{-1 + \sqrt{1+e^4} \approx 6.456\right\}$.

17. $2^{2x} + 2^x - 12 = 0$
$\left(2^x\right)^2 + 2^x - 12 = 0$
$\left(2^x - 3\right)\left(2^x + 4\right) = 0$
$2^x - 3 = 0$ or $2^x + 4 = 0$
$2^x = 3$ or $2^x = -4$
$\log(2^x) = \log 3$ No solution
$x\log 2 = \log 3$
$x = \dfrac{\log 3}{\log 2} \approx 1.585$
The solution set is $\left\{\dfrac{\log 3}{\log 2} \approx 1.585\right\}$.

19. $3^{2x} + 3^{x+1} - 4 = 0$
$\left(3^x\right)^2 + 3 \cdot 3^x - 4 = 0$
$\left(3^x - 1\right)\left(3^x + 4\right) = 0$
$3^x - 1 = 0$ or $3^x + 4 = 0$
$3^x = 1$ or $3^x = -4$
$x = 0$ No solution
The solution set is $\{0\}$.

21.
$$2^x = 10$$
$$\log(2^x) = \log 10$$
$$x \log 2 = 1$$
$$x = \frac{1}{\log 2} \approx 3.322$$
The solution set is $\left\{\dfrac{1}{\log 2} \approx 3.322\right\}$

23.
$$8^{-x} = 1.2$$
$$\log(8^{-x}) = \log(1.2)$$
$$-x \log 8 = \log(1.2)$$
$$x = \frac{\log(1.2)}{-\log 8} \approx -0.088$$
The solution set is $\left\{\dfrac{\log(1.2)}{-\log 8} \approx -0.088\right\}$.

25.
$$3^{1-2x} = 4^x$$
$$\log(3^{1-2x}) = \log(4^x)$$
$$(1-2x)\log 3 = x \log 4$$
$$\log 3 - 2x \log 3 = x \log 4$$
$$\log 3 = x \log 4 + 2x \log 3$$
$$\log 3 = x(\log 4 + 2 \log 3)$$
$$x = \frac{\log 3}{\log 4 + 2 \log 3} \approx 0.307$$
The solution set is $\left\{\dfrac{\log 3}{\log 4 + 2 \log 3} \approx 0.307\right\}$.

27.
$$\left(\frac{3}{5}\right)^x = 7^{1-x}$$
$$\log\left(\frac{3}{5}\right)^x = \log(7^{1-x})$$
$$x \log(3/5) = (1-x)\log 7$$
$$x \log(3/5) = \log 7 - x \log 7$$
$$x \log(3/5) + x \log 7 = \log 7$$
$$x(\log(3/5) + \log 7) = \log 7$$
$$x = \frac{\log 7}{\log(3/5) + \log 7} \approx 1.356$$
The solution set is $\left\{\dfrac{\log 7}{\log(3/5) + \log 7} \approx 1.356\right\}$.

29.
$$1.2^x = (0.5)^{-x}$$
$$\log 1.2^x = \log(0.5)^{-x}$$
$$x \log(1.2) = -x \log(0.5)$$
$$x \log(1.2) + x \log(0.5) = 0$$
$$x(\log(1.2) + \log(0.5)) = 0$$
$$x = 0$$
The solution set is $\{0\}$.

31.
$$\pi^{1-x} = e^x$$
$$\ln \pi^{1-x} = \ln e^x$$
$$(1-x)\ln \pi = x$$
$$\ln \pi - x \ln \pi = x$$
$$\ln \pi = x + x \ln \pi$$
$$\ln \pi = x(1 + \ln \pi)$$
$$x = \frac{\ln \pi}{1 + \ln \pi} \approx 0.534$$
The solution set is $\left\{\dfrac{\ln \pi}{1 + \ln \pi} \approx 0.534\right\}$.

33.
$$5(2^{3x}) = 8$$
$$2^{3x} = \tfrac{8}{5}$$
$$\log 2^{3x} = \log\left(\tfrac{8}{5}\right)$$
$$3x \log 2 = \log(8/5)$$
$$x = \frac{\log(8/5)}{3 \log 2} \approx 0.226$$
The solution set is $\left\{\dfrac{\log(8/5)}{3 \log 2} \approx 0.226\right\}$.

35. $\log_a(x-1) - \log_a(x+6) = \log_a(x-2) - \log_a(x+3)$
$$\log_a\left(\frac{x-1}{x+6}\right) = \log_a\left(\frac{x-2}{x+3}\right)$$
$$\frac{x-1}{x+6} = \frac{x-2}{x+3}$$
$$(x-1)(x+3) = (x-2)(x+6)$$
$$x^2 + 2x - 3 = x^2 + 4x - 12$$
$$2x - 3 = 4x - 12$$
$$9 = 2x$$
$$x = \tfrac{9}{2}$$
Since each of the original logarithms is defined for $x = \tfrac{9}{2}$, the solution set is $\left\{\tfrac{9}{2}\right\}$.

Chapter 4: Exponential and Logarithmic Functions

37. $\log_{1/3}(x^2+x) - \log_{1/3}(x^2-x) = -1$

$$\log_{1/3}\left(\frac{x^2+x}{x^2-x}\right) = -1$$

$$\frac{x^2+x}{x^2-x} = \left(\frac{1}{3}\right)^{-1}$$

$$\frac{x^2+x}{x^2-x} = 3$$

$$x^2 + x = 3(x^2 - x)$$

$$x^2 + x = 3x^2 - 3x$$

$$-2x^2 + 4x = 0$$

$$-2x(x-2) = 0$$

$$-2x = 0 \quad \text{or} \quad x - 2 = 0$$

$$x = 0 \quad \text{or} \quad x = 2$$

Since each of the original logarithms are not defined for $x=0$, but are defined for $x=2$, the solution set is $\{2\}$.

39. $\log_2(x+1) - \log_4 x = 1$

$$\log_2(x+1) - \frac{\log_2 x}{\log_2 4} = 1$$

$$\log_2(x+1) - \frac{\log_2 x}{2} = 1$$

$$2\log_2(x+1) - \log_2 x = 2$$

$$\log_2(x+1)^2 - \log_2 x = 2$$

$$\log_2\left(\frac{(x+1)^2}{x}\right) = 2$$

$$\frac{(x+1)^2}{x} = 2^2$$

$$x^2 + 2x + 1 = 4x$$

$$x^2 - 2x + 1 = 0$$

$$(x-1)^2 = 0$$

$$x - 1 = 0$$

$$x = 1$$

Since each of the original logarithms is defined for $x=1$, the solution set is $\{1\}$.

41. $\log_{16} x + \log_4 x + \log_2 x = 7$

$$\frac{\log_2 x}{\log_2 16} + \frac{\log_2 x}{\log_2 4} + \log_2 x = 7$$

$$\frac{\log_2 x}{4} + \frac{\log_2 x}{2} + \log_2 x = 7$$

$$\log_2 x + 2\log_2 x + 4\log_2 x = 28$$

$$7\log_2 x = 28$$

$$\log_2 x = 4$$

$$x = 2^4 = 16$$

Since each of the original logarithms is defined for $x=16$, the solution set is $\{16\}$.

43. $\left(\sqrt[3]{2}\right)^{2-x} = 2^{x^2}$

$$\left(2^{1/3}\right)^{2-x} = 2^{x^2}$$

$$2^{\frac{1}{3}(2-x)} = 2^{x^2}$$

$$\frac{1}{3}(2-x) = x^2$$

$$2 - x = 3x^2$$

$$3x^2 + x - 2 = 0$$

$$(3x-2)(x+1) = 0$$

$$x = \frac{2}{3} \quad \text{or} \quad x = -1$$

The solution set is $\left\{-1, \frac{2}{3}\right\}$.

45. $\frac{e^x + e^{-x}}{2} = 1$

$$e^x + e^{-x} = 2$$

$$e^x\left(e^x + e^{-x}\right) = 2e^x$$

$$e^{2x} + 1 = 2e^x$$

$$(e^x)^2 - 2e^x + 1 = 0$$

$$\left(e^x - 1\right)^2 = 0$$

$$e^x - 1 = 0$$

$$e^x = 1$$

$$x = 0$$

The solution set is $\{0\}$.

47.
$$\frac{e^x - e^{-x}}{2} = 2$$
$$e^x - e^{-x} = 4$$
$$e^x(e^x - e^{-x}) = 4e^x$$
$$e^{2x} - 1 = 4e^x$$
$$(e^x)^2 - 4e^x - 1 = 0$$
$$e^x = \frac{-(-4) \pm \sqrt{(-4)^2 - 4(1)(-1)}}{2(1)}$$
$$= \frac{4 \pm \sqrt{20}}{2}$$
$$= \frac{4 \pm 2\sqrt{5}}{2}$$
$$= 2 \pm \sqrt{5}$$
$$x = \ln(2 - \sqrt{5}) \quad \text{or} \quad x = \ln(2 + \sqrt{5})$$
$$x \approx \ln(-0.236) \quad \text{or} \quad x \approx 1.444$$

Since $\ln(-0.236)$ is undefined, the solution set is $\left\{\ln(2+\sqrt{5}) \approx 1.444\right\}$.

49.
$$\log_5 x + \log_3 x = 1$$
$$\frac{\log x}{\log 5} + \frac{\log x}{\log 3} = 1$$
$$(\log x)\left(\frac{1}{\log 5} + \frac{1}{\log 3}\right) = 1$$
$$\log x = \frac{1}{\frac{1}{\log 5} + \frac{1}{\log 3}}$$
$$x = 10^{\left(\frac{1}{\frac{1}{\log 5} + \frac{1}{\log 3}}\right)} \approx 1.921$$

The solution set is $\left\{10^{\frac{1}{\frac{1}{\log 5} + \frac{1}{\log 3}}} \approx 1.921\right\}$.

51. $\log_5(x+1) - \log_4(x-2) = 1$
Using INTERSECT to solve:
$y_1 = \ln(x+1)/\ln(5) - \ln(x-2)/\ln(4)$
$y_2 = 1$

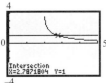

Thus, $x \approx 2.79$, so the solution set is $\{2.79\}$.

53. $e^x = -x$
Using INTERSECT to solve: $y_1 = e^x$; $y_2 = -x$

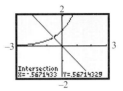

Thus, $x \approx -0.57$, so the solution set is $\{-0.57\}$.

55. $e^x = x^2$
Using INTERSECT to solve:
$y_1 = e^x$; $y_2 = x^2$

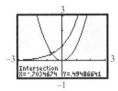

Thus, $x \approx -0.70$, so the solution set is $\{-0.70\}$.

57. $\ln x = -x$
Using INTERSECT to solve:
$y_1 = \ln x$; $y_2 = -x$

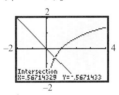

Thus, $x \approx 0.57$, so the solution set is $\{0.57\}$.

59. $\ln x = x^3 - 1$
Using INTERSECT to solve:
$y_1 = \ln x$; $y_2 = x^3 - 1$

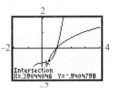 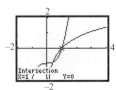

Thus, $x \approx 0.39$ or $x = 1$, so the solution set is $\{0.39, 1\}$.

61. $e^x + \ln x = 4$

Using INTERSECT to solve:

$y_1 = e^x + \ln x;\ y_2 = 4$

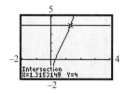

Thus, $x \approx 1.32$, so the solution set is $\{1.32\}$.

63. $e^{-x} = \ln x$

Using INTERSECT to solve:

$y_1 = e^{-x};\ y_2 = \ln x$

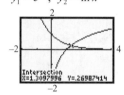

Thus, $x \approx 1.31$, so the solution set is $\{1.31\}$.

65. a. $282(1.011)^{t-2000} = 303$

$(1.011)^{t-2000} = \dfrac{303}{282}$

$\log(1.011)^{t-2000} = \log\left(\dfrac{101}{94}\right)$

$(t-2000)\log(1.011) = \log\left(\dfrac{101}{94}\right)$

$t - 2000 = \dfrac{\log(101/94)}{\log(1.011)}$

$t = \dfrac{\log(101/94)}{\log(1.011)} + 2000$

≈ 2006.57

According to the model, the population of the U.S. will reach 303 million people around the middle of the year 2006.

b. $282(1.011)^{t-2000} = 355$

$(1.011)^{t-2000} = \dfrac{355}{282}$

$\log(1.011)^{t-2000} = \log\left(\dfrac{355}{282}\right)$

$(t-2000)\log(1.011) = \log\left(\dfrac{355}{282}\right)$

$t - 2000 = \dfrac{\log(355/282)}{\log(1.011)}$

$t = \dfrac{\log(355/282)}{\log(1.011)} + 2000$

≈ 2021.04

According to the model, the population of the U.S. will reach 355 million people in the beginning of the middle of the year 2021.

67. a. $14{,}512(0.82)^t = 9{,}000$

$(0.82)^t = \dfrac{9{,}000}{14{,}512}$

$\log(0.82)^t = \log\left(\dfrac{9{,}000}{14{,}512}\right)$

$t\log(0.82) = \log\left(\dfrac{9{,}000}{14{,}512}\right)$

$t = \dfrac{\log(9{,}000/14{,}512)}{\log(0.82)}$

≈ 2.4

According to the model, the car will be worth $9,000 after about 2.4 years.

b. $14{,}512(0.82)^t = 4{,}000$

$(0.82)^t = \dfrac{4{,}000}{14{,}512}$

$\log(0.82)^t = \log\left(\dfrac{4{,}000}{14{,}512}\right)$

$t\log(0.82) = \log\left(\dfrac{4{,}000}{14{,}512}\right)$

$t = \dfrac{\log(4{,}000/14{,}512)}{\log(0.82)}$

≈ 6.5

According to the model, the car will be worth $4,000 after about 6.5 years.

c. $14{,}512(0.82)^t = 2{,}000$

$$(0.82)^t = \frac{2{,}000}{14{,}512}$$

$$\log(0.82)^t = \log\left(\frac{2{,}000}{14{,}512}\right)$$

$$t\log(0.82) = \log\left(\frac{2{,}000}{14{,}512}\right)$$

$$t = \frac{\log(2{,}000/14{,}512)}{\log(0.82)}$$

$$\approx 10.0$$

According to the model, the car will be worth $2,000 after about 10 years.

69. Solution A: change to exponential expression; square root method; meaning of $\pm$; solve.

Solution B: $\log_a M^r = r\log_a M$; divide by 2; change to exponential expression; solve.

The power rule $\log_a M^r = r\log_a M$ only applies when $M > 0$. In this equation, $M = x - 1$. Now, $x = -2$ causes $M = -2 - 1 = -3$. Thus, if we use the power rule, we lose the valid solution $x = -2$.

Section 4.7

1. $P = \$500$, $r = 0.06$, $t = 6$ months $= 0.5$ year
$I = Prt = (500)(0.06)(0.5) = \$15.00$

3. $P = \$100$, $r = 0.04$, $n = 4$, $t = 2$
$A = P\left(1 + \dfrac{r}{n}\right)^{nt} = 100\left(1 + \dfrac{0.04}{4}\right)^{(4)(2)} \approx \$108.29$

5. $P = \$500$, $r = 0.08$, $n = 4$, $t = 2.5$
$A = P\left(1 + \dfrac{r}{n}\right)^{nt} = 500\left(1 + \dfrac{0.08}{4}\right)^{(4)(2.5)} \approx \$609.50$

7. $P = \$600$, $r = 0.05$, $n = 365$, $t = 3$
$A = P\left(1 + \dfrac{r}{n}\right)^{nt} = 600\left(1 + \dfrac{0.05}{365}\right)^{(365)(3)} \approx \$697.09$

9. $P = \$10$, $r = 0.11$, $t = 2$
$A = Pe^{rt} = 10e^{(0.11)(2)} \approx \$12.46$

11. $P = \$100$, $r = 0.10$, $t = 2.25$
$A = Pe^{rt} = 100e^{(0.10)(2.25)} \approx \$125.23$

13. $A = \$100$, $r = 0.06$, $n = 12$, $t = 2$
$P = A\left(1 + \dfrac{r}{n}\right)^{-nt} = 100\left(1 + \dfrac{0.06}{12}\right)^{(-12)(2)} \approx \$88.72$

15. $A = \$1000$, $r = 0.06$, $n = 365$, $t = 2.5$
$P = A\left(1 + \dfrac{r}{n}\right)^{-nt}$
$= 1000\left(1 + \dfrac{0.06}{365}\right)^{(-365)(2.5)}$
$\approx \$860.72$

17. $A = \$600$, $r = 0.04$, $n = 4$, $t = 2$
$P = A\left(1 + \dfrac{r}{n}\right)^{-nt} = 600\left(1 + \dfrac{0.04}{4}\right)^{(-4)(2)} \approx \$554.09$

19. $A = \$80$, $r = 0.09$, $t = 3.25$
$P = Ae^{-rt} = 80e^{(-0.09)(3.25)} \approx \$59.71$

21. $A = \$400$, $r = 0.10$, $t = 1$
$P = Ae^{-rt} = 400e^{(-0.10)(1)} \approx \$361.93$

23. $\$1000$ invested for 1 year at $5\tfrac{1}{4}\%$

Compounded quarterly yields:
$$100\left(1 + \dfrac{0.0525}{4}\right)^{(4)(1)} = \$1053.54.$$

The interest earned is
$\$1053.54 - \$1000.00 = \$53.54$
Thus, $I = Prt$
$5.354 = 1000 \cdot r \cdot 1$
$r = \dfrac{5.354}{1000} = .05354$

The effective interest rate is 5.354%.

25. $2P = P(1+r)^3$
$2 = (1+r)^3$
$\sqrt[3]{2} = 1 + r$
$r = \sqrt[3]{2} - 1$
$\approx 1.26 - 1$
$= 0.26$
$r \approx 26\%$

27. 6% compounded quarterly:
$$A = 10,000\left(1+\frac{0.06}{4}\right)^{(4)(1)} = \$10,613.64$$

$6\frac{1}{4}\%$ compounded annually:
$$A = 10,000(1+0.0625)^1 = \$10,625$$

$6\frac{1}{4}\%$ compounded annually yields the larger amount.

29. 9% compounded monthly:
$$A = 10,000\left(1+\frac{0.09}{12}\right)^{(12)(1)} = \$10,938.07$$

8.8% compounded daily:
$$A = 10,000\left(1+\frac{0.088}{365}\right)^{365} = \$10,919.77$$

9% compounded monthly yields the larger amount.

31. $2P = P\left(1+\frac{0.08}{12}\right)^{12t}$

$2 \approx (1.006667)^{12t}$

$\ln 2 \approx 12t \ln(1.006667)$

$t \approx \dfrac{\ln 2}{12 \ln(1.006667)} \approx 8.69$

Compounded monthly, it will take about 8.69 years (or 104.32 months) to double.

$2P = Pe^{0.08t}$

$2 = e^{0.08t}$

$\ln 2 = 0.08t$

$t = \dfrac{\ln 2}{0.08} \approx 8.66$

Compounded continuously, it will take about 8.66 years (or 103.97 months) to double.

33. $150 = 100\left(1+\frac{0.08}{12}\right)^{12t}$

$1.5 \approx (1.006667)^{12t}$

$\ln 1.5 \approx 12t \ln(1.006667)$

$t \approx \dfrac{\ln 1.5}{12 \ln(1.006667)} \approx 5.09$

Compounded monthly, it will take about 5.09 years (or 61.02 months).

$150 = 100e^{0.08t}$

$1.5 = e^{0.08t}$

$\ln 1.5 = 0.08t$

$t = \dfrac{\ln 1.5}{0.08} \approx 5.07$

Compounded continuously, it will take about 5.07 years (or 60.82 months).

35. $25,000 = 10,000e^{0.06t}$

$2.5 = e^{0.06t}$

$\ln 2.5 = 0.06t$

$t = \dfrac{\ln 2.5}{0.06} \approx 15.27$

It will take about 15.27 years (or 15 years, 4 months).

37. $A = 90,000(1+0.03)^5 = \$104,335$

39. $P = 15,000e^{(-0.05)(3)} \approx \$12,910.62$

41. $A = 15(1+0.15)^5 = 15(1.15)^5 \approx \$30.17$ per share for a total of about $3017.

43. $850,000 = 650,000(1+r)^3$

$\dfrac{85}{65} = (1+r)^3$

$\sqrt[3]{\dfrac{85}{65}} = 1+r$

$r \approx \sqrt[3]{1.3077} - 1 \approx 0.0935$

The annual return is approximately 9.35%.

45. 5.6% compounded continuously:
$A = 1000e^{(0.056)(1)} = \$1057.60$
Jim will not have enough money to buy the computer.

5.9% compounded monthly:
$$A = 1000\left(1+\frac{0.059}{12}\right)^{12} = \$1060.62$$
The second bank offers the better deal.

47. Will: 9% compounded semiannually:
$$A = 2000\left(1+\frac{0.09}{2}\right)^{(2)(20)} = \$11,632.73$$

Henry: 8.5% compounded continuously:
$A = 2000e^{(0.085)(20)} = \$10,947.89$

Will has more money after 20 years.

49. $P = 50{,}000;\ t = 5$

  a. Simple interest at 12% per annum:
  $A = 50{,}000 + 50{,}000(0.12)(5) = \$80{,}000$
  $I = \$80{,}000 - \$50{,}000 = \$30{,}000$

  b. 11.5% compounded monthly:
  $A = 50{,}000\left(1 + \dfrac{0.115}{12}\right)^{(12)(5)} = \$88{,}613.59$
  $I = \$88{,}613.59 - \$50{,}000 = \$38{,}613.59$

  c. 11.25% compounded continuously:
  $A = 50{,}000 e^{(0.1125)(5)} = \$87{,}752.73$
  $I = \$87{,}752.73 - \$50{,}000 = \$37{,}752.73$

  Thus, simple interest at 12% is the best option since it results in the least interest.

51. **a.** $A = \$10{,}000,\ r = 0.10,\ n = 12,\ t = 20$
  $P = 10{,}000\left(1 + \dfrac{0.10}{12}\right)^{(-12)(20)} \approx \$1364.62$

  b. $A = \$10{,}000,\ r = 0.10,\ t = 20$
  $P = 10{,}000 e^{(-0.10)(20)} \approx \$1353.35$

53. $A = \$10{,}000,\ r = 0.08,\ n = 1,\ t = 10$
  $P = 10{,}000\left(1 + \dfrac{0.08}{1}\right)^{(-1)(10)} \approx \$4631.93$

55. **a.** $t = \dfrac{\ln 2}{1 \cdot \ln\left(1 + \dfrac{0.12}{1}\right)} = \dfrac{\ln 2}{\ln(1.12)} \approx 6.12$ years

  b. $t = \dfrac{\ln 3}{4 \cdot \ln\left(1 + \dfrac{0.06}{4}\right)} = \dfrac{\ln 3}{4\ln(1.015)} \approx 18.45$ years

  c. $mP = P\left(1 + \dfrac{r}{n}\right)^{nt}$
  $m = \left(1 + \dfrac{r}{n}\right)^{nt}$
  $\ln m = nt \cdot \ln\left(1 + \dfrac{r}{n}\right)$
  $t = \dfrac{\ln m}{n \cdot \ln\left(1 + \dfrac{r}{n}\right)}$

57. Answers will vary.

59. Answers will vary.

Section 4.8

1. $P(t) = 500 e^{0.02t}$

  a. $P(0) = 500 e^{(0.02)(0)} = 500$ insects

  b. growth rate = 2 %

  c.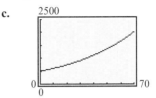

  d. $P(10) = 500 e^{(0.02)(10)} \approx 611$ insects

  e. Find t when $P = 800$:
  $800 = 500 e^{0.02t}$
  $1.6 = e^{0.02t}$
  $\ln 1.6 = 0.02 t$
  $t = \dfrac{\ln 1.6}{0.02} \approx 23.5$ days

  f. Find t when $P = 1000$:
  $1000 = 500 e^{0.02t}$
  $2 = e^{0.02t}$
  $\ln 2 = 0.02 t$
  $t = \dfrac{\ln 2}{0.02} \approx 34.7$ days

3. $A(t) = A_0 e^{-0.0244t} = 500 e^{-0.0244t}$

  a. decay rate = 2.44 %

  b.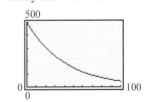

  c. $A(10) = 500 e^{(-0.0244)(10)} \approx 391.7$ grams

  d. Find t when $A = 400$:
  $400 = 500 e^{-0.0244t}$
  $0.8 = e^{-0.0244t}$
  $\ln 0.8 = -0.0244 t$
  $t = \dfrac{\ln 0.8}{-0.0244} \approx 9.1$ years

e. Find t when $A = 250$:
$$250 = 500e^{-0.0244t}$$
$$0.5 = e^{-0.0244t}$$
$$\ln 0.5 = -0.0244t$$
$$t = \frac{\ln 0.5}{-0.0244} \approx 28.4 \text{ years}$$

5. Use $N(t) = N_0 e^{kt}$ and solve for k:
$$1800 = 1000e^{k(1)}$$
$$1.8 = e^k$$
$$k = \ln 1.8$$
When $t = 3$:
$$N(3) = 1000e^{(\ln 1.8)(3)} = 5832 \text{ mosquitos}$$
Find t when $N(t) = 10,000$:
$$10,000 = 1000e^{(\ln 1.8)t}$$
$$10 = e^{(\ln 1.8)t}$$
$$\ln 10 = (\ln 1.8)t$$
$$t = \frac{\ln 10}{\ln 1.8} \approx 3.9 \text{ days}$$

7. Use $P(t) = P_0 e^{kt}$ and solve for k:
$$2P_0 = P_0 e^{k(1.5)}$$
$$2 = e^{1.5k}$$
$$\ln 2 = 1.5k$$
$$k = \frac{\ln 2}{1.5}$$
When $t = 2$:
$$P(2) = 10,000 e^{\left(\frac{\ln 2}{1.5}\right)(2)} \approx 25,198 \text{ will be the population 2 years from now.}$$

9. Use $A = A_0 e^{kt}$ and solve for k:
$$0.5 A_0 = A_0 e^{k(1690)}$$
$$0.5 = e^{1690k}$$
$$\ln 0.5 = 1690k$$
$$k = \frac{\ln 0.5}{1690}$$
When $A_0 = 10$ and $t = 50$:
$$A = 10 e^{\left(\frac{\ln 0.5}{1690}\right)(50)} \approx 9.797 \text{ grams}$$

11. a. Use $A = A_0 e^{kt}$ and solve for k:
   half-life $= 5600$ years
   $$0.5 A_0 = A_0 e^{k(5600)}$$
   $$0.5 = e^{5600k}$$
   $$\ln 0.5 = 5600k$$
   $$k = \frac{\ln 0.5}{5600}$$
   Solve for t when $A = 0.3 A_0$:
   $$0.3 A_0 = A_0 e^{\left(\frac{\ln 0.5}{5600}\right)t}$$
   $$0.3 = e^{\left(\frac{\ln 0.5}{5600}\right)t}$$
   $$\ln 0.3 = \left(\frac{\ln 0.5}{5600}\right)t$$
   $$t = \frac{\ln 0.3}{\left(\frac{\ln 0.5}{5600}\right)} \approx 9727$$
   The tree died approximately 9727 years ago.

   b. $Y_1 = e^{\left(\frac{\ln 0.5}{5600}\right)t}$
   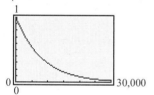

   c. $Y_1 = e^{\left(\frac{\ln 0.5}{5600}\right)t}$; $Y_2 = 0.5$
   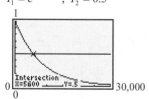
   Thus, 5600 years will elapse until half of the carbon 14 remains.

   d. $Y_1 = e^{\left(\frac{\ln 0.5}{5600}\right)t}$; $Y_2 = 0.3$
   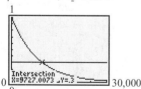
   This verifies that the tree died approximately 9727 years ago.

13. a. Using $u = T + (u_0 - T)e^{kt}$ with $t = 5$, $T = 70$, $u_0 = 450$, and $u = 300$:

$$300 = 70 + (450 - 70)e^{k(5)}$$
$$230 = 380e^{5k}$$
$$\frac{230}{380} = e^{5k}$$
$$\ln\left(\frac{23}{38}\right) = 5k$$
$$k = \frac{\ln\left(\frac{23}{38}\right)}{5} \approx -0.1004$$

$T = 70$, $u_0 = 450$, $u = 135$:

$$135 = 70 + (450 - 70)e^{\frac{\ln(23/38)}{5}t}$$
$$65 = 380e^{\frac{\ln(23/38)}{5}t}$$
$$\frac{65}{380} = e^{\frac{\ln(23/38)}{5}t}$$
$$\ln\left(\frac{65}{380}\right) = \frac{\ln(23/38)}{5}t$$
$$t = \frac{\ln(65/380)}{\left(\frac{\ln(23/38)}{5}\right)} \approx 18 \text{ minutes}$$

The pizza will be cool enough to eat at about 5:18 PM.

b. $Y_1 = 70 + (450 - 70)e^{\frac{\ln(23/38)}{5}x}$

c. $Y_1 = 70 + (450 - 70)e^{\frac{\ln(23/38)}{5}x}$; $Y_2 = 160$

Intersection X=14.343604, Y=160

The pizza will be 160°F after about 14.3 minutes.

d. As time passes, the temperature gets closer to 70°F.

15. a. Using $u = T + (u_0 - T)e^{kt}$ with $t = 3$, $T = 35$, $u_0 = 8$, and $u = 15$:

$$15 = 35 + (8 - 35)e^{k(3)}$$
$$-20 = -27e^{3k}$$
$$\frac{20}{27} = e^{3k}$$
$$\ln\left(\frac{20}{27}\right) = 3k$$
$$k = \frac{\ln(20/27)}{3}$$

At $t = 5$: $u = 35 + (8 - 35)e^{\left(\frac{\ln(20/27)}{3}\right)(5)} \approx 18.63°C$

After 5 minutes, the thermometer will read approximately $18.63°C$.

At $t = 10$: $u = 35 + (8 - 35)e^{\left(\frac{\ln(20/27)}{3}\right)(10)} \approx 25.1°C$

After 10 minutes, the thermometer will read approximately 25.1°C

b. $Y_1 = 35 + (8 - 35)e^{\left(\frac{\ln(20/27)}{3}\right)x}$

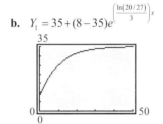

17. Use $A = A_0 e^{kt}$ and solve for k:

$$15 = 25e^{k(10)}$$
$$0.6 = e^{10k}$$
$$\ln 0.6 = 10k$$
$$k = \frac{\ln 0.6}{10}$$

When $A_0 = 25$ and $t = 24$:

$$A = 25e^{\left(\frac{\ln 0.6}{10}\right)(24)} \approx 7.34$$

There will be about 7.34 kilograms of salt left after 1 day.

Find t when $A = 0.5A_0$:

$$0.5 = 25e^{\left(\frac{\ln 0.6}{10}\right)t}$$
$$0.02 = e^{\left(\frac{\ln 0.6}{10}\right)t}$$
$$\ln 0.02 = \left(\frac{\ln 0.6}{10}\right)t$$
$$t = \frac{\ln 0.02}{\left(\frac{\ln 0.6}{10}\right)} \approx 76.6$$

It will take about 76.6 hours until ½ kilogram of salt is left.

19. Use $A = A_0 e^{kt}$ and solve for k:

$$0.5A_0 = A_0 e^{k(8)}$$
$$0.5 = e^{8k}$$
$$\ln 0.5 = 8k$$
$$k = \frac{\ln 0.5}{8}$$

Find t when $A = 0.1A_0$:

$$0.1A_0 = A_0 e^{\left(\frac{\ln 0.5}{8}\right)t}$$
$$0.1 = e^{\left(\frac{\ln 0.5}{8}\right)t}$$
$$\ln 0.1 = \left(\frac{\ln 0.5}{8}\right)t$$
$$t = \frac{\ln 0.1}{\left(\frac{\ln 0.5}{8}\right)} \approx 26.6$$

The farmers need to wait about 26.6 days before using the hay.

21. **a.** The maximum proportion is the carrying capacity, $c = 0.9 = 90\%$.

b. $P(0) = \dfrac{0.9}{1 + 6e^{-0.32(0)}} = \dfrac{0.9}{1 + 6 \cdot 1} = \dfrac{0.9}{7} = 0.1286$

In 1984, about 12.86% of U.S. households owned a VCR.

c. $Y_1 = \dfrac{0.9}{1 + 6e^{-0.32x}}$

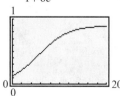

d. $t = 1999 - 1984 = 15$

$P(15) = \dfrac{0.9}{1 + 6e^{-0.32(15)}} = \dfrac{0.9}{1 + 6e^{-4.8}} \approx 0.8577$

In 1999, about 85.77% of U.S. households owned a VCR.

e. We need to find t such that $P = 0.8$.

$Y_1 = \dfrac{0.9}{1 + 6e^{-0.32x}}$; $Y_2 = 0.8$:

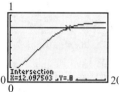

Thus, $t \approx 12.1$. Since $1984 + 12.1 = 1996.1$, 80% of households owned a VCR in 1996.

f. We need to find t such that $P = 0.45$.

$Y_1 = \dfrac{0.9}{1 + 6e^{-0.32x}}$; $Y_2 = 0.45$:

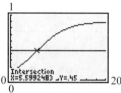

Thus, $t \approx 5.6$. About 5.6 years after 1984 (i.e. in about mid 1989), 45% of the population owned a VCR.

23. **a.** As $t \to \infty$, $e^{-0.439t} \to 0$. Thus, $P(t) \to 1000$.

The carrying capacity is 1000 grams of bacteria.

b. Growth rate $= 0.439 = 43.9\%$.

c. $P(0) = \dfrac{1000}{1 + 32.33e^{-0.439(0)}} = \dfrac{1000}{33.33} = 30$

The initial population was 30 grams of bacteria.

d. $Y_1 = \dfrac{1000}{1 + 32.33e^{-0.439x}}$

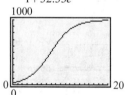

e. $P(9) = \dfrac{1000}{1+32.33e^{-0.439(9)}} \approx 616.8$

After 9 hours, the population of bacteria will be about 616.8 grams.

f. We need to find t such that $P = 700$:

$Y_1 = \dfrac{1000}{1+32.33e^{-0.439x}}$; $Y_2 = 700$

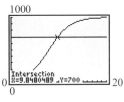

Thus, $t \approx 9.85$. The population of bacteria will be 700 grams after about 9.85 hours.

g. We need to find t such that

$P = \dfrac{1}{2}(1000) = 500$:

$Y_1 = \dfrac{1000}{1+32.33e^{-0.439x}}$; $Y_2 = 500$

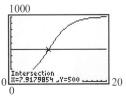

Thus, $t \approx 7.9$. The population of bacteria will reach one-half of is carrying capacity after about 7.9 hours.

25. a. $y = \dfrac{6}{1+e^{-(5.085-0.1156(100))}} \approx 0.0092$

At $100°F$, the predicted number of eroded or leaky primary O-rings will be about 0.

b. $y = \dfrac{6}{1+e^{-(5.085-0.1156(60))}} \approx 0.81$

At $60°F$, the predicted number of eroded or leaky primary O-rings will be about 1.

c. $y = \dfrac{6}{1+e^{-(5.085-0.1156(30))}} \approx 5.01$

At $30°F$, the predicted number of eroded or leaky primary O-rings will be about 5.

d. $Y_1 = \dfrac{6}{1+e^{-(5.085-0.1156x)}}$

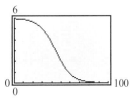

Use INTERSECT with $Y_2 = 1, 3,$ and 5:

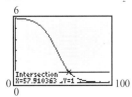

The predicted number of eroded or leaky O-rings is 1 when the temperature is about $57.91°F$.

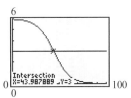

The predicted number of eroded or leaky O-rings is 3 when the temperature is about $43.99°F$.

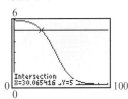

The predicted number of eroded or leaky O-rings is 5 when the temperature is about $37.07°F$.

27. a. $P(0) = \dfrac{95.4993}{1+0.0405e^{0.1968(0)}} = \dfrac{95.4993}{1.0405} \approx 91.8$

In 1984, about 91.8% of households did not own a personal computer.

b. $Y_1 = \dfrac{95.4993}{1+0.0405e^{0.1968x}}$

c. $t = 1995 - 1984 = 11$

$$P(11) = \frac{95.4993}{1 + 0.0405e^{0.1968(11)}} \approx 70.6$$

In 1995, about 70.6% of households did not own a personal computer.

d. We need to find t such that $P = 20$

$$Y_1 = \frac{95.4993}{1 + 0.0405e^{0.1968x}}; \; Y_2 = 20$$

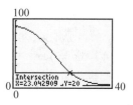

Thus, $t \approx 23$. Now, $1984 + 23 = 2007$. The percentage of households that do not own a personal computer will reach 20% during 2007.

29. a. $Y_1 = \dfrac{113.3198}{1 + 0.115e^{0.0912x}}$

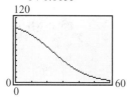

b. $P(15) = \dfrac{113.3198}{1 + 0.115e^{0.0912(15)}} \approx 78$

In a room of 15 people, the probability that no two people share the same birthday is about 78% or 0.78.

c. We need to find n such that $P = 10$.

$$Y_1 = \frac{113.3198}{1 + 0.115e^{0.0912x}}; \; Y_2 = 10$$

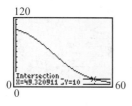

Thus, $t \approx 49.3$. The probability is 10% when about 50 people are in the room.

d. As $n \to \infty$, $1 + 0.115e^{0.0912n} \to \infty$. Thus, $P(n) \to 0$. This means that as the number of people in the room increases, the more likely it will be that two will share the same birthday.

Section 4.9

1. a.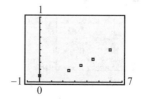

b. Using EXPonential REGression on the data yields: $y = 0.0903(1.3384)^x$

c. $y = 0.0903(1.3384)^x$

$= 0.0903\left(e^{\ln(1.3384)}\right)^x$

$= 0.0903 e^{\ln(1.3384)x}$

$N(t) = 0.0903 e^{0.2915t}$

d. $Y_1 = 0.0903 e^{0.2915x}$

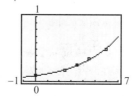

e. $N(7) = 0.0903 e^{(0.2915) \cdot 7} \approx 0.69$ bacteria

f. We need to find t when $N = 0.75$:

$0.0903 e^{(0.2915) \cdot t} = 0.75$

$e^{(0.2915) \cdot t} = \dfrac{0.75}{0.0903}$

$0.2915t = \ln\left(\dfrac{0.75}{0.0903}\right)$

$t \approx \dfrac{\ln\left(\dfrac{0.75}{0.0903}\right)}{0.2915} \approx 7.26$ hours

3. a.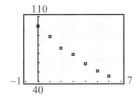

b. Using EXPonential REGression on the data yields: $y = 100.3263(0.8769)^x$

c. $y = 100.3263(0.8769)^x$
$= 100.3263\left(e^{\ln(0.8769)}\right)^x$
$= 100.3263 e^{\ln(0.8769)x}$
$A(t) = 100.3263 e^{(-0.1314)t}$

d. $Y_1 = 100.3263 e^{(-0.1314)x}$

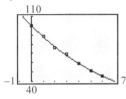

e. We need to find t when $A(t) = 0.5 \cdot A_0$
$100.3263 e^{(-0.1314)t} = (0.5)(100.3263)$
$e^{(-0.1314)t} = 0.5$
$-0.1314 t = \ln 0.5$
$t = \dfrac{\ln 0.5}{-0.1314} \approx 5.3$ weeks

f. $A(50) = 100.3263 e^{(-0.1314)\cdot 50} \approx 0.14$ grams

g. We need to find t when $A(t) = 20$.
$100.3263 e^{(-0.1314)t} = 20$
$e^{(-0.1314)t} = \dfrac{20}{100.3263}$
$-0.1314 t = \ln\left(\dfrac{20}{100.3263}\right)$
$t = \dfrac{\ln\left(\dfrac{20}{100.3263}\right)}{-0.1314}$
≈ 12.3 weeks

5. a. Let $x = 1$ correspond to 1994, $x = 2$ correspond to 1995, etc.

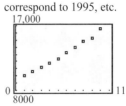

b. Using EXPonential REGression on the data yields: $y = 9478.4453(1.056554737)^x$

c. The average annual rate of return over the 10 years is
$1.056554737 - 1 \approx .0566 = 5.66\%$.

d. The year 2021 corresponds to $x = 28$, so
$y = 9478.4453(1.056554737)^{28} \approx 44{,}229.61$.
The value of the account in 2021 is about $\$44{,}229.61$.

e. We need to find x when $y = 50{,}000$:
$9478.4453(1.056554737)^x = 50{,}000$
$(1.056554737)^x = \dfrac{50{,}000}{9478.4453}$
$x \ln 1.056554737 = \ln\left(\dfrac{50{,}000}{9478.4453}\right)$
$x = \dfrac{\ln\left(\dfrac{50{,}000}{9478.4453}\right)}{\ln 1.056554737}$
≈ 30.23
The account will be worth $\$50{,}000$ early in the year 2023.

7. a.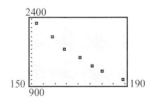

b. Using LnREGression on the data yields:
$y = 32{,}741.02 - 6070.96 \ln x$

c. $Y_1 = 32{,}741.02 - 6070.96 \ln x$

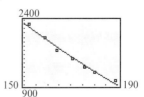

d. We need to find x when $y = 1650$:
$$1650 = 32,741.02 - 6070.96 \ln x$$
$$-31,091.02 = -6070.96 \ln x$$
$$\frac{-31,091.02}{-6070.96} = \ln x$$
$$5.1213 \approx \ln x$$
$$e^{5.1213} \approx x$$
$$x \approx 168$$

If the price were $1650, then approximately 168 computers would be demanded.

9. a. Let $x = 0$ correspond to 1900, $x = 10$ correspond to 1910, $x = 20$ correspond to 1920, etc.

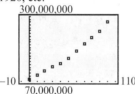

b. Using LOGISTIC REGression on the data yields:
$$y = \frac{799,475,916.5}{1 + 9.1968 e^{-0.01603x}}$$

c. $Y_1 = \frac{799,475,916.5}{1 + 9.1968 e^{-0.01603x}}$:

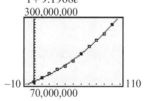

d. As $x \to \infty$, $9.1968 e^{-0.01603x} \to 0$, which means $1 + 9.1968 e^{-0.01603x} \to 1$, so
$$y = \frac{799,475,916.5}{1 + 9.1968 e^{-0.01603x}} \to 799,475,916.5$$

Therefore, the carrying capacity of the United States is approximately 799,475,917 people.

e. The year 2004 corresponds to $x = 104$, so
$$y = \frac{799,475,916.5}{1 + 9.1968 e^{-0.01603(104)}}$$
$$\approx 292,183,980 \text{ people}$$

f. Find x when $y = 300,000,000$
$$\frac{799,475,916.5}{1 + 9.1968 e^{-0.01603x}} = 300,000,000$$
$$799,475,916.5 = 300,000,000 \left(1 + 9.1968 e^{-0.01603x}\right)$$
$$\frac{799,475,916.5}{300,000,000} = 1 + 9.1968 e^{-0.01603x}$$
$$\frac{799,475,916.5}{300,000,000} - 1 = 9.1968 e^{-0.01603x}$$
$$1.6649 \approx 9.1968 e^{-0.01603x}$$
$$\frac{1.6649}{9.1968} \approx e^{-0.01603x}$$
$$\ln\left(\frac{1.6649}{9.1968}\right) \approx -0.01603x$$
$$\frac{\ln\left(\frac{1.6649}{9.1968}\right)}{-0.01603} \approx x$$
$$x \approx 107$$

Therefore, the United States population will be 300,000,000 around the year 2007.

11. a. Let $x = 0$ correspond to 1900, $x = 10$ correspond to 1910, $x = 20$ correspond to 1920, etc.

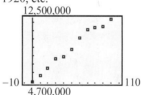

b. Using LOGISTIC REGression on the data yields: $y = \frac{14,471,245.24}{1 + 2.01527 e^{-0.02458x}}$

c. $Y_1 = \frac{14,471,245.24}{1 + 2.01527 e^{-0.02458x}}$

d. As $x \to \infty$, $2.01527 e^{-0.02458x} \to 0$, which means $1 + 2.01527 e^{-0.02458x} \to 1$, so
$$y = \frac{14,471,245.24}{1 + 2.01527 e^{-0.02458x}} \to 14,471,245.24$$

Therefore, the carrying capacity of Illinois is approximately 14,471,245 people.

e. The year 2010 corresponds to $x = 110$, so
$$y = \frac{14,471,245.24}{1 + 2.01527e^{-0.02458(110)}} \approx 12,750,854.$$
In 2010, the population of Illinois will be approximately 12,750,854 people.

Chapter 4 Review

1. $f(x) = 3x - 5 \qquad g(x) = 1 - 2x^2$

    a. $(f \circ g)(2) = f(g(2))$
    $= f(1 - 2(2)^2)$
    $= f(-7)$
    $= 3(-7) - 5$
    $= -26$

    b. $(g \circ f)(-2) = g(f(-2))$
    $= g(3(-2) - 5)$
    $= g(-11)$
    $= 1 - 2(-11)^2$
    $= -241$

    c. $(f \circ f)(4) = f(f(4))$
    $= f(3(4) - 5)$
    $= f(7)$
    $= 3(7) - 5$
    $= 16$

    d. $(g \circ g)(-1) = g(g(-1))$
    $= g(1 - 2(-1)^2)$
    $= g(-1)$
    $= 1 - 2(-1)^2$
    $= -1$

3. $f(x) = \sqrt{x+2} \qquad g(x) = 2x^2 + 1$

    a. $(f \circ g)(2) = f(g(2))$
    $= f(2(2)^2 + 1)$
    $= f(9)$
    $= \sqrt{9+2}$
    $= \sqrt{11}$

    b. $(g \circ f)(-2) = g(f(-2))$
    $= g(\sqrt{-2+2})$
    $= g(0)$
    $= 2(0)^2 + 1$
    $= 1$

    c. $(f \circ f)(4) = f(f(4))$
    $= f(\sqrt{4+2})$
    $= f(\sqrt{6})$
    $= \sqrt{\sqrt{6}+2}$

    d. $(g \circ g)(-1) = g(g(-1))$
    $= g(2(-1)^2 + 1)$
    $= g(3)$
    $= 2(3)^2 + 1$
    $= 19$

5. $f(x) = e^x \qquad g(x) = 3x - 2$

    a. $(f \circ g)(2) = f(g(2))$
    $= f(3(2) - 2)$
    $= f(4)$
    $= e^4$

    b. $(g \circ f)(-2) = g(f(-2))$
    $= g(e^{-2})$
    $= 3e^{-2} - 2$
    $= \frac{3}{e^2} - 2$

    c. $(f \circ f)(4) = f(f(4)) = f(e^4) = e^{e^4}$

    d. $(g \circ g)(-1) = g(g(-1))$
    $= g(3(-1) - 2)$
    $= g(-5)$
    $= 3(-5) - 2$
    $= -17$

Chapter 4: Exponential and Logarithmic Functions

7. $f(x) = 2 - x \qquad g(x) = 3x + 1$

    The domain of f is $\{x \mid x \text{ is any real number}\}$.

    The domain of g is $\{x \mid x \text{ is any real number}\}$.

    $(f \circ g)(x) = f(g(x))$
    $= f(3x + 1)$
    $= 2 - (3x + 1)$
    $= 2 - 3x - 1$
    $= 1 - 3x$

    Domain: $\{x \mid x \text{ is any real number}\}$.

    $(g \circ f)(x) = g(f(x))$
    $= g(2 - x)$
    $= 3(2 - x) + 1$
    $= 6 - 3x + 1$
    $= 7 - 3x$

    Domain: $\{x \mid x \text{ is any real number}\}$.

    $(f \circ f)(x) = f(f(x))$
    $= f(2 - x)$
    $= 2 - (2 - x)$
    $= 2 - 2 + x$
    $= x$

    Domain: $\{x \mid x \text{ is any real number}\}$.

    $(g \circ g)(x) = g(g(x))$
    $= g(3x + 1)$
    $= 3(3x + 1) + 1$
    $= 9x + 3 + 1$
    $= 9x + 4$

    Domain: $\{x \mid x \text{ is any real number}\}$.

9. $f(x) = 3x^2 + x + 1 \qquad g(x) = |3x|$

    The domain of f is $\{x \mid x \text{ is any real number}\}$.

    The domain of g is $\{x \mid x \text{ is any real number}\}$.

    $(f \circ g)(x) = f(g(x))$
    $= f(|3x|)$
    $= 3(|3x|)^2 + (|3x|) + 1$
    $= 27x^2 + 3|x| + 1$

    Domain: $\{x \mid x \text{ is any real number}\}$.

    $(g \circ f)(x) = g(f(x))$
    $= g(3x^2 + x + 1)$
    $= |3(3x^2 + x + 1)|$
    $= |3(3x^2 + x + 1)|$
    $= 3|3x^2 + x + 1|$

    Domain: $\{x \mid x \text{ is any real number}\}$.

    $(f \circ f)(x) = f(f(x))$
    $= f(3x^2 + x + 1)$
    $= 3(3x^2 + x + 1)^2 + (3x^2 + x + 1) + 1$
    $= 3(9x^4 + 6x^3 + 7x^2 + 2x + 1) + 3x^2 + x + 1 + 1$
    $= 27x^4 + 18x^3 + 24x^2 + 7x + 5$

    Domain: $\{x \mid x \text{ is any real number}\}$.

    $(g \circ g)(x) = g(g(x))$
    $= g(|3x|)$
    $= |3|3x||$
    $= 9|x|$

    Domain: $\{x \mid x \text{ is any real number}\}$.

11. $f(x) = \dfrac{x+1}{x-1} \qquad g(x) = \dfrac{1}{x}$

    The domain of f is $\{x \mid x \neq 1\}$.

    The domain of g is $\{x \mid x \neq 0\}$.

    $(f \circ g)(x) = f(g(x))$
    $= f\left(\dfrac{1}{x}\right)$
    $= \dfrac{\dfrac{1}{x} + 1}{\dfrac{1}{x} - 1}$
    $= \dfrac{\left(\dfrac{1}{x} + 1\right)x}{\left(\dfrac{1}{x} - 1\right)x}$
    $= \dfrac{1 + x}{1 - x}$

    Domain $\{x \mid x \neq 0, x \neq 1\}$.

$(g \circ f)(x) = g(f(x))$
$= g\left(\dfrac{x+1}{x-1}\right)$
$= \dfrac{1}{\left(\dfrac{x+1}{x-1}\right)}$
$= \dfrac{x-1}{x+1}$

Domain $\{x \mid x \neq -1,\, x \neq 1\}$

$(f \circ f)(x) = f(f(x))$
$= f\left(\dfrac{x+1}{x-1}\right)$
$= \dfrac{\dfrac{x+1}{x-1}+1}{\dfrac{x+1}{x-1}-1}$
$= \dfrac{\left(\dfrac{x+1}{x-1}+1\right)(x-1)}{\left(\dfrac{x+1}{x-1}-1\right)(x-1)}$
$= \dfrac{x+1+x-1}{x+1-(x-1)}$
$= \dfrac{2x}{2}$
$= x$

Domain $\{x \mid x \neq 1\}$.

$(g \circ g)(x) = g(g(x)) = g\left(\dfrac{1}{x}\right) = \dfrac{1}{\left(\dfrac{1}{x}\right)} = x$

Domain $\{x \mid x \neq 0\}$.

13. **a.** The function is one-to-one because there are no two distinct inputs that correspond to the same output.

   b. The inverse is $\{(2,1),(5,3),(8,5),(10,6)\}$.

15. The function f is one-to-one because every horizontal line intersects the graph at exactly one point.

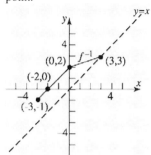

17. $f(x) = \dfrac{2x+3}{5x-2}$

$y = \dfrac{2x+3}{5x-2}$

$x = \dfrac{2y+3}{5y-2}$ Inverse

$x(5y-2) = 2y+3$

$5xy - 2x = 2y + 3$

$5xy - 2y = 2x + 3$

$y(5x-2) = 2x+3$

$y = \dfrac{2x+3}{5x-2}$

$f^{-1}(x) = \dfrac{2x+3}{5x-2}$

Domain of f = Range of f^{-1}
 = All real numbers except $\dfrac{2}{5}$.

Range of f = Domain of f^{-1}
 = All real numbers except $\dfrac{2}{5}$.

19. $f(x) = \dfrac{1}{x-1}$

$y = \dfrac{1}{x-1}$

$x = \dfrac{1}{y-1}$ Inverse

$x(y-1) = 1$

$xy - x = 1$

$xy = x + 1$

$y = \dfrac{x+1}{x}$

$f^{-1}(x) = \dfrac{x+1}{x}$

Domain of f = Range of f^{-1}
= All real numbers except 1
Range of f = Domain of f^{-1}
= All real numbers except 0

21. $f(x) = \dfrac{3}{x^{1/3}}$

$y = \dfrac{3}{x^{1/3}}$

$x = \dfrac{3}{y^{1/3}}$ Inverse

$xy^{1/3} = 3$

$y^{1/3} = \dfrac{3}{x}$

$y = \left(\dfrac{3}{x}\right)^3 = \dfrac{27}{x^3}$

$f^{-1}(x) = \dfrac{27}{x^3}$

Domain of f = Range of f^{-1}
= All real numbers except 0
Range of f = Domain of f^{-1}
= All real numbers except 0

23. a. $f(4) = 3^4 = 81$

b. $g(9) = \log_3(9) = \log_3(3^2) = 2$

c. $f(-2) = 3^{-2} = \dfrac{1}{9}$

d. $g\left(\dfrac{1}{27}\right) = \log_3\left(\dfrac{1}{27}\right) = \log_3(3^{-3}) = -3$

25. $5^2 = z$ is equivalent to $2 = \log_5 z$

27. $\log_5 u = 13$ is equivalent to $5^{13} = u$

29. $f(x) = \log(3x - 2)$ requires:
$3x - 2 > 0$

$x > \dfrac{2}{3}$

Domain: $\left\{x \,\middle|\, x > \dfrac{2}{3}\right\}$

31. $H(x) = \log_2(x^2 - 3x + 2)$ requires
$p(x) = x^2 - 3x + 2 > 0$
$(x-2)(x-1) > 0$
$x = 2$ and $x = 1$ are the zeros of p.

Interval	$(-\infty, 1)$	$(1, 2)$	$(2, \infty)$
Test Value	0	$\dfrac{3}{2}$	3
Value of p	2	$-\dfrac{1}{4}$	2
Conclusion	positive	negative	positive

Thus, the domain of $H(x) = \log_2(x^2 - 3x + 2)$ is $\{x \,|\, x < 1 \text{ or } x > 2\}$.

33. $\log_2\left(\dfrac{1}{8}\right) = \log_2 2^{-3} = -3\log_2 2 = -3$

35. $\ln e^{\sqrt{2}} = \sqrt{2}$

37. $2^{\log_2 0.4} = 0.4$

39. $\log_3\left(\dfrac{uv^2}{w}\right) = \log_3 uv^2 - \log_3 w$

$= \log_3 u + \log_3 v^2 - \log_3 w$

$= \log_3 u + 2\log_3 v - \log_3 w$

41. $\log\left(x^2\sqrt{x^3+1}\right) = \log x^2 + \log\left(x^3+1\right)^{1/2}$

$= 2\log x + \dfrac{1}{2}\log\left(x^3+1\right)$

43. $\ln\left(\dfrac{x\sqrt[3]{x^2+1}}{x-3}\right) = \ln\left(x\sqrt[3]{x^2+1}\right) - \ln(x-3)$

$\qquad = \ln x + \ln\left(x^2+1\right)^{1/3} - \ln(x-3)$

$\qquad = \ln x + \dfrac{1}{3}\ln\left(x^2+1\right) - \ln(x-3)$

45. $3\log_4 x^2 + \dfrac{1}{2}\log_4 \sqrt{x} = \log_4\left(x^2\right)^3 + \log_4\left(x^{1/2}\right)^{1/2}$

$\qquad = \log_4 x^6 + \log_4 x^{1/4}$

$\qquad = \log_4\left(x^6 \cdot x^{1/4}\right)$

$\qquad = \log_4 x^{25/4}$

$\qquad = \dfrac{25}{4}\log_4 x$

47. $\ln\left(\dfrac{x-1}{x}\right) + \ln\left(\dfrac{x}{x+1}\right) - \ln\left(x^2-1\right)$

$\qquad = \ln\left(\dfrac{x-1}{x} \cdot \dfrac{x}{x+1}\right) - \ln\left(x^2-1\right)$

$\qquad = \ln\left[\dfrac{\dfrac{x-1}{x+1}}{x^2-1}\right]$

$\qquad = \ln\left(\dfrac{x-1}{x+1} \cdot \dfrac{1}{(x-1)(x+1)}\right)$

$\qquad = \ln\dfrac{1}{(x+1)^2}$

$\qquad = \ln(x+1)^{-2}$

$\qquad = -2\ln(x+1)$

49. $2\log 2 + 3\log x - \dfrac{1}{2}\left[\log(x+3) + \log(x-2)\right]$

$\qquad = \log 2^2 + \log x^3 - \dfrac{1}{2}\log\left[(x+3)(x-2)\right]$

$\qquad = \log\left(4x^3\right) - \log\left((x+3)(x-2)\right)^{1/2}$

$\qquad = \log\left(\dfrac{4x^3}{\left[(x+3)(x-2)\right]^{1/2}}\right)$

51. $\log_4 19 = \dfrac{\log 19}{\log 4} \approx 2.124$

53. $Y_1 = \log_3 x = \dfrac{\ln x}{\ln 3}$

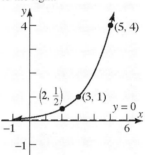

55. $f(x) = 2^{x-3}$

Using the graph of $y = 2^x$, shift the graph 3 units to the right.

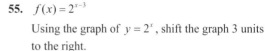

Domain: $(-\infty, \infty)$
Range: $(0, \infty)$
Horizontal Asymptote: $y = 0$

57. $f(x) = \dfrac{1}{2}\left(3^{-x}\right)$

Using the graph of $y = 3^x$, reflect the graph about the y-axis, and shrink vertically by a factor of $\dfrac{1}{2}$.

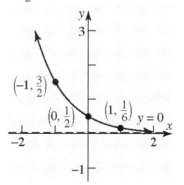

Domain: $(-\infty, \infty)$
Range: $(0, \infty)$
Horizontal Asymptote: $y = 0$

59. $f(x) = 1 - e^x$

Using the graph of $y = e^x$, reflect about the x-axis, and shift up 1 unit.

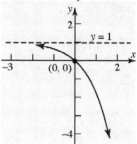

Domain: $(-\infty, \infty)$
Range: $(-\infty, 1)$
Horizontal Asymptote: $y = 1$

61. $f(x) = \dfrac{1}{2} \ln x$

Using the graph of $y = \ln x$, shrink vertically by a factor of $\dfrac{1}{2}$.

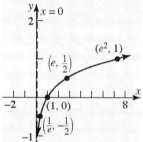

Domain: $(0, \infty)$
Range: $(-\infty, \infty)$
Vertical Asymptote: $x = 0$

63. $f(x) = 3 - e^{-x}$

Using the graph of $y = e^x$, reflect the graph about the y-axis, reflect about the x-axis, and shift up 3 units.

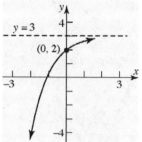

Domain: $(-\infty, \infty)$
Range: $(-\infty, 3)$
Horizontal Asymptote: $y = 3$

65.
$$4^{1-2x} = 2$$
$$\left(2^2\right)^{1-2x} = 2$$
$$2^{2-4x} = 2^1$$
$$2 - 4x = 1$$
$$-4x = -1$$
$$x = \dfrac{1}{4}$$

The solution set is $\left\{\dfrac{1}{4}\right\}$.

67.
$$3^{x^2+x} = \sqrt{3}$$
$$3^{x^2+x} = 3^{1/2}$$
$$x^2 + x = \dfrac{1}{2}$$
$$2x^2 + 2x - 1 = 0$$
$$x = \dfrac{-2 \pm \sqrt{2^2 - 4(2)(-1)}}{2(2)}$$
$$= \dfrac{-2 \pm \sqrt{12}}{4}$$
$$= \dfrac{-2 \pm 2\sqrt{3}}{4}$$
$$= \dfrac{-1 \pm \sqrt{3}}{2}$$

The solution set is $\left\{\dfrac{-1-\sqrt{3}}{2}, \dfrac{-1+\sqrt{3}}{2}\right\}$.

69. $\log_x 64 = -3$

$x^{-3} = 64$

$(x^{-3})^{-1/3} = 64^{-1/3}$

$x = \dfrac{1}{\sqrt[3]{64}} = \dfrac{1}{4}$

The solution set is $\left\{\dfrac{1}{4}\right\}$.

71. $5^x = 3^{x+2}$

$\log(5^x) = \log(3^{x+2})$

$x \log 5 = (x+2) \log 3$

$x \log 5 = x \log 3 + 2 \log 3$

$x \log 5 - x \log 3 = 2 \log 3$

$x(\log 5 - \log 3) = 2 \log 3$

$x = \dfrac{2 \log 3}{\log 5 - \log 3} \approx 4.301$

The solution set is $\left\{\dfrac{2 \log 3}{\log 5 - \log 3} \approx 4.301\right\}$.

73. $9^{2x} = 27^{3x-4}$

$(3^2)^{2x} = (3^3)^{3x-4}$

$3^{4x} = 3^{9x-12}$

$4x = 9x - 12$

$-5x = -12$

$x = \dfrac{12}{5}$

The solution set is $\left\{\dfrac{12}{5}\right\}$.

75. $\log_3 \sqrt{x-2} = 2$

$\sqrt{x-2} = 3^2$

$\sqrt{x-2} = 9$

$x - 2 = 9^2$

$x - 2 = 81$

$x = 83$

Check: $\log_3 \sqrt{83-2} = \log_3 \sqrt{81}$

$= \log_3 9$

$= 2$

The solution set is $\{83\}$.

77. $8 = 4^{x^2} \cdot 2^{5x}$

$2^3 = (2^2)^{x^2} \cdot 2^{5x}$

$2^3 = 2^{2x^2 + 5x}$

$3 = 2x^2 + 5x$

$0 = 2x^2 + 5x - 3$

$0 = (2x - 1)(x + 3)$

$x = \dfrac{1}{2}$ or $x = -3$

The solution set is $\left\{-3, \dfrac{1}{2}\right\}$.

79. $\log_6(x+3) + \log_6(x+4) = 1$

$\log_6((x+3)(x+4)) = 1$

$(x+3)(x+4) = 6^1$

$x^2 + 7x + 12 = 6$

$x^2 + 7x + 6 = 0$

$(x+6)(x+1) = 0$

$x = -6$ or $x = -1$

Since $\log_6(-6+3) = \log_6(-3)$ is undefined, the solution set is $\{-1\}$.

81. $e^{1-x} = 5$

$1 - x = \ln 5$

$-x = -1 + \ln 5$

$x = 1 - \ln 5 \approx -0.609$

The solution set is $\{1 - \ln 5 \approx -0.609\}$.

83. $2^{3x} = 3^{2x+1}$

$\ln 2^{3x} = \ln 3^{2x+1}$

$3x \ln 2 = (2x+1) \ln 3$

$3x \ln 2 = 2x \ln 3 + \ln 3$

$3x \ln 2 - 2x \ln 3 = \ln 3$

$x(3 \ln 2 - 2 \ln 3) = \ln 3$

$x = \dfrac{\ln 3}{3 \ln 2 - 2 \ln 3} \approx -9.327$

The solution set is $\left\{\dfrac{\ln 3}{3 \ln 2 - 2 \ln 3} \approx -9.327\right\}$.

85. $h(300) = (30(0) + 8000) \log\left(\dfrac{760}{300}\right)$

≈ 3229.5 meters

87. $P = 25e^{0.1d}$

   a. $P = 25e^{0.1(4)} = 25e^{0.4} \approx 37.3$ watts

   b. $50 = 25e^{0.1d}$
   $2 = e^{0.1d}$
   $\ln 2 = 0.1d$
   $d = \dfrac{\ln 2}{0.1} \approx 6.9$ decibels

89. a. $n = \dfrac{\log 10{,}000 - \log 90{,}000}{\log(1 - 0.20)} \approx 9.85$ years

   b. $n = \dfrac{\log(0.5i) - \log(i)}{\log(1 - 0.15)}$
   $= \dfrac{\log\left(\dfrac{0.5i}{i}\right)}{\log 0.85}$
   $= \dfrac{\log 0.5}{\log 0.85}$
   ≈ 4.27 years

91. $P = A\left(1 + \dfrac{r}{n}\right)^{-nt}$
   $= 85{,}000\left(1 + \dfrac{0.04}{2}\right)^{-2(18)}$
   $\approx \$41{,}668.97$

93. $A = A_0 e^{kt}$
   $0.5 A_0 = A_0 e^{k(5600)}$
   $0.5 = e^{5600k}$
   $\ln 0.5 = 5600k$
   $k = \dfrac{\ln 0.5}{5600}$
   $0.05 A_0 = A_0 e^{\left(\frac{\ln 0.5}{5600}\right)t}$
   $0.05 = e^{\left(\frac{\ln 0.5}{5600}\right)t}$
   $\ln 0.05 = \left(\dfrac{\ln 0.5}{5600}\right)t$
   $t = \dfrac{\ln 0.05}{\left(\dfrac{\ln 0.5}{5600}\right)} \approx 24{,}203$

The man died approximately 24,203 years ago.

95. $P = P_0 e^{kt}$
   $= 6{,}302{,}486{,}693 e^{0.0167(7)}$
   $\approx 6{,}835{,}600{,}129$ people

97. a. $P(0) = \dfrac{0.8}{1 + 1.67 e^{-0.16(0)}} = \dfrac{0.8}{1 + 1.67} \approx 0.3$
   In 2003, about 30% of cars had a GPS.

   b. The maximum proportion is the carrying capacity, $c = 0.8 = 80\%$.

   c. $Y_1 = \dfrac{0.8}{1 + 1.67 e^{-0.16x}}$

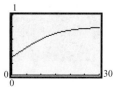

   d. Find t such that $P(t) = 0.75$.
   $\dfrac{0.8}{1 + 1.67 e^{-0.16t}} = 0.75$
   $0.8 = 0.75\left(1 + 1.67 e^{-0.16t}\right)$
   $\dfrac{0.8}{0.75} = 1 + 1.67 e^{-0.16t}$
   $\dfrac{0.8}{0.75} - 1 = 1.67 e^{-0.16t}$
   $\dfrac{\dfrac{0.8}{0.75} - 1}{1.67} = e^{-0.16t}$
   $\ln\left(\dfrac{\dfrac{0.8}{0.75} - 1}{1.67}\right) = -0.16t$
   $t = \dfrac{\ln\left(\dfrac{\dfrac{0.8}{0.75} - 1}{1.67}\right)}{-0.16} \approx 20.13$

Note that $2003 + 20.13 = 2023.13$, so 75% of new cars will have GPS in 2023.

99. a.

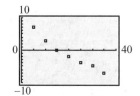

   b. Using LnREGression on the data yields: $y = 18.9028 - 7.0963 \ln x$

c. $Y_1 = 18.9028 - 7.0963 \ln x$

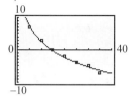

d. If $x = 23$, then
$y = 18.9028 - 7.0963 \ln 23 \approx -3°\text{F}$.

Chapter 4 Test

1. $f(x) = \dfrac{x+2}{x-2}$ $g(x) = 2x+5$

The domain of f is $\{x \mid x \neq 2\}$.
The domain of g is all real numbers.

a. $(f \circ g)(x) = f(g(x))$
$= f(2x+5)$
$= \dfrac{(2x+5)+2}{(2x+5)-2}$
$= \dfrac{2x+7}{2x+3}$

Domain $\left\{x \mid x \neq -\dfrac{3}{2}\right\}$.

b. $(g \circ f)(x) = g(f(-2))$
$= g\left(\dfrac{-2+2}{-2-2}\right)$
$= g(0)$
$= 2(0)+5$
$= 5$

c. $(f \circ g)(x) = f(g(-2))$
$= f(2(-2)+5)$
$= f(1)$
$= \dfrac{1+2}{1-2}$
$= \dfrac{3}{-1}$
$= -3$

2. a. Graph $y = 4x^2 + 3$:

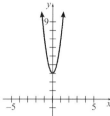

The function is not one-to-one because it fails the horizontal line test. A horizontal line (for example, $y = 4$) intersects the graph twice.

b. Graph $y = \sqrt{x+3} - 5$:

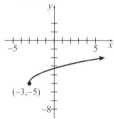

The function is one-to-one because it passes the horizontal line test. Every horizontal line intersects the graph at most once.

3. $f(x) = \dfrac{2}{3x-5}$

$y = \dfrac{2}{3x-5}$

$x = \dfrac{2}{3y-5}$ Inverse

$x(3y-5) = 2$
$3xy - 5x = 2$
$3xy = 5x + 2$
$y = \dfrac{5x+2}{3x}$
$f^{-1}(x) = \dfrac{5x+2}{3x}$

Domain of f = Range of f^{-1}
= All real numbers except $\dfrac{5}{3}$.

Range of f = Domain of f^{-1}
= All real numbers except 0.

4. If the point $(3, -5)$ is on the graph of f, then the point $(-5, 3)$ must be on the graph of f^{-1}.

5. $3^x = 243$
   $3^x = 3^5$
   $x = 5$

6. $\log_b 16 = 2$
   $b^2 = 16$
   $b = \pm\sqrt{16} = \pm 4$
   Since the base of a logarithm must be positive, the only viable solution is $b = 4$.

7. $\log_5 x = 4$
   $x = 5^4$
   $x = 625$

8. $e^3 + 2 \approx 22.086$

9. $\log 20 \approx 1.301$

10. $\log_3 21 = \dfrac{\log 21}{\log 3} \approx 2.771$

11. $\ln 133 \approx 4.890$

12. $f(x) = 4^{x+1} - 2$
    Using the graph of $y = 4^x$, shift the graph to the left 1 unit and down 2 units.

    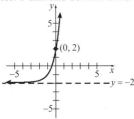

    Domain: $(-\infty, \infty)$
    Range: $(-2, \infty)$
    Horizontal Asymptote: $y = -2$

13. $g(x) = 1 - \log_5(x - 2)$
    Using the graph of $y = \log_5 x$, shift the graph to the right 2 units, reflect about the x-axis, and shift up 1 unit.

    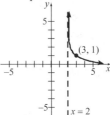

    Domain: $(2, \infty)$
    Range: $(-\infty, \infty)$
    Vertical Asymptote: $x = 2$

14. $5^{x+2} = 125$
    $5^{x+2} = 5^3$
    $x + 2 = 3$
    $x = 1$
    The solution set is $\{1\}$.

15. $\log(x + 9) = 2$
    $x + 9 = 10^2$
    $x + 9 = 100$
    $x = 91$
    The solution set is $\{91\}$.

16. $8 - 2e^{-x} = 4$
    $-2e^{-x} = -4$
    $e^{-x} = 2$
    $-x = \ln 2$
    $x = -\ln 2 \approx -0.693$
    The solution set is $\{-\ln 2 \approx -0.693\}$.

17. $\log(x^2 + 3) = \log(x + 6)$
    $x^2 + 3 = x + 6$
    $x^2 - x - 3 = 0$
    $x = \dfrac{-(-1) \pm \sqrt{(-1)^2 - 4(1)(-3)}}{2(1)}$
    $= \dfrac{1 \pm \sqrt{13}}{2}$
    The solution set is $\left\{\dfrac{1 - \sqrt{13}}{2}, \dfrac{1 + \sqrt{13}}{2}\right\}$.

18.
$$7^{x+3} = e^x$$
$$\ln 7^{x+3} = \ln e^x$$
$$(x+3)\ln 7 = x$$
$$x\ln 7 + 3\ln 7 = x$$
$$x\ln 7 - x = -3\ln 7$$
$$x(\ln 7 - 1) = -3\ln 7$$
$$x = \frac{-3\ln 7}{\ln 7 - 1} = \frac{3\ln 7}{1 - \ln 7} \approx -6.172$$

The solution set is $\left\{\dfrac{3\ln 7}{1-\ln 7} \approx -6.172\right\}$.

19. $\log_2(x-4) + \log_2(x+4) = 3$
$$\log_2[(x-4)(x+4)] = 3$$
$$\log_2(x^2 - 16) = 3$$
$$x^2 - 16 = 2^3$$
$$x^2 - 16 = 8$$
$$x^2 = 24$$
$$x = \pm\sqrt{24} = \pm 2\sqrt{6}$$

Because $x = -2\sqrt{6}$ results in a negative arguments for the original logarithms, the only viable solution is $x = 2\sqrt{6}$. That is, the solution set is $\{2\sqrt{6}\}$.

20. $\log_2\left(\dfrac{4x^3}{x^2 - 3x - 18}\right)$
$$= \log_2\left(\frac{2^2 x^3}{(x+3)(x-6)}\right)$$
$$= \log_2(2^2 x^3) - \log_2[(x+3)(x-6)]$$
$$= \log_2 2^2 + \log_2 x^3 - [\log_2(x+3) + \log_2(x-6)]$$
$$= 2 + 3\log_2 x - \log_2(x+3) - \log_2(x-6)$$

21.
$$A = A_0 e^{kt}$$
$$34 = 50 e^{k(30)}$$
$$0.68 = e^{30k}$$
$$\ln 0.68 = 30k$$
$$k = \frac{\ln 0.68}{30}$$

Thus, the decay model is $A = 50 e^{\left(\frac{\ln 0.68}{30}\right)t}$.
We need to find t when $A = 2$:

$$2 = 50 e^{\left(\frac{\ln 0.68}{30}\right)t}$$
$$0.04 = e^{\left(\frac{\ln 0.68}{30}\right)t}$$
$$\ln 0.04 = \left(\frac{\ln 0.68}{30}\right)t$$
$$t = \frac{\ln 0.04}{\left(\frac{\ln 0.68}{30}\right)} \approx 250.39$$

There will be 2 mg of the substance remaining after about 259 days.

22. a. The 2013 – 2014 academic year is $t = 10$ years after the 2003 – 2004 academic year. We use the formula $A = P\left(1 + \dfrac{r}{n}\right)^{nt}$, with $P = 19,710$, $r = 0.06$, $n = 1$, and $t = 10$.

$$A = 19,710\left(1 + \frac{0.06}{1}\right)^{(1)(10)}$$
$$= 19,710(1.06)^{10}$$
$$\approx \$35,297.61$$

In 2013 – 2014, the average cost of college at 4-year private colleges will be about $35,297.61.

b. We use the formula $A = Pe^{rt}$, with $A = 35,297.61$, $r = 0.06$, and $t = 10$.
$$35,297.61 = Pe^{0.05(10)}$$
$$35,297.61 = Pe^{0.5}$$
$$P = \frac{35,297.61}{e^{0.5}} \approx \$21,409.08$$

Angie will need to put about $21,409.08 into the savings plan.

23. a. $80 = 10\log\left(\dfrac{I}{10^{-12}}\right)$
$$8 = \log\left(\frac{I}{10^{-12}}\right)$$
$$8 = \log I - \log 10^{-12}$$
$$8 = \log I - (-12)$$
$$8 = \log I + 12$$
$$-4 = \log I$$
$$I = 10^{-4} = 0.0001$$

If one person shouts, the intensity is 10^{-4} watts per square meter. Thus, if two people shout at the same time, the intensity will be 2×10^{-4} watts per square meter. Thus, the loudness will be

$$D = 10\log\left(\frac{2\times 10^{-4}}{10^{-12}}\right) = 10\log(2\times 10^8) \approx 83.01$$

decibels

b. Let n represent the number of people who must shout. Then the intensity will be $n \times 10^{-4}$. If $D = 125$, then

$$125 = 10\log\left(\frac{n\times 10^{-4}}{10^{-12}}\right)$$
$$125 = 10\log(n\times 10^8)$$
$$12.5 = \log(n\times 10^8)$$
$$n\times 10^8 = 10^{12.5}$$
$$n = 10^{4.5} \approx 31,623$$

About 31,623 people would have to shout at the same time in order for the resulting sound level to meet the pain threshold.

24. Let x = the number of years since 1985.

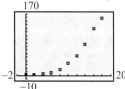

Using LOGISTIC REGression on the data yields:

$$y = \frac{213.005}{1+205.859e^{-0.3564x}}$$

Note that $2007 - 1985 = 22$. Substituting $x = 22$, we find that

$$y = \frac{213.005}{1+205.859e^{-0.3564(22)}} \approx 197.05.$$

Thus, we predict that in 2007 there will be about 197.05 million U.S. cell phone subscribers.

Cumulative Review R-4

1. The graph represents a function since it passes the Vertical Line Test. The function is not a one-to-one function since the graph fails the Horizontal Line Test.

3. $x^2 + y^2 = 1$

   a. $\left(\frac{1}{2}\right)^2 + \left(\frac{1}{2}\right)^2 = \frac{1}{4} + \frac{1}{4} = \frac{1}{2} \neq 1$; $\left(\frac{1}{2},\frac{1}{2}\right)$ is not on the graph.

   b. $\left(\frac{1}{2}\right)^2 + \left(\frac{\sqrt{3}}{2}\right)^2 = \frac{1}{4} + \frac{3}{4} = 1$; $\left(\frac{1}{2},\frac{\sqrt{3}}{2}\right)$ is on the graph.

5. $2x - 4y = 16$

   x-intercept:
   $2x - 4(0) = 16$
   $2x = 16$
   $x = 8$

   y-intercept:
   $2(0) - 4y = 16$
   $-4y = 16$
   $y = -4$

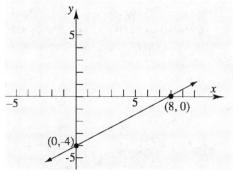

7. Given that the graph of $f(x) = ax^2 + bx + c$ has vertex $(4,-8)$ and passes through the point $(0,24)$, we can conclude $-\frac{b}{2a} = 4$, $f(4) = -8$, and $f(0) = 24$.
Notice that
$$f(0) = 24$$
$$a(0)^2 + b(0) + c = 24$$
$$c = 24$$
Therefore $f(x) = ax^2 + bx + c = ax^2 + bx + 24$.
Furthermore, $-\frac{b}{2a} = 4$, so that $b = -8a$, and

$$f(4) = -8$$
$$a(4)^2 + b(4) + 24 = -8$$
$$16a + 4b + 24 = -8$$
$$16a + 4b = -32$$
$$4a + b = -8$$
Replacing b with $-8a$ in this equation yields
$$4a - 8a = -8$$
$$-4a = -8$$
$$a = 2$$
So $b = -8a = -8(2) = -16$.
Therefore, we have the function
$$f(x) = 2x^2 - 16x + 24.$$

9. $f(x) = x^2 + 2 \qquad g(x) = \dfrac{2}{x-3}$

$$f(g(x)) = f\left(\dfrac{2}{x-3}\right)$$
$$= \left(\dfrac{2}{x-3}\right)^2 + 2$$
$$= \dfrac{4}{(x-3)^2} + 2$$

The domain of f is $\{x \mid x \text{ is any real number}\}$.
The domain of g is $\{x \mid x \neq 3\}$.
So, the domain of $f(g(x))$ is $\{x \mid x \neq 3\}$.

$$f(g(5)) = \dfrac{4}{(5-3)^2} + 2 = \dfrac{4}{2^2} + 2 = \dfrac{4}{4} + 2 = 3$$

11. **a.** $g(x) = 3^x + 2$
   Using the graph of $y = 3^x$, shift up 2 units.
   Domain: $(-\infty, \infty)$
   Range: $(2, \infty)$
   Horizontal Asymptote: $y = 2$

   b. $g(x) = 3^x + 2$
   $$y = 3^x + 2$$
   $$x = 3^y + 2 \quad \text{Inverse}$$
   $$x - 2 = 3^y$$
   $$y = \log_3(x-2)$$
   $$g^{-1}(x) = \log_3(x-2)$$

Domain: $(2, \infty)$
Range: $(-\infty, \infty)$
Vertical Asymptote: $x = 2$

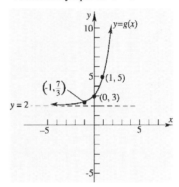

c.

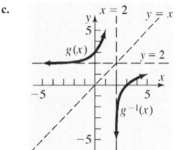

13. $\log_3(x+1) + \log_3(2x-3) = \log_9 9$
$$\log_3((x+1)(2x-3)) = 1$$
$$(x+1)(2x-3) = 3^1$$
$$2x^2 - x - 3 = 3$$
$$2x^2 - x - 6 = 0$$
$$(2x+3)(x-2) = 0$$
$$x = -\dfrac{3}{2} \text{ or } x = 2$$

Since $\log_3\left(-\dfrac{3}{2}+1\right) = \log_3\left(-\dfrac{1}{2}\right)$ is undefined the solution set is $\{2\}$.

15. **a.**

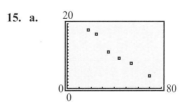

   b. Answers will vary.
   c. Answers will vary.

Chapter 5
Systems of Equations and Inequalities

Section 5.1

1. $3x + 4 = 8 - x$
   $4x = 4$
   $x = 1$
   The solution set is $\{1\}$.

3. inconsistent

5. False

7. $\begin{cases} 2x - y = 5 \\ 5x + 2y = 8 \end{cases}$
   Substituting the values of the variables:
   $\begin{cases} 2(2) - (-1) = 4 + 1 = 5 \\ 5(2) + 2(-1) = 10 - 2 = 8 \end{cases}$
   Each equation is satisfied, so $x = 2$, $y = -1$ is a solution of the system of equations.

9. $\begin{cases} 3x - 4y = 4 \\ \dfrac{1}{2}x - 3y = -\dfrac{1}{2} \end{cases}$
   Substituting the values of the variables:
   $\begin{cases} 3(2) - 4\left(\dfrac{1}{2}\right) = 6 - 2 = 4 \\ \dfrac{1}{2}(2) - 3\left(\dfrac{1}{2}\right) = 1 - \dfrac{3}{2} = -\dfrac{1}{2} \end{cases}$
   Each equation is satisfied, so $x = 2$, $y = \dfrac{1}{2}$ is a solution of the system of equations.

11. $\begin{cases} x - y = 3 \\ \dfrac{1}{2}x + y = 3 \end{cases}$
    Substituting the values of the variables, we obtain:
    $\begin{cases} 4 - 1 = 3 \\ \dfrac{1}{2}(4) + 1 = 2 + 1 = 3 \end{cases}$
    Each equation is satisfied, so $x = 4$, $y = 1$ is a solution of the system of equations.

13. $\begin{cases} 3x + 3y + 3z = 4 \\ x - y - z = 0 \\ 2y - 3z = -8 \end{cases}$
    Substituting the values of the variables:
    $\begin{cases} 3(1) + 3(-1) + 2(2) = 3 - 3 + 4 = 4 \\ 1 - (-1) - 2 = 1 + 1 - 2 = 0 \\ 2(-1) - 3(2) = -2 - 6 = -8 \end{cases}$
    Each equation is satisfied, so $x = 1$, $y = -1$, $z = 2$ is a solution of the system of equations.

15. $\begin{cases} 3x + 3y + 2z = 4 \\ x - 3y + z = 10 \\ 5x - 2y - 3z = 8 \end{cases}$
    Substituting the values of the variables:
    $\begin{cases} 3(2) + 3(-2) + 2(2) = 6 - 6 + 4 = 4 \\ 2 - 3(-2) + 2 = 2 + 6 + 2 = 10 \\ 5(2) - 2(-2) - 3(2) = 10 + 4 - 6 = 8 \end{cases}$
    Each equation is satisfied, so $x = 2$, $y = -2$, $z = 2$ is a solution of the system of equations.

17. $\begin{cases} x + y = 8 \\ x - y = 4 \end{cases}$
    Solve the first equation for y, substitute into the second equation and solve:
    $\begin{cases} y = 8 - x \\ x - y = 4 \end{cases}$
    $x - (8 - x) = 4$
    $x - 8 + x = 4$
    $2x = 12$
    $x = 6$
    Since $x = 6$, $y = 8 - 6 = 2$. The solution of the system is $x = 6$, $y = 2$.

19. $\begin{cases} 5x - y = 13 \\ 2x + 3y = 12 \end{cases}$

Multiply each side of the first equation by 3 and add the equations to eliminate y:

$\begin{cases} 15x - 3y = 39 \\ 2x + 3y = 12 \end{cases}$
$17x = 51$
$x = 3$

Substitute and solve for y:
$5(3) - y = 13$
$15 - y = 13$
$-y = -2$
$y = 2$

The solution of the system is $x = 3, y = 2$.

21. $\begin{cases} 3x = 24 \\ x + 2y = 0 \end{cases}$

Solve the first equation for x and substitute into the second equation:

$\begin{cases} x = 8 \\ x + 2y = 0 \end{cases}$

$8 + 2y = 0$
$2y = -8$
$y = -4$

The solution of the system is $x = 8, y = -4$.

23. $\begin{cases} 3x - 6y = 2 \\ 5x + 4y = 1 \end{cases}$

Multiply each side of the first equation by 2 and each side of the second equation by 3, then add to eliminate y:

$\begin{cases} 6x - 12y = 4 \\ 15x + 12y = 3 \end{cases}$
$21x = 7$
$x = \dfrac{1}{3}$

Substitute and solve for y:
$3(1/3) - 6y = 2$
$1 - 6y = 2$
$-6y = 1$
$y = -\dfrac{1}{6}$

The solution of the system is $x = \dfrac{1}{3}, y = -\dfrac{1}{6}$.

25. $\begin{cases} 2x + y = 1 \\ 4x + 2y = 3 \end{cases}$

Solve the first equation for y, substitute into the second equation and solve:

$\begin{cases} y = 1 - 2x \\ 4x + 2y = 3 \end{cases}$

$4x + 2(1 - 2x) = 3$
$4x + 2 - 4x = 3$
$0x = 1$

This has no solution, so the system is inconsistent.

27. $\begin{cases} 2x - y = 0 \\ 3x + 2y = 7 \end{cases}$

Solve the first equation for y, substitute into the second equation and solve:

$\begin{cases} y = 2x \\ 3x + 2y = 7 \end{cases}$

$3x + 2(2x) = 7$
$3x + 4x = 7$
$7x = 7$
$x = 1$

Since $x = 1, y = 2(1) = 2$
The solution of the system is $x = 1, y = 2$.

29. $\begin{cases} x + 2y = 4 \\ 2x + 4y = 8 \end{cases}$

Solve the first equation for x, substitute into the second equation and solve:

$\begin{cases} x = 4 - 2y \\ 2x + 4y = 8 \end{cases}$

$2(4 - 2y) + 4y = 8$
$8 - 4y + 4y = 8$
$0y = 0$

These equations are dependent. Any real number is a solution for y. The solution of the system is $x = 4 - 2y$, where y is any real number.

31. $\begin{cases} 2x - 3y = -1 \\ 10x + y = 11 \end{cases}$

Multiply each side of the first equation by –5, and add the equations to eliminate x:

$\begin{cases} -10x + 15y = 5 \\ 10x + y = 11 \end{cases}$
$16y = 16$
$y = 1$

Substitute and solve for x:
$2x - 3(1) = -1$
$2x - 3 = -1$
$2x = 2$
$x = 1$

The solution of the system is $x = 1$, $y = 1$.

33. $\begin{cases} 2x + 3y = 6 \\ x - y = \dfrac{1}{2} \end{cases}$

Solve the second equation for x, substitute into the first equation and solve:

$\begin{cases} 2x + 3y = 6 \\ x = y + \dfrac{1}{2} \end{cases}$

$2\left(y + \dfrac{1}{2}\right) + 3y = 6$
$2y + 1 + 3y = 6$
$5y = 5$
$y = 1$

Since $y = 1$, $x = 1 + \dfrac{1}{2} = \dfrac{3}{2}$. The solution of the system is $x = \dfrac{3}{2}$, $y = 1$.

35. $\begin{cases} \dfrac{1}{2}x + \dfrac{1}{3}y = 3 \\ \dfrac{1}{4}x - \dfrac{2}{3}y = -1 \end{cases}$

Multiply each side of the first equation by –6 and each side of the second equation by 12, then add to eliminate x:

$\begin{cases} -3x - 2y = -18 \\ 3x - 8y = -12 \end{cases}$
$-10y = -30$
$y = 3$

Substitute and solve for x:
$\dfrac{1}{2}x + \dfrac{1}{3}(3) = 3$
$\dfrac{1}{2}x + 1 = 3$
$\dfrac{1}{2}x = 2$
$x = 4$

The solution of the system is $x = 4$, $y = 3$.

37. $\begin{cases} 3x - 5y = 3 \\ 15x + 5y = 21 \end{cases}$

Add the equations to eliminate y:

$\begin{cases} 3x - 5y = 3 \\ 15x + 5y = 21 \end{cases}$
$18x = 24$
$x = \dfrac{4}{3}$

Substitute and solve for y:
$3(4/3) - 5y = 3$
$4 - 5y = 3$
$-5y = -1$
$y = \dfrac{1}{5}$

The solution of the system is $x = \dfrac{4}{3}$, $y = \dfrac{1}{5}$.

39. $\begin{cases} \dfrac{1}{x}+\dfrac{1}{y}=8 \\ \dfrac{3}{x}-\dfrac{5}{y}=0 \end{cases}$

Rewrite letting $u=\dfrac{1}{x},\ v=\dfrac{1}{y}$:

$\begin{cases} u+v=8 \\ 3u-5v=0 \end{cases}$

Solve the first equation for u, substitute into the second equation and solve:

$\begin{cases} u=8-v \\ 3u-5v=0 \end{cases}$

$3(8-v)-5v=0$
$24-3v-5v=0$
$\quad\quad\quad -8v=-24$
$\quad\quad\quad\quad v=3$

Since $v=3$, $u=8-3=5$. Thus, $x=\dfrac{1}{u}=\dfrac{1}{5}$,

$y=\dfrac{1}{v}=\dfrac{1}{3}$. The solution of the system is

$x=\dfrac{1}{5},\ y=\dfrac{1}{3}$.

41. $\begin{cases} x-y=6 \\ 2x-3z=16 \\ 2y+z=4 \end{cases}$

Multiply each side of the first equation by –2 and add to the second equation to eliminate x:
$\quad -2x+2y\quad\quad =-12$
$\quad\ \ 2x\quad\quad -3z=\ \ 16$
$\quad\overline{\quad\quad\ 2y-3z=\ \ \ 4}$

Multiply each side of the result by –1 and add to the original third equation to eliminate y:
$\quad -2y+3z=-4$
$\quad\ \ 2y+\ z=\ \ 4$
$\quad\overline{\quad\quad\quad 4z=\ \ 0}$
$\quad\quad\quad\quad z=0$

Substituting and solving for the other variables:
$\quad 2y+0=4\quad\quad\quad 2x-3(0)=16$
$\quad\quad 2y=4\quad\quad\quad\quad\quad 2x=16$
$\quad\quad\ y=2\quad\quad\quad\quad\quad\quad x=8$

The solution is $x=8,\ y=2,\ z=0$.

43. $\begin{cases} x-2y+3z=\ \ 7 \\ 2x+y+z=\ \ 4 \\ -3x+2y-2z=-10 \end{cases}$

Multiply each side of the first equation by –2 and add to the second equation to eliminate x; and multiply each side of the first equation by 3 and add to the third equation to eliminate x:
$\quad -2x+4y-6z=-14$
$\quad\ \ 2x+\ y+\ z=\ \ 4$
$\quad\overline{\quad\quad\ 5y-5z=-10}$

$\quad 3x-6y+9z=\ \ 21$
$\quad -3x+2y-2z=-10$
$\quad\overline{\quad\quad -4y+7z=\ \ 11}$

Multiply each side of the first result by $\dfrac{4}{5}$ and add to the second result to eliminate y:
$\quad 4y-4z=-8$
$\quad -4y+7z=11$
$\quad\overline{\quad\quad\ 3z=3}$
$\quad\quad\quad z=1$

Substituting and solving for the other variables:
$\quad y-1=-2\quad\quad\quad x-2(-1)+3(1)=7$
$\quad\quad y=-1\quad\quad\quad\quad\quad x+2+3=7$
$\quad\quad\quad\quad\quad\quad\quad\quad\quad\quad\quad\quad x=2$

The solution is $x=2,\ y=-1,\ z=1$.

45. $\begin{cases} x-y-z=1 \\ 2x+3y+z=2 \\ 3x+2y=0 \end{cases}$

Add the first and second equations to eliminate z:
$\quad x-\ y-z=1$
$\quad 2x+3y+z=2$
$\quad\overline{\quad 3x+2y=3}$

Multiply each side of the result by –1 and add to the original third equation to eliminate y:
$\quad -3x-2y=-3$
$\quad\ \ 3x+2y=\ \ 0$
$\quad\overline{\quad\quad\quad 0=-3}$

This result has no solution, so the system is inconsistent.

47. $\begin{cases} x - y - z = 1 \\ -x + 2y - 3z = -4 \\ 3x - 2y - 7z = 0 \end{cases}$

Add the first and second equations to eliminate x; multiply the first equation by -3 and add to the third equation to eliminate x:

$\begin{array}{r} x - y - z = 1 \\ \underline{-x + 2y - 3z = -4} \\ y - 4z = -3 \end{array}$

$\begin{array}{r} -3x + 3y + 3z = -3 \\ \underline{3x - 2y - 7z = 0} \\ y - 4z = -3 \end{array}$

Multiply each side of the first result by -1 and add to the second result to eliminate y:

$\begin{array}{r} -y + 4z = 3 \\ \underline{y - 4z = -3} \\ 0 = 0 \end{array}$

The system is dependent. If z is any real number, then $y = 4z - 3$.

Solving for x in terms of z in the first equation:
$x - (4z - 3) - z = 1$
$x - 4z + 3 - z = 1$
$x - 5z + 3 = 1$
$x = 5z - 2$

The solution is $x = 5z - 2$, $y = 4z - 3$, z is any real number.

49. $\begin{cases} 2x - 2y + 3z = 6 \\ 4x - 3y + 2z = 0 \\ -2x + 3y - 7z = 1 \end{cases}$

Multiply the first equation by -2 and add to the second equation to eliminate x; add the first and third equations to eliminate x:

$\begin{array}{r} -4x + 4y - 6z = -12 \\ \underline{4x - 3y + 2z = 0} \\ y - 4z = -12 \end{array}$

$\begin{array}{r} 2x - 2y + 3z = 6 \\ \underline{-2x + 3y - 7z = 1} \\ y - 4z = 7 \end{array}$

Multiply each side of the first result by -1 and add to the second result to eliminate y:

$\begin{array}{r} -y + 4z = 12 \\ \underline{y - 4z = 7} \\ 0 = 19 \end{array}$

This result has no solution, so the system is inconsistent.

51. $\begin{cases} x + y - z = 6 \\ 3x - 2y + z = -5 \\ x + 3y - 2z = 14 \end{cases}$

Add the first and second equations to eliminate z; multiply the second equation by 2 and add to the third equation to eliminate z:

$\begin{array}{r} x + y - z = 6 \\ \underline{3x - 2y + z = -5} \\ 4x - y = 1 \end{array}$

$\begin{array}{r} 6x - 4y + 2z = -10 \\ \underline{x + 3y - 2z = 14} \\ 7x - y = 4 \end{array}$

Multiply each side of the first result by -1 and add to the second result to eliminate y:

$\begin{array}{r} -4x + y = -1 \\ \underline{7x - y = 4} \\ 3x = 3 \\ x = 1 \end{array}$

Substituting and solving for the other variables:
$\begin{array}{rr} 4(1) - y = 1 & 3(1) - 2(3) + z = -5 \\ -y = -3 & 3 - 6 + z = -5 \\ y = 3 & z = -2 \end{array}$

The solution is $x = 1$, $y = 3$, $z = -2$.

53. $\begin{cases} x + 2y - z = -3 \\ 2x - 4y + z = -7 \\ -2x + 2y - 3z = 4 \end{cases}$

Add the first and second equations to eliminate z; multiply the second equation by 3 and add to the third equation to eliminate z:

$\begin{array}{r} x + 2y - z = -3 \\ \underline{2x - 4y + z = -7} \\ 3x - 2y = -10 \end{array}$

$\begin{array}{r} 6x - 12y + 3z = -21 \\ \underline{-2x + 2y - 3z = 4} \\ 4x - 10y = -17 \end{array}$

Multiply each side of the first result by -5 and add to the second result to eliminate y:

$$-15x + 10y = 50$$
$$\underline{4x - 10y = -17}$$
$$-11x = 33$$
$$x = -3$$

Substituting and solving for the other variables:
$$3(-3) - 2y = -10$$
$$-9 - 2y = -10$$
$$-2y = -1$$
$$y = \frac{1}{2}$$

$$-3 + 2\left(\frac{1}{2}\right) - z = -3$$
$$-3 + 1 - z = -3$$
$$-z = -1$$
$$z = 1$$

The solution is $x = -3$, $y = \frac{1}{2}$, $z = 1$.

55. Let l be the length of the rectangle and w be the width of the rectangle. Then:
$l = 2w$ and $2l + 2w = 90$

Solve by substitution:
$$2(2w) + 2w = 90$$
$$4w + 2w = 90$$
$$6w = 90$$
$$w = 15 \text{ feet}$$
$$l = 2(15) = 30 \text{ feet}$$

The dimensions of the floor are 15 feet by 30 feet.

57. Let $x =$ the cost of one cheeseburger and $y =$ the cost of one shake. Then:
$4x + 2y = 790$ and $2y = x + 15$

Solve by substitution:
$$4x + x + 15 = 790 \qquad 2y = 155 + 15$$
$$5x = 775 \qquad\qquad 2y = 170$$
$$x = 155 \qquad\qquad y = 85$$

A cheeseburger cost $1.55 and a shake costs $0.85.

59. Let $x =$ the number of pounds of cashews.
Let $y =$ is the number of pounds in the mixture.
The value of the cashews is $5x$.
The value of the peanuts is $1.50(30) = 45$.
The value of the mixture is $3y$.
Then $x + 30 = y$ represent the amount of mixture.
$5x + 45 = 3y$ represents the value of the mixture.

Solve by substitution:
$$5x + 45 = 3(x + 30)$$
$$2x = 45$$
$$x = 22.5$$

So, 22.5 pounds of cashews should be used in the mixture.

61. Let $x =$ the plane's air speed and $y =$ the wind speed.

	Rate	Time	Distance
With Wind	$x + y$	3	600
Against	$x - y$	4	600

$$\begin{cases} (x+y)(3) = 600 \\ (x-y)(4) = 600 \end{cases}$$

Multiply each side of the first equation by $\frac{1}{3}$, multiply each side of the second equation by $\frac{1}{4}$, and add the result to eliminate y

$$x + y = 200$$
$$\underline{x - y = 150}$$
$$2x = 350$$
$$x = 175$$

$$175 + y = 200$$
$$y = 25$$

The airspeed of the plane is 175 mph, and the wind speed is 25 mph.

63. Let x = the number of $25-design.
    Let y = the number of $45-design.
    Then $x + y$ = the total number of sets of dishes.
    $25x + 45y$ = the cost of the dishes.
    Setting up the equations and solving by substitution:
    $$\begin{cases} x + y = 200 \\ 25x + 45y = 7400 \end{cases}$$
    Solve the first equation for y, the solve by substitution: $y = 200 - x$
    $25x + 45(200 - x) = 7400$
    $25x + 9000 - 45x = 7400$
    $-20x = -1600$
    $x = 80$
    $y = 200 - 80 = 120$
    Thus, 80 sets of the $25 dishes and 120 sets of the $45 dishes should be ordered.

65. Let x = the cost per package of bacon.
    Let y = the cost of a carton of eggs.
    Set up a system of equations for the problem:
    $$\begin{cases} 3x + 2y = 7.45 \\ 2x + 3y = 6.45 \end{cases}$$
    Multiply each side of the first equation by 3 and each side of the second equation by –2 and solve by elimination:
    $9x + 6y = 22.35$
    $-4x - 6y = -12.90$
    _____
    $5x\quad = 9.45$
    $x = 1.89$
    Substitute and solve for y:
    $3(1.89) + 2y = 7.45$
    $5.67 + 2y = 7.45$
    $2y = 1.78$
    $y = 0.89$
    A package of bacon costs $1.89 and a carton of eggs cost $0.89. The refund for 2 packages of bacon and 2 cartons of eggs will be
    2($1.89) + 2($0.89) = $5.56.

67. Let x = the # of mg of compound 1.
    Let y = the # of mg of compound 2.
    Setting up the equations and solving by substitution:
    $$\begin{cases} 0.2x + 0.4y = 40 \quad \text{vitamin C} \\ 0.3x + 0.2y = 30 \quad \text{vitamin D} \end{cases}$$
    Multiplying each equation by 10 yields
    $$\begin{cases} 2x + 4y = 400 \\ 6x + 4y = 600 \end{cases}$$
    Subtracting the bottom equation from the top equation yields
    $2x + 4y - (6x + 4y) = 400 - 600$
    $2x - 6x = -200$
    $-4x = -200$
    $x = 50$
    $2(50) + 4y = 400$
    $100 + 4y = 400$
    $4y = 300$
    $y = \dfrac{300}{4} = 75$
    So 50 mg of compound 1 should be mixed with 75 mg of compound 2.

69. $y = ax^2 + bx + c$
    At (–1, 4) the equation becomes:
    $4 = a(-1)^2 + b(-1) + c$
    $4 = a - b + c$
    At (2, 3) the equation becomes:
    $3 = a(2)^2 + b(2) + c$
    $3 = 4a + 2b + c$
    At (0, 1) the equation becomes:
    $1 = a(0)^2 + b(0) + c$
    $1 = c$
    The system of equations is:
    $$\begin{cases} a - b + c = 4 \\ 4a + 2b + c = 3 \\ c = 1 \end{cases}$$
    Substitute $c = 1$ into the first and second equations and simplify:
    $a - b + 1 = 4 \qquad 4a + 2b + 1 = 3$
    $a - b = 3 \qquad\quad 4a + 2b = 2$
    $a = b + 3$

Solve the first result for a, substitute into the second result and solve:
$$4(b+3) + 2b = 2$$
$$4b + 12 + 2b = 2$$
$$6b = -10$$
$$b = -\frac{5}{3}$$
$$a = -\frac{5}{3} + 3 = \frac{4}{3}$$

The solution is $a = \frac{4}{3}$, $b = -\frac{5}{3}$, $c = 1$. The equation is $y = \frac{4}{3}x^2 - \frac{5}{3}x + 1$.

71. $\begin{cases} 0.06Y - 5000r = 240 \\ 0.06Y + 6000r = 900 \end{cases}$

Multiply the first equation by -1, the add the result to the second equation to eliminate Y.
$$-0.06Y + 5000r = -240$$
$$0.06Y + 6000r = 900$$
$$\overline{11000r = 660}$$
$$r = 0.06$$

Substitute this result into the first equation to find Y.
$$0.06Y - 5000(0.06) = 240$$
$$0.06Y - 300 = 240$$
$$0.06Y = 540$$
$$Y = 9000$$

The equilibrium level of income and interest rates is $9000 million and 6%.

73. $\begin{cases} I_2 = I_1 + I_3 \\ 5 - 3I_1 - 5I_2 = 0 \\ 10 - 5I_2 - 7I_3 = 0 \end{cases}$

Substitute the expression for I_2 into the second and third equations and simplify:
$$5 - 3I_1 - 5(I_1 + I_3) = 0$$
$$-8I_1 - 5I_3 = -5$$

$$10 - 5(I_1 + I_3) - 7I_3 = 0$$
$$-5I_1 - 12I_3 = -10$$

Multiply both sides of the first result by 5 and multiply both sides of the second result by –8 to eliminate I_1:

$$-40I_1 - 25I_3 = -25$$
$$40I_1 + 96I_3 = 80$$
$$\overline{71I_3 = 55}$$
$$I_3 = \frac{55}{71}$$

Substituting and solving for the other variables:
$$-8I_1 - 5\left(\frac{55}{71}\right) = -5$$
$$-8I_1 - \frac{275}{71} = -5$$
$$-8I_1 = -\frac{80}{71}$$
$$I_1 = \frac{10}{71}$$

$$I_2 = \left(\frac{10}{71}\right) + \frac{55}{71}$$
$$I_2 = \frac{65}{71}$$

The solution is $I_1 = \frac{10}{71}$, $I_2 = \frac{65}{71}$, $I_3 = \frac{55}{71}$.

75. Let x = the number of orchestra seats.
Let y = the number of main seats.
Let z = the number of balcony seats.
Since the total number of seats is 500, $x + y + z = 500$.
Since the total revenue is $17,100 if all seats are sold, $50x + 35y + 25z = 17,100$.
If only half of the orchestra seats are sold, the revenue is $14,600.

So, $50\left(\frac{1}{2}x\right) + 35y + 25z = 14,600$.

Thus, we have the following system:
$$\begin{cases} x + y + z = 500 \\ 50x + 35y + 25z = 17,100 \\ 25x + 35y + 25z = 14,600 \end{cases}$$

Multiply each side of the first equation by –25 and add to the second equation to eliminate z; multiply each side of the third equation by –1 and add to the second equation to eliminate z:
$$-25x - 25y - 25z = -12,500$$
$$50x + 35y + 25z = 17,100$$
$$\overline{25x + 10y = 4600}$$

$50x + 35y + 25z = 17,100$
$\underline{-25x - 35y - 25z = -14,600}$
$25x = 2500$
$x = 100$

Substituting and solving for the other variables:
$25(100) + 10y = 4600 \qquad 100 + 210 + z = 500$
$2500 + 10y = 4600 \qquad 310 + z = 500$
$10y = 2100 \qquad z = 190$
$y = 210$

There are 100 orchestra seats, 210 main seats, and 190 balcony seats.

77. Let x = the number of servings of chicken.
Let y = the number of servings of corn.
Let z = the number of servings of 2% milk.

Protein equation: $30x + 3y + 9z = 66$
Carbohydrate equation: $35x + 16y + 13z = 94.5$
Calcium equation: $200x + 10y + 300z = 910$

Multiply each side of the first equation by –16 and multiply each side of the second equation by 3 and add them to eliminate y; multiply each side of the second equation by –5 and multiply each side of the third equation by 8 and add to eliminate y:
$-480x - 48y - 144z = -1056$
$\underline{105x + 48y + 39z = 283.5}$
$-375x - 105z = -772.5$

$-175x - 80y - 65z = -472.5$
$\underline{1600x + 80y + 2400z = 7280}$
$1425x + 2335z = 6807.5$

Multiply each side of the first result by 19 and multiply each side of the second result by 5 to eliminate x:
$-7125x - 1995z = -14,677.5$
$\underline{7125x + 11,675z = 34,037.5}$
$9680z = 19,360$
$z = 2$

Substituting and solving for the other variables:
$-375x - 105(2) = -772.5$
$-375x - 210 = -772.5$
$-375x = -562.5$
$x = 1.5$

$30(1.5) + 3y + 9(2) = 66$
$45 + 3y + 18 = 66$
$3y = 3$
$y = 1$

The dietitian should serve 1.5 servings of chicken, 1 serving of corn, and 2 servings of 2% milk.

79. Let x = the price of 1 hamburger.
Let y = the price of 1 order of fries.
Let z = the price of 1 drink.
We can construct the system
$\begin{cases} 8x + 6y + 6z = 26.10 \\ 10x + 6y + 8z = 31.60 \end{cases}$

A system involving only 2 equations that contain 3 or more unknowns cannot be solved uniquely.

Multiply the first equation by $-\dfrac{1}{2}$ and the second equation by $\dfrac{1}{2}$, then add to eliminate y:
$-4x - 3y - 3z = -13.05$
$\underline{5x + 3y + 4z = 15.80}$
$x + z = 2.75$
$x = 2.75 - z$

Substitute and solve for y in terms of z:
$5(2.75 - z) + 3y + 4z = 15.80$
$13.75 + 3y - z = 15.80$
$3y = z + 2.05$
$y = \dfrac{1}{3}z + \dfrac{41}{60}$

Solutions of the system are: $x = 2.75 - z$, $y = \dfrac{1}{3}z + \dfrac{41}{60}$.

Since we are given that $0.60 \le z \le 0.90$, we choose values of z that give two-decimal-place values of x and y with $1.75 \le x \le 2.25$ and $0.75 \le y \le 1.00$.

The possible values of x, y, and z are shown in the table.

x	y	z
2.13	0.89	0.62
2.10	0.90	0.65

SSM: College Algebra EGU   Chapter 5: Systems of Equations and Inequalities

2.07	.091	0.68
2.04	.092	0.71
2.01	.093	0.74
1.98	.094	0.77
1.95	0.95	0.80
1.92	0.96	0.83
1.89	0.97	0.86
1.86	0.98	0.89
1.83	0.99	0.92
1.80	1.00	0.95

81. Let x = Beth's time working alone.
    Let y = Bill's time working alone.
    Let z = Edie's time working alone.

    We can use the following tables to organize our work:

	Beth	Bill	Edie
Hours to do job	x	y	z
Part of job done in 1 hour	$\frac{1}{x}$	$\frac{1}{y}$	$\frac{1}{z}$

    In 10 hours they complete 1 entire job, so
    $10\left(\frac{1}{x}+\frac{1}{y}+\frac{1}{z}\right)=1$
    $\frac{1}{x}+\frac{1}{y}+\frac{1}{z}=\frac{1}{10}$

	Bill	Edie
Hours to do job	y	z
Part of job done in 1 hour	$\frac{1}{y}$	$\frac{1}{z}$

    In 15 hours they complete 1 entire job, so
    $15\left(\frac{1}{y}+\frac{1}{z}\right)=1$.
    $\frac{1}{y}+\frac{1}{z}=\frac{1}{15}$

	Beth	Bill	Edie
Hours to do job	x	y	z
Part of job done in 1 hour	$\frac{1}{x}$	$\frac{1}{y}$	$\frac{1}{z}$

    With all 3 working for 4 hours and Beth and Bill working for an additional 8 hours, they complete 1 entire job, so $4\left(\frac{1}{x}+\frac{1}{y}+\frac{1}{z}\right)+8\left(\frac{1}{x}+\frac{1}{y}\right)=1$
    $\frac{12}{x}+\frac{12}{y}+\frac{4}{z}=1$

    We have the system
    $$\begin{cases} \frac{1}{x}+\frac{1}{y}+\frac{1}{z}=\frac{1}{10} \\ \frac{1}{y}+\frac{1}{z}=\frac{1}{15} \\ \frac{12}{x}+\frac{12}{y}+\frac{4}{z}=1 \end{cases}$$

    Subtract the second equation from the first equation:
    $\frac{1}{x}+\frac{1}{y}+\frac{1}{z}=\frac{1}{10}$
    $\phantom{\frac{1}{x}+}\frac{1}{y}+\frac{1}{z}=\frac{1}{15}$
    _____
    $\frac{1}{x}\phantom{+\frac{1}{y}+\frac{1}{z}}=\frac{1}{30}$
    $x=30$

    Substitute $x=30$ into the third equation:
    $\frac{12}{30}+\frac{12}{y}+\frac{4}{z}=1$
    $\frac{12}{y}+\frac{4}{z}=\frac{3}{5}$.

    Now consider the system consisting of the last result and the second original equation. Multiply the second original equation by –12 and add it to the last result to eliminate y:

343

$$\frac{-12}{y} + \frac{-12}{z} = \frac{-12}{15}$$

$$\frac{12}{y} + \frac{4}{z} = \frac{3}{5}$$

$$-\frac{8}{z} = -\frac{3}{15}$$

$$z = 40$$

Plugging $z = 40$ to find y:

$$\frac{12}{y} + \frac{4}{z} = \frac{3}{5}$$

$$\frac{12}{y} + \frac{4}{40} = \frac{3}{5}$$

$$\frac{12}{y} = \frac{1}{2}$$

$$y = 24$$

Working alone, it would take Beth 30 hours, Bill 24 hours, and Edie 40 hours to complete the job.

83. Answers will vary.

Section 5.2

1. matrix

3. True

5. Writing the augmented matrix for the system of equations:
$$\begin{cases} x - 5y = 5 \\ 4x + 3y = 6 \end{cases} \rightarrow \begin{bmatrix} 1 & -5 & | & 5 \\ 4 & 3 & | & 6 \end{bmatrix}$$

7. $\begin{cases} 2x + 3y - 6 = 0 \\ 4x - 6y + 2 = 0 \end{cases}$

Write the system in standard form and then write the augmented matrix for the system of equations:
$$\begin{cases} 2x + 3y = 6 \\ 4x - 6y = -2 \end{cases} \rightarrow \begin{bmatrix} 2 & 3 & | & 6 \\ 4 & -6 & | & -2 \end{bmatrix}$$

9. Writing the augmented matrix for the system of equations:
$$\begin{cases} 0.01x - 0.03y = 0.06 \\ 0.13x + 0.10y = 0.20 \end{cases} \rightarrow \begin{bmatrix} 0.01 & -0.03 & | & 0.06 \\ 0.13 & 0.10 & | & 0.20 \end{bmatrix}$$

11. Writing the augmented matrix for the system of equations:
$$\begin{cases} x - y + z = 10 \\ 3x + 3y = 5 \\ x + y + 2z = 2 \end{cases} \rightarrow \begin{bmatrix} 1 & -1 & 1 & | & 10 \\ 3 & 3 & 0 & | & 5 \\ 1 & 1 & 2 & | & 2 \end{bmatrix}$$

13. Writing the augmented matrix for the system of equations:
$$\begin{cases} x + y - z = 2 \\ 3x - 2y = 2 \\ 5x + 3y - z = 1 \end{cases} \rightarrow \begin{bmatrix} 1 & 1 & -1 & | & 2 \\ 3 & -2 & 0 & | & 2 \\ 5 & 3 & -1 & | & 1 \end{bmatrix}$$

15. Writing the augmented matrix for the system of equations:
$$\begin{cases} x - y - z = 10 \\ 2x + y + 2z = -1 \\ -3x + 4y = 5 \\ 4x - 5y + z = 0 \end{cases} \rightarrow \begin{bmatrix} 1 & -1 & -1 & | & 10 \\ 2 & 1 & 2 & | & -1 \\ -3 & 4 & 0 & | & 5 \\ 4 & -5 & 1 & | & 0 \end{bmatrix}$$

17. $R_2 = -2r_1 + r_2$
$$\begin{bmatrix} 1 & -3 & | & -2 \\ 2 & -5 & | & 5 \end{bmatrix} \rightarrow \begin{bmatrix} 1 & -3 & | & -2 \\ -2(1)+2 & 2(-3)-5 & | & -2(-2)+5 \end{bmatrix}$$
$$\rightarrow \begin{bmatrix} 1 & -3 & | & -2 \\ 0 & 1 & | & 9 \end{bmatrix}$$

19. a. $R_2 = -2r_1 + r_2$
$$\begin{bmatrix} 1 & -3 & 4 & | & 3 \\ 2 & -5 & 6 & | & 6 \\ -3 & 3 & 4 & | & 6 \end{bmatrix}$$
$$\rightarrow \begin{bmatrix} 1 & -3 & 4 & | & 3 \\ -2(1)+2 & -2(-3)-5 & -2(4)+6 & | & -2(3)+6 \\ -3 & 3 & 4 & | & 6 \end{bmatrix}$$
$$\rightarrow \begin{bmatrix} 1 & -3 & 4 & | & 3 \\ 0 & 1 & -2 & | & 0 \\ -3 & 3 & 4 & | & 6 \end{bmatrix}$$

b. $R_3 = 3r_1 + r_3$

$$\begin{bmatrix} 1 & -3 & 4 & | & 3 \\ 2 & -5 & 6 & | & 6 \\ -3 & 3 & 4 & | & 6 \end{bmatrix}$$

$$\rightarrow \begin{bmatrix} 1 & -3 & 4 & | & 3 \\ 2 & -5 & 6 & | & 6 \\ 3(1)-3 & 3(-3)+3 & 3(4)+4 & | & 3(3)+6 \end{bmatrix}$$

$$\rightarrow \begin{bmatrix} 1 & -3 & 4 & | & 3 \\ 2 & -5 & 6 & | & 6 \\ 0 & -6 & 16 & | & 15 \end{bmatrix}$$

21. a. $R_2 = -2r_1 + r_2$

$$\begin{bmatrix} 1 & -3 & 2 & | & -6 \\ 2 & -5 & 3 & | & -4 \\ -3 & -6 & 4 & | & 6 \end{bmatrix}$$

$$\rightarrow \begin{bmatrix} 1 & -3 & 2 & | & -6 \\ -2(1)+2 & -2(-3)-5 & -2(2)+3 & | & -2(-6)-4 \\ -3 & -6 & 4 & | & 6 \end{bmatrix}$$

$$\rightarrow \begin{bmatrix} 1 & -3 & 2 & | & -6 \\ 0 & 1 & -1 & | & 8 \\ -3 & -6 & 4 & | & 6 \end{bmatrix}$$

b. $R_3 = 3r_1 + r_3$

$$\begin{bmatrix} 1 & -3 & 2 & | & -6 \\ 2 & -5 & 3 & | & -4 \\ -3 & -6 & 4 & | & 6 \end{bmatrix}$$

$$\rightarrow \begin{bmatrix} 1 & -3 & 2 & | & -6 \\ 2 & -5 & 3 & | & -4 \\ 3(1)-3 & 3(-3)-6 & 3(2)+4 & | & 3(-6)+6 \end{bmatrix}$$

$$\rightarrow \begin{bmatrix} 1 & -3 & 2 & | & -6 \\ 2 & -5 & 3 & | & -4 \\ 0 & -15 & 10 & | & -12 \end{bmatrix}$$

23. a. $R_2 = -2r_1 + r_2$

$$\begin{bmatrix} 1 & -3 & 1 & | & -2 \\ 2 & -5 & 6 & | & -2 \\ -3 & 1 & 4 & | & 6 \end{bmatrix}$$

$$\rightarrow \begin{bmatrix} 1 & -3 & 1 & | & -2 \\ -2(1)+2 & -2(-3)-5 & -2(1)+6 & | & -2(-2)-2 \\ -3 & 1 & 4 & | & 6 \end{bmatrix}$$

$$\rightarrow \begin{bmatrix} 1 & -3 & 1 & | & -2 \\ 0 & 1 & 4 & | & 2 \\ -3 & 1 & 4 & | & 6 \end{bmatrix}$$

b. $R_3 = 3r_1 + r_3$

$$\begin{bmatrix} 1 & -3 & 1 & | & -2 \\ 2 & -5 & 6 & | & -2 \\ -3 & 1 & 4 & | & 6 \end{bmatrix}$$

$$\rightarrow \begin{bmatrix} 1 & -3 & 1 & | & -2 \\ 2 & -5 & 6 & | & -2 \\ 3(1)-3 & 3(-3)+1 & 3(1)+4 & | & 3(-2)+6 \end{bmatrix}$$

$$\rightarrow \begin{bmatrix} 1 & -3 & 1 & | & -2 \\ 2 & -5 & 6 & | & -2 \\ 0 & -8 & 7 & | & 0 \end{bmatrix}$$

25. $\begin{cases} x = 5 \\ y = -1 \end{cases}$ consistent; $x = 5, y = -1$

27. $\begin{cases} x = 1 \\ y = 2 \\ 0 = 3 \end{cases}$ inconsistent

29. $\begin{cases} x + 2z = -1 \\ y - 4z = -2 \\ 0 = 0 \end{cases}$ consistent; $x = -1 - 2z$, $y = -2 + 4z$, z is any real number

31. $\begin{cases} x_1 = 1 \\ x_2 + x_4 = 2 \\ x_3 + 2x_4 = 3 \end{cases}$ consistent; $x_1 = 1$, $x_2 = 2 - x_4$, $x_3 = 3 - 2x_4$, x_4 is any real number

33. $\begin{cases} x_1 + 4x_4 = 2 \\ x_2 + x_3 + 3x_4 = 3 \\ 0 = 0 \end{cases}$ consistent; $x_1 = 2 - 4x_4$, $x_2 = 3 - x_3 - 3x_4$, x_3, x_4 are any real numbers

35. $\begin{cases} x_1 + x_4 = -2 \\ x_2 + 2x_4 = 2 \\ x_3 - x_4 = 0 \\ 0 = 0 \end{cases}$ consistent; $x_1 = -2 - x_4$, $x_2 = 2 - 2x_4$, $x_3 = x_4$, x_4 is any real number

37. $\begin{cases} x+y=8 \\ x-y=4 \end{cases}$

Write the augmented matrix:
$\begin{bmatrix} 1 & 1 & | & 8 \\ 1 & -1 & | & 4 \end{bmatrix} \to \begin{bmatrix} 1 & 1 & | & 8 \\ 0 & -2 & | & -4 \end{bmatrix}$ $(R_2 = -r_1 + r_2)$

$\to \begin{bmatrix} 1 & 1 & | & 8 \\ 0 & 1 & | & 2 \end{bmatrix}$ $(R_2 = -\tfrac{1}{2}r_2)$

$\to \begin{bmatrix} 1 & 0 & | & 6 \\ 0 & 1 & | & 2 \end{bmatrix}$ $(R_1 = -r_2 + r_1)$

The solution is $x=6, y=2$.

39. $\begin{cases} 2x-4y=-2 \\ 3x+2y=3 \end{cases}$

Write the augmented matrix:
$\begin{bmatrix} 2 & -4 & | & -2 \\ 3 & 2 & | & 3 \end{bmatrix} \to \begin{bmatrix} 1 & -2 & | & -1 \\ 3 & 2 & | & 3 \end{bmatrix}$ $(R_1 = \tfrac{1}{2}r_1)$

$\to \begin{bmatrix} 1 & -2 & | & -1 \\ 0 & 8 & | & 6 \end{bmatrix}$ $(R_2 = -3r_1 + r_2)$

$\to \begin{bmatrix} 1 & -2 & | & -1 \\ 0 & 1 & | & \tfrac{3}{4} \end{bmatrix}$ $(R_2 = \tfrac{1}{8}r_2)$

$\to \begin{bmatrix} 1 & 0 & | & \tfrac{1}{2} \\ 0 & 1 & | & \tfrac{3}{4} \end{bmatrix}$ $(R_1 = 2r_2 + r_1)$

The solution is $x=\tfrac{1}{2}, y=\tfrac{3}{4}$.

41. $\begin{cases} x+2y=4 \\ 2x+4y=8 \end{cases}$

Write the augmented matrix:
$\begin{bmatrix} 1 & 2 & | & 4 \\ 2 & 4 & | & 8 \end{bmatrix} \to \begin{bmatrix} 1 & 2 & | & 4 \\ 0 & 0 & | & 0 \end{bmatrix}$ $(R_2 = -2r_1 + r_2)$

This is a dependent system and the solution is $x = 4 - 2y$, y is any real number.

43. $\begin{cases} 2x+3y=6 \\ x-y=\tfrac{1}{2} \end{cases}$

Write the augmented matrix:
$\begin{bmatrix} 2 & 3 & | & 6 \\ 1 & -1 & | & \tfrac{1}{2} \end{bmatrix} \to \begin{bmatrix} 1 & \tfrac{3}{2} & | & 3 \\ 1 & -1 & | & \tfrac{1}{2} \end{bmatrix}$ $(R_1 = \tfrac{1}{2}r_1)$

$\to \begin{bmatrix} 1 & \tfrac{3}{2} & | & 3 \\ 0 & -\tfrac{5}{2} & | & -\tfrac{5}{2} \end{bmatrix}$ $(R_2 = -r_1 + r_2)$

$\to \begin{bmatrix} 1 & \tfrac{3}{2} & | & 3 \\ 0 & 1 & | & 1 \end{bmatrix}$ $(R_2 = -\tfrac{2}{5}r_2)$

$\to \begin{bmatrix} 1 & 0 & | & \tfrac{3}{2} \\ 0 & 1 & | & 1 \end{bmatrix}$ $(R_1 = -\tfrac{3}{2}r_2 + r_1)$

The solution is $x=\tfrac{3}{2}, y=1$.

45. $\begin{cases} 3x-5y=3 \\ 15x+5y=21 \end{cases}$

Write the augmented matrix:
$\begin{bmatrix} 3 & -5 & | & 3 \\ 15 & 5 & | & 21 \end{bmatrix} \to \begin{bmatrix} 1 & -\tfrac{5}{3} & | & 1 \\ 15 & 5 & | & 21 \end{bmatrix}$ $(R_1 = \tfrac{1}{3}r_1)$

$\to \begin{bmatrix} 1 & -\tfrac{5}{3} & | & 1 \\ 0 & 30 & | & 6 \end{bmatrix}$ $(R_2 = -15r_1 + r_2)$

$\to \begin{bmatrix} 1 & -\tfrac{5}{3} & | & 1 \\ 0 & 1 & | & \tfrac{1}{5} \end{bmatrix}$ $(R_2 = \tfrac{1}{30}r_2)$

$\to \begin{bmatrix} 1 & 0 & | & \tfrac{4}{3} \\ 0 & 1 & | & \tfrac{1}{5} \end{bmatrix}$ $(R_1 = \tfrac{5}{3}r_2 + r_1)$

The solution is $x=\tfrac{4}{3}, y=\tfrac{1}{5}$.

SSM: College Algebra EGU

Chapter 5: Systems of Equations and Inequalities

47. $\begin{cases} x - y = 6 \\ 2x - 3z = 16 \\ 2y + z = 4 \end{cases}$

Write the augmented matrix:
$$\begin{bmatrix} 1 & -1 & 0 & | & 6 \\ 2 & 0 & -3 & | & 16 \\ 0 & 2 & 1 & | & 4 \end{bmatrix}$$

$\rightarrow \begin{bmatrix} 1 & -1 & 0 & | & 6 \\ 0 & 2 & -3 & | & 4 \\ 0 & 2 & 1 & | & 4 \end{bmatrix} \quad (R_2 = -2r_1 + r_2)$

$\rightarrow \begin{bmatrix} 1 & -1 & 0 & | & 6 \\ 0 & 1 & -\frac{3}{2} & | & 2 \\ 0 & 2 & 1 & | & 4 \end{bmatrix} \quad (R_2 = \frac{1}{2} r_2)$

$\rightarrow \begin{bmatrix} 1 & 0 & -\frac{3}{2} & | & 8 \\ 0 & 1 & -\frac{3}{2} & | & 2 \\ 0 & 0 & 4 & | & 0 \end{bmatrix} \quad \begin{pmatrix} R_1 = r_2 + r_1 \\ R_3 = -2r_2 + r_3 \end{pmatrix}$

$\rightarrow \begin{bmatrix} 1 & 0 & -\frac{3}{2} & | & 8 \\ 0 & 1 & -\frac{3}{2} & | & 2 \\ 0 & 0 & 1 & | & 0 \end{bmatrix} \quad (R_3 = \frac{1}{4} r_3)$

$\rightarrow \begin{bmatrix} 1 & 0 & 0 & | & 8 \\ 0 & 1 & 0 & | & 2 \\ 0 & 0 & 1 & | & 0 \end{bmatrix} \quad \begin{pmatrix} R_1 = \frac{3}{2} r_3 + r_1 \\ R_2 = \frac{3}{2} r_3 + r_2 \end{pmatrix}$

The solution is $x = 8, y = 2, z = 0$.

49. $\begin{cases} x - 2y + 3z = 7 \\ 2x + y + z = 4 \\ -3x + 2y - 2z = -10 \end{cases}$

Write the augmented matrix:
$$\begin{bmatrix} 1 & -2 & 3 & | & 7 \\ 2 & 1 & 1 & | & 4 \\ -3 & 2 & -2 & | & -10 \end{bmatrix}$$

$\rightarrow \begin{bmatrix} 1 & -2 & 3 & | & 7 \\ 0 & 5 & -5 & | & -10 \\ 0 & -4 & 7 & | & 11 \end{bmatrix} \quad \begin{pmatrix} R_2 = -2r_1 + r_2 \\ R_3 = 3r_1 + r_3 \end{pmatrix}$

$\rightarrow \begin{bmatrix} 1 & -2 & 3 & | & 7 \\ 0 & 1 & -1 & | & -2 \\ 0 & -4 & 7 & | & 11 \end{bmatrix} \quad (R_2 = \frac{1}{5} r_2)$

$\rightarrow \begin{bmatrix} 1 & 0 & 1 & | & 3 \\ 0 & 1 & -1 & | & -2 \\ 0 & 0 & 3 & | & 3 \end{bmatrix} \quad \begin{pmatrix} R_1 = 2r_2 + r_1 \\ R_3 = 4r_2 + r_3 \end{pmatrix}$

$\rightarrow \begin{bmatrix} 1 & 0 & 1 & | & 3 \\ 0 & 1 & -1 & | & -2 \\ 0 & 0 & 1 & | & 1 \end{bmatrix} \quad (R_3 = \frac{1}{3} r_3)$

$\rightarrow \begin{bmatrix} 1 & 0 & 0 & | & 2 \\ 0 & 1 & 0 & | & -1 \\ 0 & 0 & 1 & | & 1 \end{bmatrix} \quad \begin{pmatrix} R_1 = -r_3 + r_1 \\ R_2 = r_3 + r_2 \end{pmatrix}$

The solution is $x = 2, y = -1, z = 1$.

51. $\begin{cases} 2x - 2y - 2z = 2 \\ 2x + 3y + z = 2 \\ 3x + 2y = 0 \end{cases}$

Write the augmented matrix:
$$\begin{bmatrix} 2 & -2 & -2 & | & 2 \\ 2 & 3 & 1 & | & 2 \\ 3 & 2 & 0 & | & 0 \end{bmatrix}$$

$\rightarrow \begin{bmatrix} 1 & -1 & -1 & | & 1 \\ 2 & 3 & 1 & | & 2 \\ 3 & 2 & 0 & | & 0 \end{bmatrix} \quad (R_1 = \frac{1}{2} r_1)$

$\rightarrow \begin{bmatrix} 1 & -1 & -1 & | & 1 \\ 0 & 5 & 3 & | & 0 \\ 0 & 5 & 3 & | & -3 \end{bmatrix} \quad \begin{pmatrix} R_2 = -2r_1 + r_2 \\ R_3 = -3r_1 + r_3 \end{pmatrix}$

$\rightarrow \begin{bmatrix} 1 & -1 & -1 & | & 1 \\ 0 & 5 & 3 & | & 0 \\ 0 & 0 & 0 & | & -3 \end{bmatrix} \quad (R_3 = -r_2 + r_3)$

There is no solution. The system is inconsistent.

53. $\begin{cases} -x+y+z=-1 \\ -x+2y-3z=-4 \\ 3x-2y-7z=0 \end{cases}$

Write the augmented matrix:
$\begin{bmatrix} -1 & 1 & 1 & | & -1 \\ -1 & 2 & -3 & | & -4 \\ 3 & -2 & -7 & | & 0 \end{bmatrix}$

$\rightarrow \begin{bmatrix} 1 & -1 & -1 & | & 1 \\ -1 & 2 & -3 & | & -4 \\ 3 & -2 & -7 & | & 0 \end{bmatrix} \quad (R_1 = -r_1)$

$\rightarrow \begin{bmatrix} 1 & -1 & -1 & | & 1 \\ 0 & 1 & -4 & | & -3 \\ 0 & 1 & -4 & | & -3 \end{bmatrix} \quad \begin{pmatrix} R_2 = r_1 + r_2 \\ R_3 = -3r_1 + r_3 \end{pmatrix}$

$\rightarrow \begin{bmatrix} 1 & 0 & -5 & | & -2 \\ 0 & 1 & -4 & | & -3 \\ 0 & 0 & 0 & | & 0 \end{bmatrix} \quad \begin{pmatrix} R_1 = r_2 + r_1 \\ R_3 = -r_2 + r_3 \end{pmatrix}$

The matrix in the last step represents the system
$\begin{cases} x - 5z = -2 \\ y - 4z = -3 \\ 0 = 0 \end{cases}$

The solution is $x = 5z - 2$, $y = 4z - 3$, z is any real number.

55. $\begin{cases} 2x - 2y + 3z = 6 \\ 4x - 3y + 2z = 0 \\ -2x + 3y - 7z = 1 \end{cases}$

Write the augmented matrix:
$\begin{bmatrix} 2 & -2 & 3 & | & 6 \\ 4 & -3 & 2 & | & 0 \\ -2 & 3 & -7 & | & 1 \end{bmatrix}$

$\rightarrow \begin{bmatrix} 1 & -1 & \frac{3}{2} & | & 3 \\ 4 & -3 & 2 & | & 0 \\ -2 & 3 & -7 & | & 1 \end{bmatrix} \quad (R_1 = \tfrac{1}{2}r_1)$

$\rightarrow \begin{bmatrix} 1 & -1 & \frac{3}{2} & | & 3 \\ 0 & 1 & -4 & | & -12 \\ 0 & 1 & -4 & | & 7 \end{bmatrix} \quad \begin{pmatrix} R_2 = -4r_1 + r_2 \\ R_3 = 2r_1 + r_3 \end{pmatrix}$

$\rightarrow \begin{bmatrix} 1 & 0 & -\frac{5}{2} & | & -9 \\ 0 & 1 & -4 & | & -12 \\ 0 & 0 & 0 & | & 19 \end{bmatrix} \quad \begin{pmatrix} R_1 = r_2 + r_1 \\ R_3 = -r_2 + r_3 \end{pmatrix}$

There is no solution. The system is inconsistent.

57. $\begin{cases} x + y - z = 6 \\ 3x - 2y + z = -5 \\ x + 3y - 2z = 14 \end{cases}$

Write the augmented matrix:
$\begin{bmatrix} 1 & 1 & -1 & | & 6 \\ 3 & -2 & 1 & | & -5 \\ 1 & 3 & -2 & | & 14 \end{bmatrix}$

$\rightarrow \begin{bmatrix} 1 & 1 & -1 & | & 6 \\ 0 & -5 & 4 & | & -23 \\ 0 & 2 & -1 & | & 8 \end{bmatrix} \quad \begin{pmatrix} R_2 = -3r_1 + r_2 \\ R_3 = -r_1 + r_3 \end{pmatrix}$

$\rightarrow \begin{bmatrix} 1 & 1 & -1 & | & 6 \\ 0 & 1 & -\frac{4}{5} & | & \frac{23}{5} \\ 0 & 2 & -1 & | & 8 \end{bmatrix} \quad (R_2 = -\tfrac{1}{5}r_2)$

$\rightarrow \begin{bmatrix} 1 & 0 & -\frac{1}{5} & | & \frac{7}{5} \\ 0 & 1 & -\frac{4}{5} & | & \frac{23}{5} \\ 0 & 0 & \frac{3}{5} & | & -\frac{6}{5} \end{bmatrix} \quad \begin{pmatrix} R_1 = -r_2 + r_1 \\ R_3 = -2r_2 + r_3 \end{pmatrix}$

$\rightarrow \begin{bmatrix} 1 & 0 & -\frac{1}{5} & | & \frac{7}{5} \\ 0 & 1 & -\frac{4}{5} & | & \frac{23}{5} \\ 0 & 0 & 1 & | & -2 \end{bmatrix} \quad (R_3 = \tfrac{5}{3}r_3)$

$\rightarrow \begin{bmatrix} 1 & 0 & 0 & | & 1 \\ 0 & 1 & 0 & | & 3 \\ 0 & 0 & 1 & | & -2 \end{bmatrix} \quad \begin{pmatrix} R_1 = \tfrac{1}{5}r_3 + r_1 \\ R_2 = \tfrac{4}{5}r_3 + r_2 \end{pmatrix}$

The solution is $x = 1$, $y = 3$, $z = -2$.

SSM: College Algebra EGU   **Chapter 5:** Systems of Equations and Inequalities

59. $\begin{cases} x+2y-z=-3 \\ 2x-4y+z=-7 \\ -2x+2y-3z=4 \end{cases}$

Write the augmented matrix:

$\begin{bmatrix} 1 & 2 & -1 & | & -3 \\ 2 & -4 & 1 & | & -7 \\ -2 & 2 & -3 & | & 4 \end{bmatrix}$

$\rightarrow \begin{bmatrix} 1 & 2 & -1 & | & -3 \\ 0 & -8 & 3 & | & -1 \\ 0 & 6 & -5 & | & -2 \end{bmatrix} \quad \begin{pmatrix} R_2 = -2r_1 + r_2 \\ R_3 = 2r_1 + r_3 \end{pmatrix}$

$\rightarrow \begin{bmatrix} 1 & 2 & -1 & | & -3 \\ 0 & 1 & -\frac{3}{8} & | & \frac{1}{8} \\ 0 & 6 & -5 & | & -2 \end{bmatrix} \quad \left(R_2 = -\frac{1}{8}r_2\right)$

$\rightarrow \begin{bmatrix} 1 & 0 & -\frac{1}{4} & | & -\frac{13}{4} \\ 0 & 1 & -\frac{3}{8} & | & \frac{1}{8} \\ 0 & 0 & -\frac{11}{4} & | & -\frac{11}{4} \end{bmatrix} \quad \begin{pmatrix} R_1 = -2r_2 + r_1 \\ R_3 = -6r_2 + r_3 \end{pmatrix}$

$\rightarrow \begin{bmatrix} 1 & 0 & -\frac{1}{4} & | & -\frac{13}{4} \\ 0 & 1 & -\frac{3}{8} & | & \frac{1}{8} \\ 0 & 0 & 1 & | & 1 \end{bmatrix} \quad \left(R_3 = -\frac{4}{11}r_3\right)$

$\rightarrow \begin{bmatrix} 1 & 0 & 0 & | & -3 \\ 0 & 1 & 0 & | & \frac{1}{2} \\ 0 & 0 & 1 & | & 1 \end{bmatrix} \quad \begin{pmatrix} R_1 = \frac{1}{4}r_3 + r_1 \\ R_2 = \frac{3}{8}r_3 + r_2 \end{pmatrix}$

The solution is $x=-3, y=\frac{1}{2}, z=1$.

61. $\begin{cases} 3x+y-z=\frac{2}{3} \\ 2x-y+z=1 \\ 4x+2y=\frac{8}{3} \end{cases}$

Write the augmented matrix:

$\begin{bmatrix} 3 & 1 & -1 & | & \frac{2}{3} \\ 2 & -1 & 1 & | & 1 \\ 4 & 2 & 0 & | & \frac{8}{3} \end{bmatrix}$

$\rightarrow \begin{bmatrix} 1 & \frac{1}{3} & -\frac{1}{3} & | & \frac{2}{9} \\ 2 & -1 & 1 & | & 1 \\ 4 & 2 & 0 & | & \frac{8}{3} \end{bmatrix} \quad \left(R_1 = \frac{1}{3}r_1\right)$

$\rightarrow \begin{bmatrix} 1 & \frac{1}{3} & -\frac{1}{3} & | & \frac{2}{9} \\ 0 & -\frac{5}{3} & \frac{5}{3} & | & \frac{5}{9} \\ 0 & \frac{2}{3} & \frac{4}{3} & | & \frac{16}{9} \end{bmatrix} \quad \begin{pmatrix} R_2 = -2r_1 + r_2 \\ R_3 = -4r_1 + r_3 \end{pmatrix}$

$\rightarrow \begin{bmatrix} 1 & \frac{1}{3} & -\frac{1}{3} & | & \frac{2}{9} \\ 0 & 1 & -1 & | & -\frac{1}{3} \\ 0 & \frac{2}{3} & \frac{4}{3} & | & \frac{16}{9} \end{bmatrix} \quad \left(R_2 = -\frac{3}{5}r_2\right)$

$\rightarrow \begin{bmatrix} 1 & 0 & 0 & | & \frac{1}{3} \\ 0 & 1 & -1 & | & -\frac{1}{3} \\ 0 & 0 & 2 & | & 2 \end{bmatrix} \quad \begin{pmatrix} R_1 = -\frac{1}{3}r_2 + r_1 \\ R_3 = -\frac{2}{3}r_2 + r_3 \end{pmatrix}$

$\rightarrow \begin{bmatrix} 1 & 0 & 0 & | & \frac{1}{3} \\ 0 & 1 & -1 & | & -\frac{1}{3} \\ 0 & 0 & 1 & | & 1 \end{bmatrix} \quad \left(R_3 = \frac{1}{2}r_3\right)$

$\rightarrow \begin{bmatrix} 1 & 0 & 0 & | & \frac{1}{3} \\ 0 & 1 & 0 & | & \frac{2}{3} \\ 0 & 0 & 1 & | & 1 \end{bmatrix} \quad (R_2 = r_3 + r_2)$

The solution is $x=\frac{1}{3}, y=\frac{2}{3}, z=1$.

63. $\begin{cases} x+y+z+w=4 \\ 2x-y+z=0 \\ 3x+2y+z-w=6 \\ x-2y-2z+2w=-1 \end{cases}$

Write the augmented matrix:

$\begin{bmatrix} 1 & 1 & 1 & 1 & | & 4 \\ 2 & -1 & 1 & 0 & | & 0 \\ 3 & 2 & 1 & -1 & | & 6 \\ 1 & -2 & -2 & 2 & | & -1 \end{bmatrix}$

$\rightarrow \begin{bmatrix} 1 & 1 & 1 & 1 & | & 4 \\ 0 & -3 & -1 & -2 & | & -8 \\ 0 & -1 & -2 & -4 & | & -6 \\ 0 & -3 & -3 & 1 & | & -5 \end{bmatrix} \quad \begin{pmatrix} R_2 = -2r_1 + r_2 \\ R_3 = -3r_1 + r_3 \\ R_4 = -r_1 + r_4 \end{pmatrix}$

$\rightarrow \begin{bmatrix} 1 & 1 & 1 & 1 & | & 4 \\ 0 & -1 & -2 & -4 & | & -6 \\ 0 & -3 & -1 & -2 & | & -8 \\ 0 & -3 & -3 & 1 & | & -5 \end{bmatrix} \quad \begin{pmatrix} \text{Interchange} \\ r_2 \text{ and } r_3 \end{pmatrix}$

$$\rightarrow \begin{bmatrix} 1 & 1 & 1 & 1 & | & 4 \\ 0 & 1 & 2 & 4 & | & 6 \\ 0 & -3 & -1 & -2 & | & -8 \\ 0 & -3 & -3 & 1 & | & -5 \end{bmatrix} (R_2 = -r_2)$$

$$\rightarrow \begin{bmatrix} 1 & 0 & -1 & -3 & | & -2 \\ 0 & 1 & 2 & 4 & | & 6 \\ 0 & 0 & 5 & 10 & | & 10 \\ 0 & 0 & 3 & 13 & | & 13 \end{bmatrix} \begin{pmatrix} R_1 = -r_2 + r_1 \\ R_3 = 3r_2 + r_3 \\ R_4 = 3r_2 + r_4 \end{pmatrix}$$

$$\rightarrow \begin{bmatrix} 1 & 0 & -1 & -3 & | & -2 \\ 0 & 1 & 2 & 4 & | & 6 \\ 0 & 0 & 1 & 2 & | & 2 \\ 0 & 0 & 3 & 13 & | & 13 \end{bmatrix} (R_3 = \tfrac{1}{5} r_3)$$

$$\rightarrow \begin{bmatrix} 1 & 0 & 0 & -1 & | & 0 \\ 0 & 1 & 0 & 0 & | & 2 \\ 0 & 0 & 1 & 2 & | & 2 \\ 0 & 0 & 0 & 7 & | & 7 \end{bmatrix} \begin{pmatrix} R_1 = r_3 + r_1 \\ R_2 = -2r_3 + r_2 \\ R_4 = -3r_3 + r_4 \end{pmatrix}$$

$$\rightarrow \begin{bmatrix} 1 & 0 & 0 & -1 & | & 0 \\ 0 & 1 & 0 & 0 & | & 2 \\ 0 & 0 & 1 & 2 & | & 2 \\ 0 & 0 & 0 & 1 & | & 1 \end{bmatrix} (R_4 = \tfrac{1}{7} r_4)$$

$$\rightarrow \begin{bmatrix} 1 & 0 & 0 & 0 & | & 1 \\ 0 & 1 & 0 & 0 & | & 2 \\ 0 & 0 & 1 & 0 & | & 0 \\ 0 & 0 & 0 & 1 & | & 1 \end{bmatrix} \begin{pmatrix} R_1 = r_4 + r_1 \\ R_3 = -2r_4 + r_3 \end{pmatrix}$$

The solution is $x = 1$, $y = 2$, $z = 0$, $w = 1$.

65. $\begin{cases} x + 2y + z = 1 \\ 2x - y + 2z = 2 \\ 3x + y + 3z = 3 \end{cases}$

Write the augmented matrix:

$$\begin{bmatrix} 1 & 2 & 1 & | & 1 \\ 2 & -1 & 2 & | & 2 \\ 3 & 1 & 3 & | & 3 \end{bmatrix}$$

$$\rightarrow \begin{bmatrix} 1 & 2 & 1 & | & 1 \\ 0 & -5 & 0 & | & 0 \\ 0 & -5 & 0 & | & 0 \end{bmatrix} \begin{pmatrix} R_2 = -2r_1 + r_2 \\ R_3 = -3r_1 + r_3 \end{pmatrix}$$

$$\rightarrow \begin{bmatrix} 1 & 2 & 1 & | & 1 \\ 0 & -5 & 0 & | & 0 \\ 0 & 0 & 0 & | & 0 \end{bmatrix} (R_3 = -r_2 + r_3)$$

The matrix in the last step represents the system

$\begin{cases} x + 2y + z = 1 \\ -5y = 0 \\ 0 = 0 \end{cases}$

Substitute and solve:
$-5y = 0 \qquad x + 2(0) + z = 1$
$y = 0 \qquad\quad x + z = 1$
$\qquad\qquad\qquad z = 1 - x$

The solution is $y = 0$, $z = 1 - x$, x is any real number.

67. $\begin{cases} x - y + z = 5 \\ 3x + 2y - 2z = 0 \end{cases}$

Write the augmented matrix:

$$\begin{bmatrix} 1 & -1 & 1 & | & 5 \\ 3 & 2 & -2 & | & 0 \end{bmatrix}$$

$$\rightarrow \begin{bmatrix} 1 & -1 & 1 & | & 5 \\ 0 & 5 & -5 & | & -15 \end{bmatrix} (R_2 = -3r_1 + r_2)$$

$$\rightarrow \begin{bmatrix} 1 & -1 & 1 & | & 5 \\ 0 & 1 & -1 & | & -3 \end{bmatrix} (R_2 = \tfrac{1}{5} r_2)$$

$$\rightarrow \begin{bmatrix} 1 & 0 & 0 & | & 2 \\ 0 & 1 & -1 & | & -3 \end{bmatrix} (R_1 = r_2 + r_1)$$

The matrix in the last step represents the system

$\begin{cases} x = 2 \\ y - z = -3 \end{cases}$

Thus, the solution is $x = 2$, $y = z - 3$, z is any real number.

69. $\begin{cases} 2x + 3y - z = 3 \\ x - y - z = 0 \\ -x + y + z = 0 \\ x + y + 3z = 5 \end{cases}$

Write the augmented matrix:

$$\begin{bmatrix} 2 & 3 & -1 & | & 3 \\ 1 & -1 & -1 & | & 0 \\ -1 & 1 & 1 & | & 0 \\ 1 & 1 & 3 & | & 5 \end{bmatrix}$$

$$\rightarrow \begin{bmatrix} 1 & -1 & -1 & | & 0 \\ 2 & 3 & -1 & | & 3 \\ -1 & 1 & 1 & | & 0 \\ 1 & 1 & 3 & | & 5 \end{bmatrix} \begin{pmatrix} \text{interchange} \\ r_1 \text{ and } r_2 \end{pmatrix}$$

$$\rightarrow \begin{bmatrix} 1 & -1 & -1 & | & 0 \\ 0 & 5 & 1 & | & 3 \\ 0 & 0 & 0 & | & 0 \\ 0 & 2 & 4 & | & 5 \end{bmatrix} \begin{pmatrix} R_2 = -2r_1 + r_2 \\ R_3 = r_1 + r_3 \\ R_4 = -r_1 + r_4 \end{pmatrix}$$

$$\rightarrow \begin{bmatrix} 1 & -1 & -1 & | & 0 \\ 0 & 5 & 1 & | & 3 \\ 0 & 2 & 4 & | & 5 \\ 0 & 0 & 0 & | & 0 \end{bmatrix} \begin{pmatrix} \text{interchange} \\ r_3 \text{ and } r_4 \end{pmatrix}$$

$$\rightarrow \begin{bmatrix} 1 & -1 & -1 & | & 0 \\ 0 & 1 & -7 & | & -7 \\ 0 & 2 & 4 & | & 5 \\ 0 & 0 & 0 & | & 0 \end{bmatrix} (R_2 = -2r_3 + r_2)$$

$$\rightarrow \begin{bmatrix} 1 & 0 & -8 & | & -7 \\ 0 & 1 & -7 & | & -7 \\ 0 & 0 & 18 & | & 19 \\ 0 & 0 & 0 & | & 0 \end{bmatrix} \begin{pmatrix} R_1 = r_2 + r_1 \\ R_3 = -2r_2 + r_3 \end{pmatrix}$$

$$\rightarrow \begin{bmatrix} 1 & 0 & -8 & | & -7 \\ 0 & 1 & -7 & | & -7 \\ 0 & 0 & 1 & | & \frac{19}{18} \\ 0 & 0 & 0 & | & 0 \end{bmatrix} (R_3 = \tfrac{1}{18} r_3)$$

The matrix in the last step represents the system
$$\begin{cases} x - 8z = -7 \\ y - 7z = -7 \\ z = \dfrac{19}{18} \end{cases}$$

Substitute and solve:
$$y - 7\left(\frac{19}{18}\right) = 7 \qquad x - 8\left(\frac{19}{18}\right) = -7$$
$$y = \frac{7}{18} \qquad\qquad x = \frac{13}{9}$$

Thus, the solution is $x = \dfrac{13}{9}$, $y = \dfrac{7}{18}$, $z = \dfrac{19}{18}$.

71. $\begin{cases} 4x + y + z - w = 4 \\ x - y + 2z + 3w = 3 \end{cases}$

Write the augmented matrix:
$$\begin{bmatrix} 4 & 1 & 1 & -1 & | & 4 \\ 1 & -1 & 2 & 3 & | & 3 \end{bmatrix}$$

$$\rightarrow \begin{bmatrix} 1 & -1 & 2 & 3 & | & 3 \\ 4 & 1 & 1 & -1 & | & 4 \end{bmatrix} \begin{pmatrix} \text{interchange} \\ r_1 \text{ and } r_2 \end{pmatrix}$$

$$\rightarrow \begin{bmatrix} 1 & -1 & 2 & 3 & | & 3 \\ 0 & 5 & -7 & -13 & | & -8 \end{bmatrix} (R_2 = -4r_1 + r_2)$$

The matrix in the last step represents the system
$$\begin{cases} x - y + 2z + 3w = 3 \\ 5y - 7z - 13w = -8 \end{cases}$$

The second equation yields
$$5y - 7z - 13w = -8$$
$$5y = 7z + 13w - 8$$
$$y = \frac{7}{5}z + \frac{13}{5}w - \frac{8}{5}$$

The first equation yields
$$x - y + 2z + 3w = 3$$
$$x = 3 + y - 2z - 3w$$

Substituting for y:
$$x = 3 + \left(-\frac{8}{5} + \frac{7}{5}z + \frac{13}{5}w\right) - 2z - 3w$$
$$x = -\frac{3}{5}z - \frac{2}{5}w + \frac{7}{5}$$

Thus, the solution is $x = -\dfrac{3}{5}z - \dfrac{2}{5}w + \dfrac{7}{5}$, $y = \dfrac{7}{5}z + \dfrac{13}{5}w - \dfrac{8}{5}$, z and w are any real numbers.

73. Each of the points must satisfy the equation $y = ax^2 + bx + c$.

(1, 2): $\quad 2 = a + b + c$
(−2, −7): $\quad -7 = 4a - 2b + c$
(2, −3): $\quad -3 = 4a + 2b + c$

Set up a matrix and solve:
$$\begin{bmatrix} 1 & 1 & 1 & | & 2 \\ 4 & -2 & 1 & | & -7 \\ 4 & 2 & 1 & | & -3 \end{bmatrix}$$

$$\rightarrow \begin{bmatrix} 1 & 1 & 1 & | & 2 \\ 0 & -6 & -3 & | & -15 \\ 0 & -2 & -3 & | & -11 \end{bmatrix} \begin{pmatrix} R_2 = -4r_1 + r_2 \\ R_3 = -4r_1 + r_3 \end{pmatrix}$$

$$\rightarrow \begin{bmatrix} 1 & 1 & 1 & | & 2 \\ 0 & 1 & \tfrac{1}{2} & | & \tfrac{5}{2} \\ 0 & -2 & -3 & | & -11 \end{bmatrix} (R_2 = -\tfrac{1}{6} r_2)$$

$$\rightarrow \begin{bmatrix} 1 & 0 & \tfrac{1}{2} & | & -\tfrac{1}{2} \\ 0 & 1 & \tfrac{1}{2} & | & \tfrac{5}{2} \\ 0 & 0 & -2 & | & -6 \end{bmatrix} \begin{pmatrix} R_1 = -r_2 + r_1 \\ R_3 = 2r_2 + r_3 \end{pmatrix}$$

$$\rightarrow \begin{bmatrix} 1 & 0 & \frac{1}{2} & -\frac{1}{2} \\ 0 & 1 & \frac{1}{2} & \frac{5}{2} \\ 0 & 0 & 1 & 3 \end{bmatrix} \rightarrow \quad \left(R_3 = -\frac{1}{2}r_3\right)$$

$$\rightarrow \begin{bmatrix} 1 & 0 & 0 & -2 \\ 0 & 1 & 0 & 1 \\ 0 & 0 & 1 & 3 \end{bmatrix} \quad \begin{pmatrix} R_1 = -\frac{1}{2}r_3 + r_1 \\ R_2 = -\frac{1}{2}r_3 + r_2 \end{pmatrix}$$

The solution is $a = -2, b = 1, c = 3$; so the equation is $y = -2x^2 + x + 3$.

75. Each of the points must satisfy the equation
$f(x) = ax^3 + bx^2 + cx + d$.
$f(-3) = -12$: $\quad -27a + 9b - 3c + d = -112$
$f(-1) = -2$: $\quad -a + b - c + d = -2$
$f(1) = 4$: $\quad a + b + c + d = 4$
$f(2) = 13$: $\quad 8a + 4b + 2c + d = 13$

Set up a matrix and solve:
$$\begin{bmatrix} -27 & 9 & -3 & 1 & -112 \\ -1 & 1 & -1 & 1 & -2 \\ 1 & 1 & 1 & 1 & 4 \\ 8 & 4 & 2 & 1 & 13 \end{bmatrix}$$

$$\rightarrow \begin{bmatrix} 1 & 1 & 1 & 1 & 4 \\ -1 & 1 & -1 & 1 & -2 \\ -27 & 9 & -3 & 1 & -112 \\ 8 & 4 & 2 & 1 & 13 \end{bmatrix} \quad \begin{pmatrix} \text{Interchange} \\ r_3 \text{ and } r_1 \end{pmatrix}$$

$$\rightarrow \begin{bmatrix} 1 & 1 & 1 & 1 & 4 \\ 0 & 2 & 0 & 2 & 2 \\ 0 & 36 & 24 & 28 & -4 \\ 0 & -4 & -6 & -7 & -19 \end{bmatrix} \quad \begin{pmatrix} R_2 = r_1 + r_2 \\ R_3 = 27r_1 + r_3 \\ R_4 = -8r_1 + r_4 \end{pmatrix}$$

$$\rightarrow \begin{bmatrix} 1 & 1 & 1 & 1 & 4 \\ 0 & 1 & 0 & 1 & 1 \\ 0 & 36 & 24 & 28 & -4 \\ 0 & -4 & -6 & -7 & -19 \end{bmatrix} \quad \left(R_2 = \frac{1}{2}r_2\right)$$

$$\rightarrow \begin{bmatrix} 1 & 0 & 1 & 0 & 3 \\ 0 & 1 & 0 & 1 & 1 \\ 0 & 0 & 24 & -8 & -40 \\ 0 & 0 & -6 & -3 & -15 \end{bmatrix} \quad \begin{pmatrix} R_1 = -r_2 + r_1 \\ R_3 = -36r_2 + r_3 \\ R_4 = 4r_2 + r_4 \end{pmatrix}$$

$$\rightarrow \begin{bmatrix} 1 & 0 & 1 & 0 & 3 \\ 0 & 1 & 0 & 1 & 1 \\ 0 & 0 & 1 & -\frac{5}{3} & -\frac{5}{3} \\ 0 & 0 & -6 & -3 & -15 \end{bmatrix} \quad \left(R_3 = \frac{1}{24}r_3\right)$$

$$\rightarrow \begin{bmatrix} 1 & 0 & 0 & \frac{1}{3} & \frac{14}{3} \\ 0 & 1 & 0 & 1 & 1 \\ 0 & 0 & 1 & -\frac{1}{3} & -\frac{5}{3} \\ 0 & 0 & 0 & -5 & -25 \end{bmatrix} \quad \begin{pmatrix} R_1 = -r_3 + r_1 \\ R_4 = 6r_3 + r_4 \end{pmatrix}$$

$$\rightarrow \begin{bmatrix} 1 & 0 & 0 & \frac{1}{3} & \frac{14}{3} \\ 0 & 1 & 0 & 1 & 1 \\ 0 & 0 & 1 & -\frac{1}{3} & -\frac{5}{3} \\ 0 & 0 & 0 & 1 & 5 \end{bmatrix} \quad \left(R_4 = -\frac{1}{5}r_4\right)$$

$$\rightarrow \begin{bmatrix} 1 & 0 & 0 & 0 & 3 \\ 0 & 1 & 0 & 0 & -4 \\ 0 & 0 & 1 & 0 & 0 \\ 0 & 0 & 0 & 1 & 5 \end{bmatrix} \quad \begin{pmatrix} R_1 = -\frac{1}{3}r_4 + r_1 \\ R_2 = -r_4 + r_2 \\ R_3 = \frac{1}{3}r_4 + r_3 \end{pmatrix}$$

The solution is $a = 3, b = -4, c = 0, d = 5$; so the equation is $f(x) = 3x^3 - 4x^2 + 5$.

77. Let x = the number of servings of salmon steak.
Let y = the number of servings of baked eggs.
Let z = the number of servings of acorn squash.
Protein equation: $30x + 15y + 3z = 78$
Carbohydrate equation: $20x + 2y + 25z = 59$
Vitamin A equation: $2x + 20y + 32z = 75$

Set up a matrix and solve:
$$\begin{bmatrix} 30 & 15 & 3 & 78 \\ 20 & 2 & 25 & 59 \\ 2 & 20 & 32 & 75 \end{bmatrix}$$

$$\rightarrow \begin{bmatrix} 2 & 20 & 32 & 75 \\ 20 & 2 & 25 & 59 \\ 30 & 15 & 3 & 78 \end{bmatrix} \quad \begin{pmatrix} \text{Interchange} \\ r_3 \text{ and } r_1 \end{pmatrix}$$

$$\rightarrow \begin{bmatrix} 1 & 10 & 16 & 37.5 \\ 20 & 2 & 25 & 59 \\ 30 & 15 & 3 & 78 \end{bmatrix} \quad \left(R_1 = \frac{1}{2}r_1\right)$$

$$\rightarrow \begin{bmatrix} 1 & 10 & 16 & 37.5 \\ 0 & -198 & -295 & -691 \\ 0 & -285 & -477 & -1047 \end{bmatrix} \quad \begin{pmatrix} R_2 = -20r_1 + r_2 \\ R_3 = -30r_1 + r_3 \end{pmatrix}$$

$$\rightarrow \begin{bmatrix} 1 & 10 & 16 & 37.5 \\ 0 & -198 & -295 & -691 \\ 0 & 0 & -\frac{3457}{66} & -\frac{3457}{66} \end{bmatrix} \quad \left(R_3 = -\frac{95}{66}r_2 + r_3\right)$$

352

$$\rightarrow \begin{bmatrix} 1 & 10 & 16 & | & 37.5 \\ 0 & -198 & -295 & | & -691 \\ 0 & 0 & 1 & | & 1 \end{bmatrix} \quad \left(R_3 = -\tfrac{66}{3457} r_3\right) \text{ Sub}$$

stitute $z = 1$ and solve:
$$-198y - 295(1) = -691$$
$$-198y = -396$$
$$y = 2$$
$$x + 10(2) + 16(1) = 37.5$$
$$x + 36 = 37.5$$
$$x = 1.5$$

The dietitian should serve 1.5 servings of salmon steak, 2 servings of baked eggs, and 1 serving of acorn squash.

79. Let $x =$ the amount invested in Treasury bills.
Let $y =$ the amount invested in Treasury bonds.
Let $z =$ the amount invested in corporate bonds.
Total investment equation: $x + y + z = 10{,}000$
Annual income equation:
$$0.06x + 0.07y + 0.08z = 680$$
Condition on investment equation:
$$z = 0.5x$$
$$x - 2z = 0$$

Set up a matrix and solve:
$$\begin{bmatrix} 1 & 1 & 1 & | & 10{,}000 \\ 0.06 & 0.07 & 0.08 & | & 680 \\ 1 & 0 & -2 & | & 0 \end{bmatrix}$$

$$\rightarrow \begin{bmatrix} 1 & 1 & 1 & | & 10{,}000 \\ 0 & 0.01 & 0.02 & | & 80 \\ 0 & -1 & -3 & | & -10{,}000 \end{bmatrix} \quad \begin{pmatrix} R_2 = -0.06 r_1 + r_2 \\ R_3 = -r_1 + r_3 \end{pmatrix}$$

$$\rightarrow \begin{bmatrix} 1 & 1 & 1 & | & 10{,}000 \\ 0 & 1 & 2 & | & 8000 \\ 0 & -1 & -3 & | & -10{,}000 \end{bmatrix} \quad (R_2 = 100 r_2)$$

$$\rightarrow \begin{bmatrix} 1 & 0 & -1 & | & 2000 \\ 0 & 1 & 2 & | & 8000 \\ 0 & 0 & -1 & | & -2000 \end{bmatrix} \quad \begin{pmatrix} R_1 = -r_2 + r_1 \\ R_3 = r_2 + r_3 \end{pmatrix}$$

$$\rightarrow \begin{bmatrix} 1 & 0 & -1 & | & 2000 \\ 0 & 1 & 2 & | & 8000 \\ 0 & 0 & 1 & | & 2000 \end{bmatrix} \quad (R_3 = -r_3)$$

$$\rightarrow \begin{bmatrix} 1 & 0 & 0 & | & 4000 \\ 0 & 1 & 0 & | & 4000 \\ 0 & 0 & 1 & | & 2000 \end{bmatrix} \quad \begin{pmatrix} R_1 = r_3 + r_1 \\ R_2 = -2r_3 + r_2 \end{pmatrix}$$

$4000 in Treasury bills, $4000 in Treasury bonds, and $2000 in corporate bonds.

81. Let $x =$ the number of Deltas produced.
Let $y =$ the number of Betas produced.
Let $z =$ the number of Sigmas produced.
Painting equation: $10x + 16y + 8z = 240$
Drying equation: $3x + 5y + 2z = 69$
Polishing equation: $2x + 3y + z = 41$
Set up a matrix and solve:
$$\begin{bmatrix} 10 & 16 & 8 & | & 240 \\ 3 & 5 & 2 & | & 69 \\ 2 & 3 & 1 & | & 41 \end{bmatrix}$$

$$\rightarrow \begin{bmatrix} 1 & 1 & 2 & | & 33 \\ 3 & 5 & 2 & | & 69 \\ 2 & 3 & 1 & | & 41 \end{bmatrix} \quad (R_1 = -3r_2 + r_1)$$

$$\rightarrow \begin{bmatrix} 1 & 1 & 2 & | & 33 \\ 0 & 2 & -4 & | & -30 \\ 0 & 1 & -3 & | & -25 \end{bmatrix} \quad \begin{pmatrix} R_2 = -3r_1 + r_2 \\ R_3 = -2r_1 + r_3 \end{pmatrix}$$

$$\rightarrow \begin{bmatrix} 1 & 1 & 2 & | & 33 \\ 0 & 1 & -2 & | & -15 \\ 0 & 1 & -3 & | & -25 \end{bmatrix} \quad \left(R_2 = \tfrac{1}{2} r_2\right)$$

$$\rightarrow \begin{bmatrix} 1 & 0 & 4 & | & 48 \\ 0 & 1 & -2 & | & -15 \\ 0 & 0 & -1 & | & -10 \end{bmatrix} \quad \begin{pmatrix} R_1 = r_1 - r_2 \\ R_3 = r_3 - r_2 \end{pmatrix}$$

$$\rightarrow \begin{bmatrix} 1 & 0 & 4 & | & 48 \\ 0 & 1 & -2 & | & -15 \\ 0 & 0 & 1 & | & 10 \end{bmatrix} \quad (R_3 = -r_3)$$

$$\rightarrow \begin{bmatrix} 1 & 0 & 0 & | & 8 \\ 0 & 1 & 0 & | & 5 \\ 0 & 0 & 1 & | & 10 \end{bmatrix} \quad \begin{pmatrix} R_1 = -4r_3 + r_1 \\ R_2 = 2r_3 + r_2 \end{pmatrix}$$

The company should produce 8 Deltas, 5 Betas, and 10 Sigmas.

83. Rewrite the system to set up the matrix and solve:

$$\begin{cases} -4+8-2I_2 = 0 \\ 8 = 5I_4 + I_1 \\ 4 = 3I_3 + I_1 \\ I_3 + I_4 = I_1 \end{cases} \rightarrow \begin{cases} 2I_2 = 4 \\ I_1 + 5I_4 = 8 \\ I_1 + 3I_3 = 4 \\ I_1 - I_3 - I_4 = 0 \end{cases}$$

$$\begin{bmatrix} 0 & 2 & 0 & 0 & | & 4 \\ 1 & 0 & 0 & 5 & | & 8 \\ 1 & 0 & 3 & 0 & | & 4 \\ 1 & 0 & -1 & -1 & | & 0 \end{bmatrix}$$

$$\rightarrow \begin{bmatrix} 1 & 0 & 0 & 5 & | & 8 \\ 0 & 2 & 0 & 0 & | & 4 \\ 1 & 0 & 3 & 0 & | & 4 \\ 1 & 0 & -1 & -1 & | & 0 \end{bmatrix} \begin{pmatrix} \text{Interchange} \\ r_2 \text{ and } r_1 \end{pmatrix}$$

$$\rightarrow \begin{bmatrix} 1 & 0 & 0 & 5 & | & 8 \\ 0 & 1 & 0 & 0 & | & 2 \\ 0 & 0 & 3 & -5 & | & -4 \\ 0 & 0 & -1 & -6 & | & -8 \end{bmatrix} \begin{pmatrix} R_2 = \frac{1}{2}r_2 \\ R_3 = -r_1 + r_3 \\ R_4 = -r_1 + r_4 \end{pmatrix}$$

$$\rightarrow \begin{bmatrix} 1 & 0 & 0 & 5 & | & 8 \\ 0 & 1 & 0 & 0 & | & 2 \\ 0 & 0 & -1 & -6 & | & -8 \\ 0 & 0 & 3 & -5 & | & -4 \end{bmatrix} \begin{pmatrix} \text{Interchange} \\ r_3 \text{ and } r_4 \end{pmatrix}$$

$$\rightarrow \begin{bmatrix} 1 & 0 & 0 & 5 & | & 8 \\ 0 & 1 & 0 & 0 & | & 2 \\ 0 & 0 & 1 & 6 & | & 8 \\ 0 & 0 & 0 & -23 & | & -28 \end{bmatrix} \begin{pmatrix} R_3 = -r_3 \\ R_4 = -3r_3 + r_4 \end{pmatrix}$$

$$\rightarrow \begin{bmatrix} 1 & 0 & 0 & 5 & | & 8 \\ 0 & 1 & 0 & 0 & | & 2 \\ 0 & 0 & 1 & 6 & | & 8 \\ 0 & 0 & 0 & 1 & | & \frac{28}{23} \end{bmatrix} \left(R_4 = -\frac{1}{23}r_4 \right)$$

$$\rightarrow \begin{bmatrix} 1 & 0 & 0 & 0 & | & \frac{44}{23} \\ 0 & 1 & 0 & 0 & | & 2 \\ 0 & 0 & 1 & 0 & | & \frac{16}{23} \\ 0 & 0 & 0 & 1 & | & \frac{28}{23} \end{bmatrix} \begin{pmatrix} R_1 = -5r_4 + r_1 \\ R_3 = -6r_4 + r_3 \end{pmatrix}$$

The solution is $I_1 = \frac{44}{23}$, $I_2 = 2$, $I_3 = \frac{16}{23}$, $I_4 = \frac{28}{23}$.

85. Let x = the amount invested in Treasury bills.
Let y = the amount invested in corporate bonds.
Let z = the amount invested in junk bonds.

a. Total investment equation:
$$x + y + z = 20{,}000$$
Annual income equation:
$$0.07x + 0.09y + 0.11z = 2000$$
Set up a matrix and solve:
$$\begin{bmatrix} 1 & 1 & 1 & | & 20{,}000 \\ 0.07 & 0.09 & 0.11 & | & 2000 \end{bmatrix}$$

$$\rightarrow \begin{bmatrix} 1 & 1 & 1 & | & 20{,}000 \\ 7 & 9 & 11 & | & 200{,}000 \end{bmatrix} \left(R_2 = 100r_2 \right)$$

$$\rightarrow \begin{bmatrix} 1 & 1 & 1 & | & 20{,}000 \\ 0 & 2 & 4 & | & 60{,}000 \end{bmatrix} \left(R_2 = r_2 - 7r_1 \right)$$

$$\rightarrow \begin{bmatrix} 1 & 1 & 1 & | & 20{,}000 \\ 0 & 1 & 2 & | & 30{,}000 \end{bmatrix} \left(R_2 = \frac{1}{2}r_2 \right)$$

$$\rightarrow \begin{bmatrix} 1 & 0 & -1 & | & -10{,}000 \\ 0 & 1 & 2 & | & 30{,}000 \end{bmatrix} \left(R_1 = r_1 - r_2 \right)$$

The matrix in the last step represents the system $\begin{cases} x - z = -10{,}000 \\ y + 2z = 30{,}000 \end{cases}$

Therefore the solution is $x = -10{,}000 + z$, $y = 30{,}000 - 2z$, z is any real number.

Possible investment strategies:

Amount Invested At

7%	9%	11%
0	10,000	10,000
1000	8000	11,000
2000	6000	12,000
3000	4000	13,000
4000	2000	14,000
5000	0	15,000

b. Total investment equation:
$$x + y + z = 25{,}000$$
Annual income equation:
$$0.07x + 0.09y + 0.11z = 2000$$
Set up a matrix and solve:

$$\begin{bmatrix} 1 & 1 & 1 & | & 25,000 \\ 0.07 & 0.09 & 0.11 & | & 2000 \end{bmatrix}$$

$$\rightarrow \begin{bmatrix} 1 & 1 & 1 & | & 25,000 \\ 7 & 9 & 11 & | & 200,000 \end{bmatrix} \quad (R_2 = 100r_2)$$

$$\rightarrow \begin{bmatrix} 1 & 1 & 1 & | & 25,000 \\ 0 & 2 & 4 & | & 25,000 \end{bmatrix} \quad (R_2 = r_2 - 7r_1)$$

$$\rightarrow \begin{bmatrix} 1 & 1 & 1 & | & 25,000 \\ 0 & 1 & 2 & | & 12,500 \end{bmatrix} \quad (R_2 = \tfrac{1}{2}r_2)$$

$$\rightarrow \begin{bmatrix} 1 & 0 & -1 & | & 12,500 \\ 0 & 1 & 2 & | & 12,500 \end{bmatrix} \quad (R_1 = r_1 - r_2)$$

The matrix in the last step represents the system $\begin{cases} x - z = 12,500 \\ y + 2z = 12,500 \end{cases}$

Thus, the solution is $x = z + 12,500$, $y = -2z + 12,500$, z is any real number.

Possible investment strategies:

Amount Invested At

7%	9%	11%
12,500	12,500	0
14,500	8500	2000
16,500	4500	4000
18,750	0	6250

c. Total investment equation:
$x + y + z = 30,000$
Annual income equation:
$0.07x + 0.09y + 0.11z = 2000$
Set up a matrix and solve:

$$\begin{bmatrix} 1 & 1 & 1 & | & 30,000 \\ 0.07 & 0.09 & 0.11 & | & 2000 \end{bmatrix}$$

$$\rightarrow \begin{bmatrix} 1 & 1 & 1 & | & 30,000 \\ 7 & 9 & 11 & | & 200,000 \end{bmatrix} \quad (R_2 = 100r_2)$$

$$\rightarrow \begin{bmatrix} 1 & 1 & 1 & | & 30,000 \\ 0 & 2 & 4 & | & -10,000 \end{bmatrix} \quad (R_1 = r_2 - 7r_1)$$

$$\rightarrow \begin{bmatrix} 1 & 1 & 1 & | & 30,000 \\ 0 & 1 & 2 & | & -5000 \end{bmatrix} \quad (R_2 = \tfrac{1}{2}r_2)$$

$$\rightarrow \begin{bmatrix} 1 & 0 & -1 & | & 35,000 \\ 0 & 1 & 2 & | & -5000 \end{bmatrix} \quad (R_1 = r_1 - r_2)$$

The matrix in the last step represents the system $\begin{cases} x - z = 35,000 \\ y + 2z = -5000 \end{cases}$

Thus, the solution is $x = z + 35,000$, $y = -2z - 5000$, z is any real number.
However, y and z cannot be negative. From $y = -2z - 5000$, we must have $y = z = 0$.

One possible investment strategy

Amount Invested At

7%	9%	11%
30,000	0	0

This will yield ($30,000)(0.07) = $2100, which is more than the required income.

d. Answers will vary.

87. Let x = the amount of supplement 1.
Let y = the amount of supplement 2.
Let z = the amount of supplement 3.

$\begin{cases} 0.20x + 0.40y + 0.30z = 40 \quad \text{Vitamin C} \\ 0.30x + 0.20y + 0.50z = 30 \quad \text{Vitamin D} \end{cases}$

Multiplying each equation by 10 yields
$\begin{cases} 2x + 4y + 3z = 400 \\ 3x + 2y + 5z = 300 \end{cases}$

Set up a matrix and solve:

$$\begin{bmatrix} 2 & 4 & 3 & | & 400 \\ 3 & 2 & 5 & | & 300 \end{bmatrix}$$

$$\rightarrow \begin{bmatrix} 1 & 2 & \tfrac{3}{2} & | & 200 \\ 3 & 2 & 5 & | & 300 \end{bmatrix} \quad (R_1 = \tfrac{1}{2}r_1)$$

$$\rightarrow \begin{bmatrix} 1 & 2 & \tfrac{3}{2} & | & 200 \\ 0 & -4 & \tfrac{1}{2} & | & -300 \end{bmatrix} \quad (R_2 = r_2 - 3r_1)$$

$$\rightarrow \begin{bmatrix} 1 & 2 & \tfrac{3}{2} & | & 200 \\ 0 & 1 & -\tfrac{1}{8} & | & 75 \end{bmatrix} \quad (R_2 = -\tfrac{1}{4}r_2)$$

$$\rightarrow \begin{bmatrix} 1 & 0 & \tfrac{7}{4} & | & 50 \\ 0 & 1 & -\tfrac{1}{8} & | & 75 \end{bmatrix} \quad (R_1 = r_1 - 2r_2)$$

The matrix in the last step represents the system
$\begin{cases} x + \tfrac{7}{4}z = 50 \\ y - \tfrac{1}{8}z = 75 \end{cases}$

Therefore the solution is $x = 50 - \frac{7}{4}z$,

$y = 75 + \frac{1}{8}z$, z is any real number.

Possible combinations:

Supplement 1	Supplement 2	Supplement 3
50mg	75mg	0mg
36mg	76mg	8mg
22mg	77mg	16mg
8mg	78mg	24mg

89–91. Answers will vary.

Section 5.3

1. determinants

3. False

5. $\begin{vmatrix} 3 & 1 \\ 4 & 2 \end{vmatrix} = 3(2) - 4(1) = 6 - 4 = 2$

7. $\begin{vmatrix} 6 & 4 \\ -1 & 3 \end{vmatrix} = 6(3) - (-1)(4) = 18 + 4 = 22$

9. $\begin{vmatrix} -3 & -1 \\ 4 & 2 \end{vmatrix} = -3(2) - 4(-1) = -6 + 4 = -2$

11. $\begin{vmatrix} 3 & 4 & 2 \\ 1 & -1 & 5 \\ 1 & 2 & -2 \end{vmatrix} = 3\begin{vmatrix} -1 & 5 \\ 2 & -2 \end{vmatrix} - 4\begin{vmatrix} 1 & 5 \\ 1 & -2 \end{vmatrix} + 2\begin{vmatrix} 1 & -1 \\ 1 & 2 \end{vmatrix}$

$= 3[(-1)(-2) - 2(5)] - 4[1(-2) - 1(5)]$
$\qquad + 2[1(2) - 1(-1)]$
$= 3(-8) - 4(-7) + 2(3)$
$= -24 + 28 + 6$
$= 10$

13. $\begin{vmatrix} 4 & -1 & 2 \\ 6 & -1 & 0 \\ 1 & -3 & 4 \end{vmatrix} = 4\begin{vmatrix} -1 & 0 \\ -3 & 4 \end{vmatrix} - (-1)\begin{vmatrix} 6 & 0 \\ 1 & 4 \end{vmatrix} + 2\begin{vmatrix} 6 & -1 \\ 1 & -3 \end{vmatrix}$

$= 4[-1(4) - 0(-3)] + 1[6(4) - 1(0)]$
$\qquad + 2[6(-3) - 1(-1)]$
$= 4(-4) + 1(24) + 2(-17)$
$= -16 + 24 - 34$
$= -26$

15. $\begin{cases} x + y = 8 \\ x - y = 4 \end{cases}$

$D = \begin{vmatrix} 1 & 1 \\ 1 & -1 \end{vmatrix} = -1 - 1 = -2$

$D_x = \begin{vmatrix} 8 & 1 \\ 4 & -1 \end{vmatrix} = -8 - 4 = -12$

$D_y = \begin{vmatrix} 1 & 8 \\ 1 & 4 \end{vmatrix} = 4 - 8 = -4$

Find the solutions by Cramer's Rule:

$x = \frac{D_x}{D} = \frac{-12}{-2} = 6 \qquad y = \frac{D_y}{D} = \frac{-4}{-2} = 2$

17. $\begin{cases} 5x - y = 13 \\ 2x + 3y = 12 \end{cases}$

$D = \begin{vmatrix} 5 & -1 \\ 2 & 3 \end{vmatrix} = 15 + 2 = 17$

$D_x = \begin{vmatrix} 13 & -1 \\ 12 & 3 \end{vmatrix} = 39 + 12 = 51$

$D_y = \begin{vmatrix} 5 & 13 \\ 2 & 12 \end{vmatrix} = 60 - 26 = 34$

Find the solutions by Cramer's Rule:

$x = \frac{D_x}{D} = \frac{51}{17} = 3 \qquad y = \frac{D_y}{D} = \frac{34}{17} = 2$

19. $\begin{cases} 3x = 24 \\ x + 2y = 0 \end{cases}$

$D = \begin{vmatrix} 3 & 0 \\ 1 & 2 \end{vmatrix} = 6 - 0 = 6$

$D_x = \begin{vmatrix} 24 & 0 \\ 0 & 2 \end{vmatrix} = 48 - 0 = 48$

$D_y = \begin{vmatrix} 3 & 24 \\ 1 & 0 \end{vmatrix} = 0 - 24 = -24$

Find the solutions by Cramer's Rule:

$x = \dfrac{D_x}{D} = \dfrac{48}{6} = 8 \qquad y = \dfrac{D_y}{D} = \dfrac{-24}{6} = -4$

21. $\begin{cases} 3x - 6y = 24 \\ 5x + 4y = 12 \end{cases}$

$D = \begin{vmatrix} 3 & -6 \\ 5 & 4 \end{vmatrix} = 12 - (-30) = 42$

$D_x = \begin{vmatrix} 24 & -6 \\ 12 & 4 \end{vmatrix} = 96 - (-72) = 168$

$D_y = \begin{vmatrix} 3 & 24 \\ 5 & 12 \end{vmatrix} = 36 - 120 = -84$

Find the solutions by Cramer's Rule:

$x = \dfrac{D_x}{D} = \dfrac{168}{42} = 4 \qquad y = \dfrac{D_y}{D} = \dfrac{-84}{42} = -2$

23. $\begin{cases} 3x - 2y = 4 \\ 6x - 4y = 0 \end{cases}$

$D = \begin{vmatrix} 3 & -2 \\ 6 & -4 \end{vmatrix} = -12 - (-12) = 0$

Since $D = 0$, Cramer's Rule does not apply.

25. $\begin{cases} 2x - 4y = -2 \\ 3x + 2y = 3 \end{cases}$

$D = \begin{vmatrix} 2 & -4 \\ 3 & 2 \end{vmatrix} = 4 + 12 = 16$

$D_x = \begin{vmatrix} -2 & -4 \\ 3 & 2 \end{vmatrix} = -4 + 12 = 8$

$D_y = \begin{vmatrix} 2 & -2 \\ 3 & 3 \end{vmatrix} = 6 + 6 = 12$

Find the solutions by Cramer's Rule:

$x = \dfrac{D_x}{D} = \dfrac{8}{16} = \dfrac{1}{2} \qquad y = \dfrac{D_y}{D} = \dfrac{12}{16} = \dfrac{3}{4}$

27. $\begin{cases} 2x - 3y = -1 \\ 10x + 10y = 5 \end{cases}$

$D = \begin{vmatrix} 2 & -3 \\ 10 & 10 \end{vmatrix} = 20 - (-30) = 50$

$D_x = \begin{vmatrix} -1 & -3 \\ 5 & 10 \end{vmatrix} = -10 - (-15) = 5$

$D_y = \begin{vmatrix} 2 & -1 \\ 10 & 5 \end{vmatrix} = 10 - (-10) = 20$

Find the solutions by Cramer's Rule:

$x = \dfrac{D_x}{D} = \dfrac{5}{50} = \dfrac{1}{10} \qquad y = \dfrac{D_y}{D} = \dfrac{20}{50} = \dfrac{2}{5}$

29. $\begin{cases} 2x + 3y = 6 \\ x - y = \dfrac{1}{2} \end{cases}$

$D = \begin{vmatrix} 2 & 3 \\ 1 & -1 \end{vmatrix} = -2 - 3 = -5$

$D_x = \begin{vmatrix} 6 & 3 \\ \frac{1}{2} & -1 \end{vmatrix} = -6 - \dfrac{3}{2} = -\dfrac{15}{2}$

$D_y = \begin{vmatrix} 2 & 6 \\ 1 & \frac{1}{2} \end{vmatrix} = 1 - 6 = -5$

Find the solutions by Cramer's Rule:

$x = \dfrac{D_x}{D} = \dfrac{-\frac{15}{2}}{-5} = \dfrac{3}{2} \qquad y = \dfrac{D_y}{D} = \dfrac{-5}{-5} = 1$

31. $\begin{cases} 3x - 5y = 3 \\ 15x + 5y = 21 \end{cases}$

$D = \begin{vmatrix} 3 & -5 \\ 15 & 5 \end{vmatrix} = 15 - (-75) = 90$

$D_x = \begin{vmatrix} 3 & -5 \\ 21 & 5 \end{vmatrix} = 15 - (-105) = 120$

$D_y = \begin{vmatrix} 3 & 3 \\ 15 & 21 \end{vmatrix} = 63 - 45 = 18$

Find the solutions by Cramer's Rule:

$x = \dfrac{D_x}{D} = \dfrac{120}{90} = \dfrac{4}{3} \qquad y = \dfrac{D_y}{D} = \dfrac{18}{90} = \dfrac{1}{5}$

Chapter 5: Systems of Equations and Inequalities

33. $\begin{cases} x + y - z = 6 \\ 3x - 2y + z = -5 \\ x + 3y - 2z = 14 \end{cases}$

$D = \begin{vmatrix} 1 & 1 & -1 \\ 3 & -2 & 1 \\ 1 & 3 & -2 \end{vmatrix}$

$= 1\begin{vmatrix} -2 & 1 \\ 3 & -2 \end{vmatrix} - 1\begin{vmatrix} 3 & 1 \\ 1 & -2 \end{vmatrix} + (-1)\begin{vmatrix} 3 & -2 \\ 1 & 3 \end{vmatrix}$

$= 1(4-3) - 1(-6-1) - 1(9+2)$

$= 1 + 7 - 11$

$= -3$

$D_x = \begin{vmatrix} 6 & 1 & -1 \\ -5 & -2 & 1 \\ 14 & 3 & -2 \end{vmatrix}$

$= 6\begin{vmatrix} -2 & 1 \\ 3 & -2 \end{vmatrix} - 1\begin{vmatrix} -5 & 1 \\ 14 & -2 \end{vmatrix} + (-1)\begin{vmatrix} -5 & -2 \\ 14 & 3 \end{vmatrix}$

$= 6(4-3) - 1(10-14) - 1(-15+28)$

$= 6 + 4 - 13$

$= -3$

$D_y = \begin{vmatrix} 1 & 6 & -1 \\ 3 & -5 & 1 \\ 1 & 14 & -2 \end{vmatrix}$

$= 1\begin{vmatrix} -5 & 1 \\ 14 & -2 \end{vmatrix} - 6\begin{vmatrix} 3 & 1 \\ 1 & -2 \end{vmatrix} + (-1)\begin{vmatrix} 3 & -5 \\ 1 & 14 \end{vmatrix}$

$= 1(10-14) - 6(-6-1) - 1(42+5)$

$= -4 + 42 - 47$

$= -9$

$D_z = \begin{vmatrix} 1 & 1 & 6 \\ 3 & -2 & -5 \\ 1 & 3 & 14 \end{vmatrix}$

$= 1\begin{vmatrix} -2 & -5 \\ 3 & 14 \end{vmatrix} - 1\begin{vmatrix} 3 & -5 \\ 1 & 14 \end{vmatrix} + 6\begin{vmatrix} 3 & -2 \\ 1 & 3 \end{vmatrix}$

$= 1(-28+15) - 1(42+5) + 6(9+2)$

$= -13 - 47 + 66$

$= 6$

Find the solutions by Cramer's Rule:

$x = \dfrac{D_x}{D} = \dfrac{-3}{-3} = 1 \qquad y = \dfrac{D_y}{D} = \dfrac{-9}{-3} = 3$

$z = \dfrac{D_z}{D} = \dfrac{6}{-3} = -2$

35. $\begin{cases} x + 2y - z = -3 \\ 2x - 4y + z = -7 \\ -2x + 2y - 3z = 4 \end{cases}$

$D = \begin{vmatrix} 1 & 2 & -1 \\ 2 & -4 & 1 \\ -2 & 2 & -3 \end{vmatrix}$

$= 1\begin{vmatrix} -4 & 1 \\ 2 & -3 \end{vmatrix} - 2\begin{vmatrix} 2 & 1 \\ -2 & -3 \end{vmatrix} + (-1)\begin{vmatrix} 2 & -4 \\ -2 & 2 \end{vmatrix}$

$= 1(12-2) - 2(-6+2) - 1(4-8)$

$= 10 + 8 + 4$

$= 22$

$D_x = \begin{vmatrix} -3 & 2 & -1 \\ -7 & -4 & 1 \\ 4 & 2 & -3 \end{vmatrix}$

$= -3\begin{vmatrix} -4 & 1 \\ 2 & -3 \end{vmatrix} - 2\begin{vmatrix} -7 & 1 \\ 4 & -3 \end{vmatrix} + (-1)\begin{vmatrix} -7 & -4 \\ 4 & 2 \end{vmatrix}$

$= -3(12-2) - 2(21-4) - 1(-14+16)$

$= -30 - 34 - 2$

$= -66$

$D_y = \begin{vmatrix} 1 & -3 & -1 \\ 2 & -7 & 1 \\ -2 & 4 & -3 \end{vmatrix}$

$= 1\begin{vmatrix} -7 & 1 \\ 4 & -3 \end{vmatrix} - (-3)\begin{vmatrix} 2 & 1 \\ -2 & -3 \end{vmatrix} + (-1)\begin{vmatrix} 2 & -7 \\ -2 & 4 \end{vmatrix}$

$= 1(21-4) + 3(-6+2) - 1(8-14)$

$= 17 - 12 + 6$

$= 11$

$D_z = \begin{vmatrix} 1 & 2 & -3 \\ 2 & -4 & -7 \\ -2 & 2 & 4 \end{vmatrix}$

$= 1\begin{vmatrix} -4 & -7 \\ 2 & 4 \end{vmatrix} - 2\begin{vmatrix} 2 & -7 \\ -2 & 4 \end{vmatrix} + (-3)\begin{vmatrix} 2 & -4 \\ -2 & 2 \end{vmatrix}$

$= 1(-16+14) - 2(8-14) - 3(4-8)$

$= -2 + 12 + 12$

$= 22$

Find the solutions by Cramer's Rule:

$x = \dfrac{D_x}{D} = \dfrac{-66}{22} = -3 \qquad y = \dfrac{D_y}{D} = \dfrac{11}{22} = \dfrac{1}{2}$

$z = \dfrac{D_z}{D} = \dfrac{22}{22} = 1$

37. $\begin{cases} x - 2y + 3z = 1 \\ 3x + y - 2z = 0 \\ 2x - 4y + 6z = 2 \end{cases}$

$D = \begin{vmatrix} 1 & -2 & 3 \\ 3 & 1 & -2 \\ 2 & -4 & 6 \end{vmatrix}$

$= 1\begin{vmatrix} 1 & -2 \\ -4 & 6 \end{vmatrix} - (-2)\begin{vmatrix} 3 & -2 \\ 2 & 6 \end{vmatrix} + 3\begin{vmatrix} 3 & 1 \\ 2 & -4 \end{vmatrix}$

$= 1(6-8) + 2(18+4) + 3(-12-2)$

$= -2 + 44 - 42$

$= 0$

Since $D = 0$, Cramer's Rule does not apply.

39. $\begin{cases} x + 2y - z = 0 \\ 2x - 4y + z = 0 \\ -2x + 2y - 3z = 0 \end{cases}$

$D = \begin{vmatrix} 1 & 2 & -1 \\ 2 & -4 & 1 \\ -2 & 2 & -3 \end{vmatrix}$

$= 1\begin{vmatrix} -4 & 1 \\ 2 & -3 \end{vmatrix} - 2\begin{vmatrix} 2 & 1 \\ -2 & -3 \end{vmatrix} + (-1)\begin{vmatrix} 2 & -4 \\ -2 & 2 \end{vmatrix}$

$= 1(12-2) - 2(-6+2) - 1(4-8)$

$= 10 + 8 + 4$

$= 22$

$D_x = \begin{vmatrix} 0 & 2 & -1 \\ 0 & -4 & 1 \\ 0 & 2 & -3 \end{vmatrix} = 0$ (By Theorem 12)

$D_y = \begin{vmatrix} 1 & 0 & -1 \\ 2 & 0 & 1 \\ -2 & 0 & -3 \end{vmatrix} = 0$ (By Theorem 12)

$D_z = \begin{vmatrix} 1 & 2 & 0 \\ 2 & -4 & 0 \\ -2 & 2 & 0 \end{vmatrix} = 0$ (By Theorem 12)

Find the solutions by Cramer's Rule:

$x = \dfrac{D_x}{D} = \dfrac{0}{22} = 0 \qquad y = \dfrac{D_y}{D} = \dfrac{0}{22} = 0$

$z = \dfrac{D_z}{D} = \dfrac{0}{22} = 0$

41. $\begin{cases} x - 2y + 3z = 0 \\ 3x + y - 2z = 0 \\ 2x - 4y + 6z = 0 \end{cases}$

$D = \begin{vmatrix} 1 & -2 & 3 \\ 3 & 1 & -2 \\ 2 & -4 & 6 \end{vmatrix}$

$= 1\begin{vmatrix} 1 & -2 \\ -4 & 6 \end{vmatrix} - (-2)\begin{vmatrix} 3 & -2 \\ 2 & 6 \end{vmatrix} + 3\begin{vmatrix} 3 & 1 \\ 2 & -4 \end{vmatrix}$

$= 1(6-8) + 2(18+4) + 3(-12-2)$

$= -2 + 44 - 42$

$= 0$

Since $D = 0$, Cramer's Rule does not apply.

43. Solve for x:

$\begin{vmatrix} x & x \\ 4 & 3 \end{vmatrix} = 5$

$3x - 4x = 5$

$-x = 5$

$x = -5$

45. Solve for x:

$\begin{vmatrix} x & 1 & 1 \\ 4 & 3 & 2 \\ -1 & 2 & 5 \end{vmatrix} = 2$

$x\begin{vmatrix} 3 & 2 \\ 2 & 5 \end{vmatrix} - 1\begin{vmatrix} 4 & 2 \\ -1 & 5 \end{vmatrix} + 1\begin{vmatrix} 4 & 3 \\ -1 & 2 \end{vmatrix} = 2$

$x(15-4) - (20+2) + (8+3) = 2$

$11x - 22 + 11 = 2$

$11x = 13$

$x = \dfrac{13}{11}$

47. Solve for x:
$$\begin{vmatrix} x & 2 & 3 \\ 1 & x & 0 \\ 6 & 1 & -2 \end{vmatrix} = 7$$

$$x\begin{vmatrix} x & 0 \\ 1 & -2 \end{vmatrix} - 2\begin{vmatrix} 1 & 0 \\ 6 & -2 \end{vmatrix} + 3\begin{vmatrix} 1 & x \\ 6 & 1 \end{vmatrix} = 7$$

$$x(-2x) - 2(-2) + 3(1-6x) = 7$$
$$-2x^2 + 4 + 3 - 18x = 7$$
$$-2x^2 - 18x = 0$$
$$-2x(x+9) = 0$$
$$x = 0 \text{ or } x = -9$$

49. $\begin{vmatrix} x & y & z \\ u & v & w \\ 1 & 2 & 3 \end{vmatrix} = 4$

By Theorem (11), the value of a determinant changes sign if any two rows are interchanged.

Thus, $\begin{vmatrix} 1 & 2 & 3 \\ u & v & w \\ x & y & z \end{vmatrix} = -4$.

51. Let $\begin{vmatrix} x & y & z \\ u & v & w \\ 1 & 2 & 3 \end{vmatrix} = 4$.

$$\begin{vmatrix} x & y & z \\ -3 & -6 & -9 \\ u & v & w \end{vmatrix} = -3\begin{vmatrix} x & y & z \\ 1 & 2 & 3 \\ u & v & w \end{vmatrix} \quad \text{Theorem (14)}$$

$$= -3(-1)\begin{vmatrix} x & y & z \\ u & v & w \\ 1 & 2 & 3 \end{vmatrix} \quad \text{Theorem (11)}$$

$$= 3(4)$$
$$= 12$$

53. Let $\begin{vmatrix} x & y & z \\ u & v & w \\ 1 & 2 & 3 \end{vmatrix} = 4$

$$\begin{vmatrix} 1 & 2 & 3 \\ x-3 & y-6 & z-9 \\ 2u & 2v & 2w \end{vmatrix}$$

$$= 2\begin{vmatrix} 1 & 2 & 3 \\ x-3 & y-6 & z-9 \\ u & v & w \end{vmatrix} \quad \text{Theorem (14)}$$

$$= 2(-1)\begin{vmatrix} x-3 & y-6 & z-9 \\ 1 & 2 & 3 \\ u & v & w \end{vmatrix} \quad \text{Theorem (11)}$$

$$= 2(-1)(-1)\begin{vmatrix} x-3 & y-6 & z-9 \\ u & v & w \\ 1 & 2 & 3 \end{vmatrix} \quad \text{Theorem (11)}$$

$$= 2(-1)(-1)\begin{vmatrix} x & y & z \\ u & v & w \\ 1 & 2 & 3 \end{vmatrix} \quad \begin{array}{l}\text{Theorem (15)}\\(R_1 = -3r_3 + r_1)\end{array}$$

$$= 2(-1)(-1)(4)$$
$$= 8$$

55. Let $\begin{vmatrix} x & y & z \\ u & v & w \\ 1 & 2 & 3 \end{vmatrix} = 4$

$$\begin{vmatrix} 1 & 2 & 3 \\ 2x & 2y & 2z \\ u-1 & v-2 & w-3 \end{vmatrix}$$

$$= 2\begin{vmatrix} 1 & 2 & 3 \\ x & y & z \\ u-1 & v-2 & w-3 \end{vmatrix} \quad \text{Theorem (14)}$$

$$= 2(-1)\begin{vmatrix} x & y & z \\ 1 & 2 & 3 \\ u-1 & v-2 & w-3 \end{vmatrix} \quad \text{Theorem (11)}$$

$$= 2(-1)(-1)\begin{vmatrix} x & y & z \\ u-1 & v-2 & w-3 \\ 1 & 2 & 3 \end{vmatrix} \quad \text{Theorem (11)}$$

$$= 2(-1)(-1)\begin{vmatrix} x & y & z \\ u & v & w \\ 1 & 2 & 3 \end{vmatrix} \quad \begin{array}{l}\text{Theorem (15)}\\(R_2 = -r_3 + r_2)\end{array}$$

$$= 2(-1)(-1)(4)$$
$$= 8$$

57. Expanding the determinant:
$$\begin{vmatrix} x & y & 1 \\ x_1 & y_1 & 1 \\ x_2 & y_2 & 1 \end{vmatrix} = 0$$

$$x\begin{vmatrix} y_1 & 1 \\ y_2 & 1 \end{vmatrix} - y\begin{vmatrix} x_1 & 1 \\ x_2 & 1 \end{vmatrix} + 1\begin{vmatrix} x_1 & y_1 \\ x_2 & y_2 \end{vmatrix} = 0$$

$$x(y_1 - y_2) - y(x_1 - x_2) + (x_1 y_2 - x_2 y_1) = 0$$

$$x(y_1 - y_2) + y(x_2 - x_1) = x_2 y_1 - x_1 y_2$$
$$y(x_2 - x_1) = x_2 y_1 - x_1 y_2 + x(y_2 - y_1)$$
$$y(x_2 - x_1) - y_1(x_2 - x_1)$$
$$= x_2 y_1 - x_1 y_2 + x(y_2 - y_1) - y_1(x_2 - x_1)$$
$$(x_2 - x_1)(y - y_1)$$
$$= x(y_2 - y_1) + x_2 y_1 - x_1 y_2 - y_1 x_2 + y_1 x_1$$
$$(x_2 - x_1)(y - y_1) = (y_2 - y_1)x - (y_2 - y_1)x_1$$
$$(x_2 - x_1)(y - y_1) = (y_2 - y_1)(x - x_1)$$
$$(y - y_1) = \frac{(y_2 - y_1)}{(x_2 - x_1)}(x - x_1)$$

This is the 2-point form of the equation for a line.

59. Expanding the determinant:
$$\begin{vmatrix} x^2 & x & 1 \\ y^2 & y & 1 \\ z^2 & z & 1 \end{vmatrix}$$
$$= x^2 \begin{vmatrix} y & 1 \\ z & 1 \end{vmatrix} - x \begin{vmatrix} y^2 & 1 \\ z^2 & 1 \end{vmatrix} + 1 \begin{vmatrix} y^2 & y \\ z^2 & z \end{vmatrix}$$
$$= x^2(y - z) - x(y^2 - z^2) + 1(y^2 z - z^2 y)$$
$$= x^2(y - z) - x(y - z)(y + z) + yz(y - z)$$
$$= (y - z)\left[x^2 - xy - xz + yz\right]$$
$$= (y - z)\left[x(x - y) - z(x - y)\right]$$
$$= (y - z)(x - y)(x - z)$$

61. Evaluating the determinant to show the relationship:
$$\begin{vmatrix} a_{13} & a_{12} & a_{11} \\ a_{23} & a_{22} & a_{21} \\ a_{33} & a_{32} & a_{31} \end{vmatrix}$$
$$= a_{13} \begin{vmatrix} a_{22} & a_{21} \\ a_{32} & a_{31} \end{vmatrix} - a_{12} \begin{vmatrix} a_{23} & a_{21} \\ a_{33} & a_{31} \end{vmatrix} + a_{11} \begin{vmatrix} a_{23} & a_{22} \\ a_{33} & a_{32} \end{vmatrix}$$
$$= a_{13}(a_{22}a_{31} - a_{21}a_{32}) - a_{12}(a_{23}a_{31} - a_{21}a_{33})$$
$$\quad + a_{11}(a_{23}a_{32} - a_{22}a_{33})$$
$$= a_{13}a_{22}a_{31} - a_{13}a_{21}a_{32} - a_{12}a_{23}a_{31} + a_{12}a_{21}a_{33}$$
$$\quad + a_{11}a_{23}a_{32} - a_{11}a_{22}a_{33}$$
$$= -a_{11}a_{22}a_{33} + a_{11}a_{23}a_{32} + a_{12}a_{21}a_{33} - a_{12}a_{23}a_{31}$$
$$\quad - a_{13}a_{21}a_{32} + a_{13}a_{22}a_{31}$$

$$= -a_{11}(a_{22}a_{33} - a_{23}a_{32}) + a_{12}(a_{21}a_{33} - a_{23}a_{31})$$
$$\quad - a_{13}(a_{21}a_{32} - a_{22}a_{31})$$
$$= -a_{11}\begin{vmatrix} a_{22} & a_{23} \\ a_{32} & a_{33} \end{vmatrix} + a_{12}\begin{vmatrix} a_{21} & a_{23} \\ a_{31} & a_{33} \end{vmatrix} - a_{13}\begin{vmatrix} a_{21} & a_{22} \\ a_{31} & a_{32} \end{vmatrix}$$
$$= -\left(a_{11}\begin{vmatrix} a_{22} & a_{23} \\ a_{32} & a_{33} \end{vmatrix} - a_{12}\begin{vmatrix} a_{21} & a_{23} \\ a_{31} & a_{33} \end{vmatrix} + a_{13}\begin{vmatrix} a_{21} & a_{22} \\ a_{31} & a_{32} \end{vmatrix}\right)$$
$$= -\begin{vmatrix} a_{11} & a_{12} & a_{13} \\ a_{21} & a_{22} & a_{23} \\ a_{31} & a_{32} & a_{33} \end{vmatrix}$$

63. Set up a 3 by 3 determinant in which the first column and third column are the same and evaluate by expanding down column 2:
$$\begin{vmatrix} a & b & a \\ c & d & c \\ e & f & e \end{vmatrix} = -b\begin{vmatrix} c & c \\ e & e \end{vmatrix} + d\begin{vmatrix} a & a \\ e & e \end{vmatrix} - f\begin{vmatrix} a & a \\ c & c \end{vmatrix}$$
$$= -b(ce - ce) + d(ae - ae) - f(ac - ac)$$
$$= -b(0) + d(0) - f(0) = 0$$

Section 5.4

1. inverse

3. identity

5. False

7. a. $A + B = \begin{bmatrix} 0 & 3 & -5 \\ 1 & 2 & 6 \end{bmatrix} + \begin{bmatrix} 4 & 1 & 0 \\ -2 & 3 & -2 \end{bmatrix}$
$$= \begin{bmatrix} 0+4 & 3+1 & -5+0 \\ 1+(-2) & 2+3 & 6+(-2) \end{bmatrix}$$
$$= \begin{bmatrix} 4 & 4 & -5 \\ -1 & 5 & 4 \end{bmatrix}$$

b. Enter the matrices into a graphing utility. The result is shown below:

9. a. $4A = 4\begin{bmatrix} 0 & 3 & -5 \\ 1 & 2 & 6 \end{bmatrix}$

$= \begin{bmatrix} 4 \cdot 0 & 4 \cdot 3 & 4(-5) \\ 4 \cdot 1 & 4 \cdot 2 & 4 \cdot 6 \end{bmatrix}$

$= \begin{bmatrix} 0 & 12 & -20 \\ 4 & 8 & 24 \end{bmatrix}$

b. Enter the matrices into a graphing utility. The result is shown below:

```
4[A]
    [[0  12  -20]
     [4   8   24]]
```

11. a. $3A - 2B = 3\begin{bmatrix} 0 & 3 & -5 \\ 1 & 2 & 6 \end{bmatrix} - 2\begin{bmatrix} 4 & 1 & 0 \\ -2 & 3 & -2 \end{bmatrix}$

$= \begin{bmatrix} 0 & 9 & -15 \\ 3 & 6 & 18 \end{bmatrix} - \begin{bmatrix} 8 & 2 & 0 \\ -4 & 6 & -4 \end{bmatrix}$

$= \begin{bmatrix} -8 & 7 & -15 \\ 7 & 0 & 22 \end{bmatrix}$

b. Enter the matrices into a graphing utility. The result is shown below:

```
3[A]-2[B]
    [[-8  7  -15]
     [ 7  0   22]]
```

13. a. $AC = \begin{bmatrix} 0 & 3 & -5 \\ 1 & 2 & 6 \end{bmatrix} \cdot \begin{bmatrix} 4 & 1 \\ 6 & 2 \\ -2 & 3 \end{bmatrix}$

$= \begin{bmatrix} 0(4) + 3(6) + (-5)(-2) & 0(1) + 3(2) + (-5)(3) \\ 1(4) + 2(6) + 6(-2) & 1(1) + 2(2) + 6(3) \end{bmatrix}$

$= \begin{bmatrix} 28 & -9 \\ 4 & 23 \end{bmatrix}$

b. Enter the matrices into a graphing utility. The result is shown below:

```
[A][C]
    [[28  -9]
     [ 4  23]]
```

15. a. $CA = \begin{bmatrix} 4 & 1 \\ 6 & 2 \\ -2 & 3 \end{bmatrix} \cdot \begin{bmatrix} 0 & 3 & -5 \\ 1 & 2 & 6 \end{bmatrix}$

$= \begin{bmatrix} 4(0) + 1(1) & 4(3) + 1(2) & 4(-5) + 1(6) \\ 6(0) + 2(1) & 6(3) + 2(2) & 6(-5) + 2(6) \\ -2(0) + 3(1) & -2(3) + 3(2) & -2(-5) + 3(6) \end{bmatrix}$

$= \begin{bmatrix} 1 & 14 & -14 \\ 2 & 22 & -18 \\ 3 & 0 & 28 \end{bmatrix}$

b. Enter the matrices into a graphing utility. The result is shown below:

```
[C][A]
    [[1  14  -14]
     [2  22  -18]
     [3   0   28]]
```

17. a. $C(A+B) = \begin{bmatrix} 4 & 1 \\ 6 & 2 \\ -2 & 3 \end{bmatrix} \left(\begin{bmatrix} 0 & 3 & -5 \\ 1 & 2 & 6 \end{bmatrix} + \begin{bmatrix} 4 & 1 & 0 \\ -2 & 3 & -2 \end{bmatrix} \right)$

$= \begin{bmatrix} 4 & 1 \\ 6 & 2 \\ -2 & 3 \end{bmatrix} \cdot \begin{bmatrix} 4 & 4 & -5 \\ -1 & 5 & 4 \end{bmatrix}$

$= \begin{bmatrix} 15 & 21 & -16 \\ 22 & 34 & -22 \\ -11 & 7 & 22 \end{bmatrix}$

b. Enter the matrices into a graphing utility. The result is shown below:

```
[C]*([A]+[B])
    [[ 15  21  -16]
     [ 22  34  -22]
     [-11   7   22]]
```

19. a. $AC - 3I_2 = \begin{bmatrix} 0 & 3 & -5 \\ 1 & 2 & 6 \end{bmatrix} \cdot \begin{bmatrix} 4 & 1 \\ 6 & 2 \\ -2 & 3 \end{bmatrix} - 3\begin{bmatrix} 1 & 0 \\ 0 & 1 \end{bmatrix}$

$= \begin{bmatrix} 28 & -9 \\ 4 & 23 \end{bmatrix} - \begin{bmatrix} 3 & 0 \\ 0 & 3 \end{bmatrix}$

$= \begin{bmatrix} 25 & -9 \\ 4 & 20 \end{bmatrix}$

b. Enter the matrices into a graphing utility. Use $I = \begin{bmatrix} 1 & 0 \\ 0 & 1 \end{bmatrix}$. The result is shown below:

21. a. $CA - CB$

$= \begin{bmatrix} 4 & 1 \\ 6 & 2 \\ -2 & 3 \end{bmatrix} \begin{bmatrix} 0 & 3 & -5 \\ 1 & 2 & 6 \end{bmatrix} - \begin{bmatrix} 4 & 1 \\ 6 & 2 \\ -2 & 3 \end{bmatrix} \begin{bmatrix} 4 & 1 & 0 \\ -2 & 3 & -2 \end{bmatrix}$

$= \begin{bmatrix} 1 & 14 & -14 \\ 2 & 22 & -18 \\ 3 & 0 & 28 \end{bmatrix} - \begin{bmatrix} 14 & 7 & -2 \\ 20 & 12 & -4 \\ -14 & 7 & -6 \end{bmatrix}$

$= \begin{bmatrix} -13 & 7 & -12 \\ -18 & 10 & -14 \\ 17 & -7 & 34 \end{bmatrix}$

b. Enter the matrices into a graphing utility. The result is shown below:

23. a. $a_{11} = 2(2) + (-2)(3) = -2$

$a_{12} = 2(1) + (-2)(-1) = 4$

$a_{13} = 2(4) + (-2)(3) = 2$

$a_{14} = 2(6) + (-2)(2) = 8$

$a_{21} = 1(2) + 0(3) = 2$

$a_{22} = 1(1) + 0(-1) = 1$

$a_{23} = 1(4) + 0(3) = 4$

$a_{24} = 1(6) + 0(2) = 6$

$\begin{bmatrix} 2 & -2 \\ 1 & 0 \end{bmatrix} \begin{bmatrix} 2 & 1 & 4 & 6 \\ 3 & -1 & 3 & 2 \end{bmatrix} = \begin{bmatrix} -2 & 4 & 2 & 8 \\ 2 & 1 & 4 & 6 \end{bmatrix}$

b. Enter the matrices into a graphing utility using $A = \begin{bmatrix} 2 & -2 \\ 1 & 0 \end{bmatrix}$ and $B = \begin{bmatrix} 2 & 1 & 4 & 6 \\ 3 & -1 & 3 & 2 \end{bmatrix}$.

The result is shown below:

25. a. $\begin{bmatrix} 1 & 2 & 3 \\ 0 & -1 & 4 \end{bmatrix} \begin{bmatrix} 1 & 2 \\ -1 & 0 \\ 2 & 4 \end{bmatrix}$

$= \begin{bmatrix} 1(1) + 2(-1) + 3(2) & 1(2) + 2(0) + 3(4) \\ 0(1) + (-1)(-1) + 4(2) & 0(2) + (-1)(0) + 4(4) \end{bmatrix}$

$= \begin{bmatrix} 5 & 14 \\ 9 & 16 \end{bmatrix}$

b. Enter the matrices into a graphing utility using $A = \begin{bmatrix} 1 & 2 & 3 \\ 0 & -1 & 4 \end{bmatrix}$ and $B = \begin{bmatrix} 1 & 2 \\ -1 & 0 \\ 2 & 4 \end{bmatrix}$.

The result is shown below:

27. a. $\begin{bmatrix} 1 & 0 & 1 \\ 2 & 4 & 1 \\ 3 & 6 & 1 \end{bmatrix} \begin{bmatrix} 1 & 3 \\ 6 & 2 \\ 8 & -1 \end{bmatrix}$

$= \begin{bmatrix} 1(1) + 0(6) + 1(8) & 1(3) + 0(2) + 1(-1) \\ 2(1) + 4(6) + 1(8) & 2(3) + 4(2) + 1(-1) \\ 3(1) + 6(6) + 1(8) & 3(3) + 6(2) + 1(-1) \end{bmatrix}$

$= \begin{bmatrix} 9 & 2 \\ 34 & 13 \\ 47 & 20 \end{bmatrix}$

b. Enter the matrices into a graphing utility using $A = \begin{bmatrix} 1 & 0 & 1 \\ 2 & 4 & 1 \\ 3 & 6 & 1 \end{bmatrix}$ and $B = \begin{bmatrix} 1 & 3 \\ 6 & 2 \\ 8 & -1 \end{bmatrix}$. The result is shown below:

29. $A = \begin{bmatrix} 2 & 1 \\ 1 & 1 \end{bmatrix}$

Augment the matrix with the identity and use row operations to find the inverse:

$\begin{bmatrix} 2 & 1 & | & 1 & 0 \\ 1 & 1 & | & 0 & 1 \end{bmatrix}$

$\rightarrow \begin{bmatrix} 1 & 1 & | & 0 & 1 \\ 2 & 1 & | & 1 & 0 \end{bmatrix}$ $\begin{pmatrix} \text{Interchange} \\ r_1 \text{ and } r_2 \end{pmatrix}$

$\rightarrow \begin{bmatrix} 1 & 1 & | & 0 & 1 \\ 0 & -1 & | & 1 & -2 \end{bmatrix}$ $(R_2 = -2r_1 + r_2)$

$\rightarrow \begin{bmatrix} 1 & 1 & | & 0 & 1 \\ 0 & 1 & | & -1 & 2 \end{bmatrix}$ $(R_2 = -r_2)$

$\rightarrow \begin{bmatrix} 1 & 0 & | & 1 & -1 \\ 0 & 1 & | & -1 & 2 \end{bmatrix}$ $(R_1 = -r_2 + r_1)$

Thus, $A^{-1} = \begin{bmatrix} 1 & -1 \\ -1 & 2 \end{bmatrix}$.

```
[A]⁻¹
     [[1  -1]
      [-1  2]]
```

31. $A = \begin{bmatrix} 6 & 5 \\ 2 & 2 \end{bmatrix}$

Augment the matrix with the identity and use row operations to find the inverse:

$\begin{bmatrix} 6 & 5 & | & 1 & 0 \\ 2 & 2 & | & 0 & 1 \end{bmatrix}$

$\rightarrow \begin{bmatrix} 2 & 2 & | & 0 & 1 \\ 6 & 5 & | & 1 & 0 \end{bmatrix}$ $\begin{pmatrix} \text{Interchange} \\ r_1 \text{ and } r_2 \end{pmatrix}$

$\rightarrow \begin{bmatrix} 2 & 2 & | & 0 & 1 \\ 0 & -1 & | & 1 & -3 \end{bmatrix}$ $(R_2 = -3r_1 + r_2)$

$\rightarrow \begin{bmatrix} 1 & 1 & | & 0 & \frac{1}{2} \\ 0 & 1 & | & -1 & 3 \end{bmatrix}$ $\begin{pmatrix} R_1 = \frac{1}{2}r_1 \\ R_2 = -r_2 \end{pmatrix}$

$\rightarrow \begin{bmatrix} 1 & 0 & | & 1 & -\frac{5}{2} \\ 0 & 1 & | & -1 & 3 \end{bmatrix}$ $(R_1 = -r_2 + r_1)$

Thus, $A^{-1} = \begin{bmatrix} 1 & -\frac{5}{2} \\ -1 & 3 \end{bmatrix}$.

```
[A]⁻¹▶Frac
     [[1   -5/2]
      [-1   3  ]]
```

33. $A = \begin{bmatrix} 2 & 1 \\ a & a \end{bmatrix}$ where $a \neq 0$.

Augment the matrix with the identity and use row operations to find the inverse:

$\begin{bmatrix} 2 & 1 & | & 1 & 0 \\ a & a & | & 0 & 1 \end{bmatrix}$

$\rightarrow \begin{bmatrix} 1 & \frac{1}{2} & | & \frac{1}{2} & 0 \\ a & a & | & 0 & 1 \end{bmatrix}$ $\left(R_1 = \frac{1}{2}r_1\right)$

$\rightarrow \begin{bmatrix} 1 & \frac{1}{2} & | & \frac{1}{2} & 0 \\ 0 & \frac{1}{2}a & | & -\frac{1}{2}a & 1 \end{bmatrix}$ $(R_2 = -ar_1 + r_2)$

$\rightarrow \begin{bmatrix} 1 & \frac{1}{2} & | & \frac{1}{2} & 0 \\ 0 & 1 & | & -1 & \frac{2}{a} \end{bmatrix}$ $\left(R_2 = \frac{2}{a}r_2\right)$

$\rightarrow \begin{bmatrix} 1 & 0 & | & 1 & -\frac{1}{a} \\ 0 & 1 & | & -1 & \frac{2}{a} \end{bmatrix}$ $\left(R_1 = -\frac{1}{2}r_2 + r_1\right)$

Thus, $A^{-1} = \begin{bmatrix} 1 & -\frac{1}{a} \\ -1 & \frac{2}{a} \end{bmatrix}$.

35. $A = \begin{bmatrix} 1 & -1 & 1 \\ 0 & -2 & 1 \\ -2 & -3 & 0 \end{bmatrix}$

Augment the matrix with the identity and use row operations to find the inverse:

$\begin{bmatrix} 1 & -1 & 1 & | & 1 & 0 & 0 \\ 0 & -2 & 1 & | & 0 & 1 & 0 \\ -2 & -3 & 0 & | & 0 & 0 & 1 \end{bmatrix}$

$\rightarrow \begin{bmatrix} 1 & -1 & 1 & | & 1 & 0 & 0 \\ 0 & -2 & 1 & | & 0 & 1 & 0 \\ 0 & -5 & 2 & | & 2 & 0 & 1 \end{bmatrix}$ $(R_3 = 2r_1 + r_3)$

$\rightarrow \begin{bmatrix} 1 & -1 & 1 & | & 1 & 0 & 0 \\ 0 & 1 & -\frac{1}{2} & | & 0 & -\frac{1}{2} & 0 \\ 0 & -5 & 2 & | & 2 & 0 & 1 \end{bmatrix}$ $\left(R_2 = -\frac{1}{2}r_2\right)$

$\rightarrow \begin{bmatrix} 1 & 0 & \frac{1}{2} & | & 1 & -\frac{1}{2} & 0 \\ 0 & 1 & -\frac{1}{2} & | & 0 & -\frac{1}{2} & 0 \\ 0 & 0 & -\frac{1}{2} & | & 2 & -\frac{5}{2} & 1 \end{bmatrix}$ $\begin{pmatrix} R_1 = r_2 + r_1 \\ R_3 = 5r_2 + r_3 \end{pmatrix}$

$\rightarrow \begin{bmatrix} 1 & 0 & \frac{1}{2} & | & 1 & -\frac{1}{2} & 0 \\ 0 & 1 & -\frac{1}{2} & | & 0 & -\frac{1}{2} & 0 \\ 0 & 0 & 1 & | & -4 & 5 & -2 \end{bmatrix}$ $(R_3 = -2r_3)$

$$\rightarrow \begin{bmatrix} 1 & 0 & 0 & | & 3 & -3 & 1 \\ 0 & 1 & 0 & | & -2 & 2 & -1 \\ 0 & 0 & 1 & | & -4 & 5 & -2 \end{bmatrix} \quad \begin{pmatrix} R_1 = -\tfrac{1}{2}r_3 + r_1 \\ R_2 = \tfrac{1}{2}r_3 + r_2 \end{pmatrix}$$

Thus, $A^{-1} = \begin{bmatrix} 3 & -3 & 1 \\ -2 & 2 & -1 \\ -4 & 5 & -2 \end{bmatrix}$.

37. $A = \begin{bmatrix} 1 & 1 & 1 \\ 3 & 2 & -1 \\ 3 & 1 & 2 \end{bmatrix}$

Augment the matrix with the identity and use row operations to find the inverse:

$$\begin{bmatrix} 1 & 1 & 1 & | & 1 & 0 & 0 \\ 3 & 2 & -1 & | & 0 & 1 & 0 \\ 3 & 1 & 2 & | & 0 & 0 & 1 \end{bmatrix}$$

$$\rightarrow \begin{bmatrix} 1 & 1 & 1 & | & 1 & 0 & 0 \\ 0 & -1 & -4 & | & -3 & 1 & 0 \\ 0 & -2 & -1 & | & -3 & 0 & 1 \end{bmatrix} \quad \begin{pmatrix} R_2 = -3r_1 + r_2 \\ R_3 = -3r_1 + r_3 \end{pmatrix}$$

$$\rightarrow \begin{bmatrix} 1 & 1 & 1 & | & 1 & 0 & 0 \\ 0 & 1 & 4 & | & 3 & -1 & 0 \\ 0 & -2 & -1 & | & -3 & 0 & 1 \end{bmatrix} \quad (R_2 = -r_2)$$

$$\rightarrow \begin{bmatrix} 1 & 0 & -3 & | & -2 & 1 & 0 \\ 0 & 1 & 4 & | & 3 & -1 & 0 \\ 0 & 0 & 1 & | & \tfrac{3}{7} & -\tfrac{2}{7} & \tfrac{1}{7} \end{bmatrix} \quad (R_3 = \tfrac{1}{7}r_3)$$

$$\rightarrow \begin{bmatrix} 1 & 0 & 0 & | & -\tfrac{5}{7} & \tfrac{1}{7} & \tfrac{3}{7} \\ 0 & 1 & 0 & | & \tfrac{9}{7} & \tfrac{1}{7} & -\tfrac{4}{7} \\ 0 & 0 & 1 & | & \tfrac{3}{7} & -\tfrac{2}{7} & \tfrac{1}{7} \end{bmatrix} \quad \begin{pmatrix} R_1 = 3r_3 + r_1 \\ R_2 = -4r_3 + r_2 \end{pmatrix}$$

Thus, $A^{-1} = \begin{bmatrix} -\tfrac{5}{7} & \tfrac{1}{7} & \tfrac{3}{7} \\ \tfrac{9}{7} & \tfrac{1}{7} & -\tfrac{4}{7} \\ \tfrac{3}{7} & -\tfrac{2}{7} & \tfrac{1}{7} \end{bmatrix}$.

39. $\begin{cases} 2x + y = 8 \\ x + y = 5 \end{cases}$

Rewrite the system of equations in matrix form:
$A = \begin{bmatrix} 2 & 1 \\ 1 & 1 \end{bmatrix}$, $X = \begin{bmatrix} x \\ y \end{bmatrix}$, $B = \begin{bmatrix} 8 \\ 5 \end{bmatrix}$

Find the inverse of A and solve $X = A^{-1}B$:

From Problem 29, $A^{-1} = \begin{bmatrix} 1 & -1 \\ -1 & 2 \end{bmatrix}$, so

$X = A^{-1}B = \begin{bmatrix} 1 & -1 \\ -1 & 2 \end{bmatrix} \begin{bmatrix} 8 \\ 5 \end{bmatrix} = \begin{bmatrix} 3 \\ 2 \end{bmatrix}$.

The solution is $x = 3$, $y = 2$.

41. $\begin{cases} 2x + y = 0 \\ x + y = 5 \end{cases}$

Rewrite the system of equations in matrix form:
$A = \begin{bmatrix} 2 & 1 \\ 1 & 1 \end{bmatrix}$, $X = \begin{bmatrix} x \\ y \end{bmatrix}$, $B = \begin{bmatrix} 0 \\ 5 \end{bmatrix}$

Find the inverse of A and solve $X = A^{-1}B$:

From Problem 29, $A^{-1} = \begin{bmatrix} 1 & -1 \\ -1 & 2 \end{bmatrix}$, so

$X = A^{-1}B = \begin{bmatrix} 1 & -1 \\ -1 & 2 \end{bmatrix} \begin{bmatrix} 0 \\ 5 \end{bmatrix} = \begin{bmatrix} -5 \\ 10 \end{bmatrix}$.

The solution is $x = -5$, $y = 10$.

43. $\begin{cases} 6x + 5y = 7 \\ 2x + 2y = 2 \end{cases}$

Rewrite the system of equations in matrix form:
$A = \begin{bmatrix} 6 & 5 \\ 2 & 2 \end{bmatrix}$, $X = \begin{bmatrix} x \\ y \end{bmatrix}$, $B = \begin{bmatrix} 7 \\ 2 \end{bmatrix}$

Find the inverse of A and solve $X = A^{-1}B$:

From Problem 31, $A^{-1} = \begin{bmatrix} 1 & -\tfrac{5}{2} \\ -1 & 3 \end{bmatrix}$, so

$X = A^{-1}B = \begin{bmatrix} 1 & -\tfrac{5}{2} \\ -1 & 3 \end{bmatrix} \begin{bmatrix} 7 \\ 2 \end{bmatrix} = \begin{bmatrix} 2 \\ -1 \end{bmatrix}$.

The solution is $x = 2$, $y = -1$.

45. $\begin{cases} 6x+5y=13 \\ 2x+2y=5 \end{cases}$

Rewrite the system of equations in matrix form:

$A = \begin{bmatrix} 6 & 5 \\ 2 & 2 \end{bmatrix}$, $X = \begin{bmatrix} x \\ y \end{bmatrix}$, $B = \begin{bmatrix} 13 \\ 5 \end{bmatrix}$

Find the inverse of A and solve $X = A^{-1}B$:

From Problem 31, $A^{-1} = \begin{bmatrix} 1 & -\frac{5}{2} \\ -1 & 3 \end{bmatrix}$, so

$X = A^{-1}B = \begin{bmatrix} 1 & -\frac{5}{2} \\ -1 & 3 \end{bmatrix}\begin{bmatrix} 13 \\ 5 \end{bmatrix} = \begin{bmatrix} \frac{1}{2} \\ 2 \end{bmatrix}$.

The solution is $x = \frac{1}{2}$, $y = 2$.

47. $\begin{cases} 2x+y=-3 \\ ax+ay=-a \end{cases} \quad a \neq 0$

Rewrite the system of equations in matrix form:

$A = \begin{bmatrix} 2 & 1 \\ a & a \end{bmatrix}$, $X = \begin{bmatrix} x \\ y \end{bmatrix}$, $B = \begin{bmatrix} -3 \\ -a \end{bmatrix}$

Find the inverse of A and solve $X = A^{-1}B$:

From Problem 33, $A^{-1} = \begin{bmatrix} 1 & -\frac{1}{a} \\ -1 & \frac{2}{a} \end{bmatrix}$, so

$X = A^{-1}B = \begin{bmatrix} 1 & -\frac{1}{a} \\ -1 & \frac{2}{a} \end{bmatrix}\begin{bmatrix} -3 \\ -a \end{bmatrix} = \begin{bmatrix} -2 \\ 1 \end{bmatrix}$.

The solution is $x = -2$, $y = 1$.

49. $\begin{cases} 2x+y=\frac{7}{a} \\ ax+ay=5 \end{cases} \quad a \neq 0$

Rewrite the system of equations in matrix form:

$A = \begin{bmatrix} 2 & 1 \\ a & a \end{bmatrix}$, $X = \begin{bmatrix} x \\ y \end{bmatrix}$, $B = \begin{bmatrix} \frac{7}{a} \\ 5 \end{bmatrix}$

Find the inverse of A and solve $X = A^{-1}B$:

From Problem 33, $A^{-1} = \begin{bmatrix} 1 & -\frac{1}{a} \\ -1 & \frac{2}{a} \end{bmatrix}$, so

$X = A^{-1}B = \begin{bmatrix} 1 & -\frac{1}{a} \\ -1 & \frac{2}{a} \end{bmatrix}\begin{bmatrix} \frac{7}{a} \\ 5 \end{bmatrix} = \begin{bmatrix} \frac{2}{a} \\ \frac{3}{a} \end{bmatrix}$.

The solution is $x = \frac{2}{a}$, $y = \frac{3}{a}$.

51. $\begin{cases} x-y+z=0 \\ -2y+z=-1 \\ -2x-3y=-5 \end{cases}$

Rewrite the system of equations in matrix form:

$A = \begin{bmatrix} 1 & -1 & 1 \\ 0 & -2 & 1 \\ -2 & -3 & 0 \end{bmatrix}$, $X = \begin{bmatrix} x \\ y \\ z \end{bmatrix}$, $B = \begin{bmatrix} 0 \\ -1 \\ -5 \end{bmatrix}$

Find the inverse of A and solve $X = A^{-1}B$:

From Problem 35, $A^{-1} = \begin{bmatrix} 3 & -3 & 1 \\ -2 & 2 & -1 \\ -4 & 5 & -2 \end{bmatrix}$, so

$X = A^{-1}B = \begin{bmatrix} 3 & -3 & 1 \\ -2 & 2 & -1 \\ -4 & 5 & -2 \end{bmatrix}\begin{bmatrix} 0 \\ -1 \\ -5 \end{bmatrix} = \begin{bmatrix} -2 \\ 3 \\ 5 \end{bmatrix}$.

The solution is $x = -2$, $y = 3$, $z = 5$.

53. $\begin{cases} x-y+z=2 \\ -2y+z=2 \\ -2x-3y=\frac{1}{2} \end{cases}$

Rewrite the system of equations in matrix form:

$A = \begin{bmatrix} 1 & -1 & 1 \\ 0 & -2 & 1 \\ -2 & -3 & 0 \end{bmatrix}$, $X = \begin{bmatrix} x \\ y \\ z \end{bmatrix}$, $B = \begin{bmatrix} 2 \\ 2 \\ \frac{1}{2} \end{bmatrix}$

Find the inverse of A and solve $X = A^{-1}B$:

From Problem 35, $A^{-1} = \begin{bmatrix} 3 & -3 & 1 \\ -2 & 2 & -1 \\ -4 & 5 & -2 \end{bmatrix}$, so

$X = A^{-1}B = \begin{bmatrix} 3 & -3 & 1 \\ -2 & 2 & -1 \\ -4 & 5 & -2 \end{bmatrix}\begin{bmatrix} 2 \\ 2 \\ \frac{1}{2} \end{bmatrix} = \begin{bmatrix} \frac{1}{2} \\ -\frac{1}{2} \\ 1 \end{bmatrix}$.

The solution is $x = \frac{1}{2}$, $y = -\frac{1}{2}$, $z = 1$.

55. $\begin{cases} x + y + z = 9 \\ 3x + 2y - z = 8 \\ 3x + y + 2z = 1 \end{cases}$

Rewrite the system of equations in matrix form:

$A = \begin{bmatrix} 1 & 1 & 1 \\ 3 & 2 & -1 \\ 3 & 1 & 2 \end{bmatrix}$, $X = \begin{bmatrix} x \\ y \\ z \end{bmatrix}$, $B = \begin{bmatrix} 9 \\ 8 \\ 1 \end{bmatrix}$

Find the inverse of A and solve $X = A^{-1}B$:

From Problem 37, $A^{-1} = \begin{bmatrix} -\frac{5}{7} & \frac{1}{7} & \frac{3}{7} \\ \frac{9}{7} & \frac{1}{7} & -\frac{4}{7} \\ \frac{3}{7} & -\frac{2}{7} & \frac{1}{7} \end{bmatrix}$, so

$X = A^{-1}B = \begin{bmatrix} -\frac{5}{7} & \frac{1}{7} & \frac{3}{7} \\ \frac{9}{7} & \frac{1}{7} & -\frac{4}{7} \\ \frac{3}{7} & -\frac{2}{7} & \frac{1}{7} \end{bmatrix} \begin{bmatrix} 9 \\ 8 \\ 1 \end{bmatrix} = \begin{bmatrix} -\frac{34}{7} \\ \frac{85}{7} \\ \frac{12}{7} \end{bmatrix}$.

The solution is $x = -\frac{34}{7}, y = \frac{85}{7}, z = \frac{12}{7}$.

57. $\begin{cases} x + y + z = 2 \\ 3x + 2y - z = \frac{7}{3} \\ 3x + y + 2z = \frac{10}{3} \end{cases}$

Rewrite the system of equations in matrix form:

$A = \begin{bmatrix} 1 & 1 & 1 \\ 3 & 2 & -1 \\ 3 & 1 & 2 \end{bmatrix}$, $X = \begin{bmatrix} x \\ y \\ z \end{bmatrix}$, $B = \begin{bmatrix} 2 \\ \frac{7}{3} \\ \frac{10}{3} \end{bmatrix}$

Find the inverse of A and solve $X = A^{-1}B$:

From Problem 37, $A^{-1} = \begin{bmatrix} -\frac{5}{7} & \frac{1}{7} & \frac{3}{7} \\ \frac{9}{7} & \frac{1}{7} & -\frac{4}{7} \\ \frac{3}{7} & -\frac{2}{7} & \frac{1}{7} \end{bmatrix}$, so

$X = A^{-1}B = \begin{bmatrix} -\frac{5}{7} & \frac{1}{7} & \frac{3}{7} \\ \frac{9}{7} & \frac{1}{7} & -\frac{4}{7} \\ \frac{3}{7} & -\frac{2}{7} & \frac{1}{7} \end{bmatrix} \begin{bmatrix} 2 \\ \frac{7}{3} \\ \frac{10}{3} \end{bmatrix} = \begin{bmatrix} \frac{1}{3} \\ 1 \\ \frac{2}{3} \end{bmatrix}$.

The solution is $x = \frac{1}{3}, y = 1, z = \frac{2}{3}$.

59. $A = \begin{bmatrix} 4 & 2 \\ 2 & 1 \end{bmatrix}$

Augment the matrix with the identity and use row operations to find the inverse:

$\begin{bmatrix} 4 & 2 & | & 1 & 0 \\ 2 & 1 & | & 0 & 1 \end{bmatrix}$

$\rightarrow \begin{bmatrix} 4 & 2 & | & 1 & 0 \\ 0 & 0 & | & -\frac{1}{2} & 1 \end{bmatrix}$ $\left(R_2 = -\frac{1}{2}r_1 + r_2 \right)$

$\rightarrow \begin{bmatrix} 1 & \frac{1}{2} & | & \frac{1}{4} & 0 \\ 0 & 0 & | & -\frac{1}{2} & 1 \end{bmatrix}$ $\left(R_1 = \frac{1}{4}r_1 \right)$

There is no way to obtain the identity matrix on the left. Thus, this matrix has no inverse.

61. $A = \begin{bmatrix} 15 & 3 \\ 10 & 2 \end{bmatrix}$

Augment the matrix with the identity and use row operations to find the inverse:

$\begin{bmatrix} 15 & 3 & | & 1 & 0 \\ 10 & 2 & | & 0 & 1 \end{bmatrix}$

$\rightarrow \begin{bmatrix} 15 & 3 & | & 1 & 0 \\ 0 & 0 & | & -\frac{2}{3} & 1 \end{bmatrix}$ $\left(R_2 = -\frac{2}{3}r_1 + r_2 \right)$

$\rightarrow \begin{bmatrix} 1 & \frac{1}{5} & | & \frac{1}{15} & 0 \\ 0 & 0 & | & -\frac{2}{3} & 1 \end{bmatrix}$ $\left(R_1 = \frac{1}{15}r_1 \right)$

There is no way to obtain the identity matrix on the left; thus, there is no inverse.

63. $A = \begin{bmatrix} -3 & 1 & -1 \\ 1 & -4 & -7 \\ 1 & 2 & 5 \end{bmatrix}$

Augment the matrix with the identity and use row operations to find the inverse:

$\begin{bmatrix} -3 & 1 & -1 & | & 1 & 0 & 0 \\ 1 & -4 & -7 & | & 0 & 1 & 0 \\ 1 & 2 & 5 & | & 0 & 0 & 1 \end{bmatrix}$

$\rightarrow \begin{bmatrix} 1 & 2 & 5 & | & 0 & 0 & 1 \\ 1 & -4 & -7 & | & 0 & 1 & 0 \\ -3 & 1 & -1 & | & 1 & 0 & 0 \end{bmatrix}$ $\begin{pmatrix} \text{Interchange} \\ r_1 \text{ and } r_3 \end{pmatrix}$

$\rightarrow \begin{bmatrix} 1 & 2 & 5 & | & 0 & 0 & 1 \\ 0 & -6 & -12 & | & 0 & 1 & -1 \\ 0 & 7 & 14 & | & 1 & 0 & 3 \end{bmatrix}$ $\begin{pmatrix} R_2 = -r_1 + r_2 \\ R_3 = 3r_1 + r_3 \end{pmatrix}$

$$\rightarrow \begin{bmatrix} 1 & 2 & 5 & | & 0 & 0 & 1 \\ 0 & 1 & 2 & | & 0 & -\frac{1}{6} & \frac{1}{6} \\ 0 & 7 & 14 & | & 1 & 0 & 3 \end{bmatrix} \quad \left(R_2 = -\frac{1}{6}r_2\right)$$

$$\rightarrow \begin{bmatrix} 1 & 0 & 1 & | & 0 & \frac{1}{3} & \frac{2}{3} \\ 0 & 1 & 2 & | & 0 & -\frac{1}{6} & \frac{1}{6} \\ 0 & 0 & 0 & | & 1 & \frac{7}{6} & \frac{11}{6} \end{bmatrix} \quad \begin{pmatrix} R_1 = -2r_2 + r_1 \\ R_3 = -7r_2 + r_3 \end{pmatrix}$$

There is no way to obtain the identity matrix on the left; thus, there is no inverse.

65. $A = \begin{bmatrix} 25 & 61 & -12 \\ 18 & -2 & 4 \\ 8 & 35 & 21 \end{bmatrix}$

[A]$^{-1}$
[[.01 .05 -.0...
 [.01 -.02 .01...
 [-.02 .01 .03...

[A]$^{-1}$
...01 .05 -.01]
...01 -.02 .01]
...02 .01 .03]]

Thus, $A^{-1} \approx \begin{bmatrix} 0.01 & 0.05 & -0.01 \\ 0.01 & -0.02 & 0.01 \\ -0.02 & 0.01 & 0.03 \end{bmatrix}$

67. $A = \begin{bmatrix} 44 & 21 & 18 & 6 \\ -2 & 10 & 15 & 5 \\ 21 & 12 & -12 & 4 \\ -8 & -16 & 4 & 9 \end{bmatrix}$

[A]$^{-1}$
[[.02 -.04 -.0...
 [-.02 .05 .03...
 [.02 .01 -.0...
 [-.02 .06 .07...

[A]$^{-1}$
... -.01 .01]
... .03 -.03]
... -.04 4.88E-4]
... .07 .06]]

Thus, $A^{-1} \approx \begin{bmatrix} 0.02 & -0.04 & -0.01 & 0.01 \\ -0.02 & 0.05 & 0.03 & -0.03 \\ 0.02 & 0.01 & -0.04 & 0.00 \\ -0.02 & 0.06 & 0.07 & 0.06 \end{bmatrix}$.

69. $A = \begin{bmatrix} 25 & 61 & -12 \\ 18 & -12 & 7 \\ 3 & 4 & -1 \end{bmatrix}; B = \begin{bmatrix} 10 \\ -9 \\ 12 \end{bmatrix}$

Enter the matrices into a graphing utility and use $A^{-1}B$ to solve the system. The result is shown below:

Thus, the solution to the system is $x \approx 4.57$, $y \approx -6.44$, $z \approx -24.07$.

71. $A = \begin{bmatrix} 25 & 61 & -12 \\ 18 & -12 & 7 \\ 3 & 4 & -1 \end{bmatrix}; B = \begin{bmatrix} 21 \\ 7 \\ -2 \end{bmatrix}$

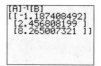

Thus, the solution to the system is $x \approx -1.19$, $y \approx 2.46$, $z \approx 8.27$.

73. a. The rows of the 2 by 3 matrix represent stainless steel and aluminum. The columns represent 10-gallon, 5-gallon, and 1-gallon.

The 2 by 3 matrix is: $\begin{bmatrix} 500 & 350 & 400 \\ 700 & 500 & 850 \end{bmatrix}$.

The 3 by 2 matrix is: $\begin{bmatrix} 500 & 700 \\ 350 & 500 \\ 400 & 850 \end{bmatrix}$.

b. The 3 by 1 matrix representing the amount of material is: $\begin{bmatrix} 15 \\ 8 \\ 3 \end{bmatrix}$.

c. The days usage of materials is:

$\begin{bmatrix} 500 & 350 & 400 \\ 700 & 500 & 850 \end{bmatrix} \cdot \begin{bmatrix} 15 \\ 8 \\ 3 \end{bmatrix} = \begin{bmatrix} 11,500 \\ 17,050 \end{bmatrix}$

Thus, 11,500 pounds of stainless steel and 17,050 pounds of aluminum were used that day.

d. The 1 by 2 matrix representing cost is: $\begin{bmatrix} 0.10 & 0.05 \end{bmatrix}$.

e. The total cost of the day's production was:

$\begin{bmatrix} 0.10 & 0.05 \end{bmatrix} \cdot \begin{bmatrix} 11,500 \\ 17,050 \end{bmatrix} = \begin{bmatrix} 2002.50 \end{bmatrix}$.

The total cost of the day's production was $2002.50.

75. $A = \begin{bmatrix} a & b \\ c & d \end{bmatrix}$

If $D = ad - bc \neq 0$, then $a \neq 0$ and $d \neq 0$, or $b \neq 0$ and $c \neq 0$. Assuming the former, then

$\begin{bmatrix} a & b & | & 1 & 0 \\ c & d & | & 0 & 1 \end{bmatrix}$

$\rightarrow \begin{bmatrix} 1 & \frac{b}{a} & | & \frac{1}{a} & 0 \\ c & d & | & 0 & 1 \end{bmatrix} \quad \left(R_1 = \frac{1}{a} \cdot r_1\right)$

$\rightarrow \begin{bmatrix} 1 & \frac{b}{a} & | & \frac{1}{a} & 0 \\ 0 & d - \frac{bc}{a} & | & -\frac{c}{a} & 1 \end{bmatrix} \quad \left(R_2 = -c \cdot r_1 + r_2\right)$

$\rightarrow \begin{bmatrix} 1 & \frac{b}{a} & | & \frac{1}{a} & 0 \\ 0 & \frac{ad-bc}{a} & | & -\frac{c}{a} & 1 \end{bmatrix}$

$\rightarrow \begin{bmatrix} 1 & \frac{b}{a} & | & \frac{1}{a} & 0 \\ 0 & 1 & | & \frac{-c}{ad-bc} & \frac{a}{ad-bc} \end{bmatrix} \quad \left(R_2 = \frac{a}{ad-bc} \cdot r_2\right)$

$\rightarrow \begin{bmatrix} 1 & 0 & | & \frac{1}{a} + \frac{bc}{a(ad-bc)} & \frac{-b}{ad-bc} \\ 0 & 1 & | & \frac{-c}{ad-bc} & \frac{a}{ad-bc} \end{bmatrix} \quad \left(R_1 = -\frac{b}{a} \cdot r_2 + r_1\right)$

$\rightarrow \begin{bmatrix} 1 & 0 & | & \frac{d}{ad-bc} & \frac{-b}{ad-bc} \\ 0 & 1 & | & \frac{-c}{ad-bc} & \frac{a}{ad-bc} \end{bmatrix}$

$\rightarrow \begin{bmatrix} 1 & 0 & | & \frac{d}{D} & \frac{-b}{D} \\ 0 & 1 & | & \frac{-c}{D} & \frac{a}{D} \end{bmatrix}$

Thus, $A^{-1} = \begin{bmatrix} \frac{d}{D} & \frac{-b}{D} \\ \frac{-c}{D} & \frac{a}{D} \end{bmatrix} = \frac{1}{D} \begin{bmatrix} d & -b \\ -c & a \end{bmatrix}$

where $D = ad - bc$.

Section 5.5

1. True

3. $3x^4 + 6x^3 + 3x^2 = 3x^2(x^2 + 2x + 1)$
   $= 3x^2(x+1)^2$

5. The rational expression $\frac{x}{x^2-1}$ is proper, since the degree of the numerator is less than the degree of the denominator.

7. The rational expression $\frac{x^2+5}{x^2-4}$ is improper, so perform the division:

$$x^2-4 \overline{\smash{\big)}\, x^2 + 5}$$
$$\underline{x^2 - 4}$$
$$9$$

The proper rational expression is:
$\frac{x^2+5}{x^2-4} = 1 + \frac{9}{x^2-4}$

9. The rational expression $\frac{5x^3+2x-1}{x^2-4}$ is improper, so perform the division:

$$x^2-4 \overline{\smash{\big)}\, 5x^3 + 2x - 1}$$
$$\underline{5x^3 - 20x}$$
$$22x - 1$$

The proper rational expression is:
$\frac{5x^3+2x-1}{x^2-4} = 5x + \frac{22x-1}{x^2-4}$

11. The rational expression
    $\frac{x(x-1)}{(x+4)(x-3)} = \frac{x^2-x}{x^2+x-12}$ is improper, so perform the division:

$$x^2+x-12 \overline{\smash{\big)}\, x^2 - x + 0}$$
$$\underline{x^2 + x - 12}$$
$$-2x + 12$$

The proper rational expression is:
$\frac{x(x-1)}{(x+4)(x-3)} = 1 + \frac{-2x+12}{x^2+x-12}$
$= 1 + \frac{-2(x-6)}{(x+4)(x-3)}$

13. Find the partial fraction decomposition:
    $\frac{4}{x(x-1)} = \frac{A}{x} + \frac{B}{x-1}$

    $x(x-1)\left(\frac{4}{x(x-1)}\right) = x(x-1)\left(\frac{A}{x} + \frac{B}{x-1}\right)$
    $4 = A(x-1) + Bx$

    Let $x = 1$, then $4 = A(0) + B \rightarrow B = 4$
    Let $x = 0$, then $4 = A(-1) + B(0) \rightarrow A = -4$

    $\frac{4}{x(x-1)} = \frac{-4}{x} + \frac{4}{x-1}$

15. Find the partial fraction decomposition:
$$\frac{1}{x(x^2+1)} = \frac{A}{x} + \frac{Bx+C}{x^2+1}$$

$$x(x^2+1)\left(\frac{1}{x(x^2+1)}\right) = x(x^2+1)\left(\frac{A}{x} + \frac{Bx+C}{x^2+1}\right)$$

$$1 = A(x^2+1) + (Bx+C)x$$

Let $x=0$, then $1 = A(0^2+1) + (B(0)+C)(0)$
$A = 1$

Let $x=1$, then $1 = A(1^2+1) + (B(1)+C)(1)$
$1 = 2A + B + C$
$1 = 2(1) + B + C$
$B + C = -1$

Let $x=-1$, then
$1 = A((-1)^2+1) + (B(-1)+C)(-1)$
$1 = A(1+1) + (-B+C)(-1)$
$1 = 2A + B - C$
$1 = 2(1) + B - C$
$B - C = -1$

Solve the system of equations:
$B + C = -1$
$B - C = -1$
$2B = -2 \qquad -1 + C = -1$
$B = -1 \qquad C = 0$

$$\frac{1}{x(x^2+1)} = \frac{1}{x} + \frac{-x}{x^2+1}$$

17. Find the partial fraction decomposition:
$$\frac{x}{(x-1)(x-2)} = \frac{A}{x-1} + \frac{B}{x-2}$$

Multiplying both sides by $(x-1)(x-2)$, we obtain: $x = A(x-2) + B(x-1)$

Let $x=1$, then $1 = A(1-2) + B(1-1)$
$1 = -A$
$A = -1$

Let $x=2$, then $2 = A(2-2) + B(2-1)$
$2 = B$

$$\frac{x}{(x-1)(x-2)} = \frac{-1}{x-1} + \frac{2}{x-2}$$

19. Find the partial fraction decomposition:
$$\frac{x^2}{(x-1)^2(x+1)} = \frac{A}{x-1} + \frac{B}{(x-1)^2} + \frac{C}{x+1}$$

Multiplying both sides by $(x-1)^2(x+1)$, we obtain: $x^2 = A(x-1)(x+1) + B(x+1) + C(x-1)^2$

Let $x=1$, then
$1^2 = A(1-1)(1+1) + B(1+1) + C(1-1)^2$
$1 = A(0)(2) + B(2) + C(0)^2$
$1 = 2B$
$B = \dfrac{1}{2}$

Let $x=-1$, then
$(-1)^2 = A(-1-1)(-1+1) + B(-1+1) + C(-1-1)^2$
$1 = A(-2)(0) + B(0) + C(-2)^2$
$1 = 4C$
$C = \dfrac{1}{4}$

Let $x=0$, then
$0^2 = A(0-1)(0+1) + B(0+1) + C(0-1)^2$
$0 = -A + B + C$
$A = B + C$
$A = \dfrac{1}{2} + \dfrac{1}{4} = \dfrac{3}{4}$

$$\frac{x^2}{(x-1)^2(x+1)} = \frac{(3/4)}{x-1} + \frac{(1/2)}{(x-1)^2} + \frac{(1/4)}{x+1}$$

21. Find the partial fraction decomposition:
$$\frac{1}{x^3-8} = \frac{1}{(x-2)(x^2+2x+4)}$$

$$\frac{1}{(x-2)(x^2+2x+4)} = \frac{A}{x-2} + \frac{Bx+C}{x^2+2x+4}$$

Multiplying both sides by $(x-2)(x^2+2x+4)$, we obtain: $1 = A(x^2+2x+4) + (Bx+C)(x-2)$

Let $x=2$, then
$1 = A(2^2+2(2)+4) + (B(2)+C)(2-2)$
$1 = 12A$
$A = \dfrac{1}{12}$

370

Let $x = 0$, then
$$1 = A(0^2 + 2(0) + 4) + (B(0) + C)(0 - 2)$$
$$1 = 4A - 2C$$
$$1 = 4(1/12) - 2C$$
$$-2C = \frac{2}{3}$$
$$C = -\frac{1}{3}$$

Let $x = 1$, then
$$1 = A(1^2 + 2(1) + 4) + (B(1) + C)(1 - 2)$$
$$1 = 7A - B - C$$
$$1 = 7(1/12) - B + \frac{1}{3}$$
$$B = -\frac{1}{12}$$

$$\frac{1}{x^3 - 8} = \frac{(1/12)}{x - 2} + \frac{-(1/12)x - 1/3}{x^2 + 2x + 4}$$
$$= \frac{(1/12)}{x - 2} + \frac{-(1/12)(x + 4)}{x^2 + 2x + 4}$$

23. Find the partial fraction decomposition:
$$\frac{x^2}{(x-1)^2(x+1)^2} = \frac{A}{x-1} + \frac{B}{(x-1)^2} + \frac{C}{x+1} + \frac{D}{(x+1)^2}$$

Multiplying both sides by $(x-1)^2(x+1)^2$, we obtain:
$$x^2 = A(x-1)(x+1)^2 + B(x+1)^2 + C(x-1)^2(x+1) + D(x-1)^2$$

Let $x = 1$, then
$$1^2 = A(1-1)(1+1)^2 + B(1+1)^2 + C(1-1)^2(1+1) + D(1-1)^2$$
$$1 = 4B$$
$$B = \frac{1}{4}$$

Let $x = -1$, then
$$(-1)^2 = A(-1-1)(-1+1)^2 + B(-1+1)^2 + C(-1-1)^2(-1+1) + D(-1-1)^2$$
$$1 = 4D$$
$$D = \frac{1}{4}$$

Let $x = 0$, then
$$0^2 = A(0-1)(0+1)^2 + B(0+1)^2 + C(0-1)^2(0+1) + D(0-1)^2$$
$$0 = -A + B + C + D$$
$$A - C = B + D$$
$$A - C = \frac{1}{4} + \frac{1}{4} = \frac{1}{2}$$

Let $x = 2$, then
$$2^2 = A(2-1)(2+1)^2 + B(2+1)^2 + C(2-1)^2(2+1) + D(2-1)^2$$
$$4 = 9A + 9B + 3C + D$$
$$9A + 3C = 4 - 9B - D$$
$$9A + 3C = 4 - 9\left(\frac{1}{4}\right) - \frac{1}{4} = \frac{3}{2}$$
$$3A + C = \frac{1}{2}$$

Solve the system of equations:
$$A - C = \frac{1}{2}$$
$$3A + C = \frac{1}{2}$$
$$4A = 1$$
$$A = \frac{1}{4}$$

$$\frac{3}{4} + C = \frac{1}{2}$$
$$C = -\frac{1}{4}$$

$$\frac{x^2}{(x-1)^2(x+1)^2} = \frac{(1/4)}{x-1} + \frac{(1/4)}{(x-1)^2} + \frac{(-1/4)}{x+1} + \frac{(1/4)}{(x+1)^2}$$

25. Find the partial fraction decomposition:
$$\frac{x-3}{(x+2)(x+1)^2} = \frac{A}{x+2} + \frac{B}{x+1} + \frac{C}{(x+1)^2}$$

Multiplying both sides by $(x+2)(x+1)^2$, we obtain:
$$x - 3 = A(x+1)^2 + B(x+2)(x+1) + C(x+2)$$

Let $x = -2$, then
$$-2 - 3 = A(-2+1)^2 + B(-2+2)(-2+1) + C(-2+2)$$
$$-5 = A$$
$$A = -5$$

Let $x = -1$, then

$-1-3 = A(-1+1)^2 + B(-1+2)(-1+1) + C(-1+2)$
$-4 = C$
$C = -4$

Let $x = 0$, then
$0 - 3 = A(0+1)^2 + B(0+2)(0+1) + C(0+2)$
$-3 = A + 2B + 2C$
$-3 = -5 + 2B + 2(-4)$
$2B = 10$
$B = 5$

$$\frac{x-3}{(x+2)(x+1)^2} = \frac{-5}{x+2} + \frac{5}{x+1} + \frac{-4}{(x+1)^2}$$

27. Find the partial fraction decomposition:
$$\frac{x+4}{x^2(x^2+4)} = \frac{A}{x} + \frac{B}{x^2} + \frac{Cx+D}{x^2+4}$$

Multiplying both sides by $x^2(x^2+4)$, we obtain:
$x + 4 = Ax(x^2+4) + B(x^2+4) + (Cx+D)x^2$

Let $x = 0$, then
$0 + 4 = A(0)(0^2+4) + B(0^2+4) + (C(0)+D)(0)^2$
$4 = 4B$
$B = 1$

Let $x = 1$, then
$1 + 4 = A(1)(1^2+4) + B(1^2+4) + (C(1)+D)(1)^2$
$5 = 5A + 5B + C + D$
$5 = 5A + 5 + C + D$
$5A + C + D = 0$

Let $x = -1$, then
$-1+4 = A(-1)((-1)^2+4) + B((-1)^2+4)$
$\qquad + (C(-1)+D)(-1)^2$
$3 = -5A + 5B - C + D$
$3 = -5A + 5 - C + D$
$-5A - C + D = -2$

Let $x = 2$, then
$2+4 = A(2)(2^2+4) + B(2^2+4) + (C(2)+D)(2)^2$
$6 = 16A + 8B + 8C + 4D$
$6 = 16A + 8 + 8C + 4D$
$16A + 8C + 4D = -2$

Solve the system of equations:
$5A + C + D = 0$
$-5A - C + D = -2$
$2D = -2$
$D = -1$

$5A + C - 1 = 0$
$C = 1 - 5A$
$16A + 8(1-5A) + 4(-1) = -2$
$16A + 8 - 40A - 4 = -2$
$-24A = -6$
$A = \frac{1}{4}$

$C = 1 - 5\left(\frac{1}{4}\right)$
$C = 1 - \frac{5}{4} = -\frac{1}{4}$

$$\frac{x+4}{x^2(x^2+4)} = \frac{(1/4)}{x} + \frac{1}{x^2} + \frac{-(1/4)x - 1}{x^2+4}$$
$$= \frac{(1/4)}{x} + \frac{1}{x^2} + \frac{-(1/4)(x+4)}{x^2+4}$$

29. Find the partial fraction decomposition:
$$\frac{x^2 + 2x + 3}{(x+1)(x^2+2x+4)} = \frac{A}{x+1} + \frac{Bx+C}{x^2+2x+4}$$

Multiplying both sides by $(x+1)(x^2+2x+4)$, we obtain:
$x^2 + 2x + 3 = A(x^2+2x+4) + (Bx+C)(x+1)$

Let $x = -1$, then
$(-1)^2 + 2(-1) + 3 = A((-1)^2 + 2(-1) + 4)$
$\qquad + (B(-1)+C)(-1+1)$
$2 = 3A$
$A = \frac{2}{3}$

Let $x = 0$, then
$0^2 + 2(0) + 3 = A(0^2 + 2(0) + 4) + (B(0)+C)(0+1)$
$3 = 4A + C$
$3 = 4(2/3) + C$
$C = \frac{1}{3}$

Let $x = 1$, then
$$1^2 + 2(1) + 3 = A(1^2 + 2(1) + 4) + (B(1) + C)(1+1)$$
$$6 = 7A + 2B + 2C$$
$$6 = 7(2/3) + 2B + 2(1/3)$$
$$2B = 6 - \frac{14}{3} - \frac{2}{3}$$
$$2B = \frac{2}{3}$$
$$B = \frac{1}{3}$$

$$\frac{x^2 + 2x + 3}{(x+1)(x^2 + 2x + 4)} = \frac{(2/3)}{x+1} + \frac{(1/3)x + 1/3}{x^2 + 2x + 4}$$
$$= \frac{(2/3)}{x+1} + \frac{(1/3)(x+1)}{x^2 + 2x + 4}$$

31. Find the partial fraction decomposition:
$$\frac{x}{(3x-2)(2x+1)} = \frac{A}{3x-2} + \frac{B}{2x+1}$$
Multiplying both sides by $(3x-2)(2x+1)$, we obtain: $x = A(2x+1) + B(3x-2)$

Let $x = -\frac{1}{2}$, then
$$-\frac{1}{2} = A(2(-1/2) + 1) + B(3(-1/2) - 2)$$
$$-\frac{1}{2} = -\frac{7}{2}B$$
$$B = \frac{1}{7}$$

Let $x = \frac{2}{3}$, then
$$\frac{2}{3} = A(2(2/3) + 1) + B(3(2/3) - 2)$$
$$\frac{2}{3} = \frac{7}{3}A$$
$$A = \frac{2}{7}$$

$$\frac{x}{(3x-2)(2x+1)} = \frac{(2/7)}{3x-2} + \frac{(1/7)}{2x+1}$$

33. Find the partial fraction decomposition:
$$\frac{x}{x^2 + 2x - 3} = \frac{x}{(x+3)(x-1)} = \frac{A}{x+3} + \frac{B}{x-1}$$
Multiplying both sides by $(x+3)(x-1)$, we obtain: $x = A(x-1) + B(x+3)$

Let $x = 1$, then $1 = A(1-1) + B(1+3)$
$$1 = 4B$$
$$B = \frac{1}{4}$$

Let $x = -3$, then $-3 = A(-3-1) + B(-3+3)$
$$-3 = -4A$$
$$A = \frac{3}{4}$$

$$\frac{x}{x^2 + 2x - 3} = \frac{(3/4)}{x+3} + \frac{(1/4)}{x-1}$$

35. Find the partial fraction decomposition:
$$\frac{x^2 + 2x + 3}{(x^2 + 4)^2} = \frac{Ax + B}{x^2 + 4} + \frac{Cx + D}{(x^2 + 4)^2}$$
Multiplying both sides by $(x^2 + 4)^2$, we obtain:
$$x^2 + 2x + 3 = (Ax + B)(x^2 + 4) + Cx + D$$
$$x^2 + 2x + 3 = Ax^3 + Bx^2 + 4Ax + 4B + Cx + D$$
$$x^2 + 2x + 3 = Ax^3 + Bx^2 + (4A + C)x + 4B + D$$

$A = 0$; $B = 1$;

$4A + C = 2$ $\qquad$ $4B + D = 3$
$4(0) + C = 2$ $\qquad$ $4(1) + D = 3$
$C = 2$ $\qquad$ $D = -1$

$$\frac{x^2 + 2x + 3}{(x^2 + 4)^2} = \frac{1}{x^2 + 4} + \frac{2x - 1}{(x^2 + 4)^2}$$

37. Find the partial fraction decomposition:
$$\frac{7x + 3}{x^3 - 2x^2 - 3x} = \frac{7x + 3}{x(x-3)(x+1)}$$
$$= \frac{A}{x} + \frac{B}{x-3} + \frac{C}{x+1}$$
Multiplying both sides by $x(x-3)(x+1)$, we obtain:
$$7x + 3 = A(x-3)(x+1) + Bx(x+1) + Cx(x-3)$$

Let $x = 0$, then
$$7(0) + 3 = A(0-3)(0+1) + B(0)(0+1) + C(0)(0-3)$$
$$3 = -3A$$
$$A = -1$$

Let $x = 3$, then
$$7(3) + 3 = A(3-3)(3+1) + B(3)(3+1) + C(3)(3-3)$$
$$24 = 12B$$
$$B = 2$$

Let $x = -1$, then
$$7(-1) + 3 = A(-1-3)(-1+1) + B(-1)(-1+1)$$
$$+ C(-1)(-1-3)$$
$$-4 = 4C$$
$$C = -1$$
$$\frac{7x+3}{x^3 - 2x^2 - 3x} = \frac{-1}{x} + \frac{2}{x-3} + \frac{-1}{x+1}$$

39. Perform synthetic division to find a factor:

$$\begin{array}{r|rrrr} 2) & 1 & -4 & 5 & -2 \\ & & 2 & -4 & 2 \\ \hline & 1 & -2 & 1 & 0 \end{array}$$

$$x^3 - 4x^2 + 5x - 2 = (x-2)(x^2 - 2x + 1)$$
$$= (x-2)(x-1)^2$$

Find the partial fraction decomposition:
$$\frac{x^2}{x^3 - 4x^2 + 5x - 2} = \frac{x^2}{(x-2)(x-1)^2}$$
$$= \frac{A}{x-2} + \frac{B}{x-1} + \frac{C}{(x-1)^2}$$

Multiplying both sides by $(x-2)(x-1)^2$, we obtain:
$$x^2 = A(x-1)^2 + B(x-2)(x-1) + C(x-2)$$

Let $x = 2$, then
$$2^2 = A(2-1)^2 + B(2-2)(2-1) + C(2-2)$$
$$4 = A$$

Let $x = 1$, then
$$1^2 = A(1-1)^2 + B(1-2)(1-1) + C(1-2)$$
$$1 = -C$$
$$C = -1$$

Let $x = 0$, then
$$0^2 = A(0-1)^2 + B(0-2)(0-1) + C(0-2)$$
$$0 = A + 2B - 2C$$
$$0 = 4 + 2B - 2(-1)$$
$$-2B = 6$$
$$B = -3$$

$$\frac{x^2}{x^3 - 4x^2 + 5x - 2} = \frac{4}{x-2} + \frac{-3}{x-1} + \frac{-1}{(x-1)^2}$$

41. Find the partial fraction decomposition:
$$\frac{x^3}{(x^2+16)^3} = \frac{Ax+B}{x^2+16} + \frac{Cx+D}{(x^2+16)^2} + \frac{Ex+F}{(x^2+16)^3}$$

Multiplying both sides by $(x^2+16)^3$, we obtain:
$$x^3 = (Ax+B)(x^2+16)^2 + (Cx+D)(x^2+16)$$
$$+ Ex + F$$
$$x^3 = (Ax+B)(x^4 + 32x^2 + 256) + Cx^3 + Dx^2$$
$$+ 16Cx + 16D + Ex + F$$
$$x^3 = Ax^5 + Bx^4 + 32Ax^3 + 32Bx^2 + 256Ax$$
$$+ 256B + Cx^3 + Dx^2$$
$$+ 16Cx + 16D + Ex + F$$
$$x^3 = Ax^5 + Bx^4 + (32A+C)x^3 + (32B+D)x^2$$
$$+ (256A + 16C + E)x$$
$$+ (256B + 16D + F)$$

$A = 0;\ B = 0;\ \ \ \ 32A + C = 1$
$\ 32(0) + C = 1$
$\ C = 1$

$32B + D = 0\ \ \ \ \ \ \ 256A + 16C + E = 0$
$32(0) + D = 0\ \ \ \ \ \ 256(0) + 16(1) + E = 0$
$\ \ \ \ \ \ \ \ D = 0\ E = -16$

$256B + 16D + F = 0$
$256(0) + 16(0) + F = 0$
$\ \ \ \ \ \ \ \ \ \ \ \ \ \ \ \ \ \ F = 0$

$$\frac{x^3}{(x^2+16)^3} = \frac{x}{(x^2+16)^2} + \frac{-16x}{(x^2+16)^3}$$

43. Find the partial fraction decomposition:
$$\frac{4}{2x^2 - 5x - 3} = \frac{4}{(x-3)(2x+1)} = \frac{A}{x-3} + \frac{B}{2x+1}$$

Multiplying both sides by $(x-3)(2x+1)$, we obtain: $4 = A(2x+1) + B(x-3)$

Let $x = -\frac{1}{2}$, then
$$4 = A(2(-1/2)+1) + B\left(-\frac{1}{2} - 3\right)$$
$$4 = -\frac{7}{2}B$$
$$B = -\frac{8}{7}$$

Let $x = 3$, then $4 = A(2(3)+1) + B(3-3)$
$$4 = 7A$$
$$A = \frac{4}{7}$$
$$\frac{4}{2x^2 - 5x - 3} = \frac{(4/7)}{x-3} + \frac{(-8/7)}{2x+1}$$

45. Find the partial fraction decomposition:
$$\frac{2x+3}{x^4 - 9x^2} = \frac{2x+3}{x^2(x-3)(x+3)}$$
$$= \frac{A}{x} + \frac{B}{x^2} + \frac{C}{x-3} + \frac{D}{x+3}$$

Multiplying both sides by $x^2(x-3)(x+3)$, we obtain:
$$2x + 3 = Ax(x-3)(x+3) + B(x-3)(x+3)$$
$$+ Cx^2(x+3) + Dx^2(x-3)$$

Let $x = 0$, then
$$2 \cdot 0 + 3 = A \cdot 0(0-3)(0+3) + B(0-3)(0+3)$$
$$+ C \cdot 0^2(0+3) + D \cdot 0^2(0-3)$$
$$3 = -9B$$
$$B = -\frac{1}{3}$$

Let $x = 3$, then
$$2 \cdot 3 + 3 = A \cdot 3(3-3)(3+3) + B(3-3)(3+3)$$
$$+ C \cdot 3^2(3+3) + D \cdot 3^2(3-3)$$
$$9 = 54C$$
$$C = \frac{1}{6}$$

Let $x = -3$, then
$$2(-3) + 3 = A(-3)(-3-3)(-3+3)$$
$$+ B(-3-3)(-3+3)$$
$$+ C(-3)^2(-3+3)$$
$$+ D(-3)^2(-3-3)$$
$$-3 = -54D$$
$$D = \frac{1}{18}$$

Let $x = 1$, then
$$2 \cdot 1 + 3 = A \cdot 1(1-3)(1+3) + B(1-3)(1+3)$$
$$+ C \cdot 1^2(1+3) + D \cdot 1^2(1-3)$$
$$5 = -8A - 8B + 4C - 2D$$

$$5 = -8A - 8(-1/3) + 4(1/6) - 2(1/18)$$
$$5 = -8A + \frac{8}{3} + \frac{2}{3} - \frac{1}{9}$$
$$-8A = \frac{16}{9}$$
$$A = -\frac{2}{9}$$

$$\frac{2x+3}{x^4 - 9x^2} = \frac{(-2/9)}{x} + \frac{(-1/3)}{x^2} + \frac{(1/6)}{x-3} + \frac{(1/18)}{x+3}$$

Section 5.6

1. $3x + 4 < 8 - x$
$$4x < 4$$
$$x < 1$$
$$\{x \mid x < 1\} \text{ or } (-\infty, 1)$$

3. $x^2 + y^2 = 9$
The graph is a circle. Center: $(0, 0)$; Radius: 3

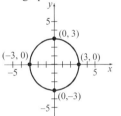

5. True

7. satisfied

9. False

11. a. $x \geq 0$

Graph the line $x = 0$. Use a solid line since the inequality uses $\geq$. Choose a test point not on the line, such as $(2, 0)$. Since $2 \geq 0$ is true, shade the side of the line containing $(2, 0)$.

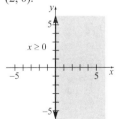

b. Set the calculator viewing WINDOW as shown below. Graph the vertical line $x = 0$ by using the SHADE commands in the calculators DRAW menu: Shade $(-5, 5, 0, 5)$

13. a. $x \geq 4$

    Graph the line $x = 4$. Use a solid line since the inequality uses $\geq$. Choose a test point not on the line, such as $(5, 0)$. Since $5 \geq 0$ is true, shade the side of the line containing $(5, 0)$.

    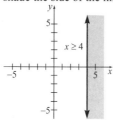

    b. Set the calculator viewing WINDOW as shown below. Graph the vertical line $x = 0$ by using the SHADE commands in the calculators DRAW menu: Shade $(-5, 5, 4, 5)$.

15. a. $2x + y \geq 6$

    Graph the line $2x + y = 6$. Use a solid line since the inequality uses $\geq$. Choose a test point not on the line, such as $(0, 0)$. Since $2(0) + 0 \geq 6$ is false, shade the opposite side of the line from $(0, 0)$.

    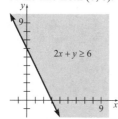

    b. Set the calculator viewing WINDOW as shown below. Graph the linear inequality $2x + y \geq 6$ $(y \geq -2x + 6)$ by entering $Y_1 = -2x + 6$ and adjusting the setting to the left of Y_1 as shown below:

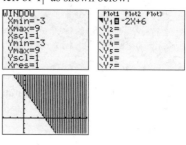

17. a. $x^2 + y^2 > 1$

    Graph the circle $x^2 + y^2 = 1$. Use a dashed line since the inequality uses $>$. Choose a test point not on the circle, such as $(0, 0)$. Since $0^2 + 0^2 > 1$ is false, shade the opposite side of the circle from $(0, 0)$.

    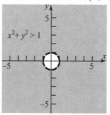

    b. Solving $x^2 + y^2 = 1$ for y, we obtain $y = \pm\sqrt{1-x^2}$. We need to shade above $y = \sqrt{1-x^2}$ and below $y = -\sqrt{1-x^2}$. Since these functions are only defined on the domain $-1 \leq x \leq 1$, we must also shade to the left of $x = -1$ and to the right of $x = 1$. To do so, first set the calculator viewing WINDOW as shown below. The SHADE commands shown below shade above, below, to the left, and to the right of the circle, respectively:

    Shade $\left(\sqrt{1-x^2}, 5, -1, 1\right)$

    Shade $\left(-5, -\sqrt{1-x^2}, -1, 1\right)$

    Shade $(-5, 5, -5, -1)$

    Shade $(-5, 5, 1, 5)$

19. a. $y \leq x^2 - 1$

   Graph the parabola $y = x^2 - 1$. Use a solid line since the inequality uses $\leq$. Choose a test point not on the parabola, such as (0, 0). Since $0 \leq 0^2 - 1$ is false, shade the opposite side of the parabola from (0, 0).

   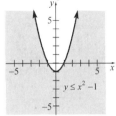

   b. Set the calculator viewing WINDOW as shown below. Enter $Y_1 = x^2 - 1$ and adjust the setting to the left of Y_1 as shown below:

21. a. $xy \geq 4$

   Graph the hyperbola $xy = 4$. Use a solid line since the inequality uses $\geq$. Choose a test point not on the hyperbola, such as (0, 0). Since $0 \cdot 0 \geq 4$ is false, shade the opposite side of the hyperbola from (0, 0).

   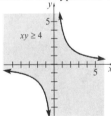

   b. Solving $xy = 4$ for y, we obtain $y = \dfrac{4}{x}$. We need to shade below $y = \dfrac{4}{x}$ when $x < 0$ and above $y = \dfrac{4}{x}$ when $x > 0$. To do so, first set the calculator viewing WINDOW as shown below. The SHADE commands shown

below shade the necessary area, respectively:

Shade $\left(-6, \dfrac{4}{x}, -6, 0\right)$

Shade $\left(\dfrac{4}{x}, 6, 0, 6\right)$

23. $\begin{cases} x + y \leq 2 \\ 2x + y \geq 4 \end{cases}$

   Graph the line $x + y = 2$. Use a solid line since the inequality uses $\leq$. Choose a test point not on the line, such as (0, 0). Since $0 + 0 \leq 2$ is true, shade the side of the line containing (0, 0). Graph the line $2x + y = 4$. Use a solid line since the inequality uses $\geq$. Choose a test point not on the line, such as (0, 0). Since $2(0) + 0 \geq 4$ is false, shade the opposite side of the line from (0, 0). The overlapping region is the solution.

   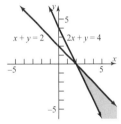

25. $\begin{cases} 2x - y \leq 4 \\ 3x + 2y \geq -6 \end{cases}$

   Graph the line $2x - y = 4$. Use a solid line since the inequality uses $\leq$. Choose a test point not on the line, such as (0, 0). Since $2(0) - 0 \leq 4$ is true, shade the side of the line containing (0, 0). Graph the line $3x + 2y = -6$. Use a solid line since the inequality uses $\geq$. Choose a test point not on the line, such as (0, 0). Since $3(0) + 2(0) \geq -6$ is true, shade the side of the line containing (0, 0). The overlapping region is the solution.

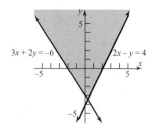

27. $\begin{cases} 2x - 3y \leq 0 \\ 3x + 2y \leq 6 \end{cases}$

Graph the line $2x - 3y = 0$. Use a solid line since the inequality uses $\leq$. Choose a test point not on the line, such as (0, 3). Since $2(0) - 3(3) \leq 0$ is true, shade the side of the line containing (0, 3). Graph the line $3x + 2y = 6$. Use a solid line since the inequality uses $\leq$. Choose a test point not on the line, such as (0, 0). Since $3(0) + 2(0) \leq 6$ is true, shade the side of the line containing (0, 0). The overlapping region is the solution.

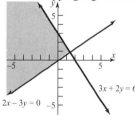

29. $\begin{cases} x - 2y \leq 6 \\ 2x - 4y \geq 0 \end{cases}$

Graph the line $x - 2y = 6$. Use a solid line since the inequality uses $\leq$. Choose a test point not on the line, such as (0, 0). Since $0 - 2(0) \leq 6$ is true, shade the side of the line containing (0, 0). Graph the line $2x - 4y = 0$. Use a solid line since the inequality uses $\geq$. Choose a test point not on the line, such as (0, 2). Since $2(0) - 4(2) \geq 0$ is false, shade the opposite side of the line from (0, 2). The overlapping region is the solution.

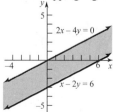

31. $\begin{cases} 2x + y \geq -2 \\ 2x + y \geq 2 \end{cases}$

Graph the line $2x + y = -2$. Use a solid line since the inequality uses $\geq$. Choose a test point not on the line, such as (0, 0). Since $2(0) + 0 \geq -2$ is true, shade the side of the line containing (0, 0). Graph the line $2x + y = 2$. Use a solid line since the inequality uses $\geq$. Choose a test point not on the line, such as (0, 0). Since $2(0) + 0 \geq 2$ is false, shade the opposite side of the line from (0, 0). The overlapping region is the solution.

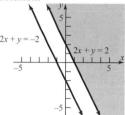

33. $\begin{cases} 2x + 3y \geq 6 \\ 2x + 3y \leq 0 \end{cases}$

Graph the line $2x + 3y = 6$. Use a solid line since the inequality uses $\geq$. Choose a test point not on the line, such as (0, 0). Since $2(0) + 3(0) \geq 6$ is false, shade the opposite side of the line from (0, 0). Graph the line $2x + 3y = 0$. Use a solid line since the inequality uses $\leq$. Choose a test point not on the line, such as (0, 2). Since $2(0) + 3(2) \leq 0$ is false, shade the opposite side of the line from (0, 2). Since the regions do not overlap, the solution is an empty set.

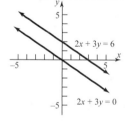

35. $\begin{cases} x \geq 0 \\ y \geq 0 \\ 2x + y \leq 6 \\ x + 2y \leq 6 \end{cases}$

Graph $x \geq 0$; $y \geq 0$. Shaded region is the first quadrant. Graph the line $2x + y = 6$. Use a solid line since the inequality uses $\leq$. Choose a test point not on the line, such as (0, 0). Since $2(0) + 0 \leq 6$ is true, shade the side of the line containing (0, 0). Graph the line $x + 2y = 6$. Use a solid line since the inequality uses $\leq$. Choose a test point not on the line, such as (0, 0). Since $0 + 2(0) \leq 6$ is true, shade the side of the line containing (0, 0). The overlapping region is the solution. The graph is bounded. Find the vertices:

The x-axis and y-axis intersect at (0, 0). The intersection of $x + 2y = 6$ and the y-axis is (0, 3). The intersection of $2x + y = 6$ and the x-axis is (3, 0). To find the intersection of $x + 2y = 6$ and $2x + y = 6$, solve the system:
$$\begin{cases} x + 2y = 6 \\ 2x + y = 6 \end{cases}$$
Solve the first equation for x: $x = 6 - 2y$.
Substitute and solve:
$$2(6 - 2y) + y = 6$$
$$12 - 4y + y = 6$$
$$12 - 3y = 6$$
$$-3y = -6$$
$$y = 2$$
$x = 6 - 2(2) = 2$

The point of intersection is (2, 2).
The four corner points are (0, 0), (0, 3), (3, 0), and (2, 2).

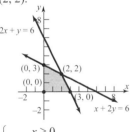

37. $\begin{cases} x \geq 0 \\ y \geq 0 \\ x + y \geq 2 \\ 2x + y \geq 4 \end{cases}$

Graph $x \geq 0$; $y \geq 0$. Shaded region is the first quadrant. Graph the line $x + y = 2$. Use a solid line since the inequality uses $\geq$. Choose a test point not on the line, such as (0, 0). Since $0 + 0 \geq 2$ is false, shade the opposite side of the line from (0, 0). Graph the line $2x + y = 4$. Use a solid line since the inequality uses $\geq$. Choose a test point not on the line, such as (0, 0). Since

$2(0) + 0 \geq 4$ is false, shade the opposite side of the line from (0, 0). The overlapping region is the solution. The graph is unbounded.

Find the vertices:
The intersection of $x + y = 2$ and the x-axis is (2, 0). The intersection of $2x + y = 4$ and the y-axis is (0, 4). The two corner points are (2, 0), and (0, 4).

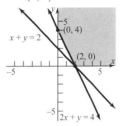

39. $\begin{cases} x \geq 0 \\ y \geq 0 \\ x + y \geq 2 \\ 2x + 3y \leq 12 \\ 3x + y \leq 12 \end{cases}$

Graph $x \geq 0$; $y \geq 0$. Shaded region is the first quadrant. Graph the line $x + y = 2$. Use a solid line since the inequality uses $\geq$. Choose a test point not on the line, such as (0, 0). Since $0 + 0 \geq 2$ is false, shade the opposite side of the line from (0, 0). Graph the line $2x + 3y = 12$. Use a solid line since the inequality uses $\leq$. Choose a test point not on the line, such as (0, 0). Since $2(0) + 3(0) \leq 12$ is true, shade the side of the line containing (0, 0). Graph the line $3x + y = 12$. Use a solid line since the inequality uses $\leq$. Choose a test point not on the line, such as (0, 0). Since $3(0) + 0 \leq 12$ is true, shade the side of the line containing (0, 0). The overlapping region is the solution. The graph is bounded.

Find the vertices:
The intersection of $x + y = 2$ and the y-axis is (0, 2). The intersection of $x + y = 2$ and the x-axis is (2, 0). The intersection of $2x + 3y = 12$ and the y-axis is (0, 4). The intersection of $3x + y = 12$ and the x-axis is (4, 0).

To find the intersection of $2x + 3y = 12$ and $3x + y = 12$, solve the system:

$\begin{cases} 2x+3y=12 \\ 3x+y=12 \end{cases}$

Solve the second equation for y: $y=12-3x$.
Substitute and solve:
$$2x+3(12-3x)=12$$
$$2x+36-9x=12$$
$$-7x=-24$$
$$x=\frac{24}{7}$$
$$y=12-3\left(\frac{24}{7}\right)=12-\frac{72}{7}=\frac{12}{7}$$

The point of intersection is $\left(\frac{24}{7},\frac{12}{7}\right)$.

The five corner points are (0, 2), (0, 4), (2, 0), (4, 0), and $\left(\frac{24}{7},\frac{12}{7}\right)$.

Find the vertices:
The intersection of $x+y=2$ and the y-axis is (0, 2). The intersection of $x+y=2$ and the x-axis is (2, 0). The intersection of $x+y=8$ and the y-axis is (0, 8). The intersection of $2x+y=10$ and the x-axis is (5, 0). To find the intersection of $x+y=8$ and $2x+y=10$, solve the system:
$\begin{cases} x+y=8 \\ 2x+y=10 \end{cases}$

Solve the first equation for y: $y=8-x$.
Substitute and solve:
$$2x+8-x=10$$
$$x=2$$
$$y=8-2=6$$

The point of intersection is (2, 6).
The five corner points are (0, 2), (0, 8), (2, 0), (5, 0), and (2, 6).

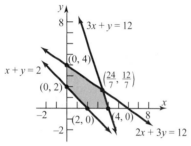

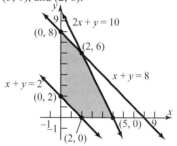

41. $\begin{cases} x\geq 0 \\ y\geq 0 \\ x+y\geq 2 \\ x+y\leq 8 \\ 2x+y\leq 10 \end{cases}$

Graph $x\geq 0$; $y\geq 0$. Shaded region is the first quadrant. Graph the line $x+y=2$. Use a solid line since the inequality uses $\geq$. Choose a test point not on the line, such as (0, 0). Since $0+0\geq 2$ is false, shade the opposite side of the line from (0, 0). Graph the line $x+y=8$. Use a solid line since the inequality uses $\leq$. Choose a test point not on the line, such as (0, 0). Since $0+0\leq 8$ is true, shade the side of the line containing (0, 0). Graph the line $2x+y=10$. Use a solid line since the inequality uses $\leq$. Choose a test point not on the line, such as (0, 0). Since $2(0)+0\leq 10$ is true, shade the side of the line containing (0, 0). The overlapping region is the solution. The graph is bounded.

43. $\begin{cases} x\geq 0 \\ y\geq 0 \\ x+2y\geq 1 \\ x+2y\leq 10 \end{cases}$

Graph $x\geq 0$; $y\geq 0$. Shaded region is the first quadrant. Graph the line $x+2y=1$. Use a solid line since the inequality uses $\geq$. Choose a test point not on the line, such as (0, 0). Since $0+2(0)\geq 1$ is false, shade the opposite side of the line from (0, 0). Graph the line $x+2y=10$. Use a solid line since the inequality uses $\leq$. Choose a test point not on the line, such as (0, 0). Since $0+2(0)\leq 10$ is true, shade the side of the line containing (0, 0). The overlapping region is the solution. The graph is bounded.

Find the vertices:
The intersection of $x+2y=1$ and the y-axis is (0, 0.5). The intersection of $x+2y=1$ and the x-axis is (1, 0). The intersection of $x+2y=10$

and the y-axis is (0, 5). The intersection of $x+2y=10$ and the x-axis is (10, 0). The four corner points are (0, 0.5), (0, 5), (1, 0), and (10, 0).

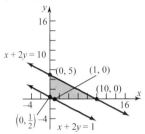

45. The system of linear inequalities is:
$$\begin{cases} x \geq 0 \\ y \geq 0 \\ x \leq 4 \\ x+y \leq 6 \end{cases}$$

47. The system of linear inequalities is:
$$\begin{cases} x \geq 0 \\ y \geq 15 \\ x \leq 20 \\ x+y \leq 50 \\ x-y \leq 0 \end{cases}$$

49. a. Let x = the amount invested in Treasury bills, and let y = the amount invested in corporate bonds.
The constraints are:
$x \geq 0$, $y \geq 0$ because a non-negative amount must be invested.
$x+y \leq 50,000$ because the total investment cannot exceed $50,000.
$y \leq 10,000$ because the amount invested in corporate bonds must not exceed $10,000.
$x \geq 35,000$ because the amount invested in Treasury bills must be at least $35,000.
The system is
$$\begin{cases} x \geq 0 \\ y \geq 0 \\ x+y \leq 50,000 \\ y \leq 10,000 \\ x \geq 35,000 \end{cases}$$

b. Graph the system.

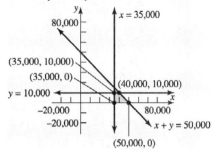

The corner points are (35,000, 0), (35,000, 10,000), (40,000, 10,000), (50,000, 0).

51. a. Let x = the # of packages of the economy blend, and let y = the # of packages of the superior blend.
The constraints are:
$x \geq 0$, $y \geq 0$ because a non-negative # of packages must be produced.
$4x+8y \leq 75 \cdot 16$ because the total amount of "A grade" coffee cannot exceed 75 pounds. (Note: 75 pounds = (75)(16) ounces.)
$12x+8y \leq 120 \cdot 16$ because the total amount of "B grade" coffee cannot exceed 120 pounds. (Note: 120 pounds = (120)(16) ounces.)
Simplifying the inequalities, we obtain:

$4x+8y \leq 75 \cdot 16$ $12x+8y \leq 120 \cdot 16$
$x+2y \leq 75 \cdot 4$ $3x+2y \leq 120 \cdot 4$
$x+2y \leq 300$ $3x+2y \leq 480$

The system is:
$$\begin{cases} x \geq 0 \\ y \geq 0 \\ x+2y \leq 300 \\ 3x+2y \leq 480 \end{cases}$$

b. Graph the system.

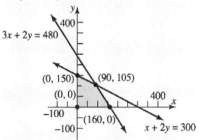

The corner points are (0, 0), (0, 150), (90, 105), (160, 0).

Chapter 5: Systems of Equations and Inequalities

53. **a.** Let x = the # of microwaves, and let y = the # of printers.

    The constraints are:
    $x \geq 0$, $y \geq 0$ because a non-negative # of items must be shipped.
    $30x + 20y \leq 1600$ because a total cargo weight cannot exceed 1600 pounds.
    $2x + 3y \leq 150$ because the total cargo volume cannot exceed 150 cubic feet. Note that the inequality $30x + 20y \leq 1600$ can be simplified: $3x + 2y \leq 160$.

    The system is:
    $$\begin{cases} x \geq 0;\ y \geq 0 \\ 3x + 2y \leq 160 \\ 2x + 3y \leq 150 \end{cases}$$

    b. Graph the system.

    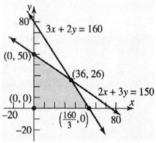

    The corner points are $(0, 0)$, $(0, 50)$, $(36, 26)$, $(160/3, 0)$.

Section 5.7

1. objective function

3. $z = x + y$

Vertex	Value of $z = x + y$
(0, 3)	$z = 0 + 3 = 3$
(0, 6)	$z = 0 + 6 = 6$
(5, 6)	$z = 5 + 6 = 11$
(5, 2)	$z = 5 + 2 = 7$
(4, 0)	$z = 4 + 0 = 4$

    The maximum value is 11 at $(5, 6)$, and the minimum value is 3 at $(0, 3)$.

5. $z = x + 10y$

Vertex	Value of $z = x + 10y$
(0, 3)	$z = 0 + 10(3) = 30$
(0, 6)	$z = 0 + 10(6) = 60$
(5, 6)	$z = 5 + 10(6) = 65$
(5, 2)	$z = 5 + 10(2) = 25$
(4, 0)	$z = 4 + 10(0) = 4$

    The maximum value is 65 at $(5, 6)$, and the minimum value is 4 at $(4, 0)$.

7. $z = 5x + 7y$

Vertex	Value of $z = 5x + 7y$
(0, 3)	$z = 5(0) + 7(3) = 21$
(0, 6)	$z = 5(0) + 7(6) = 42$
(5, 6)	$z = 5(5) + 7(6) = 67$
(5, 2)	$z = 5(5) + 7(2) = 39$
(4, 0)	$z = 5(4) + 7(0) = 20$

    The maximum value is 67 at $(5, 6)$, and the minimum value is 20 at $(4, 0)$.

9. Maximize $z = 2x + y$ subject to $x \geq 0$, $y \geq 0$, $x + y \leq 6$, $x + y \geq 1$. Graph the constraints.

    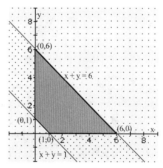

    The corner points are $(0, 1)$, $(1, 0)$, $(0, 6)$, $(6, 0)$.
    Evaluate the objective function:

Vertex	Value of $z = 2x + y$
(0, 1)	$z = 2(0) + 1 = 1$
(0, 6)	$z = 2(0) + 6 = 6$
(1, 0)	$z = 2(1) + 0 = 2$
(6, 0)	$z = 2(6) + 0 = 12$

    The maximum value is 12 at $(6, 0)$.

11. Minimize $z = 2x + 5y$ subject to $x \geq 0$, $y \geq 0$, $x + y \geq 2$, $x \leq 5$, $y \leq 3$. Graph the constraints.

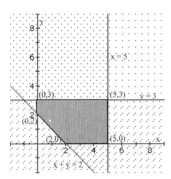

The corner points are (0, 2), (2, 0), (0, 3), (5, 0), (5, 3). Evaluate the objective function:

Vertex	Value of $z = 2x + 5y$
(0, 2)	$z = 2(0) + 5(2) = 10$
(0, 3)	$z = 2(0) + 5(3) = 15$
(2, 0)	$z = 2(2) + 5(0) = 4$
(5, 0)	$z = 2(5) + 5(0) = 10$
(5, 3)	$z = 2(5) + 5(3) = 25$

The minimum value is 4 at (2, 0).

13. Maximize $z = 3x + 5y$ subject to $x \geq 0$, $y \geq 0$, $x + y \geq 2$, $2x + 3y \leq 12$, $3x + 2y \leq 12$. Graph the constraints.

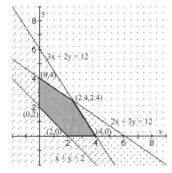

To find the intersection of $2x + 3y = 12$ and $3x + 2y = 12$, solve the system:
$$\begin{cases} 2x + 3y = 12 \\ 3x + 2y = 12 \end{cases}$$

Solve the second equation for y: $y = 6 - \frac{3}{2}x$

Substitute and solve:

$2x + 3\left(6 - \frac{3}{2}x\right) = 12$

$2x + 18 - \frac{9}{2}x = 12$

$-\frac{5}{2}x = -6$

$x = \frac{12}{5}$

$y = 6 - \frac{3}{2}\left(\frac{12}{5}\right) = 6 - \frac{18}{5} = \frac{12}{5}$

The point of intersection is $(2.4, 2.4)$.

The corner points are (0, 2), (2, 0), (0, 4), (4, 0), (2.4, 2.4). Evaluate the objective function:

Vertex	Value of $z = 3x + 5y$
(0, 2)	$z = 3(0) + 5(2) = 10$
(0, 4)	$z = 3(0) + 5(4) = 20$
(2, 0)	$z = 3(2) + 5(0) = 6$
(4, 0)	$z = 3(4) + 5(0) = 12$
(2.4, 2.4)	$z = 3(2.4) + 5(2.4) = 19.2$

The maximum value is 20 at (0, 4).

15. Minimize $z = 5x + 4y$ subject to $x \geq 0$, $y \geq 0$, $x + y \geq 2$, $2x + 3y \leq 12$, $3x + y \leq 12$. Graph the constraints.

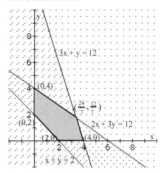

To find the intersection of $2x + 3y = 12$ and $3x + y = 12$, solve the system:
$$\begin{cases} 2x + 3y = 12 \\ 3x + y = 12 \end{cases}$$

Solve the second equation for y: $y = 12 - 3x$
Substitute and solve:

$2x+3(12-3x)=12$
$2x+36-9x=12$
$-7x=-24$
$x=\dfrac{24}{7}$
$y=12-3\left(\dfrac{24}{7}\right)=12-\dfrac{72}{7}=\dfrac{12}{7}$

The point of intersection is $\left(\dfrac{24}{7},\dfrac{12}{7}\right)$.

The corner points are (0, 2), (2, 0), (0, 4), (4, 0), $\left(\dfrac{24}{7},\dfrac{12}{7}\right)$. Evaluate the objective function:

Vertex	Value of $z=5x+4y$
(0, 2)	$z=5(0)+4(2)=8$
(0, 4)	$z=5(0)+4(4)=16$
(2, 0)	$z=5(2)+4(0)=10$
(4, 0)	$z=5(4)+4(0)=20$
$\left(\dfrac{24}{7},\dfrac{12}{7}\right)$	$z=5\left(\dfrac{24}{7}\right)+4\left(\dfrac{12}{7}\right)=24$

The minimum value is 8 at (0, 2).

17. Maximize $z=5x+2y$ subject to $x\geq 0$, $y\geq 0$, $x+y\leq 10$, $2x+y\geq 10$, $x+2y\geq 10$.
Graph the constraints.

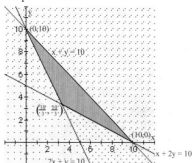

To find the intersection of $2x+y=10$ and $x+2y=10$, solve the system:
$\begin{cases} 2x+y=10 \\ x+2y=10 \end{cases}$
Solve the first equation for y: $y=10-2x$.
Substitute and solve:

$x+2(10-2x)=10$
$x+20-4x=10$
$-3x=-10$
$x=\dfrac{10}{3}$
$y=10-2\left(\dfrac{10}{3}\right)=10-\dfrac{20}{3}=\dfrac{10}{3}$

The point of intersection is (10/3 10/3).
The corner points are (0, 10), (10, 0), (10/3, 10/3).
Evaluate the objective function:

Vertex	Value of $z=5x+2y$
(0, 10)	$z=5(0)+2(10)=20$
(10, 0)	$z=5(10)+2(0)=50$
$\left(\dfrac{10}{3},\dfrac{10}{3}\right)$	$z=5\left(\dfrac{10}{3}\right)+2\left(\dfrac{10}{3}\right)=\dfrac{70}{3}=23\dfrac{1}{3}$

The maximum value is 50 at (10, 0).

19. Let x = the number of downhill skis produced, and let y = the number of cross-country skis produced. The total profit is: $P=70x+50y$. Profit is to be maximized, so this is the objective function. The constraints are:
$x\geq 0$, $y\geq 0$ A positive number of skis must be produced.
$2x+y\leq 40$ Manufacturing time available.
$x+y\leq 32$ Finishing time available.
Graph the constraints.

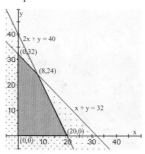

To find the intersection of $x+y=32$ and $2x+y=40$, solve the system:
$\begin{cases} x+y=32 \\ 2x+y=40 \end{cases}$
Solve the first equation for y: $y=32-x$.
Substitute and solve:
$2x+(32-x)=40$
$x=8$
$y=32-8=24$
The point of intersection is (8, 24).

The corner points are (0, 0), (0, 32), (20, 0), (8, 24). Evaluate the objective function:

Vertex	Value of $P = 70x + 50y$
(0, 0)	$P = 70(0) + 50(0) = 0$
(0, 32)	$P = 70(0) + 50(32) = 1600$
(20, 0)	$P = 70(20) + 50(0) = 1400$
(8, 24)	$P = 70(8) + 50(24) = 1760$

The maximum profit is $1760, when 8 downhill skis and 24 cross-country skis are produced.

With the increase of the manufacturing time to 48 hours, we do the following:
The constraints are:

$x \geq 0, \ y \geq 0$ A positive number of skis must be produced.
$2x + y \leq 48$ Manufacturing time available.
$x + y \leq 32$ Finishing time available.

Graph the constraints.

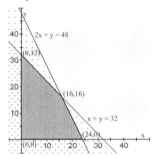

To find the intersection of $x + y = 32$ and $2x + y = 48$, solve the system:
$$\begin{cases} x + y = 32 \\ 2x + y = 48 \end{cases}$$

Solve the first equation for y: $y = 32 - x$.
Substitute and solve:
$2x + (32 - x) = 48$
$\qquad x = 16$
$y = 32 - 16 = 16$

The point of intersection is (16, 16).
The corner points are (0, 0), (0, 32), (24, 0), (16, 16). Evaluate the objective function:

Vertex	Value of $P = 70x + 50y$
(0, 0)	$P = 70(0) + 50(0) = 0$
(0, 32)	$P = 70(0) + 50(32) = 1600$
(24, 0)	$P = 70(24) + 50(0) = 1680$
(16, 16)	$P = 70(16) + 50(16) = 1920$

The maximum profit is $1920, when 16 downhill skis and 16 cross-country skis are produced.

21. Let $x =$ the number of acres of corn planted, and let $y =$ the number of acres of soybeans planted. The total profit is: $P = 250x + 200y$. Profit is to be maximized, so this is the objective function.
The constraints are:

$x \geq 0, \ y \geq 0$ A non-negative number of acres must be planted.
$x + y \leq 100$ Acres available to plant.
$60x + 40y \leq 1800$ Money available for cultivation costs.
$60x + 60y \leq 2400$ Money available for labor costs.

Graph the constraints.

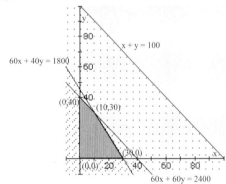

To find the intersection of $60x + 40y = 1800$ and $60x + 60y = 2400$, solve the system:
$$\begin{cases} 60x + 40y = 1800 \\ 60x + 60y = 2400 \end{cases}$$

Subtract multiply the first equation by -1 and add the result to the second equation:
$-60x - 40y = -1800$
$\underline{60x + 60y = 2400}$
$\qquad 20y = 600$
$\qquad\ \ y = 30$

$60x + 40(30) = 1800$
$60x + 1200 = 1800$
$\qquad 60x = 600$
$\qquad\ \ x = 10$

The point of intersection is (10, 30).
The corner points are (0, 0), (0, 40), (30, 0), (10, 30). Evaluate the objective function:

Vertex	Value of $P = 250x + 200y$
(0, 0)	$P = 250(0) + 200(0) = 0$
(0, 40)	$P = 250(0) + 200(40) = 8000$
(30, 0)	$P = 250(30) + 200(0) = 7500$
(10, 30)	$P = 250(10) + 200(30) = 8500$

The maximum profit is $8500, when 10 acres of corn and 30 acres of soybeans are planted.

23. Let $x =$ the number of hours that machine 1 is operated, and let $y =$ the number of hours that machine 2 is operated. The total cost is: $C = 50x + 30y$. Cost is to be minimized, so this is the objective function.
The constraints are:
$x \geq 0, \ y \geq 0$ A positive number of hours must be used.
$x \leq 10$ Time used on machine 1.
$y \leq 10$ Time used on machine 2.
$60x + 40y \geq 240$ 8-inch pliers to be produced.
$70x + 20y \geq 140$ 6-inch pliers to be produced.
Graph the constraints.

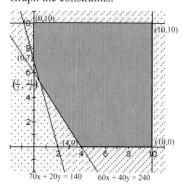

To find the intersection of $60x + 40y = 240$ and $70x + 20y = 140$, solve the system:
$$\begin{cases} 60x + 40y = 240 \\ 70x + 20y = 140 \end{cases}$$
Divide the first equation by -2 and add the result to the second equation:
$$\begin{aligned} -30x - 20y &= -120 \\ 70x + 20y &= 140 \\ \hline 40x &= 20 \\ x &= 0.5 \end{aligned}$$
Substitute and solve:

$60(0.5) + 40y = 240$
$30 + 40y = 240$
$40y = 210$
$y = 5.25$

The point of intersection is $(0.5, 5.25)$.
The corner points are (0, 7), (0, 10), (4, 0), (10, 0), (10, 10), $(0.5, 5.25)$. Evaluate the objective function:

Vertex	Value of $C = 50x + 30y$
(0, 7)	$C = 50(0) + 30(7) = 210$
(0, 10)	$C = 50(0) + 30(10) = 300$
(4, 0)	$C = 50(4) + 30(0) = 200$
(10, 0)	$C = 50(10) + 30(0) = 500$
(10, 10)	$C = 50(10) + 30(10) = 800$
(0.5, 5.25)	$P = 50(0.5) + 30(5.25) = 182.5$

The minimum cost is $182.50, when machine 1 is used for 0.5 hour and machine 2 is used for 5.25 hours.

25. Let $x =$ the number of pounds of ground beef, and let $y =$ the number of pounds of ground pork. The total cost is: $C = 0.75x + 0.45y$. Cost is to be minimized, so this is the objective function. The constraints are:
$x \geq 0, \ y \geq 0$ A positive number of pounds must be used.
$x \leq 200$ Only 200 pounds of ground beef are available.
$y \geq 50$ At least 50 pounds of ground pork must be used.
$0.75x + 0.60y \geq 0.70(x + y)$ Leanness condition
(Note that the last equation will simplify to $y \leq \frac{1}{2}x$.) Graph the constraints.

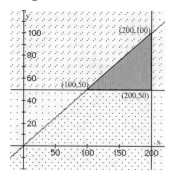

The corner points are (100, 50), (200, 50), (200, 100). Evaluate the objective function:

Vertex	Value of $C = 0.75x + 0.45y$
(100, 50)	$C = 0.75(100) + 0.45(50)$ $= 97.50$
(200, 50)	$C = 0.75(200) + 0.45(50)$ $= 172.50$
(200, 100)	$C = 0.75(200) + 0.45(100)$ $= 195$

The minimum cost is $97.50, when 100 pounds of ground beef and 50 pounds of ground pork are used.

27. Let x = the number of racing skates manufactured, and let y = the number of figure skates manufactured. The total profit is: $P = 10x + 12y$. Profit is to be maximized, so this is the objective function. The constraints are:
$x \geq 0, \ y \geq 0$ A positive number of skates must be manufactured.
$6x + 4y \leq 120$ Only 120 hours are available for fabrication.
$x + 2y \leq 40$ Only 40 hours are available for finishing.
Graph the constraints.

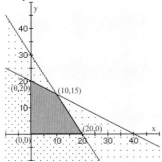

To find the intersection of $6x + 4y = 120$ and $x + 2y = 40$, solve the system:
$$\begin{cases} 6x + 4y = 120 \\ x + 2y = 40 \end{cases}$$
Solve the second equation for x: $x = 40 - 2y$
Substitute and solve:
$6(40 - 2y) + 4y = 120$
$240 - 12y + 4y = 120$
$-8y = -120$
$y = 15$
$x = 40 - 2(15) = 10$

The point of intersection is (10, 15).
The corner points are (0, 0), (0, 20), (20, 0), (10, 15). Evaluate the objective function:

Vertex	Value of $P = 10x + 12y$
(0, 0)	$P = 10(0) + 12(0) = 0$
(0, 20)	$P = 10(0) + 12(20) = 240$
(20, 0)	$P = 10(20) + 12(0) = 200$
(10, 15)	$P = 10(10) + 12(15) = 280$

The maximum profit is $280, when 10 racing skates and 15 figure skates are produced.

29. Let x = the number of metal fasteners, and let y = the number of plastic fasteners. The total cost is: $C = 9x + 4y$. Cost is to be minimized, so this is the objective function. The constraints are:
$x \geq 2, \ y \geq 2$ At least 2 of each fastener must be made.
$x + y \geq 6$ At least 6 fasteners are needed.
$4x + 2y \leq 24$ Only 24 hours are available.
Graph the constraints.

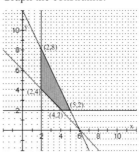

The corner points are (2, 4), (2, 8), (4, 2), (5, 2). Evaluate the objective function:

Vertex	Value of $C = 9x + 4y$
(2, 4)	$C = 9(2) + 4(4) = 34$
(2, 8)	$C = 9(2) + 4(8) = 50$
(4, 2)	$C = 9(4) + 4(2) = 44$
(5, 2)	$C = 9(5) + 4(2) = 53$

The minimum cost is $34, when 2 metal fasteners and 4 plastic fasteners are ordered.

31. Let x = the number of first class seats, and let y = the number of coach seats. Using the hint, the revenue from x first class seats and y coach seats is $Fx + Cy$, where $F > C > 0$. Thus, $R = Fx + Cy$ is the objective function to be maximized. The constraints are:
$8 \leq x \leq 16$ Restriction on first class seats.
$80 \leq y \leq 120$ Restriction on coach seats.

a. $\dfrac{x}{y} \leq \dfrac{1}{12}$ Ratio of seats.

The constraints are:
$8 \leq x \leq 16$
$80 \leq y \leq 120$
$12x \leq y$
Graph the constraints.

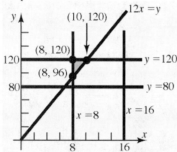

The corner points are (8, 96), (8, 120), and (10, 120). Evaluate the objective function:

Vertex	Value of $R = Fx + Cy$
(8, 96)	$R = 8F + 96C$
(8, 120)	$R = 8F + 120C$
(10, 120)	$R = 10F + 120C$

Since $C > 0$, $120C > 96C$, so
$8F + 120C > 8F + 96C$.
Since $F > 0$, $10F > 8F$, so
$10F + 120C > 8F + 120C$.
Thus, the maximum revenue occurs when the aircraft is configured with 10 first class seats and 120 coach seats.

b. $\dfrac{x}{y} \leq \dfrac{1}{8}$

The constraints are:
$8 \leq x \leq 16$
$80 \leq y \leq 120$
$8x \leq y$
Graph the constraints.

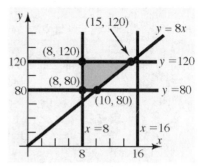

The corner points are (8, 80), (8, 120), (15, 120), and (10, 80).
Evaluate the objective function:

Vertex	Value of $R = Fx + Cy$
(8, 80)	$R = 8F + 80C$
(8, 120)	$R = 8F + 120C$
(15, 120)	$R = 15F + 120C$
(10, 80)	$R = 10F + 80C$

Since $F > 0$ and $C > 0$, $120C > 96C$, the maximum value of R occurs at (15, 120). The maximum revenue occurs when the aircraft is configured with 15 first class seats and 120 coach seats.

c. Answers will vary.

33. Answers will vary.

Chapter 5 Review

1. $\begin{cases} 2x - y = 5 \\ 5x + 2y = 8 \end{cases}$

   Solve the first equation for y: $y = 2x - 5$.
   Substitute and solve:
   $5x + 2(2x - 5) = 8$
   $5x + 4x - 10 = 8$
   $9x = 18$
   $x = 2$
   $y = 2(2) - 5 = 4 - 5 = -1$
   The solution is $x = 2$, $y = -1$.

3. $\begin{cases} 3x - 4y = 4 \\ x - 3y = \dfrac{1}{2} \end{cases}$

   Solve the second equation for x: $x = 3y + \dfrac{1}{2}$
   Substitute into the first equation and solve:
   $3\left(3y + \dfrac{1}{2}\right) - 4y = 4$
   $9y + \dfrac{3}{2} - 4y = 4$
   $5y = \dfrac{5}{2}$
   $y = \dfrac{1}{2}$
   $x = 3\left(\dfrac{1}{2}\right) + \dfrac{1}{2} = 2$
   The solution is $x = 2$, $y = \dfrac{1}{2}$.

5. $\begin{cases} x - 2y - 4 = 0 \\ 3x + 2y - 4 = 0 \end{cases}$

   Solve the first equation for x: $x = 2y + 4$
   Substitute into the second equation and solve:
   $3(2y + 4) + 2y - 4 = 0$
   $6y + 12 + 2y - 4 = 0$
   $8y = -8$
   $y = -1$
   $x = 2(-1) + 4 = 2$
   The solution is $x = 2$, $y = -1$.

7. $\begin{cases} y = 2x - 5 \\ x = 3y + 4 \end{cases}$

   Substitute the first equation into the second equation and solve:
   $x = 3(2x - 5) + 4$
   $x = 6x - 15 + 4$
   $-5x = -11$
   $x = \dfrac{11}{5}$
   $y = 2\left(\dfrac{11}{5}\right) - 5 = -\dfrac{3}{5}$
   The solution is $x = \dfrac{11}{5}$, $y = -\dfrac{3}{5}$.

9. $\begin{cases} x - 3y + 4 = 0 \\ \dfrac{1}{2}x - \dfrac{3}{2}y + \dfrac{4}{3} = 0 \end{cases}$

   Multiply each side of the first equation by 3 and each side of the second equation by -6 and add:
   $\begin{cases} 3x - 9y + 12 = 0 \\ -3x + 9y - 8 = 0 \end{cases}$
   $ 4 = 0$
   There is no solution to the system. The system is inconsistent.

11. $\begin{cases} 2x + 3y - 13 = 0 \\ 3x - 2y = 0 \end{cases}$

    Multiply each side of the first equation by 2 and each side of the second equation by 3, and add to eliminate y:
    $\begin{cases} 4x + 6y - 26 = 0 \\ 9x - 6y = 0 \end{cases}$
    $13x - 26 = 0$
    $13x = 26$
    $x = 2$
    Substitute and solve for y:
    $3(2) - 2y = 0$
    $-2y = -6$
    $y = 3$
    The solution of the system is $x = 2$, $y = 3$.

13. $\begin{cases} 3x - 2y = 8 \\ x - \dfrac{2}{3}y = 12 \end{cases}$

Multiply each side of the second equation by −3 and add to eliminate x:

$\begin{cases} 3x - 2y = 8 \\ -3x + 2y = -36 \end{cases}$
$\ 0 = -28$

The system has no solution, so the system is inconsistent.

15. $\begin{cases} x + 2y - z = 6 \\ 2x - y + 3z = -13 \\ 3x - 2y + 3z = -16 \end{cases}$

Multiply each side of the first equation by −2 and add to the second equation to eliminate x;

$\begin{cases} -2x - 4y + 2z = -12 \\ 2x - y + 3z = -13 \end{cases}$
$\ -5y + 5z = -25$
$\ y - z = 5$

Multiply each side of the first equation by −3 and add to the third equation to eliminate x:
$-3x - 6y + 3z = -18$
$\ \ 3x - 2y + 3z = -16$
$ -8y + 6z = -34$

Multiply each side of the first result by 8 and add to the second result to eliminate y:
$8y - 8z = 40$
$-8y + 6z = -34$
$ -2z = 6$
$ z = -3$

Substituting and solving for the other variables:
$y - (-3) = 5 \quad x + 2(2) - (-3) = 6$
$\ y = 2 x + 4 + 3 = 6$
$\ x = -1$

The solution is $x = -1$, $y = 2$, $z = -3$.

17. $\begin{cases} 2x - 4y + z = -15 \\ x + 2y - 4z = 27 \\ 5x - 6y - 2z = -3 \end{cases}$

Multiply the first equation by −1 and the second equation by 2, and then add to eliminate x:

$\begin{cases} -2x + 4y - z = 15 \\ 2x + 4y - 8z = 54 \end{cases}$
$\ 8y - 9z = 69$

Multiply the second equation by −5 and add to the third equation to eliminate x:
$-5x - 10y + 20z = -135$
$\ \ 5x - 6y - 2z = -3$
$ -16y + 18z = -138$

Multiply both sides of the first result by 2 and add to the second result to eliminate y:
$16y - 18z = 138$
$-16y + 18z = -138$
$ 0 = 0$

The system is dependent.
$-16y + 18z = -138$
$18z + 138 = 16y$
$y = \dfrac{9}{8}z + \dfrac{69}{8}$

Substituting into the second equation and solving for x:
$x + 2\left(\dfrac{9}{8}z + \dfrac{69}{8}\right) - 4z = 27$
$x + \dfrac{9}{4}z + \dfrac{69}{4} - 4z = 27$
$x = \dfrac{7}{4}z + \dfrac{39}{4}$

The solution is $x = \dfrac{7}{4}z + \dfrac{39}{4}$, $y = \dfrac{9}{8}z + \dfrac{69}{8}$, z is any real number.

19. $\begin{cases} 3x + 2y = 8 \\ x + 4y = -1 \end{cases}$

21. $A + C = \begin{bmatrix} 1 & 0 \\ 2 & 4 \\ -1 & 2 \end{bmatrix} + \begin{bmatrix} 3 & -4 \\ 1 & 5 \\ 5 & 2 \end{bmatrix}$

$= \begin{bmatrix} 1+3 & 0+(-4) \\ 2+1 & 4+5 \\ -1+5 & 2+2 \end{bmatrix}$

$= \begin{bmatrix} 4 & -4 \\ 3 & 9 \\ 4 & 4 \end{bmatrix}$

23. $6A = 6 \cdot \begin{bmatrix} 1 & 0 \\ 2 & 4 \\ -1 & 2 \end{bmatrix} = \begin{bmatrix} 6 \cdot 1 & 6 \cdot 0 \\ 6 \cdot 2 & 6 \cdot 4 \\ 6(-1) & 6 \cdot 2 \end{bmatrix} = \begin{bmatrix} 6 & 0 \\ 12 & 24 \\ -6 & 12 \end{bmatrix}$

25. $AB = \begin{bmatrix} 1 & 0 \\ 2 & 4 \\ -1 & 2 \end{bmatrix} \cdot \begin{bmatrix} 4 & -3 & 0 \\ 1 & 1 & -2 \end{bmatrix}$

$= \begin{bmatrix} 1(4)+0(1) & 1(-3)+0(1) & 1(0)+2(-2) \\ 2(4)+2(1) & 2(-3)+4(1) & 2(0)+4(-2) \\ -1(4)+2(1) & -1(-3)+2(1) & -1(0)+2(-2) \end{bmatrix}$

$= \begin{bmatrix} 4 & -3 & 0 \\ 12 & -2 & -8 \\ -2 & 5 & -4 \end{bmatrix}$

27. $CB = \begin{bmatrix} 3 & -4 \\ 1 & 5 \\ 5 & 2 \end{bmatrix} \cdot \begin{bmatrix} 4 & -3 & 0 \\ 1 & 1 & -2 \end{bmatrix}$

$= \begin{bmatrix} 3(4)-4(1) & 3(-3)-4(1) & 3(0)-4(-2) \\ 1(4)+5(1) & 1(-3)+5(1) & 1(0)+5(-2) \\ 5(4)+2(1) & 5(-3)+2(1) & 5(0)+2(-2) \end{bmatrix}$

$= \begin{bmatrix} 8 & -13 & 8 \\ 9 & 2 & -10 \\ 22 & -13 & -4 \end{bmatrix}$

29. $A = \begin{bmatrix} 4 & 6 \\ 1 & 3 \end{bmatrix}$

Augment the matrix with the identity and use row operations to find the inverse:

$\begin{bmatrix} 4 & 6 & | & 1 & 0 \\ 1 & 3 & | & 0 & 1 \end{bmatrix}$

$\to \begin{bmatrix} 1 & 3 & | & 0 & 1 \\ 4 & 6 & | & 1 & 0 \end{bmatrix}$ $\begin{pmatrix} \text{Interchange} \\ r_1 \text{ and } r_2 \end{pmatrix}$

$\to \begin{bmatrix} 1 & 3 & | & 0 & 1 \\ 0 & -6 & | & 1 & -4 \end{bmatrix}$ $(R_2 = -4r_1 + r_2)$

$\to \begin{bmatrix} 1 & 3 & | & 0 & 1 \\ 0 & 1 & | & -\frac{1}{6} & \frac{2}{3} \end{bmatrix}$ $(R_2 = -\frac{1}{6}r_2)$

$\to \begin{bmatrix} 1 & 0 & | & \frac{1}{2} & -1 \\ 0 & 1 & | & -\frac{1}{6} & \frac{2}{3} \end{bmatrix}$ $(R_1 = -3r_2 + r_1)$

Thus, $A^{-1} = \begin{bmatrix} \frac{1}{2} & -1 \\ -\frac{1}{6} & \frac{2}{3} \end{bmatrix}$.

31. $A = \begin{bmatrix} 1 & 3 & 3 \\ 1 & 2 & 1 \\ 1 & -1 & 2 \end{bmatrix}$

Augment the matrix with the identity and use row operations to find the inverse:

$\begin{bmatrix} 1 & 3 & 3 & | & 1 & 0 & 0 \\ 1 & 2 & 1 & | & 0 & 1 & 0 \\ 1 & -1 & 2 & | & 0 & 0 & 1 \end{bmatrix}$

$\to \begin{bmatrix} 1 & 3 & 3 & | & 1 & 0 & 0 \\ 0 & -1 & -2 & | & -1 & 1 & 0 \\ 0 & -4 & -1 & | & -1 & 0 & 1 \end{bmatrix}$ $\begin{pmatrix} R_2 = -r_1 + r_2 \\ R_3 = -r_1 + r_3 \end{pmatrix}$

$\to \begin{bmatrix} 1 & 3 & 3 & | & 1 & 0 & 0 \\ 0 & 1 & 2 & | & 1 & -1 & 0 \\ 0 & -4 & -1 & | & -1 & 0 & 1 \end{bmatrix}$ $(R_2 = -r_2)$

$\to \begin{bmatrix} 1 & 0 & -3 & | & -2 & 3 & 0 \\ 0 & 1 & 2 & | & 1 & -1 & 0 \\ 0 & 0 & 7 & | & 3 & -4 & 1 \end{bmatrix}$ $\begin{pmatrix} R_1 = -3r_2 + r_1 \\ R_3 = 4r_2 + r_3 \end{pmatrix}$

$\to \begin{bmatrix} 1 & 0 & -3 & | & -2 & 3 & 0 \\ 0 & 1 & 2 & | & 1 & -1 & 0 \\ 0 & 0 & 1 & | & \frac{3}{7} & -\frac{4}{7} & \frac{1}{7} \end{bmatrix}$ $(R_3 = \frac{1}{7}r_3)$

$\to \begin{bmatrix} 1 & 0 & 0 & | & -\frac{5}{7} & \frac{9}{7} & \frac{3}{7} \\ 0 & 1 & 0 & | & \frac{1}{7} & \frac{1}{7} & -\frac{2}{7} \\ 0 & 0 & 1 & | & \frac{3}{7} & -\frac{4}{7} & \frac{1}{7} \end{bmatrix}$ $\begin{pmatrix} R_1 = 3r_3 + r_1 \\ R_2 = -2r_3 + r_2 \end{pmatrix}$

Thus, $A^{-1} = \begin{bmatrix} -\frac{5}{7} & \frac{9}{7} & \frac{3}{7} \\ \frac{1}{7} & \frac{1}{7} & -\frac{2}{7} \\ \frac{3}{7} & -\frac{4}{7} & \frac{1}{7} \end{bmatrix}$.

33. $A = \begin{bmatrix} 4 & -8 \\ -1 & 2 \end{bmatrix}$

Augment the matrix with the identity and use row operations to find the inverse:

$\begin{bmatrix} 4 & -8 & | & 1 & 0 \\ -1 & 2 & | & 0 & 1 \end{bmatrix}$

$\rightarrow \begin{bmatrix} -1 & 2 & | & 0 & 1 \\ 4 & -8 & | & 1 & 0 \end{bmatrix}$ $\begin{pmatrix} \text{Interchange} \\ r_1 \text{ and } r_2 \end{pmatrix}$

$\rightarrow \begin{bmatrix} -1 & 2 & | & 0 & 1 \\ 0 & 0 & | & 1 & 4 \end{bmatrix}$ $(R_2 = 4r_1 + r_2)$

$\rightarrow \begin{bmatrix} 1 & -2 & | & 0 & -1 \\ 0 & 0 & | & 1 & 4 \end{bmatrix}$ $(R_1 = -r_1)$

There is no inverse because there is no way to obtain the identity on the left side. The matrix is singular.

35. $\begin{cases} 3x - 2y = 1 \\ 10x + 10y = 5 \end{cases}$

Write the augmented matrix:

$\begin{bmatrix} 3 & -2 & | & 1 \\ 10 & 10 & | & 5 \end{bmatrix}$

$\rightarrow \begin{bmatrix} 3 & -2 & | & 1 \\ 1 & 16 & | & 2 \end{bmatrix}$ $(R_2 = -3r_1 + r_2)$

$\rightarrow \begin{bmatrix} 1 & 16 & | & 2 \\ 3 & -2 & | & 1 \end{bmatrix}$ $\begin{pmatrix} \text{Interchange} \\ r_1 \text{ and } r_2 \end{pmatrix}$

$\rightarrow \begin{bmatrix} 1 & 16 & | & 2 \\ 0 & -50 & | & -5 \end{bmatrix}$ $(R_2 = -3r_1 + r_2)$

$\rightarrow \begin{bmatrix} 1 & 16 & | & 2 \\ 0 & 1 & | & \frac{1}{10} \end{bmatrix}$ $(R_2 = -\frac{1}{50}r_2)$

$\rightarrow \begin{bmatrix} 1 & 0 & | & \frac{2}{5} \\ 0 & 1 & | & \frac{1}{10} \end{bmatrix}$ $(R_1 = -16r_2 + r_1)$

The solution is $x = \frac{2}{5}, y = \frac{1}{10}$.

37. $\begin{cases} 5x + 6y - 3z = 6 \\ 4x - 7y - 2z = -3 \\ 3x + y - 7z = 1 \end{cases}$

Write the augmented matrix:

$\begin{bmatrix} 5 & 6 & -3 & | & 6 \\ 4 & -7 & -2 & | & -3 \\ 3 & 1 & -7 & | & 1 \end{bmatrix}$

$\rightarrow \begin{bmatrix} 1 & 13 & -1 & | & 9 \\ 4 & -7 & -2 & | & -3 \\ 3 & 1 & -7 & | & 1 \end{bmatrix}$ $(R_1 = -r_2 + r_1)$

$\rightarrow \begin{bmatrix} 1 & 13 & -1 & | & 9 \\ 0 & -59 & 2 & | & -39 \\ 0 & -38 & -4 & | & -26 \end{bmatrix}$ $\begin{pmatrix} R_2 = -4r_1 + r_2 \\ R_3 = -3r_1 + r_3 \end{pmatrix}$

$\rightarrow \begin{bmatrix} 1 & 13 & -1 & | & 9 \\ 0 & 1 & -\frac{2}{59} & | & \frac{39}{59} \\ 0 & -38 & -4 & | & -26 \end{bmatrix}$ $(R_2 = -\frac{1}{59}r_2)$

$\rightarrow \begin{bmatrix} 1 & 0 & -\frac{33}{59} & | & \frac{24}{59} \\ 0 & 1 & -\frac{2}{59} & | & \frac{39}{59} \\ 0 & 0 & -\frac{312}{59} & | & -\frac{52}{59} \end{bmatrix}$ $\begin{pmatrix} R_1 = -13r_2 + r_1 \\ R_3 = 38r_2 + r_3 \end{pmatrix}$

$\rightarrow \begin{bmatrix} 1 & 0 & -\frac{33}{59} & | & \frac{24}{59} \\ 0 & 1 & -\frac{2}{59} & | & \frac{39}{59} \\ 0 & 0 & 1 & | & \frac{1}{6} \end{bmatrix}$ $(R_3 = -\frac{59}{312}r_3)$

$\rightarrow \begin{bmatrix} 1 & 0 & 0 & | & \frac{1}{2} \\ 0 & 1 & 0 & | & \frac{2}{3} \\ 0 & 0 & 1 & | & \frac{1}{6} \end{bmatrix}$ $\begin{pmatrix} R_1 = \frac{33}{59}r_3 + r_1 \\ R_2 = \frac{2}{59}r_3 + r_2 \end{pmatrix}$

The solution is $x = \frac{1}{2}, y = \frac{2}{3}, z = \frac{1}{6}$.

39. $\begin{cases} x - 2z = 1 \\ 2x + 3y = -3 \\ 4x - 3y - 4z = 3 \end{cases}$

Write the augmented matrix:

$\begin{bmatrix} 1 & 0 & -2 & | & 1 \\ 2 & 3 & 0 & | & -3 \\ 4 & -3 & -4 & | & 3 \end{bmatrix}$

$\rightarrow \begin{bmatrix} 1 & 0 & -2 & | & 1 \\ 0 & 3 & 4 & | & -5 \\ 0 & -3 & 4 & | & -1 \end{bmatrix}$ $\begin{pmatrix} R_2 = -2r_1 + r_2 \\ R_3 = -4r_1 + r_3 \end{pmatrix}$

392

$$\rightarrow \begin{bmatrix} 1 & 0 & -2 & | & 1 \\ 0 & 1 & \frac{4}{3} & | & -\frac{5}{3} \\ 0 & -3 & 4 & | & -1 \end{bmatrix} \quad (R_2 = \tfrac{1}{3}r_2)$$

$$\rightarrow \begin{bmatrix} 1 & 0 & -2 & | & 1 \\ 0 & 1 & \frac{4}{3} & | & -\frac{5}{3} \\ 0 & 0 & 8 & | & -6 \end{bmatrix} \quad (R_3 = 3r_2 + r_3)$$

$$\rightarrow \begin{bmatrix} 1 & 0 & -2 & | & 1 \\ 0 & 1 & \frac{4}{3} & | & -\frac{5}{3} \\ 0 & 0 & 1 & | & -\frac{3}{4} \end{bmatrix} \quad (R_3 = \tfrac{1}{8}r_3)$$

$$\rightarrow \begin{bmatrix} 1 & 0 & 0 & | & -\frac{1}{2} \\ 0 & 1 & 0 & | & -\frac{2}{3} \\ 0 & 0 & 1 & | & -\frac{3}{4} \end{bmatrix} \quad \begin{pmatrix} R_1 = 2r_3 + r_1 \\ R_2 = -\tfrac{4}{3}r_3 + r_2 \end{pmatrix}$$

The solution is $x = -\dfrac{1}{2}$, $y = -\dfrac{2}{3}$, $z = -\dfrac{3}{4}$.

41. $\begin{cases} x - y + z = 0 \\ x - y - 5z = 6 \\ 2x - 2y + z = 1 \end{cases}$

Write the augmented matrix:
$$\begin{bmatrix} 1 & -1 & 1 & | & 0 \\ 1 & -1 & -5 & | & 6 \\ 2 & -2 & 1 & | & 1 \end{bmatrix}$$

$$\rightarrow \begin{bmatrix} 1 & -1 & 1 & | & 0 \\ 0 & 0 & -6 & | & 6 \\ 0 & 0 & -1 & | & 1 \end{bmatrix} \quad \begin{pmatrix} R_2 = -r_1 + r_2 \\ R_3 = -2r_1 + r_3 \end{pmatrix}$$

$$\rightarrow \begin{bmatrix} 1 & -1 & 1 & | & 0 \\ 0 & 0 & 1 & | & -1 \\ 0 & 0 & -1 & | & 1 \end{bmatrix} \quad (R_2 = -\tfrac{1}{6}r_2)$$

$$\rightarrow \begin{bmatrix} 1 & -1 & 0 & | & 1 \\ 0 & 0 & 1 & | & -1 \\ 0 & 0 & 0 & | & 0 \end{bmatrix} \quad \begin{pmatrix} R_1 = -r_2 + r_1 \\ R_3 = r_2 + r_3 \end{pmatrix}$$

The system is dependent.
$\begin{cases} x = y + 1 \\ z = -1 \end{cases}$

The solution is $x = y + 1$, $z = -1$, y is any real number.

43. $\begin{cases} x - y - z - t = 1 \\ 2x + y - z + 2t = 3 \\ x - 2y - 2z - 3t = 0 \\ 3x - 4y + z + 5t = -3 \end{cases}$

Write the augmented matrix:
$$\begin{bmatrix} 1 & -1 & -1 & -1 & | & 1 \\ 2 & 1 & -1 & 2 & | & 3 \\ 1 & -2 & -2 & -3 & | & 0 \\ 3 & -4 & 1 & 5 & | & -3 \end{bmatrix}$$

$$\rightarrow \begin{bmatrix} 1 & -1 & -1 & -1 & | & 1 \\ 0 & 3 & 1 & 4 & | & 1 \\ 0 & -1 & -1 & -2 & | & -1 \\ 0 & -1 & 4 & 8 & | & -6 \end{bmatrix} \quad \begin{pmatrix} R_2 = -2r_1 + r_2 \\ R_3 = -r_1 + r_3 \\ R_4 = -3r_1 + r_4 \end{pmatrix}$$

$$\rightarrow \begin{bmatrix} 1 & -1 & -1 & -1 & | & 1 \\ 0 & -1 & -1 & -2 & | & -1 \\ 0 & 3 & 1 & 4 & | & 1 \\ 0 & -1 & 4 & 8 & | & -6 \end{bmatrix} \quad \begin{pmatrix} \text{Interchange} \\ r_2 \text{ and } r_3 \end{pmatrix}$$

$$\rightarrow \begin{bmatrix} 1 & -1 & -1 & -1 & | & 1 \\ 0 & 1 & 1 & 2 & | & 1 \\ 0 & 3 & 1 & 4 & | & 1 \\ 0 & -1 & 4 & 8 & | & -6 \end{bmatrix} \quad (R_2 = -r_2)$$

$$\rightarrow \begin{bmatrix} 1 & 0 & 0 & 1 & | & 2 \\ 0 & 1 & 1 & 2 & | & 1 \\ 0 & 0 & -2 & -2 & | & -2 \\ 0 & 0 & 5 & 10 & | & -5 \end{bmatrix} \quad \begin{pmatrix} R_1 = r_2 + r_1 \\ R_3 = -3r_2 + r_3 \\ R_4 = r_2 + r_4 \end{pmatrix}$$

$$\rightarrow \begin{bmatrix} 1 & 0 & 0 & 1 & | & 2 \\ 0 & 1 & 1 & 2 & | & 1 \\ 0 & 0 & 1 & 1 & | & 1 \\ 0 & 0 & 1 & 2 & | & -1 \end{bmatrix} \quad \begin{pmatrix} R_3 = -\tfrac{1}{2}r_3 \\ R_4 = \tfrac{1}{5}r_4 \end{pmatrix}$$

$$\rightarrow \begin{bmatrix} 1 & 0 & 0 & 1 & | & 2 \\ 0 & 1 & 0 & 1 & | & 0 \\ 0 & 0 & 1 & 1 & | & 1 \\ 0 & 0 & 0 & 1 & | & -2 \end{bmatrix} \quad \begin{pmatrix} R_2 = -r_3 + r_2 \\ R_4 = -r_3 + r_4 \end{pmatrix}$$

$$\rightarrow \begin{bmatrix} 1 & 0 & 0 & 0 & | & 4 \\ 0 & 1 & 0 & 0 & | & 2 \\ 0 & 0 & 1 & 0 & | & 3 \\ 0 & 0 & 0 & 1 & | & -2 \end{bmatrix} \quad \begin{pmatrix} R_1 = -r_4 + r_1 \\ R_2 = -r_4 + r_2 \\ R_3 = -r_4 + r_3 \end{pmatrix}$$

The solution is $x = 4$, $y = 2$, $z = 3$, $t = -2$.

45. $\begin{vmatrix} 3 & 4 \\ 1 & 3 \end{vmatrix} = 3(3) - 4(1) = 9 - 4 = 5$

47. $\begin{vmatrix} 1 & 4 & 0 \\ -1 & 2 & 6 \\ 4 & 1 & 3 \end{vmatrix} = 1\begin{vmatrix} 2 & 6 \\ 1 & 3 \end{vmatrix} - 4\begin{vmatrix} -1 & 6 \\ 4 & 3 \end{vmatrix} + 0\begin{vmatrix} -1 & 2 \\ 4 & 1 \end{vmatrix}$
$= 1(6-6) - 4(-3-24) + 0(-1-8)$
$= 1(0) - 4(-27) + 0(-9) = 0 + 108 + 0$
$= 108$

49. $\begin{vmatrix} 2 & 1 & -3 \\ 5 & 0 & 1 \\ 2 & 6 & 0 \end{vmatrix} = 2\begin{vmatrix} 0 & 1 \\ 6 & 0 \end{vmatrix} - 1\begin{vmatrix} 5 & 1 \\ 2 & 0 \end{vmatrix} + (-3)\begin{vmatrix} 5 & 0 \\ 2 & 6 \end{vmatrix}$
$= 2(0-6) - 1(0-2) - 3(30-0)$
$= 2(-6) - 1(-2) - 3(30)$
$= -12 + 2 - 90$
$= -100$

51. $\begin{cases} x - 2y = 4 \\ 3x + 2y = 4 \end{cases}$

Set up and evaluate the determinants to use Cramer's Rule:

$D = \begin{vmatrix} 1 & -2 \\ 3 & 2 \end{vmatrix} = 1(2) - 3(-2) = 2 + 6 = 8$

$D_x = \begin{vmatrix} 4 & -2 \\ 4 & 2 \end{vmatrix} = 4(2) - 4(-2) = 8 + 8 = 16$

$D_y = \begin{vmatrix} 1 & 4 \\ 3 & 4 \end{vmatrix} = 1(4) - 3(4) = 4 - 12 = -8$

The solution is $x = \dfrac{D_x}{D} = \dfrac{16}{8} = 2$,

$y = \dfrac{D_y}{D} = \dfrac{-8}{8} = -1$

53. $\begin{cases} 2x + 3y - 13 = 0 \\ 3x - 2y = 0 \end{cases}$

Write the system is standard form:
$\begin{cases} 2x + 3y = 13 \\ 3x - 2y = 0 \end{cases}$

Set up and evaluate the determinants to use Cramer's Rule:

$D = \begin{vmatrix} 2 & 3 \\ 3 & -2 \end{vmatrix} = -4 - 9 = -13$

$D_x = \begin{vmatrix} 13 & 3 \\ 0 & -2 \end{vmatrix} = -26 - 0 = -26$

$D_y = \begin{vmatrix} 2 & 13 \\ 3 & 0 \end{vmatrix} = 0 - 39 = -39$

The solution is $x = \dfrac{D_x}{D} = \dfrac{-26}{-13} = 2$,

$y = \dfrac{D_y}{D} = \dfrac{-39}{-13} = 3$.

55. $\begin{cases} x + 2y - z = 6 \\ 2x - y + 3z = -13 \\ 3x - 2y + 3z = -16 \end{cases}$

Set up and evaluate the determinants to use Cramer's Rule:

$D = \begin{vmatrix} 1 & 2 & -1 \\ 2 & -1 & 3 \\ 3 & -2 & 3 \end{vmatrix}$

$= 1\begin{vmatrix} -1 & 3 \\ -2 & 3 \end{vmatrix} - 2\begin{vmatrix} -1 & 3 \\ -2 & 3 \end{vmatrix} + (-1)\begin{vmatrix} 2 & -1 \\ 3 & -2 \end{vmatrix}$

$= 1(-3+6) - 2(-3+6) + (-1)(-4+3)$
$= 3 + 6 + 1 = 10$

$D_x = \begin{vmatrix} 6 & 2 & -1 \\ -13 & -1 & 3 \\ -16 & -2 & 3 \end{vmatrix}$

$= 6\begin{vmatrix} -1 & 3 \\ -2 & 3 \end{vmatrix} - 2\begin{vmatrix} -13 & 3 \\ -16 & 3 \end{vmatrix} + (-1)\begin{vmatrix} -13 & -1 \\ -16 & -2 \end{vmatrix}$

$= 6(-3+6) - 2(-39+48) + (-1)(26-16)$
$= 18 - 18 - 10 = -10$

$D_y = \begin{vmatrix} 1 & 6 & -1 \\ 2 & -13 & 3 \\ 3 & -16 & 3 \end{vmatrix}$

$= 1\begin{vmatrix} -13 & 3 \\ -16 & 3 \end{vmatrix} - 6\begin{vmatrix} 2 & 3 \\ 3 & 3 \end{vmatrix} + (-1)\begin{vmatrix} 2 & -13 \\ 3 & -16 \end{vmatrix}$

$= 1(-39+48) - 6(6-9) + (-1)(-32+39)$
$= 9 + 18 - 7 = 20$

$D_z = \begin{vmatrix} 1 & 2 & 6 \\ 2 & -1 & -13 \\ 3 & -2 & -16 \end{vmatrix}$

$= 1\begin{vmatrix} -1 & -13 \\ -2 & -16 \end{vmatrix} - 2\begin{vmatrix} 2 & -13 \\ 3 & -16 \end{vmatrix} + 6\begin{vmatrix} 2 & -1 \\ 3 & -2 \end{vmatrix}$

$= 1(16-26) - 2(-32+39) + 6(-4+3)$
$= -10 - 14 - 6 = -30$

The solution is $x = \dfrac{D_x}{D} = \dfrac{-10}{10} = -1$,

$y = \dfrac{D_y}{D} = \dfrac{20}{10} = 2$, $z = \dfrac{D_z}{D} = \dfrac{-30}{10} = -3$.

57. Let $\begin{vmatrix} x & y \\ a & b \end{vmatrix} = 8$.

Then $\begin{vmatrix} 2x & y \\ 2a & b \end{vmatrix} = 2(8) = 16$ by Theorem (14).

The value of the determinant is multiplied by k when the elements of a column are multiplied by k.

59. Find the partial fraction decomposition:

$$x(x-4)\left(\frac{6}{x(x-4)}\right) = x(x-4)\left(\frac{A}{x} + \frac{B}{x-4}\right)$$

$$6 = A(x-4) + Bx$$

Let $x = 4$, then $6 = A(4-4) + B(4)$

$$4B = 6$$

$$B = \frac{3}{2}$$

Let $x = 0$, then $6 = A(0-4) + B(0)$

$$-4A = 6$$

$$A = -\frac{3}{2}$$

$$\frac{6}{x(x-4)} = \frac{(-3/2)}{x} + \frac{(3/2)}{x-4}$$

61. Find the partial fraction decomposition:

$$\frac{x-4}{x^2(x-1)} = \frac{A}{x} + \frac{B}{x^2} + \frac{C}{x-1}$$

Multiply both sides by $x^2(x-1)$

$$x - 4 = Ax(x-1) + B(x-1) + Cx^2$$

Let $x = 1$, then

$$1 - 4 = A(1)(1-1) + B(1-1) + C(1)^2$$

$$-3 = C$$

$$C = -3$$

Let $x = 0$, then

$$0 - 4 = A(0)(0-1) + B(0-1) + C(0)^2$$

$$-4 = -B$$

$$B = 4$$

Let $x = 2$, then

$$2 - 4 = A(2)(2-1) + B(2-1) + C(2)^2$$

$$-2 = 2A + B + 4C$$

$$2A = -2 - 4 - 4(-3)$$

$$2A = 6$$

$$A = 3$$

$$\frac{x-4}{x^2(x-1)} = \frac{3}{x} + \frac{4}{x^2} + \frac{-3}{x-1}$$

63. Find the partial fraction decomposition:

$$\frac{x}{(x^2+9)(x+1)} = \frac{A}{x+1} + \frac{Bx+C}{x^2+9}$$

Multiply both sides by $(x+1)(x^2+9)$.

$$x = A(x^2+9) + (Bx+C)(x+1)$$

Let $x = -1$, then

$$-1 = A\left((-1)^2 + 9\right) + \left(B(-1) + C\right)(-1+1)$$

$$-1 = A(10) + (-B+C)(0)$$

$$-1 = 10A$$

$$A = -\frac{1}{10}$$

Let $x = 0$, then

$$0 = A(0^2 + 9) + (B(0) + C)(0+1)$$

$$0 = 9A + C$$

$$0 = 9\left(-\frac{1}{10}\right) + C$$

$$C = \frac{9}{10}$$

Let $x = 1$, then $1 = A(1^2 + 9) + (B(1) + C)(1+1)$

$$1 = A(10) + (B+C)(2)$$

$$1 = 10A + 2B + 2C$$

$$1 = 10\left(-\frac{1}{10}\right) + 2B + 2\left(\frac{9}{10}\right)$$

$$1 = -1 + 2B + \frac{9}{5}$$

$$2B = \frac{1}{5}$$

$$B = \frac{1}{10}$$

$$\frac{x}{(x^2+9)(x+1)} = \frac{-(1/10)}{x+1} + \frac{(1/10)x + 9/10}{x^2+9}$$

65. Find the partial fraction decomposition:

$$\frac{x^3}{(x^2+4)^2} = \frac{Ax+B}{x^2+4} + \frac{Cx+D}{(x^2+4)^2}$$

Multiply both sides by $(x^2+4)^2$.

$x^3 = (Ax+B)(x^2+4) + Cx+D$
$x^3 = Ax^3 + Bx^2 + 4Ax + 4B + Cx + D$
$x^3 = Ax^3 + Bx^2 + (4A+C)x + 4B + D$
$A = 1;\ B = 0$
$4A + C = 0$
$4(1) + C = 0$
$\quad C = -4$
$4B + D = 0$
$4(0) + D = 0$
$\quad D = 0$

$$\frac{x^3}{(x^2+4)^2} = \frac{x}{x^2+4} + \frac{-4x}{(x^2+4)^2}$$

67. Find the partial fraction decomposition:

$$\frac{x^2}{(x^2+1)(x^2-1)} = \frac{x^2}{(x^2+1)(x-1)(x+1)}$$

$$= \frac{A}{x-1} + \frac{B}{x+1} + \frac{Cx+D}{x^2+1}$$

Multiply both sides by $(x-1)(x+1)(x^2+1)$.

$x^2 = A(x+1)(x^2+1) + B(x-1)(x^2+1)$
$\qquad + (Cx+D)(x-1)(x+1)$

Let $x = 1$, then
$1^2 = A(1+1)(1^2+1) + B(1-1)(1^2+1)$
$\qquad + (C(1)+D)(1-1)(1+1)$
$1 = 4A$
$A = \dfrac{1}{4}$

Let $x = -1$, then
$(-1)^2 = A(-1+1)((-1)^2+1)$
$\qquad + B(-1-1)((-1)^2+1)$
$\qquad + (C(-1)+D)(-1-1)(-1+1)$
$1 = -4B$
$B = -\dfrac{1}{4}$

Let $x = 0$, then

$0^2 = A(0+1)(0^2+1) + B(0-1)(0^2+1)$
$\qquad + (C(0)+D)(0-1)(0+1)$
$0 = A - B - D$
$0 = \dfrac{1}{4} - \left(-\dfrac{1}{4}\right) - D$
$D = \dfrac{1}{2}$

Let $x = 2$, then
$2^2 = A(2+1)(2^2+1) + B(2-1)(2^2+1)$
$\qquad + (C(2)+D)(2-1)(2+1)$
$4 = 15A + 5B + 6C + 3D$
$4 = 15\left(\dfrac{1}{4}\right) + 5\left(-\dfrac{1}{4}\right) + 6C + 3\left(\dfrac{1}{2}\right)$
$6C = 4 - \dfrac{15}{4} + \dfrac{5}{4} - \dfrac{3}{2}$
$6C = 0$
$C = 0$

$$\frac{x^2}{(x^2+1)(x^2-1)} = \frac{(1/4)}{x-1} + \frac{-(1/4)}{x+1} + \frac{(1/2)}{x^2+1}$$

69. a. $3x + 4y \leq 12$

Graph the line $3x + 4y = 12$. Use a solid line since the inequality uses $\leq$. Choose a test point not on the line, such as (0, 0). Since $3(0) + 4(0) \leq 12$ is true, shade the side of the line containing (0, 0).

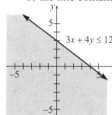

b. Solve the inequality for y:
$3x + 4y \leq 12$
$4y \leq -3x + 12$
$y \leq -\dfrac{3}{4}x + 3$

Set the calculator viewing WINDOW as shown below. Graph the linear inequality $y \leq -\dfrac{3}{4}x + 3$ by entering $Y_1 = -\dfrac{3}{4}x + 3$ and adjusting the setting to the left of Y_1 as shown below:

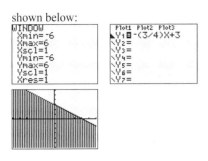

71. a. $y \leq x^2$

Graph the parabola $y = x^2$. Use a solid curve since the inequality uses $\leq$. Choose a test point not on the parabola, such as (0, 1). Since $0 \leq 1^2$ is false, shade the opposite side of the parabola from (0, 1).

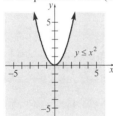

b. Set the calculator viewing WINDOW as shown below. Enter $Y_1 = x^2$ and adjust the setting to the left of Y_1 as shown below:

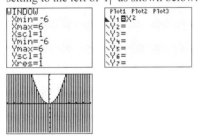

73. $\begin{cases} -2x + y \leq 2 \\ x + y \geq 2 \end{cases}$

Graph the line $-2x + y = 2$. Use a solid line since the inequality uses $\leq$. Choose a test point not on the line, such as (0, 0). Since $-2(0) + 0 \leq 2$ is true, shade the side of the line containing (0, 0). Graph the line $x + y = 2$. Use a solid line since the inequality uses $\geq$. Choose a test point not on the line, such as (0, 0). Since $0 + 0 \geq 2$ is false, shade the opposite side of the line from (0, 0). The overlapping region is the solution.

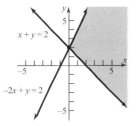

The graph is unbounded. Find the vertices: To find the intersection of $x + y = 2$ and $-2x + y = 2$, solve the system:

$$\begin{cases} x + y = 2 \\ -2x + y = 2 \end{cases}$$

Solve the first equation for x: $x = 2 - y$.
Substitute and solve:
$-2(2 - y) + y = 2$
$-4 + 2y + y = 2$
$3y = 6$
$y = 2$
$x = 2 - 2 = 0$

The point of intersection is (0, 2).
The corner point is (0, 2).

75. $\begin{cases} x \geq 0 \\ y \geq 0 \\ x + y \leq 4 \\ 2x + 3y \leq 6 \end{cases}$

Graph $x \geq 0$; $y \geq 0$. Shaded region is the first quadrant. Graph the line $x + y = 4$. Use a solid line since the inequality uses $\leq$. Choose a test point not on the line, such as (0, 0). Since $0 + 0 \leq 4$ is true, shade the side of the line containing (0, 0). Graph the line $2x + 3y = 6$. Use a solid line since the inequality uses $\leq$. Choose a test point not on the line, such as (0, 0). Since $2(0) + 3(0) \leq 6$ is true, shade the side of the line containing (0, 0).

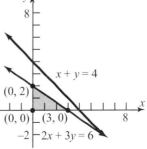

The overlapping region is the solution. The graph

is bounded. Find the vertices: The x-axis and y-axis intersect at (0, 0). The intersection of $2x + 3y = 6$ and the y-axis is (0, 2). The intersection of $2x + 3y = 6$ and the x-axis is (3, 0). The three corner points are (0, 0), (0, 2), and (3, 0).

77. $\begin{cases} x \geq 0 \\ y \geq 0 \\ 2x + y \leq 8 \\ x + 2y \geq 2 \end{cases}$

Graph $x \geq 0$; $y \geq 0$. Shaded region is the first quadrant. Graph the line $2x + y = 8$. Use a solid line since the inequality uses $\leq$. Choose a test point not on the line, such as (0, 0). Since $2(0) + 0 \leq 8$ is true, shade the side of the line containing (0, 0). Graph the line $x + 2y = 2$. Use a solid line since the inequality uses $\geq$. Choose a test point not on the line, such as (0, 0). Since $0 + 2(0) \geq 2$ is false, shade the opposite side of the line from (0, 0).

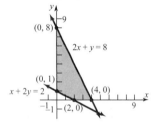

The overlapping region is the solution. The graph is bounded. Find the vertices: The intersection of $x + 2y = 2$ and the y-axis is (0, 1). The intersection of $x + 2y = 2$ and the x-axis is (2, 0). The intersection of $2x + y = 8$ and the y-axis is (0, 8). The intersection of $2x + y = 8$ and the x-axis is (4, 0). The four corner points are (0, 1), (0, 8), (2, 0), and (4, 0).

79. Maximize $z = 3x + 4y$ subject to $x \geq 0$, $y \geq 0$, $3x + 2y \geq 6$, $x + y \leq 8$. Graph the constraints.

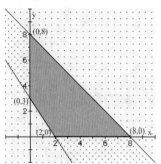

The corner points are (0, 3), (2, 0), (0, 8), (8, 0). Evaluate the objective function:

Vertex	Value of $z = 3x + 4y$
(0, 3)	$z = 3(0) + 4(3) = 12$
(0, 8)	$z = 3(0) + 4(8) = 32$
(2, 0)	$z = 3(2) + 4(0) = 6$
(8, 0)	$z = 3(8) + 4(0) = 24$

The maximum value is 32 at (0, 8).

81. Minimize $z = 3x + 5y$ subject to $x \geq 0$, $y \geq 0$, $x + y \geq 1$, $3x + 2y \leq 12$, $x + 3y \leq 12$.
Graph the constraints.

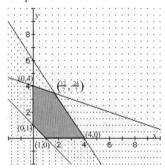

To find the intersection of $3x + 2y = 12$ and $x + 3y = 12$, solve the system:
$\begin{cases} 3x + 2y = 12 \\ x + 3y = 12 \end{cases}$
Solve the second equation for x: $x = 12 - 3y$
Substitute and solve:
$3(12 - 3y) + 2y = 12$
$36 - 9y + 2y = 12$
$-7y = -24$
$y = \dfrac{24}{7}$

$x = 12 - 3\left(\dfrac{24}{7}\right) = 12 - \dfrac{72}{7} = \dfrac{12}{7}$

The point of intersection is $\left(\dfrac{12}{7}, \dfrac{24}{7}\right)$.

The corner points are (0, 1), (1, 0), (0, 4), (4, 0), $\left(\dfrac{12}{7}, \dfrac{24}{7}\right)$.

Evaluate the objective function:

Vertex	Value of $z = 3x + 5y$
(0, 1)	$z = 3(0) + 5(1) = 5$
(0, 4)	$z = 3(0) + 5(4) = 20$
(1, 0)	$z = 3(1) + 5(0) = 3$
(4, 0)	$z = 3(4) + 5(0) = 12$
$\left(\dfrac{12}{7}, \dfrac{24}{7}\right)$	$z = 3\left(\dfrac{12}{7}\right) + 5\left(\dfrac{24}{7}\right) = \dfrac{156}{7}$

The minimum value is 3 at (1, 0).

83. $\begin{cases} 2x + 5y = 5 \\ 4x + 10y = A \end{cases}$

Multiply each side of the first equation by –2 and eliminate x:

$\begin{cases} -4x - 10y = -10 \\ \underline{4x + 10y = A} \\ 0 = A - 10 \end{cases}$

If there are to be infinitely many solutions, the result of elimination should be 0 = 0. Therefore, $A - 10 = 0$ or $A = 10$.

85. $y = ax^2 + bx + c$

At (0, 1) the equation becomes:
$1 = a(0)^2 + b(0) + c$
$c = 1$

At (1, 0) the equation becomes:
$0 = a(1)^2 + b(1) + c$
$0 = a + b + c$
$a + b + c = 0$

At (–2, 1) the equation becomes:
$1 = a(-2)^2 + b(-2) + c$
$1 = 4a - 2b + c$
$4a - 2b + c = 1$

The system of equations is:
$\begin{cases} a + b + c = 0 \\ 4a - 2b + c = 1 \\ c = 1 \end{cases}$

Substitute $c = 1$ into the first and second equations and simplify:
$a + b + 1 = 0 \qquad 4a - 2b + 1 = 1$
$a + b = -1 \qquad 4a - 2b = 0$
$a = -b - 1$

Solve the first equation for a, substitute into the second equation and solve:
$4(-b - 1) - 2b = 0$
$-4b - 4 - 2b = 0$
$-6b = 4$
$b = -\dfrac{2}{3}$

$a = \dfrac{2}{3} - 1 = -\dfrac{1}{3}$

The quadratic function is $y = -\dfrac{1}{3}x^2 - \dfrac{2}{3}x + 1$.

87. Let x = the number of pounds of coffee that costs $3.00 per pound, and let y = the number of pounds of coffee that costs $6.00 per pound. Then $x + y = 100$ represents the total amount of coffee in the blend. The value of the blend will be represented by the equation:
$3x + 6y = 3.90(100)$. Solve the system of equations:

$\begin{cases} x + y = 100 \\ 3x + 6y = 390 \end{cases}$

Solve the first equation for y: $y = 100 - x$.
Solve by substitution:
$3x + 6(100 - x) = 390$
$3x + 600 - 6x = 390$
$-3x = -210$
$x = 70$
$y = 100 - 70 = 30$

The blend is made up of 70 pounds of the $3 per pound coffee and 30 pounds of the $6-per pound coffee.

89. Let x = the number of small boxes, let y = the number of medium boxes, and let z = the number of large boxes.
Oatmeal raisin equation: $x + 2y + 2z = 15$
Chocolate chip equation: $x + y + 2z = 10$
Shortbread equation: $y + 3z = 11$

$\begin{cases} x + 2y + 2z = 15 \\ x + y + 2z = 10 \\ y + 3z = 11 \end{cases}$

Chapter 5: Systems of Equations and Inequalities   SSM: College Algebra EGU

Multiply each side of the second equation by -1 and add to the first equation to eliminate x:
$$\begin{cases} x+2y+2z = 15 \\ -x-y-2z = -10 \end{cases}$$
$$y+3z = 11$$
$$y = 5$$
Substituting and solving for the other variables:
$5+3z = 11$ $x+5+2(2) = 10$
$3z = 6$ $x+9 = 10$
$z = 2$ $x = 1$
Thus, 1 small box, 5 medium boxes, and 2 large boxes of cookies should be purchased.

91. Let x = the speed of the boat in still water, and let y = the speed of the river current. The distance from Chiritza to the Flotel Orellana is 100 kilometers.

	Rate	Time	Distance
trip downstream	$x+y$	5/2	100
trip downstream	$x-y$	3	100

The system of equations is:
$$\begin{cases} \frac{5}{2}(x+y) = 100 \\ 3(x-y) = 100 \end{cases}$$
Multiply both sides of the first equation by 6, multiply both sides of the second equation by 5, and add the results.
$15x+15y = 600$
$15x-15y = 500$
$30x = 1100$
$x = \frac{1100}{30} = \frac{110}{3}$
$3\left(\frac{110}{3}\right) - 3y = 100$
$110 - 3y = 100$
$10 = 3y$
$y = \frac{10}{3}$
The speed of the boat is $110/3 \approx 36.67$ km/hr; the speed of the current is $10/3 \approx 3.33$ km/hr.

93. Let x = the number of hours for Bruce to do the job alone, let y = the number of hours for Bryce to do the job alone, and let z = the number of hours for Marty to do the job alone.
Then $1/x$ represents the fraction of the job that Bruce does in one hour.

$1/y$ represents the fraction of the job that Bryce does in one hour.
$1/z$ represents the fraction of the job that Marty does in one hour.
The equation representing Bruce and Bryce working together is:
$$\frac{1}{x}+\frac{1}{y} = \frac{1}{(4/3)} = \frac{3}{4} = 0.75$$
The equation representing Bryce and Marty working together is:
$$\frac{1}{y}+\frac{1}{z} = \frac{1}{(8/5)} = \frac{5}{6} = 0.625$$
The equation representing Bruce and Marty working together is:
$$\frac{1}{x}+\frac{1}{z} = \frac{1}{(8/3)} = \frac{3}{8} = 0.375$$
Solve the system of equations:
$$\begin{cases} x^{-1}+y^{-1} = 0.75 \\ y^{-1}+z^{-1} = 0.625 \\ x^{-1}+z^{-1} = 0.375 \end{cases}$$
Let $u = x^{-1}$, $v = y^{-1}$, $w = z^{-1}$
$$\begin{cases} u+v = 0.75 \\ v+w = 0.625 \\ u+w = 0.375 \end{cases}$$
Solve the first equation for u: $u = 0.75 - v$.
Solve the second equation for w: $w = 0.625 - v$.
Substi into the third equation and solve:
$(0.75-v)+(0.625-v) = 0.375$
$-2v = -1$
$v = 0.5$
$u = 0.75 - 0.5 = 0.25$
$w = 0.625 - 0.5 = 0.125$
Solve for
x, y, and z: $x = 4$, $y = 2$, $z = 8$ (reciprocals)
Bruce can do the job in 4 hours, Bryce in 2 hours, and Marty in 8 hours.

95. Let x = the number of gasoline engines produced each week, and let y = the number of diesel engines produced each week. The total cost is: $C = 450x + 550y$. Cost is to be minimized; thus, this is the objective function. The constraints are:
$20 \leq x \leq 60$ number of gasoline engines needed and capacity each week.
$15 \leq y \leq 40$ number of diesel engines needed and capacity each week.

$x + y \geq 50$ number of engines produced to prevent layoffs.

Graph the constraints.

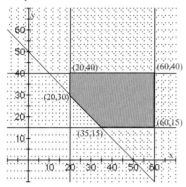

The corner points are (20, 30), (20, 40), (35, 15), (60, 15), (60, 40)

Evaluate the objective function:

Vertex	Value of $C = 450x + 550y$
(20, 30)	$C = 450(20) + 550(30) = 25,500$
(20, 40)	$C = 450(35) + 550(40) = 31,000$
(35, 15)	$C = 450(35) + 550(15) = 24,000$
(60, 15)	$C = 450(60) + 550(15) = 35,250$
(60, 40)	$C = 450(60) + 550(40) = 49,000$

The minimum cost is $24,000, when 35 gasoline engines and 15 diesel engines are produced. The excess capacity is 15 gasoline engines, since only 20 gasoline engines had to be delivered.

Chapter 5 Test

1. $\begin{cases} -2x + y = -7 \\ 4x + 3y = 9 \end{cases}$

   Substitution:
   We solve the first equation for y, obtaining
   $y = 2x - 7$
   Next we substitute this result for y in the second equation and solve for x.
   $$4x + 3y = 9$$
   $$4x + 3(2x - 7) = 9$$
   $$4x + 6x - 21 = 9$$
   $$10x = 30$$
   $$x = \frac{30}{10} = 3$$
   We can now obtain the value for y by letting $x = 3$ in our substitution for y.

   $y = 2x - 7$
   $y = 2(3) - 7$
   $ = 6 - 7$
   $ = -1$
   The solution of the system is $x = 3$, $y = -1$.

   Elimination:
   Multiply each side of the first equation by 2 so that the coefficients of x in the two equations are negatives of each other. The result is the equivalent system
   $\begin{cases} -4x + 2y = -14 \\ 4x + 3y = 9 \end{cases}$
   We can replace the second equation of this system by the sum of the two equations. The result is the equivalent system
   $\begin{cases} -4x + 2y = -14 \\ 5y = -5 \end{cases}$
   Now we solve the second equation for y.
   $5y = -5$
   $$y = \frac{-5}{5} = -1$$
   We back-substitute this value for y into the original first equation and solve for x.
   $$-2x + y = -7$$
   $$-2x + (-1) = -7$$
   $$-2x = -6$$
   $$x = \frac{-6}{-2} = 3$$
   The solution of the system is $x = 3$, $y = -1$.

   Check:
   To check our solution graphically, we need to solve each equation for y by putting them in the form $y = mx + b$.
   $\begin{array}{ll} -2x + y = -7 & 4x + 3y = 9 \\ y = 2x - 7 & 3y = -4x + 9 \\ & y = -\frac{4}{3}x + 3 \end{array}$
   Since the slopes of the two lines are not the same, we know that the two lines will intersect. We enter both equations into our calculator and find the intersection point.

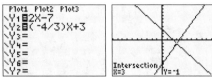

The solution of the system is $x = 3$, $y = -1$.

2. $\begin{cases} \dfrac{1}{3}x - 2y = 1 \\ 5x - 30y = 18 \end{cases}$

We choose to use the method of elimination and multiply the first equation by -15 to obtain the equivalent system
$\begin{cases} -5x + 30y = -15 \\ 5x - 30y = 18 \end{cases}$

We replace the second equation by the sum of the two equations to obtain the equivalent system
$\begin{cases} -5x + 30y = -15 \\ 0 = 3 \end{cases}$

The second equation is a contradiction and has no solution. This means that the system itself has no solution and is therefore inconsistent.

Check:
To check our solution graphically, we first solve each equation for y by putting them in the form $y = mx + b$.

$\dfrac{1}{3}x - 2y = 1$ $\qquad$ $5x - 30y = 18$

$-2y = -\dfrac{1}{3}x + 1$ $\qquad$ $-30y = -5x + 18$

$y = \dfrac{1}{6}x - \dfrac{1}{2}$ $\qquad$ $y = \dfrac{1}{6}x - \dfrac{3}{5}$

We can see that the slopes are the same, but the y-intercepts are different. Therefore, the two lines are parallel and will not intersect. The system has no solution and is therefore inconsistent.

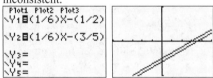

3. $\begin{cases} x - y + 2z = 5 & (1) \\ 3x + 4y - z = -2 & (2) \\ 5x + 2y + 3z = 8 & (3) \end{cases}$

We use the method of elimination and begin by eliminating the variable y from equation (2). Multiply each side of equation (1) by 4 and add the result to equation (2). This result becomes our new equation (2).

$\begin{array}{l} x - y + 2z = 5 \\ 3x + 4y - z = -2 \end{array}$ $\qquad$ $\begin{array}{l} 4x - 4y + 8z = 20 \\ 3x + 4y - z = -2 \\ \hline 7x + 7z = 18 \quad (2) \end{array}$

We now eliminate the variable y from equation (3) by multiplying each side of equation (1) by 2 and adding the result to equation (3). The result becomes our new equation (3).

$\begin{array}{l} x - y + 2z = 5 \\ 5x + 2y + 3z = 8 \end{array}$ $\qquad$ $\begin{array}{l} 2x - 2y + 4z = 10 \\ 5x + 2y + 3z = 8 \\ \hline 7x + 7z = 18 \quad (3) \end{array}$

Our (equivalent) system now looks like
$\begin{cases} x - y + 2z = 5 & (1) \\ 7x + 7z = 18 & (2) \\ 7x + 7z = 18 & (3) \end{cases}$

Treat equations (2) and (3) as a system of two equations containing two variables, and eliminate the x variable by multiplying each side of equation (2) by -1 and adding the result to equation (3). The result becomes our new equation (3).

$\begin{array}{l} 7x + 7z = 18 \\ 7x + 7z = 18 \end{array}$ $\qquad$ $\begin{array}{l} -7x - 7z = -18 \\ 7x + 7z = 18 \\ \hline 0 = 0 \quad (3) \end{array}$

We now have the equivalent system
$\begin{cases} x - y + 2z = 5 & (1) \\ 7x + 7z = 18 & (2) \\ \phantom{7x + 7z =\ } 0 = 0 & (3) \end{cases}$

This is equivalent to a system of two equations with three variables. Since one of the equations contains three variables and one contains only two variables, the system will be dependent. There are infinitely many solutions.
We solve equation (2) for x and determine that
$x = -z + \dfrac{18}{7}$. Substitute this expression into equation (1) to obtain y in terms of z.

$$x - y + 2z = 5$$
$$\left(-z + \frac{18}{7}\right) - y + 2z = 5$$
$$-z + \frac{18}{7} - y + 2z = 5$$
$$-y + z = \frac{17}{7}$$
$$y = z - \frac{17}{7}$$

We can write the solution as
$$\begin{cases} x = -z + \dfrac{18}{7} \\ y = z - \dfrac{17}{7} \end{cases}$$
where z can be any real number.

Check:
To check our solution on a graphing utility, we enter the augmented matrix and use the **rref** command.

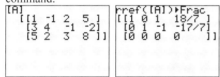

4. $\begin{cases} 3x + 2y - 8z = -3 \quad (1) \\ -x - \frac{2}{3}y + z = 1 \quad (2) \\ 6x - 3y + 15z = 8 \quad (3) \end{cases}$

We start by clearing the fraction in equation (2) by multiplying both sides of the equation by 3.
$$\begin{cases} 3x + 2y - 8z = -3 \quad (1) \\ -3x - 2y + 3z = 3 \quad (2) \\ 6x - 3y + 15z = 8 \quad (3) \end{cases}$$

We use the method of elimination and begin by eliminating the variable x from equation (2). The coefficients on x in equations (1) and (2) are negatives of each other so we simply add the two equations together. This result becomes our new equation (2).
$$3x + 2y - 8z = -3$$
$$\underline{-3x - 2y + 3z = 3}$$
$$-5z = 0 \quad (2)$$

We now eliminate the variable x from equation (3) by multiplying each side of equation (1) by -2 and adding the result to equation (3). The result becomes our new equation (3).
$$3x + 2y - 8z = -3 \qquad -6x - 4y + 16z = 6$$
$$6x - 3y + 15z = 8 \qquad \underline{6x - 3y + 15z = 8}$$
$$ \qquad -7y + 31z = 14 \quad (3)$$

Our (equivalent) system now looks like
$$\begin{cases} 3x + 2y - 8z = -3 \quad (1) \\ -5z = 0 \quad (2) \\ -7y + 31z = 14 \quad (3) \end{cases}$$

We solve equation (2) for z by dividing both sides of the equation by -5.
$$-5z = 0$$
$$z = 0$$

Back-substitute $z = 0$ into equation (3) and solve for y.
$$-7y + 31z = 14$$
$$-7y + 31(0) = 14$$
$$-7y = 14$$
$$y = -2$$

Finally, back-substitute $y = -2$ and $z = 0$ into equation (1) and solve for x.
$$3x + 2y - 8z = -3$$
$$3x + 2(-2) - 8(0) = -3$$
$$3x - 4 = -3$$
$$3x = 1$$
$$x = \frac{1}{3}$$

The solution of the original system is $x = \dfrac{1}{3}$, $y = -2$, and $z = 0$.

Check:
To check our solution with a graphing utility, we enter the augmented matrix and use the **rref** command.

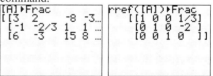

The solution of the system is $x = \dfrac{1}{3}$, $y = -2$, and $z = 0$.

5. We first check the equations to make sure that all variable terms are on the left side of the equation and the constants are on the right side. If a variable is missing, we put it in with a coefficient of 0. Our system can be rewritten as
$$\begin{cases} 4x - 5y + z = 0 \\ -2x - y + 0z = -25 \\ x + 5y - 5z = 10 \end{cases}$$
The augmented matrix is
$$\begin{bmatrix} 4 & -5 & 1 & | & 0 \\ -2 & -1 & 0 & | & -25 \\ 1 & 5 & -5 & | & 10 \end{bmatrix}$$

6. The matrix has three rows and represents a system with three equations. The three columns to the left of the vertical bar indicate that the system has three variables. We can let x, y, and z denote these variables. The column to the right of the vertical bar represents the constants on the right side of the equations. The system is
$$\begin{cases} 3x + 2y + 4z = -6 \\ 1x + 0y + 8z = 2 \\ -2x + 1y + 3z = -11 \end{cases} \text{ or } \begin{cases} 3x + 2y + 4z = -6 \\ x + 8z = 2 \\ -2x + y + 3z = -11 \end{cases}$$

7. $2A + C = 2\begin{bmatrix} 1 & -1 \\ 0 & -4 \\ 3 & 2 \end{bmatrix} + \begin{bmatrix} 4 & 6 \\ 1 & -3 \\ -1 & 8 \end{bmatrix}$

$= \begin{bmatrix} 2 & -2 \\ 0 & -8 \\ 6 & 4 \end{bmatrix} + \begin{bmatrix} 4 & 6 \\ 1 & -3 \\ -1 & 8 \end{bmatrix}$

$= \begin{bmatrix} 6 & 4 \\ 1 & -11 \\ 5 & 12 \end{bmatrix}$

```
2[A]+[C]
 [[6  4  ]
  [1  -11]
  [5  12 ]]
```

8. $A - 3C = \begin{bmatrix} 1 & -1 \\ 0 & -4 \\ 3 & 2 \end{bmatrix} - 3\begin{bmatrix} 4 & 6 \\ 1 & -3 \\ -1 & 8 \end{bmatrix}$

$= \begin{bmatrix} 1 & -1 \\ 0 & -4 \\ 3 & 2 \end{bmatrix} - \begin{bmatrix} 12 & 18 \\ 3 & -9 \\ -3 & 24 \end{bmatrix}$

$= \begin{bmatrix} -11 & -19 \\ -3 & 5 \\ 6 & -22 \end{bmatrix}$

```
[A]-3[C]
 [[-11  -19]
  [-3   5  ]
  [6    -22]]
```

9. AC cannot be computed because the dimensions are mismatched. To multiply two matrices, we need the number of columns in the first matrix to be the same as the number of rows in the second matrix. Matrix A has 2 columns, but matrix C has 3 rows. Therefore, the operation cannot be performed.

10. Here we are taking the product of a 2×3 matrix and a 3×2 matrix. Since the number of columns in the first matrix is the same as the number of rows in the second matrix (3 in both cases), the operation can be performed and will result in a 2×2 matrix.

$BA = \begin{bmatrix} 1 & -2 & 5 \\ 0 & 3 & 1 \end{bmatrix} \begin{bmatrix} 1 & -1 \\ 0 & -4 \\ 3 & 2 \end{bmatrix}$

$= \begin{bmatrix} 1 \cdot 1 + (-2) \cdot 0 + 5 \cdot 3 & 1 \cdot (-1) + (-2) \cdot (-4) + 5 \cdot 2 \\ 0 \cdot 1 + 3 \cdot 0 + 1 \cdot 3 & 0 \cdot (-1) + 3(-4) + 1 \cdot 2 \end{bmatrix}$

$= \begin{bmatrix} 16 & 17 \\ 3 & -10 \end{bmatrix}$

```
[B][A]
 [[16  17 ]
  [3   -10]]
```

11. We first form the matrix
$$[A \mid I_2] = \begin{bmatrix} 3 & 2 & | & 1 & 0 \\ 5 & 4 & | & 0 & 1 \end{bmatrix}$$
Next we use row operations to transform $[A \mid I_2]$ into reduced row echelon form.

$$\begin{bmatrix} 3 & 2 & | & 1 & 0 \\ 5 & 4 & | & 0 & 1 \end{bmatrix} R_1 = \tfrac{1}{3} r_1$$

$$\begin{bmatrix} 1 & \tfrac{2}{3} & | & \tfrac{1}{3} & 0 \\ 5 & 4 & | & 0 & 1 \end{bmatrix} R_2 = -5r_1 + r_2$$

$$\begin{bmatrix} 1 & \tfrac{2}{3} & | & \tfrac{1}{3} & 0 \\ 0 & \tfrac{2}{3} & | & -\tfrac{5}{3} & 1 \end{bmatrix} R_2 = \tfrac{3}{2} r_2$$

$$\begin{bmatrix} 1 & \tfrac{2}{3} & | & \tfrac{1}{3} & 0 \\ 0 & 1 & | & -\tfrac{5}{2} & \tfrac{3}{2} \end{bmatrix} R_1 = -\tfrac{2}{3} r_2 + r_1$$

$$\begin{bmatrix} 1 & 0 & | & 2 & -1 \\ 0 & 1 & | & -\tfrac{5}{2} & \tfrac{3}{2} \end{bmatrix}$$

Therefore, $A^{-1} = \begin{bmatrix} 2 & -1 \\ -\tfrac{5}{2} & \tfrac{3}{2} \end{bmatrix}$.

12. We first form the matrix
$$[B \mid I_3] = \begin{bmatrix} 1 & -1 & 1 & | & 1 & 0 & 0 \\ 2 & 5 & -1 & | & 0 & 1 & 0 \\ 2 & 3 & 0 & | & 0 & 0 & 1 \end{bmatrix}$$
Next we use row operations to transform $[B \mid I_3]$ into reduced row echelon form.

$$\begin{bmatrix} 1 & -1 & 1 & | & 1 & 0 & 0 \\ 2 & 5 & -1 & | & 0 & 1 & 0 \\ 2 & 3 & 0 & | & 0 & 0 & 1 \end{bmatrix} \begin{matrix} R_2 = -2r_1 + r_2 \\ R_3 = -2r_1 + r_3 \end{matrix}$$

$$= \begin{bmatrix} 1 & -1 & 1 & | & 1 & 0 & 0 \\ 0 & 7 & -3 & | & -2 & 1 & 0 \\ 0 & 5 & -2 & | & -2 & 0 & 1 \end{bmatrix} R_2 = \tfrac{1}{7} r_2$$

$$= \begin{bmatrix} 1 & -1 & 1 & | & 1 & 0 & 0 \\ 0 & 1 & -\tfrac{3}{7} & | & -\tfrac{2}{7} & \tfrac{1}{7} & 0 \\ 0 & 5 & -2 & | & -2 & 0 & 1 \end{bmatrix} \begin{matrix} R_1 = r_2 + r_1 \\ R_3 = -5r_2 + r_3 \end{matrix}$$

$$= \begin{bmatrix} 1 & 0 & \tfrac{4}{7} & | & \tfrac{5}{7} & \tfrac{1}{7} & 0 \\ 0 & 1 & -\tfrac{3}{7} & | & -\tfrac{2}{7} & \tfrac{1}{7} & 0 \\ 0 & 0 & \tfrac{1}{7} & | & -\tfrac{4}{7} & -\tfrac{5}{7} & 1 \end{bmatrix} R_3 = 7r_3$$

$$= \begin{bmatrix} 1 & 0 & \tfrac{4}{7} & | & \tfrac{5}{7} & \tfrac{1}{7} & 0 \\ 0 & 1 & -\tfrac{3}{7} & | & -\tfrac{2}{7} & \tfrac{1}{7} & 0 \\ 0 & 0 & 1 & | & -4 & -5 & 7 \end{bmatrix} \begin{matrix} R_1 = -\tfrac{4}{7} r_3 + r_1 \\ R_2 = \tfrac{3}{7} r_3 + r_2 \end{matrix}$$

$$= \begin{bmatrix} 1 & 0 & 0 & | & 3 & 3 & -4 \\ 0 & 1 & 0 & | & -2 & -2 & 3 \\ 0 & 0 & 1 & | & -4 & -5 & 7 \end{bmatrix}$$

Thus, $B^{-1} = \begin{bmatrix} 3 & 3 & -4 \\ -2 & -2 & 3 \\ -4 & -5 & 7 \end{bmatrix}$

13. $\begin{cases} 6x + 3y = 12 \\ 2x - y = -2 \end{cases}$

We start by writing the augmented matrix for the system.
$$\begin{bmatrix} 6 & 3 & | & 12 \\ 2 & -1 & | & -2 \end{bmatrix}$$

Next we use row operations to transform the augmented matrix into row echelon form.

$$\begin{bmatrix} 6 & 3 & | & 12 \\ 2 & -1 & | & -2 \end{bmatrix} \to \begin{bmatrix} 2 & -1 & | & -2 \\ 6 & 3 & | & 12 \end{bmatrix} \begin{pmatrix} R_1 = r_2 \\ R_2 = r_1 \end{pmatrix}$$

$$\to \begin{bmatrix} 1 & -\tfrac{1}{2} & | & -1 \\ 6 & 3 & | & 12 \end{bmatrix} \left(R_1 = \tfrac{1}{2} r_1\right)$$

$$\to \begin{bmatrix} 1 & -\tfrac{1}{2} & | & -1 \\ 0 & 6 & | & 18 \end{bmatrix} \left(R_2 = -6r_1 + r_2\right)$$

$$\to \begin{bmatrix} 1 & -\tfrac{1}{2} & | & -1 \\ 0 & 1 & | & 3 \end{bmatrix} \left(R_2 = \tfrac{1}{6} r_2\right)$$

$$\to \begin{bmatrix} 1 & 0 & | & -\tfrac{1}{2} \\ 0 & 1 & | & 3 \end{bmatrix} \left(R_2 = \tfrac{1}{2} r_2 + r_1\right)$$

The solution of the system is $x = \tfrac{1}{2}$, $y = 3$.

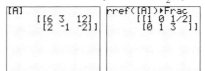

14. $\begin{cases} x + \dfrac{1}{4}y = 7 \\ 8x + 2y = 56 \end{cases}$

We start by writing the augmented matrix for the system.
$\begin{bmatrix} 1 & \frac{1}{4} & | & 7 \\ 8 & 2 & | & 56 \end{bmatrix}$

Next we use row operations to transform the augmented matrix into row echelon form.
$\begin{bmatrix} 1 & \frac{1}{4} & | & 7 \\ 8 & 2 & | & 56 \end{bmatrix} R_2 = -8R_1 + r_2$

$\begin{bmatrix} 1 & \frac{1}{4} & | & 7 \\ 0 & 0 & | & 0 \end{bmatrix}$

The augmented matrix is now in row echelon form. Because the bottom row consists entirely of 0's, the system actually consists of one equation in two variables. The system is dependent and therefore has an infinite number of solutions. Any ordered pair satisfying the equation $x + \dfrac{1}{4}y = 7$, or $y = -4x + 28$, is a solution to the system.

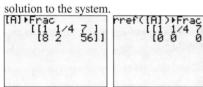

15. $\begin{cases} x + 2y + 4z = -3 \\ 2x + 7y + 15z = -12 \\ 4x + 7y + 13z = -10 \end{cases}$

We start by writing the augmented matrix for the system.
$\begin{bmatrix} 1 & 2 & 4 & | & -3 \\ 2 & 7 & 15 & | & -12 \\ 4 & 7 & 13 & | & -10 \end{bmatrix}$

Next we use row operations to transform the augmented matrix into row echelon form.
$\begin{bmatrix} 1 & 2 & 4 & | & -3 \\ 2 & 7 & 15 & | & -12 \\ 4 & 7 & 13 & | & -10 \end{bmatrix}$

$\rightarrow \begin{bmatrix} 1 & 2 & 4 & | & -3 \\ 0 & 3 & 7 & | & -6 \\ 0 & -1 & -3 & | & 2 \end{bmatrix} \begin{pmatrix} R_2 = -2r_1 + r_2 \\ R_3 = -4r_1 + r_3 \end{pmatrix}$

$\rightarrow \begin{bmatrix} 1 & 2 & 4 & | & -3 \\ 0 & 1 & 3 & | & -2 \\ 0 & 3 & 7 & | & -6 \end{bmatrix} \begin{pmatrix} R_2 = -r_3 \\ R_3 = r_2 \end{pmatrix}$

$\rightarrow \begin{bmatrix} 1 & 2 & 4 & | & -3 \\ 0 & 1 & 3 & | & -2 \\ 0 & 0 & -2 & | & 0 \end{bmatrix} (R_3 = -3r_2 + r_3)$

$\rightarrow \begin{bmatrix} 1 & 2 & 4 & | & -3 \\ 0 & 1 & 3 & | & -2 \\ 0 & 0 & 1 & | & 0 \end{bmatrix} (R_3 = -\tfrac{1}{2}r_3)$

The matrix is now in row echelon form. The last row represents the equation $z = 0$. Using $z = 0$ we back-substitute into the equation $y + 3z = -2$ (from the second row) and obtain
$y + 3z = -2$
$y + 3(0) = -2$
$y = -2$

Using $y = -2$ and $z = 0$, we back-substitute into the equation $x + 2y + 4z = -3$ (from the first row) and obtain
$x + 2y + 4z = -3$
$x + 2(-2) + 4(0) = -3$
$x = 1$

The solution of the system is $x = 1$, $y = -2$, $z = 0$.

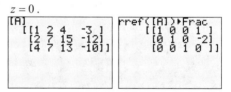

16. $\begin{cases} 2x + 2y - 3z = 5 \\ x - y + 2z = 8 \\ 3x + 5y - 8z = -2 \end{cases}$

We start by writing the augmented matrix for the system.
$\begin{bmatrix} 2 & 2 & -3 & | & 5 \\ 1 & -1 & 2 & | & 8 \\ 3 & 5 & -8 & | & -2 \end{bmatrix}$

Next we use row operations to transform the augmented matrix into row echelon form.

$$\begin{bmatrix} 2 & 2 & -3 & | & 5 \\ 1 & -1 & 2 & | & 8 \\ 3 & 5 & -8 & | & -2 \end{bmatrix}$$

$$= \begin{bmatrix} 1 & -1 & 2 & | & 8 \\ 2 & 2 & -3 & | & 5 \\ 3 & 5 & -8 & | & -2 \end{bmatrix} \quad \begin{pmatrix} R_1 = r_2 \\ R_2 = r_1 \end{pmatrix}$$

$$= \begin{bmatrix} 1 & -1 & 2 & | & 8 \\ 0 & 4 & -7 & | & -11 \\ 0 & 8 & -14 & | & -26 \end{bmatrix} \quad \begin{pmatrix} R_2 = -2r_1 + r_2 \\ R_3 = -3r_1 + r_3 \end{pmatrix}$$

$$= \begin{bmatrix} 1 & -1 & 2 & | & 8 \\ 0 & 1 & -\frac{7}{4} & | & -\frac{11}{4} \\ 0 & 8 & -14 & | & -26 \end{bmatrix} \quad (R_2 = \tfrac{1}{4} r_2)$$

$$= \begin{bmatrix} 1 & -1 & 2 & | & 8 \\ 0 & 1 & -\frac{7}{4} & | & -\frac{11}{4} \\ 0 & 0 & 0 & | & -4 \end{bmatrix} \quad (R_3 = -8r_2 + r_3)$$

The matrix is now in row echelon form. The last row represents the equation $0 = -4$ which is a contradiction. Therefore, the system has no solution and is said to be inconsistent.

17. $\begin{vmatrix} -2 & 5 \\ 3 & 7 \end{vmatrix} = (-2)(7) - (5)(3)$

$\qquad = -14 - 15$

$\qquad = -29$

18. We choose to expand across row 2 because it contains a 0.

$\begin{vmatrix} 2 & -4 & 6 \\ 1 & 4 & 0 \\ -1 & 2 & -4 \end{vmatrix}$

$= (-1)^{2+1}(1)\begin{vmatrix} -4 & 6 \\ 2 & -4 \end{vmatrix} + (-1)^{2+2}(4)\begin{vmatrix} 2 & 6 \\ -1 & -4 \end{vmatrix}$

$\qquad + (-1)^{2+3}(0)\begin{vmatrix} 2 & -4 \\ -1 & 2 \end{vmatrix}$

$= -1\begin{vmatrix} -4 & 6 \\ 2 & -4 \end{vmatrix} + 4\begin{vmatrix} 2 & 6 \\ -1 & -4 \end{vmatrix} - 0\begin{vmatrix} 2 & -4 \\ -1 & 2 \end{vmatrix}$

$= -1(16-12) + 4(-8+6) - 0(4-4)$

$= -1(4) + 4(-2)$

$= -4 - 8$

$= -12$

19. $\begin{cases} 4x + 3y = -23 \\ 3x - 5y = 19 \end{cases}$

The determinant D of the coefficients of the variables is

$D = \begin{vmatrix} 4 & 3 \\ 3 & -5 \end{vmatrix} = (4)(-5) - (3)(3) = -20 - 9 = -29$

Since $D \neq 0$, Cramer's Rule can be applied.

$D_x = \begin{vmatrix} -23 & 3 \\ 19 & -5 \end{vmatrix} = (-23)(-5) - (3)(19) = 58$

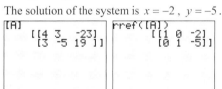

$D_y = \begin{vmatrix} 4 & -23 \\ 3 & 19 \end{vmatrix} = (4)(19) - (-23)(3) = 145$

$x = \dfrac{D_x}{D} = \dfrac{58}{-29} = -2$

$y = \dfrac{D_y}{D} = \dfrac{145}{-29} = -5$

The solution of the system is $x = -2$, $y = -5$.

20. $\begin{cases} 4x - 3y + 2z = 15 \\ -2x + y - 3z = -15 \\ 5x - 5y + 2z = 18 \end{cases}$

The determinant D of the coefficients of the variables is

$D = \begin{vmatrix} 4 & -3 & 2 \\ -2 & 1 & -3 \\ 5 & -5 & 2 \end{vmatrix}$

$= 4\begin{vmatrix} 1 & -3 \\ -5 & 2 \end{vmatrix} - (-3)\begin{vmatrix} -2 & -3 \\ 5 & 2 \end{vmatrix} + 2\begin{vmatrix} -2 & 1 \\ 5 & -5 \end{vmatrix}$

$= 4(2-15) + 3(-4+15) + 2(10-5)$

$= 4(-13) + 3(11) + 2(5)$

$= -52 + 33 + 10$

$= -9$

Since $D \neq 0$, Cramer's Rule can be applied.

$D_x = \begin{vmatrix} 15 & -3 & 2 \\ -15 & 1 & -3 \\ 18 & -5 & 2 \end{vmatrix}$

$= 15\begin{vmatrix} 1 & -3 \\ -5 & 2 \end{vmatrix} - (-3)\begin{vmatrix} -15 & -3 \\ 18 & 2 \end{vmatrix} + 2\begin{vmatrix} -15 & 1 \\ 18 & -5 \end{vmatrix}$

$= 15(2-15) + 3(-30+54) + 2(75-18)$

$= 15(-13) + 3(24) + 2(57)$

$= -9$

$D_y = \begin{vmatrix} 4 & 15 & 2 \\ -2 & -15 & -3 \\ 5 & 18 & 2 \end{vmatrix}$

$= 4\begin{vmatrix} -15 & -3 \\ 18 & 2 \end{vmatrix} - 15\begin{vmatrix} -2 & -3 \\ 5 & 2 \end{vmatrix} + 2\begin{vmatrix} -2 & -15 \\ 5 & 18 \end{vmatrix}$

$= 4(-30+54) - 15(-4+15) + 2(-36+75)$

$= 4(24) - 15(11) + 2(39)$

$= -9$

$D_z = \begin{vmatrix} 4 & -3 & 15 \\ -2 & 1 & -15 \\ 5 & -5 & 18 \end{vmatrix}$

$= 4\begin{vmatrix} 1 & -15 \\ -5 & 18 \end{vmatrix} - (-3)\begin{vmatrix} -2 & -15 \\ 5 & 18 \end{vmatrix} + 15\begin{vmatrix} -2 & 1 \\ 5 & -5 \end{vmatrix}$

$= 4(18-75) + 3(-36+75) + 15(10-5)$

$= 4(-57) + 3(39) + 15(5)$

$= -36$

$x = \dfrac{D_x}{D} = \dfrac{-9}{-9} = 1$, $y = \dfrac{D_y}{D} = \dfrac{9}{-9} = -1$,

$z = \dfrac{D_z}{D} = \dfrac{-36}{-9} = 4$

The solution of the system is $x = 1$, $y = -1$, $z = 4$.

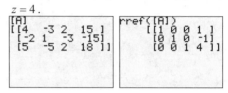

21. $\dfrac{3x+7}{(x+3)^2}$

The denominator contains the repeated linear factor $x + 3$. Thus, the partial fraction decomposition takes on the form

$\dfrac{3x+7}{(x+3)^2} = \dfrac{A}{x+3} + \dfrac{B}{(x+3)^2}$

Clear the fractions by multiplying both sides by $(x+3)^2$. The result is the identity

$3x + 7 = A(x+3) + B$

or

$3x + 7 = Ax + (3A + B)$

We equate coefficients of like powers of x to obtain the system

$\begin{cases} 3 = A \\ 7 = 3A + B \end{cases}$

Therefore, we have $A = 3$. Substituting this result into the second equation gives

$7 = 3A + B$

$7 = 3(3) + B$

$-2 = B$

Thus, the partial fraction decomposition is

$\dfrac{3x+7}{(x+3)^2} = \dfrac{3}{x+3} + \dfrac{-2}{(x+3)^2}$.

22. $\dfrac{4x^2-3}{x(x^2+3)^2}$

The denominator contains the linear factor x and the repeated irreducible quadratic factor $x^2 + 3$. The partial fraction decomposition takes on the form

$\dfrac{4x^2-3}{x(x^2+3)^2} = \dfrac{A}{x} + \dfrac{Bx+C}{x^2+3} + \dfrac{Dx+E}{(x^2+3)^2}$

We clear the fractions by multiplying both sides by $x(x^2+3)^2$ to obtain the identity
$$4x^2 - 3 = A(x^2+3)^2 + x(x^2+3)(Bx+C) + x(Dx+E)$$
Collecting like terms yields
$$4x^2 - 3 = (A+B)x^4 + Cx^3 + (6A+3B+D)x^2 + (3C+E)x + (9A)$$
Equating coefficients, we obtain the system
$$\begin{cases} A+B=0 \\ C=0 \\ 6A+3B+D=4 \\ 3C+E=0 \\ 9A=-3 \end{cases}$$
From the last equation we get $A = -\frac{1}{3}$.
Substituting this value into the first equation gives $B = \frac{1}{3}$. From the second equation, we know $C = 0$. Substituting this value into the fourth equation yields $E = 0$.
Substituting $A = -\frac{1}{3}$ and $B = \frac{1}{3}$ into the third equation gives us
$$6\left(-\tfrac{1}{3}\right) + 3\left(\tfrac{1}{3}\right) + D = 4$$
$$-2 + 1 + D = 4$$
$$D = 5$$
Therefore, the partial fraction decomposition is
$$\frac{4x^2-3}{x(x^2+3)^2} = \frac{-\frac{1}{3}}{x} + \frac{\frac{1}{3}x}{(x^2+3)} + \frac{5x}{(x^2+3)^2}$$

23. $\begin{cases} x \geq 0 \\ x+2y \geq 8 \\ 2x-3y \geq 2 \end{cases}$

The inequalities $x \geq 0$ and $y \geq 0$ require that the graph be in quadrant I.
$$x + 2y \geq 8$$
$$2y \geq -x + 8$$
$$y \geq -\frac{1}{2}x + 4$$
Test the point $(0,0)$.

$$x + 2y \geq 8$$
$$0 + 2(0) \geq 8 \ ?$$
$$0 \geq 8 \text{ false}$$
The point $(0,0)$ is not a solution. Thus, the graph of the inequality $x+2y \geq 8$ includes the half-plane above the line $y = -\frac{1}{2}x + 4$. Because the inequality is non-strict, the line is also part of the graph of the solution.
$$2x - 3y \geq 2$$
$$-3y \geq -2x + 2$$
$$y \leq \frac{2}{3}x - \frac{2}{3}$$
Test the point $(0,0)$.
$$2x - 3y \geq 2$$
$$2(0) - 3(0) \geq 2 \ ?$$
$$0 \geq 2 \text{ false}$$
The point $(0,0)$ is not a solution. Thus, the graph of the inequality $2x - 3y \geq 2$ includes the half-plane below the line $y = \frac{2}{3}x - \frac{2}{3}$.
Because the inequality is non-strict, the line is also part of the graph of the solution.
The overlapping shaded region (that is, the shaded region in the graph below) is the solution to the system of linear inequalities.

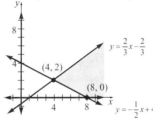

The graph is unbounded. The corner points are $(4,2)$ and $(8,0)$.

24. The objective function is $z = 5x + 8y$. We seek the largest value of z that can occur if x and y are solutions of the system of linear inequalities
$$\begin{cases} x \geq 0 \\ y \geq 0 \\ 2x+y \leq 8 \\ x-3y \leq -3 \end{cases}$$

$2x+y=8$ $x-3y=-3$
$y=-2x+8$ $-3y=-x-3$
              $y=\dfrac{1}{3}x+1$

The graph of this system (the feasible points) is shown as the shaded region in the figure below. The corner points of the feasible region are $(0,1)$, $(3,2)$, and $(0,8)$.

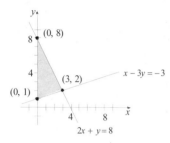

Corner point, (x,y)	Value of obj. function, z
$(0,1)$	$z=5(0)+8(1)=8$
$(3,2)$	$z=5(3)+8(2)=31$
$(0,8)$	$z=5(0)+8(8)=64$

From the table, we can see that the maximum value of z is 64, and it occurs at the point $(0,8)$.

25. Let j = unit price for flare jeans, c = unit price for camisoles, and t = unit price for t-shirts. The given information yields a system of equations with each of the three women yielding an equation.

$$\begin{cases} 2j+2c+4t=90 & \text{(Megan)} \\ j+3t=42.5 & \text{(Paige)} \\ j+3c+2t=62 & \text{(Kara)} \end{cases}$$

We can solve this system by using matrices.

$\begin{bmatrix} 2 & 2 & 4 & | & 90 \\ 1 & 0 & 3 & | & 42.5 \\ 1 & 3 & 2 & | & 62 \end{bmatrix}$

$=\begin{bmatrix} 1 & 1 & 2 & | & 45 \\ 1 & 0 & 3 & | & 42.5 \\ 1 & 3 & 2 & | & 62 \end{bmatrix}$ $\left(R_1=\tfrac{1}{2}r_1\right)$

$=\begin{bmatrix} 1 & 1 & 2 & | & 45 \\ 0 & -1 & 1 & | & -2.5 \\ 0 & 2 & 0 & | & 17 \end{bmatrix}$ $\left(\begin{array}{l}R_2=-r_1+r_2 \\ R_3=-r_1+r_3\end{array}\right)$

$=\begin{bmatrix} 1 & 1 & 2 & | & 45 \\ 0 & 1 & -1 & | & 2.5 \\ 0 & 2 & 0 & | & 17 \end{bmatrix}$ $\left(R_2=-r_2\right)$

$=\begin{bmatrix} 1 & 0 & 3 & | & 42.5 \\ 0 & 1 & -1 & | & 2.5 \\ 0 & 0 & 2 & | & 12 \end{bmatrix}$ $\left(\begin{array}{l}R_1=-r_2+r_1 \\ R_3=-2r_2+r_3\end{array}\right)$

$=\begin{bmatrix} 1 & 0 & 3 & | & 42.5 \\ 0 & 1 & -1 & | & 2.5 \\ 0 & 0 & 1 & | & 6 \end{bmatrix}$ $\left(R_3=\tfrac{1}{2}r_3\right)$

The last row represents the equation $z=6$. Substituting this result into $y-z=2.5$ (from the second row) gives
$y-z=2.5$
$y-6=2.5$
$y=8.5$

Substituting $z=6$ into $x+3z=42.5$ (from the first row) gives
$x+3z=42.5$
$x+3(6)=42.5$
$x=24.5$

Thus, flare jeans cost $24.50, camisoles cost $8.50, and t-shirts cost $6.00.

Cumulative Review R-5

1. $2x^2-x=0$
$x(2x-1)=0$
$x=0$ or $2x-1=0$
$\phantom{x=0\text{ or }}2x=1$
$\phantom{x=0\text{ or }}x=\dfrac{1}{2}$

The solution set is $\left\{0,\dfrac{1}{2}\right\}$.

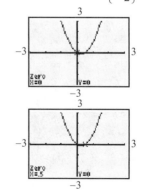

3. $2x^3 - 3x^2 - 8x - 3 = 0$

The graph of $Y_1 = 2x^3 - 3x^2 - 8x - 3$ appears to have an x-intercept at $x = 3$.

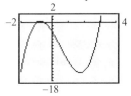

Using synthetic division:

$$\begin{array}{r|rrrr} 3) & 2 & -3 & -8 & -3 \\ & & 6 & 9 & 3 \\ \hline & 2 & 3 & 1 & 0 \end{array}$$

Therefore, $2x^3 - 3x^2 - 8x - 3 = 0$
$(x-3)(2x^2 + 3x + 1) = 0$
$(x-3)(2x+1)(x+1) = 0$
$x = 3$ or $x = -\dfrac{1}{2}$ or $x = -1$

The solution set is $\left\{-1, -\dfrac{1}{2}, 3\right\}$.

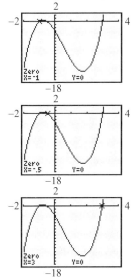

5. $\log_3(x-1) + \log_3(2x+1) = 2$
$\log_3\big((x-1)(2x+1)\big) = 2$
$(x-1)(2x+1) = 3^2$
$2x^2 - x - 1 = 9$
$2x^2 - x - 10 = 0$
$(2x-5)(x+2) = 0$
$x = \dfrac{5}{2}$ or $x = -2$

Since $x = -2$ makes the original logarithms undefined, the solution set is $\left\{\dfrac{5}{2}\right\}$.

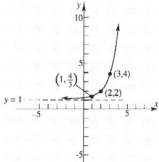

7. $g(x) = \dfrac{2x^3}{x^4 + 1}$

$g(-x) = \dfrac{2(-x)^3}{(-x)^4 + 1} = \dfrac{-2x^3}{x^4 + 1} = -g(x)$

Thus, g is an odd function and its graph is symmetric with respect to the origin.

9. $f(x) = 3^{x-2} + 1$

Using the graph of $y = 3^x$, shift the graph horizontally 2 units to the right, then shift the graph vertically upward 1 unit.

Domain: $(-\infty, \infty)$ Range: $(1, \infty)$
Horizontal Asymptote: $y = 1$

11. a. $y = 3x + 6$

The graph is a line.

x-intercept:
$0 = 3x + 6$
$3x = -6$
$x = -2$

y-intercept:
$y = 3(0) + 6$
$= 6$

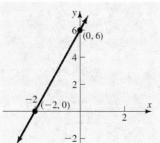

b. $x^2 + y^2 = 4$

The graph is a circle with center $(0, 0)$ and radius 2.

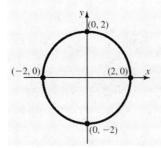

c. $y = x^3$

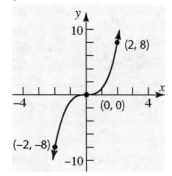

d. $y = \dfrac{1}{x}$

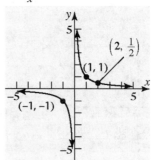

e. $y = \sqrt{x}$

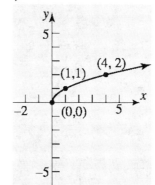

f. $y = e^x$

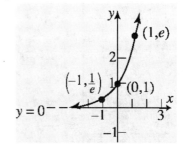

g. $y = \ln x$

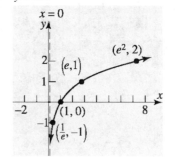

Chapter 6
Analytic Geometry

Section 6.1

Not applicable

Section 6.2

1. $\sqrt{(x_2-x_1)^2+(y_2-y_1)^2}$

3. $(x+4)^2 = 9$
   $x+4 = \pm 3$
   $x+4 = 3 \Rightarrow x = -1$
   or $x+4 = -3 \Rightarrow x = -7$
   The solution set is $\{-7,-1\}$.

5. 3, up

7. paraboloid of revolution

9. True

11. B; the graph has a vertex $(h,k) = (0,0)$ and opens up. Therefore, the equation of the graph has the form $x^2 = 4ay$. The graph passes through the point $(2,1)$ so we have
    $(2)^2 = 4a(1)$
    $4 = 4a$
    $1 = a$
    Thus, the equation of the graph is $x^2 = 4y$.

13. E; the graph has vertex $(h,k) = (1,1)$ and opens to the right. Therefore, the equation of the graph has the form $(y-1)^2 = 4a(x-1)$.

15. H; the graph has vertex $(h,k) = (-1,-1)$ and opens down. Therefore, the equation of the graph has the form $(x+1)^2 = -4a(y+1)$.

17. C; the graph has vertex $(h,k) = (0,0)$ and opens to the left. Therefore, the equation of the graph has the form $y^2 = -4ax$. The graph passes through the point $(-1,-2)$ so we have
    $(-2)^2 = -4a(-1)$
    $4 = 4a$
    $1 = a$
    Thus, the equation of the graph is $y^2 = -4x$.

19. The focus is (4, 0) and the vertex is (0, 0). Both lie on the horizontal line $y = 0$. $a = 4$ and since (4, 0) is to the right of (0, 0), the parabola opens to the right. The equation of the parabola is:
    $y^2 = 4ax$
    $y^2 = 4 \cdot 4 \cdot x$
    $y^2 = 16x$
    Letting $x = 4$, we find $y^2 = 64$ or $y = \pm 8$.
    The points (4, 8) and (4, –8) define the latus rectum.

    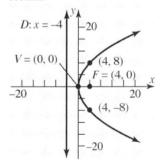

21. The focus is (0, –3) and the vertex is (0, 0). Both lie on the vertical line $x = 0$. $a = 3$ and since (0, –3) is below (0, 0), the parabola opens down. The equation of the parabola is:
    $x^2 = -4ay$
    $x^2 = -4 \cdot 3 \cdot y$
    $x^2 = -12y$
    Letting $y = -3$, we find $x^2 = 36$ or $x = \pm 6$.
    The points $(-6,-3)$ and $(6,-3)$ define the latus rectum.

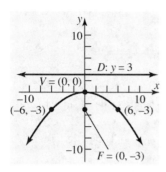

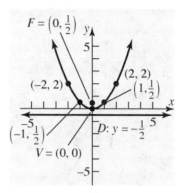

23. The focus is (–2, 0) and the directrix is $x = 2$. The vertex is (0, 0). $a = 2$ and since (–2, 0) is to the left of (0, 0), the parabola opens to the left. The equation of the parabola is:
$y^2 = -4ax$
$y^2 = -4 \cdot 2 \cdot x$
$y^2 = -8x$

Letting $x = -2$, we find $y^2 = 16$ or $y = \pm 4$. The points (–2, 4) and (–2, –4) define the latus rectum.

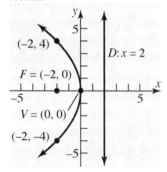

25. The directrix is $y = -\frac{1}{2}$ and the vertex is (0, 0). The focus is $\left(0, \frac{1}{2}\right)$. $a = \frac{1}{2}$ and since $\left(0, \frac{1}{2}\right)$ is above (0, 0), the parabola opens up. The equation of the parabola is:
$x^2 = 4ay$
$x^2 = 4 \cdot \frac{1}{2} \cdot y \Rightarrow x^2 = 2y$

Letting $y = \frac{1}{2}$, we find $x^2 = 1$ or $x = \pm 1$.

The points $\left(1, \frac{1}{2}\right)$ and $\left(-1, \frac{1}{2}\right)$ define the latus rectum.

27. Vertex: (0, 0). Since the axis of symmetry is vertical, the parabola opens up or down. Since (2, 3) is above (0, 0), the parabola opens up. The equation has the form $x^2 = 4ay$. Substitute the coordinates of (2, 3) into the equation to find a:
$2^2 = 4a \cdot 3 \Rightarrow 4 = 12a \Rightarrow a = \frac{1}{3}$

The equation of the parabola is: $x^2 = \frac{4}{3}y$. The focus is $\left(0, \frac{1}{3}\right)$. Letting $y = \frac{1}{3}$, we find $x^2 = \frac{4}{9}$ or $x = \pm\frac{2}{3}$. The points $\left(\frac{2}{3}, \frac{1}{3}\right)$ and $\left(-\frac{2}{3}, \frac{1}{3}\right)$ define the latus rectum.

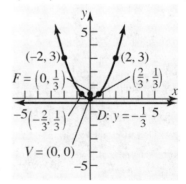

29. The vertex is (2, –3) and the focus is (2, –5). Both lie on the vertical line $x = 2$.
$a = |-5-(-3)| = 2$ and since (2, –5) is below (2, –3), the parabola opens down. The equation of the parabola is:
$$(x-h)^2 = -4a(y-k)$$
$$(x-2)^2 = -4(2)(y-(-3))$$
$$(x-2)^2 = -8(y+3)$$
Letting $y = -5$, we find
$$(x-2)^2 = 16$$
$x - 2 = \pm 4 \Rightarrow x = -2$ or $x = 6$
The points (–2, –5) and (6, –5) define the latus rectum.

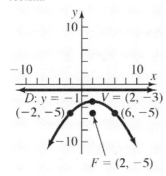

31. The vertex is (–1, –2) and the focus is (0, –2). Both lie on the horizontal line $y = -2$.
$a = |-1-0| = 1$ and since (0, –2) is to the right of (–1, –2), the parabola opens to the right. The equation of the parabola is:
$$(y-k)^2 = 4a(x-h)$$
$$(y-(-2))^2 = 4(1)(x-(-1))$$
$$(y+2)^2 = 4(x+1)$$
Letting $x = 0$, we find
$$(y+2)^2 = 4$$
$y + 2 = \pm 2 \Rightarrow y = -4$ or $y = 0$
The points (0, –4) and (0, 0) define the latus rectum.

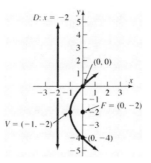

33. The directrix is $y = 2$ and the focus is (–3, 4). This is a vertical case, so the vertex is (–3, 3). $a = 1$ and since (–3, 4) is above $y = 2$, the parabola opens up. The equation of the parabola is: $(x-h)^2 = 4a(y-k)$
$$(x-(-3))^2 = 4 \cdot 1 \cdot (y-3)$$
$$(x+3)^2 = 4(y-3)$$
Letting $y = 4$, we find $(x+3)^2 = 4$ or $x + 3 = \pm 2$. So, $x = -1$ or $x = -5$. The points (–1, 4) and (–5, 4) define the latus rectum.

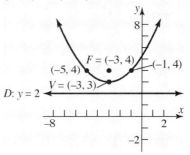

35. The directrix is $x = 1$ and the focus is (–3, –2). This is a horizontal case, so the vertex is (–1, –2). $a = 2$ and since (–3, –2) is to the left of $x = 1$, the parabola opens to the left. The equation of the parabola is:
$$(y-k)^2 = -4a(x-h)$$
$$(y-(-2))^2 = -4 \cdot 2 \cdot (x-(-1))$$
$$(y+2)^2 = -8(x+1)$$
Letting $x = -3$, we find $(y+2)^2 = 16$ or $y + 2 = \pm 4$. So, $y = 2$ or $y = -6$. The points (–3, 2) and (–3, –6) define the latus rectum.

37. a. The equation $x^2 = 4y$ is in the form $x^2 = 4ay$ where $4a = 4$ or $a = 1$.
Thus, we have:
Vertex: $(0, 0)$
Focus: $(0, 1)$
Directrix: $y = -1$

b.
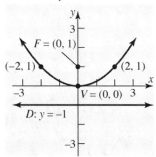

39. a. The equation $y^2 = -16x$ is in the form $y^2 = -4ax$ where $-4a = -16$ or $a = 4$. Thus, we have:
Vertex: $(0, 0)$
Focus: $(-4, 0)$
Directrix: $x = 4$

b.
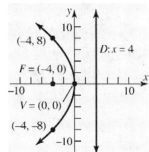

41. a. The equation $(y-2)^2 = 8(x+1)$ is in the form $(y-k)^2 = 4a(x-h)$ where $4a = 8$ or $a = 2$, $h = -1$, and $k = 2$. Thus, we have:
Vertex: $(-1, 2)$
Focus: $(1, 2)$
Directrix: $x = -3$

b.
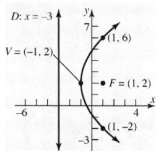

43. a. The equation $(x-3)^2 = -(y+1)$ is in the form $(x-h)^2 = -4a(y-k)$ where $-4a = -1$ or $a = \frac{1}{4}$, $h = 3$, and $k = -1$.
Thus, we have:
Vertex: $(3, -1)$; Focus: $\left(3, -\frac{5}{4}\right)$;
Directrix: $y = -\frac{3}{4}$

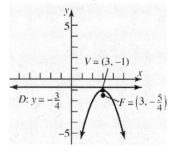

b.

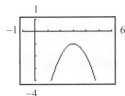

45. a. The equation $(y+3)^2 = 8(x-2)$ is in the form $(y-k)^2 = 4a(x-h)$ where $4a = 8$ or $a = 2$, $h = 2$, and $k = -3$. Thus, we have:
Vertex: $(2, -3)$
Focus: $(4, -3)$
Directrix: $x = 0$

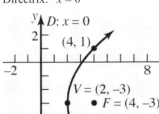

b.

47. a. Complete the square to put in standard form:
$y^2 - 4y + 4x + 4 = 0$
$y^2 - 4y + 4 = -4x$
$(y-2)^2 = -4x$

The equation is in the form $(y-k)^2 = -4a(x-h)$ where $-4a = -4$ or $a = 1$, $h = 0$, and $k = 2$. Thus, we have:
Vertex: $(0, 2)$
Focus: $(-1, 2)$
Directrix: $x = 1$

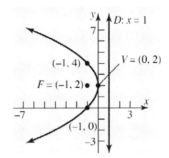

b.

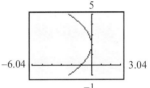

49. a. Complete the square to put in standard form:
$x^2 + 8x = 4y - 8$
$x^2 + 8x + 16 = 4y - 8 + 16$
$(x+4)^2 = 4(y+2)$

The equation is in the form $(x-h)^2 = 4a(y-k)$ where $4a = 4$ or $a = 1$, $h = -4$, and $k = -2$. Thus, we have:
Vertex: $(-4, -2)$
Focus: $(-4, -1)$
Directrix: $y = -3$

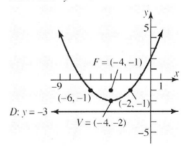

b.

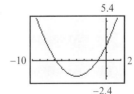

51. a. Complete the square to put in standard form:
$$y^2 + 2y - x = 0$$
$$y^2 + 2y + 1 = x + 1$$
$$(y+1)^2 = x+1$$
The equation is in the form
$(y-k)^2 = 4a(x-h)$ where $4a = 1$ or $a = \frac{1}{4}$,
$h = -1$, and $k = -1$. Thus, we have:
Vertex: $(-1, -1)$
Focus: $\left(-\frac{3}{4}, -1\right)$
Directrix: $x = -\frac{5}{4}$

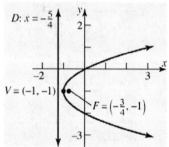

b.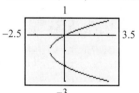

53. a. Complete the square to put in standard form:
$$x^2 - 4x = y + 4$$
$$x^2 - 4x + 4 = y + 4 + 4$$
$$(x-2)^2 = y + 8$$
The equation is in the form
$(x-h)^2 = 4a(y-k)$ where $4a = 1$ or $a = \frac{1}{4}$,
$h = 2$, and $k = -8$. Thus, we have:
Vertex: $(2, -8)$
Focus: $\left(2, -\frac{31}{4}\right)$
Directrix: $y = -\frac{33}{4}$

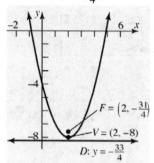

b.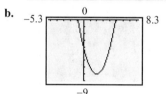

55. $(y-1)^2 = c(x-0)$
$(y-1)^2 = cx$
$(2-1)^2 = c(1) \Rightarrow 1 = c$
$(y-1)^2 = x$

57. $(y-1)^2 = c(x-2)$
$(0-1)^2 = c(1-2)$
$1 = -c \Rightarrow c = -1$
$(y-1)^2 = -(x-2)$

59. $(x-0)^2 = c(y-1)$
$x^2 = c(y-1)$
$2^2 = c(2-1)$
$4 = c$
$x^2 = 4(y-1)$

61. $(y-0)^2 = c(x-(-2))$
$y^2 = c(x+2)$
$1^2 = c(0+2) \Rightarrow 1 = 2c \Rightarrow c = \frac{1}{2}$
$y^2 = \frac{1}{2}(x+2)$

63. Set up the problem so that the vertex of the parabola is at (0, 0) and it opens up. Then the equation of the parabola has the form: $x^2 = 4ay$. Since the parabola is 10 feet across and 4 feet deep, the points (5, 4) and (–5, 4) are on the parabola. Substitute and solve for a:
$$5^2 = 4a(4) \Rightarrow 25 = 16a \Rightarrow a = \frac{25}{16}$$
a is the distance from the vertex to the focus. Thus, the receiver (located at the focus) is $\frac{25}{16} = 1.5625$ feet, or 18.75 inches from the base of the dish, along the axis of the parabola.

65. Set up the problem so that the vertex of the parabola is at (0, 0) and it opens up. Then the equation of the parabola has the form: $x^2 = 4ay$. Since the parabola is 4 inches across and 1 inch deep, the points (2, 1) and (–2, 1) are on the parabola. Substitute and solve for a:
$$2^2 = 4a(1) \Rightarrow 4 = 4a \Rightarrow a = 1$$
a is the distance from the vertex to the focus. Thus, the bulb (located at the focus) should be 1 inch from the vertex.

67. Set up the problem so that the vertex of the parabola is at (0, 0) and it opens up. Then the equation of the parabola has the form: $x^2 = cy$. The point (300, 80) is a point on the parabola. Solve for c and find the equation:
$$300^2 = c(80) \Rightarrow c = 1125$$
$$x^2 = 1125y$$

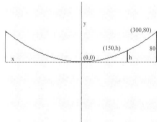

Since the height of the cable 150 feet from the center is to be found, the point (150, h) is a point on the parabola. Solve for h:
$$150^2 = 1125h$$
$$22{,}500 = 1125h$$
$$20 = h$$
The height of the cable 150 feet from the center is 20 feet.

69. Set up the problem so that the vertex of the parabola is at (0, 0) and it opens up. Then the equation of the parabola has the form: $x^2 = 4ay$. a is the distance from the vertex to the focus (where the source is located), so $a = 2$. Since the opening is 5 feet across, there is a point (2.5, y) on the parabola.
Solve for y: $\quad x^2 = 8y$
$$2.5^2 = 8y$$
$$6.25 = 8y$$
$$y = 0.78125 \text{ feet}$$
The depth of the searchlight should be 0.78125 feet.

71. Set up the problem so that the vertex of the parabola is at (0, 0) and it opens up. Then the equation of the parabola has the form: $x^2 = 4ay$. Since the parabola is 20 feet across and 6 feet deep, the points (10, 6) and (–10, 6) are on the parabola. Substitute and solve for a:
$$10^2 = 4a(6) \Rightarrow 100 = 24a \Rightarrow a \approx 4.17 \text{ feet}$$
The heat will be concentrated about 4.17 feet from the base, along the axis of symmetry.

73. Set up the problem so that the vertex of the parabola is at (0, 0) and it opens down. Then the equation of the parabola has the form: $x^2 = cy$. The point (60, –25) is a point on the parabola. Solve for c and find the equation:
$$60^2 = c(-25) \Rightarrow c = -144$$
$$x^2 = -144y$$

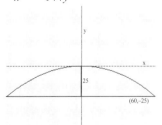

To find the height of the bridge 10 feet from the center the point (10, y) is a point on the parabola. Solve for y:
$$10^2 = -144y$$
$$100 = -144y$$
$$-0.69 \approx y$$
The height of the bridge 10 feet from the center is about $25 - 0.69 = 24.31$ feet. To find the height of the bridge 30 feet from the center the

point $(30, y)$ is a point on the parabola.
Solve for y:
$30^2 = -144y$
$900 = -144y$
$-6.25 = y$
The height of the bridge 30 feet from the center is $25 - 6.25 = 18.75$ feet. To find the height of the bridge, 50 feet from the center, the point $(50, y)$ is a point on the parabola. Solve for y:
$50^2 = -144y \Rightarrow 2500 = -144y \Rightarrow y = -17.36$
The height of the bridge 50 feet from the center is about $25 - 17.36 = 7.64$ feet.

75. $Ax^2 + Ey = 0 \quad A \neq 0, \ E \neq 0$

$Ax^2 = -Ey \Rightarrow x^2 = -\dfrac{E}{A}y$

This is the equation of a parabola with vertex at $(0, 0)$ and axis of symmetry being the y-axis.

The focus is $\left(0, -\dfrac{E}{4A}\right)$. The directrix is $y = \dfrac{E}{4A}$.

The parabola opens up if $-\dfrac{E}{A} > 0$ and down if $-\dfrac{E}{A} < 0$.

77. $Ax^2 + Dx + Ey + F = 0 \quad A \neq 0$

   a. If $E \neq 0$, then:
   $Ax^2 + Dx = -Ey - F$
   $A\left(x^2 + \dfrac{D}{A}x + \dfrac{D^2}{4A^2}\right) = -Ey - F + \dfrac{D^2}{4A}$
   $\left(x + \dfrac{D}{2A}\right)^2 = \dfrac{1}{A}\left(-Ey - F + \dfrac{D^2}{4A}\right)$
   $\left(x + \dfrac{D}{2A}\right)^2 = \dfrac{-E}{A}\left(y + \dfrac{F}{E} - \dfrac{D^2}{4AE}\right)$
   $\left(x + \dfrac{D}{2A}\right)^2 = \dfrac{-E}{A}\left(y - \dfrac{D^2 - 4AF}{4AE}\right)$

   This is the equation of a parabola whose vertex is $\left(-\dfrac{D}{2A}, \dfrac{D^2 - 4AF}{4AE}\right)$ and whose axis of symmetry is parallel to the y-axis.

   b. If $E = 0$, then
   $Ax^2 + Dx + F = 0 \Rightarrow x = \dfrac{-D \pm \sqrt{D^2 - 4AF}}{2A}$
   If $D^2 - 4AF = 0$, then $x = -\dfrac{D}{2A}$ is a single vertical line.

   c. If $E = 0$, then
   $Ax^2 + Dx + F = 0 \Rightarrow x = \dfrac{-D \pm \sqrt{D^2 - 4AF}}{2A}$
   If $D^2 - 4AF > 0$, then
   $x = \dfrac{-D + \sqrt{D^2 - 4AF}}{2A}$ and
   $x = \dfrac{-D - \sqrt{D^2 - 4AF}}{2A}$ are two vertical lines.

   d. If $E = 0$, then
   $Ax^2 + Dx + F = 0 \Rightarrow x = \dfrac{-D \pm \sqrt{D^2 - 4AF}}{2A}$
   If $D^2 - 4AF < 0$, there is no real solution. The graph contains no points.

Section 6.3

1. $d = \sqrt{(4-2)^2 + (-2-(-5))^2} = \sqrt{2^2 + 3^2} = \sqrt{13}$

3. x-intercepts: $0^2 = 16 - 4x^2$
   $4x^2 = 16$
   $x^2 = 4$
   $x = \pm 2 \to (-2, 0), (2, 0)$

   y-intercepts: $y^2 = 16 - 4(0)^2$
   $y^2 = 16$
   $y = \pm 4 \to (0, -4), (0, 4)$

   The intercepts are $(-2, 0)$, $(2, 0)$, $(0, -4)$, and $(0, 4)$.

5. left 1; down 4

7. ellipse

9. $(0, -5)$ and $(0, 5)$

11. True

13. C; the major axis is along the x-axis and the vertices are at $(-4, 0)$ and $(4, 0)$.

15. B; the major axis is along the y-axis and the vertices are at $(0, -2)$ and $(0, 2)$.

17. $\dfrac{x^2}{25}+\dfrac{y^2}{4}=1$

The center of the ellipse is at the origin.
$a=5$, $b=2$. The vertices are (5, 0) and (–5, 0).
Find the value of c:
$c^2=a^2-b^2=25-4=21 \to c=\sqrt{21}$
The foci are $\left(\sqrt{21},0\right)$ and $\left(-\sqrt{21},0\right)$.

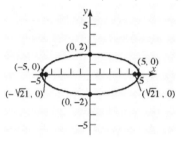

19. $\dfrac{x^2}{9}+\dfrac{y^2}{25}=1$

The center of the ellipse is at the origin.
$a=5$, $b=3$. The vertices are (0, 5) and (0, –5).
Find the value of c:
$c^2=a^2-b^2=25-9=16$
$c=4$
The foci are (0, 4) and (0, –4).

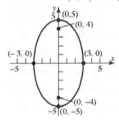

21. $4x^2+y^2=16$

Divide by 16 to put in standard form:
$\dfrac{4x^2}{16}+\dfrac{y^2}{16}=\dfrac{16}{16} \to \dfrac{x^2}{4}+\dfrac{y^2}{16}=1$
The center of the ellipse is at the origin.
$a=4$, $b=2$.
The vertices are (0, 4) and (0, –4). Find the value of c:
$c^2=a^2-b^2=16-4=12$
$c=\sqrt{12}=2\sqrt{3}$

The foci are $\left(0, 2\sqrt{3}\right)$ and $\left(0, -2\sqrt{3}\right)$.

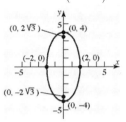

23. $4y^2+x^2=8$

Divide by 8 to put in standard form:
$\dfrac{4y^2}{8}+\dfrac{x^2}{8}=\dfrac{8}{8} \to \dfrac{x^2}{8}+\dfrac{y^2}{2}=1$
The center of the ellipse is at the origin.
$a=\sqrt{8}=2\sqrt{2}$, $b=\sqrt{2}$.
The vertices are $\left(2\sqrt{2},0\right)$ and $\left(-2\sqrt{2},0\right)$. Find the value of c:
$c^2=a^2-b^2=8-2=6$
$c=\sqrt{6}$
The foci are $\left(\sqrt{6},0\right)$ and $\left(-\sqrt{6},0\right)$.

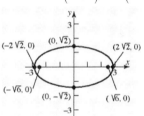

25. $x^2+y^2=16$

This is the equation of a circle whose center is at (0, 0) and radius = 4.

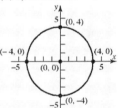

27. Center: (0, 0); Focus: (3, 0); Vertex: (5, 0);
Major axis is the x-axis; $a = 5$; $c = 3$. Find b:
$b^2 = a^2 - c^2 = 25 - 9 = 16$
$b = 4$
Write the equation: $\dfrac{x^2}{25} + \dfrac{y^2}{16} = 1$

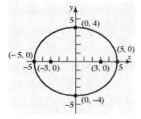

29. Center: (0, 0); Focus: (0, –4); Vertex: (0, 5);
Major axis is the y-axis; $a = 5$; $c = 4$. Find b:
$b^2 = a^2 - c^2 = 25 - 16 = 9$
$b = 3$
Write the equation: $\dfrac{x^2}{9} + \dfrac{y^2}{25} = 1$

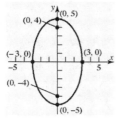

31. Foci: (±2, 0); Length of major axis is 6.
Center: (0, 0); Major axis is the x-axis;
$a = 3$; $c = 2$. Find b:
$b^2 = a^2 - c^2 = 9 - 4 = 5$ → $b = \sqrt{5}$
Write the equation: $\dfrac{x^2}{9} + \dfrac{y^2}{5} = 1$

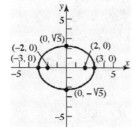

33. Focus: $(-4, 0)$; Vertices: $(-5, 0)$ and $(5, 0)$;
Center: $(0, 0)$; Major axis is the x-axis.
$a = 5$; $c = 4$. Find b:
$b^2 = a^2 - c^2 = 25 - 16 = 9$ → $b = 3$
Write the equation: $\dfrac{x^2}{25} + \dfrac{y^2}{9} = 1$

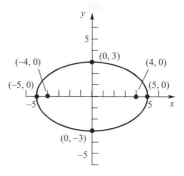

35. Foci: (0, ±3); x-intercepts are ±2.
Center: (0, 0); Major axis is the y-axis;
$c = 3$; $b = 2$. Find a:
$a^2 = b^2 + c^2 = 4 + 9 = 13$ → $a = \sqrt{13}$
Write the equation: $\dfrac{x^2}{4} + \dfrac{y^2}{13} = 1$

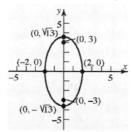

37. Center: (0, 0); Vertex: (0, 4); $b = 1$; Major
axis is the y-axis; $a = 4$; $b = 1$.
Write the equation: $\dfrac{x^2}{1} + \dfrac{y^2}{16} = 1$

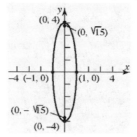

39. Center: $(-1, 1)$
Major axis: parallel to x-axis
Length of major axis: $4 = 2a \to a = 2$
Length of minor axis: $2 = 2b \to b = 1$
$$\frac{(x+1)^2}{4} + \frac{(y-1)^2}{1} = 1$$

41. Center: $(1, 0)$
Major axis: parallel to y-axis
Length of major axis: $4 = 2a \to a = 2$
Length of minor axis: $2 = 2b \to b = 1$
$$\frac{(x-1)^2}{1} + \frac{y^2}{4} = 1$$

43. a. The equation $\frac{(x-3)^2}{4} + \frac{(y+1)^2}{9} = 1$ is in the form $\frac{(x-h)^2}{b^2} + \frac{(y-k)^2}{a^2} = 1$ (major axis parallel to the y-axis) where $a = 3$, $b = 2$, $h = 3$, and $k = -1$. Solving for c:
$c^2 = a^2 - b^2 = 9 - 4 = 5 \to c = \sqrt{5}$
Thus, we have:
Center: $(3, -1)$
Foci: $(3, -1 + \sqrt{5})$, $(3, -1 - \sqrt{5})$
Vertices: $(3, 2)$, $(3, -4)$

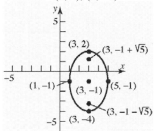

b. To graph, enter:
$y_1 = -1 + 3\sqrt{1 - (x-3)^2 / 4}$;
$y_2 = -1 - 3\sqrt{1 - (x-3)^2 / 4}$

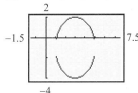

45. a. Divide by 16 to put the equation in standard form:
$(x+5)^2 + 4(y-4)^2 = 16$
$$\frac{(x+5)^2}{16} + \frac{4(y-4)^2}{16} = \frac{16}{16}$$
$$\frac{(x+5)^2}{16} + \frac{(y-4)^2}{4} = 1$$
The equation is in the form
$\frac{(x-h)^2}{a^2} + \frac{(y-k)^2}{b^2} = 1$ (major axis parallel to the x-axis) where $a = 4$, $b = 2$, $h = -5$, and $k = 4$.
Solving for c:
$c^2 = a^2 - b^2 = 16 - 4 = 12 \to c = \sqrt{12} = 2\sqrt{3}$
Thus, we have:
Center: $(-5, 4)$
Foci: $(-5 - 2\sqrt{3}, 4)$, $(-5 + 2\sqrt{3}, 4)$
Vertices: $(-9, 4)$, $(-1, 4)$

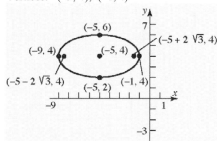

b. To graph, enter:
$y_1 = 4 + 2\sqrt{1 - (x+5)^2 / 16}$;
$y_2 = 4 - 2\sqrt{1 - (x+5)^2 / 16}$

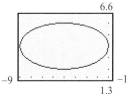

47. a. Complete the square to put the equation in standard form:
$$x^2 + 4x + 4y^2 - 8y + 4 = 0$$
$$(x^2 + 4x + 4) + 4(y^2 - 2y + 1) = -4 + 4 + 4$$
$$(x+2)^2 + 4(y-1)^2 = 4$$
$$\frac{(x+2)^2}{4} + \frac{4(y-1)^2}{4} = \frac{4}{4}$$
$$\frac{(x+2)^2}{4} + \frac{(y-1)^2}{1} = 1$$

The equation is in the form
$$\frac{(x-h)^2}{a^2} + \frac{(y-k)^2}{b^2} = 1 \text{ (major axis parallel}$$
to the x-axis) where
$a = 2$, $b = 1$, $h = -2$, and $k = 1$.
Solving for c:
$c^2 = a^2 - b^2 = 4 - 1 = 3 \rightarrow c = \sqrt{3}$
Thus, we have:
Center: $(-2, 1)$
Foci: $\left(-2 - \sqrt{3}, 1\right)$, $\left(-2 + \sqrt{3}, 1\right)$
Vertices: $(-4, 1)$, $(0, 1)$

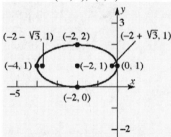

b. To graph, enter:
$y_1 = 1 + \sqrt{1 - (x+2)^2/4}$;
$y_2 = 1 - \sqrt{1 - (x+2)^2/4}$

49. a. Complete the square to put the equation in standard form:
$$2x^2 + 3y^2 - 8x + 6y + 5 = 0$$
$$2(x^2 - 4x) + 3(y^2 + 2y) = -5$$
$$2(x^2 - 4x + 4) + 3(y^2 + 2y + 1) = -5 + 8 + 3$$
$$2(x-2)^2 + 3(y+1)^2 = 6$$
$$\frac{2(x-2)^2}{6} + \frac{3(y+1)^2}{6} = \frac{6}{6}$$
$$\frac{(x-2)^2}{3} + \frac{(y+1)^2}{2} = 1$$

The equation is in the form
$$\frac{(x-h)^2}{a^2} + \frac{(y-k)^2}{b^2} = 1 \text{ (major axis parallel}$$
to the x-axis) where
$a = \sqrt{3}$, $b = \sqrt{2}$, $h = 2$, and $k = -1$.
Solving for c:
$c^2 = a^2 - b^2 = 3 - 2 = 1 \rightarrow c = 1$
Thus, we have:
Center: $(2, -1)$
Foci: $(1, -1)$, $(3, -1)$
Vertices: $\left(2 - \sqrt{3}, -1\right)$, $\left(2 + \sqrt{3}, -1\right)$

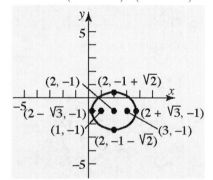

b. To graph, enter:
$y_1 = -1 + \sqrt{2 - 2(x-2)^2/3}$;
$y_2 = -1 - \sqrt{2 - 2(x-2)^2/3}$

51. a. Complete the square to put the equation in standard form:
$$9x^2 + 4y^2 - 18x + 16y - 11 = 0$$
$$9(x^2 - 2x) + 4(y^2 + 4y) = 11$$
$$9(x^2 - 2x + 1) + 4(y^2 + 4y + 4) = 11 + 9 + 16$$
$$9(x-1)^2 + 4(y+2)^2 = 36$$
$$\frac{9(x-1)^2}{36} + \frac{4(y+2)^2}{36} = \frac{36}{36}$$
$$\frac{(x-1)^2}{4} + \frac{(y+2)^2}{9} = 1$$

The equation is in the form
$$\frac{(x-h)^2}{b^2} + \frac{(y-k)^2}{a^2} = 1 \text{ (major axis parallel}$$
to the y-axis) where
$a = 3$, $b = 2$, $h = 1$, and $k = -2$.
Solving for c:
$c^2 = a^2 - b^2 = 9 - 4 = 5 \rightarrow c = \sqrt{5}$
Thus, we have:
Center: $(1, -2)$
Foci: $\left(1, -2 + \sqrt{5}\right)$, $\left(1, -2 - \sqrt{5}\right)$
Vertices: $(1, 1)$, $(1, -5)$

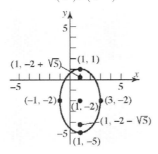

b. To graph, enter:
$$y_1 = -2 + 3\sqrt{1 - (x-1)^2/4};$$
$$y_2 = -2 - 3\sqrt{1 - (x-1)^2/4}$$

53. a. Complete the square to put the equation in standard form:
$$4x^2 + y^2 + 4y = 0$$
$$4x^2 + y^2 + 4y + 4 = 4$$
$$4x^2 + (y+2)^2 = 4$$
$$\frac{4x^2}{4} + \frac{(y+2)^2}{4} = \frac{4}{4}$$
$$\frac{x^2}{1} + \frac{(y+2)^2}{4} = 1$$

The equation is in the form
$$\frac{(x-h)^2}{b^2} + \frac{(y-k)^2}{a^2} = 1 \text{ (major axis parallel}$$
to the y-axis) where
$a = 2$, $b = 1$, $h = 0$, and $k = -2$.
Solving for c:
$c^2 = a^2 - b^2 = 4 - 1 = 3 \rightarrow c = \sqrt{3}$
Thus, we have:
Center: $(0, -2)$
Foci: $\left(0, -2 + \sqrt{3}\right)$, $\left(0, -2 - \sqrt{3}\right)$
Vertices: $(0, 0)$, $(0, -4)$

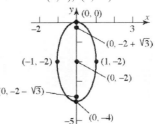

b. To graph, enter:
$$y_1 = -2 + 2\sqrt{1 - x^2};$$
$$y_2 = -2 - 2\sqrt{1 - x^2}$$

55. Center: $(2, -2)$; Vertex: $(7, -2)$; Focus: $(4, -2)$;
Major axis parallel to the x-axis; $a = 5$; $c = 2$.
Find b:
$b^2 = a^2 - c^2 = 25 - 4 = 21 \rightarrow b = \sqrt{21}$
Write the equation: $\dfrac{(x-2)^2}{25} + \dfrac{(y+2)^2}{21} = 1$

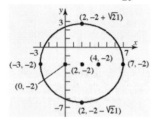

57. Vertices: $(4, 3), (4, 9)$; Focus: $(4, 8)$;
Center: $(4, 6)$; Major axis parallel to the y-axis;
$a = 3$; $c = 2$. Find b:
$b^2 = a^2 - c^2 = 9 - 4 = 5 \rightarrow b = \sqrt{5}$
Write the equation: $\dfrac{(x-4)^2}{5} + \dfrac{(y-6)^2}{9} = 1$

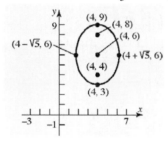

59. Foci: $(5, 1), (-1, 1)$;
Length of the major axis = 8; Center: $(2, 1)$;
Major axis parallel to the x-axis; $a = 4$; $c = 3$.
Find b:
$b^2 = a^2 - c^2 = 16 - 9 = 7 \rightarrow b = \sqrt{7}$
Write the equation: $\dfrac{(x-2)^2}{16} + \dfrac{(y-1)^2}{7} = 1$

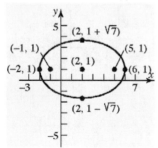

61. Center: $(1, 2)$; Focus: $(4, 2)$; contains the point $(1, 3)$; Major axis parallel to the x-axis; $c = 3$.
The equation has the form:
$\dfrac{(x-1)^2}{a^2} + \dfrac{(y-2)^2}{b^2} = 1$
Since the point $(1, 3)$ is on the curve:
$\dfrac{0}{a^2} + \dfrac{1}{b^2} = 1$
$\dfrac{1}{b^2} = 1 \rightarrow b^2 = 1 \rightarrow b = 1$
Find a:
$a^2 = b^2 + c^2 = 1 + 9 = 10 \rightarrow a = \sqrt{10}$
Write the equation: $\dfrac{(x-1)^2}{10} + \dfrac{(y-2)^2}{1} = 1$

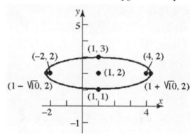

63. Center: $(1, 2)$; Vertex: $(4, 2)$; contains the point $(1, 3)$; Major axis parallel to the x-axis; $a = 3$.
The equation has the form:
$\dfrac{(x-1)^2}{a^2} + \dfrac{(y-2)^2}{b^2} = 1$
Since the point $(1, 3)$ is on the curve:
$\dfrac{0}{9} + \dfrac{1}{b^2} = 1$
$\dfrac{1}{b^2} = 1 \rightarrow b^2 = 1 \rightarrow b = 1$
Solve for c:
$c^2 = a^2 - b^2 = 9 - 1 = 8$
Thus, $c = \pm 2\sqrt{2}$.
Write the equation: $\dfrac{(x-1)^2}{9} + \dfrac{(y-2)^2}{1} = 1$

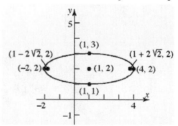

65. Rewrite the equation:
$$y = \sqrt{16-4x^2}$$
$$y^2 = 16-4x^2, \quad y \geq 0$$
$$4x^2 + y^2 = 16, \quad y \geq 0$$
$$\frac{x^2}{4} + \frac{y^2}{16} = 1, \quad y \geq 0$$

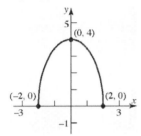

67. Rewrite the equation:
$$y = -\sqrt{64-16x^2}$$
$$y^2 = 64-16x^2, \quad y \leq 0$$
$$16x^2 + y^2 = 64, \quad y \leq 0$$
$$\frac{x^2}{4} + \frac{y^2}{64} = 1, \quad y \leq 0$$

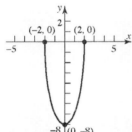

69. The center of the ellipse is (0, 0). The length of the major axis is 20, so $a = 10$. The length of half the minor axis is 6, so $b = 6$. The ellipse is situated with its major axis on the x-axis. The equation is: $\frac{x^2}{100} + \frac{y^2}{36} = 1$.

71. Assume that the half ellipse formed by the gallery is centered at (0, 0). Since the hall is 100 feet long, $2a = 100$ or $a = 50$. The distance from the center to the foci is 25 feet, so $c = 25$. Find the height of the gallery which is b:
$$b^2 = a^2 - c^2 = 2500 - 625 = 1875$$
$$b = \sqrt{1875} \approx 43.3$$
The ceiling will be 43.3 feet high in the center.

73. Place the semi-elliptical arch so that the x-axis coincides with the water and the y-axis passes through the center of the arch. Since the bridge has a span of 120 feet, the length of the major axis is 120, or $2a = 120$ or $a = 60$. The maximum height of the bridge is 25 feet, so $b = 25$. The equation is: $\frac{x^2}{3600} + \frac{y^2}{625} = 1$.

The height 10 feet from the center:
$$\frac{10^2}{3600} + \frac{y^2}{625} = 1$$
$$\frac{y^2}{625} = 1 - \frac{100}{3600}$$
$$y^2 = 625 \cdot \frac{3500}{3600}$$
$$y \approx 24.65 \text{ feet}$$

The height 30 feet from the center:
$$\frac{30^2}{3600} + \frac{y^2}{625} = 1$$
$$\frac{y^2}{625} = 1 - \frac{900}{3600}$$
$$y^2 = 625 \cdot \frac{2700}{3600}$$
$$y \approx 21.65 \text{ feet}$$

The height 50 feet from the center:
$$\frac{50^2}{3600} + \frac{y^2}{625} = 1$$
$$\frac{y^2}{625} = 1 - \frac{2500}{3600}$$
$$y^2 = 625 \cdot \frac{1100}{3600}$$
$$y \approx 13.82 \text{ feet}$$

75. Place the semi-elliptical arch so that the x-axis coincides with the major axis and the y-axis passes through the center of the arch. Since the ellipse is 40 feet wide, the length of the major axis is 40, or $2a = 40$ or $a = 20$. The height is 15 feet at the center, so $b = 15$. The equation is:
$$\frac{x^2}{400} + \frac{y^2}{225} = 1.$$

The height a distance of 10 feet on either side of the center:
$$\frac{10^2}{400} + \frac{y^2}{225} = 1$$
$$\frac{y^2}{225} = 1 - \frac{100}{400}$$
$$y^2 = 225 \cdot \frac{3}{4}$$
$$y \approx 12.99 \text{ feet}$$

The height a distance of 20 feet on either side of the center:
$$\frac{20^2}{400} + \frac{y^2}{225} = 1$$
$$\frac{y^2}{225} = 1 - \frac{400}{400}$$
$$y^2 = 225 \cdot 0$$
$$y \approx 0 \text{ feet}$$

Heights: 0 ft, 12.99 ft, 15 ft, 12.99 ft, 0 ft.

77. Since the mean distance is 93 million miles, $a = 93$ million. The length of the major axis is 186 million. The perihelion is 186 million – 94.5 million = 91.5 million miles.

The distance from the center of the ellipse to the sun (focus) is 93 million – 91.5 million = 1.5 million miles. Therefore, $c = 1.5$ million. Find b:
$$b^2 = a^2 - c^2 = (93 \times 10^6)^2 - (1.5 \times 10^6)^2$$
$$= 8.64675 \times 10^{15} = 8646.75 \times 10^{12}$$
$$b = 92.99 \times 10^6$$

The equation of the orbit is:
$$\frac{x^2}{(93 \times 10^6)^2} + \frac{y^2}{(92.99 \times 10^6)^2} = 1$$

We can simplify the equation by letting our units for x and y be millions of miles. The equation then becomes:
$$\frac{x^2}{8649} + \frac{y^2}{8646.75} = 1$$

79. The mean distance is 507 million – 23.2 million = 483.8 million miles.

The perihelion is 483.8 million – 23.2 million = 460.6 million miles.

Since $a = 483.8 \times 10^6$ and $c = 23.2 \times 10^6$, we can find b:
$$b^2 = a^2 - c^2 = (483.8 \times 10^6)^2 - (23.2 \times 10^6)^2$$
$$= 2.335242 \times 10^{17}$$
$$b = 483.2 \times 10^6$$

The equation of the orbit of Jupiter is:
$$\frac{x^2}{(483.8 \times 10^6)^2} + \frac{y^2}{(483.2 \times 10^6)^2} = 1$$

We can simplify the equation by letting our units for x and y be millions of miles. The equation then becomes:
$$\frac{x^2}{234{,}062.44} + \frac{y^2}{233{,}524.2} = 1$$

81. If the x-axis is placed along the 100 foot length and the y-axis is placed along the 50 foot length, the equation for the ellipse is: $\frac{x^2}{50^2} + \frac{y^2}{25^2} = 1$.

Find y when x = 40:
$$\frac{40^2}{50^2} + \frac{y^2}{25^2} = 1$$
$$\frac{y^2}{625} = 1 - \frac{1600}{2500}$$
$$y^2 = 625 \cdot \frac{9}{25}$$
$$y \approx 15 \text{ feet}$$

To get the width of the ellipse at $x = 40$, we need to double the y value. Thus, the width 10 feet from a vertex is 30 feet.

83. a. Put the equation in standard ellipse form:
$$Ax^2 + Cy^2 + F = 0$$
$$Ax^2 + Cy^2 = -F$$
$$\frac{Ax^2}{-F} + \frac{Cy^2}{-F} = 1$$
$$\frac{x^2}{(-F/A)} + \frac{y^2}{(-F/C)} = 1$$
where $A \neq 0, C \neq 0, F \neq 0$, and $-F/A$ and $-F/C$ are positive.

If $A \neq C$, then $-\frac{F}{A} \neq -\frac{F}{C}$. So, this is the equation of an ellipse with center at (0, 0).

b. If $A = C$, the equation becomes:
$$Ax^2 + Ay^2 = -F \rightarrow x^2 + y^2 = \frac{-F}{A}$$
This is the equation of a circle with center at (0, 0) and radius of $\sqrt{\frac{-F}{A}}$.

85. Answers will vary.

Section 6.4

1. $d = \sqrt{(-2-3)^2 + (1-(-4))^2}$
$= \sqrt{(-5)^2 + (5)^2} = \sqrt{25 + 25}$
$= \sqrt{50} = 5\sqrt{2}$

3. x-intercepts: $0^2 = 9 + 4x^2$
$4x^2 = -9$
$x^2 = -\frac{9}{4}$ (no real solution)

y-intercepts: $y^2 = 9 + 4(0)^2$
$y^2 = 9$
$y = \pm 3 \rightarrow (0, -3), (0, 3)$

The intercepts are $(0, -3)$ and $(0, 3)$.

5. right 5 units; down 4 units

7. hyperbola

9. $y = \frac{3}{2}x$, $y = -\frac{3}{2}x$

11. True; hyperbolas always have two asymptotes.

13. B; the hyperbola opens to the left and right, and has vertices at $(\pm 1, 0)$. Thus, the graph has an equation of the form $x^2 - \frac{y^2}{b^2} = 1$.

15. A; the hyperbola opens to the left and right, and has vertices at $(\pm 2, 0)$. Thus, the graph has an equation of the form $\frac{x^2}{4} - \frac{y^2}{b^2} = 1$.

17. Center: (0, 0); Focus: (3, 0); Vertex: (1, 0); Transverse axis is the x-axis; $a = 1$; $c = 3$. Find the value of b:
$b^2 = c^2 - a^2 = 9 - 1 = 8$
$b = \sqrt{8} = 2\sqrt{2}$

Write the equation: $\frac{x^2}{1} - \frac{y^2}{8} = 1$.

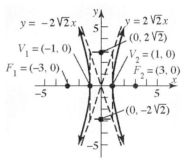

19. Center: (0, 0); Focus: (0, –6); Vertex: (0, 4) Transverse axis is the y-axis; $a = 4$; $c = 6$. Find the value of b:
$b^2 = c^2 - a^2 = 36 - 16 = 20$
$b = \sqrt{20} = 2\sqrt{5}$

Write the equation: $\frac{y^2}{16} - \frac{x^2}{20} = 1$.

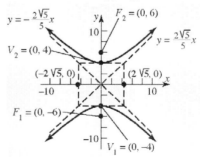

21. Foci: $(-5, 0), (5, 0)$; Vertex: $(3, 0)$
Center: $(0, 0)$; Transverse axis is the x-axis;
$a = 3$; $c = 5$.
Find the value of b:
$b^2 = c^2 - a^2 = 25 - 9 = 16 \Rightarrow b = 4$
Write the equation: $\dfrac{x^2}{9} - \dfrac{y^2}{16} = 1$.

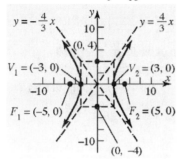

23. Vertices: $(0, -6), (0, 6)$; asymptote: $y = 2x$;
Center: $(0, 0)$; Transverse axis is the y-axis;
$a = 6$. Find the value of b using the slope of the asymptote: $\dfrac{a}{b} = \dfrac{6}{b} = 2 \Rightarrow 2b = 6 \Rightarrow b = 3$
Find the value of c:
$c^2 = a^2 + b^2 = 36 + 9 = 45$
$c = 3\sqrt{5}$
Write the equation: $\dfrac{y^2}{36} - \dfrac{x^2}{9} = 1$.

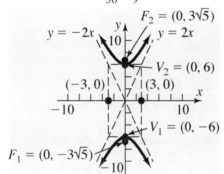

25. Foci: $(-4, 0), (4, 0)$; asymptote: $y = -x$;
Center: $(0, 0)$; Transverse axis is the x-axis;
$c = 4$. Using the slope of the asymptote:
$-\dfrac{b}{a} = -1 \Rightarrow -b = -a \Rightarrow b = a$.
Find the value of b:
$b^2 = c^2 - a^2 \Rightarrow a^2 + b^2 = c^2$ $(c = 4)$
$b^2 + b^2 = 16 \Rightarrow 2b^2 = 16 \Rightarrow b^2 = 8$
$b = \sqrt{8} = 2\sqrt{2}$
$a = \sqrt{8} = 2\sqrt{2}$ $(a = b)$
Write the equation: $\dfrac{x^2}{8} - \dfrac{y^2}{8} = 1$.

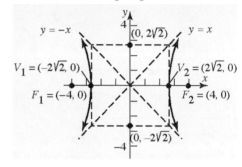

27. a. $\dfrac{x^2}{25} - \dfrac{y^2}{9} = 1$
The center of the hyperbola is at $(0, 0)$.
$a = 5$, $b = 3$. The vertices are $(5, 0)$ and $(-5, 0)$. Find the value of c:
$c^2 = a^2 + b^2 = 25 + 9 = 34 \Rightarrow c = \sqrt{34}$
The foci are $(\sqrt{34}, 0)$ and $(-\sqrt{34}, 0)$.
The transverse axis is the x-axis. The asymptotes are $y = \dfrac{3}{5}x$; $y = -\dfrac{3}{5}x$.

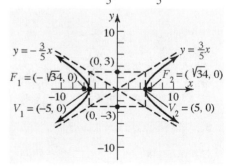

b. $\dfrac{x^2}{25} - \dfrac{y^2}{9} = 1$

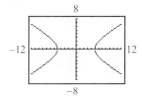

29. a. $4x^2 - y^2 = 16$

Divide both sides by 16 to put in standard form: $\dfrac{4x^2}{16} - \dfrac{y^2}{16} = \dfrac{16}{16} \Rightarrow \dfrac{x^2}{4} - \dfrac{y^2}{16} = 1$

. The center of the hyperbola is at (0, 0).
$a = 2,\ b = 4$.
The vertices are (2, 0) and (–2, 0).
Find the value of c:
$c^2 = a^2 + b^2 = 4 + 16 = 20$
$c = \sqrt{20} = 2\sqrt{5}$
The foci are $\left(2\sqrt{5}, 0\right)$ and $\left(-2\sqrt{5}, 0\right)$.
The transverse axis is the x-axis. The asymptotes are $y = 2x;\ y = -2x$.

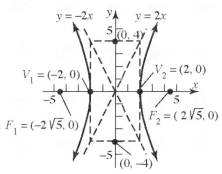

b. $4x^2 - y^2 = 16$

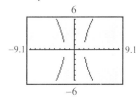

31. a. $y^2 - 9x^2 = 9$

Divide both sides by 9 to put in standard form: $\dfrac{y^2}{9} - \dfrac{9x^2}{9} = \dfrac{9}{9} \Rightarrow \dfrac{y^2}{9} - \dfrac{x^2}{1} = 1$

The center of the hyperbola is at (0, 0).
$a = 3,\ b = 1$.
The vertices are (0, 3) and (0, –3).
Find the value of c:
$c^2 = a^2 + b^2 = 9 + 1 = 10$
$c = \sqrt{10}$
The foci are $\left(0, \sqrt{10}\right)$ and $\left(0, -\sqrt{10}\right)$.
The transverse axis is the y-axis.
The asymptotes are $y = 3x;\ y = -3x$.

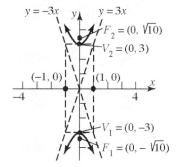

b. $y^2 - 9x^2 = 9$

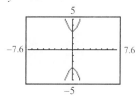

33. a. $y^2 - x^2 = 25$

Divide both sides by 25 to put in standard form: $\dfrac{y^2}{25} - \dfrac{x^2}{25} = 1$.

The center of the hyperbola is at (0, 0).
$a = 5,\ b = 5$. The vertices are $(0, 5)$ and $(0, -5)$. Find the value of c:
$c^2 = a^2 + b^2 = 25 + 25 = 50$
$c = \sqrt{50} = 5\sqrt{2}$
The foci are $\left(0, 5\sqrt{2}\right)$ and $\left(0, -5\sqrt{2}\right)$.
The transverse axis is the y-axis.

The asymptotes are $y = x$; $y = -x$.

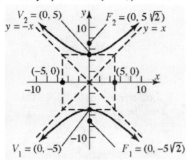

b. $y^2 - x^2 = 25$

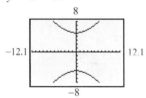

35. The center of the hyperbola is at $(0, 0)$.
$a = 1$, $b = 1$. The vertices are $(1, 0)$ and $(-1, 0)$.
Find the value of c:
$c^2 = a^2 + b^2 = 1 + 1 = 2$
$c = \sqrt{2}$
The foci are $(\sqrt{2}, 0)$ and $(-\sqrt{2}, 0)$.
The transverse axis is the x-axis.
The asymptotes are $y = x$; $y = -x$.

37. The center of the hyperbola is at $(0, 0)$.
$a = 6$, $b = 3$.
The vertices are $(0, -6)$ and $(0, 6)$. Find the value of c:
$c^2 = a^2 + b^2 = 36 + 9 = 45$
$c = \sqrt{45} = 3\sqrt{5}$
The foci are $(0, -3\sqrt{5})$ and $(0, 3\sqrt{5})$.
The transverse axis is the y-axis.
The asymptotes are $y = 2x$; $y = -2x$. The equation is: $\dfrac{y^2}{36} - \dfrac{x^2}{9} = 1$.

39. Center: $(4, -1)$; Focus: $(7, -1)$; Vertex: $(6, -1)$;
Transverse axis is parallel to the x-axis;
$a = 2$; $c = 3$.
Find the value of b:
$b^2 = c^2 - a^2 = 9 - 4 = 5 \Rightarrow b = \sqrt{5}$

Write the equation: $\dfrac{(x-4)^2}{4} - \dfrac{(y+1)^2}{5} = 1$.

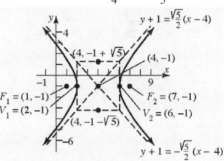

41. Center: $(-3, -4)$; Focus: $(-3, -8)$;
Vertex: $(-3, -2)$;
Transverse axis is parallel to the y-axis;
$a = 2$; $c = 4$.
Find the value of b:
$b^2 = c^2 - a^2 = 16 - 4 = 12$
$b = \sqrt{12} = 2\sqrt{3}$

Write the equation: $\dfrac{(y+4)^2}{4} - \dfrac{(x+3)^2}{12} = 1$.

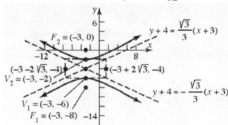

43. Foci: $(3, 7)$, $(7, 7)$; Vertex: $(6, 7)$;
Center: $(5, 7)$; Transverse axis is parallel to the x-axis; $a = 1$; $c = 2$.
Find the value of b:
$b^2 = c^2 - a^2 = 4 - 1 = 3$
$b = \sqrt{3}$

Write the equation: $\dfrac{(x-5)^2}{1} - \dfrac{(y-7)^2}{3} = 1$.

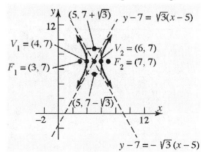

45. Vertices: $(-1,-1)$, $(3,-1)$; Center: $(1,-1)$;
Transverse axis is parallel to the x-axis; $a = 2$.
Asymptote: $y+1 = \frac{3}{2}(x-1)$

Using the slope of the asymptote, find the value of b:
$\frac{b}{a} = \frac{b}{2} = \frac{3}{2} \Rightarrow b = 3$

Find the value of c:
$c^2 = a^2 + b^2 = 4 + 9 = 13$
$c = \sqrt{13}$

Write the equation: $\frac{(x-1)^2}{4} - \frac{(y+1)^2}{9} = 1$.

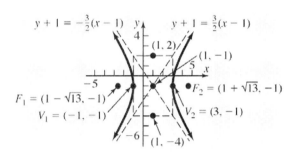

47. a. $\frac{(x-2)^2}{4} - \frac{(y+3)^2}{9} = 1$

The center of the hyperbola is at $(2,-3)$.
$a = 2$, $b = 3$.
The vertices are $(0,-3)$ and $(4,-3)$.
Find the value of c:
$c^2 = a^2 + b^2 = 4 + 9 = 13 \Rightarrow c = \sqrt{13}$
Foci: $\left(2-\sqrt{13},-3\right)$ and $\left(2+\sqrt{13},-3\right)$.
Transverse axis: $y = -3$, parallel to x-axis.
Asymptotes: $y+3 = \frac{3}{2}(x-2)$;
$y+3 = -\frac{3}{2}(x-2)$

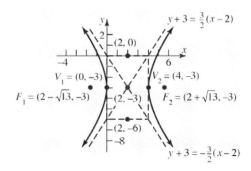

b. $\frac{(x-2)^2}{4} - \frac{(y+3)^2}{9} = 1$

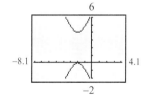

49. a. $(y-2)^2 - 4(x+2)^2 = 4$

Divide both sides by 4 to put in standard form: $\frac{(y-2)^2}{4} - \frac{(x+2)^2}{1} = 1$.

The center of the hyperbola is at $(-2, 2)$.
$a = 2$, $b = 1$.

The vertices are $(-2, 4)$ and $(-2, 0)$. Find the value of c:
$c^2 = a^2 + b^2 = 4 + 1 = 5 \Rightarrow c = \sqrt{5}$
Foci: $\left(-2, 2-\sqrt{5}\right)$ and $\left(-2, 2+\sqrt{5}\right)$.

Transverse axis: $x = -2$, parallel to the y-axis.

Asymptotes:
$y-2 = 2(x+2)$; $y-2 = -2(x+2)$.

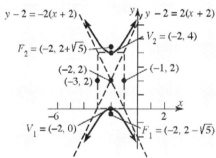

b. $(y-2)^2 - 4(x+2)^2 = 4$

51. a. $(x+1)^2 - (y+2)^2 = 4$
Divide both sides by 4 to put in standard form: $\dfrac{(x+1)^2}{4} - \dfrac{(y+2)^2}{4} = 1$.
The center of the hyperbola is $(-1, -2)$.
$a = 2$, $b = 2$.
The vertices are $(-3, -2)$ and $(1, -2)$.
Find the value of c:
$c^2 = a^2 + b^2 = 4 + 4 = 8$
$c = \sqrt{8} = 2\sqrt{2}$
Foci: $\left(-1 - 2\sqrt{2}, -2\right)$ and $\left(-1 + 2\sqrt{2}, -2\right)$
Transverse axis: $y = -2$, parallel to the x-axis.
Asymptotes: $y + 2 = x + 1$;
$y + 2 = -(x + 1)$

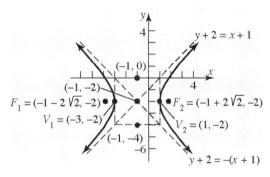

b. $(x+1)^2 - (y+2)^2 = 4$

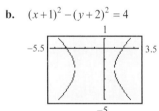

53. a. Complete the squares to put in standard form:
$x^2 - y^2 - 2x - 2y - 1 = 0$
$(x^2 - 2x + 1) - (y^2 + 2y + 1) = 1 + 1 - 1$
$(x - 1)^2 - (y + 1)^2 = 1$
The center of the hyperbola is $(1, -1)$.
$a = 1$, $b = 1$. The vertices are $(0, -1)$ and $(2, -1)$. Find the value of c:
$c^2 = a^2 + b^2 = 1 + 1 = 2 \Rightarrow c = \sqrt{2}$
Foci: $\left(1 - \sqrt{2}, -1\right)$ and $\left(1 + \sqrt{2}, -1\right)$.
Transverse axis: $y = -1$, parallel to x-axis.

Asymptotes: $y + 1 = x - 1$; $y + 1 = -(x - 1)$.

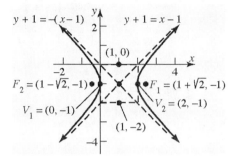

b. $(x-1)^2 - (y+1)^2 = 1$

55. a. Complete the squares to put in standard form:
$y^2 - 4x^2 - 4y - 8x - 4 = 0$
$(y^2 - 4y + 4) - 4(x^2 + 2x + 1) = 4 + 4 - 4$
$(y - 2)^2 - 4(x + 1)^2 = 4$
$\dfrac{(y-2)^2}{4} - \dfrac{(x+1)^2}{1} = 1$
The center of the hyperbola is $(-1, 2)$.
$a = 2$, $b = 1$.
The vertices are $(-1, 4)$ and $(-1, 0)$.
Find the value of c:
$c^2 = a^2 + b^2 = 4 + 1 = 5 \Rightarrow c = \sqrt{5}$
Foci: $\left(-1, 2 - \sqrt{5}\right)$ and $\left(-1, 2 + \sqrt{5}\right)$.
Transverse axis: $x = -1$, parallel to the y-axis.
Asymptotes:
$y - 2 = 2(x + 1)$; $y - 2 = -2(x + 1)$.

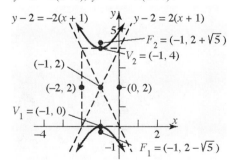

b. $\dfrac{(y-2)^2}{4}-(x+1)^2=1$

59. a. Complete the squares to put in standard form:
$$y^2-4x^2-16x-2y-19=0$$
$$(y^2-2y+1)-4(x^2+4x+4)=19+1-16$$
$$(y-1)^2-4(x+2)^2=4$$
$$\dfrac{(y-1)^2}{4}-\dfrac{(x+2)^2}{1}=1$$
The center of the hyperbola is $(-2, 1)$.
$a=2,\ b=1$.
The vertices are $(-2, 3)$ and $(-2, -1)$. Find the value of c:
$$c^2=a^2+b^2=4+1=5$$
$$c=\sqrt{5}$$
Foci: $\left(-2, 1-\sqrt{5}\right)$ and $\left(-2, 1+\sqrt{5}\right)$.
Transverse axis: $x=-2$, parallel to the y-axis.
Asymptotes:
$y-1=2(x+2);\ y-1=-2(x+2)$.

57. a. Complete the squares to put in standard form:
$$4x^2-y^2-24x-4y+16=0$$
$$4(x^2-6x+9)-(y^2+4y+4)=-16+36-4$$
$$4(x-3)^2-(y+2)^2=16$$
$$\dfrac{(x-3)^2}{4}-\dfrac{(y+2)^2}{16}=1$$
The center of the hyperbola is $(3, -2)$.
$a=2,\ b=4$.
The vertices are $(1, -2)$ and $(5, -2)$. Find the value of c:
$$c^2=a^2+b^2=4+16=20$$
$$c=\sqrt{20}=2\sqrt{5}$$
Foci: $\left(3-2\sqrt{5}, -2\right)$ and $\left(3+2\sqrt{5}, -2\right)$.
Transverse axis: $y=-2$, parallel to x-axis.
Asymptotes: $y+2=2(x-3);$
$$y+2=-2(x-3)$$

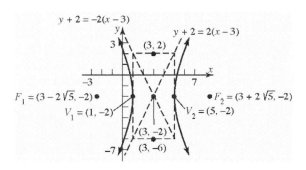

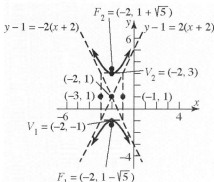

b. $\dfrac{(y-1)^2}{4}-\dfrac{(x+2)^2}{1}=1$

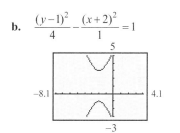

b. $\dfrac{(x-3)^2}{4}-\dfrac{(y+2)^2}{16}=1$

61. Rewrite the equation:
$$y = \sqrt{16 + 4x^2}$$
$$y^2 = 16 + 4x^2, \quad y \geq 0$$
$$y^2 - 4x^2 = 16, \quad y \geq 0$$
$$\frac{y^2}{16} - \frac{x^2}{4} = 1, \quad y \geq 0$$

63. Rewrite the equation:
$$y = -\sqrt{-25 + x^2}$$
$$y^2 = -25 + x^2, \quad y \leq 0$$
$$x^2 - y^2 = 25, \quad y \leq 0$$
$$\frac{x^2}{25} - \frac{y^2}{25} = 1, \quad y \leq 0$$

65. First note that all points where a burst could take place, such that the time difference would be the same as that for the first burst, would form a hyperbola with A and B as the foci.
Start with a diagram:

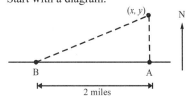

The ordered pair (x, y) represents the location of the fireworks. We know that sound travels at 1100 feet per second, so the person at point A is 1100 feet closer to the fireworks display than the person at point B. Since the difference of the distance from (x, y) to A and from (x, y) to B is the constant 1100, the point (x, y) lies on a hyperbola whose foci are at A and B. The hyperbola has the equation
$$\frac{x^2}{a^2} - \frac{y^2}{b^2} = 1$$
where $2a = 1100$, so $a = 550$. Because the distance between the two people is 2 miles (10,560 feet) and each person is at a focus of the hyperbola, we have
$$2c = 10,560$$
$$c = 5280$$
$$b^2 = c^2 - a^2 = 5280^2 - 550^2 = 27,575,900$$
The equation of the hyperbola that describes the location of the fireworks display is
$$\frac{x^2}{550^2} - \frac{y^2}{27,575,900} = 1$$
Since the fireworks display is due north of the individual at A, we let $x = 5280$ and solve the equation for y.
$$\frac{5280^2}{550^2} - \frac{y^2}{27,575,900} = 1$$
$$-\frac{y^2}{27,575,900} = -91.16$$
$$y^2 = 2,513,819,044$$
$$y = 50,138$$
Therefore, the fireworks display was 50,138 feet (approximately 9.5 miles) due north of the person at A.

67. a. Since the particles are deflected at a $45°$ angle, the asymptotes will be $y = \pm x$.

b. Since the vertex is 10 cm from the center of the hyperbola, we know that $a = 10$. The slope of the asymptotes is given by $\pm\frac{b}{a}$. Therefore, we have
$$\frac{b}{a} = 1$$
$$\frac{b}{10} = 1 \rightarrow b = 10$$
Using the origin as the center of the hyperbola, the equation of the particle path would be
$$\frac{x^2}{100} - \frac{y^2}{100} = 1$$

69. Assume $\dfrac{x^2}{a^2} - \dfrac{y^2}{b^2} = 1$.

If the eccentricity is close to 1, then $c \approx a$ and $b \approx 0$. When b is close to 0, the hyperbola is very narrow, because the slopes of the asymptotes are close to 0.

If the eccentricity is very large, then c is much larger than a and b is very large. The result is a hyperbola that is very wide, because the slopes of the asymptotes are very large.

71. $\dfrac{x^2}{4} - y^2 = 1 \quad (a = 2,\ b = 1)$

This is a hyperbola with horizontal transverse axis, centered at (0, 0) and has asymptotes:

$y = \pm \dfrac{1}{2} x$

$y^2 - \dfrac{x^2}{4} = 1 \quad (a = 1,\ b = 2)$

This is a hyperbola with vertical transverse axis, centered at (0, 0) and has asymptotes:

$y = \pm \dfrac{1}{2} x$.

Since the two hyperbolas have the same asymptotes, they are conjugates.

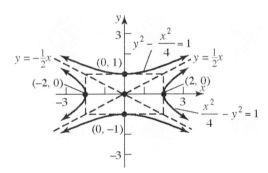

73. Put the equation in standard hyperbola form:

$Ax^2 + Cy^2 + F = 0 \qquad A \cdot C < 0,\ F \neq 0$

$Ax^2 + Cy^2 = -F$

$\dfrac{Ax^2}{-F} + \dfrac{Cy^2}{-F} = 1$

$\dfrac{x^2}{\left(-\dfrac{F}{A}\right)} + \dfrac{y^2}{\left(-\dfrac{F}{C}\right)} = 1$

Since $-F/A$ and $-F/C$ have opposite signs, this is a hyperbola with center (0, 0).
The transverse axis is the x-axis if $-F/A > 0$.
The transverse axis is the y-axis if $-F/A < 0$.

Section 6.5

1. $y = 3x + 2$
The graph is a line.
x-intercept:
$0 = 3x + 2$
$3x = -2$
$x = -\dfrac{2}{3}$

y-intercept: $y = 3(0) + 2 = 2$

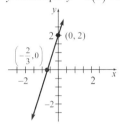

3. $2x + 3y < 6$
$2x + 3y < 6$
$3y < -2x + 6$
$y < -\dfrac{2}{3} x + 2$

The inequality is strict so we graph the line $y = -\dfrac{2}{3} x + 2$ with a dashed line. Test a point not on the line, such as $(0, 0)$.

$2(0) + 3(0) < 6$
$0 < 6$

This is a true statement so we shade the half-plane containing the point $(0, 0)$.

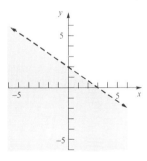

5. $\begin{cases} y = 2x - 3 \\ 2x - 3y = -7 \end{cases}$

Substitute the expression $2x - 3$ for y in the second equation.
$$2x - 3y = -7$$
$$2x - 3(2x - 3) = -7$$
$$2x - 6x + 9 = -7$$
$$-4x = -16$$
$$x = 4$$
Substitute this value for x in the first equation and solve for y.
$$y = 2x - 3$$
$$y = 2(4) - 3 = 8 - 3 = 5$$
The solution to the system is $(4, 5)$.

7. $\begin{cases} y = x^2 + 1 \\ y = x + 1 \end{cases}$

Solve by substitution:
$$x^2 + 1 = x + 1$$
$$x^2 - x = 0$$
$$x(x - 1) = 0$$
$$x = 0 \text{ or } x = 1$$
$$y = 1 \quad y = 2$$
Solutions: $(0, 1)$ and $(1, 2)$

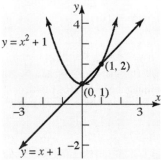

$(0, 1)$ and $(1, 2)$ are the intersection points.

9. $\begin{cases} y = \sqrt{36 - x^2} \\ y = 8 - x \end{cases}$

Solve by substitution:
$$\sqrt{36 - x^2} = 8 - x$$
$$36 - x^2 = 64 - 16x + x^2$$
$$2x^2 - 16x + 28 = 0$$
$$x^2 - 8x + 14 = 0$$
$$x = \frac{8 \pm \sqrt{64 - 56}}{2}$$
$$= \frac{8 \pm 2\sqrt{2}}{2}$$
$$= 4 \pm \sqrt{2}$$

If $x = 4 + \sqrt{2}$, $y = 8 - (4 + \sqrt{2}) = 4 - \sqrt{2}$
If $x = 4 - \sqrt{2}$, $y = 8 - (4 - \sqrt{2}) = 4 + \sqrt{2}$
Solutions:
$(4 + \sqrt{2}, 4 - \sqrt{2})$ and $(4 - \sqrt{2}, 4 + \sqrt{2})$

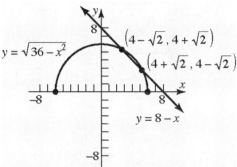

$(2.59, 5.41)$ and $(5.41, 2.59)$ are the intersection points.

11. $\begin{cases} y = \sqrt{x} \\ y = 2 - x \end{cases}$

Solve by substitution:
$$\sqrt{x} = 2 - x$$
$$x = 4 - 4x + x^2$$
$$x^2 - 5x + 4 = 0$$
$$(x - 4)(x - 1) = 0$$
$$x = 4 \quad \text{or} \quad x = 1$$
$$y = -2 \quad \text{or} \quad y = 1$$
Eliminate $(4, -2)$; we must have $y \geq 0$.
Solution: $(1, 1)$

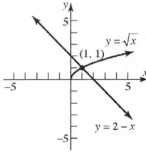

(1, 1) is the intersection point.

13. $\begin{cases} x = 2y \\ x = y^2 - 2y \end{cases}$

Solve by substitution:
$2y = y^2 - 2y$
$y^2 - 4y = 0$
$y(y-4) = 0$
$y = 0$ or $y = 4$
$x = 0$ or $x = 8$
Solutions: $(0, 0)$ and $(8, 4)$

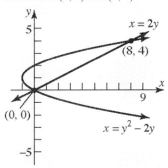

$(0, 0)$ and $(8, 4)$ are the intersection points.

15. $\begin{cases} x^2 + y^2 = 4 \\ x^2 + 2x + y^2 = 0 \end{cases}$

Substitute 4 for $x^2 + y^2$ in the second equation.
$2x + 4 = 0$
$2x = -4$
$x = -2$
$y = \sqrt{4 - (-2)^2} = 0$
Solution: $(-2, 0)$

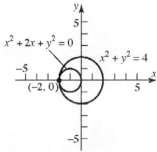

$(-2, 0)$ is the intersection point.

17. $\begin{cases} y = 3x - 5 \\ x^2 + y^2 = 5 \end{cases}$

Solve by substitution:
$x^2 + (3x - 5)^2 = 5$
$x^2 + 9x^2 - 30x + 25 = 5$
$10x^2 - 30x + 20 = 0$
$x^2 - 3x + 2 = 0$
$(x-1)(x-2) = 0$
$x = 1$ or $x = 2$
$y = -2$ $\quad y = 1$
Solutions: $(1, -2)$ and $(2, 1)$

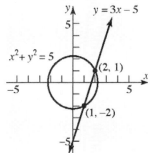

$(1, -2)$ and $(2, 1)$ are the intersection points.

19. $\begin{cases} x^2 + y^2 = 4 \\ y^2 - x = 4 \end{cases}$

Substitute $x + 4$ for y^2 in the first equation:
$x^2 + x + 4 = 4$
$x^2 + x = 0$
$x(x+1) = 0$
$x = 0$ or $x = -1$
$y^2 = 4$ $\quad y^2 = 3$
$y = \pm 2$ $\quad y = \pm\sqrt{3}$

Solutions: $(0, -2), (0, 2), (-1, \sqrt{3}), (-1, -\sqrt{3})$

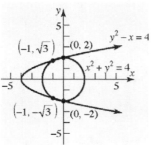

$(-1, 1.73), (-1, -1.73), (0, 2),$ and $(0, -2)$ are the intersection points.

21. $\begin{cases} xy = 4 \\ x^2 + y^2 = 8 \end{cases}$

Solve by substitution:
$$x^2 + \left(\frac{4}{x}\right)^2 = 8$$
$$x^2 + \frac{16}{x^2} = 8$$
$$x^4 + 16 = 8x^2$$
$$x^4 - 8x^2 + 16 = 0$$
$$(x^2 - 4)^2 = 0$$
$$x^2 - 4 = 0$$
$$x^2 = 4$$
$$x = 2 \text{ or } x = -2$$
$$y = 2 \quad\quad y = -2$$

Solutions: $(-2, -2)$ and $(2, 2)$

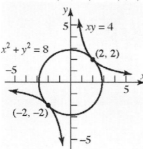

$(-2, -2)$ and $(2, 2)$ are the intersection points.

23. $\begin{cases} x^2 + y^2 = 4 \\ y = x^2 - 9 \end{cases}$

Solve by substitution:
$$x^2 + (x^2 - 9)^2 = 4$$
$$x^2 + x^4 - 18x^2 + 81 = 4$$
$$x^4 - 17x^2 + 77 = 0$$
$$x^2 = \frac{17 \pm \sqrt{289 - 4(77)}}{2}$$
$$= \frac{17 \pm \sqrt{-19}}{2}$$

There are no real solutions to this expression. Inconsistent.

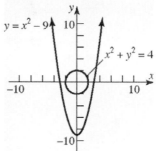

No solution; Inconsistent.

25. $\begin{cases} y = x^2 - 4 \\ y = 6x - 13 \end{cases}$

Solve by substitution:
$$x^2 - 4 = 6x - 13$$
$$x^2 - 6x + 9 = 0$$
$$(x - 3)^2 = 0$$
$$x - 3 = 0$$
$$x = 3 \implies y = (3)^2 - 4 = 5$$

Solution: $(3, 5)$

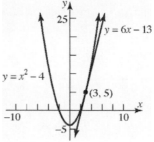

$(3, 5)$ is the intersection point.

27. Solve the second equation for y, substitute into the first equation and solve:
$$\begin{cases} 2x^2 + y^2 = 18 \\ xy = 4 \implies y = \dfrac{4}{x} \end{cases}$$
$$2x^2 + \left(\dfrac{4}{x}\right)^2 = 18$$
$$2x^2 + \dfrac{16}{x^2} = 18$$
$$2x^4 + 16 = 18x^2$$
$$2x^4 - 18x^2 + 16 = 0$$
$$x^4 - 9x^2 + 8 = 0$$
$$(x^2 - 8)(x^2 - 1) = 0$$
$$x^2 = 8 \implies x = \pm\sqrt{8} = \pm 2\sqrt{2}$$
or $x^2 = 1 \implies x = \pm 1$

If $x = 2\sqrt{2}$: $y = \dfrac{4}{2\sqrt{2}} = \sqrt{2}$

If $x = -2\sqrt{2}$: $y = \dfrac{4}{-2\sqrt{2}} = -\sqrt{2}$

If $x = 1$: $y = \dfrac{4}{1} = 4$

If $x = -1$: $y = \dfrac{4}{-1} = -4$

Solutions:
$(2\sqrt{2}, \sqrt{2}), (-2\sqrt{2}, -\sqrt{2}), (1, 4), (-1, -4)$

29. Substitute the first equation into the second equation and solve:
$$\begin{cases} y = 2x + 1 \\ 2x^2 + y^2 = 1 \end{cases}$$
$$2x^2 + (2x+1)^2 = 1$$
$$2x^2 + 4x^2 + 4x + 1 = 1$$
$$6x^2 + 4x = 0$$
$$2x(3x + 2) = 0$$
$$2x = 0 \text{ or } 3x + 2 = 0$$
$$x = 0 \text{ or } x = -\dfrac{2}{3}$$

If $x = 0$: $y = 2(0) + 1 = 1$

If $x = -\dfrac{2}{3}$: $y = 2\left(-\dfrac{2}{3}\right) + 1 = -\dfrac{4}{3} + 1 = -\dfrac{1}{3}$

Solutions: $(0, 1), \left(-\dfrac{2}{3}, -\dfrac{1}{3}\right)$

31. Solve the first equation for y, substitute into the second equation and solve:
$$\begin{cases} x + y + 1 = 0 \implies y = -x - 1 \\ x^2 + y^2 + 6y - x = -5 \end{cases}$$
$$x^2 + (-x-1)^2 + 6(-x-1) - x = -5$$
$$x^2 + x^2 + 2x + 1 - 6x - 6 - x = -5$$
$$2x^2 - 5x = 0$$
$$x(2x - 5) = 0$$
$$x = 0 \text{ or } x = \dfrac{5}{2}$$

If $x = 0$: $y = -(0) - 1 = -1$

If $x = \dfrac{5}{2}$: $y = -\dfrac{5}{2} - 1 = -\dfrac{7}{2}$

Solutions: $(0, -1), \left(\dfrac{5}{2}, -\dfrac{7}{2}\right)$

33. Solve the second equation for y, substitute into the first equation and solve:
$$\begin{cases} 4x^2 - 3xy + 9y^2 = 15 \\ 2x + 3y = 5 \implies y = -\dfrac{2}{3}x + \dfrac{5}{3} \end{cases}$$
$$4x^2 - 3x\left(-\dfrac{2}{3}x + \dfrac{5}{3}\right) + 9\left(-\dfrac{2}{3}x + \dfrac{5}{3}\right)^2 = 15$$
$$4x^2 + 2x^2 - 5x + 4x^2 - 20x + 25 = 15$$
$$10x^2 - 25x + 10 = 0$$
$$2x^2 - 5x + 2 = 0$$
$$(2x - 1)(x - 2) = 0$$
$$x = \dfrac{1}{2} \text{ or } x = 2$$

If $x = \dfrac{1}{2}$: $y = -\dfrac{2}{3}\left(\dfrac{1}{2}\right) + \dfrac{5}{3} = \dfrac{4}{3}$

If $x = 2$: $y = -\dfrac{2}{3}(2) + \dfrac{5}{3} = \dfrac{1}{3}$

Solutions: $\left(\dfrac{1}{2}, \dfrac{4}{3}\right), \left(2, \dfrac{1}{3}\right)$

35. Multiply each side of the second equation by 4 and add the equations to eliminate y:

$$\begin{cases} x^2 - 4y^2 = -7 \\ 3x^2 + y^2 = 31 \end{cases} \xrightarrow{4} \begin{array}{r} x^2 - 4y^2 = -7 \\ 12x^2 + 4y^2 = 124 \\ \hline 13x^2 = 117 \end{array}$$

$$x^2 = 9$$
$$x = \pm 3$$

If $x = 3$:
$3(3)^2 + y^2 = 31 \Rightarrow y^2 = 4 \Rightarrow y = \pm 2$
If $x = -3$:
$3(-3)^2 + y^2 = 31 \Rightarrow y^2 = 4 \Rightarrow y = \pm 2$
Solutions: $(3, 2), (3, -2), (-3, 2), (-3, -2)$

37. $\begin{cases} 7x^2 - 3y^2 + 5 = 0 \\ 3x^2 + 5y^2 = 12 \end{cases}$

$\begin{cases} 7x^2 - 3y^2 = -5 \\ 3x^2 + 5y^2 = 12 \end{cases}$

Multiply each side of the first equation by 5 and each side of the second equation by 3 and add the equations to eliminate y:

$$\begin{array}{r} 35x^2 - 15y^2 = -25 \\ 9x^2 + 15y^2 = 36 \\ \hline 44x^2 = 11 \end{array}$$

$$x^2 = \frac{1}{4}$$
$$x = \pm \frac{1}{2}$$

If $x = \frac{1}{2}$:
$3\left(\frac{1}{2}\right)^2 + 5y^2 = 12 \Rightarrow y^2 = \frac{9}{4} \Rightarrow y = \pm \frac{3}{2}$
If $x = -\frac{1}{2}$:
$3\left(-\frac{1}{2}\right)^2 + 5y^2 = 12 \Rightarrow y^2 = \frac{9}{4} \Rightarrow y = \pm \frac{3}{2}$
Solutions:
$\left(\frac{1}{2}, \frac{3}{2}\right), \left(\frac{1}{2}, -\frac{3}{2}\right), \left(-\frac{1}{2}, \frac{3}{2}\right), \left(-\frac{1}{2}, -\frac{3}{2}\right)$

39. Multiply each side of the second equation by 2 and add the equations to eliminate xy:

$$\begin{cases} x^2 + 2xy = 10 \\ 3x^2 - xy = 2 \end{cases} \xrightarrow{2} \begin{array}{r} x^2 + 2xy = 10 \\ 6x^2 - 2xy = 4 \\ \hline 7x^2 = 14 \end{array}$$

$$x^2 = 2$$
$$x = \pm \sqrt{2}$$

If $x = \sqrt{2}$:
$3(\sqrt{2})^2 - \sqrt{2} \cdot y = 2$
$\Rightarrow -\sqrt{2} \cdot y = -4 \Rightarrow y = \frac{4}{\sqrt{2}} \Rightarrow y = 2\sqrt{2}$
If $x = -\sqrt{2}$:
$3(-\sqrt{2})^2 - (-\sqrt{2})y = 2$
$\Rightarrow \sqrt{2} \cdot y = -4 \Rightarrow y = \frac{-4}{\sqrt{2}} \Rightarrow y = -2\sqrt{2}$
Solutions: $(\sqrt{2}, 2\sqrt{2}), (-\sqrt{2}, -2\sqrt{2})$

41. $\begin{cases} 2x^2 + y^2 = 2 \\ x^2 - 2y^2 + 8 = 0 \end{cases}$

$\begin{cases} 2x^2 + y^2 = 2 \\ x^2 - 2y^2 = -8 \end{cases}$

Multiply each side of the first equation by 2 and add the equations to eliminate y:

$$\begin{array}{r} 4x^2 + 2y^2 = 4 \\ x^2 - 2y^2 = -8 \\ \hline 5x^2 = -4 \end{array}$$

$$x^2 = -\frac{4}{5}$$

No real solution. The system is inconsistent.

43. $\begin{cases} x^2 + 2y^2 = 16 \\ 4x^2 - y^2 = 24 \end{cases}$

Multiply each side of the second equation by 2 and add the equations to eliminate y:

$x^2 + 2y^2 = 16$
$\underline{8x^2 - 2y^2 = 48}$
$9x^2 = 64$
$x^2 = \dfrac{64}{9}$
$x = \pm \dfrac{8}{3}$

If $x = \dfrac{8}{3}$:

$\left(\dfrac{8}{3}\right)^2 + 2y^2 = 16 \Rightarrow 2y^2 = \dfrac{80}{9}$
$\Rightarrow y^2 = \dfrac{40}{9} \Rightarrow y = \pm \dfrac{2\sqrt{10}}{3}$

If $x = -\dfrac{8}{3}$:

$\left(-\dfrac{8}{3}\right)^2 + 2y^2 = 16 \Rightarrow 2y^2 = \dfrac{80}{9}$
$\Rightarrow y^2 = \dfrac{40}{9} \Rightarrow y = \pm \dfrac{2\sqrt{10}}{3}$

Solutions:

$\left(\dfrac{8}{3}, \dfrac{2\sqrt{10}}{3}\right), \left(\dfrac{8}{3}, -\dfrac{2\sqrt{10}}{3}\right), \left(-\dfrac{8}{3}, \dfrac{2\sqrt{10}}{3}\right),$
$\left(-\dfrac{8}{3}, -\dfrac{2\sqrt{10}}{3}\right)$

45. $\begin{cases} \dfrac{5}{x^2} - \dfrac{2}{y^2} + 3 = 0 \\ \dfrac{3}{x^2} + \dfrac{1}{y^2} = 7 \end{cases}$

$\begin{cases} \dfrac{5}{x^2} - \dfrac{2}{y^2} = -3 \\ \dfrac{3}{x^2} + \dfrac{1}{y^2} = 7 \end{cases}$

Multiply each side of the second equation by 2 and add the equations to eliminate y:

$\dfrac{5}{x^2} - \dfrac{2}{y^2} = -3$
$\underline{\dfrac{6}{x^2} + \dfrac{2}{y^2} = 14}$
$\dfrac{11}{x^2} = 11$
$x^2 = 1$
$x = \pm 1$

If $x = 1$:

$\dfrac{3}{(1)^2} + \dfrac{1}{y^2} = 7 \Rightarrow \dfrac{1}{y^2} = 4 \Rightarrow y^2 = \dfrac{1}{4}$
$\Rightarrow y = \pm \dfrac{1}{2}$

If $x = -1$:

$\dfrac{3}{(-1)^2} + \dfrac{1}{y^2} = 7 \Rightarrow \dfrac{1}{y^2} = 4 \Rightarrow y^2 = \dfrac{1}{4}$
$\Rightarrow y = \pm \dfrac{1}{2}$

Solutions: $\left(1, \dfrac{1}{2}\right), \left(1, -\dfrac{1}{2}\right), \left(-1, \dfrac{1}{2}\right), \left(-1, -\dfrac{1}{2}\right)$

47. $\begin{cases} \dfrac{1}{x^4} + \dfrac{6}{y^4} = 6 \\ \dfrac{2}{x^4} - \dfrac{2}{y^4} = 19 \end{cases}$

Multiply each side of the first equation by -2 and add the equations to eliminate x:

$\dfrac{-2}{x^4} - \dfrac{12}{y^4} = -12$
$\underline{\dfrac{2}{x^4} - \dfrac{2}{y^4} = 19}$
$-\dfrac{14}{y^4} = 7$
$y^4 = -2$

There are no real solutions. The system is inconsistent.

49. $\begin{cases} x^2 - 3xy + 2y^2 = 0 \\ x^2 + xy = 6 \end{cases}$

Subtract the second equation from the first to eliminate the x^2 term.
$$-4xy + 2y^2 = -6$$
$$2xy - y^2 = 3$$

Since $y \neq 0$, we can solve for x in this equation to get
$$x = \frac{y^2 + 3}{2y}, \quad y \neq 0$$

Now substitute for x in the second equation and solve for y.
$$x^2 + xy = 6$$
$$\left(\frac{y^2 + 3}{2y}\right)^2 + \left(\frac{y^2 + 3}{2y}\right)y = 6$$
$$\frac{y^4 + 6y^2 + 9}{4y^2} + \frac{y^2 + 3}{2} = 6$$
$$y^4 + 6y^2 + 9 + 2y^4 + 6y^2 = 24y^2$$
$$3y^4 - 12y^2 + 9 = 0$$
$$y^4 - 4y^2 + 3 = 0$$
$$(y^2 - 3)(y^2 - 1) = 0$$

Thus, $y = \pm\sqrt{3}$ or $y = \pm 1$.
If $y = 1$: $x = 2 \cdot 1 = 2$
If $y = -1$: $x = 2(-1) = -2$
If $y = \sqrt{3}$: $x = \sqrt{3}$
If $y = -\sqrt{3}$: $x = -\sqrt{3}$
Solutions:
$(2, 1), (-2, -1), (\sqrt{3}, \sqrt{3}), (-\sqrt{3}, -\sqrt{3})$

51. $\begin{cases} y^2 + y + x^2 - x - 2 = 0 \\ y + 1 + \dfrac{x-2}{y} = 0 \end{cases}$

Multiply each side of the second equation by $-y$ and add the equations to eliminate y:
$$y^2 + y + x^2 - x - 2 = 0$$
$$\underline{-y^2 - y \quad\quad - x + 2 = 0}$$
$$x^2 - 2x = 0$$
$$x(x - 2) = 0$$
$$x = 0 \text{ or } x = 2$$

If $x = 0$:
$$y^2 + y + 0^2 - 0 - 2 = 0 \Rightarrow y^2 + y - 2 = 0$$
$$\Rightarrow (y + 2)(y - 1) = 0 \Rightarrow y = -2 \text{ or } y = 1$$

If $x = 2$:
$$y^2 + y + 2^2 - 2 - 2 = 0 \Rightarrow y^2 + y = 0$$
$$\Rightarrow y(y + 1) = 0 \Rightarrow y = 0 \text{ or } y = -1$$

Note: $y \neq 0$ because of division by zero.
Solutions: $(0, -2), (0, 1), (2, -1)$

53. Rewrite each equation in exponential form:
$$\begin{cases} \log_x y = 3 \;\rightarrow\; y = x^3 \\ \log_x(4y) = 5 \;\rightarrow\; 4y = x^5 \end{cases}$$

Substitute the first equation into the second and solve:
$$4x^3 = x^5$$
$$x^5 - 4x^3 = 0$$
$$x^3(x^2 - 4) = 0$$
$$x^3 = 0 \text{ or } x^2 = 4 \Rightarrow x = 0 \text{ or } x = \pm 2$$

The base of a logarithm must be positive, thus $x \neq 0$ and $x \neq -2$.
If $x = 2$: $y = 2^3 = 8$
Solution: $(2, 8)$

55. Rewrite each equation in exponential form:
$$\ln x = 4 \ln y \Rightarrow x = e^{4 \ln y} = e^{\ln y^4} = y^4$$
$$\log_3 x = 2 + 2 \log_3 y$$
$$x = 3^{2 + 2\log_3 y} = 3^2 \cdot 3^{2\log_3 y} = 3^2 \cdot 3^{\log_3 y^2} = 9y^2$$

So we have the system $\begin{cases} x = y^4 \\ x = 9y^2 \end{cases}$

Therefore we have:
$$9y^2 = y^4 \Rightarrow 9y^2 - y^4 = 0 \Rightarrow y^2(9 - y^2) = 0$$
$$y^2(3 + y)(3 - y) = 0$$
$$y = 0 \text{ or } y = -3 \text{ or } y = 3$$

Since $\ln y$ is undefined when $y \leq 0$, the only solution is $y = 3$.
If $y = 3$: $x = y^4 \Rightarrow x = 3^4 = 81$
Solution: $(81, 3)$

57. Solve the first equation for x, substitute into the second equation and solve:
$$\begin{cases} x+2y=0 \Rightarrow x=-2y \\ (x-1)^2+(y-1)^2=5 \end{cases}$$
$$(-2y-1)^2+(y-1)^2=5$$
$$4y^2+4y+1+y^2-2y+1=5 \Rightarrow 5y^2+2y-3=0$$
$$(5y-3)(y+1)=0$$
$$y=\frac{3}{5}=0.6 \quad \text{or} \quad y=-1$$
$$x=-\frac{6}{5}=-1.2 \quad \text{or} \quad x=2$$
The points of intersection are $(-1.2, 0.6), (2, -1)$.

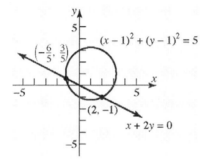

59. Complete the square on the second equation.
$$y^2+4y+4=x-1+4$$
$$(y+2)^2=x+3$$
Substitute this result into the first equation.
$$(x-1)^2+x+3=4$$
$$x^2-2x+1+x+3=4$$
$$x^2-x=0$$
$$x(x-1)=0$$
$$x=0 \quad \text{or} \quad x=1$$
If $x=0$: $(y+2)^2=0+3$
$$y+2=\pm\sqrt{3} \Rightarrow y=-2\pm\sqrt{3}$$
If $x=1$: $(y+2)^2=1+3$
$$y+2=\pm 2 \Rightarrow y=-2\pm 2$$
The points of intersection are:
$(0, -2-\sqrt{3}), (0, -2+\sqrt{3}), (1, -4), (1, 0)$.

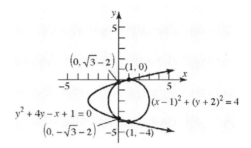

61. Solve the first equation for x, substitute into the second equation and solve:
$$\begin{cases} y=\dfrac{4}{x-3} \\ x^2-6x+y^2+1=0 \end{cases}$$
$$y=\frac{4}{x-3}$$
$$x-3=\frac{4}{y}$$
$$x=\frac{4}{y}+3$$
$$\left(\frac{4}{y}+3\right)^2-6\left(\frac{4}{y}+3\right)+y^2+1=0$$
$$\frac{16}{y^2}+\frac{24}{y}+9-\frac{24}{y}-18+y^2+1=0$$
$$\frac{16}{y^2}+y^2-8=0$$
$$16+y^4-8y^2=0$$
$$y^4-8y^2+16=0$$
$$(y^2-4)^2=0$$
$$y^2-4=0$$
$$y^2=4 \rightarrow y=\pm 2$$
If $y=2$: $x=\dfrac{4}{2}+3=5$
If $y=-2$: $x=\dfrac{4}{-2}+3=1$
The points of intersection are: $(1, -2), (5, 2)$.

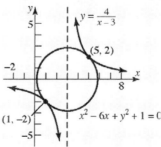

63. Graph: $y_1 = x \wedge (2/3)$; $y_2 = e \wedge (-x)$
Use INTERSECT to solve:

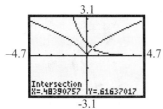

Solution: (0.48, 0.62)

65. Graph: $y_1 = \sqrt[3]{2 - x^2}$; $y_2 = 4/x^3$
Use INTERSECT to solve:

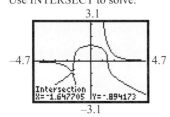

Solution: (−1.65, −0.89)

67. Graph: $y_1 = \sqrt[4]{12 - x^4}$; $y_2 = -\sqrt[4]{12 - x^4}$;
$y_3 = \sqrt{2/x}$; $y_4 = -\sqrt{2/x}$
Use INTERSECT to solve:

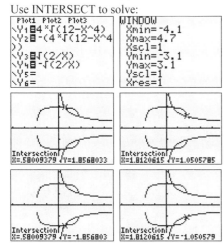

Solutions: (0.58, 1.86), (1.81, 1.05),
(1.81, −1.05), (0.58, −1.86)

69. Graph: $y_1 = 2/x$; $y_2 = \ln x$
Use INTERSECT to solve:

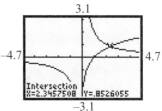

Solution: (2.35, 0.85)

71. $\begin{cases} x^2 + y^2 \leq 9 \\ x + y \geq 3 \end{cases}$

Graph the circle $x^2 + y^2 = 9$. Use a solid line since the inequality uses $\leq$. Choose a test point not on the circle, such as (0, 0). Since $0^2 + 0^2 \leq 9$ is true, shade the same side of the circle as (0, 0).
Graph the line $x + y = 3$. Use a solid line since the inequality uses $\geq$. Choose a test point not on the line, such as (0, 0). Since $0 + 0 \geq 3$ is false, shade the opposite side of the line from (0, 0). The overlapping region is the solution.

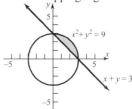

73. $\begin{cases} y \geq x^2 - 4 \\ y \leq x - 2 \end{cases}$

Graph the parabola $y = x^2 - 4$. Use a solid line since the inequality uses $\geq$. Choose a test point not on the parabola, such as (0, 0). Since $0 \geq 0^2 - 4$ is true, shade the same side of the parabola as (0, 0). Graph the line $y = x - 2$. Use a solid line since the inequality uses $\leq$. Choose a test point not on the line, such as (0, 0). Since $0 \leq 0 - 2$ is false, shade the opposite side of the line from (0, 0). The overlapping region is the solution.

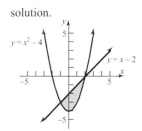

overlapping region is the solution.

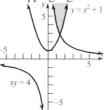

75. $\begin{cases} x^2 + y^2 \leq 16 \\ y \geq x^2 - 4 \end{cases}$

Graph the circle $x^2 + y^2 = 16$. Use a sold line since the inequality is not strict. Choose a test point not on the circle, such as $(0,0)$. Since $0^2 + 0^2 \leq 16$ is true, shade the side of the circle containing $(0,0)$. Graph the parabola $y = x^2 - 4$. Use a solid line since the inequality is not strict. Choose a test point not on the parabola, such as $(0,0)$. Since $0 \geq 0^2 - 4$ is true, shade the side of the parabola that contains $(0,0)$. The overlapping region is the solution.

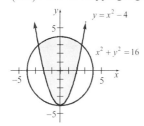

77. $\begin{cases} xy \geq 4 \\ y \geq x^2 + 1 \end{cases}$

Graph the hyperbola $xy = 4$. Use a solid line since the inequality uses $\geq$. Choose a test point not on the parabola, such as $(0, 0)$. Since $0 \cdot 0 \geq 4$ is false, shade the opposite side of the hyperbola from $(0, 0)$. Graph the parabola $y = x^2 + 1$. Use a solid line since the inequality uses $\geq$. Choose a test point not on the parabola, such as $(0, 0)$. Since $0 \geq 0^2 + 1$ is false, shade the opposite side of the parabola from $(0, 0)$. The

79. Let x and y be the two numbers. The system of equations is:
$\begin{cases} x - y = 2 & \Rightarrow x = y + 2 \\ x^2 + y^2 = 10 \end{cases}$

Solve the first equation for x, substitute into the second equation and solve:
$(y+2)^2 + y^2 = 10$
$y^2 + 4y + 4 + y^2 = 10$
$y^2 + 2y - 3 = 0$
$(y+3)(y-1) = 0 \Rightarrow y = -3$ or $y = 1$
If $y = -3$: $x = -3 + 2 = -1$
If $y = 1$: $x = 1 + 2 = 3$
The two numbers are 1 and 3 or –1 and –3.

81. Let x and y be the two numbers. The system of equations is:
$\begin{cases} xy = 4 \Rightarrow x = \dfrac{4}{y} \\ x^2 + y^2 = 8 \end{cases}$

Solve the first equation for x, substitute into the second equation and solve:
$\left(\dfrac{4}{y}\right)^2 + y^2 = 8$
$\dfrac{16}{y^2} + y^2 = 8$
$16 + y^4 = 8y^2$
$y^4 - 8y^2 + 16 = 0$
$(y^2 - 4) = 0$
$y^2 = 4$
$y = \pm 2$
If $y = 2$: $x = \dfrac{4}{2} = 2$;   If $y = -2$: $x = \dfrac{4}{-2} = -2$

The two numbers are 2 and 2 or –2 and –2.

83. Let x and y be the two numbers. The system of equations is:
$$\begin{cases} x - y = xy \\ \dfrac{1}{x} + \dfrac{1}{y} = 5 \end{cases}$$
Solve the first equation for x, substitute into the second equation and solve:
$$x - xy = y$$
$$x(1-y) = y \Rightarrow x = \dfrac{y}{1-y}$$
$$\dfrac{1}{\frac{y}{1-y}} + \dfrac{1}{y} = 5$$
$$\dfrac{1-y}{y} + \dfrac{1}{y} = 5$$
$$\dfrac{2-y}{y} = 5$$
$$2 - y = 5y$$
$$6y = 2$$
$$y = \dfrac{1}{3} \Rightarrow x = \dfrac{\frac{1}{3}}{1-\frac{1}{3}} = \dfrac{\frac{1}{3}}{\frac{2}{3}} = \dfrac{1}{2}$$
The two numbers are $\dfrac{1}{2}$ and $\dfrac{1}{3}$.

85. $\begin{cases} \dfrac{a}{b} = \dfrac{2}{3} \\ a + b = 10 \Rightarrow a = 10 - b \end{cases}$

Solve the second equation for a, substitute into the first equation and solve:
$$\dfrac{10-b}{b} = \dfrac{2}{3}$$
$$3(10-b) = 2b$$
$$30 - 3b = 2b$$
$$30 = 5b$$
$$b = 6 \Rightarrow a = 4$$
$a + b = 10;\ b - a = 2$
The ratio of $a+b$ to $b-a$ is $\dfrac{10}{2} = 5$.

87. Let x = the width of the rectangle.
Let y = the length of the rectangle.
$$\begin{cases} 2x + 2y = 16 \\ xy = 15 \end{cases}$$
Solve the first equation for y, substitute into the second equation and solve.
$$2x + 2y = 16$$
$$2y = 16 - 2x$$
$$y = 8 - x$$
$$x(8-x) = 15$$
$$8x - x^2 = 15$$
$$x^2 - 8x + 15 = 0$$
$$(x-5)(x-3) = 0$$
$$x = 5 \quad \text{or} \quad x = 3$$
The dimensions of the rectangle are 3 inches by 5 inches.

89. Let x = the radius of the first circle.
Let y = the radius of the second circle.
$$\begin{cases} 2\pi x + 2\pi y = 12\pi \\ \pi x^2 + \pi y^2 = 20\pi \end{cases}$$
Solve the first equation for y, substitute into the second equation and solve:
$$2\pi x + 2\pi y = 12\pi$$
$$x + y = 6$$
$$y = 6 - x$$

$$\pi x^2 + \pi y^2 = 20\pi$$
$$x^2 + y^2 = 20$$
$$x^2 + (6-x)^2 = 20$$
$$x^2 + 36 - 12x + x^2 = 20 \Rightarrow 2x^2 - 12x + 16 = 0$$
$$x^2 - 6x + 8 = 0 \Rightarrow (x-4)(x-2) = 0$$
$$x = 4 \quad \text{or} \quad x = 2$$
$$y = 2 \qquad y = 4$$
The radii of the circles are 2 centimeters and 4 centimeters.

91. The tortoise takes $9 + 3 = 12$ minutes or 0.2 hour longer to complete the race than the hare.
Let r = the rate of the hare.
Let t = the time for the hare to complete the race. Then $t + 0.2$ = the time for the tortoise and $r - 0.5$ = the rate for the tortoise. Since the length of the race is 21 meters, the distance equations are:
$$\begin{cases} rt = 21 \Rightarrow r = \dfrac{21}{t} \\ (r - 0.5)(t + 0.2) = 21 \end{cases}$$
Solve the first equation for r, substitute into the second equation and solve:
$$\left(\dfrac{21}{t} - 0.5\right)(t + 0.2) = 21$$
$$21 + \dfrac{4.2}{t} - 0.5t - 0.1 = 21$$
$$10t\left(21 + \dfrac{4.2}{t} - 0.5t - 0.1\right) = 10t \cdot (21)$$
$$210t + 42 - 5t^2 - t = 210t$$
$$5t^2 + t - 42 = 0$$
$$(5t - 14)(t + 3) = 0$$
$5t - 14 = 0$ or $t + 3 = 0$
$5t = 14$ $t = -3$
$t = \dfrac{14}{5} = 2.8$

$t = -3$ makes no sense, since time cannot be negative.
Solve for r:
$r = \dfrac{21}{2.8} = 7.5$

The average speed of the hare is 7.5 meters per hour, and the average speed for the tortoise is 7 meters per hour.

93. Let x = the width of the cardboard. Let y = the length of the cardboard. The width of the box will be $x - 4$, the length of the box will be $y - 4$, and the height is 2. The volume is $V = (x - 4)(y - 4)(2)$.
Solve the system of equations:
$$\begin{cases} xy = 216 \quad \Rightarrow y = \dfrac{216}{x} \\ 2(x - 4)(y - 4) = 224 \end{cases}$$
Solve the first equation for y, substitute into the second equation and solve.

$$(2x - 8)\left(\dfrac{216}{x} - 4\right) = 224$$
$$432 - 8x - \dfrac{1728}{x} + 32 = 224$$
$$432x - 8x^2 - 1728 + 32x = 224x$$
$$8x^2 - 240x + 1728 = 0$$
$$x^2 - 30x + 216 = 0$$
$$(x - 12)(x - 18) = 0$$
$x - 12 = 0$ or $x - 18 = 0$
$x = 12$ $x = 18$

The cardboard should be 12 centimeters by 18 centimeters.

95. Find equations relating area and perimeter:
$$\begin{cases} x^2 + y^2 = 4500 \\ 3x + 3y + (x - y) = 300 \end{cases}$$
Solve the second equation for y, substitute into the first equation and solve:
$$4x + 2y = 300$$
$$2y = 300 - 4x$$
$$y = 150 - 2x$$
$$x^2 + (150 - 2x)^2 = 4500$$
$$x^2 + 22{,}500 - 600x + 4x^2 = 4500$$
$$5x^2 - 600x + 18{,}000 = 0$$
$$x^2 - 120x + 3600 = 0$$
$$(x - 60)^2 = 0$$
$$x - 60 = 0$$
$$x = 60$$
$$y = 150 - 2(60) = 30$$
The sides of the squares are 30 feet and 60 feet.

97. Solve the system for l and w:
$$\begin{cases} 2l + 2w = P \\ lw = A \end{cases}$$
Solve the first equation for l, substitute into the second equation and solve.
$$2l = P - 2w$$
$$l = \frac{P}{2} - w$$
$$\left(\frac{P}{2} - w\right)w = A$$
$$\frac{P}{2}w - w^2 = A$$
$$w^2 - \frac{P}{2}w + A = 0$$
$$w = \frac{\frac{P}{2} \pm \sqrt{\frac{P^2}{4} - 4A}}{2} = \frac{\frac{P}{2} \pm \sqrt{\frac{P^2}{4} - \frac{16A}{4}}}{2}$$
$$= \frac{\frac{P}{2} \pm \frac{\sqrt{P^2 - 16A}}{2}}{2} = \frac{P \pm \sqrt{P^2 - 16A}}{4}$$
If $w = \frac{P + \sqrt{P^2 - 16A}}{4}$ then
$$l = \frac{P}{2} - \frac{P + \sqrt{P^2 - 16A}}{4} = \frac{P - \sqrt{P^2 - 16A}}{4}$$
If $w = \frac{P - \sqrt{P^2 - 16A}}{4}$ then
$$l = \frac{P}{2} - \frac{P - \sqrt{P^2 - 16A}}{4} = \frac{P + \sqrt{P^2 - 16A}}{4}$$
If it is required that length be greater than width, then the solution is:
$$w = \frac{P - \sqrt{P^2 - 16A}}{4} \text{ and } l = \frac{P + \sqrt{P^2 - 16A}}{4}$$

99. Solve the equation: $m^2 - 4(2m - 4) = 0$
$$m^2 - 8m + 16 = 0$$
$$(m - 4)^2 = 0$$
$$m = 4$$
Use the point-slope equation with slope 4 and the point (2, 4) to obtain the equation of the tangent line:
$$y - 4 = 4(x - 2) \Rightarrow y - 4 = 4x - 8 \Rightarrow y = 4x - 4$$

101. Solve the system:
$$\begin{cases} y = x^2 + 2 \\ y = mx + b \end{cases}$$
Solve the system by substitution:
$$x^2 + 2 = mx + b \Rightarrow x^2 - mx + 2 - b = 0$$
Note that the tangent line passes through (1, 3).
Find the relation between m and b:
$$3 = m(1) + b \Rightarrow b = 3 - m$$
Substitute into the quadratic to eliminate b:
$$x^2 - mx + 2 - (3 - m) = 0 \Rightarrow x^2 - mx + (m - 1) = 0$$
Find when the discriminant equals 0:
$$(-m)^2 - 4(1)(m - 1) = 0$$
$$m^2 - 4m + 4 = 0$$
$$(m - 2)^2 = 0$$
$$m - 2 = 0$$
$$m = 2$$
$$b = 3 - m = 3 - 2 = 1$$
The equation of the tangent line is $y = 2x + 1$.

103. Solve the system:
$$\begin{cases} 2x^2 + 3y^2 = 14 \\ y = mx + b \end{cases}$$
Solve the system by substitution:
$$2x^2 + 3(mx + b)^2 = 14$$
$$2x^2 + 3m^2x^2 + 6mbx + 3b^2 = 14$$
$$(3m^2 + 2)x^2 + 6mbx + 3b^2 - 14 = 0$$
Note that the tangent line passes through (1, 2).
Find the relation between m and b:
$$2 = m(1) + b \Rightarrow b = 2 - m$$
Substitute into the quadratic to eliminate b:
$$(3m^2 + 2)x^2 + 6m(2 - m)x + 3(2 - m)^2 - 14 = 0$$
$$(3m^2 + 2)x^2 + (12m - 6m^2)x + (3m^2 - 12m - 2) = 0$$
Find when the discriminant equals 0:
$$(12m - 6m^2)^2 - 4(3m^2 + 2)(3m^2 - 12m - 2) = 0$$
$$144m^2 + 96m + 16 = 0$$
$$9m^2 + 6m + 1 = 0$$
$$(3m + 1)^2 = 0$$
$$3m + 1 = 0$$
$$m = -\frac{1}{3}$$
$$b = 2 - m = 2 - \left(-\frac{1}{3}\right) = \frac{7}{3}$$
The equation of the tangent line is $y = -\frac{1}{3}x + \frac{7}{3}$.

105. Solve the system:
$$\begin{cases} x^2 - y^2 = 3 \\ y = mx + b \end{cases}$$
Solve the system by substitution:
$$x^2 - (mx+b)^2 = 3$$
$$x^2 - m^2x^2 - 2mbx - b^2 = 3$$
$$(1-m^2)x^2 - 2mbx - b^2 - 3 = 0$$
Note that the tangent line passes through (2, 1). Find the relation between m and b:
$$1 = m(2) + b \Rightarrow b = 1 - 2m$$
Substitute into the quadratic to eliminate b:
$$(1-m^2)x^2 - 2m(1-2m)x - (1-2m)^2 - 3 = 0$$
$$(1-m^2)x^2 + (-2m+4m^2)x - 1 + 4m - 4m^2 - 3 = 0$$
$$(1-m^2)x^2 + (-2m+4m^2)x + (-4m^2+4m-4) = 0$$
Find when the discriminant equals 0:
$$(-2m+4m^2)^2 - 4(1-m^2)(-4m^2+4m-4) = 0$$
$$4m^2 - 16m^3 + 16m^4 - 16m^4 + 16m^3 - 16m + 16 = 0$$
$$4m^2 - 16m + 16 = 0$$
$$m^2 - 4m + 4 = 0$$
$$(m-2)^2 = 0$$
$$m = 2$$
The equation of the tangent line is $y = 2x - 3$.

107. Solve for r_1 and r_2:
$$\begin{cases} r_1 + r_2 = -\dfrac{b}{a} \\ r_1 r_2 = \dfrac{c}{a} \end{cases}$$
Substitute and solve:
$$r_1 = -r_2 - \frac{b}{a}$$
$$\left(-r_2 - \frac{b}{a}\right) r_2 = \frac{c}{a}$$
$$-r_2^2 - \frac{b}{a}r_2 - \frac{c}{a} = 0$$
$$ar_2^2 + br_2 + c = 0$$

$$r_2 = \frac{-b \pm \sqrt{b^2 - 4ac}}{2a}$$
$$r_1 = -r_2 - \frac{b}{a} =$$
$$= -\left(\frac{-b \pm \sqrt{b^2 - 4ac}}{2a}\right) - \frac{b}{a}$$
$$= \frac{-b \mp \sqrt{b^2 - 4ac}}{2a}$$
The solutions are:
$$\frac{-b + \sqrt{b^2-4ac}}{2a} \text{ and } \frac{-b - \sqrt{b^2-4ac}}{2a}.$$

109. Since the area of the square piece of sheet metal is 100 square feet, the sheet's dimensions are 10 feet by 10 feet. Let $x =$ the length of the cut.

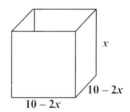

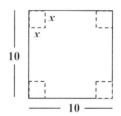

The dimensions of the box are
length $= 10 - 2x$; width $= 10 - 2x$; height $= x$
Note that each of these expressions must be positive. So we must have
$x > 0$ and $10 - 2x > 0 \Rightarrow x < 5$, that is, $0 < x < 5$.
So the volume of the box is given by
$$V = (\text{length}) \cdot (\text{width}) \cdot (\text{height})$$
$$= (10-2x)(10-2x)(x)$$
$$= (10-2x)^2 (x)$$

a. In order to get a volume equal to 9 cubic feet, we solve $(10-2x)^2(x) = 9$.

$(10-2x)^2(x) = 9$

$(100-40x+4x^2)x = 9$

$100x - 40x^2 + 4x^3 = 9$

So we need to solve the equation
$4x^3 - 40x^2 + 100x - 9 = 0$.

Graphing the function
$y_1 = 4x^3 - 40x^2 + 100x - 9$ on a calculator yields the graph

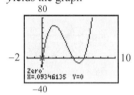

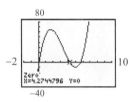

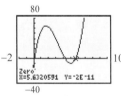

The graph indicates that there are three real zeros on the interval [0, 6].
Using the ZERO feature of a graphing calculator, we find that the three roots shown occur at $x \approx 0.093$, $x \approx 4.274$ and $x \approx 5.632$.
But we've already noted that we must have 0<x<5, so the only practical values for the sides of the square base are $x \approx 0.093$ feet and $x \approx 4.274$ feet.

b. Answers will vary.

Chapter 6 Review

1. $y^2 = -16x$
This is a parabola.
$a = 4$
Vertex: (0, 0)
Focus: (–4, 0)
Directrix: $x = 4$

3. $\dfrac{x^2}{25} - y^2 = 1$
This is a hyperbola.
$a = 5$, $b = 1$.
Find the value of c:
$c^2 = a^2 + b^2 = 25 + 1 = 26$
$c = \sqrt{26}$
Center: (0, 0)
Vertices: (5, 0), (–5, 0)
Foci: $(\sqrt{26}, 0)$, $(-\sqrt{26}, 0)$
Asymptotes: $y = \dfrac{1}{5}x$; $y = -\dfrac{1}{5}x$

5. $\dfrac{y^2}{25} + \dfrac{x^2}{16} = 1$
This is an ellipse.
$a = 5$, $b = 4$.
Find the value of c:
$c^2 = a^2 - b^2 = 25 - 16 = 9$
$c = 3$
Center: (0, 0)
Vertices: (0, 5), (0, –5)
Foci: (0, 3), (0, –3)

7. $x^2 + 4y = 4$
This is a parabola.
Write in standard form:
$x^2 = -4y + 4$
$x^2 = -4(y-1)$
$a = 1$
Vertex: (0, 1)
Focus: (0, 0)
Directrix: $y = 2$

9. $4x^2 - y^2 = 8$
   This is a hyperbola.
   Write in standard form:
   $$\frac{x^2}{2} - \frac{y^2}{8} = 1$$
   $a = \sqrt{2}, \ b = \sqrt{8} = 2\sqrt{2}$.
   Find the value of c:
   $c^2 = a^2 + b^2 = 2 + 8 = 10$
   $c = \sqrt{10}$
   Center: (0, 0)
   Vertices: $\left(-\sqrt{2}, 0\right), \left(\sqrt{2}, 0\right)$
   Foci: $\left(-\sqrt{10}, 0\right), \left(\sqrt{10}, 0\right)$
   Asymptotes: $y = 2x; \ y = -2x$

11. $x^2 - 4x = 2y$
    This is a parabola.
    Write in standard form:
    $x^2 - 4x + 4 = 2y + 4$
    $(x-2)^2 = 2(y+2)$
    $a = \frac{1}{2}$
    Vertex: (2, −2)
    Focus: $\left(2, -\frac{3}{2}\right)$
    Directrix: $y = -\frac{5}{2}$

13. $y^2 - 4y - 4x^2 + 8x = 4$
    This is a hyperbola.
    Write in standard form:
    $(y^2 - 4y + 4) - 4(x^2 - 2x + 1) = 4 + 4 - 4$
    $(y-2)^2 - 4(x-1)^2 = 4$
    $$\frac{(y-2)^2}{4} - \frac{(x-1)^2}{1} = 1$$
    $a = 2, \ b = 1$.
    Find the value of c:
    $c^2 = a^2 + b^2 = 4 + 1 = 5$
    $c = \sqrt{5}$
    Center: (1, 2)
    Vertices: (1, 0), (1, 4)
    Foci: $\left(1, 2-\sqrt{5}\right), \left(1, 2+\sqrt{5}\right)$
    Asymptotes: $y - 2 = 2(x-1); \ y - 2 = -2(x-1)$

15. $4x^2 + 9y^2 - 16x - 18y = 11$
    This is an ellipse.
    Write in standard form:
    $4x^2 + 9y^2 - 16x - 18y = 11$
    $4(x^2 - 4x + 4) + 9(y^2 - 2y + 1) = 11 + 16 + 9$
    $4(x-2)^2 + 9(y-1)^2 = 36$
    $$\frac{(x-2)^2}{9} + \frac{(y-1)^2}{4} = 1$$
    $a = 3, \ b = 2$.
    Find the value of c:
    $c^2 = a^2 - b^2 = 9 - 4 = 5$
    $c = \sqrt{5}$
    Center: (2, 1); Vertices: (−1, 1), (5, 1)
    Foci: $\left(2-\sqrt{5}, 1\right), \left(2+\sqrt{5}, 1\right)$

17. $4x^2 - 16x + 16y + 32 = 0$
    This is a parabola.
    Write in standard form:
    $4(x^2 - 4x + 4) = -16y - 32 + 16$
    $4(x-2)^2 = -16(y+1)$
    $(x-2)^2 = -4(y+1)$
    $a = 1$
    Vertex: (2, −1); Focus: (2, −2);
    Directrix: $y = 0$

19. $9x^2 + 4y^2 - 18x + 8y = 23$
    This is an ellipse.
    Write in standard form:
    $9(x^2 - 2x + 1) + 4(y^2 + 2y + 1) = 23 + 9 + 4$
    $9(x-1)^2 + 4(y+1)^2 = 36$
    $$\frac{(x-1)^2}{4} + \frac{(y+1)^2}{9} = 1$$
    $a = 3, \ b = 2$.
    Find the value of c:
    $c^2 = a^2 - b^2 = 9 - 4 = 5$
    $c = \sqrt{5}$
    Center: (1, −1)
    Vertices: (1, −4), (1, 2)
    Foci: $\left(1, -1-\sqrt{5}\right), \left(1, -1+\sqrt{5}\right)$

21. Parabola: The focus is (–2, 0) and the directrix is $x = 2$. The vertex is (0, 0). $a = 2$ and since (–2, 0) is to the left of (0, 0), the parabola opens to the left. The equation of the parabola is:
$y^2 = -4ax$
$y^2 = -4 \cdot 2 \cdot x$
$y^2 = -8x$

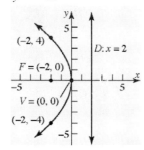

23. Hyperbola: Center: (0, 0); Focus: (0, 4); Vertex: (0, –2); Transverse axis is the y-axis; $a = 2$; $c = 4$.
Find b:
$b^2 = c^2 - a^2 = 16 - 4 = 12$
$b = \sqrt{12} = 2\sqrt{3}$
Write the equation: $\dfrac{y^2}{4} - \dfrac{x^2}{12} = 1$

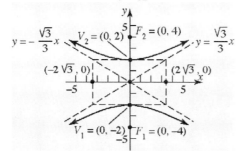

25. Ellipse: Foci: (–3, 0), (3, 0); Vertex: (4, 0); Center: (0, 0); Major axis is the x-axis; $a = 4$; $c = 3$. Find b:
$b^2 = a^2 - c^2 = 16 - 9 = 7$
$b = \sqrt{7}$

Write the equation: $\dfrac{x^2}{16} + \dfrac{y^2}{7} = 1$

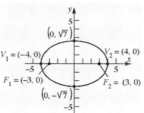

27. Parabola: The focus is (2, –4) and the vertex is (2, –3). Both lie on the vertical line $x = 2$. $a = 1$ and since (2, –4) is below (2, –3), the parabola opens down. The equation of the parabola is:
$(x-h)^2 = -4a(y-k)$
$(x-2)^2 = -4 \cdot 1 \cdot (y-(-3))$
$(x-2)^2 = -4(y+3)$

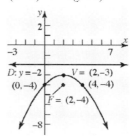

29. Hyperbola: Center: (–2, –3); Focus: (–4, –3); Vertex: (–3, –3); Transverse axis is parallel to the x-axis; $a = 1$; $c = 2$. Find b:
$b^2 = c^2 - a^2 = 4 - 1 = 3$
$b = \sqrt{3}$
Write the equation: $\dfrac{(x+2)^2}{1} - \dfrac{(y+3)^2}{3} = 1$

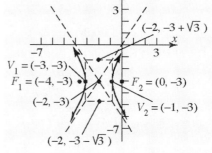

31. Ellipse: Foci: $(-4, 2), (-4, 8)$; Vertex: $(-4, 10)$; Center: $(-4, 5)$; Major axis is parallel to the y-axis; $a = 5$; $c = 3$. Find b:
$b^2 = a^2 - c^2 = 25 - 9 = 16 \rightarrow b = 4$
Write the equation: $\dfrac{(x+4)^2}{16} + \dfrac{(y-5)^2}{25} = 1$

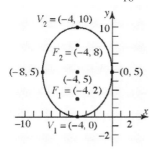

33. Hyperbola: Center: $(-1, 2)$; $a = 3$; $c = 4$; Transverse axis parallel to the x-axis; Find b:
$b^2 = c^2 - a^2 = 16 - 9 = 7$
$b = \sqrt{7}$
Write the equation: $\dfrac{(x+1)^2}{9} - \dfrac{(y-2)^2}{7} = 1$

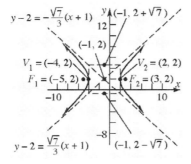

35. Hyperbola: Vertices: $(0, 1), (6, 1)$; Asymptote: $3y + 2x - 9 = 0$; Center: $(3, 1)$; Transverse axis is parallel to the x-axis; $a = 3$; The slope of the asymptote is $-\dfrac{2}{3}$; Find b:
$\dfrac{-b}{a} = \dfrac{-b}{3} = \dfrac{-2}{3} \rightarrow -3b = -6 \rightarrow b = 2$

Write the equation: $\dfrac{(x-3)^2}{9} - \dfrac{(y-1)^2}{4} = 1$

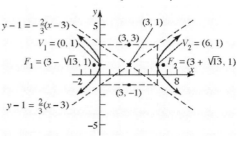

37. Solve the first equation for y, substitute into the second equation and solve:
$\begin{cases} 2x + y + 3 = 0 \rightarrow y = -2x - 3 \\ x^2 + y^2 = 5 \end{cases}$
$x^2 + (-2x-3)^2 = 5 \rightarrow x^2 + 4x^2 + 12x + 9 = 5$
$5x^2 + 12x + 4 = 0 \rightarrow (5x+2)(x+2) = 0$
$\Rightarrow x = -\dfrac{2}{5}$ or $x = -2$
$y = -\dfrac{11}{5} \quad y = 1$
Solutions: $\left(-\dfrac{2}{5}, -\dfrac{11}{5}\right), (-2, 1)$.

39. Multiply each side of the second equation by 2 and add the equations to eliminate xy:
$\begin{cases} 2xy + y^2 = 10 \longrightarrow 2xy + y^2 = 10 \\ -xy + 3y^2 = 2 \xrightarrow{2} -2xy + 6y^2 = 4 \end{cases}$
$7y^2 = 14$
$y^2 = 2$
$y = \pm\sqrt{2}$

If $y = \sqrt{2}$:
$2x(\sqrt{2}) + (\sqrt{2})^2 = 10 \rightarrow 2\sqrt{2}x = 8 \rightarrow x = 2\sqrt{2}$

If $y = -\sqrt{2}$:
$2x(-\sqrt{2}) + (-\sqrt{2})^2 = 10 \rightarrow -2\sqrt{2}x = 8$
$\rightarrow x = -2\sqrt{2}$

Solutions: $(2\sqrt{2}, \sqrt{2}), (-2\sqrt{2}, -\sqrt{2})$

41. Substitute into the second equation into the first equation and solve:
$$\begin{cases} x^2 + y^2 = 6y \\ x^2 = 3y \end{cases}$$
$3y + y^2 = 6y$
$y^2 - 3y = 0$
$y(y-3) = 0 \to y = 0$ or $y = 3$
If $y = 0$: $x^2 = 3(0) \to x^2 = 0 \to x = 0$
If $y = 3$: $x^2 = 3(3) \to x^2 = 9 \to x = \pm 3$
Solutions: $(0, 0), (-3, 3), (3, 3)$.

43. Factor the second equation, solve for x, substitute into the first equation and solve:
$$\begin{cases} 3x^2 + 4xy + 5y^2 = 8 \\ x^2 + 3xy + 2y^2 = 0 \end{cases}$$
$x^2 + 3xy + 2y^2 = 0$
$(x + 2y)(x + y) = 0 \to x = -2y$ or $x = -y$
Substitute $x = -2y$ and solve:
$3x^2 + 4xy + 5y^2 = 8$
$3(-2y)^2 + 4(-2y)y + 5y^2 = 8$
$12y^2 - 8y^2 + 5y^2 = 8$
$9y^2 = 8$
$y^2 = \frac{8}{9} \Rightarrow y = \pm \frac{2\sqrt{2}}{3}$

Substitute $x = -y$ and solve:
$3x^2 + 4xy + 5y^2 = 8$
$3(-y)^2 + 4(-y)y + 5y^2 = 8$
$3y^2 - 4y^2 + 5y^2 = 8$
$4y^2 = 8$
$y^2 = 2 \Rightarrow y = \pm\sqrt{2}$

If $y = \frac{2\sqrt{2}}{3}$: $x = -2\left(\frac{2\sqrt{2}}{3}\right) = \frac{-4\sqrt{2}}{3}$
If $y = \frac{-2\sqrt{2}}{3}$: $x = -2\left(\frac{-2\sqrt{2}}{3}\right) = \frac{4\sqrt{2}}{3}$
If $y = \sqrt{2}$: $x = -\sqrt{2}$
If $y = -\sqrt{2}$: $x = \sqrt{2}$

Solutions:
$\left(\frac{-4\sqrt{2}}{3}, \frac{2\sqrt{2}}{3}\right), \left(\frac{4\sqrt{2}}{3}, \frac{-2\sqrt{2}}{3}\right), \left(-\sqrt{2}, \sqrt{2}\right),$
$\left(\sqrt{2}, -\sqrt{2}\right)$

45. $$\begin{cases} x^2 - 3x + y^2 + y = -2 \\ \dfrac{x^2 - x}{y} + y + 1 = 0 \end{cases}$$
Multiply each side of the second equation by $-y$ and add the equations to eliminate y:
$x^2 - 3x + y^2 + y = -2$
$-x^2 + x - y^2 - y = 0$
$-2x = -2 \Rightarrow x = 1$
If $x = 1$: $1^2 - 3(1) + y^2 + y = -2$
$y^2 + y = 0$
$y(y + 1) = 0$
$y = 0$ or $y = -1$
Note that $y \neq 0$ because that would cause division by zero in the original system.
Solution: $(1, -1)$

47. Graph the system of inequalities:
$$\begin{cases} x^2 + y^2 \leq 16 \\ x + y \geq 2 \end{cases}$$
Graph the circle $x^2 + y^2 = 16$. Use a solid line since the inequality uses $\leq$. Choose a test point not on the circle, such as $(0, 0)$. Since $0^2 + 0^2 \leq 16$ is true, shade the side of the circle containing $(0, 0)$.
Graph the line $x + y = 2$. Use a solid line since the inequality uses $\geq$. Choose a test point not on the line, such as $(0, 0)$. Since $0 + 0 \geq 2$ is false, shade the opposite side of the line from $(0, 0)$. The overlapping region is the solution.

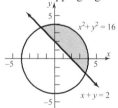

49. Graph the system of inequalities:
$$\begin{cases} y \le x^2 \\ xy \le 4 \end{cases}$$

Graph the parabola $y = x^2$. Use a solid line since the inequality uses $\le$. Choose a test point not on the parabola, such as (1, 2). Since $2 \le 1^2$ is false, shade the opposite side of the parabola from (1, 2).
Graph the hyperbola $xy = 4$. Use a solid line since the inequality uses $\le$. Choose a test point not on the hyperbola, such as (1, 2). Since $1 \cdot 2 \le 4$ is true, shade the same side of the hyperbola as (1, 2). The overlapping region is the solution.

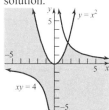

51. Write the equation in standard form:
$$4x^2 + 9y^2 = 36 \;\to\; \frac{x^2}{9} + \frac{y^2}{4} = 1$$
The center of the ellipse is (0, 0). The major axis is the x-axis.
$a = 3;\; b = 2;$
$c^2 = a^2 - b^2 = 9 - 4 = 5 \;\to\; c = \sqrt{5}.$
For the ellipse:
Vertices: $(-3, 0), (3, 0);$
Foci: $\left(-\sqrt{5}, 0\right), \left(\sqrt{5}, 0\right)$

For the hyperbola:
Foci: $(-3, 0), (3, 0);$
Vertices: $\left(-\sqrt{5}, 0\right), \left(\sqrt{5}, 0\right);$
Center: $(0, 0)$
$a = \sqrt{5};\; c = 3;$
$b^2 = c^2 - a^2 = 9 - 5 = 4 \;\to\; b = 2$
The equation of the hyperbola is: $\dfrac{x^2}{5} - \dfrac{y^2}{4} = 1$

53. Let (x, y) be any point in the collection of points.
The distance from
(x, y) to $(3, 0) = \sqrt{(x-3)^2 + y^2}$.
The distance from
(x, y) to the line $x = \dfrac{16}{3}$ is $\left| x - \dfrac{16}{3} \right|$.
Relating the distances, we have:
$$\sqrt{(x-3)^2 + y^2} = \frac{3}{4}\left| x - \frac{16}{3} \right|$$
$$(x-3)^2 + y^2 = \frac{9}{16}\left(x - \frac{16}{3} \right)^2$$
$$x^2 - 6x + 9 + y^2 = \frac{9}{16}\left(x^2 - \frac{32}{3}x + \frac{256}{9} \right)$$
$$16x^2 - 96x + 144 + 16y^2 = 9x^2 - 96x + 256$$
$$7x^2 + 16y^2 = 112$$
$$\frac{7x^2}{112} + \frac{16y^2}{112} = 1$$
$$\frac{x^2}{16} + \frac{y^2}{7} = 1$$
The set of points is an ellipse.

55. Locate the parabola so that the vertex is at (0, 0) and opens up. It then has the equation:
$x^2 = 4ay$. Since the light source is located at the focus and is 1 foot from the base, $a = 1$. Thus,
$x^2 = 4y$. The diameter is 2, so the point (1, y) is located on the parabola. Solve for y:
$1^2 = 4y \to 1 = 4y \to y = 0.25$ feet
The mirror is 0.25 feet, or 3 inches, deep.

57. Place the semi-elliptical arch so that the x-axis coincides with the water and the y-axis passes through the center of the arch. Since the bridge has a span of 60 feet, the length of the major axis is 60, or $2a = 60$ or $a = 30$. The maximum height of the bridge is 20 feet, so $b = 20$. The equation is: $\dfrac{x^2}{900} + \dfrac{y^2}{400} = 1$.
The height 5 feet from the center:

$$\frac{5^2}{900}+\frac{y^2}{400}=1$$
$$\frac{y^2}{400}=1-\frac{25}{900}$$
$$y^2=400\cdot\frac{875}{900}\rightarrow y\approx 19.72 \text{ feet}$$

The height 10 feet from the center:
$$\frac{10^2}{900}+\frac{y^2}{400}=1$$
$$\frac{y^2}{400}=1-\frac{100}{900}$$
$$y^2=400\cdot\frac{800}{900}\rightarrow y\approx 18.86 \text{ feet}$$

The height 20 feet from the center:
$$\frac{20^2}{900}+\frac{y^2}{400}=1$$
$$\frac{y^2}{400}=1-\frac{400}{900}$$
$$y^2=400\cdot\frac{500}{900}\rightarrow y\approx 14.91 \text{ feet}$$

59. First note that all points where an explosion could take place, such that the time difference would be the same as that for the first detonation, would form a hyperbola with A and B as the foci.
Start with a diagram:

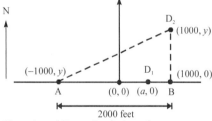

Since A and B are the foci, we have
$2c = 2000$
$c = 1000$

Since D_1 is on the transverse axis and is on the hyperbola, then it must be a vertex of the hyperbola. Since it is 200 feet from B, we have $a = 800$. Finally,
$b^2 = c^2 - a^2 = 1000^2 - 800^2 = 360,000$

Thus, the equation of the hyperbola is
$$\frac{x^2}{640,000}-\frac{y^2}{360,000}=1$$

The point $(1000, y)$ needs to lie on the graph of the hyperbola. Thus, we have
$$\frac{(1000)^2}{640,000}-\frac{y^2}{360,000}=1$$
$$-\frac{y^2}{360,000}=-\frac{9}{16}$$
$$y^2=202,500$$
$$y=450$$

The second explosion should be set off 450 feet due north of point B.

Chapter 6 Test

1. $\dfrac{(x+1)^2}{4}-\dfrac{y^2}{9}=1$

Rewriting the equation as
$$\frac{(x-(-1))^2}{2^2}-\frac{(y-0)^2}{3^2}=1\text{, we see that this is the}$$
equation of a hyperbola in the form
$\dfrac{(x-h)^2}{a^2}-\dfrac{(y-k)^2}{b^2}=1$. Therefore, we have
$h=-1$, $k=0$, $a=2$, and $b=3$. Since $a^2=4$ and $b^2=9$, we get $c^2=a^2+b^2=4+9=13$, or $c=\sqrt{13}$. The center is at $(-1,0)$ and the transverse axis is the x-axis. The vertices are at $(h\pm a,k)=(-1\pm 2,0)$, or $(-3,0)$ and $(1,0)$.
The foci are at $(h\pm c,k)=(-1\pm\sqrt{13},0)$, or $(-1-\sqrt{13},0)$ and $(-1+\sqrt{13},0)$. The asymptotes are $y-0=\pm\dfrac{3}{2}(x-(-1))$, or $y=-\dfrac{3}{2}(x+1)$ and $y=\dfrac{3}{2}(x+1)$.

2. $8y=(x-1)^2-4$
Rewriting gives
$(x-1)^2=8y+4$
$(x-1)^2=8\left(y-\left(-\dfrac{1}{2}\right)\right)$
$(x-1)^2=4(2)\left(y-\left(-\dfrac{1}{2}\right)\right)$

This is the equation of a parabola in the form $(x-h)^2=4a(y-k)$. Therefore, the axis of symmetry is parallel to the y-axis and we have $(h,k) = \left(1, -\dfrac{1}{2}\right)$ and $a = 2$. The vertex is at $(h,k) = \left(1, -\dfrac{1}{2}\right)$, the axis of symmetry is $x = 1$, the focus is at $(h, k+a) = \left(1, -\dfrac{1}{2} + 2\right) = \left(1, \dfrac{3}{2}\right)$, and the directrix is given by the line $y = k - a$, or $y = -\dfrac{5}{2}$.

3. $2x^2 + 3y^2 + 4x - 6y = 13$

Rewrite the equation by completing the square in x and y.

$$2x^2 + 3y^2 + 4x - 6y = 13$$
$$2x^2 + 4x + 3y^2 - 6y = 13$$
$$2(x^2 + 2x) + 3(y^2 - 2y) = 13$$
$$2(x^2 + 2x + 1) + 3(y^2 - 2y + 1) = 13 + 2 + 3$$
$$2(x+1)^2 + 3(y-1)^2 = 18$$
$$\dfrac{(x-(-1))^2}{9} + \dfrac{(y-1)^2}{6} = 1$$

This is the equation of an ellipse with center at $(-1, 1)$ and major axis parallel to the x-axis.

Since $a^2 = 9$ and $b^2 = 6$, we have $c^2 = a^2 - b^2 = 9 - 6 = 3$, or $c = \sqrt{3}$. The foci are $(h \pm c, k) = (-1 \pm \sqrt{3}, 1)$ or $(-1 - \sqrt{3}, 1)$ and $(-1 + \sqrt{3}, 1)$. The vertices are at $(h \pm a, k) = (-1 \pm 3, 1)$, or $(-4, 1)$ and $(2, 1)$.

4. The vertex $(-1, 3)$ and the focus $(-1, 4.5)$ both lie on the vertical line $x = -1$ (the axis of symmetry). The distance a from the vertex to the focus is $a = 1.5$. Because the focus lies above the vertex, we know the parabola opens upward. As a result, the form of the equation is

$$(x - h)^2 = 4a(y - k)$$

where $(h, k) = (-1, 3)$ and $a = 1.5$. Therefore, the equation is

$$(x + 1)^2 = 4(1.5)(y - 3)$$
$$(x + 1)^2 = 6(y - 3)$$

The points $(h \pm 2a, k)$, that is $(-4, 4.5)$ and $(2, 4.5)$, define the lattice rectum; the line $y = 1.5$ is the directrix.

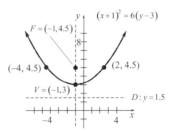

5. The center is $(h, k) = (0, 0)$ so $h = 0$ and $k = 0$. Since the center, focus, and vertex all lie on the line $x = 0$, the major axis is the y-axis. The distance from the center $(0, 0)$ to a focus $(0, 3)$ is $c = 3$. The distance from the center $(0, 0)$ to a vertex $(0, -4)$ is $a = 4$. Then,

$$b^2 = a^2 - c^2 = 4^2 - 3^2 = 16 - 9 = 7$$

The form of the equation is

$$\dfrac{(x-h)^2}{b^2} + \dfrac{(y-k)^2}{a^2} = 1$$

where $h = 0$, $k = 0$, $a = 4$, and $b = \sqrt{7}$. Thus, we get

$$\dfrac{x^2}{7} + \dfrac{y^2}{16} = 1$$

To graph the equation, we use the center $(h, k) = (0, 0)$ to locate the vertices. The major axis is the y-axis, so the vertices are $a = 4$ units above and below the center. Therefore, the vertices are $V_1 = (0, 4)$ and $V_2 = (0, -4)$. Since $c = 3$ and the major axis is the y-axis, the foci are 3 units above and below the center. Therefore, the foci are $F_1 = (0, 3)$ and $F_2 = (0, -3)$. Finally, we use the value $b = \sqrt{7}$ to find the two points left and right of the center: $(-\sqrt{7}, 0)$ and $(\sqrt{7}, 0)$.

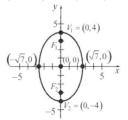

6. The center $(h,k)=(2,2)$ and vertex $(2,4)$ both lie on the line $x=2$, the transverse axis is parallel to the y-axis. The distance from the center $(2,2)$ to the vertex $(2,4)$ is $a=2$, so the other vertex must be $(2,0)$. The form of the equation is
$$\frac{(y-k)^2}{a^2}-\frac{(x-h)^2}{b^2}=1$$
where $h=2$, $k=2$, and $a=2$. This gives us
$$\frac{(y-2)^2}{4}-\frac{(x-2)^2}{b^2}=1$$
Since the graph contains the point $(x,y)=(2+\sqrt{10},5)$, we can use this point to determine the value for b.
$$\frac{(5-2)^2}{4}-\frac{(2+\sqrt{10}-2)^2}{b^2}=1$$
$$\frac{9}{4}-\frac{10}{b^2}=1$$
$$\frac{5}{4}=\frac{10}{b^2}$$
$$b^2=8$$
$$b=2\sqrt{2}$$
Therefore, the equation becomes
$$\frac{(y-2)^2}{4}-\frac{(x-2)^2}{8}=1$$
Since $c^2=a^2+b^2=4+8=12$, the distance from the center to either focus is $c=2\sqrt{3}$. Therefore, the foci are $c=2\sqrt{3}$ units above and below the center. The foci are $F_1=(2,2+2\sqrt{3})$ and $F_2=(2,2-2\sqrt{3})$. The asymptotes are given by the lines $y-k=\pm\frac{a}{b}(x-h)$. Therefore, the asymptotes are
$$y-2=\pm\frac{2}{2\sqrt{2}}(x-2)$$
$$y=\pm\frac{\sqrt{2}}{2}(x-2)+2$$

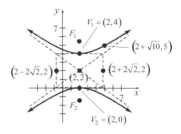

7. $\begin{cases} 3x^2+y^2=12 & (1) \\ y^2=9x & (2) \end{cases}$

We will use substitution to solve the system. Since Equation (2) is already solved for y^2, we substitute the expression $y^2=9x$ for y^2 in equation (1) and solve the equation for x.
$$3x^2+y^2=12$$
$$3x^2+9x=12$$
$$3x^2+9x-12=0$$
$$x^2+3x-4=0$$
$$(x+4)(x-1)=0$$
$$x=-4 \text{ or } x=1$$
Using these values for x in $y^2=9x$, we find
$$y^2=9(-4)=-36 \quad \text{or} \quad y^2=9(1)=9$$
The first equation, $y^2=-36$, yields no real solutions. From the second equation, we get $y^2=9$ or $y=\pm 3$.
The solutions are $(1,-3)$ and $(1,3)$.

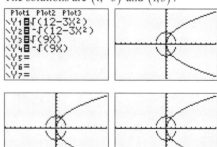

460

8. $\begin{cases} 2y^2 - 3x^2 = 5 & (1) \\ y - x = 1 & (2) \end{cases}$

We will solve the system using substitution. Solve equation (2) for y.
$y - x = 1$
$y = x + 1$

Substitute this expression for y in equation (1) and solve for x.
$$2y^2 - 3x^2 = 5$$
$$2(x+1)^2 - 3x^2 = 5$$
$$2(x^2 + 2x + 1) - 3x^2 = 5$$
$$2x^2 + 4x + 2 - 3x^2 = 5$$
$$-x^2 + 4x - 3 = 0$$
$$x^2 - 4x + 3 = 0$$
$$(x-3)(x-1) = 0$$
$$x = 3 \quad \text{or} \quad x = 1$$

Use these values for x in equation (2) to find y.
$y - (3) = 1$ or $y - (1) = 1$
$y = 4 \qquad y = 2$

The solutions are $(3,4)$ and $(1,2)$.

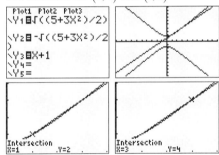

9. $\begin{cases} x^2 + y^2 \le 100 \\ 4x - 3y \ge 0 \end{cases}$

Write the second inequality in standard form:
$4x - 3y \ge 0$
$\qquad -3y \ge -4x$
$\qquad y \le \dfrac{4}{3}x$

Graph the circle $x^2 + y^2 = 100$. Use a solid line since the inequality is not strict. Choose a test point not on the circle, such as $(0, 0)$. Since $0^2 + 0^2 \le 100$ is true, shade the side of the circle containing $(0, 0)$.

Graph the line $y = \dfrac{4}{3}x$. Use a solid line since the inequality is not strict. Choose a test point not on the line, such as $(0, 1)$. Since $1 \le 0$ is false, shade the side of the line that does not contain $(0, 1)$. The overlapping region is the solution.

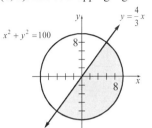

10. We can draw the parabola used to form the reflector on a rectangular coordinate system so that the vertex of the parabola is at the origin and its focus is on the positive y-axis.

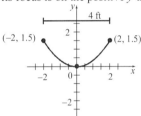

The form of the equation of the parabola is $x^2 = 4ay$ and its focus is at $(0, a)$. Since the point $(2, 1.5)$ is on the graph, we have
$$2^2 = 4a(1.5)$$
$$4 = 6a$$
$$a = \frac{2}{3}$$

The microphone should be located $\dfrac{2}{3}$ feet (or 8 inches) from the base of the reflector, along its axis of symmetry.

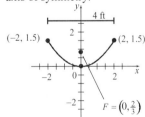

Cumulative Review R-6

1. $\dfrac{f(x+h)-f(x)}{h}$

   $= \dfrac{-3(x+h)^2 + 5(x+h) - 2 - (-3x^2 + 5x - 2)}{h}$

   $= \dfrac{-3(x^2 + 2xh + h^2) + 5x + 5h - 2 + 3x^2 - 5x + 2}{h}$

   $= \dfrac{-3x^2 - 6xh - 3h^2 + 5x + 5h - 2 + 3x^2 - 5x + 2}{h}$

   $= \dfrac{-6xh - 3h^2 + 5h}{h} = \dfrac{h(-6x - 3h + 5)}{h}$

   $= -6x - 3h + 5$

3. $9x^4 + 33x^3 - 71x^2 - 57x - 10 = 0$
   There are at most 4 real zeros.
   Possible rational zeros:
   $p = \pm 1, \pm 2, \pm 5, \pm 10; \quad q = \pm 1, \pm 3, \pm 9;$
   $\dfrac{p}{q} = \pm 1, \pm \dfrac{1}{3}, \pm \dfrac{1}{9}, \pm 2, \pm \dfrac{2}{3}, \pm \dfrac{2}{9}, \pm 5,$
   $\pm \dfrac{5}{3}, \pm \dfrac{5}{9}, \pm 10, \pm \dfrac{10}{3}, \pm \dfrac{10}{9}$

   Graphing $y_1 = 9x^4 + 33x^3 - 71x^2 - 57x - 10$ indicates that there appear to be zeros at $x = -5$ and at $x = 2$.
   Using synthetic division with $x = -5$:

   $\begin{array}{r|rrrrr} -5) & 9 & 33 & -71 & -57 & -10 \\ & & -45 & 60 & 55 & 10 \\ \hline & 9 & -12 & -11 & -2 & 0 \end{array}$

   Since the remainder is 0, –5 is a zero for f. So $x - (-5) = x + 5$ is a factor.
   The other factor is the quotient:
   $9x^3 - 12x^2 - 11x - 2$.
   Thus, $f(x) = (x+5)(9x^3 - 12x^2 - 11x - 2)$.
   Using synthetic division on the quotient and $x = 2$:

   $\begin{array}{r|rrrr} 2) & 9 & -12 & -11 & -2 \\ & & 18 & 12 & 2 \\ \hline & 9 & 6 & 1 & 0 \end{array}$

   Since the remainder is 0, 2 is a zero for f. So $x - 2$ is a factor; thus,
   $f(x) = (x+5)(x-2)(9x^2 + 6x + 1)$
   $= (x+5)(x-2)(3x+1)(3x+1)$

   Therefore, $x = -\dfrac{1}{3}$ is also a zero for f (with multiplicity 2).
   Solution set: $\left\{-5, -\dfrac{1}{3}, 2\right\}$.

5. a. This graph is a line containing points $(0, -2)$ and $(1, 0)$.
   slope $= \dfrac{\Delta y}{\Delta x} = \dfrac{0 - (-2)}{1 - 0} = \dfrac{2}{1} = 2$
   using $y - y_1 = m(x - x_1)$
   $y - 0 = 2(x - 1)$
   $y = 2x - 2$ or $2x - y - 2 = 0$

   b. This graph is a circle with center point $(2, 0)$ and radius 2.
   $(x-h)^2 + (y-k)^2 = r^2$
   $(x-2)^2 + (y-0)^2 = 2^2$
   $(x-2)^2 + y^2 = 4$

   c. This graph is an ellipse with center point $(0, 0)$; vertices $(\pm 3, 0)$ and y-intercepts $(0, \pm 2)$.
   $\dfrac{(x-h)^2}{a^2} + \dfrac{(y-k)^2}{b^2} = 1$
   $\dfrac{(x-0)^2}{3^2} + \dfrac{(y-0)^2}{2^2} = 1 \Rightarrow \dfrac{x^2}{9} + \dfrac{y^2}{4} = 1$

   d. This graph is a parabola with vertex $(1, 0)$ and y-intercept $(0, 2)$.
   $(x-h)^2 = 4a(y-k)$
   $(x-1)^2 = 4ay$
   $(0-1)^2 = 4a(2)$
   $1 = 8a$
   $a = \dfrac{1}{8}$
   $(x-1)^2 = \dfrac{1}{2}y$ or $y = 2(x-1)^2$

e. This graph is a hyperbola with center $(0,0)$ and vertices $(0,\pm 1)$, containing the point $(3,2)$.

$$\frac{(y-k)^2}{a^2} - \frac{(x-h)^2}{b^2} = 1$$

$$\frac{y^2}{1} - \frac{x^2}{b^2} = 1$$

$$\frac{(2)^2}{1} - \frac{(3)^2}{b^2} = 1$$

$$\frac{4}{1} - \frac{9}{b^2} = 1$$

$$4 - \frac{9}{b^2} = 1$$

$$3 = \frac{9}{b^2}$$

$$3b^2 = 9 \quad \rightarrow \quad b^2 = 3$$

The equation of the hyperbola is:

$$\frac{y^2}{1} - \frac{x^2}{3} = 1$$

f. This is the graph of an exponential function with y-intercept $(0,1)$, containing the point $(1,4)$.

$y = A \cdot b^x$

y-intercept $(0,1) \Rightarrow 1 = A \cdot b^0 = A \cdot 1 \Rightarrow A = 1$

point $(1,4) \Rightarrow 4 = b^1 = b$

Therefore, $y = 4^x$.

Chapter 7
Sequences; Induction; the Binomial Theorem

Section 7.1

1. $f(2) = \dfrac{2-1}{2} = \dfrac{1}{2}; \quad f(3) = \dfrac{3-1}{3} = \dfrac{2}{3}$

3. $A = P\left(1 + \dfrac{r}{n}\right)^{n \cdot t}$

   $= 1000\left(1 + \dfrac{0.04}{2}\right)^{2 \cdot 2}$

   $= 1000(1.02)^4$

   $= 1082.43$

   After two years, the account will contain $1082.43.

5. sequence

7. $\sum_{k=1}^{4}(2k) = 2\sum_{k=1}^{4} k = 2 \cdot \dfrac{4(4+1)}{2} = 4(5) = 20$

9. True; a sequence is a function whose domain is the set of positive integers.

11. $10! = 10 \cdot 9 \cdot 8 \cdot 7 \cdot 6 \cdot 5 \cdot 4 \cdot 3 \cdot 2 \cdot 1 = 3{,}628{,}800$

13. $\dfrac{9!}{6!} = \dfrac{9 \cdot 8 \cdot 7 \cdot 6!}{6!} = 9 \cdot 8 \cdot 7 = 504$

15. $\dfrac{3! \cdot 7!}{4!} = \dfrac{3 \cdot 2 \cdot 1 \cdot 7 \cdot 6 \cdot 5 \cdot 4!}{4!}$

    $= 3 \cdot 2 \cdot 1 \cdot 7 \cdot 6 \cdot 5 = 1{,}260$

17. $a_1 = 1,\ a_2 = 2,\ a_3 = 3,\ a_4 = 4,\ a_5 = 5$

19. $a_1 = \dfrac{1}{1+2} = \dfrac{1}{3},\ a_2 = \dfrac{2}{2+2} = \dfrac{2}{4} = \dfrac{1}{2},$

    $a_3 = \dfrac{3}{3+2} = \dfrac{3}{5},\ a_4 = \dfrac{4}{4+2} = \dfrac{4}{6} = \dfrac{2}{3},$

    $a_5 = \dfrac{5}{5+2} = \dfrac{5}{7}$

21. $a_1 = (-1)^{1+1}(1^2) = 1,\ a_2 = (-1)^{2+1}(2^2) = -4,$

    $a_3 = (-1)^{3+1}(3^2) = 9,\ a_4 = (-1)^{4+1}(4^2) = -16,$

    $a_5 = (-1)^{5+1}(5^2) = 25$

23. $a_1 = \dfrac{2^1}{3^1+1} = \dfrac{2}{4} = \dfrac{1}{2},\ a_2 = \dfrac{2^2}{3^2+1} = \dfrac{4}{10} = \dfrac{2}{5},$

    $a_3 = \dfrac{2^3}{3^3+1} = \dfrac{8}{28} = \dfrac{2}{7},\ a_4 = \dfrac{2^4}{3^4+1} = \dfrac{16}{82} = \dfrac{8}{41},$

    $a_5 = \dfrac{2^5}{3^5+1} = \dfrac{32}{244} = \dfrac{8}{61}$

25. $a_1 = \dfrac{(-1)^1}{(1+1)(1+2)} = \dfrac{-1}{2 \cdot 3} = -\dfrac{1}{6},$

    $a_2 = \dfrac{(-1)^2}{(2+1)(2+2)} = \dfrac{1}{3 \cdot 4} = \dfrac{1}{12},$

    $a_3 = \dfrac{(-1)^3}{(3+1)(3+2)} = \dfrac{-1}{4 \cdot 5} = -\dfrac{1}{20},$

    $a_4 = \dfrac{(-1)^4}{(4+1)(4+2)} = \dfrac{1}{5 \cdot 6} = \dfrac{1}{30},$

    $a_5 = \dfrac{(-1)^5}{(5+1)(5+2)} = \dfrac{-1}{6 \cdot 7} = -\dfrac{1}{42}$

27. $a_1 = \dfrac{1}{e^1} = \dfrac{1}{e},\ a_2 = \dfrac{2}{e^2},\ a_3 = \dfrac{3}{e^3},$

    $a_4 = \dfrac{4}{e^4},\ a_5 = \dfrac{5}{e^5}$

29. Each term is a fraction with the numerator equal to the term number and the denominator equal to one more than the term number.

    $a_n = \dfrac{n}{n+1}$

31. Each term is a fraction with the numerator equal to 1 and the denominator equal to a power of 2. The power is equal to one less than the term number.

    $a_n = \dfrac{1}{2^{n-1}}$

33. The terms form an alternating sequence. Ignoring the sign, each term always contains a 1. The sign alternates by raising -1 to a power. Since the first term is positive, we use $n+1$ as the power.

    $a_n = (-1)^{n+1}$

464

35. The terms (ignoring the sign) are equal to the term number. The alternating sign is obtained by using $(-1)^{n+1}$.
$a_n = (-1)^{n+1} \cdot n$

37. $a_1 = 2$, $a_2 = 3 + 2 = 5$, $a_3 = 3 + 5 = 8$,
$a_4 = 3 + 8 = 11$, $a_5 = 3 + 11 = 14$

39. $a_1 = -2$, $a_2 = 2 + (-2) = 0$, $a_3 = 3 + 0 = 3$,
$a_4 = 4 + 3 = 7$, $a_5 = 5 + 7 = 12$

41. $a_1 = 5$, $a_2 = 2 \cdot 5 = 10$, $a_3 = 2 \cdot 10 = 20$,
$a_4 = 2 \cdot 20 = 40$, $a_5 = 2 \cdot 40 = 80$

43. $a_1 = 3$, $a_2 = \dfrac{3}{2}$, $a_3 = \dfrac{\frac{3}{2}}{3} = \dfrac{1}{2}$,
$a_4 = \dfrac{\frac{1}{2}}{4} = \dfrac{1}{8}$, $a_5 = \dfrac{\frac{1}{8}}{5} = \dfrac{1}{40}$

45. $a_1 = 1$, $a_2 = 2$, $a_3 = 2 \cdot 1 = 2$, $a_4 = 2 \cdot 2 = 4$,
$a_5 = 4 \cdot 2 = 8$

47. $a_1 = A$, $a_2 = A + d$, $a_3 = (A+d) + d = A + 2d$,
$a_4 = (A+2d) + d = A + 3d$,
$a_5 = (A+3d) + d = A + 4d$

49. $a_1 = \sqrt{2}$, $a_2 = \sqrt{2+\sqrt{2}}$, $a_3 = \sqrt{2+\sqrt{2+\sqrt{2}}}$,
$a_4 = \sqrt{2+\sqrt{2+\sqrt{2+\sqrt{2}}}}$,
$a_5 = \sqrt{2+\sqrt{2+\sqrt{2+\sqrt{2+\sqrt{2}}}}}$

51. $\sum_{k=1}^{n}(k+2) = 3 + 4 + 5 + 6 + 7 + \cdots + (n+2)$

53. $\sum_{k=1}^{n}\dfrac{k^2}{2} = \dfrac{1}{2} + 2 + \dfrac{9}{2} + 8 + \dfrac{25}{2} + 18 + \dfrac{49}{2} + 32 + \cdots + \dfrac{n^2}{2}$

55. $\sum_{k=0}^{n}\dfrac{1}{3^k} = 1 + \dfrac{1}{3} + \dfrac{1}{9} + \dfrac{1}{27} + \cdots + \dfrac{1}{3^n}$

57. $\sum_{k=0}^{n-1}\dfrac{1}{3^{k+1}} = \dfrac{1}{3} + \dfrac{1}{9} + \dfrac{1}{27} + \cdots + \dfrac{1}{3^n}$

59. $\sum_{k=2}^{n}(-1)^k \ln k = \ln 2 - \ln 3 + \ln 4 + \cdots + (-1)^n \ln n$

61. $1 + 2 + 3 + \cdots + 20 = \sum_{k=1}^{20} k$

63. $\dfrac{1}{2} + \dfrac{2}{3} + \dfrac{3}{4} + \cdots + \dfrac{13}{13+1} = \sum_{k=1}^{13}\dfrac{k}{k+1}$

65. $1 - \dfrac{1}{3} + \dfrac{1}{9} - \dfrac{1}{27} + \cdots + (-1)^6 \left(\dfrac{1}{3^6}\right) = \sum_{k=0}^{6}(-1)^k\left(\dfrac{1}{3^k}\right)$

67. $3 + \dfrac{3^2}{2} + \dfrac{3^3}{3} + \cdots + \dfrac{3^n}{n} = \sum_{k=1}^{n}\dfrac{3^k}{k}$

69. $a + (a+d) + (a+2d) + \cdots + (a+nd) = \sum_{k=0}^{n}(a+kd)$

71. a. $\sum_{k=1}^{10} 5 = \underbrace{5 + 5 + 5 + \cdots + 5}_{10 \text{ times}} = 10(5) = 50$

b. Verify:

73. a. $\sum_{k=1}^{6} k = 1 + 2 + 3 + 4 + 5 + 6 = 21$

b. Verify:
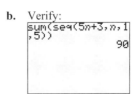

75. a. $\sum_{k=1}^{5}(5k+3) = (5 \cdot 1 + 3) + (5 \cdot 2 + 3) + (5 \cdot 3 + 3)$
$+ (5 \cdot 4 + 3) + (5 \cdot 5 + 3)$
$= 8 + 13 + 18 + 23 + 28 = 90$

b. Verify:

Chapter 7: Sequences; Induction; the Binomial Theorem

77. a. $\sum_{k=1}^{3}(k^2+4) = (1^2+4)+(2^2+4)+(3^2+4)$
$= 5+8+13 = 26$

 b. Verify:

79. a. $\sum_{k=1}^{6}(-1)^k 2^k = (-1)^1 \cdot 2^1 + (-1)^2 \cdot 2^2 + (-1)^3 \cdot 2^3$
$+(-1)^4 \cdot 2^4 + (-1)^5 \cdot 2^5 + (-1)^6 \cdot 2^6$
$= -2+4-8+16-32+64 = 42$

 b. Verify:

81. a. $\sum_{k=1}^{4}(k^3-1)$
$= (1^3-1)+(2^3-1)+(3^3-1)+(4^3-1)$
$= 0+7+26+63 = 96$

 b. Verify:

83. a. $B_2 = 1.01(3000) - 100 = \$2930$

 b. Put the graphing utility in SEQuence mode. Enter Y= as follows, then examine the TABLE:

From the table we see that the balance is below $2000 after 15 payments have been made. The balance then is $1953.70.

 c. Scrolling down the table, we find that balance is paid off in the 37th month. The last payment is $83.78. There are 36 payments of $100 and the last payment of $83.78. The total amount paid is: 36(100) + 83.78(1.01) = $3684.62. (we have to add the interest for the last month).

 d. The interest expense is:
3684.62 − 3000.00 = $684.62

85. a. $p_1 = 1.03(2000) + 20 = 2080$;
$p_2 = 1.03(2080) + 20 = 2162.4$

 b. Scrolling down the table, we find the trout population exceeds 5000 at the end of the 26th month when the population is 5084.

87. a. Since the fund returns 8% compound annually, this is equivalent to a return of 2% each quarter. Defining a recursive sequence, we have:
$a_1 = 500$, $a_n = 1.02a_{n-1} + 500$

b. Insert the formulas in your graphing utility and use the table feature to find when the value of the account will exceed $100,000:

In the 82nd quarter (approximately May 2019) the value of the account will exceed $100,000 with a value of $101,810.

c. Find the value of the account in 25 years or 100 quarters:

The value of the account will be $156,116.

89. a. Since the interest rate is 6% per annum compounded monthly, this is equivalent to a rate of 0.5% each month. Defining a recursive sequence, we have:
$a_0 = 150,000$, $a_n = 1.005a_{n-1} - 899.33$

b. $1.005(150,000) - 899.33 = \$149,850.67$

c. Enter the recursive formula in Y= and create the table:

d. Scroll through the table:

After 58 payments have been made, the balance is below $140,000. The balance is about $139,981.

e. Scroll through the table:

The loan will be paid off at the end of 360 months or 30 years.
Total amount paid = $(359)(\$899.33) + \$890.65(1.005) = \$323,754.57$.

f. The total interest expense is the difference of the total of the payments and the original loan: $323,754.57 - 150,000 = \$173,754.57$

g. (a) Since the interest rate is 6% per annum compounded monthly, this is equivalent to a rate of 0.5% each month. Defining a recursive sequence, we have:
$a_0 = 150,000$, $a_n = 1.005a_{n-1} - 999.33$

(b) $1.005(150,000) - 999.33 = \$149,750.67$

(c) Enter the recursive formula in Y= and create the table:

(d) Scroll through the table:

After 37 payments have been made, the balance is below $140,000. The balance is $139,894.

(e) Scroll through the table:

n	u(n)
278	353.69
279	-643.9
280	-1646
281	-2654
282	-3667
283	-4684
284	-5707

n=279

The loan will be paid off at the end of 279 months or 23 years and 3 months. Total amount paid = (278)($999.33) + 353.69(1.005) = $278,169.20

(f) The total interest expense is the difference of the total of the payments and the original loan:
278,169.20 − 150,000 = $128,169.20

h. Yes, if they can afford the additional monthly payment. They would save $44,586.07 in interest payments by paying the loan off sooner.

91. $a_1 = 1$, $a_2 = 1$, $a_3 = 2$, $a_4 = 3$, $a_5 = 5$,
$a_6 = 8$, $a_7 = 13$, $a_8 = 21$, $a_n = a_{n-1} + a_{n-2}$
$a_8 = a_7 + a_6 = 13 + 8 = 21$
After 7 months there are 21 mature pairs of rabbits.

93. 1, 1, 2, 3, 5, 8, 13
This is the Fibonacci sequence.

95. a. $e^{1.3} \approx \sum_{k=0}^{4} \frac{1.3^k}{k!} = \frac{1.3^0}{0!} + \frac{1.3^1}{1!} + ... + \frac{1.3^4}{4!}$
≈ 3.630170833

b. $e^{1.3} \approx \sum_{k=0}^{7} \frac{1.3^k}{k!} = \frac{1.3^0}{0!} + \frac{1.3^1}{1!} + ... + \frac{1.3^7}{7!}$
≈ 3.669060828

c. $e^{1.3} \approx 3.669296668$

e^(1.3)
        3.669296668

d. It will take $n = 12$ to approximate $e^{1.3}$ correct to 8 decimal places.

3.669296614
sum(seq(1.3^n/n!
,n,0,13))
        3.669296667
sum(seq(1.3^n/n!
,n,0,12))
        3.669296662

97. To show that $1 + 2 + 3 + ... + (n-1) + n = \frac{n(n+1)}{2}$

Let $S = 1 + 2 + 3 + ... + (n-1) + n$, we can reverse the order to get
$+S = n + (n-1) + (n-2) + ... + 2 + 1$, now add these two lines to get
$2S = [1+n] + [2+(n-1)] + [3+(n-2)] + ... + [(n-1)+2] + [n+1]$

So we have $2S = [1+n] + [1+n] + [1+n] + ... + [n+1] + [n+1] = n \cdot [n+1]$

$\therefore 2S = n(n+1) \rightarrow S = \frac{n \cdot (n+1)}{2}$

Section 7.2

1. arithmetic

3. $d = a_{n+1} - a_n$
$= (n+1+4) - (n+4)$
$= n+5-n-4$
$= 1$
The difference between consecutive terms is constant, therefore the sequence is arithmetic.
$a_1 = 1+4 = 5$, $a_2 = 2+4 = 6$, $a_3 = 3+4 = 7$,
$a_4 = 4+4 = 8$

5. $d = a_{n+1} - a_n$
$= (2(n+1) - 5) - (2n-5)$
$= 2n+2-5-2n+5$
$= 2$
The difference between consecutive terms is constant, therefore the sequence is arithmetic.
$a_1 = 2 \cdot 1 - 5 = -3$, $a_2 = 2 \cdot 2 - 5 = -1$,
$a_3 = 2 \cdot 3 - 5 = 1$, $a_4 = 2 \cdot 4 - 5 = 3$

7. $d = a_{n+1} - a_n$
$= (6 - 2(n+1)) - (6-2n)$
$= 6-2n-2-6+2n$
$= -2$
The difference between consecutive terms is constant, therefore the sequence is arithmetic.
$a_1 = 6 - 2 \cdot 1 = 4$, $a_2 = 6 - 2 \cdot 2 = 2$,
$a_3 = 6 - 2 \cdot 3 = 0$, $a_4 = 6 - 2 \cdot 4 = -2$

9. $d = a_{n+1} - a_n$
$= \left(\frac{1}{2} - \frac{1}{3}(n+1)\right) - \left(\frac{1}{2} - \frac{1}{3}n\right)$
$= \frac{1}{2} - \frac{1}{3}n - \frac{1}{3} - \frac{1}{2} + \frac{1}{3}n$
$= -\frac{1}{3}$
The difference between consecutive terms is constant, therefore the sequence is arithmetic.
$a_1 = \frac{1}{2} - \frac{1}{3} \cdot 1 = \frac{1}{6}$, $a_2 = \frac{1}{2} - \frac{1}{3} \cdot 2 = -\frac{1}{6}$,
$a_3 = \frac{1}{2} - \frac{1}{3} \cdot 3 = -\frac{1}{2}$, $a_4 = \frac{1}{2} - \frac{1}{3} \cdot 4 = -\frac{5}{6}$

11. $d = a_{n+1} - a_n$
$= \ln(3^{n+1}) - \ln(3^n)$
$= (n+1)\ln(3) - n\ln(3)$
$= \ln 3(n+1-n)$
$= \ln(3)$
The difference between consecutive terms is constant, therefore the sequence is arithmetic.
$a_1 = \ln(3^1) = \ln(3)$, $a_2 = \ln(3^2) = 2\ln(3)$,
$a_3 = \ln(3^3) = 3\ln(3)$, $a_4 = \ln(3^4) = 4\ln(3)$

13. $a_n = a + (n-1)d$
$= 2 + (n-1)3$
$= 2 + 3n - 3$
$= 3n - 1$
$a_5 = 3 \cdot 5 - 1 = 14$

15. $a_n = a + (n-1)d$
$= 5 + (n-1)(-3)$
$= 5 - 3n + 3$
$= 8 - 3n$
$a_5 = 8 - 3 \cdot 5 = -7$

17. $a_n = a + (n-1)d$
$= 0 + (n-1)\frac{1}{2}$
$= \frac{1}{2}n - \frac{1}{2}$
$= \frac{1}{2}(n-1)$
$a_5 = \frac{1}{2}\left(5 - \frac{1}{2}\right) = 2$

19. $a_n = a + (n-1)d$
$= \sqrt{2} + (n-1)\sqrt{2}$
$= \sqrt{2} + \sqrt{2}n - \sqrt{2}$
$= \sqrt{2}n$
$a_5 = 5\sqrt{2}$

21. $a_1 = 2$, $d = 2$, $a_n = a + (n-1)d$
$a_{12} = 2 + (12-1)2 = 2 + 11(2) = 2 + 22 = 24$

23. $a_1 = 1$, $d = -2 - 1 = -3$, $a_n = a + (n-1)d$
$a_{10} = 1 + (10-1)(-3) = 1 + 9(-3) = 1 - 27 = -26$

25. $a_1 = a$, $d = (a+b) - a = b$, $a_n = a + (n-1)d$
$a_8 = a + (8-1)b = a + 7b$

27. $a_8 = a + 7d = 8$ $a_{20} = a + 19d = 44$
Solve the system of equations by subtracting the first equation from the second:
$12d = 36 \Rightarrow d = 3$
$a = 8 - 7(3) = 8 - 21 = -13$
Recursive formula: $a_1 = -13$ $a_n = a_{n-1} + 3$
nth term: $a_n = a + (n-1)d$
$= -13 + (n-1)(3)$
$= -13 + 3n - 3$
$= 3n - 16$

29. $a_9 = a + 8d = -5$ $a_{15} = a + 14d = 31$
Solve the system of equations by subtracting the first equation from the second:
$6d = 36 \Rightarrow d = 6$
$a = -5 - 8(6) = -5 - 48 = -53$
Recursive formula: $a_1 = -53$ $a_n = a_{n-1} + 6$
nth term: $a_n = a + (n-1)d$
$= -53 + (n-1)(6)$
$= -53 + 6n - 6$
$= 6n - 59$

31. $a_{15} = a + 14d = 0$ $a_{40} = a + 39d = -50$
Solve the system of equations by subtracting the first equation from the second:
$25d = -50 \Rightarrow d = -2$
$a = -14(-2) = 28$
Recursive formula: $a_1 = 28$ $a_n = a_{n-1} - 2$
nth term: $a_n = a + (n-1)d$
$= 28 + (n-1)(-2)$
$= 28 - 2n + 2$
$= 30 - 2n$

33. $a_{14} = a + 13d = -1$ $a_{18} = a + 17d = -9$
Solve the system of equations by subtracting the first equation from the second:
$4d = -8 \Rightarrow d = -2$
$a = -1 - 13(-2) = -1 + 26 = 25$
Recursive formula: $a_1 = 25$ $a_n = a_{n-1} - 2$
nth term: $a_n = a + (n-1)d$
$= 25 + (n-1)(-2)$
$= 25 - 2n + 2$
$= 27 - 2n$

35. $S_n = \dfrac{n}{2}(a + a_n) = \dfrac{n}{2}(1 + (2n-1)) = \dfrac{n}{2}(2n) = n^2$

37. $S_n = \dfrac{n}{2}(a + a_n) = \dfrac{n}{2}(7 + (2 + 5n)) = \dfrac{n}{2}(9 + 5n)$

39. $a_1 = 2$, $d = 4 - 2 = 2$, $a_n = a + (n-1)d$
$70 = 2 + (n-1)2$
$70 = 2 + 2n - 2$
$70 = 2n$
$n = 35$
$S_n = \dfrac{n}{2}(a + a_n) = \dfrac{35}{2}(2 + 70)$
$= \dfrac{35}{2}(72) = 35(36)$
$= 1260$

41. $a_1 = 5$, $d = 9 - 5 = 4$, $a_n = a + (n-1)d$
$49 = 5 + (n-1)4$
$49 = 5 + 4n - 4$
$48 = 4n$
$n = 12$
$S_n = \dfrac{n}{2}(a + a_n) = \dfrac{12}{2}(5 + 49) = 6(54) = 324$

43. Using the sum of the sequence feature:

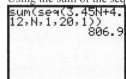

45.
$d = 5.2 - 2.8 = 2.4$
$a = 2.8$
$36.4 = 2.8 + (n-1)2.4$
$36.4 = 2.8 + 2.4n - 2.4$
$36 = 2.4n$
$n = 15$
$a_n = 2.8 + (n-1)2.4$
$= 2.8 + 2.4n - 2.4$
$= 2.4n + 0.4$

```
sum(seq(2.4N+.4,
N,1,15,1))
                294
```

47.
$d = 7.48 - 4.9 = 2.58$
$a = 4.9$
$66.82 = 4.9 + (n-1)2.58$
$66.82 = 4.9 + 2.58n - 2.58$
$64.5 = 2.58n$
$n = 25$
$a_n = 4.9 + (n-1)2.58 = 4.9 + 2.58n - 2.58$
$a_n = 2.58n + 2.32$

```
sum(seq(2.58N+2.
32,N,1,25,1))
                896.5
```

49. Find the common difference of the terms and solve the system of equations:
$(2x+1) - (x+3) = d \Rightarrow x - 2 = d$
$(5x+2) - (2x+1) = d \Rightarrow 3x + 1 = d$
$3x + 1 = x - 2$
$2x = -3$
$x = -\dfrac{3}{2}$

51. The total number of seats is:
$S = 25 + 26 + 27 + \cdots + (25 + 29(1))$

This is the sum of an arithmetic sequence with $d = 1$, $a = 25$, and $n = 30$.
Find the sum of the sequence:

$S_{30} = \dfrac{30}{2}[2(25) + (30-1)(1)]$
$= 15(50 + 29) = 15(79)$
$= 1185$

There are 1185 seats in the theater.

53. The lighter colored tiles have 20 tiles in the bottom row and 1 tile in the top row. The number decreases by 1 as we move up the triangle. This is an arithmetic sequence with $a_1 = 20$, $d = -1$, and $n = 20$. Find the sum:

$S = \dfrac{20}{2}[2(20) + (20-1)(-1)]$
$= 10(40 - 19) = 10(21)$
$= 210$

There are 210 lighter tiles.

The darker colored tiles have 19 tiles in the bottom row and 1 tile in the top row. The number decreases by 1 as we move up the triangle. This is an arithmetic sequence with $a_1 = 19$, $d = -1$, and $n = 19$. Find the sum:

$S = \dfrac{19}{2}[2(19) + (19-1)(-1)]$
$= \dfrac{19}{2}(38 - 18) = \dfrac{19}{2}(20) = 190$

There are 190 darker tiles.

55. Find n in an arithmetic sequence with $a_1 = 10$, $d = 4$, $S_n = 2040$.

$S_n = \dfrac{n}{2}[2a_1 + (n-1)d]$
$2040 = \dfrac{n}{2}[2(10) + (n-1)4]$
$4080 = n[20 + 4n - 4]$
$4080 = n(4n + 16)$
$4080 = 4n^2 + 16n$
$1020 = n^2 + 4n$
$n^2 + 4n - 1020 = 0$
$(n+34)(n-30) = 0 \Rightarrow n = -34$ or $n = 30$

There are 30 rows in the corner section of the stadium.

57. Answers will vary.

Section 7.3

1. geometric

3. True

5. $r = \dfrac{3^{n+1}}{3^n} = 3^{n+1-n} = 3$

   The ratio of consecutive terms is constant, therefore the sequence is geometric.
   $a_1 = 3^1 = 3$, $a_2 = 3^2 = 9$,
   $a_3 = 3^3 = 27$, $a_4 = 3^4 = 81$

7. $r = \dfrac{-3\left(\dfrac{1}{2}\right)^{n+1}}{-3\left(\dfrac{1}{2}\right)^n} = \left(\dfrac{1}{2}\right)^{n+1-n} = \dfrac{1}{2}$

   The ratio of consecutive terms is constant, therefore the sequence is geometric.
   $a_1 = -3\left(\dfrac{1}{2}\right)^1 = -\dfrac{3}{2}$, $a_2 = -3\left(\dfrac{1}{2}\right)^2 = -\dfrac{3}{4}$,
   $a_3 = -3\left(\dfrac{1}{2}\right)^3 = -\dfrac{3}{8}$, $a_4 = -3\left(\dfrac{1}{2}\right)^4 = -\dfrac{3}{16}$

9. $r = \dfrac{\left(\dfrac{2^{n+1-1}}{4}\right)}{\left(\dfrac{2^{n-1}}{4}\right)} = \dfrac{2^n}{2^{n-1}} = 2^{n-(n-1)} = 2$

   The ratio of consecutive terms is constant, therefore the sequence is geometric.
   $a_1 = \dfrac{2^{1-1}}{4} = \dfrac{2^0}{2^2} = 2^{-2} = \dfrac{1}{4}$,
   $a_2 = \dfrac{2^{2-1}}{4} = \dfrac{2^1}{2^2} = 2^{-1} = \dfrac{1}{2}$,
   $a_3 = \dfrac{2^{3-1}}{4} = \dfrac{2^2}{2^2} = 1$,
   $a_4 = \dfrac{2^{4-1}}{4} = \dfrac{2^3}{2^2} = 2$

11. $r = \dfrac{2^{\left(\frac{n+1}{3}\right)}}{2^{\left(\frac{n}{3}\right)}} = 2^{\left(\frac{n+1}{3} - \frac{n}{3}\right)} = 2^{1/3}$

    The ratio of consecutive terms is constant, therefore the sequence is geometric.
    $a_1 = 2^{1/3}$, $a_2 = 2^{2/3}$, $a_3 = 2^{3/3} = 2$, $a_4 = 2^{4/3}$

13. $r = \dfrac{\left(\dfrac{3^{n+1-1}}{2^{n+1}}\right)}{\left(\dfrac{3^{n-1}}{2^n}\right)} = \dfrac{3^n}{3^{n-1}} \cdot \dfrac{2^n}{2^{n+1}}$

    $= 3^{n-(n-1)} \cdot 2^{n-(n+1)} = 3 \cdot 2^{-1} = \dfrac{3}{2}$

    The ratio of consecutive terms is constant, therefore the sequence is geometric.
    $a_1 = \dfrac{3^{1-1}}{2^1} = \dfrac{3^0}{2} = \dfrac{1}{2}$, $a_2 = \dfrac{3^{2-1}}{2^2} = \dfrac{3^1}{2^2} = \dfrac{3}{4}$,
    $a_3 = \dfrac{3^{3-1}}{2^3} = \dfrac{3^2}{2^3} = \dfrac{9}{8}$, $a_4 = \dfrac{3^{4-1}}{2^4} = \dfrac{3^3}{2^4} = \dfrac{27}{16}$

15. $\{n+2\}$
    $d = (n+1+2) - (n+2) = n + 3 - n - 2 = 1$
    The difference between consecutive terms is constant, therefore the sequence is arithmetic.

17. $\{4n^2\}$ Examine the terms of the sequence: 4, 16, 36, 64, 100, ...
    There is no common difference; there is no common ratio; neither.

19. $\left\{3 - \dfrac{2}{3}n\right\}$
    $d = \left(3 - \dfrac{2}{3}(n+1)\right) - \left(3 - \dfrac{2}{3}n\right)$
    $= 3 - \dfrac{2}{3}n - \dfrac{2}{3} - 3 + \dfrac{2}{3}n = -\dfrac{2}{3}$
    The difference between consecutive terms is constant, therefore the sequence is arithmetic.

21. 1, 3, 6, 10, ...   Neither
    There is no common difference or common ratio.

23. $\left\{\left(\dfrac{2}{3}\right)^n\right\}$

$r = \dfrac{\left(\dfrac{2}{3}\right)^{n+1}}{\left(\dfrac{2}{3}\right)^n} = \left(\dfrac{2}{3}\right)^{n+1-n} = \dfrac{2}{3}$

The ratio of consecutive terms is constant, therefore the sequence is geometric.

25. $-1, -2, -4, -8, \ldots$

$r = \dfrac{-2}{-1} = \dfrac{-4}{-2} = \dfrac{-8}{-4} = 2$

The ratio of consecutive terms is constant, therefore the sequence is geometric.

27. $\left\{3^{n/2}\right\}$

$r = \dfrac{3^{\left(\frac{n+1}{2}\right)}}{3^{\left(\frac{n}{2}\right)}} = 3^{\left(\frac{n+1}{2} - \frac{n}{2}\right)} = 3^{1/2}$

The ratio of consecutive terms is constant, therefore the sequence is geometric.

29. $a_5 = 2 \cdot 3^{5-1} = 2 \cdot 3^4 = 2 \cdot 81 = 162$

$a_n = 2 \cdot 3^{n-1}$

31. $a_5 = 5(-1)^{5-1} = 5(-1)^4 = 5 \cdot 1 = 5$

$a_n = 5 \cdot (-1)^{n-1}$

33. $a_5 = 0 \cdot \left(\dfrac{1}{2}\right)^{5-1} = 0 \cdot \left(\dfrac{1}{2}\right)^4 = 0$

$a_n = 0 \cdot \left(\dfrac{1}{2}\right)^{n-1} = 0$

35. $a_5 = \sqrt{2} \cdot \left(\sqrt{2}\right)^{5-1} = \sqrt{2} \cdot \left(\sqrt{2}\right)^4 = \sqrt{2} \cdot 4 = 4\sqrt{2}$

$a_n = \sqrt{2} \cdot \left(\sqrt{2}\right)^{n-1} = \left(\sqrt{2}\right)^n$

37. $a = 1, \ r = \dfrac{1}{2}, \ n = 7$

$a_7 = 1 \cdot \left(\dfrac{1}{2}\right)^{7-1} = \left(\dfrac{1}{2}\right)^6 = \dfrac{1}{64}$

39. $a = 1, \ r = -1, \ n = 9$

$a_9 = 1 \cdot (-1)^{9-1} = (-1)^8 = 1$

41. $a = 0.4, \ r = 0.1, \ n = 8$

$a_8 = 0.4 \cdot (0.1)^{8-1} = 0.4(0.1)^7 = 0.00000004$

43. $a = \dfrac{1}{4}, \ r = 2$

$S_n = a\left(\dfrac{1-r^n}{1-r}\right) = \dfrac{1}{4}\left(\dfrac{1-2^n}{1-2}\right) = \dfrac{1}{4}\left(2^n - 1\right)$

45. $a = \dfrac{2}{3}, \ r = \dfrac{2}{3}$

$S_n = a\left(\dfrac{1-r^n}{1-r}\right) = \dfrac{2}{3}\left(\dfrac{1-\left(\dfrac{2}{3}\right)^n}{1-\dfrac{2}{3}}\right)$

$= \dfrac{2}{3}\left(\dfrac{1-\left(\dfrac{2}{3}\right)^n}{\dfrac{1}{3}}\right) = 2\left(1 - \left(\dfrac{2}{3}\right)^n\right)$

47. $a = -1, \ r = 2$

$S_n = a\left(\dfrac{1-r^n}{1-r}\right) = -1\left(\dfrac{1-2^n}{1-2}\right) = 1 - 2^n$

49. Using the sum of the sequence feature:

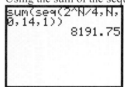

51. Using the sum of the sequence feature:

```
sum(seq((2/3)^N,
N,1,15,1))
            1.995432683
```

53. Using the sum of the sequence feature:

55. $a=1,\ r=\dfrac{1}{3}$ Since $|r|<1$,

$$S_\infty = \dfrac{a}{1-r} = \dfrac{1}{\left(1-\dfrac{1}{3}\right)} = \dfrac{1}{\left(\dfrac{2}{3}\right)} = \dfrac{3}{2}$$

57. $a=8,\ r=\dfrac{1}{2}$ Since $|r|<1$,

$$S_\infty = \dfrac{a}{1-r} = \dfrac{8}{\left(1-\dfrac{1}{2}\right)} = \dfrac{8}{\left(\dfrac{1}{2}\right)} = 16$$

59. $a=2,\ r=-\dfrac{1}{4}$ Since $|r|<1$,

$$S_\infty = \dfrac{a}{1-r} = \dfrac{2}{\left(1-\left(-\dfrac{1}{4}\right)\right)} = \dfrac{2}{\left(\dfrac{5}{4}\right)} = \dfrac{8}{5}$$

61. $a=5,\ r=\dfrac{1}{4}$ Since $|r|<1$,

$$S_\infty = \dfrac{a}{1-r} = \dfrac{5}{\left(1-\dfrac{1}{4}\right)} = \dfrac{5}{\left(\dfrac{3}{4}\right)} = \dfrac{20}{3}$$

63. $a=6,\ r=-\dfrac{2}{3}$ Since $|r|<1$,

$$S_\infty = \dfrac{a}{1-r} = \dfrac{6}{\left(1-\left(-\dfrac{2}{3}\right)\right)} = \dfrac{6}{\left(\dfrac{5}{3}\right)} = \dfrac{18}{5}$$

65. Find the common ratio of the terms and solve the system of equations:

$$\dfrac{x+2}{x} = r;\quad \dfrac{x+3}{x+2} = r$$

$$\dfrac{x+2}{x} = \dfrac{x+3}{x+2} \rightarrow x^2+4x+4 = x^2+3x \rightarrow x=-4$$

67. This is a geometric series with $a=\$18{,}000,\ r=1.05,\ n=5$. Find the 5th term:
$$a_5 = 18000(1.05)^{5-1} = 18000(1.05)^4$$
$$= \$21{,}879.11$$

69. **a.** Find the 10th term of the geometric sequence:
$a=2,\ r=0.9,\ n=10$
$$a_{10} = 2(0.9)^{10-1} = 2(0.9)^9 = 0.775 \text{ feet}$$

b. Find n when $a_n < 1$:
$$2(0.9)^{n-1} < 1$$
$$(0.9)^{n-1} < 0.5$$
$$(n-1)\log(0.9) < \log(0.5)$$
$$n-1 > \dfrac{\log(0.5)}{\log(0.9)}$$
$$n > \dfrac{\log(0.5)}{\log(0.9)} + 1 \approx 7.58$$

On the 8th swing the arc is less than 1 foot.

c. Find the sum of the first 15 swings:
$$S_{15} = 2\left(\dfrac{1-(0.9)^{15}}{1-0.9}\right) = 2\left(\dfrac{1-(0.9)^{15}}{0.1}\right)$$
$$= 20\left(1-(0.9)^{15}\right) = 15.88 \text{ feet}$$

d. Find the infinite sum of the geometric series:
$$S_\infty = \dfrac{2}{1-0.9} = \dfrac{2}{0.1} = 20 \text{ feet}$$

71. This is a geometric sequence with $a=1,\ r=2,\ n=64$.
Find the sum of the geometric series:
$$S_{64} = 1\left(\dfrac{1-2^{64}}{1-2}\right) = \dfrac{1-2^{64}}{-1} = 2^{64}-1$$
$$= 1.845 \times 10^{19} \text{ grains}$$

73. The common ratio, $r = 0.90 < 1$. The sum is:
$$S = \frac{1}{1-0.9} = \frac{1}{0.10} = 10.$$
The multiplier is 10.

75. This is an infinite geometric series with $a = 4$, and $r = \frac{1.03}{1.09}$.

Find the sum: Price $= \dfrac{4}{\left(1-\dfrac{1.03}{1.09}\right)} \approx \$72.67$.

77. Given: $a = 1000$, $r = 0.9$
Find n when $a_n < 0.01$:
$$1000(0.9)^{n-1} < 0.01$$
$$(0.9)^{n-1} < 0.00001$$
$$(n-1)\log(0.9) < \log(0.00001)$$
$$n-1 > \frac{\log(0.00001)}{\log(0.9)}$$
$$n > \frac{\log(0.00001)}{\log(0.9)} + 1 \approx 110.27$$

On the 111th day or December 20, 2001, the amount will be less than $0.01.

Find the sum of the geometric series:
$$S_{110} = a\left(\frac{1-r^n}{1-r}\right) = 1000\left(\frac{1-(0.9)^{110}}{1-0.9}\right)$$
$$= 1000\left(\frac{1-(0.9)^{110}}{0.1}\right) = \$9999.91$$

79. Find the sum of each sequence:
A: Arithmetic series with:
$a = \$1000$, $d = -1$, $n = 1000$
Find the sum of the arithmetic series:
$$S_{1000} = \frac{1000}{2}(1000+1) = 500(1001) = \$500{,}500$$

B: This is a geometric sequence with $a = 1$, $r = 2$, $n = 19$.
Find the sum of the geometric series:
$$S_{19} = 1\left(\frac{1-2^{19}}{1-2}\right) = \frac{1-2^{19}}{-1} = 2^{19}-1 = \$524{,}287$$

B results in more money.

81. Yes, a sequence can be both arithmetic and geometric. For example, the constant sequence $3,3,3,3,\ldots$ can be viewed as an arithmetic sequence with $a = 3$ and $d = 0$. Alternatively, the same sequence can be viewed as a geometric sequence with $a = 3$ and $r = 1$.

83. Answers will vary.

Section 7.4

1. I: $n = 1$: $2 \cdot 1 = 2$ and $1(1+1) = 2$

   II: If $2+4+6+\cdots+2k = k(k+1)$, then
   $$2+4+6+\cdots+2k+2(k+1)$$
   $$= [2+4+6+\cdots+2k] + 2(k+1)$$
   $$= k(k+1) + 2(k+1)$$
   $$= (k+1)(k+2)$$
   $$= (k+1)((k+1)+1)$$

   Conditions I and II are satisfied; the statement is true.

3. I: $n = 1$: $1+2 = 3$ and $\frac{1}{2} \cdot 1(1+5) = 3$

   II: If $3+4+5+\cdots+(k+2) = \frac{1}{2} \cdot k(k+5)$, then
   $$3+4+5+\cdots+(k+2)+[(k+1)+2]$$
   $$= [3+4+5+\cdots+(k+2)] + (k+3)$$
   $$= \frac{1}{2} \cdot k(k+5) + (k+3)$$
   $$= \frac{1}{2}k^2 + \frac{5}{2}k + k + 3$$
   $$= \frac{1}{2}k^2 + \frac{7}{2}k + 3$$
   $$= \frac{1}{2} \cdot (k^2 + 7k + 6)$$
   $$= \frac{1}{2} \cdot (k+1)(k+6)$$
   $$= \frac{1}{2} \cdot (k+1)((k+1)+5)$$

   Conditions I and II are satisfied; the statement is true.

5. I: $n=1$: $3\cdot 1-1=2$ and $\dfrac{1}{2}\cdot 1(3\cdot 1+1)=2$

II: If $2+5+8+\cdots+(3k-1)=\dfrac{1}{2}\cdot k(3k+1)$,
then
$2+5+8+\cdots+(3k-1)+[3(k+1)-1]$
$=[2+5+8+\cdots+(3k-1)]+(3k+2)$
$=\dfrac{1}{2}\cdot k(3k+1)+(3k+2)=\dfrac{3}{2}k^2+\dfrac{1}{2}k+3k+2$
$=\dfrac{3}{2}k^2+\dfrac{7}{2}k+2=\dfrac{1}{2}\cdot(3k^2+7k+4)$
$=\dfrac{1}{2}\cdot(k+1)(3k+4)$
$=\dfrac{1}{2}\cdot(k+1)(3(k+1)+1)$

Conditions I and II are satisfied; the statement is true.

7. I: $n=1$: $2^{1-1}=1$ and $2^1-1=1$

II: If $1+2+2^2+\cdots+2^{k-1}=2^k-1$, then
$1+2+2^2+\cdots+2^{k-1}+2^{k+1-1}$
$=\left[1+2+2^2+\cdots+2^{k-1}\right]+2^k$
$=2^k-1+2^k=2\cdot 2^k-1$
$=2^{k+1}-1$

Conditions I and II are satisfied; the statement is true.

9. I: $n=1$: $4^{1-1}=1$ and $\dfrac{1}{3}\cdot(4^1-1)=1$

II: If $1+4+4^2+\cdots+4^{k-1}=\dfrac{1}{3}\cdot(4^k-1)$, then
$1+4+4^2+\cdots+4^{k-1}+4^{k+1-1}$
$=\left[1+4+4^2+\cdots+4^{k-1}\right]+4^k$
$=\dfrac{1}{3}\cdot(4^k-1)+4^k=\dfrac{1}{3}\cdot 4^k-\dfrac{1}{3}+4^k$
$=\dfrac{4}{3}\cdot 4^k-\dfrac{1}{3}=\dfrac{1}{3}(4\cdot 4^k-1)$
$=\dfrac{1}{3}\cdot(4^{k+1}-1)$

Conditions I and II are satisfied; the statement is true.

11. I: $n=1$: $\dfrac{1}{1(1+1)}=\dfrac{1}{2}$ and $\dfrac{1}{1+1}=\dfrac{1}{2}$

II: If $\dfrac{1}{1\cdot 2}+\dfrac{1}{2\cdot 3}+\dfrac{1}{3\cdot 4}+\cdots+\dfrac{1}{k(k+1)}=\dfrac{k}{k+1}$, then
$\dfrac{1}{1\cdot 2}+\dfrac{1}{2\cdot 3}+\dfrac{1}{3\cdot 4}+\cdots+\dfrac{1}{k(k+1)}+\dfrac{1}{(k+1)(k+1+1)}=\left[\dfrac{1}{1\cdot 2}+\dfrac{1}{2\cdot 3}+\dfrac{1}{3\cdot 4}+\cdots+\dfrac{1}{k(k+1)}\right]+\dfrac{1}{(k+1)(k+2)}$
$=\dfrac{k}{k+1}+\dfrac{1}{(k+1)(k+2)}=\dfrac{k}{k+1}\cdot\dfrac{k+2}{k+2}+\dfrac{1}{(k+1)(k+2)}$
$=\dfrac{k^2+2k+1}{(k+1)(k+2)}=\dfrac{(k+1)(k+1)}{(k+1)(k+2)}=\dfrac{k+1}{k+2}=\dfrac{k+1}{(k+1)+1}$

Conditions I and II are satisfied; the statement is true.

13. I: $n=1$: $1^2 = 1$ and $\dfrac{1}{6} \cdot 1(1+1)(2 \cdot 1+1) = 1$

II: If $1^2 + 2^2 + 3^2 + \cdots + k^2 = \dfrac{1}{6} \cdot k(k+1)(2k+1)$, then

$1^2 + 2^2 + 3^2 + \cdots + k^2 + (k+1)^2 = \left[1^2 + 2^2 + 3^2 + \cdots + k^2\right] + (k+1)^2 = \dfrac{1}{6} k(k+1)(2k+1) + (k+1)^2$

$= (k+1)\left[\dfrac{1}{6}k(2k+1) + k + 1\right] = (k+1)\left[\dfrac{1}{3}k^2 + \dfrac{1}{6}k + k + 1\right] = (k+1)\left[\dfrac{1}{3}k^2 + \dfrac{7}{6}k + 1\right] = \dfrac{1}{6}(k+1)\left[2k^2 + 7k + 6\right]$

$= \dfrac{1}{6} \cdot (k+1)(k+2)(2k+3)$

$= \dfrac{1}{6} \cdot (k+1)\big((k+1)+1\big)\big(2(k+1)+1\big)$

Conditions I and II are satisfied; the statement is true.

15. I: $n=1$: $5-1=4$ and $\dfrac{1}{2} \cdot 1(9-1) = 4$

II: If $4+3+2+\cdots+(5-k) = \dfrac{1}{2} \cdot k(9-k)$, then

$4+3+2+\cdots+(5-k)+(5-(k+1)) = \left[4+3+2+\cdots+(5-k)\right]+(4-k) = \dfrac{1}{2}k(9-k)+(4-k)$

$= \dfrac{9}{2}k - \dfrac{1}{2}k^2 + 4 - k = -\dfrac{1}{2}k^2 + \dfrac{7}{2}k + 4 = -\dfrac{1}{2} \cdot \left[k^2 - 7k - 8\right]$

$= -\dfrac{1}{2} \cdot (k+1)(k-8) = \dfrac{1}{2} \cdot (k+1)(8-k) = \dfrac{1}{2} \cdot (k+1)\left[9-(k+1)\right]$

Conditions I and II are satisfied; the statement is true.

17. I: $n=1$: $1(1+1) = 2$ and $\dfrac{1}{3} \cdot 1(1+1)(1+2) = 2$

II: If $1 \cdot 2 + 2 \cdot 3 + 3 \cdot 4 + \cdots + k(k+1) = \dfrac{1}{3} \cdot k(k+1)(k+2)$, then

$1 \cdot 2 + 2 \cdot 3 + 3 \cdot 4 + \cdots + k(k+1) + (k+1)(k+1+1) = \left[1 \cdot 2 + 2 \cdot 3 + 3 \cdot 4 + \cdots + k(k+1)\right] + (k+1)(k+2)$

$= \dfrac{1}{3} \cdot k(k+1)(k+2) + (k+1)(k+2) = (k+1)(k+2)\left[\dfrac{1}{3}k + 1\right]$

$= \dfrac{1}{3} \cdot (k+1)(k+2)(k+3)$

$= \dfrac{1}{3} \cdot (k+1)((k+1)+1)((k+1)+2)$

Conditions I and II are satisfied; the statement is true.

19. I: $n=1$: $1^2+1=2$ is divisible by 2

    II: If k^2+k is divisible by 2, then
    $$(k+1)^2+(k+1)=k^2+2k+1+k+1$$
    $$=(k^2+k)+(2k+2)$$
    Since k^2+k is divisible by 2 and $2k+2$ is divisible by 2, then $(k+1)^2+(k+1)$ is divisible by 2.

    Conditions I and II are satisfied; the statement is true.

21. I: $n=1$: $1^2-1+2=2$ is divisible by 2

    II: If k^2-k+2 is divisible by 2, then
    $$(k+1)^2-(k+1)+2=k^2+2k+1-k-1+2$$
    $$=(k^2-k+2)+(2k)$$
    Since k^2-k+2 is divisible by 2 and $2k$ is divisible by 2, then $(k+1)^2-(k+1)+2$ is divisible by 2.

    Conditions I and II are satisfied; the statement is true.

23. I: $n=1$: If $x>1$ then $x^1=x>1$.

    II: Assume, for some natural number k, that if $x>1$, then $x^k>1$.
    Then $x^{k+1}>1$, for $x>1$,
    $$x^{k+1}=x^k\cdot x>1\cdot x=x>1$$
    $$\uparrow$$
    $$(x^k>1)$$
    Conditions I and II are satisfied; the statement is true.

25. I: $n=1$: $a-b$ is a factor of $a^1-b^1=a-b$.

    II: If $a-b$ is a factor of a^k-b^k, show that $a-b$ is a factor of $a^{k+1}-b^{k+1}$.
    $$a^{k+1}-b^{k+1}=a\cdot a^k-b\cdot b^k$$
    $$=a\cdot a^k-a\cdot b^k+a\cdot b^k-b\cdot b^k$$
    $$=a\left(a^k-b^k\right)+b^k(a-b)$$
    Since $a-b$ is a factor of a^k-b^k and $a-b$ is a factor of $a-b$, then $a-b$ is a factor of $a^{k+1}-b^{k+1}$.

    Conditions I and II are satisfied; the statement is true.

27. $n=1$:
    $1^2-1+41=41$ is a prime number.
    $n=41$:
    $41^2-41+41=41^2$ is not a prime number.

29. I: $n=1$: $ar^{1-1}=a$ and $a\left(\dfrac{1-r^1}{1-r}\right)=a$

    II: If $a+ar+ar^2+\cdots+ar^{k-1}=a\left(\dfrac{1-r^k}{1-r}\right)$, then
    $$a+ar+ar^2+\cdots+ar^{k-1}+ar^{k+1-1}$$
    $$=\left[a+ar+ar^2+\cdots+ar^{k-1}\right]+ar^k$$
    $$=a\left(\dfrac{1-r^k}{1-r}\right)+ar^k$$
    $$=\dfrac{a(1-r^k)+ar^k(1-r)}{1-r}$$
    $$=\dfrac{a-ar^k+ar^k-ar^{k+1}}{1-r}$$
    $$=a\left(\dfrac{1-r^{k+1}}{1-r}\right)$$

    Conditions I and II are satisfied; the statement is true.

31. I: $n=4$: The number of diagonals of a quadrilateral is $\dfrac{1}{2}\cdot 4(4-3)=2$.

    II: Assume that for any integer k, the number of diagonals of a convex polygon with k sides (k vertices) is $\dfrac{1}{2}\cdot k(k-3)$. A convex polygon with $k+1$ sides ($k+1$ vertices) consists of a convex polygon with k sides (k vertices) plus a triangle, for a total of ($k+1$) vertices. The diagonals of this $k+1$-sided convex polygon consist of the diagonals of the k-sided polygon plus $k-1$ additional diagonals. For example, consider the following diagrams.

Section 7.5

1. Pascal Triangle

3. False; $\binom{n}{j} = \dfrac{n!}{j!(n-j)!}$

5. $\binom{5}{3} = \dfrac{5!}{3!\,2!} = \dfrac{5\cdot 4\cdot 3\cdot 2\cdot 1}{3\cdot 2\cdot 1\cdot 2\cdot 1} = \dfrac{5\cdot 4}{2\cdot 1} = 10$

7. $\binom{7}{5} = \dfrac{7!}{5!\,2!} = \dfrac{7\cdot 6\cdot 5\cdot 4\cdot 3\cdot 2\cdot 1}{5\cdot 4\cdot 3\cdot 2\cdot 1\cdot 2\cdot 1} = \dfrac{7\cdot 6}{2\cdot 1} = 21$

9. $\binom{50}{49} = \dfrac{50!}{49!\,1!} = \dfrac{50\cdot 49!}{49!\cdot 1} = \dfrac{50}{1} = 50$

11. $\binom{1000}{1000} = \dfrac{1000!}{1000!\,0!} = \dfrac{1}{1} = 1$

13. $\binom{55}{23} = \dfrac{55!}{23!\,32!} \approx 1.866442159 \times 10^{15}$

15. $\binom{47}{25} = \dfrac{47!}{25!\,22!} \approx 1.483389769 \times 10^{13}$

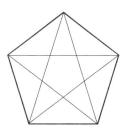

$k = 5$ sides

$k+1 = 6$ sides
$k-1 = 4$ new diagonals

Thus, we have the equation:
$$\tfrac{1}{2}\cdot k(k-3) + (k-1) = \tfrac{1}{2}k^2 - \tfrac{3}{2}k + k - 1$$
$$= \tfrac{1}{2}k^2 - \tfrac{1}{2}k - 1$$
$$= \tfrac{1}{2}\cdot(k^2 - k - 2)$$
$$= \tfrac{1}{2}\cdot(k+1)(k-2)$$
$$= \tfrac{1}{2}\cdot(k+1)((k+1)-3)$$

Conditions I and II are satisfied; the statement is true.

33. Answers will vary.

17. $(x+1)^5 = \binom{5}{0}x^5 + \binom{5}{1}x^4 + \binom{5}{2}x^3 + \binom{5}{3}x^2 + \binom{5}{4}x^1 + \binom{5}{5}x^0 = x^5 + 5x^4 + 10x^3 + 10x^2 + 5x + 1$

19. $(x-2)^6 = \binom{6}{0}x^6 + \binom{6}{1}x^5(-2) + \binom{6}{2}x^4(-2)^2 + \binom{6}{3}x^3(-2)^3 + \binom{6}{4}x^2(-2)^4 + \binom{6}{5}x(-2)^5 + \binom{6}{6}x^0(-2)^6$
$= x^6 + 6x^5(-2) + 15x^4 \cdot 4 + 20x^3(-8) + 15x^2 \cdot 16 + 6x\cdot(-32) + 64$
$= x^6 - 12x^5 + 60x^4 - 160x^3 + 240x^2 - 192x + 64$

21. $(3x+1)^4 = \binom{4}{0}(3x)^4 + \binom{4}{1}(3x)^3 + \binom{4}{2}(3x)^2 + \binom{4}{3}(3x) + \binom{4}{4}$
$= 81x^4 + 4\cdot 27x^3 + 6\cdot 9x^2 + 4\cdot 3x + 1 = 81x^4 + 108x^3 + 54x^2 + 12x + 1$

23. $(x^2+y^2)^5 = \binom{5}{0}(x^2)^5 + \binom{5}{1}(x^2)^4 y^2 + \binom{5}{2}(x^2)^3(y^2)^2 + \binom{5}{3}(x^2)^2(y^2)^3 + \binom{5}{4}x^2(y^2)^4 + \binom{5}{5}(y^2)^5$

$= x^{10} + 5x^8 y^2 + 10x^6 y^4 + 10x^4 y^6 + 5x^2 y^8 + y^{10}$

25. $(\sqrt{x}+\sqrt{2})^6 = \binom{6}{0}(\sqrt{x})^6 + \binom{6}{1}(\sqrt{x})^5(\sqrt{2})^1 + \binom{6}{2}(\sqrt{x})^4(\sqrt{2})^2 + \binom{6}{3}(\sqrt{x})^3(\sqrt{2})^3$

$\qquad + \binom{6}{4}(\sqrt{x})^2(\sqrt{2})^4 + \binom{6}{5}(\sqrt{x})(\sqrt{2})^5 + \binom{6}{6}(\sqrt{2})^6$

$= x^3 + 6\sqrt{2}x^{5/2} + 15 \cdot 2x^2 + 20 \cdot 2\sqrt{2}x^{3/2} + 15 \cdot 4x + 6 \cdot 4\sqrt{2}x^{1/2} + 8$

$= x^3 + 6\sqrt{2}x^{5/2} + 30x^2 + 40\sqrt{2}x^{3/2} + 60x + 24\sqrt{2}x^{1/2} + 8$

27. $(ax+by)^5 = \binom{5}{0}(ax)^5 + \binom{5}{1}(ax)^4 \cdot by + \binom{5}{2}(ax)^3(by)^2 + \binom{5}{3}(ax)^2(by)^3 + \binom{5}{4}ax(by)^4 + \binom{5}{5}(by)^5$

$= a^5 x^5 + 5a^4 x^4 by + 10a^3 x^3 b^2 y^2 + 10a^2 x^2 b^3 y^3 + 5axb^4 y^4 + b^5 y^5$

29. $n=10, \ j=4, \ x=x, \ a=3$

$\binom{10}{4}x^6 \cdot 3^4 = \frac{10!}{4!6!} \cdot 81x^6 = \frac{10 \cdot 9 \cdot 8 \cdot 7}{4 \cdot 3 \cdot 2 \cdot 1} \cdot 81x^6$

$= 17{,}010x^6$

The coefficient of x^6 is 17,010.

31. $n=12, \ j=5, \ x=2x, \ a=-1$

$\binom{12}{5}(2x)^7 \cdot (-1)^5 = \frac{12!}{5!7!} \cdot 128x^7(-1)$

$= \frac{12 \cdot 11 \cdot 10 \cdot 9 \cdot 8}{5 \cdot 4 \cdot 3 \cdot 2 \cdot 1} \cdot (-128)x^7$

$= -101{,}376x^7$

The coefficient of x^7 is $-101{,}376$.

33. $n=9, \ j=2, \ x=2x, \ a=3$

$\binom{9}{2}(2x)^7 \cdot 3^2 = \frac{9!}{2!7!} \cdot 128x^7(9)$

$= \frac{9 \cdot 8}{2 \cdot 1} \cdot 128x^7 \cdot 9$

$= 41{,}472x^7$

The coefficient of x^7 is 41,472.

35. $n=7, \ j=4, \ x=x, \ a=3$

$\binom{7}{4}x^3 \cdot 3^4 = \frac{7!}{4!3!} \cdot 81x^3 = \frac{7 \cdot 6 \cdot 5}{3 \cdot 2 \cdot 1} \cdot 81x^3 = 2835x^3$

37. $n=9, \ j=2, \ x=3x, \ a=-2$

$\binom{9}{2}(3x)^7 \cdot (-2)^2 = \frac{9!}{2!7!} \cdot 2187x^7 \cdot 4$

$= \frac{9 \cdot 8}{2 \cdot 1} \cdot 8748x^7 = 314{,}928x^7$

39. The x^0 term in

$\sum_{j=0}^{12}\binom{12}{j}(x^2)^{12-j}\left(\frac{1}{x}\right)^j = \sum_{j=0}^{12}\binom{12}{j}x^{24-3j}$

occurs when:

$24 - 3j = 0$

$24 = 3j \ \to \ j = 8$

The coefficient is

$\binom{12}{8} = \frac{12!}{8!4!} = \frac{12 \cdot 11 \cdot 10 \cdot 9}{4 \cdot 3 \cdot 2 \cdot 1} = 495$

41. The x^4 term in

$\sum_{j=0}^{10}\binom{10}{j}(x)^{10-j}\left(\frac{-2}{\sqrt{x}}\right)^j = \sum_{j=0}^{10}\binom{10}{j}(-2)^j x^{10-\frac{3}{2}j}$

occurs when:

$10 - \frac{3}{2}j = 4$

$-\frac{3}{2}j = -6 \ \to \ j=4$

The coefficient is

$\binom{10}{4}(-2)^4 = \frac{10!}{6!4!} \cdot 16 = \frac{10 \cdot 9 \cdot 8 \cdot 7}{4 \cdot 3 \cdot 2 \cdot 1} \cdot 16 = 3360$

43. $(1.001)^5 = (1+10^{-3})^5 = \binom{5}{0}1^5 + \binom{5}{1}1^4 \cdot 10^{-3} + \binom{5}{2}1^3 \cdot (10^{-3})^2 + \binom{5}{3}1^2 \cdot (10^{-3})^3 + \cdots$

$= 1 + 5(0.001) + 10(0.000001) + 10(0.000000001) + \cdots$

$= 1 + 0.005 + 0.000010 + 0.000000010 + \cdots$

$= 1.00501$ (correct to 5 decimal places)

45. $\binom{n}{n-1} = \dfrac{n!}{(n-1)!(n-(n-1))!} = \dfrac{n!}{(n-1)!(1)!} = \dfrac{n(n-1)!}{(n-1)!} = n$

$\binom{n}{n} = \dfrac{n!}{n!(n-n)!} = \dfrac{n!}{n!0!} = \dfrac{n!}{n! \cdot 1} = \dfrac{n!}{n!} = 1$

47. Show that $\binom{n}{0} + \binom{n}{1} + \cdots + \binom{n}{n} = 2^n$

$2^n = (1+1)^n$

$= \binom{n}{0} \cdot 1^n + \binom{n}{1} \cdot 1^{n-1} \cdot 1 + \binom{n}{2} \cdot 1^{n-2} \cdot 1^2 + \cdots + \binom{n}{n} \cdot 1^{n-n} \cdot 1^n$

$= \binom{n}{0} + \binom{n}{1} + \cdots + \binom{n}{n}$

49. $\binom{5}{0}\left(\dfrac{1}{4}\right)^5 + \binom{5}{1}\left(\dfrac{1}{4}\right)^4\left(\dfrac{3}{4}\right) + \binom{5}{2}\left(\dfrac{1}{4}\right)^3\left(\dfrac{3}{4}\right)^2 + \binom{5}{3}\left(\dfrac{1}{4}\right)^2\left(\dfrac{3}{4}\right)^3 + \binom{5}{4}\left(\dfrac{1}{4}\right)\left(\dfrac{3}{4}\right)^4 + \binom{5}{5}\left(\dfrac{3}{4}\right)^5 = \left(\dfrac{1}{4} + \dfrac{3}{4}\right)^5 = (1)^5 = 1$

Chapter 7 Review

1. $a_1 = (-1)^1 \dfrac{1+3}{1+2} = -\dfrac{4}{3},\ a_2 = (-1)^2 \dfrac{2+3}{2+2} = \dfrac{5}{4},\ a_3 = (-1)^3 \dfrac{3+3}{3+2} = -\dfrac{6}{5},\ a_4 = (-1)^4 \dfrac{4+3}{4+2} = \dfrac{7}{6},\ a_5 = (-1)^5 \dfrac{5+3}{5+2} = -\dfrac{8}{7}$

3. $a_1 = \dfrac{2^1}{1^2} = \dfrac{2}{1} = 2,\ a_2 = \dfrac{2^2}{2^2} = \dfrac{4}{4} = 1,\ a_3 = \dfrac{2^3}{3^2} = \dfrac{8}{9},\ a_4 = \dfrac{2^4}{4^2} = \dfrac{16}{16} = 1,\ a_5 = \dfrac{2^5}{5^2} = \dfrac{32}{25}$

5. $a_1 = 3,\ a_2 = \dfrac{2}{3} \cdot 3 = 2,\ a_3 = \dfrac{2}{3} \cdot 2 = \dfrac{4}{3},\ a_4 = \dfrac{2}{3} \cdot \dfrac{4}{3} = \dfrac{8}{9},\ a_5 = \dfrac{2}{3} \cdot \dfrac{8}{9} = \dfrac{16}{27}$

7. $a_1 = 2,\ a_2 = 2 - 2 = 0,\ a_3 = 2 - 0 = 2,\ a_4 = 2 - 2 = 0,\ a_5 = 2 - 0 = 2$

9. $\displaystyle\sum_{k=1}^{4}(4k+2) = (4 \cdot 1 + 2) + (4 \cdot 2 + 2) + (4 \cdot 3 + 2) + (4 \cdot 4 + 2) = (6) + (10) + (14) + (18) = 48$

11. $1 - \dfrac{1}{2} + \dfrac{1}{3} - \dfrac{1}{4} + \cdots + \dfrac{1}{13} = \displaystyle\sum_{k=1}^{13}(-1)^{k+1}\left(\dfrac{1}{k}\right)$

13. $\{n+5\}$ Arithmetic
    $d = (n+1+5)-(n+5) = n+6-n-5 = 1$
    $S_n = \dfrac{n}{2}[6+n+5] = \dfrac{n}{2}(n+11)$

15. $\{2n^3\}$ Examine the terms of the sequence: 2, 16, 54, 128, 250, ...
    There is no common difference; there is no common ratio; neither.

17. $\{2^{3n}\}$ Geometric
    $r = \dfrac{2^{3(n+1)}}{2^{3n}} = \dfrac{2^{3n+3}}{2^{3n}} = 2^{3n+3-3n} = 2^3 = 8$
    $S_n = 8\left(\dfrac{1-8^n}{1-8}\right) = 8\left(\dfrac{1-8^n}{-7}\right) = \dfrac{8}{7}(8^n-1)$

19. 0, 4, 8, 12, ... Arithmetic $d = 4-0 = 4$
    $S_n = \dfrac{n}{2}(2(0)+(n-1)4) = \dfrac{n}{2}(4(n-1)) = 2n(n-1)$

21. $3, \dfrac{3}{2}, \dfrac{3}{4}, \dfrac{3}{8}, \dfrac{3}{16}, ...$ Geometric
    $r = \dfrac{\left(\dfrac{3}{2}\right)}{3} = \dfrac{3}{2}\cdot\dfrac{1}{3} = \dfrac{1}{2}$
    $S_n = 3\left(\dfrac{1-\left(\dfrac{1}{2}\right)^n}{1-\dfrac{1}{2}}\right) = 3\left(\dfrac{1-\left(\dfrac{1}{2}\right)^n}{\left(\dfrac{1}{2}\right)}\right) = 6\left(1-\left(\dfrac{1}{2}\right)^n\right)$

23. Neither. There is no common difference or common ratio.

25. a. $\displaystyle\sum_{k=1}^{5}(k^2+12) = 13+16+21+28+37 = 115$

    b. use the sum(seq) feature:

    ```
    sum(seq(K²+12,K,
    1,5))
                115
    ```

27. a. $\displaystyle\sum_{k=1}^{10}(3k-9) = \sum_{k=1}^{10}3k - \sum_{k=1}^{10}9 = 3\sum_{k=1}^{10}k - \sum_{k=1}^{10}9$
    $= 3\left(\dfrac{10(10+1)}{2}\right) - 10(9) = 165-90$
    $= 75$

    b. use the sum(seq) feature:

    ```
    sum(seq(3K-9,K,1
    ,10))
                 75
    ```

29. a. $\displaystyle\sum_{k=1}^{7}\left(\dfrac{1}{3}\right)^k = \dfrac{1}{3}\left(\dfrac{1-\left(\dfrac{1}{3}\right)^7}{1-\dfrac{1}{3}}\right) = \dfrac{1}{3}\left(\dfrac{1-\left(\dfrac{1}{3}\right)^7}{\left(\dfrac{2}{3}\right)}\right)$
    $= \dfrac{1}{2}\left(1-\dfrac{1}{2187}\right)$
    $= \dfrac{1}{2}\cdot\dfrac{2186}{2187} = \dfrac{1093}{2187} \approx 0.49977$

    b. use the sum(seq) feature:

    ```
    sum(seq((1/3)^K,
    K,1,7))
          .4997713763
    ```

31. a. Arithmetic $a = 3$, $d = 4$, $a_n = a+(n-1)d$
    $a_9 = 3+(9-1)4 = 3+8(4) = 3+32 = 35$

    b. scroll to the end of the list:

    ```
    seq(3+(A-1)*4,A,
    1,9,1)→L₁
    ...19 23 27 31 35}
    ```

33. a. Geometric
    $a = 1$, $r = \dfrac{1}{10}$, $n = 11$; $a_n = ar^{n-1}$
    $a_{11} = 1\cdot\left(\dfrac{1}{10}\right)^{11-1} = \left(\dfrac{1}{10}\right)^{10}$
    $= \dfrac{1}{10,000,000,000}$

b. scroll to the end of the list:

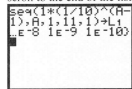

35. a. Arithmetic

$a = \sqrt{2}, \ d = \sqrt{2}, \ n = 9, \ a_n = a + (n-1)d$

$a_9 = \sqrt{2} + (9-1)\sqrt{2} = \sqrt{2} + 8\sqrt{2}$

$= 9\sqrt{2} \approx 12.7279$

b. scroll to the end of the list:

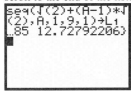

37. $a_7 = a + 6d = 31 \quad a_{20} = a + 19d = 96$;

Solve the system of equations:

$a + 6d = 31$

$a + 19d = 96$

Subtract the second equation from the first equation and solve for d.

$-13d = -65$

$d = 5$

$a = 31 - 6(5) = 31 - 30 = 1$

$a_n = a + (n-1)d$

$= 1 + (n-1)(5)$

$= 1 + 5n - 5$

$= 5n - 4$

General formula: $\{5n - 4\}$

39. $a_{10} = a + 9d = 0 \quad a_{18} = a + 17d = 8$;

Solve the system of equations:

$a + 9d = 0$

$a + 17d = 8$

Subtract the second equation from the first equation and solve for d.

$-8d = -8$

$d = 1$

$a = -9(1) = -9$

$a_n = a + (n-1)d$

$= -9 + (n-1)(1)$

$= -9 + n - 1$

$= n - 10$

General formula: $\{n - 10\}$

41. $a = 3, \ r = \dfrac{1}{3}$

Since $|r| < 1$, $\ S_n = \dfrac{a}{1-r} = \dfrac{3}{\left(1 - \dfrac{1}{3}\right)} = \dfrac{3}{\left(\dfrac{2}{3}\right)} = \dfrac{9}{2}$

43. $a = 2, \ r = -\dfrac{1}{2}$

Since $|r| < 1$, $\ S_n = \dfrac{a}{1-r} = \dfrac{2}{\left(1 - \left(-\dfrac{1}{2}\right)\right)}$

$= \dfrac{2}{\left(\dfrac{3}{2}\right)} = \dfrac{4}{3}$

45. $a = 4, \ r = \dfrac{1}{2}$

Since $|r| < 1$, $\ S_n = \dfrac{a}{1-r} = \dfrac{4}{\left(1 - \dfrac{1}{2}\right)} = \dfrac{4}{\left(\dfrac{1}{2}\right)} = 8$

47. I: $\quad n = 1: \quad 3 \cdot 1 = 3$ and $\dfrac{3 \cdot 1}{2}(1+1) = 3$

II: If $3 + 6 + 9 + \cdots + 3k = \dfrac{3k}{2}(k+1)$, then

$3 + 6 + 9 + \cdots + 3k + 3(k+1)$

$= [3 + 6 + 9 + \cdots + 3k] + 3(k+1)$

$= \dfrac{3k}{2}(k+1) + 3(k+1)$

$= (k+1)\left(\dfrac{3k}{2} + 3\right) = \dfrac{3(k+1)}{2}((k+1)+1)$

Conditions I and II are satisfied; the statement is true.

49. I: $n = 1$: $2 \cdot 3^{1-1} = 2$ and $3^1 - 1 = 2$

II: If $2 + 6 + 18 + \cdots + 2 \cdot 3^{k-1} = 3^k - 1$, then
$$2 + 6 + 18 + \cdots + 2 \cdot 3^{k-1} + 2 \cdot 3^{k+1-1}$$
$$= \left[2 + 6 + 18 + \cdots + 2 \cdot 3^{k-1}\right] + 2 \cdot 3^k$$
$$= 3^k - 1 + 2 \cdot 3^k = 3 \cdot 3^k - 1 = 3^{k+1} - 1$$

Conditions I and II are satisfied; the statement is true.

51. I: $n = 1$:
$$(3 \cdot 1 - 2)^2 = 1 \text{ and } \frac{1}{2} \cdot 1(6 \cdot 1^2 - 3 \cdot 1 - 1) = 1$$

II: If
$$1^2 + 4^2 + \cdots + (3k - 2)^2 = \frac{1}{2} \cdot k\left(6k^2 - 3k - 1\right),$$
then
$$1^2 + 4^2 + 7^2 + \cdots + (3k - 2)^2 + (3(k+1) - 2)^2$$
$$= \left[1^2 + 4^2 + 7^2 + \cdots + (3k - 2)^2\right] + (3k + 1)^2$$
$$= \frac{1}{2} \cdot k\left(6k^2 - 3k - 1\right) + (3k + 1)^2$$
$$= \frac{1}{2} \cdot \left[6k^3 - 3k^2 - k + 18k^2 + 12k + 2\right]$$
$$= \frac{1}{2} \cdot \left[6k^3 + 15k^2 + 11k + 2\right]$$
$$= \frac{1}{2} \cdot (k+1)\left[6k^2 + 9k + 2\right]$$
$$= \frac{1}{2} \cdot \left[6k^3 + 6k^2 + 9k^2 + 9k + 2k + 2\right]$$
$$= \frac{1}{2} \cdot \left[6k^2(k+1) + 9k(k+1) + 2(k+1)\right]$$
$$= \frac{1}{2} \cdot (k+1)\left[6k^2 + 12k + 6 - 3k - 3 - 1\right]$$
$$= \frac{1}{2} \cdot (k+1)\left[6(k^2 + 2k + 1) - 3(k+1) - 1\right]$$
$$= \frac{1}{2} \cdot (k+1)\left[6(k+1)^2 - 3(k+1) - 1\right]$$

Conditions I and II are satisfied; the statement is true.

53. $\binom{5}{2} = \dfrac{5!}{2! \, 3!} = \dfrac{5 \cdot 4 \cdot 3 \cdot 2 \cdot 1}{2 \cdot 1 \cdot 3 \cdot 2 \cdot 1} = \dfrac{5 \cdot 4}{2 \cdot 1} = 10$

SSM: College Algebra EGU

Chapter 7: Sequences; Induction; the Binomial Theorem

55. $(x+2)^5 = \binom{5}{0}x^5 + \binom{5}{1}x^4 \cdot 2 + \binom{5}{2}x^3 \cdot 2^2 + \binom{5}{3}x^2 \cdot 2^3 + \binom{5}{4}x^1 \cdot 2^4 + \binom{5}{5} \cdot 2^5$
$= x^5 + 5 \cdot 2x^4 + 10 \cdot 4x^3 + 10 \cdot 8x^2 + 5 \cdot 16x + 1 \cdot 32$
$= x^5 + 10x^4 + 40x^3 + 80x^2 + 80x + 32$

57. $(2x+3)^5 = \binom{5}{0}(2x)^5 + \binom{5}{1}(2x)^4 \cdot 3 + \binom{5}{2}(2x)^3 \cdot 3^2 + \binom{5}{3}(2x)^2 \cdot 3^3 + \binom{5}{4}(2x)^1 \cdot 3^4 + \binom{5}{5} \cdot 3^5$
$= 32x^5 + 5 \cdot 16x^4 \cdot 3 + 10 \cdot 8x^3 \cdot 9 + 10 \cdot 4x^2 \cdot 27 + 5 \cdot 2x \cdot 81 + 1 \cdot 243$
$= 32x^5 + 240x^4 + 720x^3 + 1080x^2 + 810x + 243$

59. $n = 9$, $j = 2$, $x = x$, $a = 2$
$\binom{9}{2}x^7 \cdot 2^2 = \frac{9!}{2!\,7!} \cdot 4x^7 = \frac{9 \cdot 8}{2 \cdot 1} \cdot 4x^7 = 144x^7$
The coefficient of x^7 is 144.

61. $n = 7$, $j = 5$, $x = 2x$, $a = 1$
$\binom{7}{5}(2x)^2 \cdot 1^5 = \frac{7!}{5!\,2!} \cdot 4x^2(1) = \frac{7 \cdot 6}{2 \cdot 1} \cdot 4x^2 = 84x^2$
The coefficient of x^2 is 84.

63. This is an arithmetic sequence with $a = 80$, $d = -3$, $n = 25$

  a. $a_{25} = 80 + (25-1)(-3) = 80 - 72 = 8$ bricks

  b. $S_{25} = \frac{25}{2}(80+8) = 25(44) = 1100$ bricks
  1100 bricks are needed to build the steps.

65. This is a geometric sequence with $a = 20$, $r = \frac{3}{4}$.

  a. After striking the ground the third time, the height is $20\left(\frac{3}{4}\right)^3 = \frac{135}{16} \approx 8.44$ feet.

  b. After striking the ground the nth time, the height is $20\left(\frac{3}{4}\right)^n$ feet.

  c If the height is less than 6 inches or 0.5 feet, then:

$0.5 \geq 20\left(\frac{3}{4}\right)^n$

$0.025 \geq \left(\frac{3}{4}\right)^n$

$\log(0.025) \geq n \log\left(\frac{3}{4}\right)$

$n \geq \frac{\log(0.025)}{\log\left(\frac{3}{4}\right)} \approx 12.82$

The height is less than 6 inches after the 13th strike.

  d. Since this is a geometric sequence with $|r| < 1$, the distance is the sum of the two infinite geometric series - the distances going down plus the distances going up.
  Distance going down:
  $S_{down} = \frac{20}{\left(1-\frac{3}{4}\right)} = \frac{20}{\left(\frac{1}{4}\right)} = 80$ feet.

  Distance going up:
  $S_{up} = \frac{15}{\left(1-\frac{3}{4}\right)} = \frac{15}{\left(\frac{1}{4}\right)} = 60$ feet.

  The total distance traveled is 140 feet.

67. a. $b_1 = 5,000$, $b_n = 1.015b_{n-1} - 100$
$b_2 = 1.015b_{(2-1)} - 100 = 1.015b_1 - 100$
$= 1.015(5000) - 100 = \$4975$

b. Enter the recursive formula in Y= and draw the graph:

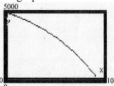

c. Scroll through the table:

At the beginning of the 33rd month, or after 32 payments have been made, the balance is below $4,000. The balance is $3982.8.

d. Scroll through the table:

The balance will be paid off at the end of 94 months or 7 years and 10 months.
Total payments =
$100(93) + 11.01(1.015) = \$9,311.18$

e. The total interest expense is the difference between the total of the payments and the original balance:
$100(93) + 11.18 - 5,000 = \$4,311.18$

Chapter 7 Test

1. $a_n = \dfrac{n^2 - 1}{n + 8}$

   $a_1 = \dfrac{1^2 - 1}{1 + 8} = \dfrac{0}{9} = 0$

   $a_2 = \dfrac{2^2 - 1}{2 + 8} = \dfrac{3}{10}$

   $a_3 = \dfrac{3^2 - 1}{3 + 8} = \dfrac{8}{11}$

   $a_4 = \dfrac{4^2 - 1}{4 + 8} = \dfrac{15}{12} = \dfrac{5}{4}$

   $a_5 = \dfrac{5^2 - 1}{5 + 8} = \dfrac{24}{13}$

   The first five terms of the sequence are 0, $\dfrac{3}{10}$, $\dfrac{8}{11}$, $\dfrac{5}{4}$, and $\dfrac{24}{13}$.

2. $a_1 = 4;\ a_n = 3a_{n-1} + 2$

   $a_2 = 3a_1 + 2 = 3(4) + 2 = 14$

   $a_3 = 3a_2 + 2 = 3(14) + 2 = 44$

   $a_4 = 3a_3 + 2 = 3(44) + 2 = 134$

   $a_5 = 3a_4 + 2 = 3(134) + 2 = 404$

   The first five terms of the sequence are 4, 14, 44, 134, and 404.

3. $\displaystyle\sum_{k=1}^{3}(-1)^{k+1}\left(\dfrac{k+1}{k^2}\right)$

   $= (-1)^{1+1}\left(\dfrac{1+1}{1^2}\right) + (-1)^{2+1}\left(\dfrac{2+1}{2^2}\right) + (-1)^{3+1}\left(\dfrac{3+1}{3^2}\right)$

   $= (-1)^2\left(\dfrac{2}{1}\right) + (-1)^3\left(\dfrac{3}{4}\right) + (-1)^4\left(\dfrac{4}{9}\right)$

   $= 2 - \dfrac{3}{4} + \dfrac{4}{9} = \dfrac{61}{36}$

4. $\displaystyle\sum_{k=1}^{4}\left[\left(\dfrac{2}{3}\right)^k - k\right]$

   $= \left[\left(\dfrac{2}{3}\right)^1 - 1\right] + \left[\left(\dfrac{2}{3}\right)^2 - 2\right] + \left[\left(\dfrac{2}{3}\right)^3 - 3\right] + \left[\left(\dfrac{2}{3}\right)^4 - 4\right]$

   $= \dfrac{2}{3} - 1 + \dfrac{4}{9} - 2 + \dfrac{8}{27} - 3 + \dfrac{16}{81} - 4$

   $= \dfrac{130}{81} - 10 = -\dfrac{680}{81}$

5. $-\dfrac{2}{5}+\dfrac{3}{6}-\dfrac{4}{7}+\ldots+\dfrac{11}{14}$

   Notice that the signs of each term alternate, with the first term being negative. This implies that the general term will include a power of -1. Also note that the numerator is always 1 more than the term number and the denominator is 4 more than the term number. Thus, each term is in the form $(-1)^k\left(\dfrac{k+1}{k+4}\right)$. The last numerator is 11 which indicates that there are 10 terms.

   $-\dfrac{2}{5}+\dfrac{3}{6}-\dfrac{4}{7}+\ldots+\dfrac{11}{14}=\sum_{k=1}^{10}(-1)^k\left(\dfrac{k+1}{k+4}\right)$

6. $\sum_{k=1}^{100}(-1)^k(k^2)=5050$

   ```
   sum(seq((-1)^X*X
   ²,X,1,100))
              5050
   ```

7. $6, 12, 36, 144, \ldots$

   $12-6=6$ and $36-12=24$

   The difference between consecutive terms is not constant. Therefore, the sequence is not arithmetic.

   $\dfrac{12}{6}=2$ and $\dfrac{36}{12}=3$

   The ratio of consecutive terms is not constant. Therefore, the sequence is not geometric.

8. $a_n=-\dfrac{1}{2}\cdot 4^n$

   $\dfrac{a_n}{a_{n-1}}=\dfrac{-\frac{1}{2}\cdot 4^n}{-\frac{1}{2}\cdot 4^{n-1}}=\dfrac{-\frac{1}{2}\cdot 4^{n-1}\cdot 4}{-\frac{1}{2}\cdot 4^{n-1}}=4$

   Since the ratio of consecutive terms is constant, the sequence is geometric with common ratio $r=4$ and first term $a_1=a=-\dfrac{1}{2}\cdot 4^1=-2$.

   The sum of the first n terms of the sequence is given by

   $S_n=a\cdot\dfrac{1-r^n}{1-r}$

   $=-2\cdot\dfrac{1-4^n}{1-4}$

   $=\dfrac{2}{3}(1-4^n)$

9. $-2, -10, -18, -26, \ldots$

   $-10-(-2)=-8$, $-18-(-10)=-8$,

   $-26-(-18)=-8$

   The difference between consecutive terms is constant. Therefore, the sequence is arithmetic with common difference $d=-8$ and first term $a_1=a=-2$.

   $a_n=a+(n-1)d$
   $=-2+(n-1)(-8)$
   $=-2-8n+8$
   $=6-8n$

   The sum of the first n terms of the sequence is given by

   $S_n=\dfrac{n}{2}(a+a_n)$

   $=\dfrac{n}{2}(-2+6-8n)$

   $=\dfrac{n}{2}(4-8n)$

   $=n(2-4n)$

10. $a_n=-\dfrac{n}{2}+7$

    $a_n-a_{n-1}=\left[-\dfrac{n}{2}+7\right]-\left[-\dfrac{(n-1)}{2}+7\right]$

    $=-\dfrac{n}{2}+7+\dfrac{n-1}{2}-7$

    $=-\dfrac{1}{2}$

    The difference between consecutive terms is constant. Therefore, the sequence is arithmetic with common difference $d=-\dfrac{1}{2}$ and first term $a_1=a=-\dfrac{1}{2}+7=\dfrac{13}{2}$.

    The sum of the first n terms of the sequence is given by

    $S_n=\dfrac{n}{2}(a+a_n)$

    $=\dfrac{n}{2}\left(\dfrac{13}{2}+\left(-\dfrac{n}{2}+7\right)\right)$

    $=\dfrac{n}{2}\left(\dfrac{27}{2}-\dfrac{n}{2}\right)$

    $=\dfrac{n}{4}(27-n)$

11. $25, 10, 4, \dfrac{8}{5}, \ldots$

$\dfrac{10}{25} = \dfrac{2}{5}, \ \dfrac{4}{10} = \dfrac{2}{5}, \ \dfrac{\frac{8}{5}}{4} = \dfrac{8}{5} \cdot \dfrac{1}{4} = \dfrac{2}{5}$

The ratio of consecutive terms is constant. Therefore, the sequence is geometric with common ratio $r = \dfrac{2}{5}$ and first term $a_1 = a = 25$.

The sum of the first n terms of the sequence is given by

$$S_n = a \cdot \dfrac{1-r^n}{1-r} = 25 \cdot \dfrac{1-\left(\dfrac{2}{5}\right)^n}{1-\dfrac{2}{5}} = 25 \cdot \dfrac{\left(1-\left(\dfrac{2}{5}\right)^n\right)}{\dfrac{3}{5}}$$

$$= 25 \cdot \dfrac{5}{3}\left(1-\left(\dfrac{2}{5}\right)^n\right) = \dfrac{125}{3}\left(1-\left(\dfrac{2}{5}\right)^n\right)$$

12. $a_n = \dfrac{2n-3}{2n+1}$

$a_n - a_{n-1} = \dfrac{2n-3}{2n+1} - \dfrac{2(n-1)-3}{2(n-1)+1} = \dfrac{2n-3}{2n+1} - \dfrac{2n-5}{2n-1}$

$= \dfrac{(2n-3)(2n-1) - (2n-5)(2n+1)}{(2n+1)(2n-1)}$

$= \dfrac{(4n^2 - 8n + 3) - (4n^2 - 8n - 5)}{4n^2 - 1}$

$= \dfrac{8}{4n^2 - 1}$

The difference of consecutive terms is not constant. Therefore, the sequence is not arithmetic.

$\dfrac{a_n}{a_{n-1}} = \dfrac{\dfrac{2n-3}{2n+1}}{\dfrac{2(n-1)-3}{2(n-1)+1}} = \dfrac{2n-3}{2n+1} \cdot \dfrac{2n-1}{2n-5}$

$= \dfrac{(2n-3)(2n-1)}{(2n+1)(2n-5)}$

The ratio of consecutive terms is not constant. Therefore, the sequence is not geometric.

13. For this geometric series we have $r = \dfrac{-64}{256} = -\dfrac{1}{4}$ and $a_1 = a = 256$. Since $|r| = \left|-\dfrac{1}{4}\right| = \dfrac{1}{4} < 1$, we get

$S_\infty = \dfrac{a}{1-r} = \dfrac{256}{1-\left(-\dfrac{1}{4}\right)} = \dfrac{256}{\dfrac{5}{4}} = \dfrac{1024}{5} = 204.8$

14. $(3m+2)^5 = \binom{5}{0}(3m)^5 + \binom{5}{1}(3m)^4(2) + \binom{5}{2}(3m)^3(2)^2 + \binom{5}{3}(3m)^2(2)^3 + \binom{5}{4}(3m)(2)^4 + \binom{5}{5}(2)^5$

$= 243m^5 + 5\cdot 81m^4 \cdot 2 + 10\cdot 27m^3 \cdot 4 + 10\cdot 9m^2 \cdot 8 + 5\cdot 3m\cdot 16 + 32$

$= 243m^5 + 810m^4 + 1080m^3 + 720m^2 + 240m + 32$

15. First we show that the statement holds for $n=1$.

$\left(1+\dfrac{1}{1}\right) = 1+1 = 2$

The equality is true for $n=1$ so Condition I holds.

Next we assume that $\left(1+\dfrac{1}{1}\right)\left(1+\dfrac{1}{2}\right)\left(1+\dfrac{1}{3}\right)\cdots\left(1+\dfrac{1}{n}\right) = n+1$ is true for some k, and we determine whether the formula then holds for $k+1$. We assume that

$\left(1+\dfrac{1}{1}\right)\left(1+\dfrac{1}{2}\right)\left(1+\dfrac{1}{3}\right)\cdots\left(1+\dfrac{1}{k}\right) = k+1$

Now we need to show that

$\left(1+\dfrac{1}{1}\right)\left(1+\dfrac{1}{2}\right)\left(1+\dfrac{1}{3}\right)\cdots\left(1+\dfrac{1}{k}\right)\left(1+\dfrac{1}{k+1}\right) = (k+1)+1 = k+2$

We do this as follows:

$\left(1+\dfrac{1}{1}\right)\left(1+\dfrac{1}{2}\right)\left(1+\dfrac{1}{3}\right)\cdots\left(1+\dfrac{1}{k}\right)\left(1+\dfrac{1}{k+1}\right) = \left[\left(1+\dfrac{1}{1}\right)\left(1+\dfrac{1}{2}\right)\left(1+\dfrac{1}{3}\right)\cdots\left(1+\dfrac{1}{k}\right)\right]\left(1+\dfrac{1}{k+1}\right)$

$= (k+1)\left(1+\dfrac{1}{k+1}\right)$ (using the induction assumption)

$= (k+1)\cdot 1 + (k+1)\cdot\dfrac{1}{k+1} = k+1+1$

$= k+2$

Condition II also holds. Thus, formula holds true for all natural numbers.

\

16. The yearly values of the Durango form a geometric sequence with first term $a_1 = a = 31,000$ and common ratio $r = 0.85$ (which represents a 15% loss in value).

$a_n = 31,000\cdot(0.85)^{n-1}$

The nth term of the sequence represents the value of the Durango at the beginning of the nth year. Since we want to know the value *after* 10 years, we are looking for the 11th term of the sequence. That is, the value of the Durango at the beginning of the 11th year.

$a_{11} = a\cdot r^{11-1}$

$= 31,000\cdot(0.85)^{10}$

$= 6,103.11$

After 10 years, the Durango will be worth $6,103.11.

17. The weights for each set form an arithmetic sequence with first term $a_1 = a = 100$ and common difference $d = 30$. If we imagine the weightlifter only performed one repetition per set, the total weight lifted in 5 sets would be the sum of the first five terms of the sequence.

$a_n = a + (n-1)d$

$a_5 = 100 + (5-1)(30) = 100 + 4(30) = 220$

$S_n = \dfrac{n}{2}(a+a_n)$

$S_5 = \dfrac{5}{2}(100+220) = \dfrac{5}{2}(320) = 800$

Since he performs 10 repetitions in each set, we multiply the sum by 10 to obtain the total weight lifted.

$10(800) = 8000$

The weightlifter will have lifted a total of 8000 pounds after 5 sets.

Cumulative Review R-7

1. $|x^2| = 9$

   $x^2 = 9$ or $x^2 = -9$

   $x = \pm 3$ or $x = \pm 3i$

3. $2e^x = 5$

   $e^x = \dfrac{5}{2}$

   $\ln(e^x) = \ln\left(\dfrac{5}{2}\right)$

   $x = \ln\left(\dfrac{5}{2}\right) \approx 0.916$

5. Given a circle with center $(-1, 2)$ and containing the point $(3, 5)$, we first use the distance formula to determine the radius.

   $r = \sqrt{(3-(-1))^2 + (5-2)^2}$
   $ = \sqrt{4^2 + 3^2} = \sqrt{16+9}$
   $ = \sqrt{25}$
   $ = 5$

   Therefore, the equation of the circle is given by

   $(x-(-1))^2 + (y-2)^2 = 5^2$
   $(x+1)^2 + (y-2)^2 = 5^2$
   $x^2 + 2x + 1 + y^2 - 4y + 4 = 25$
   $x^2 + y^2 + 2x - 4y - 20 = 0$

7. Center: $(0, 0)$; Focus: $(0, 3)$; Vertex: $(0, 4)$;
   Major axis is the y-axis; $a = 4$; $c = 3$.
   Find b: $b^2 = a^2 - c^2 = 16 - 9 = 7 \Rightarrow b = \sqrt{7}$
   Write the equation using rectangular coordinates:
   $\dfrac{x^2}{7} + \dfrac{y^2}{16} = 1$

Chapter 8
Counting and Probability

Section 8.1

1. union

3. True; the union of two sets includes those elements that are in one *or* both of the sets. The intersection consists of the elements that are in *both* sets. Thus, the intersection is a subset of the union.

5. $A \cup B = \{1, 3, 5, 7, 9\} \cup \{1, 5, 6, 7\}$
$= \{1, 3, 5, 6, 7, 9\}$

7. $A \cap B = \{1, 3, 5, 7, 9\} \cap \{1, 5, 6, 7\} = \{1, 5, 7\}$

9. $(A \cup B) \cap C$
$= (\{1, 3, 5, 7, 9\} \cup \{1, 5, 6, 7\}) \cap \{1, 2, 4, 6, 8, 9\}$
$= \{1, 3, 5, 6, 7, 9\} \cap \{1, 2, 4, 6, 8, 9\}$
$= \{1, 6, 9\}$

11. $(A \cap B) \cup C$
$= (\{1, 3, 5, 7, 9\} \cap \{1, 5, 6, 7\}) \cup \{1, 2, 4, 6, 8, 9\}$
$= \{1, 5, 7\} \cup \{1, 2, 4, 6, 8, 9\}$
$= \{1, 2, 4, 5, 6, 7, 8, 9\}$

13. $(A \cup C) \cap (B \cup C)$
$= (\{1, 3, 5, 7, 9\} \cup \{1, 2, 4, 6, 8, 9\})$
$\cap (\{1, 5, 6, 7\} \cup \{1, 2, 4, 6, 8, 9\})$
$= \{1, 2, 3, 4, 5, 6, 7, 8, 9\} \cap \{1, 2, 4, 5, 6, 7, 8, 9\}$
$= \{1, 2, 4, 5, 6, 7, 8, 9\}$

15. $\overline{A} = \{0, 2, 6, 7, 8\}$

17. $\overline{A \cap B} = \overline{\{1, 3, 4, 5, 9\} \cap \{2, 4, 6, 7, 8\}}$
$= \overline{\{4\}} = \{0, 1, 2, 3, 5, 6, 7, 8, 9\}$

19. $\overline{A} \cup \overline{B} = \{0, 2, 6, 7, 8\} \cup \{0, 1, 3, 5, 9\}$
$= \{0, 1, 2, 3, 5, 6, 7, 8, 9\}$

21. $\overline{A \cap \overline{C}} = \overline{\{1, 3, 4, 5, 9\} \cap \{0, 2, 5, 7, 8, 9\}}$
$= \overline{\{5, 9\}} = \{0, 1, 2, 3, 4, 6, 7, 8\}$

23. $\overline{A \cup B \cup C}$
$= \overline{\{1, 3, 4, 5, 9\} \cup \{2, 4, 6, 7, 8\} \cup \{1, 3, 4, 6\}}$
$= \overline{\{1, 2, 3, 4, 5, 6, 7, 8, 9\}}$
$= \{0\}$

25. $\{a\}, \{b\}, \{c\}, \{d\}, \{a, b\}, \{a, c\}, \{a, d\},$
$\{b, c\}, \{b, d\}, \{c, d\}, \{a, b, c\}, \{a, b, d\},$
$\{a, c, d\}, \{b, c, d\}, \{a, b, c, d\}, \varnothing$

27. $n(A) = 15, n(B) = 20, n(A \cap B) = 10$
$n(A \cup B) = n(A) + n(B) - n(A \cap B)$
$= 15 + 20 - 10 = 25$

29. $n(A \cup B) = 50, n(A \cap B) = 10, n(B) = 20$
$n(A \cup B) = n(A) + n(B) - n(A \cap B)$
$50 = n(A) + 20 - 10$
$40 = n(A)$

31. From the figure: $n(A) = 15 + 3 + 5 + 2 = 25$

33. From the figure:
$n(A \text{ or } B) = n(A \cup B)$
$= 15 + 2 + 5 + 3 + 10 + 2 = 37$

35. From the figure:
$n(A \text{ but not } C) = n(A) - n(A \cap C) = 25 - 7 = 18$

37. From the figure:
$n(A \text{ and } B \text{ and } C) = n(A \cap B \cap C) = 5$

491

39. Let $A = \{\text{those who will purchase a major appliance}\}$ and
$B = \{\text{those who will buy a car}\}$
$n(U) = 500,\ n(A) = 200,$
$n(B) = 150,\ n(A \cap B) = 25$
$n(A \cup B) = n(A) + n(B) - n(A \cap B)$
$\qquad = 200 + 150 - 25 = 325$
$n(\text{purchase neither}) = n(U) - n(A \cup B)$
$\qquad = -500 - 325 = 175$
$n(\text{purchase only a car}) = n(B) - n(A \cup B)$
$\qquad = 150 - 25 = 125$

41. Construct a Venn diagram:

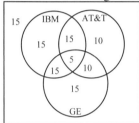

(a) 15   (b) 15
(c) 15   (d) 25
(e) 40

43. a. $n(\text{widowed or divorced}) = n(\text{widowed}) + n(\text{divorced})$
$\qquad = 2{,}632 + 8{,}659$
$\qquad = 11{,}291$
There were 11,291 thousand males 18 years old and older who were widowed or divorced.

b. $n(\text{married, widowed or divorced})$
$= n(\text{married}) + n(\text{widowed}) + n(\text{divorced})$
$= 61{,}212 + 2{,}632 + 8{,}659 = 72{,}503$
There were 72,503 thousand males 18 years old and older who were married, widowed, or divorced.

45. Answers will vary.

Section 8.2

1. 1; 1

3. permutation

5. True

7. $P(6, 2) = \dfrac{6!}{(6-2)!} = \dfrac{6!}{4!} = \dfrac{6 \cdot 5 \cdot 4!}{4!} = 30$

9. $P(4, 4) = \dfrac{4!}{(4-4)!} = \dfrac{4!}{0!} = \dfrac{4 \cdot 3 \cdot 2 \cdot 1}{1} = 24$

11. $P(7, 0) = \dfrac{7!}{(7-0)!} = \dfrac{7!}{7!} = 1$

13. $P(8, 4) = \dfrac{8!}{(8-4)!} = \dfrac{8!}{4!} = \dfrac{8 \cdot 7 \cdot 6 \cdot 5 \cdot 4!}{4!} = 1680$

15. $C(8, 2) = \dfrac{8!}{(8-2)!\,2!} = \dfrac{8!}{6!\,2!} = \dfrac{8 \cdot 7 \cdot 6!}{6! \cdot 2 \cdot 1} = 28$

17. $C(7, 4) = \dfrac{7!}{(7-4)!\,4!} = \dfrac{7!}{3!\,4!} = \dfrac{7 \cdot 6 \cdot 5 \cdot 4!}{4! \cdot 3 \cdot 2 \cdot 1} = 35$

19. $C(15, 15) = \dfrac{15!}{(15-15)!\,15!} = \dfrac{15!}{0!\,15!} = \dfrac{15!}{15! \cdot 1} = 1$

21. $C(26, 13) = \dfrac{26!}{(26-13)!\,13!} = \dfrac{26!}{13!\,13!} = 10{,}400{,}600$

23. $\{abc, abd, abe, acb, acd, ace, adb, adc,$
$ade, aeb, aec, aed, bac, bad, bae, bca,$
$bcd, bce, bda, bdc, bde, bea, bec, bed,$
$cab, cad, cae, cba, cbd, cbe, cda, cdb,$
$cde, cea, ceb, ced, dab, dac, dae, dba,$
$dbc, dbe, dca, dcb, dce, dea, deb, dec,$
$eab, eac, ead, eba, ebc, ebd, eca, ecb,$
$ecd, eda, edb, edc\}$
$P(5, 3) = \dfrac{5!}{(5-3)!} = \dfrac{5!}{2!} = \dfrac{5 \cdot 4 \cdot 3 \cdot 2!}{2!} = 60$

25. {123, 124, 132, 134, 142, 143, 213, 214, 231, 234, 241, 243, 312, 314, 321, 324, 341, 342, 412, 413, 421, 423, 431, 432}

$$P(4,3) = \frac{4!}{(4-3)!} = \frac{4!}{1!} = \frac{4 \cdot 3 \cdot 2 \cdot 1}{1} = 24$$

27. {abc, abd, abe, acd, ace, ade, bcd, bce, bde, cde}

$$C(5,3) = \frac{5!}{(5-3)!3!} = \frac{5 \cdot 4 \cdot 3!}{2 \cdot 1 \cdot 3!} = 10$$

29. {123, 124, 134, 234}

$$C(4,3) = \frac{4!}{(4-3)!3!} = \frac{4 \cdot 3!}{1!3!} = 4$$

31. There are 5 choices of shirts and 3 choices of ties; there are (5)(3) = 15 combinations.

33. There are 4 choices for the first letter in the code and 4 choices for the second letter in the code; there are (4)(4) = 16 possible two-letter codes.

35. There are two choices for each of three positions; there are (2)(2)(2) = 8 possible three-digit numbers.

37. To line up the four people, there are 4 choices for the first position, 3 choices for the second position, 2 choices for the third position, and 1 choice for the fourth position. Thus there are (4)(3)(2)(1) = 24 possible ways four people can be lined up.

39. Since no letter can be repeated, there are 5 choices for the first letter, 4 choices for the second letter, and 3 choices for the third letter. Thus, there are (5)(4)(3) = 60 possible three-letter codes.

41. There are 26 possible one-letter names. There are (26)(26) = 676 possible two-letter names. There are (26)(26)(26) = 17,576 possible three-letter names. Thus, there are 26 + 676 + 17,576 = 18,278 possible companies that can be listed on the New York Stock Exchange.

43. A committee of 4 from a total of 7 students is given by:
$$C(7,4) = \frac{7!}{(7-4)!\,4!} = \frac{7!}{3!\,4!} = \frac{7 \cdot 6 \cdot 5 \cdot 4!}{3 \cdot 2 \cdot 1 \cdot 4!} = 35$$
35 committees are possible.

45. There are 2 possible answers for each question. Therefore, there are $2^{10} = 1024$ different possible arrangements of the answers.

47. There are 9 choices for the first digit, and 10 choices for each of the other three digits. Thus, there are (9)(10)(10)(10) = 9000 possible four-digit numbers.

49. There are 5 choices for the first position, 4 choices for the second position, 3 choices for the third position, 2 choices for the fourth position, and 1 choice for the fifth position. Thus, there are (5)(4)(3)(2)(1) = 120 possible arrangements of the books.

51. There are 8 choices for the DOW stocks, 15 choices for the NASDAQ stocks, and 4 choices for the global stocks. Thus, there are (8)(15)(4) = 480 different portfolios.

53. The 1st person can have any of 365 days, the 2nd person can have any of the remaining 364 days. Thus, there are (365)(364) = 132,860 possible ways two people can have different birthdays.

55. Choosing 2 boys from the 4 boys can be done C(4,2) ways, and choosing 3 girls from the 8 girls can be done in C(8,3) ways. Thus, there are a total of:

$$C(4,2) \cdot C(8,3) = \frac{4!}{(4-2)!\,2!} \cdot \frac{8!}{(8-3)!\,3!}$$
$$= \frac{4!}{2!\,2!} \cdot \frac{8!}{5!\,3!}$$
$$= \frac{4 \cdot 3 \cdot 2!}{2 \cdot 1 \cdot 2!} \cdot \frac{8 \cdot 7 \cdot 6 \cdot 5!}{5!\,3!}$$
$$= 6 \cdot 56 = 336$$

57. This is a permutation with repetition. There are
$\dfrac{9!}{2!\,2!} = 90{,}720$ different words.

59. a. $C(7,2) \cdot C(3,1) = 21 \cdot 3 = 63$

   b. $C(7,3) \cdot C(3,0) = 35 \cdot 1 = 35$

   c. $C(3,3) \cdot C(7,0) = 1 \cdot 1 = 1$

61. There are $C(100, 22)$ ways to form the first committee. There are 78 senators left, so there are $C(78, 13)$ ways to form the second committee. There are $C(65, 10)$ ways to form the third committee. There are $C(55, 5)$ ways to form the fourth committee. There are $C(50, 16)$ ways to form the fifth committee. There are $C(34, 17)$ ways to form the sixth committee. There are $C(17, 17)$ ways to form the seventh committee.
The total number of committees
$= C(100,22) \cdot C(78,13) \cdot C(65,10) \cdot C(55,5)$
$\quad \cdot C(50,16) \cdot C(34,17) \cdot C(17,17)$
$\approx 1.1568 \times 10^{76}$

63. There are 9 choices for the first position, 8 choices for the second position, 7 for the third position, etc. There are
$9 \cdot 8 \cdot 7 \cdot 6 \cdot 5 \cdot 4 \cdot 3 \cdot 2 \cdot 1 = 9! = 362{,}800$ possible batting orders.

65. The team must have 1 pitcher and 8 position players (non-pitchers). For pitcher, choose 1 player from a group of 4 players, i.e., $C(4, 1)$. For position players, choose 8 players from a group of 11 players, i.e., $C(11, 8)$. Therefore, the number different teams possible is
$C(4,1) \cdot C(11,8) = 4 \cdot 165 = 660$.

67. Choose 2 players from a group of 6 players. Therefore, there are $C(6,2) = 15$ different teams possible.

69. Answers will vary.

71. Answers will vary.

Section 8.3

1. equally likely

3. False; probability may equal 0. In such cases, the corresponding event will never happen.

5. Probabilities must be between 0 and 1, inclusive. Thus, 0, 0.01, 0.35, and 1 could be probabilities.

7. All the probabilities are between 0 and 1.
The sum of the probabilities is
$0.2 + 0.3 + 0.1 + 0.4 = 1$.
This is a probability model.

9. All the probabilities are between 0 and 1.
The sum of the probabilities is
$0.3 + 0.2 + 0.1 + 0.3 = 0.9$.
This is not a probability model.

11. The sample space is: $S = \{HH, HT, TH, TT\}$.
Each outcome is equally likely to occur; so
$P(E) = \dfrac{n(E)}{n(S)}$.
The probabilities are:
$P(HH) = \dfrac{1}{4}$, $P(HT) = \dfrac{1}{4}$, $P(TH) = \dfrac{1}{4}$, $P(TT) = \dfrac{1}{4}$

13. The sample space of tossing two fair coins and a fair die is:
$S = \{HH1, HH2, HH3, HH4, HH5, HH6,$
$HT1, HT2, HT3, HT4, HT5, HT6, TH1,$
$TH2, TH3, TH4, TH5, TH6, TT1, TT2,$
$TT3, TT4, TT5, TT6\}$
There are 24 equally likely outcomes and the probability of each is $\dfrac{1}{24}$.

15. The sample space for tossing three fair coins is:
$S = \{HHH, HHT, HTH, THH, HTT, THT,$
$TTH, TTT\}$
There are 8 equally likely outcomes and the probability of each is $\dfrac{1}{8}$.

17. The sample space is:
$S = \{1$ Yellow, 1 Red, 1 Green, 2 Yellow, 2 Red, 2 Green, 3 Yellow, 3 Red, 3 Green, 4 Yellow, 4 Red, 4 Green$\}$
There are 12 equally likely events and the probability of each is $\dfrac{1}{12}$. The probability of getting a 2 or 4 followed by a Red is
$P(2\ \text{Red}) + P(4\ \text{Red}) = \dfrac{1}{12} + \dfrac{1}{12} = \dfrac{1}{6}$.

19. The sample space is:
    S = {1 Yellow Forward, 1 Yellow Backward, 1 Red Forward, 1 Red Backward, 1 Green Forward, 1 Green Backward, 2 Yellow Forward, 2 Yellow Backward, 2 Red Forward, 2 Red Backward, 2 Green Forward, 2 Green Backward, 3 Yellow Forward, 3 Yellow Backward, 3 Red Forward, 3 Red Backward, 3 Green Forward, 3 Green Backward, 4 Yellow Forward, 4 Yellow Backward, 4 Red Forward, 4 Red Backward, 4 Green Forward, 4 Green Backward}
    There are 24 equally likely events and the probability of each is $\frac{1}{24}$. The probability of getting a 1, followed by a Red or Green, followed by a Backward is
    $P(1\text{ Red Backward}) + P(1\text{ Green Backward}) = \frac{1}{24} + \frac{1}{24}$
    $= \frac{1}{12}$

21. The sample space is:
    S = {1 1 Yellow, 1 1 Red, 1 1 Green, 1 2 Yellow, 1 2 Red, 1 2 Green, 1 3 Yellow, 1 3 Red, 1 3 Green, 1 4 Yellow, 1 4 Red, 1 4 Green, 2 1 Yellow, 2 1 Red, 2 1 Green, 2 2 Yellow, 2 2 Red, 2 2 Green, 2 3 Yellow, 2 3 Red, 2 3 Green, 2 4 Yellow, 2 4 Red, 2 4 Green, 3 1 Yellow, 3 1 Red, 3 1 Green, 3 2 Yellow, 3 2 Red, 3 2 Green, 3 3 Yellow, 3 3 Red, 3 3 Green, 3 4 Yellow, 3 4 Red, 3 4 Green, 4 1 Yellow, 4 1 Red, 4 1 Green, 4 2 Yellow, 4 2 Red, 4 2 Green, 4 3 Yellow, 4 3 Red, 4 3 Green, 4 4 Yellow, 4 4 Red, 4 4 Green}
    There are 48 equally likely events and the probability of each is $\frac{1}{48}$. The probability of getting a 2, followed by a 2 or 4, followed by a Red or Green is
    $P(2\,2\text{ Red}) + P(2\,4\text{ Red}) + P(2\,2\text{ Green}) + P(2\,4\text{ Green})$
    $= \frac{1}{48} + \frac{1}{48} + \frac{1}{48} + \frac{1}{48} = \frac{1}{12}$

23. A, B, C, F

25. B

27. Let $P(\text{tails}) = x$, then $P(\text{heads}) = 4x$
    $x + 4x = 1$
    $5x = 1$
    $x = \frac{1}{5}$
    $P(\text{tails}) = \frac{1}{5}, \quad P(\text{heads}) = \frac{4}{5}$

29. $P(2) = P(4) = P(6) = x$
    $P(1) = P(3) = P(5) = 2x$
    $P(1) + P(2) + P(3) + P(4) + P(5) + P(6) = 1$
    $2x + x + 2x + x + 2x + x = 1$
    $9x = 1$
    $x = \frac{1}{9}$
    $P(2) = P(4) = P(6) = \frac{1}{9}$
    $P(1) = P(3) = P(5) = \frac{2}{9}$

31. $P(E) = \frac{n(E)}{n(S)} = \frac{n\{1,2,3\}}{10} = \frac{3}{10}$

33. $P(E) = \frac{n(E)}{n(S)} = \frac{n\{2,4,6,8,10\}}{10} = \frac{5}{10} = \frac{1}{2}$

35. $P(\text{white}) = \frac{n(\text{white})}{n(S)} = \frac{5}{5+10+8+7} = \frac{5}{30} = \frac{1}{6}$

37. The sample space is: S = {BBB, BBG, BGB, GBB, BGG, GBG, GGB, GGG}
    $P(3\text{ boys}) = \frac{n(3\text{ boys})}{n(S)} = \frac{1}{8}$

39. The sample space is:
    S = {BBBB, BBBG, BBGB, BGBB, GBBB, BBGG, BGBG, GBBG, BGGB, GBGB, GGBB, BGGG, GBGG, GGBG, GGGB, GGGG}
    $P(1\text{ girl, }3\text{ boys}) = \frac{n(1\text{ girl, }3\text{ boys})}{n(S)} = \frac{4}{16} = \frac{1}{4}$

41. $P(\text{sum of two dice is 7})$

$= \dfrac{n(\text{sum of two dice is 7})}{n(S)}$

$= \dfrac{n\{1,6 \text{ or } 2,5 \text{ or } 3,4 \text{ or } 4,3 \text{ or } 5,2 \text{ or } 6,1\}}{n(S)}$

$= \dfrac{6}{36} = \dfrac{1}{6}$

43. $P(\text{sum of two dice is 3}) = \dfrac{n(\text{sum of two dice is 3})}{n(S)}$

$= \dfrac{n\{1,2 \text{ or } 2,1\}}{n(S)} = \dfrac{2}{36} = \dfrac{1}{18}$

45. $P(A \cup B) = P(A) + P(B) - P(A \cap B)$
$= 0.25 + 0.45 - 0.15 = 0.55$

47. $P(A \cup B) = P(A) + P(B) = 0.25 + 0.45 = 0.70$

49. $P(A \cup B) = P(A) + P(B) - P(A \cap B)$
$0.85 = 0.60 + P(B) - 0.05$
$P(B) = 0.85 - 0.60 + 0.05 = 0.30$

51. $P(\text{not victim}) = 1 - P(\text{victim}) = 1 - 0.265 = 0.735$

53. $P(\text{not in 70's}) = 1 - P(\text{in 70's}) = 1 - 0.3 = 0.7$

55. $P(\text{white or green}) = P(\text{white}) + P(\text{green})$

$= \dfrac{n(\text{white}) + n(\text{green})}{n(S)}$

$= \dfrac{9+8}{9+8+3} = \dfrac{17}{20}$

57. $P(\text{not white}) = 1 - P(\text{white})$

$= 1 - \dfrac{n(\text{white})}{n(S)}$

$= 1 - \dfrac{9}{20} = \dfrac{11}{20}$

59. $P(\text{strike or one}) = P(\text{strike}) + P(\text{one})$

$= \dfrac{n(\text{strike}) + n(\text{one})}{n(S)}$

$= \dfrac{3+1}{8} = \dfrac{4}{8}$

$= \dfrac{1}{2}$

61. There are 30 households out of 100 with an income of $30,000 or more.

$P(E) = \dfrac{n(E)}{n(S)} = \dfrac{n(30,000 \text{ or more})}{n(\text{total households})}$

$= \dfrac{30}{100} = \dfrac{3}{10} = 0.30$

63. There are 40 households out of 100 with an income of less than $20,000.

$P(E) = \dfrac{n(E)}{n(S)} = \dfrac{n(\text{less than } \$20,000)}{n(\text{total households})}$

$= \dfrac{40}{100} = \dfrac{2}{5} = 0.40$

65. a. $P(1 \text{ or } 2) = P(1) + P(2) = 0.24 + 0.33 = 0.57$

b. $P(1 \text{ or more}) = 1 - P(\text{none})$
$= 1 - 0.05 = 0.95$

c. $P(3 \text{ or fewer}) = 1 - P(4 \text{ or more})$
$= 1 - 0.17 = 0.83$

d. $P(3 \text{ or more}) = P(3) + P(4 \text{ or more})$
$= 0.21 + 0.17 = 0.38$

e. $P(\text{fewer than 2}) = P(0) + P(1)$
$= 0.05 + 0.24 = 0.29$

f. $P(\text{fewer than 1}) = P(0) = 0.05$

g. $P(1, 2, \text{ or } 3) = P(1) + P(2) + P(3)$
$= 0.24 + 0.33 + 0.21 = 0.78$

h. $P(2 \text{ or more}) = P(2) + P(3) + P(4 \text{ or more})$
$= 0.33 + 0.21 + 0.17 = 0.71$

67. a. $P(\text{freshman or female})$
$= P(\text{freshman}) + P(\text{female}) - P(\text{freshman and female})$

$= \dfrac{n(\text{freshman}) + n(\text{female}) - n(\text{freshman and female})}{n(S)}$

$= \dfrac{18 + 15 - 8}{33} = \dfrac{25}{33}$

b. $P(\text{sophomore or male})$
$= P(\text{sophomore}) + P(\text{male}) - P(\text{sophomore and male})$

$= \dfrac{n(\text{sophomore}) + n(\text{male}) - n(\text{sophomore and male})}{n(S)}$

$= \dfrac{15 + 18 - 8}{33} = \dfrac{25}{33}$

69. $P(\text{at least 2 with same birthday})$
$= 1 - P(\text{none with same birthday})$
$= 1 - \dfrac{n(\text{different birthdays})}{n(S)}$
$= 1 - \dfrac{365 \cdot 364 \cdot 363 \cdot 362 \cdot 361 \cdot 360 \cdots 354}{365^{12}}$
$\approx 1 - 0.833$
$= 0.167$

71. The sample space for picking 5 out of 10 numbers in a particular order contains
$P(10,5) = \dfrac{10!}{(10-5)!} = \dfrac{10!}{5!} = 30,240$ possible outcomes.
One of these is the desired outcome. Thus, the probability of winning is:
$P(E) = \dfrac{n(E)}{n(S)} = \dfrac{n(\text{winning})}{n(\text{total possible outcomes})}$
$= \dfrac{1}{30,240} \approx 0.000033069$

Chapter 8 Review

1. $\{\text{Dave}\}, \{\text{Joanne}\}, \{\text{Erica}\},$
$\{\text{Dave, Joanne}\}, \{\text{Dave, Erica}\}, \{\text{Joanne, Erica}\},$
$\{\text{Dave, Joanne, Erica}\}, \varnothing$

3. $A \cup B = \{1, 3, 5, 7\} \cup \{3, 5, 6, 7, 8\}$
$= \{1, 3, 5, 6, 7, 8\}$

5. $A \cap C = \{1, 3, 5, 7\} \cap \{2, 3, 7, 8, 9\} = \{3, 7\}$

7. $\overline{A} \cup \overline{B} = \overline{\{1, 3, 5, 7\}} \cup \overline{\{3, 5, 6, 7, 8\}}$
$= \{2, 4, 6, 8, 9\} \cup \{1, 2, 4, 9\}$
$= \{1, 2, 4, 6, 8, 9\}$

9. $\overline{B \cap C} = \overline{\{3, 5, 6, 7, 8\} \cap \{2, 3, 7, 8, 9\}}$
$= \overline{\{3, 7, 8\}}$
$= \{1, 2, 4, 5, 6, 9\}$

11. $n(A) = 8, n(B) = 12, n(A \cap B) = 3$
$n(A \cup B) = n(A) + n(B) - n(A \cap B)$
$= 8 + 12 - 3 = 17$

13. From the figure: $n(A) = 20 + 2 + 6 + 1 = 29$

15. From the figure:
$n(A \text{ and } C) = n(A \cap C) = 1 + 6 = 7$

17. From the figure:
$n(\text{neither in } A \text{ nor in } C) = n(\overline{A \cup C}) = 20 + 5 = 25$

19. $P(8,3) = \dfrac{8!}{(8-3)!} = \dfrac{8!}{5!} = \dfrac{8 \cdot 7 \cdot 6 \cdot 5!}{5!} = 336$

21. $C(8,3) = \dfrac{8!}{(8-3)!\,3!} = \dfrac{8!}{5!\,3!} = \dfrac{8 \cdot 7 \cdot 6 \cdot 5!}{5! \cdot 3 \cdot 2 \cdot 1} = 56$

23. There are 2 choices of material, 3 choices of color, and 10 choices of size. The complete assortment would have: $2 \cdot 3 \cdot 10 = 60$ suits.

25. There are two possible outcomes for each game or $2 \cdot 2 \cdot 2 \cdot 2 \cdot 2 \cdot 2 \cdot 2 = 2^7 = 128$ outcomes for 7

27. Since order is significant, this is a permutation.
$P(9,4) = \dfrac{9!}{(9-4)!} = \dfrac{9!}{5!} = \dfrac{9 \cdot 8 \cdot 7 \cdot 6 \cdot 5!}{5!} = 3024$
ways to seat 4 people in 9 seats.

29. Choose 4 runners – order is significant:
$P(8,4) = \dfrac{8!}{(8-4)!} = \dfrac{8!}{4!} = \dfrac{8 \cdot 7 \cdot 6 \cdot 5 \cdot 4!}{4!} = 1680$
ways a squad can be chosen.

31. Choose 2 teams from 14 – order is not significant:
$C(14,2) = \dfrac{14!}{(14-2)!\,2!} = \dfrac{14!}{12!\,2!} = \dfrac{14 \cdot 13 \cdot 12!}{12! \cdot 2 \cdot 1} = 91$
ways to choose 2 teams.

33. There are $8 \cdot 10 \cdot 10 \cdot 10 \cdot 10 \cdot 10 \cdot 2 = 1,600,000$ possible phone numbers.

35. There are $24 \cdot 9 \cdot 10 \cdot 10 \cdot 10 = 216,000$ possible license plates.

37. Since there are repeated letters:
$\dfrac{7!}{2!\,2!} = \dfrac{7 \cdot 6 \cdot 5 \cdot 4 \cdot 3 \cdot 2 \cdot 1}{2 \cdot 1 \cdot 2 \cdot 1} = 1260$ different words can be formed.

39. a. $C(9,4) \cdot C(9,3) \cdot C(9,2) = 126 \cdot 84 \cdot 36$
$= 381,024$
committees can be formed.

b. $C(9,4) \cdot C(5,3) \cdot C(2,2) = 126 \cdot 10 \cdot 1 = 1260$
committees can be formed.

41. a. $365 \cdot 364 \cdot 363 \cdots 348 = 8.634628387 \times 10^{45}$

b. $P(\text{no one has same birthday})$
$= \dfrac{365 \cdot 364 \cdot 363 \cdots 348}{365^{18}} \approx 0.6531$

c. $P(\text{at least 2 have same birthday})$
$= 1 - P(\text{no one has same birthday})$
$= 1 - 0.6531 = 0.3469$

43. a. $P(\text{unemployed}) = 0.058$

b. $P(\text{not unemployed}) = 1 - P(\text{unemployed})$
$= 1 - 0.058 = 0.942$

45. $P(\$1 \text{ bill}) = \dfrac{n(\$1 \text{ bill})}{n(S)} = \dfrac{4}{9}$

47. Let S be all possible selections, let D be a card that is divisible by 5, and let PN be a card that is 1 or a prime number.
$n(S) = 100$
$n(D) = 20$ (There are 20 numbers divisible by 5 between 1 and 100.)
$n(PN) = 26$ (There are 25 prime numbers less than or equal to 100.)
$P(D) = \dfrac{n(D)}{n(S)} = \dfrac{20}{100} = \dfrac{1}{5} = 0.2$
$P(PN) = \dfrac{n(PN)}{n(S)} = \dfrac{26}{100} = \dfrac{13}{50} = 0.26$

Chapter 8 Test

1. $A \cap B = \{0,1,4,9\} \cap \{2,4,6,8\} = \{4\}$

2. $A \cup C = \{0,1,4,9\} \cup \{1,3,5,7,9\}$
$= \{0,1,3,4,5,7,9\}$

3. $(A \cup B) \cap C$
$= (\{0,1,4,9\} \cup \{2,4,6,8\}) \cap \{1,3,5,7,9\}$
$= \{0,1,2,4,6,8,9\} \cap \{1,3,5,7,9\}$
$= \{1,9\}$

4. $A \cup B = \{0,1,4,9\} \cup \{2,4,6,8\}$
$= \{0,1,2,4,6,8,9\}$
Since the universal set is $\{0,1,2,3,4,5,6,7,8,9\}$,
$\overline{A \cup B} = \{3,5,7\}$.

5. Since the universal set is $\{0,1,2,3,4,5,6,7,8,9\}$,
   $\overline{C} = \{0,2,4,6,8\}$.

6. $\overline{C}$ can be obtained from the previous problem.
   $A \cap \overline{C} = \{0,1,4,9\} \cap \{0,2,4,6,8\} = \{0,4\}$
   Therefore,
   $\overline{A \cap \overline{C}} = \{1,2,3,5,6,7,8,9\}$.

7. From the figure:
   $n(\text{physics}) = 4 + 2 + 7 + 9 = 22$

8. From the figure:
   $n(\text{biology } or \text{ chemistry } or \text{ physics})$
   $= 22 + 8 + 2 + 4 + 9 + 7 + 15$
   $= 67$
   Therefore,
   $n(\text{none of the three}) = 70 - 67 = 3$

9. From the figure:
   $n(\text{only biology } and \text{ chemistry})$
   $= n(\text{biol. } and \text{ chem.}) - n(\text{biol. } and \text{ chem. } and \text{ phys.})$
   $= (8+2) - 2$
   $= 8$

10. From the figure:
    $n(\text{physics } or \text{ chemistry}) = 4+2+7+9+15+8$
    $= 45$

11. $7! = 7 \cdot 6 \cdot 5 \cdot 4 \cdot 3 \cdot 2 \cdot 1 = 5040$

12. $P(10,6) = \dfrac{10!}{(10-6)!} = \dfrac{10!}{4!}$
    $= \dfrac{10 \cdot 9 \cdot 8 \cdot 7 \cdot 6 \cdot 5 \cdot 4!}{4!}$
    $= 10 \cdot 9 \cdot 8 \cdot 7 \cdot 6 \cdot 5$
    $= 151,200$

13. $C(11,5) = \dfrac{11!}{5!(11-5)!}$
    $= \dfrac{11!}{5!6!}$
    $= \dfrac{11 \cdot 10 \cdot 9 \cdot 8 \cdot 7 \cdot 6!}{5 \cdot 4 \cdot 3 \cdot 2 \cdot 1 \cdot 6!}$
    $= \dfrac{11 \cdot 10 \cdot 9 \cdot 8 \cdot 7}{5 \cdot 4 \cdot 3 \cdot 2 \cdot 1}$
    $= 462$

14. Since the order in which the colors are selected doesn't matter, this is a combination problem. We have $n = 21$ colors and we wish to select $r = 6$ of them.
    $C(21,6) = \dfrac{21!}{6!(21-6)!} = \dfrac{21!}{6!15!}$
    $= \dfrac{21 \cdot 20 \cdot 19 \cdot 18 \cdot 17 \cdot 16 \cdot 15!}{6!15!}$
    $= \dfrac{21 \cdot 20 \cdot 19 \cdot 18 \cdot 17 \cdot 16}{6 \cdot 5 \cdot 4 \cdot 3 \cdot 2 \cdot 1}$
    $= 54,264$
    There are 54,264 ways to choose 6 colors from the 21 available colors.

15. Because the letters are not distinct and order matters, we use the permutation formula for non-distinct objects. We have four different letters, two of which are repeated (E four times and D two times).
    $\dfrac{n!}{n_1!n_2!n_3!n_4!} = \dfrac{8!}{4!2!1!1!}$
    $= \dfrac{8 \cdot 7 \cdot 6 \cdot 5 \cdot 4!}{4! \cdot 2 \cdot 1}$
    $= \dfrac{8 \cdot 7 \cdot 6 \cdot 5}{2}$
    $= 4 \cdot 7 \cdot 6 \cdot 5$
    $= 840$
    There are 840 distinct arrangements of the letters in the word REDEEMED.

16. Since the order of the horses matters and all the horses are distinct, we use the permutation formula for distinct objects.
    $P(8,2) = \dfrac{8!}{(8-2)!} = \dfrac{8!}{6!} = \dfrac{8 \cdot 7 \cdot 6!}{6!} = 8 \cdot 7 = 56$
    There are 56 different exacta bets for an 8-horse race.

17. We are choosing 3 letters from 26 distinct letters and 4 digits from 10 distinct digits. The letters and numbers are placed in order following the format LLL DDDD with repetitions being allowed. Using the Multiplication Principle, we get
$26 \cdot 26 \cdot 23 \cdot 10 \cdot 10 \cdot 10 \cdot 10 = 155,480,000$
Note that there are only 23 possibilities for the third letter.
There are 155,480,000 possible license plates using the new format.

18. Let A = Kiersten accepted at USC, and B = Kiersten accepted at FSU. Then, we get $P(A) = 0.6$, $P(B) = 0.7$, and $P(A \cap B) = 0.35$.

    a. Here we need to use the Addition Rule.
    $P(A \cup B) = P(A) + P(B) - P(A \cap B)$
    $= 0.6 + 0.7 - 0.35$
    $= 0.95$
    Kiersten has a 95% chance of being admitted to at least one of the universities.

    b. Here we need the Complement of an event.
    $P(\overline{B}) = 1 - P(B) = 1 - 0.7 = 0.3$
    Kiersten has a 30% chance of not being admitted to FSU.

19. a. Since the bottle is chosen at random, all bottles are equally likely to be selected. Thus,
    $P(\text{Coke}) = \dfrac{5}{8+5+4+3} = \dfrac{5}{20} = \dfrac{1}{4} = 0.25$
    There is a 25% chance that the selected bottle contains Coke.

    b. $P(\text{Pepsi} \cup \text{IBC}) = \dfrac{8+3}{8+5+4+3}$
    $= \dfrac{11}{20} = 0.55$
    There is a 55% chance that the selected bottle contains either Pepsi or IBC.

20. Since the ages cover all possibilities and the age groups are mutually exclusive, the sum of all the probabilities must equal 1.
    $0.03 + 0.23 + 0.29 + 0.25 + 0.01 = 0.81$
    $1 - 0.81 = 0.19$
    The given probabilities sum to 0.81. This means the missing probability (for 18-20) must be 0.19.

21. The number of different selections of 6 numbers is the number of ways we can choose 5 white balls and 1 red ball, where the order of the white balls is not important. This requires the use of the Multiplication Principle and the combination formula. Thus, the total number of distinct ways to pick the 6 numbers is given by
    $n(\text{white balls}) \cdot n(\text{red ball})$
    $= C(53,5) \cdot C(42,1)$
    $= \dfrac{53!}{5!(53-5)!} \cdot \dfrac{42!}{1!(42-1)!}$
    $= \dfrac{53!}{5! \cdot 48!} \cdot \dfrac{42!}{1! \cdot 41!}$
    $= \left(\dfrac{53 \cdot 52 \cdot 51 \cdot 50 \cdot 49}{5 \cdot 4 \cdot 3 \cdot 2 \cdot 1} \right) \left(\dfrac{42 \cdot 41!}{41!} \right)$
    $= \dfrac{53 \cdot 52 \cdot 51 \cdot 50 \cdot 49 \cdot 42}{5 \cdot 4 \cdot 3 \cdot 2}$
    $= 120,526,770$
    Since each possible combination is equally likely, the probability of winning on a $1 play is
    $P(\text{win on \$1 play}) = \dfrac{1}{120,526,770}$
    ≈ 0.0000000083

22. The number of elements in the sample space can be obtained by using the Multiplication Principle:
    $6 \cdot 6 \cdot 6 \cdot 6 \cdot 6 = 7,776$
    Consider the rolls as a sequence of 5 slots. The number of ways to position 2 fours in 5 slots is $C(5,2)$. The remaining three slots can be filled with any of the five remaining numbers from the die. Repetitions are allowed so this can be done in $5 \cdot 5 \cdot 5 = 125$ different ways.
    Therefore, the total number of ways to get exactly 2 fours is
    $C(5,2) \cdot 125 = \dfrac{5!}{2! \cdot 3!} \cdot 125 = \dfrac{5 \cdot 4 \cdot 125}{2} = 1250$
    The probability of getting exactly 2 fours on 5 rolls of a die is given by
    $P(\text{exactly 2 fours}) = \dfrac{1250}{7776} \approx 0.1608$.

Cumulative Review R-8

1. $3x^2 - 2x = -1 \Rightarrow 3x^2 - 2x + 1 = 0$

$$x = \frac{-b \pm \sqrt{b^2-4ac}}{2a} = \frac{-(-2) \pm \sqrt{(-2)^2 - 4(3)(1)}}{2(3)}$$

$$= \frac{2 \pm \sqrt{4-12}}{6} = \frac{2 \pm \sqrt{-8}}{6} = \frac{2 \pm 2\sqrt{2}i}{6}$$

$$= \frac{1 \pm \sqrt{2}i}{3}$$

The solution set is $\left\{ \dfrac{1-\sqrt{2}i}{3}, \dfrac{1+\sqrt{2}i}{3} \right\}$.

3. $y = 2(x+1)^2 - 4$

Using the graph of $y = x^2$, horizontally shift to the left 1 unit, vertically stretch by a factor of 2, and vertically shift down 4 units.

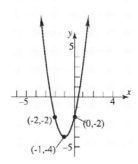

5. $f(x) = 5x^4 - 9x^3 - 7x^2 - 31x - 6$

Step 1: $f(x)$ has at most 4 real zeros.

Step 2:   Possible rational zeros:
$p = \pm 1, \pm 2, \pm 3, \pm 6; \quad q = \pm 1, \pm 5;$

$\dfrac{p}{q} = \pm 1, \pm \dfrac{1}{5}, \pm 2, \pm \dfrac{2}{5}, \pm 3, \pm \dfrac{3}{5}, \pm 6, \pm \dfrac{6}{5}$

Step 3:   Using the Bounds on Zeros Theorem:

$f(x) = 5\left(x^4 - 1.8x^3 - 1.4x^2 - 6.2x - 1.2 \right)$

$a_3 = -1.8,\ a_2 = -1.4,\ a_1 = -6.2,\ a_0 = -1.2$

Max $\{1, |-1.2|+|-6.2|+|-1.4|+|-1.8|\}$
$=$ Max $\{1, 10.6\} = 10.6$

$1 + \text{Max}\ \{|-1.2|,|-6.2|,|-1.4|,|-1.8|\}$
$= 1 + 6.2 = 7.2$

The smaller of the two numbers is 7.2. Thus, every zero of f lies between -7.2 and 7.2.
Graphing using the bounds: (Second graph has a better window.)

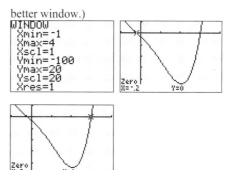

Step 4:   From the graph we see that there are x-intercepts at -0.2 and 3. Using synthetic division with 3:

$\begin{array}{r|rrrrr} 3) & 5 & -9 & -7 & -31 & -6 \\ & & 15 & 18 & 33 & 6 \\ \hline & 5 & 6 & 11 & 2 & 0 \end{array}$

Since the remainder is 0, $x - 3$ is a factor. The other factor is the quotient: $5x^3 + 6x^2 + 11x + 2$.
Using synthetic division with 2 on the quotient:

$\begin{array}{r|rrrr} -0.2) & 5 & 6 & 11 & 2 \\ & & -1 & -1 & -2 \\ \hline & 5 & 5 & 10 & 0 \end{array}$

Since the remainder is 0, $x - (-0.2) = x + 0.2$ is a factor. The other factor is the quotient:
$5x^2 + 5x + 10 = 5(x^2 + x + 2)$.

Factoring, $f(x) = 5(x^2 + x + 2)(x-3)(x+0.2)$
The real zeros are 3 and -0.2.
The complex zeros come from solving
$x^2 + x + 2 = 0$.

$$x = \frac{-b \pm \sqrt{b^2-4ac}}{2a} = \frac{-1 \pm \sqrt{1^2 - 4(1)(2)}}{2(1)}$$

$$= \frac{-1 \pm \sqrt{1-8}}{2} = \frac{-1 \pm \sqrt{-7}}{2}$$

$$= \frac{-1 \pm \sqrt{7}i}{2}$$

Therefore, over the set of complex numbers, $f(x) = 5x^4 - 9x^3 - 7x^2 - 31x - 6$ has zeros

$\left\{ -\dfrac{1}{2} + \dfrac{\sqrt{7}}{2}i, -\dfrac{1}{2} - \dfrac{\sqrt{7}}{2}i, -\dfrac{1}{5}, 3 \right\}$.

7. $\log_3(9) = \log_3(3^2) = 2$

9. Multiply each side of the first equation by –3 and add to the second equation to eliminate x; multiply each side of the first equation by 2 and add to the third equation to eliminate x:
$$\begin{cases} x - 2y + z = 15 \\ 3x + y - 3z = -8 \\ -2x + 4y - z = -27 \end{cases}$$

$$\begin{aligned} -3x + 6y - 3z &= -45 \\ 3x + y - 3z &= -8 \\ \hline 7y - 6z &= -53 \end{aligned}$$

$$\begin{aligned} x - 2y + z &= 15 \\ -2x + 4y - z &= -27 \end{aligned} \qquad \begin{aligned} 2x - 4y + 2z &= 30 \\ -2x + 4y - z &= -27 \\ \hline z &= 3 \end{aligned}$$

Substituting and solving for the other variables:
$$z = 3 \Rightarrow 7y - 6(3) = -53$$
$$7y = -35$$
$$y = -5$$
$$z = 3, y = -5 \Rightarrow x - 2(-5) + 3 = 15$$
$$x + 10 + 3 = 15 \Rightarrow x = 2$$

The solution is $x = 2$, $y = -5$, $z = 3$.